Lecture Notes in Mathematics 1960

Editors:
J.-M. Morel, Cachan
F. Takens, Groningen
B. Teissier, Paris

Joseph Lipman · Mitsuyasu Hashimoto

Foundations of Grothendieck Duality for Diagrams of Schemes

 Springer

Joseph Lipman
Mathematics Department
Purdue University
West Lafayette, IN 47907
USA
lipman@math.purdue.edu

Mitsuyasu Hashimoto
Graduate School of Mathematics
Nagoya University
Chikusa-ku, Nagoya 464-8602
Japan
hasimoto@math.nagoya-u.ac.jp

ISBN: 978-3-540-85419-7   e-ISBN: 978-3-540-85420-3
DOI: 10.1007/978-3-540-85420-3

Lecture Notes in Mathematics ISSN print edition: 0075-8434
                              ISSN electronic edition: 1617-9692

Library of Congress Control Number: 2008935627

Mathematics Subject Classification (2000): 14A20, 18E30, 14F99, 18A99, 18F99, 14L30

Cover design: SPi Publishing Services

Printed on acid-free paper

9 8 7 6 5 4 3 2 1

springer.com

Preface

This volume contains two related, though independently written, monographs.

In *Notes on Derived Functors and Grothendieck Duality* the first three chapters treat the basics of derived categories and functors, and of the rich formalism, over ringed spaces, of the derived functors, for unbounded complexes, of the sheaf functors $\otimes$, $\mathcal{H}om$, f_* and f^* where f is a ringed-space map. Included are some enhancements, for concentrated (i.e., quasi-compact and quasi-separated) schemes, of classical results such as the projection and Künneth isomorphisms. The fourth chapter presents the abstract foundations of Grothendieck Duality—existence and tor-independent base change for the right adjoint of the derived functor $\mathbf{R}f_*$ when f is a quasi-proper map of concentrated schemes, the twisted inverse image pseudofunctor for separated finite-type maps of noetherian schemes, refinements for maps of finite tor-dimension, and a brief discussion of dualizing complexes.

In *Equivariant Twisted Inverses* the theory is extended to the context of diagrams of schemes, and in particular, to schemes with a group-scheme action. An equivariant version of the twisted inverse-image pseudofunctor is defined, and equivariant versions of some of its important properties are proved, including Grothendieck duality for proper morphisms, and flat base change. Also, equivariant dualizing complexes are dealt with. As an application, a generalized version of Watanabe's theorem on the Gorenstein property of rings of invariants is proved.

More detailed overviews are given in the respective Introductions.

Contents

Part II Mitsuyasu Hashimoto: Equivariant Twisted Inverses

Part I
Joseph Lipman: Notes on Derived Functors and Grothendieck Duality

Abstract

This is a polished version of notes begun in the late 1980s, largely available from my home page since then, meant to be accessible to mid-level graduate students. The first three chapters treat the basics of derived categories and functors, and of the rich formalism, over ringed spaces, of the derived functors, for unbounded complexes, of the sheaf functors $\otimes$, $\mathcal{H}om$, f_* and f^* (where f is a ringed-space map). Included are some enhancements, for concentrated (= quasi-compact and quasi-separated) schemes, of classical results such as the projection and Künneth isomorphisms. The fourth chapter presents the abstract foundations of Grothendieck Duality—existence and tor-independent base change for the right adjoint of the derived functor $\mathbf{R}f_*$ when f is a quasi-proper map of concentrated schemes, the twisted inverse image pseudofunctor for separated finite-type maps of noetherian schemes, some refinements for maps of finite tor-dimension, and a brief discussion of dualizing complexes.

Introduction

(0.1) The first three chapters of these notes[1] treat the basics of derived categories and functors, and of the formalism of four of Grothendieck's "six operations" ([**Ay**], [**Mb**]), over, say, the category of ringed spaces (topological spaces equipped with a sheaf of rings)—namely the derived functors, for complexes which need not be bounded, of the sheaf functors $\otimes$, $\mathcal{H}om$, and of the direct and inverse image functors f_* and f^* relative to a map f. The axioms of this formalism are summarized in §3.6 under the rubric *adjoint monoidal Δ-pseudofunctors,* with values in closed categories (§3.5).

Chapter 4 develops the abstract theory of the *twisted inverse image* functor $f^!$ associated to a finite-type separated map of schemes $f \colon X \to Y$. (Suppose for now that Y is noetherian and separated, though for much of what we do, weaker hypotheses will suffice.) This functor maps the derived category of cohomologically bounded-below $\mathcal{O}_Y$-complexes with quasi-coherent homology to the analogous category over X. Its characterizing properties are:

- *Duality.* If f is proper then $f^!$ is right-adjoint to the derived direct image functor $\mathbf{R}f_*$.
- *Localization.* If f is an open immersion (or even étale), then $f^!$ is the usual inverse image functor f^*.
- *Pseudofunctoriality* (or *2-functoriality*). To each composition $X \xrightarrow{f} Y \xrightarrow{g} Z$ we can assign a natural functorial isomorphism $(gf)^! \xrightarrow{\sim} f^!g^!$, in such a way that a kind of associativity holds with respect to any composition of three maps, see §(3.6.5).

---

[1] That are a polished version of notes written largely in the late 1980s, available in part since then from < www.math.purdue.edu/~lipman >. I am grateful to Bradley Lucier for his patient instruction in some of the finer points of TEX, and for setting up the appearance macros in those days when canned style files were not common—and when compilation was several thousand times slower than nowadays.

J. Lipman, M. Hashimoto, *Foundations of Grothendieck Duality for Diagrams of Schemes*, Lecture Notes in Mathematics 1960,
© Springer-Verlag Berlin Heidelberg 2009

Additional basic properties of $f^!$ are its compatibility with *flat base change* (Theorems (4.4.3), (4.8.3)), and the existence of canonical functorial maps, for $\mathcal{O}_Y$-complexes E and F having quasi-coherent homology:

$$\mathbf{R}\mathcal{H}om(\mathbf{L}f^*E, f^!F) \rightarrow f^!\mathbf{R}\mathcal{H}om(E, F)$$

$$\mathbf{L}f^*E \underset{\equiv}{\otimes} f^!F \rightarrow f^!(E \underset{\equiv}{\otimes} F)$$

(where $\underset{\equiv}{\otimes}$ denotes the left-derived tensor product), of which the first is an isomorphism when E is cohomologically bounded above, with coherent homology, and F is cohomologically bounded below, (Exercise (4.9.3)(b)), and the second is an isomorphism whenever f has finite tor-dimension (Theorem (4.9.4)) or E is a bounded flat complex (Exercise (4.9.6)(a)).

The existence and uniqueness, up to isomorphism, of the twisted inverse image pseudofunctor is given by Theorem (4.8.1), and compatibility with flat base change by Theorem (4.8.3). These are culminating results in the notes. Various approximations to these theorems have been known for decades, see, e.g., [**H**, p. 383, 3.4]. At present, however, the proofs of the theorems, as stated here, seem to need, among other things, a compactification theorem of Nagata, that any finite-type separable map of noetherian schemes factors as an open immersion followed by a proper map, a fact whose proof was barely accessible before the appearance of [**Lt**] and [**C′**] (see also [**Vj**]); and even with that compactification theorem, I am not aware of any complete, detailed exposition of the proofs in print prior to the recent one by Nayak [**Nk**].[2] There must be a more illuminating treatment of this awesome pseudofunctor in the Plato-Erdös Book!

(**0.2**) The theory of $f^!$ was conceived by Grothendieck [**Gr′**, pp. 112–115], as a generalization of Serre's duality theorems for smooth projective varieties over fields. Grothendieck also applied his ideas in the context of étale cohomology. The fundamental technique of derived categories was developed by Verdier, who used it in establishing a duality theorem for locally compact spaces that generalizes classical duality theorems for topological manifolds. Deligne further developed the methods of Grothendieck and Verdier (cf. [**De′**] and its references).

Hartshorne gave an account of the theory in [**H**]. The method there is to treat separately several distinctive special situations, such as smooth maps, finite maps, and regular immersions (local complete intersections), where $f^!$ has a nice explicit description; and then to do the general case by pasting together special ones (locally, a general f can be factored as smooth ∘ finite). The fact that this approach works is indicative of considerable depth in the underlying structure, in that the special cases, that don't *a priori* have to

---

[2] In fact Nayak's methods, which are less dependent on compactifications, apply to other contexts as well, for example flat finitely-presentable separated maps of not-necessarily-noetherian schemes, or separated maps of noetherian formal schemes, see [**Nk**, §7]. See also the summary of Nayak's work in [**S′**, §§3.1–3.3].

be related at all, can in fact be melded; and in that the reduction from general to special involves several choices (for example, in the just-mentioned factorization) of which the final results turn out to be independent. Proving the existence of $f^!$ and its basic properties in this manner involves many compatibilities among those properties in their various epiphanies, a notable example being the "Residue Isomorphism" [**H**, p. 185]. The proof in [**H**] also makes essential use of a pseudofunctorial theory of dualizing complexes,[3] so that it does not apply, e.g., to arbitrary separated noetherian schemes.

On first acquaintance, [**De′**] appears to offer a neat way to cut through the complexity—a direct abstract proof of the existence of $f^!$, with indications about how to derive the concrete special situations (which, after all, motivate and enliven the abstract formalism). Such an impression is bolstered by Verdier's paper [**V′**]. Verdier gives a reasonably short proof of the flat base change theorem, sketches some corollaries (for example, the finite tor-dimension case is treated in half a page [*ibid.*, p. 396], as is the smooth case [*ibid.*, pp. 397–398]), and states in conclusion that "all the results of [**H**], except the theory of dualizing and residual complexes, are easy consequences of the existence theorem." In short, Verdier's concise summary of the main features, together with some background from [**H**] and a little patience, should suffice for most users of the duality machine.

Personally speaking, it was in this spirit—not unlike that in which many scientists use mathematics—that I worked on algebraic and geometric applications in the late 1970s and early 1980s. But eventually I wanted to gain a better understanding of the foundations, and began digging beneath the surface. The present notes are part of the result. They show, I believe, that there is more to the abstract theory than first meets the eye.

(**0.3**) There are a number of treatments of Grothendieck duality for the Zariski topology (not to mention other contexts, see e.g., [**Gl′**], [**De**], [**LO**]), for example, Neeman's approach via Brown representability [**N**], Hashimoto's treatment of duality for *diagrams* of schemes (in particular, schemes with group actions) [**Hsh**], duality for *formal* schemes [**AJL′**], as well as various substantial enhancements of material in Hartshorne's classic [**H**], such as [**C**], [**S**], [**LNS**] and [**YZ**]. Still, some basic results in these notes, such as Theorem (3.10.3) and Theorem (4.4.1) are difficult, if not impossible, to find elsewhere, at least in the present generality and detail. And, as indicated below, there are in these notes some significant differences in emphasis.

It should be clarified that the word "Notes" in the title indicates that the present exposition is neither entirely self-contained nor completely polished. The goal is, basically, to guide the willing reader along one path to an understanding of all that needs to be done to prove the fundamental Theorems (4.8.1) and (4.8.3), and of how to go about doing it. The intent is to provide enough in the way of foundations, yoga, and references so that the reader can,

---

[3] This enlightening theory—touched on in §4.10 below—is generalized to Cousin complexes over formal schemes in [**LNS**]. A novel approach, via "rigidity," is given in [**YZ**], at least for schemes of finite type over a fixed regular one.

more or less mechanically, fill in as much of what is missing as motivation
and patience allow.

So what is meant by "foundations and yoga"?

There are innumerable interconnections among the various properties
of the twisted inverse image, often expressible via commutativity of some
diagram of natural maps. In this way one can encode, within a formal functorial language, relationships involving higher direct images of quasi-coherent
sheaves, or, more generally, of complexes with quasi-coherent homology,
relationships whose treatment might otherwise, on the whole, prove discour-
agingly complicated.

As a strategy for coping with duality theory, disengaging the underlying
category-theoretic skeleton from the algebra and geometry which it supports
has the usual advantages of simplification, clarification, and generality. Nevertheless, the resulting fertile formalism of adjoint monoidal pseudofunctors
soon sprouts a thicket of rather complicated diagrams whose commutativity
is an essential part of the development—as may be seen, for example, in
the later parts of Chapters 3 and 4. Verifying such commutativities, fun
to begin with, soon becomes a tedious, time-consuming, chore. Such chores
must, eventually, be attended to.[4]

Thus, these notes emphasize purely formal considerations, and attention
to detail. On the whole, statements are made, whenever possible, in precise
category-theoretic terms, canonical isomorphisms are not usually treated as
equalities, and commutativity of diagrams of natural maps—a matter of
paramount importance—is not taken for granted unless explicitly proved
or straightforward to verify. The desire is to lay down transparently secure
foundations for the main results. A perusal of §2.6, which treats the basic
relation "adjoint associativity" between the derived functors $\otimes$ and $\mathsf{R}\mathcal{H}\mathrm{om}$,
and of §3.10, which treats various avatars of the tor-independence condition
on squares of quasi-compact maps of quasi-separated schemes, will illustrate
the point. (In both cases, total understanding requires a good deal of preceding material.)

> *Computer-aided proofs are often more convincing*
> *than many standard proofs based on*
> *diagrams which are claimed to commute,*
> *arrows which are supposed to be the same,*
> *and arguments which are left to the reader.*
>
> —J.-P. Serre [**R**, pp. 212–213].

In practice, the techniques used to decompose diagrams successively into
simpler ones until one reaches those whose commutativity is axiomatic do not
seem to be too varied or difficult, suggesting that sooner or later a computer
might be trained to become a skilled assistant in this exhausting task. (For
the general idea, see e.g., [**Sm**].) Or, there might be found a theorem in

---

[4] Cf. [**H**, pp. 117–119], which takes note of the problem, but entices readers to relax
their guard so as to make feasible a hike over the seemingly solid crust of a glacier.

the vein of "coherence in categories" which would help even more.[5] Though I have been saying this publicly for a long time, I have not yet made a serious enough effort to pursue the matter, but do hope that someone else will find it worthwhile to try.

(0.4) Finally, the present exposition is incomplete in that it does not include that part of the "Ideal Theorem" of [**H**, pp. 6–7] involving concrete realizations of the twisted inverse image, particularly through differential forms. Such interpretations are clearly important for applications. Moreover, connections between different such realizations—isomorphisms forced by the uniqueness properties of the twisted inverse image—give rise to fascinating maps, such as residues, with subtle properties reflecting pseudofunctoriality and base change (see [**H**, pp. 197–199], [**L′**]).

Indeed, the theory as a whole has two complementary aspects. Without the enlivening concrete interpretations, the abstract functorial approach can be rather austere—though when it comes to treating complex relationships, it can be quite advantageous. While the theory can be based on either aspect (see e.g., [**H**] and [**C**] for the concrete foundations), bridging the concrete and abstract aspects is not a trivial matter. For a simple example (recommended as an exercise), over the category of open-and-closed immersions f, it is easily seen that the functor $f^!$ is naturally isomorphic to the inverse image functor f^*; but making this isomorphism pseudofunctorial, and proving that the flat base-change isomorphism is the "obvious one," though not difficult, requires some effort.

More generally, consider smooth maps, say with d-dimensional fibers. For such $f\colon X \to Y$, and a complex $A^\bullet$ of $\mathcal{O}_Y$-modules, there is a natural isomorphism

$$f^*A^\bullet \otimes_{\mathcal{O}_X} \Omega^d_{X/Y}[d] \xrightarrow{\ \sim\ } f^!A^\bullet$$

where $\Omega^d_{X/Y}[d]$ is the complex vanishing in all degrees except $-d$, at which it is the sheaf of relative d-forms (Kähler differentials).[6] For proper such f, where $f^!$ is right-adjoint to $\mathbf{R}f_*$, there is, correspondingly, a natural map $\int(A^\bullet)\colon \mathbf{R}f_*f^!A^\bullet \to A^\bullet$. In particular, when $Y = \mathrm{Spec}(k)$, k a field, these data give *Serre Duality*, i.e., the existence of natural isomorphisms

$$\mathrm{Hom}_k(H^i(X,F), k) \xrightarrow{\ \sim\ } \mathrm{Ext}^{d-i}_X(F, \Omega^d_{X/Y})$$

for quasi-coherent $\mathcal{O}_X$-modules F.

Pseudofunctoriality of $^!$ corresponds here to the standard isomorphism

$$\Omega^d_{X/Y} \otimes_{\mathcal{O}_X} f^*\Omega^e_{Y/Z} \xrightarrow{\ \sim\ } \Omega^{d+e}_{X/Z}$$

---

[5] Warning: see Exercise (3.4.4.1) below.

[6] A striking definition of this isomorphism was given by Verdier [**V′**, p. 397, Thm. 3]. See also [**S′**, §5.1] for a generalization to formal schemes.

attached to a pair of smooth maps $X \xrightarrow{f} Y \xrightarrow{g} Z$ of respective relative dimensions d, e. For a map $h\colon Y' \to Y$, and $p_X\colon X' := X \times_Y Y' \to X$ the projection, the abstractly defined base change isomorphism ((4.4.3) below) corresponds to the natural isomorphism

$$\Omega^d_{X'/Y'} \xrightarrow{\sim} p_X^* \Omega^d_{X/Y}.$$

The proofs of these down-to-earth statements are not easy, and will not appear in these notes.

Thus, there is a *canonical* dualizing pair $(f^!, \smallint\colon \mathbf{R}f_* f^! \to 1)$ when f is smooth; and there are explicit descriptions of its basic properties in terms of differential forms. But it is not at all clear that there is a canonical such pair for all f, let alone one which restricts to the preceding one on smooth maps. At the (homology) level of dualizing sheaves the case of varieties over a fixed perfect field is dealt with in [**Lp**, §10], and this treatment is generalized in [**HS**, §4] to generically smooth equidimensional maps of noetherian schemes without embedded components.

All these facts should fit into a general theory of the *fundamental class* of an arbitrary separated finite-type flat map $f\colon X \to Y$ with d-dimensional fibers, a canonical derived-category map $\Omega^d_{X/Y}[d] \to f^! \mathcal{O}_Y$ which globalizes the local residue map, and expresses the basic relation between differentials and duality. It is hoped that a "Residue Theorem" dealing with these questions in full generality will appear not too many years after these notes do.

Chapter 1
Derived and Triangulated Categories

In this chapter we review foundational material from [**H**, Chap. 1][1] (see also [**De**, §1]) insofar as seems necessary for understanding what follows. The main points are summarized in (1.9.1).

Why derived categories? We postulate an interest in various homology objects and their functorial behavior. Homology is defined by means of complexes in appropriate abelian categories; and we can often best understand relations among homology objects as shadows of relations among their defining complexes. Derived categories provide a supple framework for doing so.

To construct the derived category $\mathbf{D}(\mathcal{A})$ of an abelian category $\mathcal{A}$, we begin with the category $\mathbf{C} = \mathbf{C}(\mathcal{A})$ of complexes in $\mathcal{A}$. Being interested basically in homology, we do not want to distinguish between homotopic maps of complexes; and we want to consider a morphism of complexes which induces homology isomorphisms (i.e., a *quasi-isomorphism*) to be an "equivalence" of complexes. So force these two considerations on $\mathbf{C}$: first *factor out* the homotopy-equivalence relation to get the category $\mathbf{K}(\mathcal{A})$ whose objects are those of $\mathbf{C}$ but whose morphisms are homotopy classes of maps of complexes; and then *localize* by formally adjoining an inverse morphism for each quasi-isomorphism. The resulting category is $\mathbf{D}(\mathcal{A})$, see §1.2 below. The category $\mathbf{D}(\mathcal{A})$ is no longer abelian; but it carries a supplementary structure given by *triangles,* which take the place of, and are functorially better-behaved than, exact sequence of complexes, see 1.4, 1.5.[2]

Restricting attention to complexes which are bounded (above, below, or both), or whose homology is bounded, or whose homology groups lie in some plump subcategory of $\mathcal{A}$, we obtain corresponding derived categories, all

---

[1] An expansion of some of [**V**], for which [**Do**] offers some motivation. See the historical notes in [**N′**, pp. 70–71]. See also [**I′**]. Some details omitted in [**H**] can be found in more recent exposés such as [**Gl**], [**Iv**, Chapter XI], [**KS**, Chapter I], [**W**, Chapter 10], [**N′**, Chapters 1 and 2], and [**Sm**].

[2] All these constructs are *Verdier quotients* with respect to the triangulated subcategory of $\mathbf{K}(\mathcal{A})$ whose objects are the exact complexes, see [**N′**, p. 74, 2.1.8].

J. Lipman, M. Hashimoto, *Foundations of Grothendieck Duality for Diagrams of Schemes*, Lecture Notes in Mathematics 1960,
© Springer-Verlag Berlin Heidelberg 2009

of which are in fact isomorphic to full triangulated subcategories of $\mathbf{D}(\mathcal{A})$, see 1.6, 1.7, and 1.9.

In 1.8 we describe some equivalences among derived categories. For example, any choice of injective resolutions, one for each homologically bounded-below complex, gives a triangle-preserving equivalence from the derived category of such complexes to its full subcategory whose objects are bounded-below injective complexes (and whose morphisms can be identified with homotopy-equivalence classes of maps of complexes). Similarly, any choice of flat resolutions gives a triangle-preserving equivalence from the derived category of homologically bounded-above complexes to its full subcategory whose objects are bounded-above flat complexes. (For flat complexes, however, quasi-isomorphisms need not have homotopy inverses). Such equivalences are useful, for example, in treating derived functors, also for unbounded complexes, see Chapter 2.

The truncation functors of 1.10 and the "way-out" lemmas of 1.11 supply repeatedly useful techniques for working with derived categories and functors. These two sections may well be skipped until needed.

1.1 The Homotopy Category K

Let $\mathcal{A}$ be an abelian category [**M**, p. 194]. $\mathbf{K} = \mathbf{K}(\mathcal{A})$ denotes the additive category [**M**, p. 192] whose objects are complexes of objects in $\mathcal{A}$:

$$C^{\bullet} \qquad \cdots \to C^{n-1} \xrightarrow{d^{n-1}} C^n \xrightarrow{d^n} C^{n+1} \to \cdots \qquad (n \in \mathbb{Z},\ d^n \circ d^{n-1} = 0)$$

and whose morphisms are homotopy-equivalence classes of maps of complexes [**H**, p. 25]. (The maps d^n are called the *differentials* in $C^{\bullet}$.)

We always assume that $\mathcal{A}$ comes equipped with a *specific choice* of the zero-object, of a kernel and cokernel for each map, and of a direct sum for any two objects. Nevertheless we will often abuse notation by allowing the symbol 0 to stand for *any* initial object in $\mathcal{A}$; thus for $A \in \mathcal{A}$, $A = 0$ means only that A is isomorphic to the zero-object.

For a complex $C^{\bullet}$ as above, since $d^n \circ d^{n-1} = 0$ therefore d^{n-1} induces a natural map

$$C^{n-1} \to (\text{kernel of } d^n),$$

the cokernel of which is defined to be the homology $H^n(C^{\bullet})$. A map of complexes $u \colon A^{\bullet} \to B^{\bullet}$ obviously induces maps

$$H^n(u) \colon H^n(A^{\bullet}) \to H^n(B^{\bullet}) \qquad (n \in \mathbb{Z}),$$

and these maps depend only on the homotopy class of u. Thus we have a family of functors

$$H^n \colon \mathbf{K} \to \mathcal{A} \qquad (n \in \mathbb{Z}).$$

We say that u (or its homotopy class $\bar{u}$, which is a morphism in $\mathbf{K}$) is a *quasi-isomorphism* if for every $n \in \mathbb{Z}$, the map $H^n(u) = H^n(\bar{u})$ is an isomorphism.

1.2 The Derived Category D

The *derived category* $\mathbf{D} = \mathbf{D}(\mathcal{A})$ is the category whose objects are the same as those of $\mathbf{K}$, but in which each morphism $A^\bullet \to B^\bullet$ is the equivalence class f/s of a pair (s, f)

$$A^\bullet \xleftarrow{s} C^\bullet \xrightarrow{f} B^\bullet$$

of morphisms in $\mathbf{K}$, with s a quasi-isomorphism, where two such pairs (s, f), (s', f') are equivalent if there is a third such pair (s'', f'') and a commutative diagram in $\mathbf{K}$:

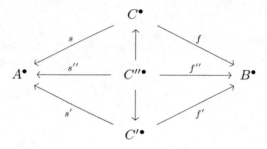

see [**H**, p. 30]. The composition of two morphisms $f/s\colon A^\bullet \to B^\bullet$, $f'/s'\colon B^\bullet \to B'^\bullet$, is $f'g/st$, where (t, g) is a pair (which always exists) such that $ft = s'g$, see [**H**, pp. 30–31, 35–36]:

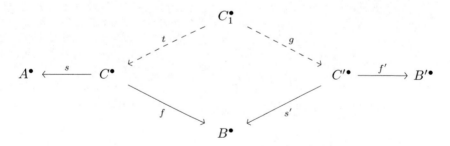

In particular, with (s, f) as above and $1_{C^\bullet}$ the homotopy class of the identity map of $C^\bullet$, we have

$$f/s = (f/1_{C^\bullet}) \circ (1_{C^\bullet}/s) = (f/1_{C^\bullet}) \circ (s/1_{C^\bullet})^{-1}.$$

There is a natural functor $Q\colon \mathbf{K} \to \mathbf{D}$ with $Q(A^\bullet) = A^\bullet$ for each complex $A^\bullet$ in $\mathbf{K}$ and $Q(f) = f/1_{A^\bullet}$ for each map $f\colon A^\bullet \to B^\bullet$ in $\mathbf{K}$. If f is a quasi-isomorphism then $Q(f) = f/1_{A^\bullet}$ is an isomorphism (with inverse $1_{A^\bullet}/f$); and in this respect, Q is universal: *any functor $Q'\colon \mathbf{K} \to \mathbf{E}$ taking quasi-isomorphisms to isomorphisms factors uniquely via Q,* i.e., there is a unique functor $\widetilde{Q}'\colon \mathbf{D} \to \mathbf{E}$ such that $Q' = \widetilde{Q}' \circ Q$ (so that $\widetilde{Q}'(A^\bullet) = Q'(A^\bullet)$ and $\widetilde{Q}'(f/s) = Q'(f) \circ Q'(s)^{-1}$).

This characterizes the pair $(\mathbf{D}, Q)$ up to canonical isomorphism.[3]

Moreover [**H**, p. 33, Prop. 3.4]: *any morphism $Q'_1 \to Q'_2$ of such functors extends uniquely to a morphism $\widetilde{Q}'_1 \to \widetilde{Q}'_2$.* In other words, composition with Q gives, for any category $\mathbf{E}$, an isomorphism of the functor category $\mathbf{Hom}(\mathbf{D}, \mathbf{E})$ onto the full subcategory of $\mathbf{Hom}(\mathbf{K}, \mathbf{E})$ whose objects are the functors $\mathbf{K} \to \mathbf{E}$ which transform quasi-isomorphisms in $\mathbf{K}$ into isomorphisms in $\mathbf{E}$.

One checks that the category $\mathbf{D}$ supports a *unique additive structure* such that the canonical functor $Q\colon \mathbf{K} \to \mathbf{D}$ is additive; and accordingly we will always regard $\mathbf{D}$ as an additive category. If the category $\mathbf{E}$ and the above functor $Q'\colon \mathbf{K} \to \mathbf{E}$ are both additive, then so is $\widetilde{Q}'$.

Remark 1.2.1. The homology functors $H^n\colon \mathbf{K} \to \mathcal{A}$ defined in (1.1) transform quasi-isomorphisms into isomorphisms, and hence may be regarded as functors on $\mathbf{D}$.

(1.2.2). *A morphism $f/s\colon A^\bullet \to B^\bullet$ in $\mathbf{D}$ is an isomorphism if and only if*

$$H^n(f/s) = H^n(f) \circ H^n(s)^{-1}\colon H^n(A^\bullet) \to H^n(B^\bullet)$$

is an isomorphism for all $n \in \mathbb{Z}$.

Indeed, if $H^n(f/s)$ is an isomorphism for all n, then so is $H^n(f)$, i.e., f is a quasi-isomorphism; and then s/f is the inverse of f/s.

(1.2.3). *There is an isomorphism of $\mathcal{A}$ onto a full subcategory of $\mathbf{D}$, taking any object $X \in \mathcal{A}$ to the complex $X^\bullet$ which is X in degree zero and 0 elsewhere, and taking a map $f\colon X \to Y$ in $\mathcal{A}$ to $f^\bullet/1_{X^\bullet}$, where $f^\bullet\colon X^\bullet \to Y^\bullet$ is the homotopy class whose sole member is the map of complexes which is f in degree zero.*

Bijectivity of the indicated map $\mathrm{Hom}_{\mathcal{A}}(X, Y) \to \mathrm{Hom}_{\mathbf{D}(\mathcal{A})}(X^\bullet, Y^\bullet)$ is a straightforward consequence of the existence of a natural functorial isomorphism $Z \xrightarrow{\sim} H^0(Z^\bullet)$ $(Z \in \mathcal{A})$.

---

[3] The set Σ of quasi-isomorphisms in $\mathbf{K}$ admits a calculus of left and of right fractions, and $\mathbf{D}$ is, up to canonical isomorphism, the *category of fractions* $\mathbf{K}[\Sigma^{-1}]$, see e.g., [**Sc**, Chapter 19.] The set-theoretic questions arising from the possibility that Σ is "too large," i.e., a class rather than a set, are dealt with in *loc. cit.* Moreover, there is often a construction of a universal pair $(\mathbf{D}, Q)$ which gets around such questions (but may need the axiom of choice), cf. (2.3.2.2) and (2.3.5) below.

1.3 Mapping Cones

An important construction is that of the *mapping cone* $C_u^\bullet$ of a map of complexes $u\colon A^\bullet \to B^\bullet$ in $\mathcal{A}$. (For this construction we need only assume that the category $\mathcal{A}$ is additive.) $C_u^\bullet$ is the complex whose degree n component is

$$C_u^n = B^n \oplus A^{n+1}$$

and whose differentials $d^n\colon C_u^n \to C_u^{n+1}$ satisfy

$$d^n|_{B^n} = d_B^n, \qquad d^n|_{A^{n+1}} = u|_{A^{n+1}} - d_A^{n+1} \qquad (n \in \mathbb{Z})$$

where the vertical bars denote "restricted to," and d_B, d_A are the differentials in $B^\bullet$, $A^\bullet$ respectively.

$$
\begin{array}{ccccc}
C_u^{n+1} & = & B^{n+1} & \oplus & A^{n+2} \\
d\uparrow & & d_B\uparrow & \ \nwarrow u & \uparrow{-d_A} \\
C_u^n & = & B^n & \oplus & A^{n+1}
\end{array}
$$

For any complex $A^\bullet$, and $m \in \mathbb{Z}$, $A^\bullet[m]$ will denote the complex having degree-n component

$$(A^\bullet[m])^n = A^{n+m} \qquad (n \in \mathbb{Z})$$

and in which the differentials $A^n[m] \to A^{n+1}[m]$ are $(-1)^m$ times the corresponding differentials $A^{n+m} \to A^{n+m+1}$ in $A^\bullet$. There is a natural "translation" functor T from the category of $\mathcal{A}$-complexes into itself satisfying $TA^\bullet = A^\bullet[1]$ for all complexes $A^\bullet$.

To any map u as above, we can then associate the sequence of maps of complexes

$$A^\bullet \xrightarrow{u} B^\bullet \xrightarrow{v} C_u^\bullet \xrightarrow{w} A^\bullet[1] \tag{1.3.1}$$

where v (resp. w) is the natural inclusion (resp. projection) map. The sequence (1.3.1) could also be represented in the form

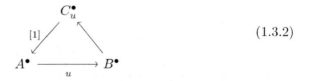

$$\tag{1.3.2}$$

and so we call such a sequence a *standard triangle*.

A commutative diagram of maps of complexes

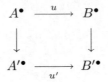

gives rise naturally to a commutative diagram of associated g triangles (each arrow representing a map of complexes):

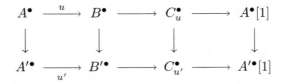

Most of the basic properties of standard triangles involve homotopy, and so are best stated in $\mathbf{K}(\mathcal{A})$. For example, the mapping cone $C_1^\bullet$ of the identity map $A^\bullet \to A^\bullet$ is homotopically equivalent to zero, a homotopy between the identity map of $C_1^\bullet$ and the zero map being as indicated:

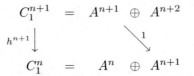

(i.e., for each $n \in \mathbb{Z}$, h^{n+1} restricts to the identity on A^{n+1} and to 0 on A^{n+2}; and $d^{n-1}h^n + h^{n+1}d^n$ is the identity of C_1^n). Other properties can be found e.g., in [**Bo**, pp. 102–105], [**Iv**, pp. 22–33]. For subsequent developments we need to axiomatize them, as follows.

1.4 Triangulated Categories (Δ-Categories)

A *triangulation* on an arbitrary additive category $\mathbf{K}$ consists of an additive automorphism T (the *translation functor*) of $\mathbf{K}$, and a collection $\mathcal{T}$ of diagrams of the form

$$A \xrightarrow{u} B \xrightarrow{v} C \xrightarrow{w} TA. \tag{1.4.1}$$

A *triangle* (*with base u and summit C*) is a diagram (1.4.1) in $\mathcal{T}$. (See (1.3.2) for a more picturesque—but typographically less convenient—representation of a triangle.) The following conditions are required to hold:

(Δ1)′   *Every diagram of the following form is a triangle:*

$$A \xrightarrow{\text{identity}} A \longrightarrow 0 \longrightarrow TA.$$

(Δ1)″   *Given a commutative diagram*

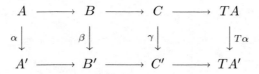

if α, β, γ are all isomorphisms and the top row is a triangle then the bottom row is a triangle.

(Δ2)   *For any triangle (1.4.1) consider the corresponding infinite diagram* (1.4.1)∞:

$$\cdots \longrightarrow T^{-1}C \xrightarrow{-T^{-1}w} A \xrightarrow{u} B \xrightarrow{v} C \xrightarrow{w} TA \xrightarrow{-Tu} TB \longrightarrow \cdots$$

in which every arrow is obtained from the third preceding one by applying −T. Then any three successive maps in (1.4.1)∞ *form a triangle.*

(Δ3)′   *Any morphism $A \xrightarrow{u} B$ in* **K** *is the base of a triangle (1.4.1).*

(Δ3)″   *For any diagram*

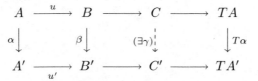

whose rows are triangles, and with maps α, β given such that βu = u′α, there exists a morphism γ: C → C′ making the entire diagram commute, i.e., making it a morphism of triangles.[4]

---

[4] (Δ3)″ is implied by a stronger "octahedral" axiom, which states that for a composition $A \xrightarrow{u} B \xrightarrow{\beta} B'$ and triangles Δ_u, $\Delta_{\beta u}$, Δ_β with respective bases u, βu, β, there exist morphisms of triangles $\Delta_u \to \Delta_{\beta u} \to \Delta_\beta$ extending the diagram

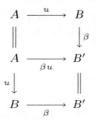

and such that the induced maps on summits $C_u \to C_{\beta u} \to C_\beta$ are themselves the sides of a triangle, whose third side is the composed map $C_\beta \to TB \to TC_u$. This axiom is incompletely stated in [**H**, p. 21], see [**V**, p. 3] or [**Iv**, pp. 453–455]. We omit it here because it plays no role in these notes (nor, as far as I can tell, in [**H**]). Thus the adjective "pre-triangulated" may be substituted for "triangulated" throughout, see [**N′**, p. 51, Definition 1.3.13 and p. 60, Remark 1.4.7].

As a consequence of these conditions we have [**H**, p. 23, Prop. 1.1 c]:

($\Delta 3$)* *If in ($\Delta 3$)'' both α and β are isomorphisms, then so is γ.*
Thus, and by ($\Delta 3$)':

Every morphism $A \xrightarrow{u} B$ *is the base of a triangle, uniquely determined up to isomorphism by u.*

Definition 1.4.2. A *triangulated category* (Δ-category for short) is an additive category together with a triangulation.

Exercise 1.4.2.1. (Cf. [**N'**, pp. 42–45].) For any triangle

$$A \xrightarrow{u} B \xrightarrow{v} C \xrightarrow{w} TA$$

in a Δ-category **K**, and any object M, the induced sequence of abelian groups

$$\mathrm{Hom}(M, A) \to \mathrm{Hom}(M, B) \to \mathrm{Hom}(M, C)$$

is *exact* [**H**, p. 23, 1.1 b)]. Using this and ($\Delta 2$) (or otherwise), show that *u is an isomorphism iff $C \cong 0$*. More generally, *the following conditions are equivalent:*
  (a) u is a monomorphism.
  (b) v is an epimorphism.
  (c) $w = 0$.
  (d) There exist maps $A \xleftarrow{t} B \xleftarrow{s} C$ such that

$$1_A = tu, \quad 1_B = sv + ut, \quad 1_C = vs$$

(so that $B \cong A \oplus C$).
    Consequently, in view of ($\Delta 3$)', any monomorphism in **K** has a left inverse and any epimorphism has a right inverse. And incidentally, the existence of finite direct sums in **K** follows from the other axioms of Δ-categories.

Examples 1.4.3. For any abelian (or just additive) category $\mathcal{A}$, the homotopy category $\mathbf{K} := \mathbf{K}(\mathcal{A})$ of (1.1) has a triangulation, with translation T such that

$$TA^\bullet = A^\bullet[1] \qquad (A^\bullet \in \mathbf{K})$$

(i.e., T is induced by the translation functor on complexes, see (1.3), a functor which respects homotopy), and with triangles all those diagrams (1.4.1) which are isomorphic (in the obvious sense, see ($\Delta 3$)*) to the image in **K** of some standard triangle, see (1.3) again. The properties ($\Delta 1$)', ($\Delta 1$)'', and ($\Delta 3$)' follow at once from the discussion in (1.3). To prove ($\Delta 3$)'' we may assume that $C = C_u^\bullet$, $C' = C_{u'}^\bullet$, and the rows of the diagram are standard triangles. By assumption, βu is homotopic to $u'\alpha$, i.e., there is a family of maps $h^n \colon A^n \to B'^{n-1}$ $(n \in \mathbb{Z})$ such that

$$\beta^n u^n - u'^n \alpha^n = d_{B'}^{n-1} h^n + h^{n+1} d_A^n \,.$$

Define γ by the family of maps

$$\gamma^n \colon C^n = B^n \oplus A^{n+1} \longrightarrow B'^n \oplus A'^{n+1} = C'^n \qquad (n \in \mathbb{Z})$$

such that for $b \in B^n$ and $a \in A^{n+1}$,

$$\gamma^n(b,a) = \left(\beta^n(b) + h^{n+1}(a),\, \alpha^{n+1}(a)\right),$$

and then check that γ is as desired.

For establishing the remaining property ($\Delta 2$), we recall some facts about cylinders of maps of complexes (see e.g., [**B**, §2.6]—modulo sign changes leading to isomorphic complexes).

Let $u \colon A^\bullet \to B^\bullet$ be a map of complexes, and let $w \colon C_u^\bullet \to A^\bullet[1]$ be the natural map, see §1.3. We define the *cylinder of u*, $\widetilde{C}_u^\bullet$, to be the complex

$$\widetilde{C}_u^\bullet := C_w^\bullet[-1].$$

($\widetilde{C}_u^\bullet$ is also the cone of the map $(-1, u) \colon A \to A \oplus B$.) One checks that there is a map of complexes $\varphi \colon \widetilde{C}_u^\bullet \to B^\bullet$ given in degree n by the map

$$\varphi^n \colon \widetilde{C}_u^n = A^n \oplus B^n \oplus A^{n+1} \to B^n$$

such that

$$\varphi^n(a, b, a') = u(a) + b.$$

The map φ is a *homotopy equivalence,* with homotopy inverse ψ given in degree n by

$$\psi^n(b) = (0, b, 0).$$

[If $d^n \colon \widetilde{C}_u^n \to \widetilde{C}_u^{n+1}$ is the differential and $h^{n+1} \colon \widetilde{C}_u^{n+1} \to \widetilde{C}_u^n$ is given by $h^{n+1}(a, b, a') = (0, 0, -a)$, then $1_{\widetilde{C}_u^n} - \psi^n\varphi^n = d^{n-1}h^n + h^{n+1}d^n \ldots$]

There results a diagram of maps of complexes

$$
\begin{array}{ccccccc}
A^\bullet & \xrightarrow{\ \tilde{u}\ } & \widetilde{C}_u^\bullet & \xrightarrow{\ \tilde{v}\ } & C_u^\bullet & \xrightarrow{\ w\ } & A^\bullet[1] \\
\Big\| & & \varphi\Big\downarrow & & \Big\| & & \Big\| \\
A^\bullet & \xrightarrow[u]{} & B^\bullet & \xrightarrow[v = \tilde{v}\psi]{} & C_u^\bullet & \xrightarrow[w]{} & A^\bullet[1]
\end{array}
\qquad (1.4.3.1)
$$

in which $\tilde{u}$ and $\tilde{v}$ are the natural maps, the bottom row is a standard triangle, the two outer squares commute, and the middle square is *homotopy-commutative,* i.e., $\tilde{v} - v\varphi = \tilde{v}(1 - \psi\varphi)$ is homotopic to 0.

Now, (1.4.3.1) implies that the diagram

$$C_u^\bullet[-1] \xrightarrow{-w[-1]} A^\bullet \xrightarrow{u} B^\bullet \xrightarrow{v} C_u^\bullet$$

is isomorphic in $\mathbf{K}$ to the diagram

$$C_u^\bullet[-1] \xrightarrow{-w[-1]} A^\bullet \xrightarrow{\tilde{u}} \widetilde{C}_u^\bullet \xrightarrow{\tilde{v}} C_u^\bullet$$

which is a standard triangle, since $\widetilde{C}_u^\bullet = C_w^\bullet[-1] = C_{-w[-1]}^\bullet$.

Hence if

$$A^\bullet \xrightarrow{u'} B^\bullet \xrightarrow{v'} C^\bullet \xrightarrow{w'} A^\bullet[1]$$

is any triangle in $\mathbf{K}$, then

$$C^\bullet[-1] \xrightarrow{-w'[-1]} A^\bullet \xrightarrow{u'} B^\bullet \xrightarrow{v'} C^\bullet$$

is a triangle, and—by the same reasoning—so is

$$B^\bullet[-1] \xrightarrow{-v'[-1]} C^\bullet[-1] \xrightarrow{-w'[-1]} A^\bullet \xrightarrow{u'} B^\bullet,$$

and consequently so is

$$B^\bullet \xrightarrow{v'} C^\bullet \xrightarrow{-w'} A^\bullet[1] \xrightarrow{u'[1]} B^\bullet[1]$$

(because if $A^\bullet \cong C_{-v'[-1]}^\bullet = C_{v'}^\bullet[-1]$, then $A^\bullet[1] \cong C_{v'}^\bullet$), as is the isomorphic diagram

$$B^\bullet \xrightarrow{v'} C^\bullet \xrightarrow{w'} A^\bullet[1] \xrightarrow{-u'[1]} B^\bullet[1].$$

Property ($\Delta 2$) for $\mathbf{K}$ results.[5]

We will always consider $\mathbf{K}$ to be a Δ-category, with this triangulation.

There is a close relation between triangles in $\mathbf{K}$ and certain exact sequences. For any exact sequence of complexes in an abelian category $\mathcal{A}$

$$0 \longrightarrow A^\bullet \xrightarrow{u} B^\bullet \xrightarrow{v} C^\bullet \longrightarrow 0 , \qquad (1.4.3.2)$$

if u_0 is the isomorphism from $A^\bullet$ onto the kernel of v induced by u, then we have a natural exact sequence of complexes

$$0 \longrightarrow C_{u_0}^\bullet \xrightarrow{\text{inclusion}} C_u^\bullet \xrightarrow{\chi} C^\bullet \longrightarrow 0 \qquad (1.4.3.3)$$

---

[5] For other treatments of ($\Delta 2$) and ($\Delta 3$)″ see [**Bo**, pp. 102–104] or [**Iv**, p. 27, 4.16; and p. 30, 4.19]. And for the octahedral axiom, use triangle (4.22) in [**Iv**, p. 32], whose vertices are the cones of two composable maps and of their composition.

where $\chi^n \colon C_u^n \to C^n$ $(n \in \mathbb{Z})$ is the composition

$$\chi^n \colon C_u^n = B^n \oplus A^{n+1} \xrightarrow{\text{natural}} B^n \xrightarrow{v} C^n$$

(see (1.3)).

It is easily checked—either directly, or because $C_{u_0}^\bullet$ is isomorphic to the cone of the identity map of $A^\bullet$—that $H^n(C_{u_0}^\bullet) = 0$ for all n; and then from the long exact cohomology sequence associated to (1.4.3.3) we conclude that χ *is a quasi-isomorphism.*

If the exact sequence (1.4.3.2) is *semi-split,* i.e., for every $n \in \mathbb{Z}$, the restriction $v^n \colon B^n \to C^n$ of v to B^n has a left inverse, say φ^n, then with

$$\Phi^n = \varphi^n \oplus (\varphi^{n+1} d_C^n - d_B^n \varphi^n) \colon C^n \to B^n \oplus A^{n+1}$$

(where A^{n+1} is identified with $\ker(v^{n+1})$ via u), the map of complexes $\Phi := (\Phi^n)_{n \in \mathbb{Z}}$ is a *homotopy inverse* for χ: $\chi \circ \Phi$ is the identity map of $C^\bullet$, and also the map $(1_{C_u^\bullet} - \Phi \circ \chi) \colon C_u^\bullet \to \ker(\chi) = C_{u_0}^\bullet \cong 0$ vanishes in $\mathbf{K}$. [More explicitly, if $h^{n+1} \colon C_u^{n+1} \to C_u^n$ is given by

$$h^{n+1}(b, a) := b - \phi^{n+1} v^{n+1} b \in A^{n+1} \subset C_u^n \qquad (b \in B^{n+1},\ a \in A^{n+2})$$

and d is the differential in $C_u^\bullet$, then $1_{C_u^n} - \Phi^n \circ \chi^n = (d^{n-1} h^n + h^{n+1} d^n)$.] Thus χ induces a *natural isomorphism in* $\mathbf{K}$

$$C_u^\bullet \xrightarrow{\ \sim\ } C^\bullet,$$

and hence by $(\Delta 1)''$ we have a triangle

$$A^\bullet \xrightarrow{\bar{u}} B^\bullet \xrightarrow{\bar{v}} C^\bullet \xrightarrow{\bar{w}} A^\bullet[1] \tag{1.4.3.4}$$

where $\bar{u}$, $\bar{v}$ are the homotopy classes of u, v respectively, and $\bar{w}$ is the homotopy class of the composed map

$$(\varphi^{n+1} d_C^n - d_B^n \varphi^n)_{n \in \mathbb{Z}} \colon C^\bullet \xrightarrow{\Phi} C_u^\bullet \xrightarrow{\text{natural}} A^\bullet[1], \tag{1.4.3.5}$$

a class independent of the choice of splitting maps φ^n, because χ does not depend on that choice, so that neither does its inverse Φ, up to homotopy. This $\bar{w}$ is called the *homotopy invariant* of (1.4.3.2) (assumed semi-split).[6]

---

[6] The category $\mathcal{A}$ need only be additive for us to define the homotopy invariant of a semi-split sequence of complexes $A^\bullet \underset{\psi}{\overset{u}{\rightleftarrows}} B^\bullet \underset{\varphi}{\overset{v}{\rightleftarrows}} C^\bullet$ (i.e., $B^n \cong A^n \oplus C^n$ for all n, and $u^n, \psi^n, v^n, \varphi^n$ are the usual maps associated with a direct sum): it's the homotopy class of the map

$$\psi(\varphi d_C - d_B \varphi) \colon C^\bullet \to A^\bullet[1],$$

a class depending, as above, only on u and v. [More directly, note that if φ' is another family of splitting maps then

$$\psi(\varphi d_C - d_B \varphi) - \psi(\varphi' d_C - d_B \varphi') = d_{A[1]} \psi(\varphi - \varphi') + \psi(\varphi - \varphi') d_C.]$$

Moreover, *any triangle in* **K** *is isomorphic to one so obtained.*

This is shown by the image in **K** of (1.4.3.1) (in which the bottom row is any standard triangle, and the homotopy equivalence φ becomes an isomorphism) as soon as one checks that the top row is in fact of the form specified by (1.4.3.4) and (1.4.3.5).

By way of illustration here is an often used fact, whose proof involves triangles. (See also [**H**, pp. 35–36].)

Lemma 1.4.3.6. *Any diagram* $A^\bullet \xleftarrow{s} C^\bullet \xrightarrow{f} B^\bullet$ *in* **K**$(\mathcal{A})$, *with* s *a quasi-isomorphism, can be embedded in a commutative diagram*

$$
\begin{array}{ccc}
C^\bullet & \xrightarrow{\ f\ } & B^\bullet \\
{\scriptstyle s}\downarrow & & \downarrow{\scriptstyle s'} \\
A^\bullet & \xrightarrow{\ f'\ } & D^\bullet
\end{array}
\qquad (1.4.3.7)
$$

with s' *a quasi-isomorphism.*

Proof. By $(\Delta 3)'$ there exists a triangle

$$
C^\bullet \xrightarrow{(s,-f)} A^\bullet \oplus B^\bullet \longrightarrow D^\bullet \longrightarrow C^\bullet[1]. \qquad (1.4.3.8)
$$

If f' is the natural composition $A^\bullet \to A^\bullet \oplus B^\bullet \to D^\bullet$, and s' is the composition $B^\bullet \to A^\bullet \oplus B^\bullet \to D^\bullet$, then commutativity of (1.4.3.7) results from the easily-verifiable fact that the composition of the first two maps in a standard triangle is homotopic to 0.[7] And if s is a quasi-isomorphism, then from (1.4.3.8) we get exact homology sequences

$$
0 \to H^n(C^\bullet) \to H^n(A^\bullet) \oplus H^n(B^\bullet) \to H^n(D^\bullet) \to 0 \qquad (n \in \mathbb{Z})
$$

(see (1.4.5) below) which quickly yield that s' is a quasi-isomorphism too.

Examples 1.4.4. The above triangulation on **K** leads naturally to one on the derived category **D** of 1.2. The translation functor $\widetilde{T}$ is determined by the relation $QT = \widetilde{T}Q$, where $Q\colon \mathbf{K} \to \mathbf{D}$ is the canonical functor, and T is the translation functor in **K** (see (1.4.3)): note that QT transforms quasi-isomorphisms into isomorphisms, and use the universal property of Q given in 1.2. In particular $\widetilde{T}(A^\bullet) = A^\bullet[1]$ for every complex $A^\bullet \in \mathbf{D}$. ($\widetilde{T}$ is additive, by the remarks just before (1.2.1).) The triangles are those diagrams which are isomorphic—in the obvious sense, see $(\Delta 3)^*$—to those coming from **K** via Q, i.e., diagrams isomorphic to natural images of standard triangles.

Conditions $(\Delta 1)'$, $(\Delta 1)''$, and $(\Delta 2)$ are easily checked.

---

[7] In fact in any Δ-category, any two successive maps in a triangle compose to 0 [**H**, p. 23, Prop. 1.1 a)].

Next, given $f/s\colon A^\bullet \to B^\bullet$ in **D**, represented by $A^\bullet \xleftarrow{s} X^\bullet \xrightarrow{f} B^\bullet$ in **K** (see 1.2), we have, by (Δ3)′ for **K**, a triangle $X^\bullet \xrightarrow{f} B^\bullet \xrightarrow{g} C^\bullet \xrightarrow{h} X^\bullet[1]$ in **K**, whose image is the top row of a commutative diagram in **D**, as follows:

$$
\begin{array}{ccccccc}
X^\bullet & \xrightarrow{Q(f)} & B^\bullet & \xrightarrow{Q(g)} & C^\bullet & \xrightarrow{Q(h)} & X^\bullet[1] \\
Q(s)\Big\downarrow \simeq & & \Big\| & & \Big\| & & \simeq \Big\downarrow \tilde{T}Q(s) \\
A^\bullet & \xrightarrow{f/s} & B^\bullet & \longrightarrow & C^\bullet & \longrightarrow & X^\bullet[1]
\end{array}
\qquad (1.4.4.1)
$$

Condition (Δ3)′ for **D** results. As for (Δ3)″, we can assume, via isomorphisms, that the rows of the diagram in question come from **K**, via Q. Then we check via definitions in 1.2 that the commutative diagram

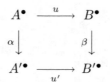

in **D** can be expanded to a commutative diagram of the form

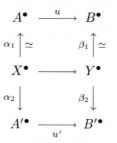

(i.e., $\alpha = \alpha_2\alpha_1^{-1}$, $\beta = \beta_2\beta_1^{-1}$), where all the arrows represent maps coming from **K**, i.e., maps of the form $Q(f)$. By (Δ3)′ and (Δ3)″ for **K**, this diagram embeds into a larger commutative one whose middle row also comes from **K**:

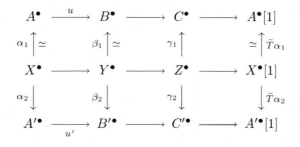

Using (1.2.2) and the exact homology sequences associated to the top two rows (see (1.4.5) below), we find that γ_1 is an isomorphism. Then $\gamma := \gamma_2 \gamma_1^{-1}$ fulfills $(\Delta 3)''$.

So we have indeed defined a triangulation; and from $(\Delta 1)''$, $(\Delta 3)^*$, and (1.4.4.1) we conclude that this is the *unique* triangulation on $\mathbf{D}$ with translation $\tilde{T}$ and such that Q transforms triangles into triangles.

We will always consider $\mathbf{D}$ to be a Δ-category, with this triangulation.

Now for any exact sequence of complexes in $\mathcal{A}$

$$0 \longrightarrow A^\bullet \overset{u}{\longrightarrow} B^\bullet \overset{v}{\longrightarrow} C^\bullet \longrightarrow 0 \qquad\qquad (1.4.4.2)$$

the quasi-isomorphism χ of (1.4.3.3) becomes an isomorphism $\tilde{\chi}$ in $\mathbf{D}$, so that in $\mathbf{D}$ there is a natural composed map

$$\tilde{w} \colon C^\bullet \overset{\tilde{\chi}^{-1}}{\longrightarrow} C_u^\bullet \longrightarrow A^\bullet[1] \,;$$

and then with $\tilde{u}$ and $\tilde{v}$ corresponding to u and v respectively, the diagram

$$A^\bullet \overset{\tilde{u}}{\longrightarrow} B^\bullet \overset{\tilde{v}}{\longrightarrow} C^\bullet \overset{\tilde{w}}{\longrightarrow} A^\bullet[1] \qquad\qquad (1.4.4.2)^\sim$$

is a triangle in $\mathbf{D}$. If the sequence (1.4.4.2) is semi-split, then $(1.4.4.2)^\sim$ is the image in $\mathbf{D}$ of the triangle (1.4.3.4) in $\mathbf{K}$. Since every triangle in $\mathbf{K}$ is isomorphic to one coming from a semi-split exact sequence (see end of example (1.4.3)), therefore *every triangle in $\mathbf{D}$ is isomorphic to one of the form $(1.4.4.2)^\sim$ arising from an exact sequence of complexes in $\mathcal{A}$ (in fact, from a semi-split such sequence).*

(1.4.5). To any triangle $A^\bullet \overset{u}{\longrightarrow} B^\bullet \overset{v}{\longrightarrow} C^\bullet \overset{w}{\longrightarrow} A^\bullet[1]$ in $\mathbf{K}$ or in $\mathbf{D}$, we can apply the homology functors H^n (see (1.2.1)) to obtain an associated *exact homology sequence*

$$\cdots \longrightarrow H^{i-1}(C^\bullet) \overset{H^{i-1}(w)}{\longrightarrow} H^i(A^\bullet) \overset{H^i(u)}{\longrightarrow} H^i(B^\bullet)$$

$$\overset{H^i(v)}{\longrightarrow} H^i(C^\bullet) \overset{H^i(w)}{\longrightarrow} H^{i+1}(A^\bullet) \longrightarrow \cdots$$
$$(1.4.5)^{\mathrm{H}}$$

Exactness is verified by reduction to the case of standard triangles.

For an exact sequence (1.4.4.2), the usual connecting homomorphism

$$H^i(C^\bullet) \to H^{i+1}(A^\bullet) \qquad (i \in \mathbb{Z})$$

is easily seen to be $-H^i(\tilde{w})$ (see $(1.4.4.2)^\sim$). Thus $(1.4.5)^{\mathrm{H}}$ (for $(1.4.4.2)^\sim$) is, except for signs, the usual homology sequence associated to (1.4.4.2).

It should now be clear why it is that we can replace exact sequences of complexes in $\mathcal{A}$ by triangles in $\mathbf{D}$. And the following notion of "Δ-functor" will eventually make it quite advantageous to do so.

1.5 Triangle-Preserving Functors (Δ-Functors)

Let $\mathbf{K_1}$, $\mathbf{K_2}$ be Δ-categories (1.4.2) with translation functors T_1, T_2 respectively. A (covariant) Δ-*functor* is defined to be a pair (F, θ) consisting of an additive functor $F\colon \mathbf{K_1} \to \mathbf{K_2}$ together with an isomorphism of functors

$$\theta\colon FT_1 \xrightarrow{\ \sim\ } T_2 F$$

such that for every triangle

$$A \xrightarrow{\ u\ } B \xrightarrow{\ v\ } C \xrightarrow{\ w\ } T_1 A$$

in $\mathbf{K_1}$, the corresponding diagram

$$FA \xrightarrow{\ Fu\ } FB \xrightarrow{\ Fu\ } FC \xrightarrow{\ \theta \circ Fw\ } T_2 FA$$

is a triangle in $\mathbf{K_2}$.

These are the exact functors of [$\mathbf{V}$, p. 4], and also the ∂-functors of [$\mathbf{H}$, p. 22]; it should be kept in mind however that θ is not always the identity transformation (see Examples (1.5.3), (1.5.4) below—but see also Exercise (1.5.5)). In practice, for given F if there is *some* θ such that (F, θ) is a Δ-functor then there will usually be a natural one, and after specifying such a θ we will simply say (abusing language) that F is a Δ-functor.

Let $\mathbf{K_3}$ be a third Δ-category, with translation T_3. If each one of $(F, \theta)\colon \mathbf{K_1} \to \mathbf{K_2}$ and $(H, \chi)\colon \mathbf{K_2} \to \mathbf{K_3}$ is a Δ-functor, then so is

$$(H \circ F,\ \chi \circ \theta)\colon \mathbf{K_1} \to \mathbf{K_3}$$

where $\chi \circ \theta$ is defined to be the composition

$$HFT_1 \xrightarrow{\ \text{via } \theta\ } HT_2 F \xrightarrow{\ \text{via } \chi\ } T_3 H F\,.$$

A *morphism* $\eta\colon (F, \theta) \to (G, \psi)$ of Δ-*functors* (from $\mathbf{K_1}$ to $\mathbf{K_2}$) is a morphism of functors $\eta\colon F \to G$ such that for all objects X in $\mathbf{K_1}$, the following diagram commutes:

$$
\begin{array}{ccc}
FT_1(X) & \xrightarrow{\ \theta(X)\ } & T_2 F(X) \\
{\scriptstyle \eta(T_1(X))}\big\downarrow & & \big\downarrow{\scriptstyle T_2(\eta(X))} \\
GT_1(X) & \xrightarrow[\ \psi(X)\ ]{} & T_2 G(X)
\end{array}
$$

The set of all such η can be made, in an obvious way, into an abelian group. If $\mu\colon (G, \psi) \to (G', \psi')$ is also a morphism of Δ-functors, then so is the composition $\mu\eta\colon (F, \theta) \to (G', \psi')$. And if $(H, \chi)\colon \mathbf{K_2} \to \mathbf{K_3}$ [respectively $(H', \chi')\colon \mathbf{K_3} \to \mathbf{K_1}$] is, as above, another Δ-functor then η naturally induces a morphism of composed Δ-functors

$$(H \circ F, \chi \circ \theta) \to (H \circ G, \chi \circ \psi)$$
$$[\text{respectively } (F \circ H', \theta \circ \chi') \to (G \circ H', \psi \circ \chi')].$$

We find then that:

Proposition. *The Δ-functors from $\mathbf{K_1}$ to $\mathbf{K_2}$, and their morphisms, form an additive category $\mathbf{Hom}_\Delta(\mathbf{K_1}, \mathbf{K_2})$; and the composition operation*

$$\mathbf{Hom}_\Delta(\mathbf{K_1}, \mathbf{K_2}) \times \mathbf{Hom}_\Delta(\mathbf{K_2}, \mathbf{K_3}) \longrightarrow \mathbf{Hom}_\Delta(\mathbf{K_1}, \mathbf{K_3})$$

is a biadditive functor.

A morphism η as above has an inverse in $\mathbf{Hom}_\Delta(\mathbf{K_1}, \mathbf{K_2})$ if and only if $\eta(X)$ is an isomorphism in $\mathbf{K_2}$ for every $X \in \mathbf{K_1}$. We call such an η a Δ-*functorial isomorphism.*

Similarly, a *contravariant Δ-functor* is a pair (F, θ) with $F \colon \mathbf{K_1} \to \mathbf{K_2}$ a contravariant additive functor and

$$\theta \colon T_2^{-1} F \xrightarrow{\ \sim\ } F T_1$$

an isomorphism of functors such that for every triangle in $\mathbf{K_1}$ as above, the corresponding diagram

$$FA \xleftarrow{Fu} FB \xleftarrow{Fv} FC \xleftarrow{-Fw \circ \theta} T_2^{-1} FA$$

is a triangle in $\mathbf{K_2}$. Composition and morphisms etc. of contravariant Δ-functors are introduced in the obvious way.

Exercise. A contravariant Δ-functor is the same thing as a covariant Δ-functor on the opposite (dual) category $\mathbf{K_1^{op}}$ [**M**, p. 33], suitably triangulated. (For example, $\mathbf{D}(\mathcal{A})^{\mathrm{op}}$ is Δ-isomorphic to $\mathbf{D}(\mathcal{A}^{\mathrm{op}})$, see (1.4.4).)

Examples. (1.5.1). (see [**H**, p. 33, Prop. 3.4]). By (1.4.4), the natural functor $Q \colon \mathbf{K} \to \mathbf{D}$ of §1.2, together with $\theta = $ identity, is a Δ-functor. Moreover, as in 1.2: *composition with Q gives, for any Δ-category $\mathbf{E}$, an isomorphism of the category of Δ-functors $\mathbf{Hom}_\Delta(\mathbf{D}, \mathbf{E})$ onto the full subcategory of $\mathbf{Hom}_\Delta(\mathbf{K}, \mathbf{E})$ whose objects are the Δ-functors (F, θ) such that F transforms quasi-isomorphisms in $\mathbf{K}$ to isomorphisms in $\mathbf{E}$.*[8]

---

[8] *Equivalently* (∗): $F(C^\bullet) \cong 0$ *for every exact complex* $C^\bullet \in \mathbf{K}$. ("$C^\bullet$ exact" means $H^i(C^\bullet) = 0$ for all i, i.e., the zero map $C^\bullet \to 0$ is a quasi-isomorphism). Exactness of the homology sequence $(1.4.5)^{\mathrm{H}}$ of a standard triangle shows that a map u in $\mathbf{K}$ is a quasi-isomorphism iff the cone $C_u^\bullet$ is exact. Also, the base of a triangle is an isomorphism iff the summit is 0, see (1.4.2.1). So since $F(C_u^\bullet)$ is the summit of a triangle with base $F(u)$, (∗) implies that if u is a quasi-isomorphism then $F(u)$ is an isomorphism.

(1.5.2). Let $F\colon \mathcal{A}_1 \to \mathcal{A}_2$ be an additive functor of abelian categories, and set $\mathbf{K_1} = \mathbf{K}(\mathcal{A}_1)$, $\mathbf{K_2} = \mathbf{K}(\mathcal{A}_2)$. Then F extends in an obvious way to an additive functor $\bar{F}\colon \mathbf{K_1} \to \mathbf{K_2}$ which commutes with translation, and which (together with $\theta = $ identity) is easily seen to be a Δ-functor, essentially because $\bar{F}$ takes cones to cones, i.e., for any map u of complexes in $\mathcal{A}_1$,

$$\bar{F}(C_u^\bullet) = C_{\bar{F}(u)}^\bullet. \tag{1.5.2.1}$$

(1.5.3) (expanding [**H**, p. 64, line 7] and illustrating [**De**, p. 265, Prop. 1.1.7]). For complexes $A^\bullet$, $B^\bullet$ in the abelian category $\mathcal{A}$, the complex of abelian groups $\mathrm{Hom}^\bullet(A^\bullet, B^\bullet)$ is given in degree n by

$$\mathrm{Hom}^n(A^\bullet, B^\bullet) = \mathrm{Hom}_{\mathrm{gr}}(A^\bullet[-n], B^\bullet) = \prod_{j \in \mathbb{Z}} \mathrm{Hom}(A^j, B^{j+n})$$

("$\mathrm{Hom}_{\mathrm{gr}}$" denotes "homomorphisms of graded groups") and the differential $d^n\colon \mathrm{Hom}^n \to \mathrm{Hom}^{n+1}$ takes $f \in \mathrm{Hom}_{\mathrm{gr}}(A^\bullet[-n], B^\bullet)$ to

$$d^n(f) := (d_B \circ f)[-1] + f \circ d_{A[-n-1]} \in \mathrm{Hom}_{\mathrm{gr}}(A^\bullet[-n-1], B^\bullet).$$

In other words, if $f = (f^j)_{j \in \mathbb{Z}}$ with $f^j \in \mathrm{Hom}(A^j, B^{j+n})$ then

$$d^n(f) = \left(d_B^{n+j} \circ f^j + (-1)^{n+1} f^{j+1} \circ d_A^j\right)_{j \in \mathbb{Z}}.^9$$

For fixed $C^\bullet$, the additive functor of complexes

$$F_1(A^\bullet) = \mathrm{Hom}^\bullet(C^\bullet, A^\bullet)$$

preserves homotopy, and so gives an additive functor (still denoted by F_1) from $\mathbf{K} = \mathbf{K}(\mathcal{A})$ into $\mathbf{K}(\mathfrak{Ab})$ (where $\mathfrak{Ab}$ is the category of abelian groups). One checks that $F_1 T = T_* F_1$, ($T = $ translation in $\mathbf{K}$, $T_* = $ translation in $\mathbf{K}(\mathfrak{Ab})$) and that F_1 takes cones to cones (cf. (1.5.2.1)); and hence F_1 (together with $\theta_1 = $ identity) is a Δ-functor.

Similarly, for fixed $D^\bullet$,

$$F_2(A^\bullet) = \mathrm{Hom}^\bullet(A^\bullet, D^\bullet)$$

gives a contravariant additive functor from $\mathbf{K}$ into $\mathbf{K}(\mathfrak{Ab})$. But now we run into sign complications: the complexes $T_*^{-1} F_2(A^\bullet)$ and $F_2 T(A^\bullet)$, while coinciding as graded objects, are *not* equal, the differential in one being the negative of the differential in the other. We define a functorial isomorphism

$$\theta_2(A^\bullet)\colon T_*^{-1} F_2(A^\bullet) \xrightarrow{\sim} F_2 T(A^\bullet)$$

to be multiplication in each degree n by $(-1)^n$, and claim that *the pair* (F_2, θ_2) *is a contravariant* Δ-functor.

---

[9] This standard d^n differs from the one in [**H**, p. 64] by a factor of $(-1)^{n+1}$.

Indeed, if $u\colon A^\bullet \to B^\bullet$ is a morphism of complexes in $\mathcal{A}$, then we check (by writing everything out explicitly) that, with $F = F_2$, $\theta = \theta_2$, the map of graded objects

$$C^\bullet_{Fu} = FA^\bullet \oplus T_* FB^\bullet \xrightarrow{T_*(\theta(A^\bullet))\oplus(-1)} T_* FTA^\bullet \oplus T_* FB^\bullet = T_* FC^\bullet_u$$

is an isomorphism of *complexes,* whence, $v\colon B^\bullet \to C^\bullet_u$ and $w\colon C^\bullet_u \to TA^\bullet$ being the canonical maps, the diagram

$$FB^\bullet \xrightarrow{Fu} FA^\bullet \xrightarrow{(T_* Fw)\circ T_*(\theta(A^\bullet))} T_* FC^\bullet_u \xrightarrow{-T_* Fv} T_* FB^\bullet$$

is a triangle in $\mathbf{K}(\mathfrak{Ab})$, i.e.,

$$T_*^{-1} FA^\bullet \xrightarrow{(-Fw)\circ\theta(A^\bullet)} F(C^\bullet_u) \xrightarrow{Fv} FB^\bullet \xrightarrow{Fu} FA^\bullet$$

is a triangle (see ($\Delta 2$) in §1.4); and the claim follows.

(1.5.4) (see again [**De**, p. 265, Prop. 1.1.7]). Let U be a topological space, $\mathcal{O}$ a sheaf of rings—say, for simplicity, commutative—and $\mathcal{A}$ the abelian category of sheaves of $\mathcal{O}$-modules. For complexes $A^\bullet$, $B^\bullet$ in $\mathcal{A}$, the complex $A^\bullet \otimes B^\bullet$ is given in degree n by

$$(A^\bullet \otimes B^\bullet)^n = \bigoplus_{p\in\mathbb{Z}}(A^p \otimes B^{n-p}) \qquad (\otimes = \otimes_{\mathcal{O}})$$

and the differential

$$d^n \colon (A^\bullet \otimes B^\bullet)^n \to (A^\bullet \otimes B^\bullet)^{n+1}$$

is the unique map whose restriction to $A^p \otimes B^{n-p}$ is

$$d^n|(A^p \otimes B^{n-p}) = d_A^p \otimes 1 + (-1)^p \otimes d_B^{n-p} \qquad (p \in \mathbb{Z}).$$

With the usual translation functor T, we have for each $i, j \in \mathbb{Z}$ a unique isomorphism of complexes

$$\theta_{ij} \colon T^i A^\bullet \otimes T^j B^\bullet \xrightarrow{\sim} T^{i+j}(A^\bullet \otimes B^\bullet)$$

satisfying, for every $p, q \in \mathbb{Z}$,

$$\theta_{ij}|(A^{p+i} \otimes B^{q+j}) = \text{multiplication by } (-1)^{pj}.$$

[Note that $A^{p+i} \otimes B^{q+j}$ is contained both in $(T^i A^\bullet \otimes T^j B^\bullet)^{p+q}$ and in $(T^{i+j}(A \otimes B))^{p+q}$.]

For fixed $A^\bullet$, we find then that the functor of complexes taking $B^\bullet$ to $B^\bullet \otimes A^\bullet$ preserves homotopy and takes cones to cones, giving an additive functor from $\mathbf{K}(\mathcal{A})$ into itself, which, together with $\theta_{10} = $ identity, is a Δ-functor.

Similarly, for fixed $A^\bullet$ the functor taking $B^\bullet$ to $A^\bullet \otimes B^\bullet$ induces a functor of $\mathbf{K}(\mathcal{A})$ into itself which, together with $\theta_{01} \neq$ identity, is a Δ-functor. And for fixed $A^\bullet$, the family of isomorphisms

$$\theta(B^\bullet)\colon A^\bullet \otimes B^\bullet \xrightarrow{\sim} B^\bullet \otimes A^\bullet \qquad (1.5.4.1)$$

defined locally by

$$\theta(B^\bullet)(a \otimes b) = (-1)^{pq}(b \otimes a) \qquad (a \in A^p, \; b \in B^q)$$

constitutes an isomorphism of Δ-functors.

Exercise 1.5.5. Let $\mathbf{K_1}$ and $\mathbf{K_2}$ be Δ-categories with respective translation functors T_1 and T_2; and let $(F, \theta)\colon \mathbf{K_1} \to \mathbf{K_2}$ be a Δ-functor. An object A in $\mathbf{K_1}$ is *periodic* if there is an integer $m > 0$ such that $T_1^m(A) = A$. Suppose that 0 is the only periodic object in $\mathbf{K_1}$. (For example, $\mathbf{K_1}$ could be any one of the Δ-categories $\mathbf{K}^*$ of §1.6 below.) Then we can choose a function $\nu\colon$ (objects of $\mathbf{K_1}) \to \mathbb{Z}$ such that $\nu(0) = 0$ and $\nu(T_1 A) = \nu(A) - 1$ for all $A \neq 0$; and using θ, we can define isomorphisms

$$\eta_A\colon F(A) \xrightarrow{\sim} T_2^{-\nu(A)} F(T_1^{\nu(A)} A) =: f(A) \qquad (A \in \mathbf{K_1}).$$

Note that $f(T_1 A) = T_2 f(A)$. Verify that there is a unique way of extending f to a functor such that the η_A form an isomorphism of Δ-functors $(F, \theta) \xrightarrow{\sim} (f, \text{identity})$.

1.6 Δ-Subcategories

A *full additive subcategory* $\mathbf{K}'$ of a Δ-category $\mathbf{K}$ carries at most one tri-angulation for which the translation is the restriction of that on $\mathbf{K}$, and such that the inclusion functor $\iota : \mathbf{K}' \hookrightarrow \mathbf{K}$ (together with the identity transformation from ιT to $T\iota$) is a Δ-functor. For the existence of such a triangulation it is necessary and sufficient that $\mathbf{K}'$ be stable under the translation automorphism and its inverse, and that the summit of any triangle in $\mathbf{K}$ with base in $\mathbf{K}'$ be isomorphic to an object in $\mathbf{K}'$; the triangles in $\mathbf{K}'$ are then precisely the triangles of $\mathbf{K}$ whose vertices are all in $\mathbf{K}'$. (Details left to the reader.) Such a $\mathbf{K}'$ is called a Δ-*subcategory* of $\mathbf{K}$.

For example, if $\mathbf{K} = \mathbf{K}(\mathcal{A})$ is as in (1.4.3), then a full additive subcategory $\mathbf{K}'$ is a Δ-subcategory if and only if:

   (i)   for every complex $A^\bullet \in \mathbf{K}$ we have $A^\bullet \in \mathbf{K}' \Leftrightarrow A^\bullet[1] \in \mathbf{K}'$, and

   (ii) the mapping cone of any $\mathcal{A}$-morphism of complexes $u\colon A^\bullet \to B^\bullet$ with $A^\bullet$ and $B^\bullet$ in $\mathbf{K}'$ is homotopically equivalent to a complex in $\mathbf{K}'$.

Examples 1.6.1. We consider various full additive subcategories $\mathbf{K}^+$, $\mathbf{K}^-$, $\mathbf{K}^b$, $\overline{\mathbf{K}}^+$, $\overline{\mathbf{K}}^-$, $\overline{\mathbf{K}}^b$, of $\mathbf{K} = \mathbf{K}(\mathcal{A})$.

The objects of $\mathbf{K}^+$ are complexes $A^\bullet$ which are *bounded below,* i.e., there is an integer n_0 (depending on $A^\bullet$) such that $A^n = 0$ for $n < n_0$. The objects of $\overline{\mathbf{K}}^+$ are complexes $B^\bullet$ whose *homology* is bounded below, i.e., $H^m(B^\bullet) = 0$ for all $m < m_0(B^\bullet)$. The objects of $\mathbf{K}^-$ and $\overline{\mathbf{K}}^-$ (respectively $\mathbf{K}^b$ and $\overline{\mathbf{K}}^b$) are specified similarly, with *"bounded above"* (resp. *"bounded above and below"*) in place of "bounded below." We have, obviously,

$$\mathbf{K}^b = \mathbf{K}^+ \cap \mathbf{K}^-, \qquad \overline{\mathbf{K}}^b = \overline{\mathbf{K}}^+ \cap \overline{\mathbf{K}}^-;$$

and if $*$ stands for any one of $^+$, $^-$, or ^{b}, then

$$\mathbf{K}^* \subset \overline{\mathbf{K}}^*.$$

Using the natural exact sequence (see (1.3))

$$0 \to B^\bullet \to C_u^\bullet \to A^\bullet[1] \to 0 \tag{1.6.2}$$

associated with a morphism $u\colon A^\bullet \to B^\bullet$ of complexes in $\mathcal{A}$, we find that if both $A^\bullet$ and $B^\bullet$ satisfy one of the above boundedness conditions then so does the cone $C_u^\bullet$, whence $\mathbf{K}^*$ *and* $\overline{\mathbf{K}}^*$ *are Δ-subcategories of* $\mathbf{K}$.

Remark 1.6.3. In (1.4.3.6) and its proof, we can replace $\mathbf{K}(\mathcal{A})$ by any Δ-subcategory.

1.7 Localizing Subcategories of K; Δ-Equivalent Categories

In the description of the derived category $\mathbf{D}$ given in §1.2, we can replace $\mathbf{K}$ by any Δ-subcategory $\mathbf{L}$, and obtain a derived category $\mathbf{D_L}$ together with a functor $Q_L\colon \mathbf{L} \to \mathbf{D_L}$ which is universal among all functors transforming quasi-isomorphisms into isomorphisms. (Here, as in 1.2, for checking details one needs [**H**, p. 35, Prop. 4.2].) Then, just as in (1.4.4), $\mathbf{D_L}$ has a unique triangulation for which the translation functor is the obvious one and for which Q_L is a Δ-functor; and (1.5.1) remains valid with Q_L in place of Q.

If $\mathbf{L}' \subset \mathbf{L}''$ are Δ-subcategories of $\mathbf{K}$ and $j\colon \mathbf{L}' \to \mathbf{L}''$ is the inclusion, then there exists a natural commutative diagram of Δ-functors

$$
\begin{array}{ccc}
\mathbf{L}' & \xrightarrow{\;j\;} & \mathbf{L}'' \\
{\scriptstyle Q':=Q_{L'}}\downarrow & & \downarrow{\scriptstyle Q_{L''}=:Q''} \\
\mathbf{D}':=\mathbf{D_{L'}} & \xrightarrow[\;\tilde{j}\;]{} & \mathbf{D_{L''}} =: \mathbf{D}''
\end{array}
$$

Note that on objects of $\mathbf{D}'$ ($=$ objects of $\mathbf{L}'$), $\tilde{j}$ is just the inclusion map to objects of $\mathbf{D}''$.

Recalling that passage to derived categories is a kind of localization in categories (§1.2, footnote), we say that $\mathbf{L}'$ localizes to a Δ-subcategory of $\mathbf{D}''$, or more briefly, that $\mathbf{L}'$ is a *localizing subcategory of* $\mathbf{L}''$, if the functor $\tilde{\jmath}$ is fully faithful, i.e., the natural map is an isomorphism

$$\mathrm{Hom}_{\mathbf{D}'}(A^{\bullet}, B^{\bullet}) \xrightarrow{\sim} \mathrm{Hom}_{\mathbf{D}''}(\tilde{\jmath}A^{\bullet}, \tilde{\jmath}B^{\bullet})$$

for all $A^{\bullet}$ and $B^{\bullet}$ in $\mathbf{D}'$.

When this condition holds, $\tilde{\jmath}$ is an additive isomorphism of $\mathbf{D}'$ onto the full subcategory $\tilde{\jmath}(\mathbf{D}')$ of $\mathbf{D}''$, so $\tilde{\jmath}$ carries the triangulation on $\mathbf{D}'$ over to a triangulation on $\tilde{\jmath}(\mathbf{D}')$; and then since $\tilde{\jmath}$ is a Δ-functor, the inclusion functor $\tilde{\jmath}(\mathbf{D}') \hookrightarrow \mathbf{D}''$, together with $\theta = $ identity, is a Δ-functor, i.e., $\tilde{\jmath}(\mathbf{D}')$ is a Δ-subcategory of $\mathbf{D}''$. Thus *if* $\mathbf{L}'$ *is localizing in* $\mathbf{L}''$, *then we can identify* $\mathbf{D}'$ *with the* Δ-*subcategory of* $\mathbf{D}''$ *whose objects are the complexes in* $\mathbf{L}'$, *and* Q' *with the restriction of* Q'' *to* $\mathbf{L}'$.

(1.7.1). From definitions in §1.2, we deduce easily the following simple sufficient condition for $\mathbf{L}'$ to be localizing in $\mathbf{L}''$:

For every quasi-isomorphism $X^{\bullet} \to B^{\bullet}$ *in* $\mathbf{L}''$ *with* $B^{\bullet}$ *in* $\mathbf{L}'$, *there exists a quasi-isomorphism* $A^{\bullet} \to X^{\bullet}$ *with* $A^{\bullet}$ *in* $\mathbf{L}'$.

(1.7.1)^{op}. A "dual" argument (see [**H**, p. 32, proof of 3.2]) yields:

The same condition with arrows reversed is also sufficient.

For example, *if the objects in* $\mathbf{L}'$ *are precisely those complexes in* $\mathbf{K}$ *which satisfy some condition on their homology* (for instance, if $\mathbf{L}'$ is any one of the categories $\overline{\mathbf{K}}^{*}$ of (1.6.1)), *then* $\mathbf{L}'$ *is localizing in* $\mathbf{L}''$.

This follows at once from (1.7.1) (take $A^{\bullet} = X^{\bullet}$).

The following results will provide a useful interpretation of various kinds of *resolutions* (injective, flat, flasque, etc.) as defining an *equivalence of* Δ-*categories.*

(1.7.2). *If for every* $X^{\bullet} \in \mathbf{L}''$ *there exists a quasi-isomorphism* $A^{\bullet} \to X^{\bullet}$ *with* $A^{\bullet} \in \mathbf{L}'$ *then* $\tilde{\jmath}$ *is an equivalence of categories, i.e., there exists a functor* $\rho \colon \mathbf{D}'' \to \mathbf{D}'$ *together with functorial isomorphisms*

$$\mathbf{1}_{\mathbf{D}''} \xrightarrow{\sim} \tilde{\jmath}\rho, \qquad \mathbf{1}_{\mathbf{D}'} \xrightarrow{\sim} \rho\tilde{\jmath} \qquad (1.7.2.1)$$

(see [**M**, p. 91]). *Moreover, for the usual translation* T *there is then a unique functorial isomorphism*

$$\theta \colon \rho T \xrightarrow{\sim} T\rho$$

such that the pair (ρ, θ) *is a* Δ-*functor and the isomorphisms* (1.7.2.1) *are isomorphisms of* Δ-*functors* (§1.5).

We say then that $\tilde{\jmath}$ and ρ—or more precisely $(\tilde{\jmath}, \text{identity})$ and (ρ, θ)—are Δ-*equivalences of categories, quasi-inverse to each other.*

(1.7.2)^{op}. *Same as* (1.7.2), *with* $A^{\bullet} \to X^{\bullet}$ *replaced by* $X^{\bullet} \to A^{\bullet}$.

To prove (1.7.2)^{op}, for example, suppose that we have a family of quasi-isomorphisms ("right $\mathbf{L}'$-resolutions")

$$\varphi_{X^{\bullet}} \colon X^{\bullet} \to A^{\bullet}_{X^{\bullet}} \in \mathbf{L}' \qquad (X^{\bullet} \in \mathbf{L}'').$$

Then by $(1.7.1)^{\mathrm{op}}$, $\mathbf{L}'$ is localizing in $\mathbf{L}''$. So finding an additive functor ρ with isomorphisms (1.7.2.1) is equivalent to finding for each object $X^\bullet$ of $\mathbf{D}''$ an isomorphism to an object in $\mathbf{D}' \subset \mathbf{D}''$ (see [$\mathbf{M}$, p. 92, (iii)$\Rightarrow$(ii)]). But $Q''(\varphi_{X^\bullet})$ is such an isomorphism. Thus we have $\rho \colon \mathbf{D}'' \to \mathbf{D}'$ with

$$\rho(X^\bullet) = A_{X^\bullet}^\bullet . \qquad (X^\bullet \in \mathbf{D}'').$$

Next, define $\theta(X^\bullet)$ to be the unique map making the following diagram (with all arrows representing *isomorphisms* in $\mathbf{D}''$) commute:

$$
\begin{array}{ccc}
 & TX^\bullet & \\
{\scriptstyle Q''(\varphi_{TX^\bullet})}\swarrow & & \searrow{\scriptstyle TQ''(\varphi_{X^\bullet})} \\
\rho TX^\bullet = A_{TX^\bullet}^\bullet \xrightarrow[\theta(X^\bullet)]{} & & TA_{X^\bullet}^\bullet = T\rho X^\bullet
\end{array}
\qquad (1.7.2.2)
$$

Then, one checks, the family $\theta(X^\bullet)$ constitutes an isomorphism of functors $\theta \colon \rho T \xrightarrow{\;\sim\;} T\rho$.

Furthermore, if

$$X^\bullet \xrightarrow{\;u\;} Y^\bullet \xrightarrow{\;v\;} Z^\bullet \xrightarrow{\;w\;} TX^\bullet$$

is a triangle in $\mathbf{D}''$, then $(\Delta 1)''$ (see §1.4) applied to the commutative diagram in $\mathbf{D}''$

$$
\begin{array}{ccccccc}
X^\bullet & \xrightarrow{\;u\;} & Y^\bullet & \xrightarrow{\;v\;} & Z^\bullet & \xrightarrow{\;w\;} & TX^\bullet \\
{\scriptstyle Q''\varphi_{X^\bullet}}\downarrow & & {\scriptstyle Q''\varphi_{Y^\bullet}}\downarrow & & \downarrow{\scriptstyle Q''\varphi_{Z^\bullet}} & & \downarrow{\scriptstyle TQ''\varphi_{X^\bullet}} \\
A_{X^\bullet}^\bullet & \xrightarrow[\rho(u)]{} & A_{Y^\bullet}^\bullet & \xrightarrow[\rho(v)]{} & A_{Z^\bullet}^\bullet & \xrightarrow[\theta(X^\bullet)\circ\rho(w)]{} & TA_{X^\bullet}^\bullet
\end{array}
$$

guarantees that the bottom row is a triangle; and so (ρ, θ) is a Δ-functor.

The fact that the isomorphisms in (1.7.2.1) (induced by the family $\varphi_{X^\bullet}$) are isomorphisms of Δ-functors is nothing but the commutativity of (1.7.2.2). Thus the family $\theta := \{\theta(X^\bullet)\}$ is the unique functorial isomorphism having the properties stated in $(1.7.2)^{\mathrm{op}}$.

Remark 1.7.2.3. It is sometimes possible to choose the functor ρ so that $\rho T = T\rho$ and $\theta = $ identity, i.e., to find a family of quasi-isomorphisms $\varphi_{X^\bullet} \colon X^\bullet \to A_{X^\bullet}^\bullet$ commuting with translation (see (1.8.1.1), (1.8.2), and (1.8.3) below).

1.8 Examples

(1.8.1). *If* $\mathbf{L'} \subset \mathbf{K}$ *is any one of the* Δ*-subcategories* $\overline{\mathbf{K}}^{\boldsymbol{*}}$ *of* (1.6.1) *and if* $\mathbf{L''}$ *is any* Δ*-subcategory of* $\mathbf{K}$ *containing* $\mathbf{L'}$*, then* $\mathbf{L'}$ *is localizing in* $\mathbf{L''}$*. The same holds for* $\mathbf{L'} = \mathbf{K}^{\boldsymbol{+}}$ *or* $\mathbf{L'} = \mathbf{K}^{\boldsymbol{-}}$*; and also for* $\mathbf{L'} = \mathbf{K}^{\mathsf{b}}$ *if* $\mathbf{L''}$ *is localizing in* $\mathbf{K}$*.*

For $\mathbf{L'} = \overline{\mathbf{K}}^{\boldsymbol{*}}$ the assertion follows at once from (1.7.1). For the rest (and for other purposes) we need the *truncation operators* τ^+, τ^-, defined as follows:
For any $B^{\bullet} \in \mathbf{K}$, set

$$i = i(B^{\bullet}) := \inf\{\, m \mid H^m(B^{\bullet}) \neq 0 \,\}$$

and let $\tau^+(B^{\bullet})$ be the complex

$$\cdots \to 0 \to 0 \to \operatorname{coker}(B^{i-1} \to B^i) \to B^{i+1} \to B^{i+2} \to \cdots \ .$$

(When $i = \infty$, i.e., when $B^{\bullet}$ is exact, this means $\tau^+(B^{\bullet}) = 0^{\bullet}$; and when $i = -\infty$, $\tau^+(B^{\bullet}) = B^{\bullet}$.) There is an obvious *quasi-isomorphism*

$$B^{\bullet} \to \tau^+(B^{\bullet})\,. \tag{1.8.1$^+$}$$

Dually, for any $C^{\bullet} \in \mathbf{K}$ set

$$s = s(C^{\bullet}) := \sup\{\, n \mid H^n(C^{\bullet}) \neq 0 \,\}$$

and let $\tau^-(C^{\bullet})$ be the complex

$$\cdots \to C^{s-2} \to C^{s-1} \to \ker(C^s \to C^{s+1}) \to 0 \to 0 \to \cdots \ .$$

There is an obvious *quasi-isomorphism*

$$\tau^-(C^{\bullet}) \to C^{\bullet}\,. \tag{1.8.1$^-$}$$

Now if $C^{\bullet} \to B^{\bullet}$ is a quasi-isomorphism in $\mathbf{L''}$ with $B^{\bullet} \in \mathbf{K}^-$ then $C^{\bullet} \in \overline{\mathbf{K}}^-$, and we have the quasi-isomorphism (1.8.1)$^-$ with $\tau^-(C^{\bullet}) \in \mathbf{K}^-$. So (1.7.1) with $\mathbf{L'} = \mathbf{K}^- \subset \mathbf{L''}$ shows that $\mathbf{K}^-$ is localizing in $\mathbf{L''}$.

Dually, via (1.8.1)$^+$, (1.7.1)^{op} implies that $\mathbf{K}^+$ is localizing in any Δ-subcategory $\mathbf{L''}$ of $\mathbf{K}$ containing $\mathbf{K}^+$.

And again via (1.8.1)$^-$, (1.7.1) shows that $\mathbf{K}^{\mathsf{b}}$ is localizing in $\mathbf{K}^+$; and since as above $\mathbf{K}^+$ is localizing in $\mathbf{K}$, the natural functors $\mathbf{D}^{\mathsf{b}} \to \mathbf{D}^+ \to \mathbf{D}$ between the corresponding derived categories are both fully faithful, whence so is their composition, i.e., $\mathbf{K}^{\mathsf{b}}$ *is localizing in* $\mathbf{K}$. It follows at once that $\mathbf{K}^{\mathsf{b}}$ is localizing in any $\mathbf{L''} \supset \mathbf{K}^{\mathsf{b}}$ such that $\mathbf{L''}$ is localizing in $\mathbf{K}$.

Consequently, as in (1.7): *the derived category* $\mathbf{D}^{\boldsymbol{*}}$ *(resp.* $\overline{\mathbf{D}}^{\boldsymbol{*}}$*) of* $\mathbf{K}^{\boldsymbol{*}}$ *(resp.* $\overline{\mathbf{K}}^{\boldsymbol{*}}$*) can be identified in a natural way with a* Δ*-subcategory of* $\mathbf{D}$*.*

Then the inclusion $\mathbf{D}^+ \hookrightarrow \overline{\mathbf{D}}^+$ is a Δ-equivalence of categories. Indeed, as in the proof of $(1.7.2)^{\mathrm{op}}$, with $\mathbf{L}' = \mathbf{K}^+$, $\mathbf{L}'' = \overline{\mathbf{K}}^+$, and $\varphi_{B^\bullet} = (1.8.1)^+$, we can see that τ^+—which commutes with translation—extends to a Δ-functor

$$(\tau^+, 1) \colon \overline{\mathbf{D}}^+ \to \mathbf{D}^+ \tag{1.8.1.1}$$

which is quasi-inverse to the inclusion.

Similarly the inclusions $\mathbf{D}^- \hookrightarrow \overline{\mathbf{D}}^-$, $\mathbf{D}^{\mathrm{b}} \hookrightarrow \overline{\mathbf{D}}^{\mathrm{b}}$ are Δ-equivalences, with respective quasi-inverses τ^- and $\tau^{\mathrm{b}} = \tau^- \circ \tau^+ = \tau^+ \circ \tau^-$. More precisely, τ^{b} is the composition

$$\overline{\mathbf{D}}^{\mathrm{b}} \xrightarrow{\tau^+} \overline{\mathbf{D}}^{\mathrm{b}} \cap \mathbf{D}^+ \xrightarrow{\tau^-} \mathbf{D}^- \cap \mathbf{D}^+ = \mathbf{D}^{\mathrm{b}}.$$

(1.8.2) Let $\mathbf{I}$ be a full additive subcategory of $\mathcal{A}$ such that every object of $\mathcal{A}$ admits a monomorphism into an object in $\mathbf{I}$. Then there exists a family of quasi-isomorphisms

$$\varphi_{B^\bullet} \colon B^\bullet \to I_{B^\bullet}^\bullet \qquad \left(B^\bullet \in \overline{\mathbf{K}}^+ = \overline{\mathbf{K}}^+(\mathcal{A})\right)$$

where each $I^\bullet = I_{B^\bullet}^\bullet$ is a bounded-below $\mathbf{I}$-complex (i.e., $I^n \in \mathbf{I}$ for all n, and $I^n = (0)$ for $n \ll 0$); and such that moreover with the usual translation functor T we have

$$I_{TB^\bullet}^\bullet = TI_{B^\bullet}^\bullet, \qquad \varphi_{TB^\bullet} = T(\varphi_{B^\bullet}). \tag{1.8.2.1}$$

To see this, first construct quasi-isomorphisms $\varphi_{B^\bullet}$ as in [**H**, p. 42, 4.6, 1)] for those $B^\bullet$ such that $H^0(B^\bullet) \neq 0$ and $B^m = 0$ for $m < 0$. Then $(1.8.2.1)$ forces the definition of $\varphi_{B^\bullet}$ for any $B^\bullet$ such that there exists $i \in \mathbb{Z}$ with $H^i(B^\bullet) \neq 0$ and $B^m = 0$ for all $m < i$ (i.e., $0^\bullet \neq B^\bullet = \tau^+ B^\bullet$, see $(1.8.1)$). Set $I_{0^\bullet} = 0^\bullet$, and finally for any $B^\bullet \in \overline{\mathbf{K}}^+$ set

$$\varphi_{B^\bullet} = (\varphi_{\tau^+ B^\bullet}) \circ (1.8.1)^+.$$

Now let $\mathbf{K}_{\mathbf{I}}^+$ be the full subcategory of $\mathbf{K}^+$ whose objects are the bounded-below $\mathbf{I}$-complexes. Since the additive subcategory $\mathbf{I} \subset \mathcal{A}$ is closed under finite direct sums, one sees that $\mathbf{K}_{\mathbf{I}}^+$ is a Δ-subcategory of $\mathbf{K}^+$. According to $(1.7.2)^{\mathrm{op}}$, the derived category $\mathbf{D}_{\mathbf{I}}^+$ of $\mathbf{K}_{\mathbf{I}}^+$ can be identified with a Δ-subcategory of $\overline{\mathbf{D}}^+$, and the above family $\varphi_{B^\bullet}$ gives rise to an $\mathbf{I}$-*resolution functor*

$$\rho \colon \overline{\mathbf{D}}^+ \to \mathbf{D}_{\mathbf{I}}^+ \tag{1.8.2.2}$$

which is, together with $\theta = \mathrm{identity}$, a Δ-*equivalence of categories*, quasi-inverse to the inclusion $\mathbf{D}_{\mathbf{I}}^+ \hookrightarrow \overline{\mathbf{D}}^+$.

For example, if $\mathbf{I}$ is the full subcategory of $\mathcal{A}$ whose objects are all the injectives in $\mathcal{A}$, then by [**H**, p. 41, Lemma 4.5] every quasi-isomorphism in $\mathbf{K}_{\mathbf{I}}^+$ is an isomorphism, so that $\mathbf{K}_{\mathbf{I}}^+$ can be identified with its derived category $\mathbf{D_I^+}$. Thus, if $\mathcal{A}$ has enough injectives (i.e., every object of $\mathcal{A}$ admits a monomorphism into an injective object), then the natural composition

$$\mathbf{D_I^+} = \mathbf{K_I^+} \hookrightarrow \overline{\mathbf{K}}^+ \to \overline{\mathbf{D}}^+$$

is a Δ-equivalence, having as quasi-inverse an injective resolution functor (1.8.2.2) (cf. [**H**, p. 46, Prop. 4.7]).

(**1.8.3**). Let $\mathbf{P}$ be a full additive subcategory of $\mathcal{A}$ such that for every object $B \in \mathcal{A}$ there exists an epimorphism $P_B \to B$ with $P_B \in \mathbf{P}$. An argument dual to that in (1.8.2) yields that *there exists a family of quasi-isomorphisms*

$$\psi_{B^\bullet} : P_{B^\bullet}^\bullet \to B^\bullet \qquad \left(B^\bullet \in \overline{\mathbf{K}}^-(\mathcal{A}) \right)$$

commuting with translation, and such that each $P_{B^\bullet}^\bullet$ is a bounded-above $\mathbf{P}$-complex.

According to (1.7.2), we have then a $\mathbf{P}$-*resolution functor* which is a Δ-*equivalence* into $\overline{\mathbf{D}}^-(\mathcal{A})$ from its Δ-subcategory whose objects are bounded-above $\mathbf{P}$-complexes.

For example, if U is a topological space, $\mathcal{O}$ is a sheaf of rings on U, and $\mathcal{A}$ is the abelian category of (sheaves of) left $\mathcal{O}$-modules, then we can take $\mathbf{P}$ to be the full subcategory of $\mathcal{A}$ whose objects are all the *flat* $\mathcal{O}$-modules [**H**, p. 86, Prop. 1.2].

1.9 Complexes with Homology in a Plump Subcategory

(**1.9.1**). Here, in brief, are some essential basic facts.

Let $\mathcal{A}^{\#}$ be a *plump* subcategory of the abelian category $\mathcal{A}$, i.e., a full subcategory containing 0 and such that for every exact sequence in $\mathcal{A}$

$$X_1 \to X_2 \to X \to X_3 \to X_4 \,,$$

if X_1, X_2, X_3, and X_4 all lie in $\mathcal{A}^{\#}$ then so does X. Then the kernel and cokernel (in $\mathcal{A}$) of any map in $\mathcal{A}^{\#}$ must lie in $\mathcal{A}^{\#}$ (whence $\mathcal{A}^{\#}$ is abelian), and any object of $\mathcal{A}$ isomorphic to an object in $\mathcal{A}^{\#}$ must itself be in $\mathcal{A}^{\#}$.

Considering only complexes in $\mathcal{A}$ whose homology objects all lie in $\mathcal{A}^{\#}$, we obtain full subcategories $\mathbf{K}_{\#}$ of $\mathbf{K}$, $\mathbf{K}_{\#}^*$ of $\mathbf{K}^*$, and $\overline{\mathbf{K}}_{\#}^*$ of $\overline{\mathbf{K}}^*$ (see (1.6.1)). Via the exact homology sequence $(1.4.5)^{\mathrm{H}}$ of a standard triangle (1.3.1), we find that these subcategories are all Δ-subcategories (see (i) and (ii) in §1.6), and indeed, by (1.7.1), *localizing subcategories*. From (1.8.1) it follows then that $\mathbf{K}_{\#}$, $\mathbf{K}_{\#}^*$, and $\overline{\mathbf{K}}_{\#}^*$ are localizing subcategories of $\mathbf{K}$, from which we derive Δ-subcategories $\mathbf{D}_{\#}$, $\mathbf{D}_{\#}^*$, and $\overline{\mathbf{D}}_{\#}^*$ of $\mathbf{D}$, with universal properties analogous to (1.5.1). As in (1.8.1) the inclusion $\mathbf{D}_{\#}^* \hookrightarrow \overline{\mathbf{D}}_{\#}^*$ is a Δ-equivalence of categories, with quasi-inverse τ^*.

(1.9.2). The following isomorphism test will be useful.

Lemma. *If $\mathcal{A}^{\#}$ is a plump subcategory of $\mathcal{A}$, and $u\colon A_1^{\bullet} \to A_2^{\bullet}$ is a map in $\overline{\mathbf{D}}_{\#}^{+}$ such that for all $B^{\bullet} \in \mathbf{D}_{\#}^{b}$ the induced map*

$$\mathrm{Hom}_{\mathbf{D}}(B^{\bullet}, A_1^{\bullet}) \to \mathrm{Hom}_{\mathbf{D}}(B^{\bullet}, A_2^{\bullet})$$

is an isomorphism, then u is an isomorphism.

Proof. Let $C^{\bullet} \in \overline{\mathbf{D}}_{\#}^{+}$ be the summit of a triangle with base u, so that by (1.4.2.1), u is an isomorphism iff $C^{\bullet} \cong 0$, i.e., iff $\tau^{+}(C^{\bullet}) = 0^{\bullet}$, see (1.8.1) and (1.2.2).

For each $m \in \mathbb{Z}$ and each object $M \in \mathcal{A}^{\#}$ we have, by (1.4.2.1) and $(\Delta 2)$ in §1.4, an exact sequence (with $\mathrm{Hom} = \mathrm{Hom}_{\mathbf{D}}$):

$$\mathrm{Hom}(M[-m], A_1^{\bullet}) \xrightarrow[\text{via } u]{\sim} \mathrm{Hom}(M[-m], A_2^{\bullet}) \longrightarrow \mathrm{Hom}(M[-m], C^{\bullet})$$

$$\longrightarrow \mathrm{Hom}(M[-m], A_1^{\bullet}[1]) \xrightarrow[\text{via } -u[1]]{\sim} \mathrm{Hom}(M[-m], A_2^{\bullet}[1]).$$

The two labeled maps are, by hypothesis, isomorphisms, and hence

$$\mathrm{Hom}(M[-m], C^{\bullet}) = 0.$$

Were $\tau^{+}(C^{\bullet}) \neq 0^{\bullet}$, then with $m := i(C^{\bullet})$ (see (1.8.1) and

$$M := H^m(C^{\bullet}) = \ker\big(\tau^{+}(C^{\bullet})^m \to \tau^{+}(C^{\bullet})^{m+1}\big) \neq 0\,,$$

the inclusion $M \hookrightarrow \tau^{+}(C^{\bullet})^m$ would lead to a map $j\colon M[-m] \to \tau^{+}(C^{\bullet})$ with $H^m(j)$ the (non-zero) identity map of M, so we'd have

$$\mathrm{Hom}\big(M[-m], C^{\bullet}\big) \xrightarrow[(1.8.1)^{+}]{\sim} \mathrm{Hom}\big(M[-m], \tau^{+}(C^{\bullet})\big) \neq 0\,,$$

a contradiction. Thus $\tau^{+}(C^{\bullet}) = 0^{\bullet}$.   Q.E.D.

1.10 Truncation Functors

Let $\mathcal{A}$ be an abelian category, and let $\mathbf{D} = \mathbf{D}(\mathcal{A})$ be the derived category. For any complex $A^{\bullet}$ in $\mathcal{A}$, and $n \in \mathbb{Z}$, we let $\tau_{\leq n}A^{\bullet}$ be the truncated complex

$$\cdots \longrightarrow A^{n-2} \longrightarrow A^{n-1} \longrightarrow \ker(A^n \to A^{n+1}) \longrightarrow 0 \longrightarrow 0 \longrightarrow \cdots \,,$$

and dually we let $\tau_{\geq n}A$ be the complex

$$\cdots \longrightarrow 0 \longrightarrow 0 \longrightarrow \mathrm{coker}(A^{n-1} \to A^n) \longrightarrow A^{n+1} \longrightarrow A^{n+2} \longrightarrow \cdots \,.$$

Note that

$$H^m(\tau_{\leq n}A^\bullet) = H^m(A^\bullet) \qquad \text{if } m \leq n,$$
$$= 0 \qquad \text{if } m > n,$$

and that

$$H^m(\tau_{\geq n}A^\bullet) = H^m(A^\bullet) \qquad \text{if } m \geq n,$$
$$= 0 \qquad \text{if } m < n.$$

One checks that $\tau_{\geq n}$ (respectively $\tau_{\leq n}$) extends naturally to an additive functor of complexes which preserves homotopy and takes quasi-isomorphisms to quasi-isomorphisms, and hence induces an additive functor $\mathbf{D} \to \mathbf{D}$, see §1.2. In fact if $\mathbf{D}_{\leq \mathbf{n}}$ (resp. $\mathbf{D}_{\geq \mathbf{n}}$) is the full subcategory of $\mathbf{D}$ whose objects are the complexes $A^\bullet$ such that $H^m(A^\bullet) = 0$ for $m > n$ (resp. $m < n$) then we have additive functors

$$\tau_{\leq n} : \mathbf{D} \longrightarrow \mathbf{D}_{\leq \mathbf{n}} \subset \mathbf{D}$$
$$\tau_{\geq n} : \mathbf{D} \longrightarrow \mathbf{D}_{\geq \mathbf{n}} \subset \mathbf{D}$$

together with obvious functorial maps

$$i_A^n : \tau_{\leq n}A^\bullet \longrightarrow A^\bullet$$
$$j_A^n : A^\bullet \longrightarrow \tau_{\geq n}A^\bullet.$$

Proposition 1.10.1. *The preceding maps i_A^n, j_A^n induce functorial isomorphisms*

$$\operatorname{Hom}_{\mathbf{D}_{\leq \mathbf{n}}}(B^\bullet, \tau_{\leq n}A^\bullet) \xrightarrow{\sim} \operatorname{Hom}_{\mathbf{D}}(B^\bullet, A^\bullet) \quad (B^\bullet \in \mathbf{D}_{\leq \mathbf{n}}), \qquad (1.10.1.1)$$

$$\operatorname{Hom}_{\mathbf{D}_{\geq \mathbf{n}}}(\tau_{\geq n}A^\bullet, C^\bullet) \xrightarrow{\sim} \operatorname{Hom}_{\mathbf{D}}(A^\bullet, C^\bullet) \quad (C^\bullet \in \mathbf{D}_{\geq \mathbf{n}}). \qquad (1.10.1.2)$$

Proof. Bijectivity of (1.10.1.1) means that any map $\varphi \colon B^\bullet \to A^\bullet$ (in $\mathbf{D}$) with $B^\bullet \in \mathbf{D}_{\leq \mathbf{n}}$ factors uniquely via $i_A := i_A^n$.

Given φ, we have a commutative diagram

$$
\begin{array}{ccc}
\tau_{\leq n}B^\bullet & \xrightarrow{\tau_{\leq n}\varphi} & \tau_{\leq n}A^\bullet \\
\downarrow{\scriptstyle i_B} & & \downarrow{\scriptstyle i_A} \\
B^\bullet & \xrightarrow{\varphi} & A^\bullet
\end{array}
$$

and since $B^\bullet \in \mathbf{D}_{\leq \mathbf{n}}$, therefore i_B is an isomorphism in $\mathbf{D}$, see (1.2.2), so we can write $\varphi = i_A \circ (\tau_{\leq n}\varphi \circ i_B^{-1})$, and thus (1.10.1.1) is *surjective*.

To prove that (1.10.1.1) is also *injective*, we assume that $i_A \circ \tau_{\leq n}\varphi = 0$ and deduce that $\tau_{\leq n}\varphi = 0$. As in §1.2, the assumption means that there is a commutative diagram in $\mathbf{K}(\mathcal{A})$

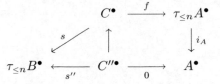

where s and s'' are quasi-isomorphisms, and $f/s = \tau_{\leq n}\varphi$.

Applying the (idempotent) functor $\tau_{\leq n}$, we get a commutative diagram

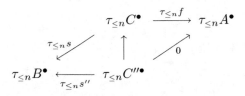

Since $\tau_{\leq n}s$ and $\tau_{\leq n}s''$ are quasi-isomorphisms, we have

$$\tau_{\leq n}\varphi = \tau_{\leq n}f/\tau_{\leq n}s = 0/\tau_{\leq n}s'' = 0 \,,$$

as desired.

A similar argument proves the bijectivity of (1.10.1.2).

Remarks 1.10.2. Let $n \in \mathbb{Z}$, $A^\bullet \in \mathbf{D}(\mathcal{A})$.

(i) There exist natural isomorphisms

$$\tau_{\leq n}\tau_{\geq n}A^\bullet \cong H^n(A^\bullet)[-n] \cong \tau_{\geq n}\tau_{\leq n}A^\bullet \,.$$

(ii) The cokernel of $i_A^{n-1} : \tau_{\leq n-1}A^\bullet \to A^\bullet$ maps quasi-isomorphically to $\tau_{\geq n}A^\bullet$; and hence there are natural triangles in $\mathbf{D}(\mathcal{A})$ (see (1.4.4.2)~):

$$\tau_{\leq n-1}A^\bullet \xrightarrow{i_A^{n-1}} A^\bullet \xrightarrow{j_A^n} \tau_{\geq n}A^\bullet \longrightarrow (\tau_{\leq n-1}A^\bullet)[1] \,, \qquad (1.10.2.1)$$

$$\tau_{\leq n-1}A^\bullet \longrightarrow \tau_{\leq n}A^\bullet \longrightarrow H^n(A^\bullet)[-n] \longrightarrow (\tau_{\leq n-1}A^\bullet)[1] \,. \qquad (1.10.2.2)$$

Details are left to the reader.

1.11 Bounded Functors; Way-Out Lemma

Many of the main results in subsequent chapters will be to the effect that some natural map or other is a functorial *isomorphism*. So we'll need isomorphism criteria. In (1.11.3) we review some commonly used ones ("Lemma on way-out functors," [**H**, p. 68, Prop. 7.1]).

Throughout this section, $\mathcal{A}$ and $\mathcal{B}$ are abelian categories, $\mathcal{A}^{\#}$ is a plump subcategory of $\mathcal{A}$, and $\overline{\mathbf{D}}_{\#}^{*}(\mathcal{A}) \subset \mathbf{D}(\mathcal{A})$ is as in (1.9.1). We identify $\mathcal{A}^{\#}$ with a full subcategory of $\overline{\mathbf{D}}_{\#}^{*}(\mathcal{A})$, see (1.2.3).

For a subcategory $\mathbf{E}$ of $\mathbf{D}(\mathcal{A})$, $\mathbf{E}_{\leq \mathbf{n}}$ (resp. $\mathbf{E}_{\geq \mathbf{n}}$) will denote the full subcategory of $\mathbf{E}$ whose objects are those complexes $A^{\bullet}$ such that $H^m(A^{\bullet}) = 0$ for $m > n$ (resp. $m < n$).

Definition 1.11.1. Let $\mathbf{E}$ be a subcategory of $\mathbf{D}(\mathcal{A})$, and let F (resp. F') be a covariant (resp. contravariant) additive functor from $\mathbf{E}$ to $\mathbf{D}(\mathcal{B})$.

The *upper dimension* $\dim^+$ and *lower dimension* $\dim^-$ of these functors are:

$$\dim^+ F := \inf\{\, d \mid F(\mathbf{E}_{\leq \mathbf{n}}) \subset \mathbf{D}_{\leq \mathbf{n+d}}(\mathcal{B}) \text{ for all } n \in \mathbb{Z} \,\},$$

$$\dim^+ F' := \inf\{\, d \mid F'(\mathbf{E}_{\geq -\mathbf{n}}) \subset \mathbf{D}_{\leq \mathbf{n+d}}(\mathcal{B}) \text{ for all } n \in \mathbb{Z} \,\},$$

$$\dim^- F := \inf\{\, d \mid F(\mathbf{E}_{\geq \mathbf{n}}) \subset \mathbf{D}_{\geq \mathbf{n-d}}(\mathcal{B}) \text{ for all } n \in \mathbb{Z} \,\},$$

$$\dim^- F' := \inf\{\, d \mid F'(\mathbf{E}_{\leq -\mathbf{n}}) \subset \mathbf{D}_{\geq \mathbf{n-d}}(\mathcal{B}) \text{ for all } n \in \mathbb{Z} \,\}.$$

The functor F is *bounded above*[10] (resp. *bounded below*)[11] if $\dim^+ F < \infty$ (resp. $\dim^- F < \infty$); and similarly for F'. F (resp. F') is *bounded* if it is both bounded-above and bounded-below.

Remarks 1.11.2. (i) Let T_1 and T_2 be the translation functors in $\mathbf{D}(\mathcal{A})$ and $\mathbf{D}(\mathcal{B})$ respectively. Suppose that $T_1 \mathbf{E} = \mathbf{E}$ and that there is a functorial isomorphism $FT_1 \xrightarrow{\sim} T_2 F$ (resp. $T_2^{-1} F' \xrightarrow{\sim} F'T_1$). (For example, $\mathbf{E}$ could be a Δ-subcategory of $\mathbf{D}(\mathcal{A})$ and F' a Δ-functor.) Then, for instance, $F'(\mathbf{E}_{\geq -\mathbf{n}}) \subset \mathbf{D}_{\leq \mathbf{n+d}}(\mathcal{B})$ holds for all $n \in \mathbb{Z}$ as soon as it holds for one single n.

(ii) If $\mathbf{E}$ is a Δ-subcategory of $\mathbf{D}(\mathcal{A})$ such that for all $n \in \mathbb{Z}$, $\tau_{\leq n}\mathbf{E} \subset \mathbf{E}$ and $\tau_{\geq n}\mathbf{E} \subset \mathbf{E}$ (e.g., $\mathbf{E} = \overline{\mathbf{D}}_{\#}^{*}(\mathcal{A})$), and if F (resp. F') is a Δ-functor, then:

$$\dim^+ F \leq d \iff H^i F(A^{\bullet}) \xrightarrow[j_A^n]{\sim} H^i F(\tau_{\geq n} A^{\bullet})$$

$$\text{for all } A^{\bullet} \in \mathbf{E}, \ n \in \mathbb{Z}, \text{ and } i \geq n+d.$$

(The display signifies that the map $H^i(j_A^n)$ (see §1.10) is an isomorphism; and as in (i), we can restrict attention to a single n.) The implication $\Rightarrow$ follows from the exact homology sequence $(1.4.5)^{\mathrm{H}}$ of the triangle gotten by applying F to (1.10.2.1); while $\Leftarrow$ is obtained by taking $A^{\bullet}$ to be an arbitrary complex in $\mathbf{E}_{\leq \mathbf{n-1}}$. An equivalent condition is that *if* $\alpha \colon A_1^{\bullet} \to A_2^{\bullet}$ *is a map in* $\mathbf{E}$ *such that* $H^i(\alpha)$ *is an isomorphism for all* $i \geq n$, (*that is, if* α *induces an isomorphism* $\tau_{\geq n} A_1^{\bullet} \xrightarrow{\sim} \tau_{\geq n} A_2^{\bullet}$), *then* $H^i(F\alpha)$ *is an isomorphism for all* $i \geq n+d$.

---

[10] *way-out left* in the terminology of [**H**, p. 68].

[11] *way-out right*.

Similarly:

$$\dim^+ F' \leq d \quad \Longleftrightarrow \quad H^i F'(A^\bullet) \xrightarrow[\;i_A^{-n}\;]{\sim} H^i F'(\tau_{\leq -n} A^\bullet) \qquad (i \geq n + d),$$

$$\dim^- F \leq d \quad \Longleftrightarrow \quad H^i F(\tau_{\leq n} A^\bullet) \xrightarrow[\;i_A^{n}\;]{\sim} H^i F(A^\bullet) \qquad (i \leq n - d),$$

$$\dim^- F' \leq d \quad \Longleftrightarrow \quad H^i F'(\tau_{\geq -n} A^\bullet) \xrightarrow[\;j_A^{-n}\;]{\sim} H^i F'(A^\bullet) \qquad (i \leq n - d).$$

(iii) If $\mathbf{E} = \mathcal{A}^\#$ (so that $\mathbf{E}_{\geq 0} = \mathbf{E} = \mathbf{E}_{\leq 0}$), then $\dim^+ F \leq d \Leftrightarrow H^j F(A) = 0$ for all $j > d$ and all $A \in \mathcal{A}^\#$. Similarly, $\dim^- F \leq d \Leftrightarrow H^j F(A) = 0$ for $j < -d$ and $A \in \mathcal{A}^\#$. These assertions remain true when F is replaced by F'.

(iv) If $\mathbf{E} = \overline{\mathbf{D}}_\#^+(\mathcal{A})$ and F is a Δ-functor, then $\dim^+ F = \dim^+ F_0$ where F_0 is the restriction $F|_{\mathcal{A}^\#}$. A similar statement holds for $\dim^- F'$; and analogous statements hold for $\dim^- F$ or $\dim^+ F'$ when $\mathbf{E} = \overline{\mathbf{D}}_\#^-(\mathcal{A})$.

Here is a typical proof: we deal with $\dim^- F'$ when $\mathbf{E} = \overline{\mathbf{D}}_\#^+(\mathcal{A})$.

Obviously $\dim^- F' \geq \dim^- F_0'$. To prove the opposite inequality, suppose that $\dim^- F_0' \leq d < \infty$, fix an $n \in \mathbb{Z}$, and let us show for any $A^\bullet \in \mathbf{E}_{\leq -n}$ that $H^j F'(A^\bullet) = 0$ whenever $j < n - d$.

We proceed by induction on the number $\nu = \nu(A^\bullet)$ of non-vanishing homology objects of $A^\bullet$, the case $\nu = 0$ being trivial. If $\nu = 1$, say $H^{-m}(A^\bullet) =: H \neq 0$ ($m \geq n$), then $A^\bullet \cong \tau^- \tau^+ A^\bullet \cong H[m]$ (see (1.8.1)), and since F' is a contravariant Δ-functor, $F'(A^\bullet) \cong F'(H)[-m]$; so by definition of $\dim^- F_0'$,

$$H^j F'(A^\bullet) \cong H^{j-m} F'(H) = 0 \quad \text{if} \quad j - m < -d,$$

whence the conclusion. When $\nu > 1$, choose any integer s such that there exist integers $p < s \leq q$ with $H^p(A^\bullet) \neq 0$, $H^q(A^\bullet) \neq 0$ (so that $\nu(\tau_{\leq s-1} A^\bullet) < \nu(A^\bullet)$ and $\nu(\tau_{\geq s} A^\bullet) < \nu(A^\bullet)$). Then apply F' to (1.10.2.1) to get a triangle

$$F'(\tau_{\leq s-1} A^\bullet) \longleftarrow F'(A^\bullet) \longleftarrow F'(\tau_{\geq s} A^\bullet) \longleftarrow F'(\tau_{\leq s-1} A^\bullet)[-1]$$

whose associated homology sequence $(1.4.5)^{\mathrm{H}}$ yields the inductive step.

Lemma 1.11.3. *Let (F, θ) and (G, ψ) be covariant Δ-functors from $\overline{\mathbf{D}}_\#^*(\mathcal{A})$ to $\mathbf{D}(\mathcal{B})$, and assume one of the following sets of conditions:*

    (i)   *$* = \mathsf{b}$.*

   (ii)   *$* = +$ and both F and G are bounded below.*

 (iii)   *$* = -$ and both F and G are bounded above.*

 (iv)   *$* = $ blank and F and G are bounded above and below.*

Then for a morphism $\eta \colon F \to G$ of Δ-functors to be an isomorphism it suffices that $\eta(X)$ be an isomorphism for all objects $X \in \mathcal{A}^\#$.

A similar assertion holds for contravariant functors if we interchange "bounded above" and "bounded below."

Complement 1.11.3.1. *Let* **I** *(resp.* **P***) be a set of objects in* $\mathcal{A}^{\#}$ *such that every object in* $\mathcal{A}^{\#}$ *admits a monomorphism into one in* **I** *(resp. is the target of an epimorphism out of one in* **P***). If* $* = +$ *and* F *and* G *are bounded below (resp.* $* = -$ *and* F *and* G *are bounded above) and if* $\eta(X)$ *is an isomorphism for all objects* $X \in$ **I** *(resp.* $X \in$ **P***), then* η *is an isomorphism.*

A similar assertion holds for contravariant functors if we interchange "bounded above" and "bounded below."

Proof. We deal first with the covariant case.

(i) Using the definition of "morphism of Δ-functors" (§1.5) we see by induction on $|n|$ that $\eta(X[-n])$ is an isomorphism for all $X \in \mathcal{A}^{\#}$ and $n \in \mathbb{Z}$. In showing that $\eta(A^{\bullet})$ is an isomorphism for all $A^{\bullet} \in \overline{\mathbf{D}}_{\#}^{\mathbf{b}}(\mathcal{A})$, we may replace $A^{\bullet}$ by the isomorphic complex $\tau^{-}(A^{\bullet}) = \tau_{\leq n}A^{\bullet}$ with $n := s(A^{\bullet})$, see (1.8.1). From (1.10.2.2), and ($\Delta 2$) of §1.4, we get a map of triangles, induced by η:

$$F(H^n(A^{\bullet})[-n-1]) \longrightarrow F(\tau_{\leq n-1}A^{\bullet}) \longrightarrow F(\tau_{\leq n}A^{\bullet}) \longrightarrow F(H^n(A^{\bullet})[-n])$$
$$\downarrow \qquad\qquad\qquad \downarrow \qquad\qquad\qquad \downarrow \qquad\qquad\qquad \downarrow$$
$$G(H^n(A^{\bullet})[-n-1]) \longrightarrow G(\tau_{\leq n-1}A^{\bullet}) \longrightarrow G(\tau_{\leq n}A^{\bullet}) \longrightarrow G(H^n(A^{\bullet})[-n])$$

and then we can conclude by $(\Delta 3)^*$ of §1.4 and induction on the number of non-vanishing homology objects of $A^{\bullet}$ (a number which is less for $\tau_{\leq n-1}A^{\bullet}$ than for $A^{\bullet}$ whenever n is finite).

(ii) By (1.2.2), it suffices to show that $\eta(A^{\bullet})$ induces an isomorphism from $H^i F(A^{\bullet})$ to $H^i G(A^{\bullet})$ for all $A^{\bullet} \in \overline{\mathbf{D}}_{\#}^{+}(\mathcal{A})$ and all $i \in \mathbb{Z}$. For this, remark (1.11.2)(ii) allows us to replace $A^{\bullet}$ by $\tau_{\leq i+d}A^{\bullet} \in \overline{\mathbf{D}}_{\#}^{\mathbf{b}}(\mathcal{A})$ for any $d \geq \max(\dim^{-} F, \dim^{-} G)$, and then (i) applies.

(iii) Similar to (ii).

(iv) As in the proof of (i), (1.10.2.1) with $n = 0$ gives rise to a map of triangles, induced by η:

$$F(\tau_{\geq 0}A^{\bullet})[-1]) \longrightarrow F(\tau_{\leq -1}A^{\bullet}) \longrightarrow F(A^{\bullet}) \longrightarrow F(\tau_{\geq 0}A^{\bullet})$$
$$\downarrow \simeq \qquad\qquad \downarrow \simeq \qquad\qquad ? \downarrow \qquad\qquad \simeq \downarrow$$
$$G((\tau_{\geq 0}A^{\bullet})[-1]) \longrightarrow G(\tau_{\leq -1}A^{\bullet}) \longrightarrow G(A^{\bullet}) \longrightarrow G(\tau_{\geq 0}A^{\bullet})$$

in which the maps other than ? are isomorphisms by (ii) and (iii), whence, by $(\Delta 3)^*$ of §1.4, so is ?.

For (1.11.3.1), it now suffices to show that $\eta(X)$ is an isomorphism for all objects $X \in \mathcal{A}^{\#}$. By a standard resolution argument (see [**H**, p. 43]), X is isomorphic in $\mathbf{D}_{\#}(\mathcal{A})$ to a bounded-below complex $I^{\bullet}$ of objects of **I** (resp. bounded-above complex $P^{\bullet}$ of objects of **P**), and so it suffices to show that $\eta(I^{\bullet})$ (resp. $\eta(P^{\bullet})$) is an isomorphism for any such $I^{\bullet}$ (resp. $P^{\bullet}$). This is done as above, except that in the inductive step in (i), say for bounded $I^{\bullet}$, one uses instead of (1.10.2.2) the triangle associated as in (1.4.3) to the natural semi-split exact sequence

$$0 \longrightarrow I^n[-n] \longrightarrow \tau'_{\leq n} I^\bullet \longrightarrow \tau'_{\leq n-1} I^\bullet \longrightarrow 0$$

where for any $A^\bullet$ and $m \in \mathbb{Z}$, $\tau'_{\leq m} A^\bullet$ is the complex

$$\cdots \longrightarrow A^{m-2} \longrightarrow A^{m-1} \longrightarrow A^m \longrightarrow 0 \longrightarrow 0 \longrightarrow \cdots ;$$

and in (ii), for example, one replaces $I^\bullet$ by the bounded complex $\tau'_{\leq i+d+1} I^\bullet$.

Similar arguments settle the contravariant case. (Or, use the exercise just before (1.5.1).)   Q.E.D.

Chapter 2
Derived Functors

Derived functors are Δ-functors out of derived categories, giving rise, upon application of homology, to functors such as Ext, Tor, and their sheaf-theoretic variants—in particular sheaf cohomology. Derived functors are characterized in §2.1 below by a universal property, and conditions for their existence are given in 2.2, leading up to the construction of right-derived functors via injective resolutions in 2.3 and, dually, of some left-derived functors via flat resolutions in 2.5. We use ideas of Spaltenstein [**Sp**] to deal throughout with unbounded complexes. The basic examples $\mathbf{R}\mathcal{H}om^{\bullet}$ and $\underset{=}{\otimes}$ are described in 2.4 and 2.5 respectively. Illustrating all that has gone before, their relation "adjoint associativity" is given in 2.6, which also includes an abbreviated discussion of what is, in all conscience, involved in constructing natural transformations of multivariate derived functors: a host of underlying category-theoretic trivialities, usually ignored, but of whose existence one should at least be aware. The last section 2.7 develops further refinements.

2.1 Definition of Derived Functors

Fix an abelian category $\mathcal{A}$, let $\mathbf{J}$ be a Δ-subcategory of $\mathbf{K}(\mathcal{A})$, let $\mathbf{D_J}$ be the corresponding derived category, and let

$$Q = Q_J \colon \mathbf{J} \to \mathbf{D_J}$$

be the canonical Δ-functor (see (1.7)). For any Δ-functors F and G from $\mathbf{J}$ to another Δ-category $\mathbf{E}$, or from $\mathbf{D_J}$ to $\mathbf{E}$, $\mathrm{Hom}(F, G)$ will denote the abelian group of Δ-functor morphisms from F to G.

Definition 2.1.1. A Δ-functor $F \colon \mathbf{J} \to \mathbf{E}$ is right-derivable if there exists a Δ-functor

$$\mathbf{R}F \colon \mathbf{D_J} \to \mathbf{E}$$

J. Lipman, M. Hashimoto, *Foundations of Grothendieck Duality for Diagrams of Schemes*, Lecture Notes in Mathematics 1960,
© Springer-Verlag Berlin Heidelberg 2009

and a morphism of Δ-functors

$$\zeta: F \to \mathbf{R}F \circ Q$$

such that for every Δ-functor $G: \mathbf{D_J} \to \mathbf{E}$ the composed map

$$\mathrm{Hom}(\mathbf{R}F,\, G) \xrightarrow{\text{natural}} \mathrm{Hom}(\mathbf{R}F \circ Q,\, G \circ Q) \xrightarrow{\text{via } \zeta} \mathrm{Hom}(F,\, G \circ Q)$$

is an isomorphism (i.e., by (1.5.1), the map "via ζ" is an isomorphism).

The Δ-functor F is left-derivable if there exists a Δ-functor

$$\mathbf{L}F: \mathbf{D_J} \to \mathbf{E}$$

and a morphism of Δ-functors

$$\xi: \mathbf{L}F \circ Q \to F$$

such that for every Δ-functor $G: \mathbf{D_J} \to \mathbf{E}$ the composed map

$$\mathrm{Hom}(G,\, \mathbf{L}F) \xrightarrow{\text{natural}} \mathrm{Hom}(G \circ Q,\, \mathbf{L}F \circ Q) \xrightarrow{\text{via } \xi} \mathrm{Hom}(G \circ Q,\, F)$$

is an isomorphism (i.e., by (1.5.1), the map "via ξ" is an isomorphism).

Such a pair $(\mathbf{R}F, \zeta)$ (respectively: $(\mathbf{L}F, \xi)$) is called a right-derived (respectively: left-derived) *functor of F*.

As in (1.5.1), composition with Q gives an embedding of Δ-functor categories

$$\mathbf{Hom_\Delta}(\mathbf{D_J},\, \mathbf{E}) \hookrightarrow \mathbf{Hom_\Delta}(\mathbf{J},\, \mathbf{E}), \qquad (2.1.1.1)$$

with image the full subcategory whose objects are the Δ-functors which transform quasi-isomorphisms into isomorphisms. Consequently we can regard a right-(left-)derived functor of F as an *initial (terminal) object* [**M**, p. 20] in the category of Δ-functor morphisms $F \to G'$ ($G' \to F$) where G' ranges over all Δ-functors from $\mathbf{J}$ to $\mathbf{E}$ which transform quasi-isomorphisms into isomorphisms. As such, the pair $(\mathbf{R}F, \zeta)$ (or $(\mathbf{L}f, \xi)$)—if it exists—is unique up to canonical isomorphism.

Complement 2.1.2. Let $\mathcal{A}'$ be another abelian category. Any additive functor $F: \mathcal{A} \to \mathcal{A}'$ extends to a Δ-functor $\bar{F}: \mathbf{K}(\mathcal{A}) \to \mathbf{K}(\mathcal{A}')$ (see (1.5.2)). $Q': \mathbf{K}(\mathcal{A}') \to \mathbf{D}(\mathcal{A}')$ being the canonical map, we will refer to derived functors of $Q'\bar{F}$, or of the restriction of $Q'\bar{F}$ to some specified Δ-subcategory $\mathbf{J}$ of $\mathbf{K}(\mathcal{A})$, as being "derived functors of F" and denote them by $\mathbf{R}F$ or $\mathbf{L}F$.

Example 2.1.3. If $F: \mathbf{J} \to \mathbf{E}$ transforms quasi-isomorphisms into isomorphisms then $F = \widetilde{F} \circ Q$ for a unique $\widetilde{F}: \mathbf{D_J} \to \mathbf{E}$; and $(\widetilde{F}, \text{identity})$ is both a right-derived and a left-derived functor of F.

Remarks 2.1.4. Let $\mathcal{A}'$ be an abelian category, and in (2.1.1) suppose that $\mathbf{E}$ is a Δ-subcategory of $\mathbf{K}(\mathcal{A}')$ or of $\mathbf{D}(\mathcal{A}')$. If $\mathbf{R}F$ exists we can set

$$\mathbf{R}^i F(A) := H^i(\mathbf{R}F(A)) \qquad (A \in \mathbf{J}, \ i \in \mathbb{Z}).$$

Since $\mathbf{R}F$ is a Δ-functor, any triangle $A \to B \to C \to A[1]$ in $\mathbf{J}$ is transformed by $\mathbf{R}F$ into a triangle in $\mathbf{E}$, and hence we have an exact homology sequence (see $(1.4.5)^{\mathrm{H}}$):

$$\cdots \to \mathbf{R}^{i-1}F(C) \to \mathbf{R}^i F(A) \to \mathbf{R}^i F(B) \to \mathbf{R}^i F(C) \to \mathbf{R}^{i+1}F(A) \to \cdots$$
$$(2.1.4)^{\mathrm{H}}$$

This applies in particular to the triangle $(1.4.4.2)^\sim$ associated to an exact sequence of $\mathcal{A}$-complexes

$$0 \to A \to B \to C \to 0 \qquad (A, B, C \in \mathbf{J}).$$

A similar remark can be made for $\mathbf{L}F$.

2.2 Existence of Derived Functors

Derivability of a given functor is often proved by reduction, via suitable Δ-equivalences of categories, to the trivial example (2.1.3), as we now explain—and summarize in (2.2.6).

We consider, as in (1.7), a diagram

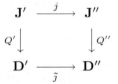

where $\mathbf{J}' \subset \mathbf{J}''$ are Δ-subcategories of $\mathbf{K}(\mathcal{A})$, $\mathbf{D}'$ and $\mathbf{D}''$ are the corresponding derived categories, Q' and Q'' are the canonical Δ-functors, j is the inclusion, and $\tilde{j}$ is the unique Δ-functor making the diagram commute; and we assume that the conditions of (1.7.2) or of $(1.7.2)^{\mathsf{op}}$ obtain. In other words we have a family of quasi-isomorphisms

$$\psi_X \colon A_X \to X, \qquad X \in \mathbf{J}'', \ A_X \in \mathbf{J}', \quad (\text{see } (1.7.2)), \qquad (2.2.1)$$

or a family of quasi-isomorphisms

$$\varphi_X \colon X \to A_X, \qquad X \in \mathbf{J}'', \ A_X \in \mathbf{J}', \quad (\text{see } (1.7.2)^{\mathsf{op}}). \qquad (2.2.1)^{\mathsf{op}}$$

In either situation, $\tilde{j}$ identifies $\mathbf{D}'$ with a Δ-subcategory of $\mathbf{D}''$; there is a Δ-functor $(\rho, \theta) \colon \mathbf{D}'' \to \mathbf{D}'$ with

$$\rho(X) = A_X \qquad (X \in \mathbf{J}'');$$

and there are isomorphisms of Δ-functors

$$1_{\mathbf{D}''} \xrightarrow{\sim} \tilde{j}\rho, \qquad 1_{\mathbf{D}'} \xrightarrow{\sim} \rho\tilde{j} \qquad\qquad (2.2.2)$$

induced by ψ or by φ.

Proposition 2.2.3. *With preceding notation, let $\mathbf{E}$ be a Δ-category, let $F \colon \mathbf{J}'' \to \mathbf{E}$ be a Δ-functor, and suppose that the restricted functor*

$$F' := F \circ j \colon \mathbf{J}' \to \mathbf{E}$$

has a right-derived functor

$$\mathbf{R}F' \colon \mathbf{D}' \to \mathbf{E}, \qquad \zeta' \colon F' \to \mathbf{R}F' \circ Q'.$$

If there exists a family $\varphi_X \colon X \to A_X$ as in $(2.2.1)^{\mathrm{op}}$, whence a functor ρ as above, then F has the right-derived functor $(\mathbf{R}F, \zeta)$ where

$$\mathbf{R}F = \mathbf{R}F' \circ \rho \colon \mathbf{D}'' \to \mathbf{E}$$

so that

$$\mathbf{R}F(X) = \mathbf{R}F'(A_X) \qquad (X \in \mathbf{J}''),$$

and where for each $X \in \mathbf{J}''$, $\zeta(X)$ is the composition

$$F(X) \xrightarrow{F(\varphi_X)} F(A_X) = F'(A_X) \xrightarrow{\zeta'(A_X)} \mathbf{R}F'(A_X) = \mathbf{R}F(X).$$

A similar statement holds for left-derived functors when there exists a family ψ_X as in (2.2).

Proof. We check first that ζ is actually a morphism of Δ-functors. Consider a map $u \colon X \to Y$ in $\mathbf{J}''$. Since $Q''(\varphi_X)$ is an isomorphism, there is a unique map $\tilde{u} \colon A_X \to A_Y$ in $\mathbf{D}''$ (and hence in the full subcategory $\mathbf{D}'$) making the following $\mathbf{D}''$-diagram commute:

$$
\begin{array}{ccc}
X & \xrightarrow{\;Q''(\varphi_X)\;} & A_X \\
{\scriptstyle Q''(u)}\big\downarrow & & \big\downarrow{\scriptstyle \tilde{u}} \\
Y & \xrightarrow[\;Q''(\varphi_Y)\;]{} & A_Y
\end{array}
$$

By the definition of the functor ρ (see proof of (1.7.2)), that ζ is a morphism of functors means that the following diagram $\mathcal{D}(u)$ commutes for all u:

$$
\begin{array}{ccccc}
F(X) & \xrightarrow{F(\varphi_X)} & F(A_X) & \xrightarrow{\zeta'(A_X)} & \mathbf{R}F'(A_X) \\
{\scriptstyle F(u)}\downarrow & & {\scriptstyle ?}\Big\downarrow & & \Big\downarrow{\scriptstyle \mathbf{R}F'(\tilde{u})} \\
F(Y) & \xrightarrow[F(\varphi_Y)]{} & F(A_Y) & \xrightarrow[\zeta'(A_Y)]{} & \mathbf{R}F'(A_Y)
\end{array}
$$

If there were a $\mathbf{J}'$-map $u' \colon A_X \to A_Y$ such that $u'\varphi_X = \varphi_Y u$, whence $Q''(u')Q''(\varphi_X) = Q''(\varphi_Y)Q''(u)$ and $\tilde{u} = Q''(u') = Q'(u')$, then the broken arrow in $\mathcal{D}(u)$ could be replaced by the map $F(u')$, making both resulting subdiagrams of $\mathcal{D}(u)$, and hence $\mathcal{D}(u)$ itself, commute. We don't know that such a u' exists; but, I claim, *there exists a quasi-isomorphism $v \colon Y \to Z$ such that (with self-explanatory notation) both v' and $(vu)'$ exist.* This being so, both diagrams $\mathcal{D}(v)$ and $\mathcal{D}(vu)$ commute; and since $\tilde{v}$ is an isomorphism (because v is a quasi-isomorphism), therefore $\mathbf{R}F'(\tilde{v})$ is an isomorphism, and it follows easily that $\mathcal{D}(u)$ also commutes, as desired.

To verify the claim, use (1.6.3) to construct in $\mathbf{J}''$ a commutative diagram

$$
\begin{array}{ccccccc}
X & \xrightarrow{\varphi_X} & A_X & & & & \\
{\scriptstyle u}\downarrow & \searrow & & \searrow {\scriptstyle w} & & & \\
Y & \xrightarrow[\varphi_Y]{} & A_Y & \xrightarrow[\varphi]{} & Z & \xrightarrow[\varphi_Z]{} & A_Z
\end{array}
$$

with φ a quasi-isomorphism, and set

$$
v := \varphi \circ \varphi_Y
$$
$$
v' := \varphi_Z \circ \varphi
$$
$$
(vu)' := \varphi_Z \circ w.
$$

Then $v'\varphi_Y = \varphi_Z v$ and $(vu)'\varphi_X = \varphi_Z(vu)$, as desired.

Thus ζ is a morphism of functors; and it is straightforward to check, via commutativity of (1.7.2.2), that ζ is in fact a morphism of Δ-functors.

Now we need to show (see (2.1.1)) that for every Δ-functor $G \colon \mathbf{D}'' \to \mathbf{E}$ the composed map

$$
\mathrm{Hom}(\mathbf{R}F, G) \xrightarrow{(1.5.1)} \mathrm{Hom}(\mathbf{R}F \circ Q'', G \circ Q'') \xrightarrow{\text{via } \zeta} \mathrm{Hom}(F, G \circ Q'')
$$

is *bijective.* For this it suffices to check that the following natural composition is an inverse map:

$$\mathrm{Hom}(F, G \circ Q'') \longrightarrow \mathrm{Hom}(F \circ j, G \circ Q'' \circ j)$$

$$=\!=\!= \mathrm{Hom}(F', G \circ \tilde{\jmath} \circ Q')$$

$$\xrightarrow{(2.1.1)} \mathrm{Hom}(\mathbf{R}F', G \circ \tilde{\jmath})$$

$$\longrightarrow \mathrm{Hom}(\mathbf{R}F' \circ \rho, G \circ \tilde{\jmath} \circ \rho)$$

$$\xrightarrow{(2.2.2)} \mathrm{Hom}(\mathbf{R}F' \circ \rho, G)$$

$$=\!=\!= \mathrm{Hom}(\mathbf{R}F, G).$$

This checking is left to the reader, as is the proof for left-derived functors.
$$\mathrm{Q.E.D.}$$

Example 2.2.4 [**H**, p. 53, Thm. 5.1]. Let $j: \mathbf{J}' \hookrightarrow \mathbf{J}''$, $F: \mathbf{J}'' \to \mathbf{E}$, and $\varphi_X: X \to A_X$ be as above, and suppose that the restricted functor $F' := F \circ j$ transforms quasi-isomorphisms into isomorphisms (or, equivalently, $F(C) \cong 0$ for every exact complex $C \in \mathbf{J}'$, see (1.5.1)). Then by (2.1.3), F' has a right-derived functor $(\mathbf{R}F', \mathbf{1})$ where $F' = \mathbf{R}F' \circ Q'$ and $\mathbf{1}$ is the identity morphism of F'.

So by (2.2.3), F has a right-derived functor $(\mathbf{R}F, \zeta)$ with

$$\mathbf{R}F(X) = F(A_X)$$

and

$$\zeta(X) = F(\varphi_X): F(X) \to F(A_X) = \mathbf{R}F(X)$$

for all $X \in \mathbf{J}''$. Note that if $X \in \mathbf{J}'$ then φ_X is a quasi-isomorphism in $\mathbf{J}'$, whence $\zeta(X)$ is an isomorphism.

The action of $\mathbf{R}F$ on maps can be described thus: if $u: X \to Y$ is a map in $\mathbf{J}''$ then with v' and $(vu)'$ as in the preceding proof,

$$\mathbf{R}F(u/1) = F(v')^{-1} \circ F((vu)');$$

and for any map f/s in $\mathbf{D}''$ (see §1.2), we have

$$\mathbf{R}F(f/s) = \mathbf{R}F(f/1) \circ \mathbf{R}F(s/1)^{-1}.$$

As for the Δ-structure on $\mathbf{R}F$, one has for each X the isomorphism

$$\theta(X): \mathbf{R}F(X[1]) = F(A_{X[1]}) \xrightarrow[F(\eta_X)]{\sim} F(A_X[1]) \xrightarrow[\theta_F]{\sim} F(A_X)[1] = \mathbf{R}F(X)[1]$$

where

$$\eta_X := Q''(\varphi_X[1]) \circ Q''(\varphi_{X[1]})^{-1}: A_{X[1]} \xrightarrow{\sim} A_X[1],$$

and where the isomorphism θ_F comes from the Δ-functoriality of F.

(2.2.5). Let $\mathcal{A}$ be an abelian category, let $\mathbf{J}$ be a Δ-subcategory of $\mathbf{K}(\mathcal{A})$, and let F be a Δ-functor from $\mathbf{J}$ to a Δ-category $\mathbf{E}$. We say that a complex X in $\mathbf{J}$ is *right-F-acyclic* if for each quasi-isomorphism $u\colon X \to Y$ in $\mathbf{J}$ there is a quasi-isomorphism $v\colon Y \to Z$ in $\mathbf{J}$ such that the map $F(vu)\colon F(X) \to F(Z)$ is an isomorphism. *Left-F-acyclicity* is defined similarly, with arrows reversed.

For example, if $\mathbf{J} := \mathbf{J}''$ in (2.2.4), then every complex $X \in \mathbf{J}'$ is right-F-acyclic—just take $Z := A_Y$ and $v := \varphi_Y$. Conversely:

Lemma 2.2.5.1. *The right-F-acyclic complexes in $\mathbf{J}$ are the objects of a localizing subcategory (§1.7). Moreover, the restriction of F to this subcategory transforms quasi-isomorphisms into isomorphisms; in other words, if the complex X is both exact and right-F-acyclic, then $F(X) \cong 0$ (see (1.5.1)).*

Proof. Since F commutes with translation—up to isomorphism—it is clear that X is right-F-acyclic iff so is $X[1]$.

Next, suppose we have a triangle $X \to X_1 \to X_2 \to X[1]$ in which X_1 and X_2 are right-F-acyclic. We will show that then X is right-F-acyclic. Any quasi-isomorphism $u\colon X \to Y$ can be embedded into a map of triangles

$$
\begin{array}{ccccccc}
X & \longrightarrow & X_1 & \longrightarrow & X_2 & \longrightarrow & X[1] \\
\downarrow{\scriptstyle u} & & \downarrow{\scriptstyle u_1} & & \downarrow{\scriptstyle u_2} & & \downarrow{\scriptstyle u[1]} \\
Y & \longrightarrow & Y_1 & \longrightarrow & Y_2 & \longrightarrow & Y[1]
\end{array}
$$

where u_1 is a quasi-isomorphism whose existence is given by (1.6.3), and where u_2 is then given by $(\Delta 3)'$ and $(\Delta 3)''$ in §1.4. Such a u_2 is also a quasi-isomorphism, as one sees by applying the five-lemma to the natural map between the homology sequences of the two triangles (see $(1.4.5)^{\mathrm{H}}$). Similarly, from the definition of right-F-acyclic we deduce a triangle-map

$$
\begin{array}{ccccccc}
Y_1 & \longrightarrow & Y_2 & \longrightarrow & Y[1] & \longrightarrow & Y_1[1] \\
\downarrow{\scriptstyle v_1} & & \downarrow{\scriptstyle v_2} & & \downarrow{\scriptstyle v[1]} & & \downarrow{\scriptstyle v_1[1]} \\
Z_1 & \longrightarrow & Z_2 & \longrightarrow & Z[1] & \longrightarrow & Z_1[1]
\end{array}
$$

where v_1, v_2, and v are quasi-isomorphisms such that $F(v_1 u_1)$ and $F(v_2 u_2)$ are isomorphisms. (Here $(\Delta 2)$ in §1.4 should be kept in mind.) We can then apply the Δ-functor F to the map of triangles

$$
\begin{array}{ccccccc}
X_1 & \longrightarrow & X_2 & \longrightarrow & X[1] & \longrightarrow & X_1[1] \\
\downarrow{\scriptstyle v_1 u_1} & & \downarrow{\scriptstyle v_2 u_2} & & \downarrow{\scriptstyle (vu)[1]} & & \downarrow{\scriptstyle (v_1 u_1)[1]} \\
Z_1 & \longrightarrow & Z_2 & \longrightarrow & Z[1] & \longrightarrow & Z_1[1]
\end{array}
$$

and deduce from $(\Delta 3)^*$ that $F((vu)[1])$, and hence $F(vu)$, is also an isomorphism. Thus X is indeed right-F-acyclic.

In particular, the direct sum of two right-F-acyclic complexes is right-F-acyclic, because the direct sum is the summit of a triangle whose base is the zero-map from one to the other, see (1.4.2.1). Also, $0 \in \mathbf{J}$ is clearly right-F-acyclic. We see then that the right-F-acyclic complexes are the objects of a Δ-subcategory of $\mathbf{J}$.

For this subcategory to be localizing it suffices, by $(1.7.1)^{\mathrm{op}}$, that if $X \to Y \to Z$ is as in the definition of right-F-acyclic, then Z is right-F-acyclic; and this follows from:

Lemma 2.2.5.2. *If X is right-F-acyclic and if there exists a quasi-isomorphism $\alpha \colon X \to Z$ such that $F(\alpha) \colon F(X) \to F(Z)$ is an epimorphism, then Z is right-F-acyclic.*

Proof. Given a quasi-isomorphism $Z \to Y'$, there exists a quasi-isomorphism $Y' \to Z'$ such that $F(X) \to F(Z) \to F(Z')$ is an isomorphism (since X is right-F-acyclic); and since $F(X) \to F(Z)$ is an epimorphism, therefore $F(Z) \to F(Z')$ is an isomorphism.   Q.E.D.

To justify the last assertion in (2.2.5.1), take $Y := 0$ in the definition of right-F-acyclicity.   Q.E.D.

We leave it to the reader to establish a corresponding statement for left-F-acyclic complexes.

In summary:

Proposition 2.2.6. *Let $\mathcal{A}$ be an abelian category, let $\mathbf{J}$ be a Δ-subcategory of $\mathbf{K}(\mathcal{A})$, and let F be a Δ-functor from $\mathbf{J}$ to a Δ-category $\mathbf{E}$. Suppose $\mathbf{J}$ contains a family of quasi-isomorphisms $\varphi_X \colon X \to A_X$ ($X \in \mathbf{J}$) such that A_X is right-F-acyclic for all X, see (2.2.5). Then F has a right-derived functor $(\mathbf{R}F, \zeta)$ such that for all $X \in \mathbf{J}$,*

$$\mathbf{R}F(X) = F(A_X) \quad \text{and} \quad \zeta(X) = F(\varphi_X) \colon F(X) \to F(A_X) = \mathbf{R}F(X).$$

Moreover, X is right-F-acyclic $\Leftrightarrow \zeta(X)$ is an isomorphism.

Proof. Everything is contained in (2.2.4) and (2.2.5), except for the fact that if $\zeta(X)$ is an isomorphism then X is right-F-acyclic, which is proved by taking, in (2.2.5), $Z := A_Y$, $v := \varphi_Y$, and noting that then $F(vu)$ is the composite isomorphism

$$F(X) \xrightarrow[\zeta(X)]{\sim} \mathbf{R}F(X) \xrightarrow{\sim} \mathbf{R}F(Y) = F(Z).$$

                                                                          Q.E.D.

Corollary 2.2.6.1. *With assumptions as in (2.2.6), if $G \colon \mathbf{E} \to \mathbf{E}'$ is any Δ-functor then $(G \circ \mathbf{R}F, G(\zeta))$ is a right-derived functor of GF.*

Proof. Clearly, right-F-acyclic complexes are right-(GF)-acyclic. It follows then from (2.2.4) and (2.2.5) that the assertion need only be proved for the restriction of F to the subcategory of right-F-acyclic complexes, in which case it follows from (2.1.3).   Q.E.D.

Corollary 2.2.7. *Let $\mathcal{A}$, $\mathcal{A}'$ be abelian categories, let $\mathbf{J} \subset \mathbf{K}(\mathcal{A})$, $\mathbf{J}' \subset \mathbf{K}(\mathcal{A}')$ be Δ-subcategories with canonical functors $Q\colon \mathbf{J} \to \mathbf{D_J}$, $Q'\colon \mathbf{J}' \to \mathbf{D_{J'}}$ to their respective derived categories, and let $F\colon \mathbf{J} \to \mathbf{J}'$ and $G\colon \mathbf{J}' \to \mathbf{E}$ be Δ-functors. Assume that G has a right-derived functor $\mathbf{R}G$ and that every complex $X \in \mathbf{J}$ admits a quasi-isomorphism into a right-$(Q'F)$-acyclic complex A_X such that $F(A_X)$ is right-G-acyclic. Then $Q'F$ and GF have right-derived functors, denoted $\mathbf{R}F$ and $\mathbf{R}(GF)$, and there is a unique Δ-functorial isomorphism*

$$\alpha\colon \mathbf{R}(GF) \xrightarrow{\;\sim\;} \mathbf{R}G\mathbf{R}F$$

such that the following natural diagram commutes for all $X \in \mathbf{J}$:

$$
\begin{array}{ccc}
GF(X) & \longrightarrow & \mathbf{R}(GF)(QX) \\
\downarrow & & \simeq \downarrow \alpha(QX) \\
\mathbf{R}GQ'F(X) & \longrightarrow & \mathbf{R}G\mathbf{R}F(QX)
\end{array}
\qquad (2.2.7.1)
$$

Proof. Derivability of $Q'F$ results from (2.2.6). Derivability of GF results similarly once we show, as follows, that A_X is right-(GF)-acyclic: note for any quasi-isomorphism $A_X \to Y$ in $\mathbf{J}$ that, by (2.2.5.1), the resulting composed map $F(A_X) \to F(Y) \to F(A_Y)$ is a quasi-isomorphism and so $GF(A_X) \xrightarrow{\sim} GF(A_Y)$. The existence of a unique Δ-functorial α making (2.2.7.1) commute follows from the definition of right-derived functor. Since A_X is right-(GF)-acyclic and right-$(Q'F)$-acyclic, and $F(A_X)$ is right-G-acyclic, (2.2.6) implies that $\alpha(QX)$ is isomorphic to the identity map of $GF(A_X)$. Thus α is an isomorphism.   Q.E.D.

We leave the corresponding statements for left-F-acyclic complexes and left-derived functors to the reader.

Incidentally, (2.2.6) generalizes in a simple way to triangulation-compatible multiplicative systems in any Δ-category (see [**H**, p. 31]). It is of course of little interest unless we can construct a family (φ_X). That matter is addressed in the following sections.

Exercises 2.2.8. (a) Verify that F transforms quasi-isomorphisms into isomorphisms iff every complex $X \in \mathbf{J}$ is right-F-acyclic.

(b) Verify that if $X \in \mathbf{J}$ is exact then X is right-F-acyclic iff $F(X) \cong 0$.

(c) Let F be a Δ-functor from $\mathbf{J}$ to a Δ-category $\mathbf{E}$. Let $\mathbf{J}'$ be the full subcategory of $\mathbf{J}$ whose objects are all the complexes in $\mathbf{J}$ admitting a quasi-isomorphism to a right-F-acyclic complex. Then $\mathbf{J}'$ is a Δ-subcategory of $\mathbf{J}$.

(d) X is right-F-acyclic iff every map $C \to X$ in $\mathbf{J}$ with C exact factors as $C \to C' \to X$ with C' exact and $F(C') \cong 0$.

(e) X is said to be "unfolded for F" if for every $Z \in \mathbf{E}$ the natural map

$$\mathrm{Hom}_{\mathbf{E}}(Z, F(X)) \to \varinjlim_{X \to Y} \mathrm{Hom}_{\mathbf{E}}(Z, F(Y))$$

is an isomorphism, where the $\varinjlim$ is taken over the category of all quasi-isomorphisms $X \to Y$ in $\mathbf{J}$ [$\mathbf{De}$, p. 274, (iv)]. Check that any right-F-acyclic X is unfolded for F; and that the converse holds under the hypotheses of (2.2.6).

(f) Show: X is unfolded for F iff every map $C \to X$ in $\mathbf{J}$ with C exact factors as $C \to C' \to X$ with C' exact and $F(C) \to F(C')$ the zero map. (For this, the octahedral axiom in $\mathbf{E}$ may be needed, see §1.4.)

2.3 Right-Derived Functors via Injective Resolutions

The basic example of a family (φ_X) as in (2.2.6) arises when $\mathcal{A}$ has enough injectives, i.e., every object of $\mathcal{A}$ admits a monomorphism into an injective object. Then every complex $X \in \overline{\mathbf{K}}^{+}(\mathcal{A})$ admits a quasi-isomorphism $\varphi_X \colon X \to I_X$ into a bounded-below complex of injectives (see (1.8.2)); and by (2.3.4) and (2.3.2.1) below, this I_X is right-F-acyclic for *every* Δ-functor $F \colon \overline{\mathbf{K}}^{+}(\mathcal{A}) \to \mathbf{E}$, whence F is right-derivable.

Later on, however, it will become important for us to be able to deal with *unbounded* complexes; and for this purpose the following more general injectivity notion is, via (2.3.5), essential.

Definition 2.3.1. Let $\mathcal{A}$ be an abelian category, and let $\mathbf{J}$ be a Δ-subcategory of $\mathbf{K}(\mathcal{A})$. A complex $I \in \mathbf{J}$ is said to be q-injective in $\mathbf{J}$ (or $\mathbf{J}$-q-injective) if for every diagram $Y \xleftarrow{s} X \xrightarrow{f} I$ in $\mathbf{J}$ with s a quasi-isomorphism, there exists $g \colon Y \to I$ such that $gs = f$.[1]

Lemma 2.3.2. $I \in \mathbf{J}$ *is* $\mathbf{J}$-*q-injective iff every quasi-isomorphism* $I \to Y$ *in* $\mathbf{J}$ *has a left inverse.*

Proof. In (2.3.1) take $X := I$ and $f :=$ identity to see that if I is q-injective then the quasi-isomorphism s has a left inverse. Conversely, by (1.6.3) any diagram $Y \xleftarrow{s} X \xrightarrow{f} I$ is part of a commutative diagram

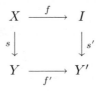

in which s' is a quasi-isomorphism; and then if t is a left inverse for s' and $g := tf'$, we have $gs = f$.   Q.E.D.

---

[1] Here "q" stands for the class of quasi-isomorphisms. The equivalent term "K-injective" in [$\mathbf{Sp}$, p. 127] seems to me less suggestive.

Corollary 2.3.2.1. $I \in \mathbf{J}$ *is* $\mathbf{J}$-q-*injective iff* I *is right-F-acyclic for every* Δ-*functor* $F\colon \mathbf{J} \to \mathbf{E}$.

Proof. If any quasi-isomorphism $I \to Y$ has a left inverse, then setting $X := I$ in (2.2.5) we see at once that I is right-F-acyclic. Conversely, if I is right-F-acyclic for the identity functor $\mathbf{J} \to \mathbf{J}$, then every quasi-isomorphism $I \to Y$ has a left inverse.   Q.E.D.

Taking $F :=$ identity in (2.2.5.1), we deduce:

Corollary 2.3.2.2. *The* $\mathbf{J}$-q-*injective complexes are the objects of a localizing subcategory* $\mathbf{I}$. *Every quasi-isomorphism in* $\mathbf{I}$ *is an isomorphism, so the pair* $(\mathbf{I}, \text{identity})$ *has the universal property of the derived category* $\mathbf{D_I}$ (§1.2), *and therefore* $\mathbf{I} \cong \mathbf{D_I}$ *can be identified with a* Δ-*subcategory of* $\mathbf{D_J}$.

Corollary 2.3.2.3. *Suppose that there exists a family of* q-*injective resolutions* $\varphi_X\colon X \to I_X$ $(X \in \mathbf{J})$, *i.e., for each* X, φ_X *is a quasi-isomorphism and* I_X *is* $\mathbf{J}$-q-*injective. Then any* Δ-*functor* $F\colon \mathbf{J} \to \mathbf{E}$ *has a right-derived functor* $(\mathbf{R}F, \zeta)^2$ *with*

$$\mathbf{R}F(X) = F(I_X) \quad \text{and} \quad \zeta(X) = F(\varphi_X)\colon F(X) \to F(I_X) = \mathbf{R}F(X),$$

and such that for any morphism $f/s\colon X_1 \xleftarrow{s} X \xrightarrow{f} X_2$ *in* $\mathbf{D_J}$,

$$\mathbf{R}F(f/s) = F(f') \circ F(s')^{-1}$$

where f' *is the unique map in* $\mathbf{I}$ *making the following square in* $\mathbf{J}$ *commute*

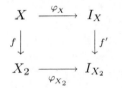

and similarly for s'.

Proof. Since φ_X becomes an isomorphism in $\mathbf{D_J}$, the map f' exists uniquely in $\mathbf{D_J}$, hence in $\mathbf{I}$ (2.3.2.2). For the rest see (2.2.4), with $\mathbf{J}' := \mathbf{I}$, $\mathbf{J}'' := \mathbf{J}$, and $v :=$ identity.   Q.E.D.

Example 2.3.3. An object I in $\mathcal{A}$ is injective iff when considered as a complex vanishing in all nonzero degrees it is q-injective in $\mathbf{K}(\mathcal{A})$ (or in $\mathbf{K}^{\mathrm{b}}(\mathcal{A})$).

Sufficiency: for any $\mathcal{A}$-diagram $Y^0 \xleftarrow{s^0} X \xrightarrow{f} I$ with s^0 a monomorphism, take Y to be the complex which looks like the natural map $Y^0 \to \mathrm{coker}(s^0)$ in degrees 0 and 1, and vanishes elsewhere, and take $s\colon X \to Y$ to be the obvious quasi-isomorphism; then deduce from (2.3.1) that if I is q-injective there exists $g^0\colon Y^0 \to I$ such that $g^0 s^0 = f$—so that I is $\mathcal{A}$-injective.

---

[2] So the embedding functor (2.1.1.1) has a *left adjoint*, taking F to $\mathbf{R}F$.

For necessity, use (2.3.2): to find a left inverse in $\mathbf{K}(\mathcal{A})$ for a quasi-isomorphism $\beta\colon I \to Y$ we may replace Y by the complex $\tau_{\geq 0}Y$, to which Y maps quasi-isomorphically (§1.10), i.e., we may assume that Y vanishes in all negative degrees; then β induces a monomorphism (in $\mathcal{A}$) $\beta^0\colon I \to Y^0$, which has a left inverse if I is $\mathcal{A}$-injective, and that gives rise, obviously, to a left inverse for β. (One could also use (iv) in (2.3.8) below.)

Example 2.3.4. Any bounded-below complex I of $\mathcal{A}$-injectives is q-injective in $\mathbf{K}(\mathcal{A})$. Indeed, by [**H**, p. 41, Lemma 4.5], I satisfies the condition in (2.3.2). (One could also use (2.3.8)(iv).) Thus (2.3.2.3) applies to $\mathbf{J} := \overline{\mathbf{K}}^+(\mathcal{A})$ whenever $\mathcal{A}$ has enough injectives (see beginning of §2.3). In that case, further, every $\overline{\mathbf{K}}^+(\mathcal{A})$-q-injective complex admits a quasi-isomorphism, hence, by (2.3.2.2), an *isomorphism,* to a bounded-below complex of $\mathcal{A}$-injectives.

Example 2.3.5. Let U be a topological space, $\mathcal{O}$ a sheaf of rings on U, and $\mathcal{A}$ the abelian category of left $\mathcal{O}$-modules. Then a theorem of Spaltenstein [**Sp**, p. 138, Theorem 4.5] asserts that *every complex in $\mathbf{K}(\mathcal{A})$ admits a q-injective resolution.* Hence by (2.3.2.3), every Δ-functor out of $\mathbf{K}(\mathcal{A})$ is right-derivable.

More generally, a q-injective resolution exists for every complex in any Grothendieck category, i.e., an abelian category with exact direct limits and having a generator [**AJS**, p. 243, Theorem 5.4]. For example, injective Cartan-Eilenberg resolutions [**EGA**, III, Chap. 0, (11.4.2)] always exist in Grothendieck categories; and their totalizations—which generally require countable direct products—give q-injective resolutions when such products of epimorphisms are epimorphisms (a condition which holds in the category of modules over a fixed ring, but fails, for instance, in most categories of sheaves on topological spaces).

Example 2.3.6. Let $\mathcal{A}_1$, $\mathcal{A}_2$ be abelian categories, $\mathcal{A}_1$ having enough injectives. As in (1.5.2) any additive functor $F\colon \mathcal{A}_1 \to \mathcal{A}_2$ extends to a Δ-functor $\bar{F}\colon \mathbf{K}^+(\mathcal{A}_1) \to \mathbf{K}^+(\mathcal{A}_2)$ which has, by (2.3.4), a right-derived functor

$$\mathbf{R}^+\bar{F}\colon \mathbf{D}^+(\mathcal{A}_1) \to \mathbf{K}^+(\mathcal{A}_2)$$

satisfying, for a given family $\varphi_X\colon X \to I_X$ of injective resolutions,

$$\mathbf{R}^+\bar{F}(X) = \bar{F}(I_X).$$

We can extend the domain of $\mathbf{R}^+\bar{F}$ to $\overline{\mathbf{D}}^+(\mathcal{A}_1)$ by composing with the equivalence τ^+ defined in (1.8.1).

Moreover, if every $\mathcal{A}_1$-complex has a q-injective resolution, then there is a further extension to a derived functor $\mathbf{R}\bar{F}\colon \mathbf{D}(\mathcal{A}_1) \to \mathbf{K}(\mathcal{A}_2)$—whose composition with the canonical map $\mathbf{K}(\mathcal{A}_2) \to \mathbf{D}(\mathcal{A}_2)$ is $\mathbf{R}F$, see (2.1.2).

With H^i the usual homology functor, let $R^i F \colon \mathcal{A}_1 \to \mathcal{A}_2$ $(i \in \mathbb{Z})$ be the composition

$$\mathcal{A}_1 \xrightarrow{\ (1.2.2)\ } \mathbf{D}^+(\mathcal{A}_1) \xrightarrow{\ \mathbf{R}^+ F\ } \mathbf{K}^+(\mathcal{A}_2) \xrightarrow{\ H^i\ } \mathcal{A}_2$$

(cf. (2.1.4)). Then $R^i F = 0$ for $i < 0$, and there is a natural map of functors $F \to R^0 F$ which is an isomorphism if and only if F is left-exact.

Example 2.3.7. Let $f \colon U_1 \to U_2$ be a continuous map of topological spaces. Let $\mathcal{A}_i$ be the category of sheaves of abelian groups on U_i $(i = 1, 2)$. Then $\mathcal{A}_i$ is abelian, and has enough injectives. The direct image functor $f_* \colon \mathcal{A}_1 \to \mathcal{A}_2$ is left-exact, and has, as in (2.3.6), a derived functor

$$\mathbf{R}^+ \overline{f_*} \colon \overline{\mathbf{D}}^+(\mathcal{A}_1) \to \mathbf{K}^+(\mathcal{A}_2) \, .$$

By (2.3.5), the composition $\mathbf{K}(\mathcal{A}_1) \xrightarrow{\overline{f_*}} \mathbf{K}(\mathcal{A}_2) \xrightarrow{Q} \mathbf{D}(\mathcal{A}_2)$ has a derived functor $\mathbf{R}f_*$, whose restriction to $\overline{\mathbf{D}}^+(\mathcal{A}_1)$ is isomorphic to $Q \circ \mathbf{R}^+ \overline{f_*}$.

In particular, when U_2 is a single point then $\mathcal{A}_2 = \mathfrak{Ab}$, the category of abelian groups, and f_* is the global section functor $\Gamma = \Gamma(U_1, -)$. In this case one usually sets, for $i \in \mathbb{Z}$, see (2.1.4),

$$\mathbf{R}f_* = \mathbf{R}\Gamma, \qquad \mathbf{R}^i f_* = \mathbf{R}^i \Gamma = \mathbf{H}^i, \qquad R^i f_*(-) = H^i(U_1, -) \, .$$

Here are some other characterizations of q-injectivity, see [**Sp**, p. 129, Prop. 1.5], [**BN**, Def. 2.6 etc.].

Proposition 2.3.8. *Let $\mathcal{A}$ be an abelian category, and let $\mathbf{J}$ be a Δ-subcategory. of $\mathbf{K}(\mathcal{A})$. The following conditions on a complex $I \in \mathbf{J}$ are equivalent:*

(i) *I is q-injective in $\mathbf{J}$.*

(i)' *For every diagram $Y \xleftarrow{s} X \xrightarrow{f} I$ in $\mathbf{J}$ with s a quasi-isomorphism there is a unique $g \colon Y \to I$ such that $gs = f$.*

(ii) *Every quasi-isomorphism $I \to Y$ in $\mathbf{J}$ has a left inverse.*

(ii)' *Every quasi-isomorphism $I \to Y$ in $\mathbf{J}$ is a monomorphism.*

(iii) *I is right-F-acyclic for every Δ-functor $F \colon \mathbf{J} \to \mathbf{E}$.*

(iii)' *I is right-F-acyclic for F the identity functor $\mathbf{J} \to \mathbf{J}$.*

(iv) *For every exact complex $X \in \mathbf{J}$, we have $\operatorname{Hom}_{\mathbf{J}}(X, I) = 0$.*

(iv)' *The Δ-functor $\operatorname{Hom}^\bullet(-, I) \colon \mathbf{J} \to \mathbf{K}(\mathfrak{Ab})$ of (1.5.3) takes quasi-isomorphisms into quasi-isomorphisms.*

(v) *For every complex $X \in \mathbf{J}$, the natural map*

$$\operatorname{Hom}_{\mathbf{J}}(X, I) \to \operatorname{Hom}_{\mathbf{D}_{\mathbf{J}}}(X, I)$$

is bijective.

Proof. The equivalence of (i), (ii), (iii) and (iii)' has already been shown (see (2.3.2) and the proof of (2.3.2.1)). For (ii) $\Leftrightarrow$ (ii)' see (1.4.2.1). Taking $Y := 0$ in (2.3.1), we see that (i) $\Rightarrow$ (iv). The equivalence of (iv) and (iv)' results from the footnote in (1.5.1) and the easily-checked relation

$$H^n\big(\mathrm{Hom}^\bullet(X,I)\big) \cong \mathrm{Hom}_{\mathbf{J}}(X[-n],I) \qquad (n \in \mathbb{Z},\ X \in \mathbf{J}). \qquad (2.3.8.1)$$

The implications (v) $\Rightarrow$ (i)$'$ $\Rightarrow$ (i) are simple to verify.

We show next that (iv) $\Rightarrow$ (ii). Let X be the summit of a triangle T in $\mathbf{J}$ whose base is a quasi-isomorphism $I \to Y$. By [**H**, p. 23, 1.1 b)], the resulting sequence

$$\mathrm{Hom}(X,I) \to \mathrm{Hom}(Y,I) \to \mathrm{Hom}(I,I) \to \mathrm{Hom}(X[-1],I)$$

is *exact.* Moreover, the exact homology sequence $(1.4.5)^{\mathrm{H}}$ of T shows that X is exact. So if (iv) holds, then $\mathrm{Hom}(Y,I) \to \mathrm{Hom}(I,I)$ is bijective, and (ii) follows.

Finally, we show (ii) $\Rightarrow$ (v). For any map $f/s\colon X \to I$ in $\mathbf{D_J}$, (1.6.3) yields a commutative diagram in $\mathbf{J}$, with s' a quasi-isomorphism:

If $ts' = $ identity, then $f/s = (s'/1)^{-1}(f'/1) = (tf')/1$, and so the map $\mathrm{Hom}_{\mathbf{J}}(X,I) \to \mathrm{Hom}_{\mathbf{D_J}}(X,I)$ is surjective. For injectivity, given $f\colon X \to I$ in $\mathbf{J}$, note that $f/1 = 0 \implies$ there exists a quasi-isomorphism $t\colon X' \to X$ such that $ft = 0$ (see §1.2) $\implies$ there exists a quasi-isomorphism $s\colon I \to Y$ such that $sf = 0$ [**H**, p. 37]; and if s has a left inverse, then $sf = 0 \implies f = 0$.
$$\text{Q.E.D.}$$

Exercise 2.3.9. Show that if $\mathcal{A}$ is a Grothendieck category then $\mathbf{D}(\mathcal{A})$ is equivalent to the homotopy category of q-injective complexes. Hence if $\mathcal{A}$ has inverse limits then so does $\mathbf{D}(\mathcal{A})$.

2.4 Derived Homomorphism Functors

Let $\mathcal{A}$ be an abelian category, and let $\mathbf{L}$ be a Δ-subcategory of $\mathbf{K}(\mathcal{A})$ in which there exists a family of quasi-isomorphisms $\varphi_X\colon X \to I_X$ $(X \in \mathbf{L})$ such that $I_X \in \mathbf{L}$ is q-injective in $\mathbf{K}(\mathcal{A})$ for every X. Then for any quasi-isomorphism $s\colon X \to Y$ with Y in $\mathbf{K}(\mathcal{A})$ there exists, by (2.3.1), a map $g\colon Y \to I_X$, necessarily a quasi-isomorphism, such that $gs = \varphi_X$; and hence by $(1.7.1)^{\mathrm{op}}$, $\mathbf{L}$ is a localizing subcategory of $\mathbf{K}(\mathcal{A})$, i.e., the derived category $\mathbf{D_L}$ identifies naturally with a Δ-subcategory of $\mathbf{D}(\mathcal{A})$.

For example, if $\mathcal{A}$ has enough injectives we could take $\mathbf{L} := \overline{\mathbf{K}}^+(\mathcal{A})$, see (2.3.4). Or, if U is a topological space with a sheaf of rings $\mathcal{O}$ and $\mathcal{A}$ is the category of left $\mathcal{O}$-modules, we could take $\mathbf{L} := \mathbf{K}(\mathcal{A})$, see (2.3.5).

By (2.3.2.3), every Δ-functor $F \colon \mathbf{L} \to \mathbf{E}$ is right-derivable. So for any fixed object $A \in \mathbf{K}(\mathcal{A})$, the Δ-functor $F_A \colon \mathbf{L} \to \mathbf{K}(\mathfrak{Ab})$ given by

$$F_A(B) = \mathrm{Hom}^\bullet(A,\, B) \qquad (B \in \mathbf{L})$$

(see (1.5.3)) has a right-derived functor

$$\mathbf{R}F_A \colon \mathbf{D_L} \to \mathbf{K}(\mathfrak{Ab})$$

with

$$\mathbf{R}F_A(B) = \mathrm{Hom}^\bullet(A,\, I_B).$$

For fixed B and variable A, $\mathrm{Hom}^\bullet(A,\, I_B)$ is a contravariant Δ-functor from $\mathbf{K}(\mathcal{A})$ to $\mathbf{K}(\mathfrak{Ab})$ (see 1.5.3), which takes quasi-isomorphisms in $\mathbf{K}(\mathcal{A})$ to quasi-isomorphisms in $\mathbf{K}(\mathfrak{Ab})$ ((2.3.8)(iv)$'$) and hence—after composition with the natural functor $Q' \colon \mathbf{K}(\mathfrak{Ab}) \to \mathbf{D}(\mathfrak{Ab})$—to isomorphisms in $\mathbf{D}(\mathfrak{Ab})$. So by (1.5.1)—and the exercise preceding it—there results a Δ-functor $\mathbf{D}(\mathcal{A})^{\mathrm{op}} \to \mathbf{D}(\mathfrak{Ab})$. Thus we obtain a functor of two variables

$$\mathbf{R}\mathrm{Hom}^\bullet(A,\, B) \colon \mathbf{D}(\mathcal{A})^{\mathrm{op}} \times \mathbf{D_L} \to \mathbf{D}(\mathfrak{Ab})$$

which, together with appropriate θ (see (1.5.3)), is a Δ-functor in each variable separately:

$$\mathbf{R}\mathrm{Hom}^\bullet(A,\, B) = Q'\mathrm{Hom}^\bullet(A,\, I_B) \tag{2.4.1}$$

for all objects $A \in \mathbf{D}(\mathcal{A})^{\mathrm{op}}$, $B \in \mathbf{D_L}$; and we leave it to the reader to make explicit the effect of $\mathbf{R}\mathrm{Hom}^\bullet$ on morphisms in $\mathbf{D}(\mathcal{A})^{\mathrm{op}}$ and $\mathbf{D_L}$ respectively.

From (2.3.8)(v) and (2.3) (with $\mathbf{J} := \mathbf{K}(\mathcal{A})$), we deduce canonical isomorphisms (*Yoneda theorem*):

$$H^n(\mathbf{R}\mathrm{Hom}^\bullet(X,\, B)) \xrightarrow{\ \sim\ } \mathrm{Hom}_{\mathbf{D}(\mathcal{A})}(X,\, B[n]) \qquad (n \in \mathbb{Z}). \tag{2.4.2}$$

This leads, in particular, to an elementary interpretation of the exact sequence $(2.1.4)^{\mathrm{H}}$ when $F := F_X$, see [**H**, p. 23, Prop. 1.1, b)].

(**2.4.3**). The variables A, B are treated quite differently in the above definition of $\mathbf{R}\mathrm{Hom}^\bullet$. But there is a more symmetric characterization of this derived functor, analogous to the one in (2.1.1). This is given in (2.4.4), after the necessary preparation.

Let $\mathbf{K_1}$, $\mathbf{K_2}$, $\mathbf{E}$ be Δ-categories, with respective translation functors T_1, T_2, T. A Δ-functor from $\mathbf{K_1} \times \mathbf{K_2}$ to $\mathbf{E}$ is defined to be a triple (F, θ_1, θ_2) with

$$F \colon \mathbf{K_1} \times \mathbf{K_2} \to \mathbf{E}$$

a functor and

$$\theta_1 \colon F \circ (T_1 \times 1) \xrightarrow{\ \sim\ } T \circ F, \qquad \theta_2 \colon F \circ (1 \times T_2) \xrightarrow{\ \sim\ } T \circ F$$

isomorphisms of functors, such that for each $B \in \mathbf{K_2}$ the functor

$$F_B(A) := F(A, B)$$

together with θ_1 is a Δ-functor from $\mathbf{K_1}$ to $\mathbf{E}$, and for each $A \in \mathbf{K_1}$ the functor

$$F_A(B) := F(A, B)$$

together with θ_2 is a Δ-functor from $\mathbf{K_2}$ to $\mathbf{E}$; and such that furthermore the composed functorial isomorphisms

$$F(T_1 \times T_2) = F(T_1 \times 1)(1 \times T_2) \xrightarrow{\text{via } \theta_1} TF(1 \times T_2) \xrightarrow{\text{via } \theta_2} TTF$$

$$F(T_1 \times T_2) = F(1 \times T_2)(T_1 \times 1) \xrightarrow{\text{via } \theta_2} TF(T_1 \times 1) \xrightarrow{\text{via } \theta_1} TTF$$

are *negatives* of each other. Similarly, we can define Δ-functors of three or more variables—with a condition indicated by the equation

$$(\text{via } \theta_i) \circ (\text{via } \theta_j) = -(\text{via } \theta_j) \circ (\text{via } \theta_i) \qquad (i \neq j).$$

Morphisms of Δ-functors are defined in the obvious way, see (1.5).

For example, let $\mathbf{L} \subset \mathbf{K} := \mathbf{K}(\mathcal{A})$ be as above, with respective derived categories $\mathbf{D_L} \subset \mathbf{D}$, and consider the functor

$$\mathrm{Hom}^\bullet \colon \mathbf{K}^{\mathrm{op}} \times \mathbf{L} \to \mathbf{K}(\mathfrak{Ab}).$$

As in the exercise preceding (1.5.1), we can consider the opposite category $\mathbf{K}^{\mathrm{op}}$ to be triangulated, with translation inverse to that in $\mathbf{K}$, in such a way that the canonical contravariant functor $\mathbf{K} \to \mathbf{K}^{\mathrm{op}}$ and its inverse, together with $\theta = $ identity, are both Δ-functors. This being so, one checks then that $\mathrm{Hom}^\bullet$ is a Δ-functor (see (1.5.3)).

Similarly

$$\mathbf{R}\mathrm{Hom}^\bullet \colon \mathbf{D}^{\mathrm{op}} \times \mathbf{D_L} \to \mathbf{D}(\mathfrak{Ab})$$

is a Δ-functor. Furthermore, the q-injective resolution maps $\varphi_B \colon B \to I_B$ induce a natural morphism of Δ-functors

$$\eta \colon Q'\mathrm{Hom}^\bullet(A, B) \to Q'\mathrm{Hom}^\bullet(A, I_B) \overset{(2.4.1)}{=} \mathbf{R}\mathrm{Hom}^\bullet(QA, QB)$$

where $Q \colon \mathbf{K} \to \mathbf{D}$ is the canonical functor. This η is, in the following sense, *universal* (hence unique up to isomorphism):

Lemma 2.4.4. *Let*

$$G \colon \mathbf{D}^{\mathrm{op}} \times \mathbf{D_L} \to \mathbf{D}(\mathfrak{Ab})$$

be a Δ-functor, and let

$$\mu \colon Q'\mathrm{Hom}^\bullet(A, B) \to G(QA, QB) \qquad (A \in \mathbf{K}^{\mathrm{op}}, \ B \in \mathbf{L})$$

be a morphism of Δ-functors. Then there exists a unique morphism of Δ-functors

$$\overline{\mu} : \mathbf{R}\mathrm{Hom}^{\bullet} \to G$$

such that $\mu = \overline{\mu}\eta$.

Proof. $\overline{\mu}$ is the composition

$$\mathbf{R}\mathrm{Hom}^{\bullet}(QA,\, QB) = Q'\mathrm{Hom}^{\bullet}(A,\, I_B) \xrightarrow{\mu} G(QA, QI_B) \xrightarrow{\sim} G(QA, QB)\,.$$

The rest is left to the reader. (See also (2.6.5) below.)

(2.4.5). Next we discuss the *sheafified version* of the above. Let U be a topological space, $\mathcal{O}$ a sheaf of commutative rings, and $\mathcal{A}$ the abelian category of (sheaves of) $\mathcal{O}$-modules. The "sheaf-hom" functor

$$\mathcal{H}om : \mathcal{A}^{\mathsf{op}} \times \mathcal{A} \to \mathcal{A}$$

extends naturally to a Δ-functor

$$\mathcal{H}om^{\bullet} : \mathbf{K}(\mathcal{A})^{\mathsf{op}} \times \mathbf{K}(\mathcal{A}) \to \mathbf{K}(\mathcal{A})$$

(essentially because everything in (1.5.3) is compatible with restriction to open subsets—details left to the reader).

Taking note of the following Lemma, we can proceed as above to derive a Δ-functor

$$\mathbf{R}\mathcal{H}om^{\bullet} : \mathbf{D}(\mathcal{A})^{\mathsf{op}} \times \mathbf{D}(\mathcal{A}) \to \mathbf{D}(\mathcal{A})\,.$$

Lemma 2.4.5.1. *If I is a q-injective complex in $\mathbf{K}(\mathcal{A})$ then the functor $\mathcal{H}om^{\bullet}(-, I)$ takes quasi-isomorphisms to quasi-isomorphisms.*

Proof. For $A \in \mathbf{K}(\mathcal{A})$ and $i \in \mathbb{Z}$, the homology $H^i(\mathcal{H}om^{\bullet}(A, I))$ is the sheaf associated to the presheaf

$$V \mapsto H^i\big(\Gamma(V, \mathcal{H}om^{\bullet}(A, I))\big) = H^i\big(\mathrm{Hom}^{\bullet}(A|V, I|V)\big) \qquad (V \text{ open in } U).$$

We can then apply (2.3.8)(iv)$'$ to the category $\mathcal{A}_V$ of $(\mathcal{O}|V)$-modules, as soon as we know:

Lemma 2.4.5.2. *Let V be an open subset of U, with inclusion map $i : V \hookrightarrow U$. Then for any q-injective complex $I \in \mathbf{K}(\mathcal{A})$, the restriction $i^*I = I|_V$ is q-injective in $\mathbf{K}(\mathcal{A}_V)$.*

Proof. The *extension by zero* of an $\mathcal{O}_V$-module M is the sheaf $i_! M$ associated to the presheaf on U which assigns $M(W)$ to any open $W \subset V$ and 0 to any open $W \nsubseteq V$. The restriction $i^* i_! M$ can be identified with M; and the stalk of $i_! M$ at any point $w \notin V$ is 0. So $i_!$ is an exact functor.

Now from any diagram $Y \xleftarrow{s} X \xrightarrow{f} i^*I$ of maps of $\mathcal{A}_V$-complexes with s a quasi-isomorphism, we get the diagram

$$i_! Y \xleftarrow{i_! s} i_! X \xrightarrow{i_! f} i_! i^* I \xrightarrow{\alpha} I$$

where $i_! s$ is a quasi-isomorphism (since $i_!$ is exact) and α is the natural map. By (2.3.1), there exists a map $g \colon i_! X \to I$ such that $g \circ i_! s = \alpha \circ i_! f$ in $\mathbf{K}(\mathcal{A})$; and then we have, in $\mathbf{K}(\mathcal{A}_V)$,

$$i^* g \circ s = i^* g \circ i^* i_! s = i^* \alpha \circ i^* i_! f = 1 \circ f = f \,.$$

Thus i^*I is indeed q-injective.   Q.E.D.

(2.4.5.3). Similarly, any functor having an exact left adjoint preserves q-injectivity.

2.5 Derived Tensor Product

Let U be a topological space, $\mathcal{O}$ a sheaf of commutative rings, and $\mathcal{A}$ the abelian category of (sheaves of) $\mathcal{O}$-modules. Recall from (1.5.4) the definition of the tensor product (over $\mathcal{O}$) of two complexes in $\mathbf{K}(\mathcal{A})$, and its Δ-functorial properties. The standard theory of the derived tensor product, via resolutions by complexes of flat modules, applies to complexes in $\overline{\mathbf{D}}^-(\mathcal{A})$, see e.g., [**H**, p. 93]. Following Spaltenstein [**Sp**] we can use direct limits to extend the theory to *arbitrary* complexes in $\mathbf{D}(\mathcal{A})$. Before defining, in (2.5.7), the derived tensor product, we need to develop an appropriate acyclicity notion, "q-flatness."

Definition 2.5.1. A complex $P \in \mathbf{K}(\mathcal{A})$ is q-flat if for every quasi-isomorphism $Q_1 \to Q_2$ in $\mathbf{K}(\mathcal{A})$, the resulting map $P \otimes Q_1 \to P \otimes Q_2$ is also a quasi-isomorphism; or equivalently (see footnote under (1.5.1)), if for every exact complex $Q \in \mathbf{K}(\mathcal{A})$, the complex $P \otimes Q$ is also exact.

Example 2.5.2. $P \in \mathbf{K}(\mathcal{A})$ is q-flat iff for each point $x \in U$, the stalk P_x is q-flat in $\mathbf{K}(\mathcal{A}_x)$, where $\mathcal{A}_x$ is the category of modules over the ring $\mathcal{O}_x$. (In verifying this statement, note that an exact $\mathcal{O}_x$-complex Q_x is the stalk at x of the exact $\mathcal{O}$-complex Q which associates Q_x to those open subsets of U which contain x, and 0 to those which don't.)

For instance, a complex P which vanishes in all degrees but one (say n) is q-flat if and only if tensoring with the degree n component P^n is an exact functor in the category of $\mathcal{O}$-modules, i.e., P^n is a flat $\mathcal{O}$-module, i.e., for each $x \in U$, P_x^n is a flat $\mathcal{O}_x$-module.

Example 2.5.3. Tensoring with a fixed complex Q is a Δ-functor, and so the exact homology sequence $(1.4.5)^{\mathrm{H}}$ of a triangle yields that the q-flat complexes are the objects of a Δ-subcategory of $\mathbf{K}(\mathcal{A})$.

A bounded complex

$$P: \quad \cdots \to 0 \to 0 \to P^m \to \cdots \to P^n \to 0 \to 0 \to \cdots$$

fits into a triangle $P' \to P \to P'' \to P'[1]$ where P' is P^n in degree n and 0 elsewhere, and where P'' is the cokernel of the obvious map $P' \to P$. So starting with (2.5.2) we see by induction on $n - m$ that any bounded complex of flat $\mathcal{O}$-modules is q-flat.

Example 2.5.4. Since (filtered) direct limits commute with both tensor product and homology, therefore any such limit of q-flat complexes is again q-flat.

A bounded-above complex

$$P: \quad \cdots \to P^m \to \cdots \to P^n \to 0 \to 0 \to \cdots$$

is the limit of the direct system $P_0 \to P_1 \to \cdots \to P_i \to \cdots$ where P_i is obtained from P by replacing all the components P^j with $j < n - i$ by 0, and the maps are the obvious ones. Hence, any bounded-above complex of flat $\mathcal{O}$-modules is q-flat.

A *q-flat resolution* of an $\mathcal{A}$-complex C is a quasi-isomorphism $P \to C$ where P is q-flat. The totality of such resolutions (with variable P and C) is the class of objects of a category, whose morphisms are the obvious ones.

Proposition 2.5.5. *Every $\mathcal{A}$-complex C is the target of a quasi-isomorphism ψ_C from a q-flat complex P_C, which can be constructed to depend functorially on C, and so that $P_{C[1]} = P_C[1]$ and $\psi_{C[1]} = \psi_C[1]$.*

Proof. Every $\mathcal{O}$-module is a quotient of a flat one; in fact there exists a functor P_0 from $\mathcal{A}$ to its full subcategory of flat $\mathcal{O}$-modules, together with a functorial epimorphism $P_0(\mathcal{F}) \twoheadrightarrow \mathcal{F}$ ($\mathcal{F} \in \mathcal{A}$). Indeed, for any open $V \subset U$ let $\mathcal{O}_V$ be the extension of $\mathcal{O}|V$ by zero, (i.e., the sheaf associated to the presheaf taking an open W to $\mathcal{O}(W)$ if $W \subset V$ and to 0 otherwise), so that $\mathcal{O}_V$ is flat, its stalk at $x \in U$ being $\mathcal{O}_x$ if $x \in V$ and 0 otherwise. There is a canonical isomorphism

$$\psi: \mathcal{F}(V) \xrightarrow{\sim} \mathrm{Hom}(\mathcal{O}_V, \mathcal{F}) \qquad (\mathcal{F} \in \mathcal{A})$$

such that $\psi(\lambda)$ takes $1 \in \mathcal{O}_V(V)$ to λ. With $\mathcal{O}_\lambda := \mathcal{O}_V$ for each $\lambda \in \mathcal{F}(V)$, the maps $\psi(\lambda)$ define an epimorphism, with flat source,

$$P_0(\mathcal{F}) := \Big(\bigoplus_{V \text{ open}} \bigoplus_{\lambda \in \mathcal{F}(V)} \mathcal{O}_\lambda \Big) \twoheadrightarrow \mathcal{F},$$

and this epimorphism depends functorially on $\mathcal{F}$.

We deduce then, for each $\mathcal{F}$, a functorial flat resolution

$$\cdots \to P_2(\mathcal{F}) \to P_1(\mathcal{F}) \to P_0(\mathcal{F}) \twoheadrightarrow \mathcal{F}$$

with $P_1(\mathcal{F}) := P_0(\ker(P_0(\mathcal{F}) \twoheadrightarrow \mathcal{F}))$, etc. Set $P_n(\mathcal{F}) = 0$ if $n < 0$. Then to a complex C we associate the *flat* complex $P = P_C$ such that $P^r := \oplus_{m-n=r} P_n(C^m)$ and the restriction of the differential $P^r \to P^{r+1}$ to $P_n(C^m)$ is $P_n(C^m \to C^{m+1}) \oplus (-1)^m(P_n(C^m) \to P_{n-1}(C^m))$, together with the natural map of complexes $P \to C$ induced by the epimorphisms $P_0(C^m) \twoheadrightarrow C^m$ ($m \in \mathbb{Z}$). Elementary arguments, with or without spectral sequences, show that for the truncations $\tau_{\leq m}C$ of §1.10, the maps $P_{\tau_{\leq m}C} \to \tau_{\leq m}C$ are *quasi-isomorphisms*. Since homology commutes with direct limits, the resulting map

$$\psi_C \colon P_C = \varinjlim_m P_{\tau_{\leq m}C} \to \varinjlim_m \tau_{\leq m} C = C,$$

(which depends functorially on C) is a quasi-isomorphism; and by (2.5.4), P_C is q-flat. That $P_{C[1]} = P_C[1]$ and $\psi_{C[1]} = \psi_C[1]$ is immediate.   Q.E.D.

Exercises 2.5.6. (a) Let P and Q be complexes of $\mathcal{O}$-modules, and suppose that for all integers s, t, u, v the complex $\tau_{\leq s}\tau_{\geq t}P \otimes_{\mathcal{O}} \tau_{\leq u}\tau_{\geq v}Q$ is exact. Then

$$P \otimes Q = \varinjlim_{s,u} \tau_{\leq s}P \otimes \tau_{\leq u}Q$$

is exact.

(b) If for all $n \in \mathbb{Z}$ the homology $H^n(P)$ is a flat $\mathcal{O}$-module and furthermore, for all n the kernel of $P^n \to P^{n+1}$ is a direct summand of P^n (or, for all n the image of $P^n \to P^{n+1}$ is a direct summand of P^{n+1}), then P is q-flat. (Use (a) to reduce to where P is bounded; then apply induction to the number of n such that $P^n \neq 0$.)

(2.5.7). Let $\mathcal{A}$ be, as above, the category of $\mathcal{O}$-modules, and let

$$\mathbf{J'} \subset \mathbf{K} := \mathbf{K}(\mathcal{A})$$

be the Δ-subcategory of $\mathbf{K}$ whose objects are all the q-flat complexes, see (2.5.3). Fix $B \in \mathbf{K}$ and consider the Δ-functor

$$F_B \colon \mathbf{K} \to \mathbf{D} := \mathbf{D}(\mathcal{A})$$

such that

$$F_B(A) = A \otimes B \qquad \text{(see (1.5.4))}.$$

If A is both q-flat and exact, then $A \otimes B$ is exact: to see this, we may replace B by any quasi-isomorphic complex B' (since A is q-flat), and by (2.5.5) we may assume that B' is q-flat, whence, by (2.5.1), $A \otimes B'$ is exact. Hence the restriction of F_B to $\mathbf{J'}$ transforms quasi-isomorphisms into isomorphisms.

There exists, by (2.5.5), a functorial family of quasi-isomorphisms

$$\psi_A \colon P_A \to A \qquad (A \in \mathbf{K}, \ P_A \in \mathbf{J}').$$

with $P_{A[1]} = P_A[1]$. An argument dual to that in (2.2.4) (with $\mathbf{J}'' := \mathbf{K}$) shows then that F_B has a left-derived Δ-functor

$$(\mathbf{L}F_B, \text{identity}) \colon \mathbf{D} \to \mathbf{D} \qquad\qquad (2.5.7.1)$$

with

$$\mathbf{L}F_B(A) = P_A \otimes B \cong P_A \otimes P_B \cong A \otimes P_B,$$

the isomorphisms being the ones induced by ψ_A and ψ_B. Alternatively, P_A is left-F_B-acyclic for all A, B (see 2.5.10(d)), so one can apply (2.2.6).

For fixed A and variable B, $P_A \otimes B$ is a Δ-functor from $\mathbf{K}$ to $\mathbf{D}$ which takes quasi-isomorphisms to isomorphisms, so by (1.5.1) there results a Δ-functor from $\mathbf{D}$ to $\mathbf{D}$. Hence there is a functor of two variables, called a *derived tensor product*,

$$\underset{=}{\otimes} \colon \mathbf{D} \times \mathbf{D} \longrightarrow \mathbf{D}$$

which together with appropriate θ (see (1.5.4)) is a Δ-functor in each variable separately (i.e., it is a Δ-functor as defined in (2.4.3)).

Though the variables A and B have been treated differently in the foregoing, their roles are essentially equivalent. Indeed, there is a universal property analogous to (the dual of) that in (2.4.4), characterizing the natural composite map of Δ-functors from $\mathbf{K} \times \mathbf{K}$ to $\mathbf{D}$:

$$QA \underset{=}{\otimes} QB \xrightarrow{\ \sim\ } Q(P_A \otimes P_B) \longrightarrow Q(A \otimes B).$$

Hence, in view of (1.5.4.1), there is a canonical Δ-bifunctorial isomorphism

$$B \underset{=}{\otimes} A \xrightarrow{\ \sim\ } A \underset{=}{\otimes} B.$$

This arises, in fact, from the natural isomorphism $P_B \otimes P_A \xrightarrow{\ \sim\ } P_A \otimes P_B$.

(2.5.8). The *local hypertor* sheaves are defined by

$$\mathcal{T}\mathrm{or}_n(A, B) = H^{-n}(A \underset{=}{\otimes} B) \qquad (n \in \mathbb{Z}; \ A, B \in \mathbf{D}).$$

As in (2.1.4), short exact sequences in either the A or B variable give rise to long exact hypertor sequences.

We remark that when U is a *scheme* and $\mathcal{O} = \mathcal{O}_U$, if the homology sheaves of the complexes A and B are all *quasi-coherent* then so are the sheaves $\mathcal{T}\mathrm{or}_n(A, B)$. This is clear, by reduction to the affine case, if A and B are quasi-coherent $\mathcal{O}_X$-modules (i.e., complexes vanishing except in degree 0). In the general case, since

$$A \otimes B = \varinjlim_{s, u} \tau_{\leq s} A \otimes \tau_{\leq u} B,$$

we may assume A and B lie in $\mathbf{D}^-$, and then argue as in [$\mathbf{H}$, p. 98, Prop. 4.3], or alternatively, use the Künneth spectral sequence

$$E^2_{pq} = \bigoplus_{i+j=q} \mathfrak{Tor}_p(H^{-i}(A),\, H^{-j}(B)) \Rightarrow \mathfrak{Tor}_\bullet(A,B)$$

(as described e.g., in [$\mathbf{B}$, p. 186, Exercise 9(b)], with *flat* resolutions replacing projective ones). Thus, with notation as in (1.9), denoting by $\mathbf{D}_{\mathsf{qc}}$ the Δ-subcategory $\mathbf{D}_\# \subset \mathbf{D}$ with $\mathcal{A}^\# \subset \mathcal{A}$ the subcategory of *quasi-coherent* $\mathcal{O}_U$-modules (which is plump, see [$\mathbf{GD}$, p. 217, (2.2.2) (iii)]), we have a Δ-functor

$$\underset{=}{\otimes}\colon \mathbf{D}_{\mathsf{qc}} \times \mathbf{D}_{\mathsf{qc}} \longrightarrow \mathbf{D}_{\mathsf{qc}}\ . \tag{2.5.8.1}$$

(2.5.9). The definitions in (1.5.4) can be extended to three (or more) variables, to give a Δ-functor $A \otimes B \otimes C$ from $\mathbf{K} \times \mathbf{K} \times \mathbf{K}$ to $\mathbf{K}$.

There exists a Δ-functor $T_3 \colon \mathbf{D} \times \mathbf{D} \times \mathbf{D} \to \mathbf{D}$ together with a Δ-functorial map

$$\eta\colon T_3(A,B,C) \longrightarrow A \otimes B \otimes C \qquad (A,B,C \in \mathbf{K})$$

such that for any Δ-functor $H \colon \mathbf{D} \times \mathbf{D} \times \mathbf{D} \to \mathbf{D}$ and any Δ-functorial map $\mu \colon H(A,B,C) \longrightarrow A \otimes B \otimes C$ there is a unique Δ-functor map $\bar{\mu} \colon H \to T_3$ such that $\mu = \eta \circ \bar{\mu}$. (The reader can fill in the missing Q's.) In fact there is such a T_3 with

$$T_3(A,B,C) = P_A \otimes P_B \otimes P_C\,.$$

We usually write

$$T_3(A,B,C) = A \underset{=}{\otimes} B \underset{=}{\otimes} C\,.$$

There are canonical Δ-functorial isomorphisms

$$(A \underset{=}{\otimes} B) \underset{=}{\otimes} C \xrightarrow{\ \sim\ } A \underset{=}{\otimes} B \underset{=}{\otimes} C \xleftarrow{\ \sim\ } A \underset{=}{\otimes} (B \underset{=}{\otimes} C)\,.$$

Similar considerations hold for $n > 3$ variables. Details are left to the reader. (See, for example, (2.6.5) below.)

Exercises 2.5.10. (a) Show that if $A \in \mathbf{K}(\mathcal{A})$ is q-flat and $B \in \mathbf{K}(\mathcal{A})$ is q-injective then $\mathcal{H}om^\bullet(A,B)$ is q-injective.

(b) Let $\Gamma \colon \mathcal{A} \to \mathfrak{Ab}$ be the global section functor. Show that there is a natural isomorphism of Δ-functors (of two variables, see (2.4.3))

$$\mathbf{R}\mathrm{Hom}^\bullet(A,B) \xrightarrow{\ \sim\ } \mathbf{R}\Gamma\mathbf{R}\mathcal{H}om^\bullet(A,B).$$

(Use (a) and (2.2.7), or [$\mathbf{Sp}$, 5.14, 5.12, 5.17].)

(c) Let (A_α) be a (small, directed) inductive system of $\mathcal{A}$-complexes. Show that for any complex $B \in \mathbf{D}(\mathcal{A})$ there are natural isomorphisms

$$\varinjlim_\alpha \mathfrak{Tor}_n(A_\alpha,\, B) \xrightarrow{\ \sim\ } \mathfrak{Tor}_n\big((\varinjlim_\alpha A_\alpha),\, B\big) \qquad (n \in \mathbb{Z}).$$

(d) Show that for P to be q-flat it is necessary that P be left-F_B-acyclic for all B (F_B as in (2.5.7)), and sufficient that P be left-F_B-acyclic for all *exact* B. (For the last part, (2.2.6) could prove helpful.) Formulate and prove an analogous statement involving q-injectivity and $\mathrm{Hom}^\bullet$. (See (2.3.8).)

2.6 Adjoint Associativity

Again let U be a topological space, $\mathcal{O}$ a sheaf of commutative rings, and $\mathcal{A}$ the abelian category of $\mathcal{O}$-modules. Set $\mathbf{K} := \mathbf{K}(\mathcal{A})$, $\mathbf{D} := \mathbf{D}(\mathcal{A})$. This section is devoted to (2.6.1)—or better, (2.6.1)* at the end—which expresses the basic adjointness relation between the Δ-functors $\mathbf{R}\mathcal{H}om^\bullet \colon \mathbf{D}^{\mathrm{op}} \times \mathbf{D} \to \mathbf{D}$ and $\underset{=}{\otimes} \colon \mathbf{D} \times \mathbf{D} \to \mathbf{D}$ defined in (2.4.5) and (2.5.7) respectively.

Proposition 2.6.1. *There is a natural isomorphism of Δ-functors*

$$\mathbf{R}\mathcal{H}om^\bullet(A \underset{=}{\otimes} B,\, C) \xrightarrow{\ \sim\ } \mathbf{R}\mathcal{H}om^\bullet(A,\, \mathbf{R}\mathcal{H}om^\bullet(B,\, C))\,.$$

Remarks. (i) In fact, the Δ-functors $\mathbf{R}\mathcal{H}om^\bullet$ and $\underset{=}{\otimes}$ are defined only up to canonical isomorphism, by universal properties, as in (2.5.9). We leave it to the reader to verify that the map in (2.6.1) (to be constructed below) is compatible, in the obvious sense, with such canonical isomorphisms.

(ii) A proof similar to the following one[3] yields a natural isomorphism

$$\mathbf{R}\mathrm{Hom}^\bullet(A \underset{=}{\otimes} B,\, C) \xrightarrow{\ \sim\ } \mathbf{R}\mathrm{Hom}^\bullet(A,\, \mathbf{R}\mathcal{H}om^\bullet(B,\, C))\,.$$

Applying homology H^0 we have, by (2.4), the *adjunction isomorphism*

$$\mathrm{Hom}_{\mathbf{D}}(A \underset{=}{\otimes} B,\, C) \xrightarrow{\ \sim\ } \mathrm{Hom}_{\mathbf{D}}(A,\, \mathbf{R}\mathcal{H}om^\bullet(B,\, C))\,. \qquad (2.6.1')$$

(iii) Prop. (2.6.1) gives a derived-category upgrade of the standard sheaf isomorphism

$$\mathcal{H}om(F \otimes G,\, H) \xrightarrow{\ \sim\ } \mathcal{H}om(F,\, \mathcal{H}om(G,\, H)) \qquad (F, G, H \in \mathcal{A})\,. \qquad (2.6.2)$$

Proof of (2.6.1). We discuss the proof at several levels of pedantry, beginning with the argument, in full, given in [**I**, p. 151, Lemme 7.4] (see also [**Sp**, p. 147, Prop. 6.6]): "Resolve C injectively and B flatly."

This argument can be expanded as follows. Choose quasi-isomorphisms

$$C \to I_C\,, \qquad P_B \to B$$

---

[3] Or application of the functor $\mathbf{R}\Gamma$ to (2.6.1), see (2.5.10),

where I_C is q-injective and P_B is q-flat. It follows from (2.3.8)(iv) that the complex of sheaves $\mathcal{H}om^\bullet(P_B, I_C)$ is *q-injective,* since for any exact complex $X \in \mathbf{K}$, the isomorphism of complexes

$$\mathrm{Hom}^\bullet(X \otimes P_B, I_C) \xrightarrow{\sim} \mathrm{Hom}^\bullet(X, \mathcal{H}om^\bullet(P_B, I_C))$$

coming out of (2.6) yields, upon application of homology H^0,

$$0 = \mathrm{Hom}_{\mathbf{K}}(X \otimes P_B, I_C) \xrightarrow{\sim} \mathrm{Hom}_{\mathbf{K}}(X, \mathcal{H}om^\bullet(P_B, I_C)).$$

Now consider the natural sequence of **D**-maps

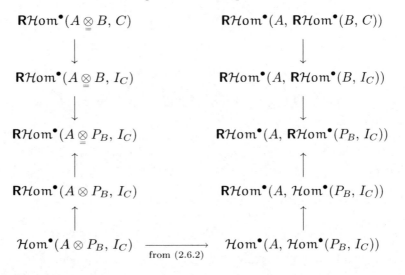

Since P_B is q-flat, and I_C and $\mathcal{H}om^\bullet(P_B, I_C)$ are q-injective, all these maps are isomorphisms (as follows, e.g., from the last assertion of (2.2.6)); so we can compose to get the isomorphism (2.6.1).

But we really should check that this isomorphism does not depend on the chosen quasi-isomorphisms, and that it is in fact Δ-functorial. This can be quite tedious. The following remarks outline a method for managing such verifications. The basic point is (2.6.4) below.

Let M be a set. An *M-category* is an additive category $\mathbf{C}$ plus a map $t \colon M \to \mathrm{Hom}(\mathbf{C}, \mathbf{C})$ from M into the set of additive functors from $\mathbf{C}$ to $\mathbf{C}$, such that with $T_m := t(m)$ it holds that $T_i \circ T_j = T_j \circ T_i$ for all $i, j \in M$. Such an M-category will be denoted $\mathbf{C}_M$, the map f—or equivalently, the commuting family $(T_m)_{m \in M}$—understood to have been specified; and when the context renders it superfluous, the subscript "M" may be omitted.

An *M-functor* $F \colon \mathbf{C}_M \to \mathbf{C}'_M$ is an additive functor $F \colon \mathbf{C} \to \mathbf{C}'$ together with isomorphisms of functors

$$\theta_i \colon F \circ T_i \xrightarrow{\sim} T'_i \circ F \qquad (i \in M)$$

(with $(T'_m)_{m \in M}$ the commuting family of functors defining the M-structure on $\mathbf{C}'$) such that for all $i \neq j$, the following diagram commutes:

$$
\begin{array}{ccccc}
F \circ T_i \circ T_j & \xrightarrow{\text{via } \theta_i} & T'_i \circ F \circ T_j & \xrightarrow{T'_i(\theta_j)} & T'_i \circ T'_j \circ F \\
\| & & & & \| \\
F \circ T_j \circ T_i & \xrightarrow[\text{via } \theta_j]{} & T'_j \circ F \circ T_i & \xrightarrow[-T'_j(\theta_i)]{} & T'_j \circ T'_i \circ F
\end{array}
$$

where, for instance, $T'_j(\theta_i)$ is the isomorphism of functors such that for each object $X \in \mathbf{C}$, $[T'_j(\theta_i)](X)$ is the $\mathbf{C}'$-isomorphism

$$
T'_j(\theta_i(X)) \colon T'_j(FT_i(X)) \overset{\sim}{\longrightarrow} T'_j(T'_i F(X)).
$$

A *morphism* $\eta \colon (F, \{\theta_i\}) \to (G, \{\psi_i\})$ of M-functors is a morphism of functors $\eta \colon F \to G$ such that for every $i \in M$ and every object X in $\mathbf{C}$, the following diagram commutes:

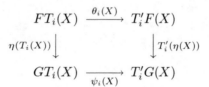

Composition of such η being defined in the obvious way, the M-functors from $\mathbf{C}$ to $\mathbf{C}'$, and their morphisms, form a category $\mathbf{H} := \mathbf{Hom}_M(\mathbf{C}, \mathbf{C}')$. If $M' \supset M$ and $\mathbf{C}'_{M'}$ is viewed as an M-category via "restriction of scalars" then $\mathbf{H}$ is itself an M'-category, with $j \in M'$ being sent to the functor $T_j^\# \colon \mathbf{H} \to \mathbf{H}$ such that on objects of $\mathbf{H}$,

$$
T_j^\#(F, \{\theta_i\}) = (T'_j \circ F, \{-T'_j(\theta_i)\}),
$$

where the isomorphism of functors

$$
T'_j(\theta_i) \colon (T'_j \circ F) \circ T_i \overset{\sim}{\longrightarrow} T'_j \circ T'_i \circ F = T'_i \circ (T'_j \circ F)
$$

is as above.[4] The definition of $T_j^\# \eta$ (η as above), and the verification that $\mathbf{H}$ is thus an M'-category, are straightforward.

Suppose given such categories $\mathbf{A}_M$, $\mathbf{B}_N$, and $\mathbf{C}_{M \cup N}$, where the sets M and N are disjoint. $\mathbf{A} \times \mathbf{B}$ is considered to be an $(M \cup N)$-category, with $i \in M$ going to the functor $T_i \times 1$ and $j \in N$ to the functor $1 \times T_j$. Also, $\mathbf{Hom}_N(\mathbf{B}, \mathbf{C})$ is considered, as above, to be an $(M \cup N)$-category.

---

[4] The reason for the minus sign in the definition of $T_j^\#$ is hidden in the details of the proof of Lemma (2.6.3) below.

Lemma 2.6.3. *With preceding notation, there is a natural isomorphism of $M \cup N$-categories*

$$\mathbf{Hom}_{M \cup N}(\mathbf{A} \times \mathbf{B}, \, \mathbf{C}) \xrightarrow{\sim} \mathbf{Hom}_M(\mathbf{A}, \mathbf{Hom}_N(\mathbf{B}, \mathbf{C}))$$

The *proof*, left to the reader, requires very little imagination, but a good deal of patience.

For any positive integer n, let $\triangle_n$ be the set $\{1, 2, \ldots, n\}$. From now on, we deal with Δ-categories, always considered to be $\triangle_1$-categories via their translation functors. If $\mathbf{C}_1, \ldots, \mathbf{C}_n$ are Δ-categories, then the product category $\mathbf{C} = \mathbf{C}_1 \times \mathbf{C}_2 \times \cdots \times \mathbf{C}_n$ becomes a $\triangle_n$-category by the product construction used in (2.6.3). A Δ-category $\mathbf{E}$ can also be made into an $\triangle_n$-category by sending each $i \in \triangle_n$ to the translation functor of $\mathbf{E}$. With these understandings, we see that the $\triangle_n$-functors from $\mathbf{C}_1 \times \mathbf{C}_2 \times \cdots \times \mathbf{C}_n$ to $\mathbf{E}$ are just the Δ-functors of (2.4.3) (categories of which we denote by $\mathbf{Hom}_\Delta$). For example, one checks that the source and target of the isomorphism in (2.6.1) are both $\triangle_3$-functors.

Now for $1 \leq i \leq n$ fix abelian categories $\mathcal{A}_i$, and let $\mathbf{L_i}$ be a Δ-subcategory of $\mathbf{K}(\mathcal{A}_i)$, with corresponding derived category $\mathbf{D_i}$ and canonical functor $Q_i \colon \mathbf{L_i} \to \mathbf{D_i}$. Let $\mathbf{E}$ be any Δ-category. We can generalize (1.5.1) as follows:

Proposition 2.6.4. *The canonical functor*

$$\mathbf{L}_1 \times \cdots \times \mathbf{L_n} \xrightarrow[\;Q_1 \times \ldots \times Q_n\;]{} \mathbf{D}_1 \times \cdots \times \mathbf{D_n}$$

induces an isomorphism from the category $\mathbf{Hom}_\Delta(\mathbf{D}_1 \times \mathbf{D}_2 \times \cdots \times \mathbf{D_n}, \mathbf{E})$ *to the full subcategory of* $\mathbf{Hom}_\Delta(\mathbf{L}_1 \times \mathbf{L}_2 \times \cdots \times \mathbf{L_n}, \mathbf{E})$ *whose objects are the Δ-functors F such that for any quasi-isomorphisms $\alpha_1, \ldots, \alpha_n$ in $\mathbf{L}_1, \ldots, \mathbf{L_n}$ respectively, $F(\alpha_1, \ldots, \alpha_n)$ is an isomorphism in $\mathbf{E}$.*

Proof. The case $n = 1$ is contained in (1.5.1). We can then proceed by induction on n, using the natural isomorphism

$$\mathbf{Hom}_{\triangle_n}(\mathbf{C}_1 \times \mathbf{C}_2 \times \cdots \times \mathbf{C_n}, \mathbf{E})$$
$$\xrightarrow{\sim} \mathbf{Hom}_{\triangle_1}(\mathbf{C}_1, \mathbf{Hom}_{\triangle_{n-1}}(\mathbf{C}_2 \times \ldots \times \mathbf{C_n}, \mathbf{E}))$$

provided by (2.6.3) (with $\mathbf{C_i} := \mathbf{D_i}$ or $\mathbf{L_i}$). Q.E.D.

Suppose next that we have pairs of Δ-subcategories $\mathbf{L}_i' \subset \mathbf{L}_i''$ in $\mathbf{K}(\mathcal{A}_i)$, with respective derived categories $\mathbf{D}_i', \mathbf{D}_i''$, and canonical functors $Q_i' \colon \mathbf{L}_i' \to \mathbf{D}_i'$, $Q_i'' \colon \mathbf{L}_i'' \to \mathbf{D}_i''$ $(1 \leq i \leq n)$. Suppose further that every complex $A \in \mathbf{L}_i''$ admits a quasi-isomorphism into a complex $I_A \in \mathbf{L}_i'$. Then as in (1.7.2) the natural Δ-functors $\tilde{\jmath}_i \colon \mathbf{D}_i' \to \mathbf{D}_i''$ are Δ-equivalences, having quasi-inverses ρ_i satisfying $\rho_i(A) = I_A$ $(A \in \mathbf{L}_i'')$. There result functors

$$\tilde{\jmath}^* : \mathbf{Hom}_\Delta(\mathbf{D}_1'' \times \cdots \times \mathbf{D}_n'', \mathbf{E}) \longrightarrow \mathbf{Hom}_\Delta(\mathbf{D}_1' \times \cdots \times \mathbf{D}_n', \mathbf{E})$$
$$\rho^* : \mathbf{Hom}_\Delta(\mathbf{D}_1' \times \cdots \times \mathbf{D}_n', \mathbf{E}) \longrightarrow \mathbf{Hom}_\Delta(\mathbf{D}_1'' \times \cdots \times \mathbf{D}_n'', \mathbf{E})$$

together with functorial isomorphisms

$$\tilde{\jmath}^*\rho^* \overset{\sim}{\longrightarrow} \text{identity}, \qquad \rho^*\tilde{\jmath}^* \overset{\sim}{\longrightarrow} \text{identity},$$

i.e., $\tilde{\jmath}^*$ and ρ^* are quasi-inverse equivalences of categories.

We deduce the following variation on the theme of (2.2.3), thereby arriving at a general method for specifying maps between Δ-functors on products of derived categories:[5]

Corollary 2.6.5. *With above notation let* $H \colon \mathbf{L}_1' \times \cdots \times \mathbf{L}_n' \to \mathbf{E}$, $F \colon \mathbf{D}_1'' \times \cdots \times \mathbf{D}_n'' \to \mathbf{E}$, *and* $G \colon \mathbf{D}_1'' \times \cdots \times \mathbf{D}_n'' \to \mathbf{E}$ *be* Δ-*functors. Let*

$$\zeta \colon H \overset{\sim}{\longrightarrow} F \circ (\tilde{\jmath}_1 Q_1' \times \ldots \times \tilde{\jmath}_n Q_n'),$$
$$\beta \colon H \longrightarrow G \circ (\tilde{\jmath}_1 Q_1' \times \ldots \times \tilde{\jmath}_n Q_n')$$

be Δ-*functorial maps, with* ζ *an isomorphism. Then:*

(i) *There exists a unique* Δ-*functorial map* $\bar{\beta} \colon F \to G$ *such that for all* $A_1 \in \mathbf{L}_1', \ldots, A_n \in \mathbf{L}_n'$, $\beta(A_1, \ldots, A_n)$ *factors as*

$$H(A_1, \ldots, A_n) \overset{\zeta}{\longrightarrow} F(A_1, \ldots, A_n) \overset{\bar{\beta}}{\longrightarrow} G(A_1, \ldots, A_n). \tag{2.6.5.1}$$

Moreover, if β *is an isomorphism then so is* $\bar{\beta}$.

(ii) *If* H *in* (i) *extends to a* Δ-*functor* $H \colon \mathbf{L}_1'' \times \ldots \times \mathbf{L}_n'' \to \mathbf{E}$, *and* ζ *(respectively* β*) to a* Δ-*functorial map* $\zeta \colon H \to F \circ (\tilde{\jmath}_1 Q_1'' \times \ldots \times \tilde{\jmath}_n Q_n'')$ *(respectively* $\beta \colon H \to G \circ (\tilde{\jmath}_1 Q_1'' \times \ldots \times \tilde{\jmath}_n Q_n'')$*), then the factorization* (2.6.5.1) *of* $\beta(A_1, \ldots, A_n)$ *holds for all* $A_1 \in \mathbf{L}_1'', \ldots, A_n \in \mathbf{L}_n''$.

Proof. (i) The assertion just means that $\bar{\beta}$ is the unique map (resp. isomorphism) $F \to G$ in the category $\mathbf{Hom}_\Delta(\mathbf{D}_1'' \times \ldots \times \mathbf{D}_n'', \mathbf{E})$ corresponding via the above equivalence $\tilde{\jmath}^*$ and (2.6.4) to the map (resp. isomorphism) $\beta\zeta^{-1}$ in the category $\mathbf{Hom}_\Delta(\mathbf{L}_1' \times \ldots \times \mathbf{L}_n', \mathbf{E})$.

(ii) Use quasi-isomorphisms $A_i \to I_{A_i}$ to map (2.6.5.1) into the corresponding diagram with $I_{A_i} \in \mathbf{L}_i'$ in place of A_i. To this latter diagram (i) applies; and as the resulting map $G(A_1, \ldots, A_n) \to G(I_{A_1}, \ldots, I_{A_n})$ is an isomorphism, the rest is clear.   Q.E.D.

---

[5] This is no more (or less) than a careful formulation of the method used, e.g., throughout [**H**, Chapter II].

We can now derive (2.6.1) as follows. Take $n = 3$, and set

$$\mathbf{L}_1' := \mathbf{K}$$

$$\mathbf{L}_2' := \begin{cases} \Delta\text{-subcategory of } \mathbf{K} \text{ whose objects are} \\ \text{the q-flat complexes (2.5.3).} \end{cases}$$

$$\mathbf{L}_3' := \begin{cases} \Delta\text{-subcategory of } \mathbf{K} \text{ whose objects are} \\ \text{the q-injective complexes (2.3.2.2).} \end{cases}$$

Let $\mathbf{D}_1'$, $\mathbf{D}_2'$, $\mathbf{D}_3'$ be the corresponding derived categories, and set

$$\mathbf{L}_i'' := \mathbf{K}, \quad \mathbf{D}_i'' := \mathbf{D} \qquad (i = 1, 2, 3),$$

so that the natural maps $j_i \colon \mathbf{D_i}' \to \mathbf{D_i}''$ are Δ-equivalences, with quasi-inverses obtained for $i = 2$ and $i = 3$ from q-flat (resp. q-injective) resolutions, i.e., from families of quasi-isomorphisms

$$P_B \to B \qquad (B \in \mathbf{K}, \; P_B \in \mathbf{L}_2'),$$

$$C \to I_C \qquad (C \in \mathbf{K}, \; I_C \in \mathbf{L}_3').$$

In Corollary (2.6.5)(ii), let $H \colon \mathbf{L}_1'' \times \mathbf{L}_2'' \times \mathbf{L}_3'' \to \mathbf{D}$ be the Δ-functor

$$H(A, B, C) := \mathcal{H}om^\bullet(A \otimes B, \, C),$$

let ζ be the natural composed Δ-functorial map

$$\mathcal{H}om^\bullet(A \otimes B, \, C) \to \mathbf{R}\mathcal{H}om^\bullet(A \otimes B, C) \to \mathbf{R}\mathcal{H}om^\bullet(A \underset{=}{\otimes} B, C),$$

and let β be the natural composed Δ-functorial map

$$\mathcal{H}om^\bullet(A \otimes B, \, C) \underset{(2.6.2)}{\overset{\sim}{\longrightarrow}} \mathcal{H}om^\bullet(A, \, \mathcal{H}om^\bullet(B, \, C))$$

$$\longrightarrow \mathbf{R}\mathcal{H}om^\bullet(A, \, \mathcal{H}om^\bullet(B, \, C))$$

$$\longrightarrow \mathbf{R}\mathcal{H}om^\bullet(A, \, \mathbf{R}\mathcal{H}om^\bullet(B, \, C)).$$

(Meticulous readers may wish to insert the missing Q's).

We saw near the beginning of the proof of (2.6.1), that for $(B, C) \in \mathbf{L}_2' \times \mathbf{L}_3'$, the complex $\mathcal{H}om^\bullet(B, \, C)$ is q-injective, and hence for such (B, C), ζ *and* β *are isomorphisms.* Modifying (2.6.5) in the obvious way to take contravariance into account, we deduce the following elaboration of (2.6.1):

Proposition (2.6.1)*. *There is a unique Δ-functorial isomorphism*

$$\alpha \colon \mathbf{R}\mathcal{H}om^\bullet(A \underset{=}{\otimes} B, \, C) \overset{\sim}{\longrightarrow} \mathbf{R}\mathcal{H}om^\bullet(A, \, \mathbf{R}\mathcal{H}om^\bullet(B, \, C))$$

such that for all $A, B, C \in \mathbf{D}$, *the following natural diagram* (*in which* $\mathcal{H}^\bullet$ *stands for* $\mathcal{H}om^\bullet$) *commutes*:

$$
\begin{array}{ccccc}
\mathcal{H}^\bullet(A \otimes B, C) & \longrightarrow & \mathbf{R}\mathcal{H}^\bullet(A \otimes B, C) & \longrightarrow & \mathbf{R}\mathcal{H}^\bullet(A \underset{\approx}{\otimes} B, C) \\
{\scriptstyle \text{via}} \downarrow {\scriptstyle (2.6.2)} & & & & {\scriptstyle \simeq} \downarrow {\scriptstyle \alpha} \\
\mathcal{H}^\bullet(A, \mathcal{H}^\bullet(B, C)) & \longrightarrow & \mathbf{R}\mathcal{H}^\bullet(A, \mathcal{H}^\bullet(B, C)) & \longrightarrow & \mathbf{R}\mathcal{H}^\bullet(A, \mathbf{R}\mathcal{H}^\bullet(B, C))
\end{array}
$$

This Δ-functorial isomorphism is the same as the one described—non-canonically, via P_B and I_C—near the beginning of this section. See also exercise (3.5.3)(e) below.

From (2.5.7.1) and (3.3.8) below (dualized), we deduce:

Corollary 2.6.6. *For fixed* A *the* Δ-*functor* $F_A(-) := \mathrm{Hom}^\bullet(A, -)$ *of* §2.4 *has a right-derived* Δ-*functor of the form* $(\mathbf{R}F_A, \mathrm{identity})$.

Exercise 2.6.7 (see [**De**, §1.2]). Define derived functors of several variables, and generalize the relevant results from §§2.2–2.3.

2.7 Acyclic Objects; Finite-Dimensional Derived Functors

This section contains additional results about acyclicity, used to get some more ways to construct derived functors, further illustrating (2.2.6). It can be skipped on first reading.

Let $\mathcal{A}$, $\mathcal{A}'$ be abelian categories, and let $\phi \colon \mathcal{A} \to \mathcal{A}'$ be an additive functor. We also denote by ϕ the composed Δ-functor

$$
\mathbf{K}(\mathcal{A}) \xrightarrow{\ \mathbf{K}(\phi)\ } \mathbf{K}(\mathcal{A}') \xrightarrow{\ Q\ } \mathbf{D}(\mathcal{A}')
$$

where $\mathbf{K}(\phi)$ is the natural extension of the original ϕ to a Δ-functor. We say then that an object in $\mathcal{A}$ is right-(or left-)ϕ-acyclic if it is so when viewed as a complex vanishing outside degree zero (see (2.2.5) with $\mathbf{J} := \mathbf{K}(\mathcal{A})$). In this section we deal mainly with the "left" context, and so we abbreviate "left-ϕ-acyclic" to "ϕ-acyclic." (The corresponding—dual—results in the "right" context are left to the reader. They are perhaps marginally less important because of the abundance of injectives in situations that we will deal with.)

If $X \in \mathcal{A}$ and $Z \to X$ is a quasi-isomorphism in $\mathbf{K}(\mathcal{A})$, then the natural map $\tau_{\leq 0} Z \to Z$ of §1.10 is a quasi-isomorphism. If furthermore the induced map $\phi(Z) \to \phi(X)$ is a quasi-isomorphism and the functor ϕ is either right exact or left exact, then, one checks, the natural composition $\phi(\tau_{\leq 0} Z) \to \phi(Z) \to \phi(X)$ is also a quasi-isomorphism.

One deduces the following characterization of ϕ-acyclicity:

Lemma 2.7.1. *If $X \in \mathcal{A}$ is such that every exact sequence*

$$\cdots \longrightarrow Y_2 \longrightarrow Y_1 \longrightarrow Y_0 \longrightarrow X \longrightarrow 0$$

embeds into a commutative diagram in $\mathcal{A}$

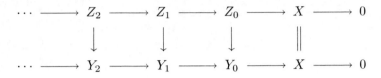

with the top row and its image under ϕ both exact, then X is ϕ-acyclic; and the converse holds whenever ϕ is either right exact or left exact.

Proposition 2.7.2. *With preceding notation, let $\mathbf{P}$ be a class of objects in $\mathcal{A}$ such that*
  (i) *every object in $\mathcal{A}$ is a quotient of (i.e., target of an epimorphism from) one in $\mathbf{P}$;*
  (ii) *if A and B are in $\mathbf{P}$ then so is $A \oplus B$; and*
  (iii) *for every exact sequence $0 \to A \to B \to C \to 0$ in $\mathcal{A}$, if B and C are in $\mathbf{P}$, then $A \in \mathbf{P}$ and moreover the corresponding sequence $0 \to \phi A \to \phi B \to \phi C \to 0$ in $\mathcal{A}'$ is also exact.*
Then every bounded-above $\mathbf{P}$-complex (i.e., complex with all components in $\mathbf{P}$)—in particular every object in $\mathbf{P}$—is ϕ-acyclic; the restriction ϕ_- of ϕ to $\overline{\mathbf{K}}^-(\mathcal{A})$ has a left-derived functor $\mathbf{L}\phi_- : \overline{\mathbf{D}}^-(\mathcal{A}) \to \mathbf{D}(\mathcal{A}')$; and if $\phi \not\cong 0$ then $\dim^+ \mathbf{L}\phi_- = 0$ (see (1.11.1)).

Proof. Since $\mathbf{P}$ is nonempty—by (i)—therefore (iii) with $B = C \in \mathbf{P}$ shows that $0 \in \mathbf{P}$. Then (ii) implies that the $\mathbf{P}$-complexes in $\mathbf{K}^-(\mathcal{A})$ are the objects of a Δ-subcategory, see (1.6). Starting from (i), an inductive argument ([**H**, p. 42, 4.6, 1)], dualized—and with assistance, if desired, from [**Iv**, p. 34, Prop. 5.2]) shows that every complex in $\mathbf{K}^-(\mathcal{A})$—and so, via (1.8.1)$^-$, in $\overline{\mathbf{K}}^-(\mathcal{A})$—is the target of a quasi-isomorphism from a bounded-above $\mathbf{P}$-complex. Hence, for the first assertion it suffices to show that ϕ transforms quasi-isomorphisms between bounded-above $\mathbf{P}$-complexes into isomorphisms, i.e., that *for any bounded-above exact $\mathbf{P}$-complex $X^\bullet$, $\phi(X^\bullet) \cong 0$* (see (1.5.1)).

Using (iii), we find by descending induction (starting with i_0 such that $X^j = 0$ for all $j > i_0$) that for every i, the kernel K^i of $X^i \to X^{i+1}$ lies in $\mathbf{P}$ and the obvious sequence

$$0 \to \phi(K^i) \to \phi(X^i) \to \phi(K^{i+1}) \to 0$$

is exact. Consequently, the complex obtained by applying ϕ to $X^\bullet$ is exact, i.e., $\phi(X^\bullet) \cong 0$ in $\mathbf{D}(\mathcal{A}')$.

Now by (2.2.4) (dualized) we see that $\mathbf{L}\phi_-$ exists and $\dim^+ \mathbf{L}\phi_- \leq 0$, with equality if $\phi(A) \not\cong 0$ for some $A \in \mathcal{A}$, because there is a natural epimorphism $H^0\mathbf{L}\phi_- A \twoheadrightarrow \phi(A)$.   Q.E.D.

Exercise 2.7.2.1. Let $\phi\colon \mathcal{A} \to \mathcal{A}'$ be as above. Let $(\Lambda_i)_{0 \leq i < \infty}$ be a "homological functor" [**Gr**, p. 140], with $\Lambda_0 = \phi$. Let $\mathbf{P}$ consist of all objects B in $\mathcal{A}$ such that $\Lambda_i(B) = 0$ for all $i > 0$, and suppose that every object $A \in \mathcal{A}$ is a quotient of one in $\mathbf{P}$. Then $\mathbf{L}\phi_-$ exists, and the homological functors (Λ_i) and $(\Lambda_i') := (H^{-i}\mathbf{L}\phi_-)$ are coeffaceable, hence universal [**Gr**, p. 141, Prop. 2.2], hence isomorphic to each other.

Example 2.7.3. A *ringed space* is a pair $(X, \mathcal{O}_X)$ with X a topological space and $\mathcal{O}_X$ a sheaf of commutative rings on X; and a *morphism* of ringed spaces $(f, \theta)\colon (X, \mathcal{O}_X) \to (Y, \mathcal{O}_Y)$ is a continuous map $f\colon X \to Y$ together with a map $\theta\colon \mathcal{O}_Y \to f_*\mathcal{O}_X$ of sheaves of rings. Any such (f, θ) gives rise to a (left-exact) *direct image* functor

$$f_*\colon \{\mathcal{O}_X\text{-modules}\} \to \{\mathcal{O}_Y\text{-modules}\}$$

such that $[f_*M](U) = M(f^{-1}U)$ for any $\mathcal{O}_X$-module M and any open set $U \subset Y$, the $\mathcal{O}_Y$-module structure on f_*M arising via θ; and also to a (right-exact) *inverse image* functor

$$f^*\colon \{\mathcal{O}_Y\text{-modules}\} \to \{\mathcal{O}_X\text{-modules}\}$$

defined up to isomorphism as being left-adjoint to f_* [**GD**, Chap. 0, §4]. For every $\mathcal{O}_Y$-module N, the stalk $(f^*N)_x$ at $x \in X$ is $\mathcal{O}_{X,x} \otimes_{\mathcal{O}_{Y,f(x)}} N_{f(x)}$.

An $\mathcal{O}_Y$-module F is *flat* if the stalk F_y is a flat $\mathcal{O}_{Y,y}$-module for all $y \in Y$. *The class* $\mathbf{P}$ *of flat* $\mathcal{O}_Y$-modules satisfies the hypotheses of (2.7.2) *when* $\phi = f^*$: (i) is given by [**H**, p. 86, Prop. 1.2], (ii) is easy, and for (iii) see [**B'**, Chap. 1, §2, no. 5]. Thus the restriction f^*_- of f^* to $\overline{\mathbf{K}}^-(Y)$ has a left-derived functor

$$\mathbf{L}f^*_-\colon \overline{\mathbf{D}}^-(Y) \to \mathbf{D}(X)$$

(where $\mathbf{D}(X)$ is the derived category of the category of $\mathcal{O}_X$-modules, etc.), defined via resolutions (on the left) by complexes of flat $\mathcal{O}_Y$-modules.

Using the family of quasi-isomorphisms $\psi_A\colon P_A \to A$ $(A \in \mathbf{D}(Y))$ with P_A q-flat (see (2.5.5)), we can, in view of (2.5.2) and (2.5.3), show as in (2.5.7) that $\mathbf{L}f^*_-$ extends to a derived Δ-functor

$$(\mathbf{L}f^*, \text{identity})\colon \mathbf{D}(Y) \to \mathbf{D}(X) \tag{2.7.3.1}$$

satisfying $\mathbf{L}f^*(A) = f^*(P_A)$.

For any $\mathcal{O}_Y$-module N, the stalk of the homology

$$L_if^*(N) := H^{-i}\mathbf{L}f^*(N) \qquad (i \geq 0)$$

at any $x \in X$ is $\operatorname{Tor}_i^{\mathcal{O}_{Y,f(x)}}(\mathcal{O}_{X,x}, N_{f(x)})$. So by the last assertion in (2.2.6) (dualized), or in (2.7.4), N is f^*-acyclic iff $\operatorname{Tor}_i^{\mathcal{O}_{Y,f(x)}}(\mathcal{O}_{X,x}, N_{f(x)}) = 0$

for all $x \in X$ and $i > 0$. (Note here that since f^* is right exact, the natural map is an isomorphism $L_0 f^*(N) \xrightarrow{\sim} f^*(N)$.) Thus—or by (2.7.2)—any flat $\mathcal{O}_Y$-module is f^*-acyclic.

Recall that an $\mathcal{O}_X$-module M is *flasque* (or *flabby*) if the restriction map $M(X) \to M(U)$ is surjective for every open subset U of X. For example, injective $\mathcal{O}_X$-modules are flasque [**G**, p. 264, 7.3.2] (with $\mathcal{L} = \mathcal{O}_X$). *The class of flasque $\mathcal{O}_X$-modules satisfies the hypotheses of* (2.7.2) (*dual version*) *when* $\phi = f_*$: for (i) see [**G**, p. 147], (ii) is easy, and (iii) follows from the fact that if

$$0 \to F \to G \to H \to 0$$

is an exact sequence of $\mathcal{O}_X$-modules, with F flasque, then for all open sets $V \subset X$ the sequence

$$0 \to F(V) \to G(V) \to H(V) \to 0$$

is still exact [**G**, p. 148, Thm. 3.1.2]. So the restriction f_*^+ of f^* to $\overline{\mathbf{K}}^{\boldsymbol{+}}(X)$ has a right-derived functor

$$\mathbf{R}f_*^+ \colon \overline{\mathbf{D}}^{\boldsymbol{+}}(X) \to \mathbf{D}(Y)$$

defined via resolutions (on the right) by complexes of flasque $\mathcal{O}_X$-modules.

Of course we already know from (2.3.4), via (somewhat less elementary) *injective* resolutions, that $\mathbf{R}f_*^+$ exists, and by (2.3.5) it extends to a derived functor $\mathbf{R}f_* \colon \mathbf{D}(X) \to \mathbf{D}(Y)$. (See also (2.3.7).) In fact, in view of (2.7.3.1), it follows from (3.2.1) and (3.3.8) (dualized) that:

(2.7.3.2). *The Δ-functor* $(f_*, \text{identity})$ *has a derived Δ-functor of the form* $(\mathbf{R}f_*, \text{identity})$.

An $\mathcal{O}_X$-module M is f_*-acyclic iff the "higher direct image" sheaves

$$R^i f_*(M) := H^i \mathbf{R}f_*(M) \qquad (i \geq 0)$$

vanish for all $i > 0$, see last assertion in (2.2.6) or in (2.7.4) (dualized). (Since f_* is left-exact, the natural map is an isomorphism $f_* \xrightarrow{\sim} R^0 f_*$.) Flasque sheaves are f_*-acyclic.

For more examples involving flasque sheaves see [**H**, p. 225, Variations 6 and 7] ("cohomology with supports").

Proposition 2.7.4. *Let $\mathcal{A}$ and $\mathcal{A}'$ be abelian categories, and let $\phi \colon \mathcal{A} \to \mathcal{A}'$ be a right-exact additive functor. If C is ϕ-acyclic, then for every exact sequence $0 \to A \to B \to C \to 0$ in $\mathcal{A}$ the corresponding sequence $0 \to \phi A \to \phi B \to \phi C \to 0$ is also exact, and A is ϕ-acyclic iff B is. So if every object in $\mathcal{A}$ is a quotient of a ϕ-acyclic one, then the conclusions of (2.7.2) hold with $\mathbf{P}$ the class of ϕ-acyclic objects; and then $D \in \mathcal{A}$ is ϕ-acyclic iff the natural map $\mathbf{L}\phi_-(D) \to \phi(D)$ is an isomorphism in $\mathbf{D}(\mathcal{A}')$, i.e., iff $H^{-i}\mathbf{L}\phi_-(D) = 0$ for all $i > 0$.*

Proof. For the first assertion, note that by (2.7.1) there exists a commutative diagram

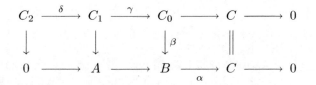

such that the top row is exact and remains so after application of ϕ. There results a commutative diagram

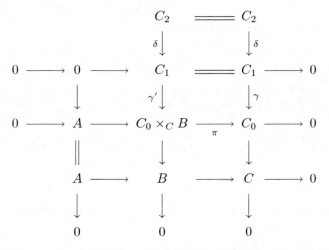

with exact columns, in which the middle row is *split exact,* a right inverse for the projection π being given by the graph of the map $\beta.$[6] (The coordinates of γ' are γ and 0.) Applying ϕ preserves split-exactness; and then, since ϕ is right-exact, so that e.g., $\phi C = \operatorname{coker}(\phi\gamma)$, the "snake lemma" yields an exact sequence

$$0 \to \ker(\phi\gamma') \to \ker(\phi\gamma) \to \phi A \to \phi B \to \phi C \to 0\,.$$

Since

$$\ker(\phi\gamma) = \operatorname{im}(\phi\delta) \subset \ker(\phi\gamma')$$

we conclude that $0 \to \phi A \to \phi B \to \phi C \to 0$ is exact, as asserted in (2.7.4).

In other words, if Z is the complex which looks like $A \to B$ in degrees -1 and 0 and which vanishes elsewhere, then the quasi-isomorphism $Z \to C$ given by the exact sequence $0 \to A \to B \to C \to 0$ becomes, upon application of ϕ, an isomorphism in $\mathbf{D}(\mathcal{A}')$; and hence, by (2.2.5.2) (dualized), Z is a ϕ-acyclic complex.

---

[6] Recall that $C_0 \times_C B$ is the kernel of the map $C_0 \oplus B \to C$ whose restriction to C_0 is $\alpha\beta$ and to B is $-\alpha$.

The natural semi-split sequence $0 \to B \to Z \to A[1] \to 0$ leads, as in (1.4.3), to a triangle

$$B \longrightarrow Z \longrightarrow A[1] \longrightarrow B[1] \, ;$$

and since the ϕ-acyclic complexes are the objects of a Δ-subcategory, see (2.2.5.1), it follows that A is ϕ-acyclic iff B is.

Since Δ-subcategories are closed under direct sum, it is clear now that (ii) and (iii) in (2.7.2) hold when $\mathbf{P}$ is the class of ϕ-acyclic objects, whence the second-last assertion in (2.7.4). In view of (2.7.2) and its proof, the last assertion of (2.7.4) is contained in (2.2.6).   Q.E.D.

The derived functor $\mathbf{L}\phi_-$ of (2.7.4) satisfies $\dim^+ \mathbf{L}\phi_- = 0$ (unless $\phi \cong 0$, see (2.7.2)). When its lower dimension satisfies $\dim^- \mathbf{L}\phi_- < \infty$, more can be said.

Proposition 2.7.5. *Let $\phi \colon \mathcal{A} \to \mathcal{A}'$ be a right-exact functor such that every object in $\mathcal{A}$ is a quotient of a ϕ-acyclic one, and let $\mathbf{L}\phi_-$ be a left-derived functor of $\phi | \overline{\mathbf{K}}^-(\mathcal{A})$, see (2.7.4). Then the following conditions on an integer $d \geq 0$ are equivalent:*

  (i) *$\dim^- \mathbf{L}\phi_- \leq d$.*
  (ii) *For any $F \in \mathcal{A}$ we have*

$$L_j \phi(F) := H^{-j} \mathbf{L}\phi_-(F) = 0 \quad \text{for all } j > d.$$

  (iii) *In any exact sequence in $\mathcal{A}$*

$$0 \to 0 \to B_d \to B_{d-1} \to \ldots \to B_0 \, ,$$

   if $B_0, B_1, \ldots, B_{d-1}$ are all ϕ-acyclic then so is B_d.[7]
  (iv) *For any $F \in \mathcal{A}$ there is an exact sequence*

$$0 \to B_d \to B_{d-1} \to \ldots \to B_0 \to F \to 0$$

   in which every B_i is ϕ-acyclic.
  (v) *For any complex $F^\bullet \in \mathbf{K}(\mathcal{A})$ and integers $m \leq n$, if $F^j = 0$ for all $j \notin [m, n]$ then there exists a quasi-isomorphism $B^\bullet \to F^\bullet$ where B^j is ϕ-acyclic for all j and $B^j = 0$ for $j \notin [m - d, n]$.*
  (vi) *For any complex $F^\bullet \in \mathbf{K}(\mathcal{A})$ and integer m, if $F^j = 0$ for all $j < m$ then there exists a quasi-isomorphism $B^\bullet \to F^\bullet$ where B^j is ϕ-acyclic for all j and $B^j = 0$ for all $j < m - d$.*

---

[7] For $d = 0$ this means that every $B \in \mathcal{A}$ is ϕ-acyclic, i.e., ϕ is an *exact* functor, see (2.7.4) (and then every $F^\bullet \in \mathbf{K}(\mathcal{A})$ is ϕ-acyclic, see (2.2.8(a))).

When there exists an integer $d \geq 0$ for which these conditions hold, then:

(a) *Every complex of ϕ-acyclic objects is a ϕ-acyclic complex.*

(b) *Every complex in $\mathcal{A}$ is the target of a quasi-isomorphism from a ϕ-acyclic complex.*

(c) *A left-derived functor $\mathbf{L}\phi\colon \mathbf{D}(\mathcal{A}) \rightarrow \mathbf{D}(\mathcal{A}')$ exists, $\dim^+ \mathbf{L}\phi = 0$ (unless $\phi \cong 0$) and $\dim^- \mathbf{L}\phi \leq d$.*

(d) *The restriction $\mathbf{L}\phi|_{\overline{\mathbf{D}}^*(\mathcal{A})}$ is a left-derived functor of $\phi|_{\overline{\mathbf{K}}^*(\mathcal{A})}$, and*

$$\mathbf{L}\phi(\overline{\mathbf{D}}^*(\mathcal{A})) \subset \overline{\mathbf{D}}^*(\mathcal{A}') \qquad (* = \texttt{+}, \texttt{-}, \text{ or } \texttt{b}).$$

Proof. (i)$\Leftrightarrow$(ii). This is given by (iii) and (iv) in (1.11.2).

(iii)$\Rightarrow$(v)$\Rightarrow$(iv). Let $F^\bullet$ and $m \leq n$ be as in (v). As in the proof of (2.7.2), there is a quasi-isomorphism $P^\bullet \rightarrow F^\bullet$ with P^j ϕ-acyclic for all j and $P^j = 0$ for $j > n$. Let B^{m-d} be the cokernel of $P^{m-d-1} \rightarrow P^{m-d}$. If (iii) holds, then B^{m-d} is ϕ-acyclic: this is trivial if $d = 0$, and otherwise follows from the exact sequence

$$0 \rightarrow B^{m-d} \rightarrow P^{m-d+1} \rightarrow \cdots \rightarrow P^{m-1} \rightarrow P^m.$$

So all components of the complex $B^\bullet = \tau_{\geq m-d}P^\bullet$ (see (1.10)) are ϕ-acyclic, and clearly $P^\bullet \rightarrow F^\bullet$ factors naturally as $P^\bullet \rightarrow B^\bullet \rightarrow F^\bullet = \tau_{\geq m-d}F^\bullet$ where both arrows represent quasi-isomorphisms. Thus (iii)$\Rightarrow$(v); and (v)$\Rightarrow$(iv) is obvious.

Recalling from (2.7.4) that $B \in \mathcal{A}$ is ϕ-acyclic iff $L_i\phi(B) = 0$ for all $i > 0$, we easily deduce the implications (iv)$\Rightarrow$(ii)$\Rightarrow$(iii) from:

Lemma 2.7.5.1. *Let*

$$0 = B_{d+1} \rightarrow B_d \rightarrow B_{d-1} \rightarrow \cdots \rightarrow B_0 \rightarrow F \rightarrow 0$$

be an exact sequence in $\mathcal{A}$ with $B_0, B_1, \ldots, B_{d-1}$ all ϕ-acyclic, and let K_j be the cokernel of $B_{j+1} \rightarrow B_j$ $(0 \leq j \leq d)$. Then for any $i > 0$, there results a natural sequence of isomorphisms

$$L_{i+d}\phi(F) = L_{i+d}\phi(K_0) \xrightarrow{\sim} L_{i+d-1}\phi(K_1) \xrightarrow{\sim} \cdots$$
$$\cdots \xrightarrow{\sim} L_{i+2}\phi(K_{d-2}) \xrightarrow{\sim} L_{i+1}\phi(K_{d-1}) \xrightarrow{\sim} L_i\phi(K_d) = L_i\phi(B_d).$$

Proof. When $d = 0$, it's obvious. If $d > 0$, apply $(2.1.4)^{\mathrm{H}}$ (dualized) to the natural exact sequences

$$0 \rightarrow K_j \rightarrow B_{j-1} \rightarrow K_{j-1} \rightarrow 0 \qquad (0 < j \leq d)$$

to obtain exact sequences

$$0 = L_{i+d-j+1}\phi(B_{j-1}) \to L_{i+d-j+1}\phi(K_{j-1})$$
$$\to L_{i+d-j}\phi(K_j) \to L_{i+d-j}\phi(B_{j-1}) = 0. \qquad\qquad \text{Q.E.D.}$$

(iii)$\Rightarrow$(vi). Condition (iii) coincides with condition (iii) of [**H**, p. 42, Lemma 4.6, 2)] (dualized, and with P the set of ϕ-acyclics in $\mathcal{A}$). Condition (i) of *loc. cit.* holds by assumption, and condition (ii) of *loc. cit.* is contained in (2.7.4). So if (iii) holds, *loc. cit.* gives the existence of a quasi-isomorphism $B^\bullet \to F^\bullet$ with B^j ϕ-acyclic for all j; and the recipe at the bottom of [**H**, p. 43] for constructing $B^\bullet$ allows us, when $F^j = 0$ for all $j < m$, to do so in such a way that $B^j = 0$ for all $j < m - d$.

(vi)$\Rightarrow$(ii). Assuming (vi), we can find for each object $F \in \mathcal{A}$ a quasi-isomorphism $B^\bullet \to F$ with all B^j ϕ-acyclic and $B^j = 0$ for $j < -d$. If K is the cokernel of $B^{-1} \to B^0$ then the natural composition

$$H^0(B^\bullet) \longrightarrow K \longrightarrow F$$

is an isomorphism, whence so are the functorially induced compositions

$$L_j\phi(H^0(B^\bullet)) \longrightarrow L_j\phi(K) \longrightarrow L_j\phi(F) \qquad (j \in \mathbb{Z}). \qquad (2.7.5.2)$$

But for every $j > d$, (2.7.5.1) with K in place of F yields $L_j\phi(K) = 0$, so that the isomorphism (2.7.5.2) is the zero-map. Thus (ii) holds.

Now suppose that (i)–(vi) hold for some $d \geq 0$. We have just seen, in proving that (iii)$\Rightarrow$(vi), that then every complex in $\mathcal{A}$ receives a quasi-isomorphism from a complex $B^\bullet$ of ϕ-acyclics; and so, as in the proof of (2.7.2), assertion (2.7.5)(a)—and hence (b)—will result if we can show that whenever such a $B^\bullet$ is exact, then so is $\phi(B^\bullet)$. But condition (iii) guarantees that when $B^\bullet$ is exact, the kernel K^i of $B^i \to B^{i+1}$ is ϕ-acyclic for all i, whence by (2.7.4) we have exact sequences

$$0 \to \phi(K^{i-1}) \to \phi(B^{i-1}) \to \phi(K^i) \to 0 \qquad (i \in \mathbb{Z})$$

which together show that $\phi(B^\bullet)$ is indeed exact.

The existence of $\mathbf{L}\phi$, via resolutions by complexes of ϕ-acyclic objects, follows now from (2.2.6); and the dimension statements follow, after application of $(1.8.1)^+$ or $(1.8.1)^-$, from (v) with $m = -\infty$ (obvious interpretation, see beginning of above proof that (iii)$\Rightarrow$(v)) and from (vi). Similar considerations yield (d).   Q.E.D.

Example 2.7.6. The *dimension* $\dim f$ of a map $f\colon X \to Y$ of ringed spaces is the upper dimension (1.11) of the functor $\mathbf{R}f_*^+\colon \overline{\mathbf{D}}^+(X) \to \mathbf{D}(Y)$ of (2.7.3):

$$\dim f := \dim^+ \mathbf{R}f_*^+,$$

a nonnegative integer unless $f_*\mathcal{O}_X \cong 0$, in which case $\dim f = -\infty$. When f has *finite* dimension, (2.7.5)(c) (dualized) gives the existence of a derived functor $\mathbf{R}f_* \colon \mathbf{D}(X) \to \mathbf{D}(Y)$ via resolutions (on the right) by complexes of f_*-acyclic objects, and we have $\infty > \dim f = \dim^+ \mathbf{R}f_*$.

The *tor-dimension* (or *flat dimension*) tor-dim f of a map $f \colon X \to Y$ of ringed spaces is defined to be the lower dimension (see (1.11)) of the functor $\mathbf{L}f_-^* \colon \overline{\mathbf{D}}^-(Y) \to \mathbf{D}(X)$ of (2.7.3):

$$\text{tor-dim } f := \dim^- \mathbf{L}f_-^* \,,$$

a nonnegative integer unless $\mathcal{O}_X \cong 0$, in which case tor-dim $f = -\infty$. When f has *finite* tor-dimension, (2.7.5)(c) gives the existence of a derived functor $\mathbf{L}f^* \colon \mathbf{D}(X) \to \mathbf{D}(Y)$ via resolutions (on the left) by complexes of f^*-acyclic objects, and we have $\infty > \text{tor-dim } f = \dim^- \mathbf{L}f^*$.

Following [**I**, p. 241, Définition 3.1] one says that an $\mathcal{O}_X$-complex E has *flat f-amplitude in $[m,n]$* if for any $\mathcal{O}_Y$-module F,

$$H^i(E \underset{\approx}{\otimes} \mathbf{L}f^*F) = 0 \ \text{ for all } i \notin [m,n],$$

or equivalently, for the functor $L_E(F) := E \underset{\approx}{\otimes} \mathbf{L}f^*F$ of $\mathcal{O}_Y$-module F,

$$\dim^+ L \leq m \ \text{ and } \ \dim^- L \leq -n.$$

This means that the stalk E_x at each $x \in X$ is $\mathbf{D}(\mathcal{O}_{Y,f(x)})$-isomorphic to a flat complex vanishing in degrees outside $[m,n]$, see [**I**, p. 242, 3.3], or argue as in (2.7.6.4) below. E has *finite flat f-amplitude* if such m and n exist.

It follows from (2.7.6.4) below and [**I**, p. 131, 5.1] that *f has finite tor-dimension* $\iff$ $\mathcal{O}_X$ *has finite flat f-amplitude*.

(2.7.6.1). If X is a compact Hausdorff space of dimension $\leq d$ (in the sense that each point has a neighborhood homeomorphic to a locally closed subspace of the Euclidean space $\mathbb{R}^d$), and $\mathcal{O}_X$ is the constant sheaf $\mathbb{Z}$, then $\dim f \leq d$.

Indeed, if $I^\bullet$ is a flasque resolution of the abelian sheaf F, then for any open $U \subset Y$ the restriction $I^\bullet|f^{-1}(U)$ is a flasque resolution of $F|f^{-1}(U)$, and $R^j f_*(F)$ is, up to isomorphism, the sheaf associated to the presheaf taking any such U to the group $H^j(\Gamma(f^{-1}(U), I^\bullet|f^{-1}(U))$, a group isomorphic to $H^j(f^{-1}(U), F|f^{-1}(U))$ [**G**, p. 181, Thm. 4.7.1(a)], and hence vanishing for $j > d$, see [**Iv**, Chap. III, §9].

More generally, if X is locally compact and we assume only that the fibers $f^{-1}y$ ($y \in Y$) are compact and have dimension $\leq d$, then $\dim f \leq d$ (because the stalk $(R^j f_* F)_y$ is the cohomology $H^j(f^{-1}y, F|f^{-1}y)$, see [**Iv**, p. 315, Thm. 1.4], whose proof does not require any assumption on Y).

(2.7.6.2). (Grothendieck, see [**H**, p. 87]). If $(X, \mathcal{O}_X)$ is a noetherian scheme of finite Krull dimension d, then $\dim f \leq d$.

(2.7.6.3). For a ringed-space map $f\colon X \to Y$ with $\mathcal{O}_X \not\cong 0$, the following conditions are equivalent:

(i)   tor-dim $f = 0$.

(i)' Every $\mathcal{O}_Y$-module is f^*-acyclic.

(i)'' The functor f^* of $\mathcal{O}_Y$-modules is exact.

(ii) f is flat (i.e., $\mathcal{O}_{X,x}$ is a flat $\mathcal{O}_{Y,f(x)}$-module for all $x \in X$).

Proof. Since every $\mathcal{O}_X$-module is a quotient of a flat one, which is f^*-acyclic (see (2.7.3)), the equivalence of (i), (i)', and (i)'' is given, e.g., by that of (i) and (iii) in (2.7.5) (for $d = 0$). The equivalence of (i) and (ii) is the case $d = 0$ of:

(2.7.6.4) Let $f\colon X \to Y$ be a ringed-space map and $d \geq 0$ an integer. Then tor-dim $f \leq d \iff$ for each $x \in X$ there exists an exact sequence of $\mathcal{O}_{Y,f(x)}$-modules

$$0 \to P_d \to P_{d-1} \to \ldots \to P_1 \to P_0 \to \mathcal{O}_{X,x} \to 0 \qquad\qquad (*)$$

with P_i flat over $\mathcal{O}_{Y,f(x)}$ $(0 \leq i \leq d)$.

Proof. ("*if*") Let F be an $\mathcal{O}_Y$-module and let $Q^\bullet \to F$ be a quasi-isomorphism with $Q^\bullet$ a flat complex (1.8.3). Then for $j \geq 0$, the homology

$$L_j f^*(F) \cong H^{-j}(f^*Q^\bullet) \qquad \text{(see (2.7.3))}$$

vanishes iff for each $x \in X$, with $y = f(x)$, $R = \mathcal{O}_{Y,y}$, and $S = \mathcal{O}_{X,x}$ we have

$$0 = H^{-j}\big((f^*Q^\bullet)_x\big) = H^{-j}\big(S \otimes_R Q_y^\bullet\big) = \operatorname{Tor}_j^R(S, F_y)$$

(where the last equality holds since $Q_y^\bullet \to F_y$ is an R-flat resolution of F_y), whence the assertion.

("*only if*") Suppose only that $L_{d+1}f^*(F) = 0$ for all F, so that (see above) $\operatorname{Tor}_{d+1}^R(S, F_y) = 0$; and let

$$\cdots \to P_2' \to P_1' \to P_0' \to S \to 0$$

be an R-flat resolution of S. Then, I claim, the module

$$P_d := \operatorname{coker}(P_{d+1}' \to P_d')$$

is R-*flat*, whence we have $(*)$ with $P_i = P_i'$ for $0 \leq i < d$.

Indeed, the flatness of P_d is equivalent to the vanishing of $\operatorname{Tor}_1^R(P_d, R/I)$ for all R-ideals I [**B'**, §4, Prop. 1]. But any such I is $\mathcal{I}_y$ where $\mathcal{I} \subset \mathcal{O}_Y$ is the $\mathcal{O}_Y$-ideal such that for any open $U \subset Y$,

$$\mathcal{I}(U) = \{\, r \in \mathcal{O}_Y(U) \mid r_y \in I \,\} \qquad \text{if } y \in U$$
$$= 0 \qquad\qquad\qquad\qquad\qquad\quad \text{if } y \notin U;$$

so that if $F = \mathcal{O}_Y/\mathcal{I}$, then $R/I = F_y$; and from the flat resolution

$$\cdots \to P_{d+2}' \to P_{d+1}' \to P_d' \to P_d \to 0$$

of P_d, we get the desired vanishing:

$$\operatorname{Tor}_1^R(P_d, R/I) = \operatorname{Tor}_1^R(P_d, F_y) = \operatorname{Tor}_{d+1}^R(S, F_y) = 0.$$

Exercise 2.7.6.5. (For amusement only.) If Y is a quasi-separated scheme, then $f: X \rightarrow Y$ satisfies tor-dim $f \leq d$ if (and only if) for every quasi-coherent $\mathcal{O}_Y$-ideal $\mathcal{I}$, we have

$$L_{d+1}f^*(\mathcal{O}_Y/\mathcal{I}) = 0.$$

If in addition Y is quasi-compact or locally noetherian, then we need only consider *finite-type* quasi-coherent $\mathcal{O}_Y$-ideals.

[The following facts in [**GD**] can be of use here: p. 111, (5.2.8); p. 313, (6.7.1); p. 294, (6.1.9) (i); p. 295, (6.1.10)(iii); p. 318, (6.9.7).]

Chapter 3
Derived Direct and Inverse Image

A *ringed space* is a pair $(X, \mathcal{O}_X)$ with X a topological space and $\mathcal{O}_X$ a sheaf of commutative rings on X; and a *morphism* (or *map*) of ringed spaces $(f, \theta): (X, \mathcal{O}_X) \to (Y, \mathcal{O}_Y)$ is a continuous map $f: X \to Y$ together with a map $\theta: \mathcal{O}_Y \to f_*\mathcal{O}_X$ of sheaves of rings. (Usually we will just denote such a morphism by $f: X \to Y$, the accompanying θ understood to be standing by.) Associated with (f, θ) are the adjoint functors

$$\mathcal{A}_X := \{\mathcal{O}_X\text{-modules}\} \underset{f_*}{\overset{f^*}{\rightleftharpoons}} \{\mathcal{O}_Y\text{-modules}\} =: \mathcal{A}_Y$$

and their respective derived functors $\mathbf{R}f_*$, $\mathbf{L}f^*$, which are also adjoint—as Δ-functors, (3.2), (3.3). In this chapter we first review the definitions and basic formal (i.e., category-theoretic) properties of these adjoint derived functors, their interactions with $\otimes$ and $\mathbf{R}\mathcal{H}om^\bullet$, and their "pseudofunctorial" behavior with respect to composition of ringed-space maps (3.6), many of the main results being packaged in (3.6.10).

A basic objective, in the spirit of Grothendieck's philosophy of the "six operations," is the categorical formalization of relations among functorial maps involving the four operations $\mathbf{R}f_*$, $\mathbf{L}f^*$, $\otimes$ and $\mathbf{R}\mathcal{H}om^\bullet$.[1]

More explicitly (details in §§3.4, 3.5), if $f: X \to Y$ is a map of ringed spaces, then the derived categories $\mathbf{D}(\mathcal{A}_X)$, $\mathbf{D}(\mathcal{A}_Y)$ have natural structures of symmetric monoidal closed categories, given by $\otimes$ and $\mathbf{R}\mathcal{H}om^\bullet$; and the adjoint Δ-functors $\mathbf{R}f_*$ and $\mathbf{L}f^*$ respect these structures, as do the conjugate isomorphisms, arising from a second map $g: Y \to Z$, $\mathbf{R}(gf)_* \xrightarrow{\sim} \mathbf{R}g_*\mathbf{R}f_*$, $\mathbf{L}f^*\mathbf{L}g^* \xrightarrow{\sim} \mathbf{L}(gf)^*$. We express all this by saying that $\mathbf{R}-_*$ and $\mathbf{L}-^*$ are *adjoint monoidal Δ-pseudofunctors*.

Thus, relations among the four operations can be worked with as instances of category-theoretic relations involving adjoint monoidal functors between

---

[1] A fifth operation, "twisted inverse image," is brought into play in Chapter 4, at least for schemes. The sixth, "direct image with proper supports" [**De′**, n°3] will not appear here, except for proper scheme-maps, where it coincides with derived direct image.

J. Lipman, M. Hashimoto, *Foundations of Grothendieck Duality for Diagrams of Schemes*, Lecture Notes in Mathematics 1960,

closed categories. This eliminates excess baggage of resolutions of complexes, which would otherwise cause intolerable tedium later on, where proofs of major results depend heavily on involved manipulations of such relations.[2] Even so, the situation is far from ideal—see the introductory remarks in §3.4, and, for example, the proof of Proposition (3.7.3), which addresses the interaction between the projection morphisms of (3.4.6) and "base change."

By way of illustration, consider the following basic functorial maps, with $A, B \in \mathbf{D}(\mathcal{A}_Y)$ and $E, F \in \mathbf{D}(\mathcal{A}_X)$:[3]

$$\mathbf{R}f_* \mathbf{R}\mathcal{H}om_X^\bullet(\mathbf{L}f^*B, E) \to \mathbf{R}\mathcal{H}om_Y^\bullet(B, \mathbf{R}f_*E), \qquad (3.2.3.2)$$

$$\mathbf{L}f^*A \underset{=}{\otimes} \mathbf{L}f^*B \leftarrow \mathbf{L}f^*(A \underset{=}{\otimes} B), \qquad (3.2.4)$$

$$\mathbf{R}f_*(E) \underset{=}{\otimes} \mathbf{R}f_*(F) \to \mathbf{R}f_*(E \underset{=}{\otimes} F), \qquad (3.2.4.2)$$

$$\mathbf{R}f_*E \underset{=}{\otimes} B \to \mathbf{R}f_*(E \underset{=}{\otimes} \mathbf{L}f^*B). \qquad (3.4.6)$$

The first two can be defined at the level of complexes, after replacing the arguments by appropriate resolutions. (The reduction is straightforward for the second, but not quite so for the first.) At that level, one sees that they are both isomorphisms. For fixed B, the source and target of the first are left-adjoint, respectively, to the target and source of the second; and it turns out that the two maps are *conjugate* (3.3.5). This is shown by reduction to the analogous statement for the ordinary direct and inverse image functors for sheaves, which can be treated concretely (3.1.10) or formally (3.5.5). So each one of these isomorphisms determines the other from a purely categorical point of view.

The second and third maps determine each other via $\mathbf{L}f^*$–$\mathbf{R}f_*$ adjunction (3.4.5), as do the third and fourth (3.4.6). When the first map is given, the second and third maps also determine each other via $\mathbf{R}\mathcal{H}om^\bullet$–$\underset{=}{\otimes}$ adjunction. (This is not obvious, see Proposition (3.2.4).)

Thus, any three of the four maps can be deduced category-theoretically from the remaining one.

In (3.9) we consider the case when our ringed spaces are *schemes*. Under mild assumptions, we note that then $\mathbf{R}f_*$ and $\mathbf{L}f^*$ "respect quasi-coherence" (3.9.1), (3.9.2). We also show that some previously introduced functorial morphisms become isomorphisms: (3.9.4) treats variants of the projection morphisms, while (3.9.5) signifies that $\mathbf{R}f_*$ behaves well—even for *unbounded* complexes—with respect to flat base change.[4] More generally, in (3.10) we see that such good behavior of $\mathbf{R}f_*$ characterizes *tor-independent* base changes,

---

[2] Cf. in this vein Hartshorne's remarks on "compatibilities" [**H**, pp. 117–119]. Note however that the formalization became fully feasible only after Spaltenstein's extension of the theory of derived functors in [**H**] to unbounded complexes [**Sp**].

[3] The first is a sheafified version of $\mathbf{L}f^*$–$\mathbf{R}f_*$ adjunction (3.2.5)(f), the second and third underly monoidality of $\mathbf{L}f^*$ and $\mathbf{R}f_*$, and the fourth is "projection."

[4] Cf. [**I**, III, 3.7 and IV, 3.1].

as does a certain Künneth map's being an isomorphism; the precise statement is given in (3.10.3), a culminating result for the chapter.

3.1 Preliminaries

For any ringed space $(X, \mathcal{O}_X)$, let $\mathcal{A}_X$ be the category of (sheaves of) $\mathcal{O}_X$-modules—which is abelian, see e.g., [**G**, Chap. II, §2.2, §2.4, and §2.6], $\mathbf{C}(X)$ the category of $\mathcal{A}_X$-complexes, $\mathbf{K}(X)$ the category of $\mathcal{A}_X$-complexes with homotopy equivalence classes of maps of complexes as morphisms, and $\mathbf{D}(X)$ the derived category gotten by "localizing" $\mathbf{K}(X)$ with respect to quasi-isomorphisms (see §§(1.1), (1.2)).

To any ringed-space map $(f, \theta) \colon (X, \mathcal{O}_X) \to (Y, \mathcal{O}_Y)$ one can associate the additive *direct image* functor

$$f_* : \mathcal{A}_X \to \mathcal{A}_Y$$

such that $[f_*M](U) = M(f^{-1}U)$ for any $\mathcal{O}_X$-module M and any open set $U \subset Y$, the $\mathcal{O}_Y$-module structure on f_*M arising via θ; and also an *inverse image* functor

$$f^* \colon \mathcal{A}_Y \to \mathcal{A}_X$$

defined up to isomorphism as a left-adjoint of f_*, see [**GD**, p. 100, (4.4.3.1)] (where $\Psi^*(\mathcal{F})$ should be $\Psi_*(\mathcal{F})$). Such an adjoint exists with, e.g.,

$$f^*A := f^{-1}A \otimes_{f^{-1}\mathcal{O}_Y} \mathcal{O}_X \qquad (A \in \mathcal{A}_Y)$$

where $f^{-1}A$ is the sheaf associated to the presheaf taking an open $V \subset X$ to $\varinjlim A(U)$ with U running through all the open neighborhoods of $f(V)$ in Y. In particular, if X is an open subset of Y, $\mathcal{O}_X$ is the restriction of $\mathcal{O}_Y$, f is the inclusion, and θ is the obvious map, then the functor "restriction to X" is left-adjoint to f_*, so it is the natural choice for f^*. Being adjoint to an additive functor, f^* is also additive.[5] From adjointness, or directly, one sees that f_* is left-exact and f^* is right-exact. (The stalk $(f^*N)_x$ at $x \in X$ is functorially isomorphic to $\mathcal{O}_{X,x} \otimes_{\mathcal{O}_{Y,f(x)}} N_{f(x)}$.)

Derived functors (see (2.1.1) and its complement)

$$\mathbf{R}f_* \colon \mathbf{D}(X) \to \mathbf{D}(Y), \qquad \mathbf{L}f^* \colon \mathbf{D}(Y) \to \mathbf{D}(X)$$

can be constructed by means of q-injective and q-flat resolutions, respectively, as follows.

---

[5] Additivity of f^* means that for any two maps $A \underset{\beta}{\overset{\alpha}{\rightrightarrows}} B$ in $\mathcal{A}_Y$ and any $E \in \mathcal{A}_X$, the sum of the induced maps $\mathrm{Hom}(f^*B, E) \rightrightarrows \mathrm{Hom}(f^*A, E)$ is the map induced by $\alpha + \beta$, a condition which follows from the additivity of f_* via the adjunction isomorphisms (of abelian groups) $\mathrm{Hom}(f^*-, E) \to \mathrm{Hom}(f_*f^*-, f_*E) \to \mathrm{Hom}(-, f_*E)$.

Assume chosen once and for all, for each ringed space X, two families of quasi-isomorphisms

$$A \to I_A, \qquad P_A \to A \qquad (A \in \mathbf{K}(X)) \tag{3.1.1}$$

with each I_A a q-injective complex and each P_A q-flat, see (2.3.5), (2.5.5), with $A \to I_A$ the identity map when A is itself q-injective, and $P_A \to A$ the identity when A is q-flat.

Then set

$$\mathbf{R}f_*(B) := f_*(I_B) \qquad (B \in \mathbf{D}(X)), \tag{3.1.2}$$

and for a map α in $\mathbf{D}(X)$ define $\mathbf{R}f_*(\alpha)$ as indicated in (2.3.2.3) (with $\mathbf{J} := \mathbf{K}(X)$). The Δ-structure on $\mathbf{R}f_*$ is specified at the end of (2.2.4). Similar considerations apply to $\mathbf{L}f^*$, once one verifies that f^* takes exact q-flat complexes to exact complexes (for which argue as in (2.5.7), keeping in mind (2.5.2)). Proceeding as in (2.2.4) (dualized, with $\mathbf{J}' \subset \mathbf{K}(Y)$ the Δ-subcategory whose objects are the q-flat complexes, and $\mathbf{J}'' := \mathbf{K}(Y)$), set

$$\mathbf{L}f^*(A) := f^*(P_A) \qquad (A \in \mathbf{D}(Y)), \tag{3.1.3}$$

etc. [See also (2.7.3).]

Proposition (3.2.1) below says in particular that *these derived functors are also adjoint*. Before getting into that we review some elementary functorial sheaf maps, and their interconnections.

For $\mathcal{O}_X$-modules E and F, there is a natural map of $\mathcal{O}_Y$-modules

$$\phi_{E,F} \colon f_* \mathcal{H}om_X(E, F) \to \mathcal{H}om_Y(f_*E, f_*F) \tag{3.1.4}$$

taking a section of $f_* \mathcal{H}om_X(E, F)$ over an open subset U of Y—i.e., a map $\alpha \colon E|_{f^{-1}U} \to F|_{f^{-1}U}$—to the section α_ϕ of $\mathcal{H}om_Y(f_*E, f_*F)$ given by the family of maps $\alpha_\phi(V) \colon (f_*E)(V) \to (f_*F)(V)$ $(V$ open $\subset U)$ with

$$\alpha_\phi(V) := \alpha(f^{-1}V) \colon E(f^{-1}V) \to F(f^{-1}V).$$

Here is another description of $\phi_{E,F}(U)$: given the commutative diagram

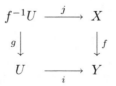

where i and j are inclusions and g is the restriction $f|_{f^{-1}U}$, and recalling that i^* and j^* are restriction functors, one verifies the functorial equalities

$$f_* j_* j^* = i_* g_* j^* = i_* i^* f_*$$

and checks then that $\phi_{E,F}(U)$ is the natural composition

$$f_* \mathcal{H}om_X(E, F)(U) \xmapsto{\text{def}} \text{Hom}(j^*E, j^*F)$$

$$\xrightarrow{\sim} \text{Hom}(E, j_*j^*F)$$

$$\longrightarrow \text{Hom}(f_*E, f_*j_*j^*F)$$

$$= \text{Hom}(f_*E, i_*i^*f_*F)$$

$$\xrightarrow{\sim} \text{Hom}(i^*f_*E, i^*f_*F) \xmapsto{\text{def}} \mathcal{H}om_Y(f_*E, f_*F)(U) .$$

Lemma 3.1.5. *Let* $f: X \to Y$ *be a ringed-space map,* $A \in \mathcal{A}_Y$, $B \in \mathcal{A}_X$, $\phi := \phi_{f^*A,B}$ *(see (3.1.4)). Let* $\eta_A: A \to f_*f^*A$ *be the map corresponding by adjunction to the identity map of* f^*A. *Then the composition*

$$f_* \mathcal{H}om_X(f^*A, B) \xrightarrow{\phi} \mathcal{H}om_Y(f_*f^*A, f_*B) \xrightarrow{\text{via } \eta_A} \mathcal{H}om_Y(A, f_*B)$$

is an isomorphism of additive bifunctors.

Proof. The preceding description of ϕ identifies (up to isomorphism) the sections over an open $U \subset Y$ of the composite map in (3.1.5) with the natural composition

$$\text{Hom}(f^*A, j_*j^*B) \longrightarrow \text{Hom}(f_*f^*A, f_*j_*j^*B) \xrightarrow{\text{via } \eta_A} \text{Hom}(A, f_*j_*j^*B)$$

which is, by adjointness of f^* and f_*, an isomorphism. Additive bifunctoriality of this isomorphism is easily verified. Q.E.D.

(3.1.6). We leave it to the reader to elaborate the foregoing to get isomorphisms of *complexes*, functorial in $A^\bullet \in \mathbf{C}(Y)$, $B^\bullet \in \mathbf{C}(X)$,

$$\text{Hom}_X^\bullet(f^*A^\bullet, B^\bullet) \xrightarrow{\sim} \text{Hom}_Y^\bullet(A^\bullet, f_*B^\bullet),$$

$$f_* \mathcal{H}om_X^\bullet(f^*A^\bullet, B^\bullet) \xrightarrow{\sim} \mathcal{H}om_Y^\bullet(A^\bullet, f_*B^\bullet).$$

(See (1.5.3) and (2.4.5) for the definitions of $\text{Hom}^\bullet$ and $\mathcal{H}om^\bullet$.)
Ditto for the maps in (3.1.7)–(3.1.9) below.

For any two $\mathcal{O}_X$-modules E, F, the tensor product $E \otimes_X F$ is by definition the sheaf associated to the presheaf $U \mapsto E(U) \otimes_{\mathcal{O}_X(U)} F(U)$ (U open $\subset X$), so there exist canonical maps

$$E(U) \otimes_{\mathcal{O}_X(U)} F(U) \to (E \otimes_X F)(U)$$

from which, taking $U = f^{-1}V$ (V open $\subset Y$), one gets a canonical map

$$f_*E \otimes_Y f_*F \to f_*(E \otimes_X F). \tag{3.1.7}$$

(3.1.8). We will abbreviate by omitting the subscripts attached to $\otimes$, and by writing $\mathcal{H}_Z(-,-)$ for $\mathcal{H}om_{\mathcal{O}_Z}(-,-)$.

The maps (3.1.4) and (3.1.7) are related via $\mathcal{H}om$-$\otimes$ adjunction (2.6.2) as follows. After taking global sections of (2.6.2) (with F, G replaced by E, F respectively) one finds, corresponding to the identity map of $E \otimes F$, a canonical map

$$E \to \mathcal{H}_X(F, E \otimes F). \tag{3.1.8.1}$$

Similarly, corresponding to the identity map of $\mathcal{H}_X(E, F)$ one has a map

$$\mathcal{H}_X(E, F) \otimes E \to F. \tag{3.1.8.2}$$

Verification of the following two assertions is left to the reader.

—The map (3.1.7) is $\mathcal{H}om$-$\otimes$ adjoint to the composition

$$f_*E \xrightarrow{\ (3.1.8.1)\ } f_*\mathcal{H}_X(F, E \otimes F) \xrightarrow{\ (3.1.4)\ } \mathcal{H}_Y(f_*F, f_*(E \otimes F)).$$

—The map (3.1.4) is $\mathcal{H}om$-$\otimes$ adjoint to the composition

$$f_*\mathcal{H}_X(E, F) \otimes f_*E \xrightarrow{\ (3.1.7)\ } f_*(\mathcal{H}_X(E, F) \otimes E) \xrightarrow{\ (3.1.8.2)\ } f_*F.$$

(3.1.9) Define the functorial map

$$f^*(A \otimes B) \xrightarrow{\ \alpha\ } f^*A \otimes f^*B \qquad (A, B \in \mathcal{A}_Y)$$

to be the adjoint of the composition

$$A \otimes B \xrightarrow{\ \text{natural}\ } f_*f^*A \otimes f_*f^*B \xrightarrow{\ (3.1.7)\ } f_*(f^*A \otimes f^*B).$$

Let $x \in X$, $y = f(x)$, so that f induces a map of local rings $\mathcal{O}_y \to \mathcal{O}_X$, where $\mathcal{O}_X$ is the stalk $\mathcal{O}_{X,x}$, and similarly for $\mathcal{O}_y$. One checks that the stalk map α_x is just the natural map

$$(A_y \otimes_{\mathcal{O}_y} B_y) \otimes_{\mathcal{O}_y} \mathcal{O}_X \to (A_y \otimes_{\mathcal{O}_y} \mathcal{O}_X) \otimes_{\mathcal{O}_X} (B_y \otimes_{\mathcal{O}_y} \mathcal{O}_X),$$

whence α coincides with the standard *isomorphism* defined, e.g., in [**GD**, p. 97, (4.3.3.1)].

Exercise 3.1.10. Show that the source and target of the map α in (3.1.9) are, as functors in the variable A, left-adjoint to the target and source (respectively) of the composed isomorphism—call it β—in (3.1.5), considered as functors in B; and that α and β are *conjugate,* see (3.3.5). (See also (3.5.5).) Work out the analog for complexes.

3.2 Adjointness of Derived Direct and Inverse Image

We begin with a direct proof of adjointness of the derived direct and inverse image functors $\mathbf{R}f_*$ and $\mathbf{L}f^*$ associated to a ringed-space map $f\colon X \to Y$.[6] A more elaborate localized formulation is given in (3.2.3). Proposition (3.2.4) introduces the basic maps connecting $\mathbf{R}f_*$ and $\mathbf{L}f^*$ to $\underline{\otimes}$. It includes derived-category versions of part of (3.1.8) and of (3.1.10), as an illustration of the basic strategy for understanding relations among maps of derived functors through purely formal considerations (see 3.5.4).

Proposition 3.2.1. *For any ringed-space map* $f\colon X \to Y$, *there is a natural bifunctorial isomorphism,*

$$\mathrm{Hom}_{\mathbf{D}(X)}(\mathbf{L}f^*A,\, B) \xrightarrow{\;\sim\;} \mathrm{Hom}_{\mathbf{D}(Y)}(A,\, \mathbf{R}f_*B)$$

$$\big(A \in \mathbf{D}(Y),\; B \in \mathbf{D}(X)\big).$$

Proof. There is a simple equivalence between giving the adjunction isomorphism (3.2.1) and giving functorial morphisms

$$\eta\colon 1 \to \mathbf{R}f_*\mathbf{L}f^*, \qquad \epsilon\colon \mathbf{L}f^*\mathbf{R}f_* \to 1 \qquad\qquad (3.2.1.0)$$

$(1:= \text{identity})$ such that the corresponding compositions

$$\mathbf{R}f_* \xrightarrow{\;\text{via }\eta\;} \mathbf{R}f_*\mathbf{L}f^*\mathbf{R}f_* \xrightarrow{\;\text{via }\epsilon\;} \mathbf{R}f_*$$

$$\mathbf{L}f^* \xrightarrow[\;\text{via }\eta\;]{} \mathbf{L}f^*\mathbf{R}f_*\mathbf{L}f^* \xrightarrow[\;\text{via }\epsilon\;]{} \mathbf{L}f^* \qquad\qquad (3.2.1.1)$$

are identity morphisms [**M**, p. 83, Thm. 2]. Indeed, $\eta(A)$ (resp. $\epsilon(B)$) corresponds under (3.2.1) to the identity map of $\mathbf{L}f^*A$ (resp. $\mathbf{R}f_*B$); and conversely, (3.2.1) can be recovered from η and ϵ thus: to a map $\alpha\colon \mathbf{L}f^*A \to B$ associate the composed map

$$A \xrightarrow{\;\eta(A)\;} \mathbf{R}f_*\mathbf{L}f^*A \xrightarrow{\;\mathbf{R}f_*\alpha\;} \mathbf{R}f_*B\,,$$

and inversely, to a map $\beta\colon A \to \mathbf{R}f_*B$ associate the composed map

$$\mathbf{L}f^*A \xrightarrow{\;\mathbf{L}f^*\beta\;} \mathbf{L}f^*\mathbf{R}f_*B \xrightarrow{\;\epsilon(B)\;} B.$$

Define ϵ to be the unique Δ-functorial map such that the following natural diagram in $\mathbf{D}(X)$ commutes for all $B \in \mathbf{K}(X)$:[7]

---

[6] An ultra-generalization of this "trivial duality formula" is given in [**De**, p. 298, Thm. 2.3.7].

[7] Here, and elsewhere, we lighten notation by omitting Qs, so that, e.g., B sometimes denotes the (physically identical) image QB of B in $\mathbf{D}(X)$. This should not cause confusion.

$$
\begin{array}{ccc}
\mathbf{L}f^*f_*B & \longrightarrow & \mathbf{L}f^*\mathbf{R}f_*B \\
\downarrow & & \downarrow{\scriptstyle\epsilon(B)} \\
f^*f_*B & \longrightarrow & B
\end{array}
\qquad (3.2.1.2)
$$

Such an ϵ exists because $\mathbf{L}f^*\mathbf{R}f_*$ is a right-derived functor of $\mathbf{L}f^*Q_Y f_*$ (where $Q_Y\colon \mathbf{K}(Y) \to \mathbf{D}(Y)$ is the canonical functor), and the natural composition $\mathbf{L}f^*Q_Y f_* \to Q_X f^*f_* \to Q_X$ is Δ-functorial, see (2.1.1) and (2.2.6.1). (Alternatively, use (2.6.5), with $n = 1$, $\mathbf{L}'' = \mathbf{K}(X)$, $\mathbf{L}' \subset \mathbf{L}''$ the Δ-subcategory whose objects are the q-injective complexes, and β the preceding Δ-functorial composition.)

Dually, define η to be the unique Δ-functorial map such that the following natural diagram commutes for all $A \in \mathbf{K}(Y)$:

$$
\begin{array}{ccc}
\mathbf{R}f_*f^*A & \longleftarrow & \mathbf{R}f_*\mathbf{L}f^*A \\
\uparrow & & \uparrow{\scriptstyle\eta(A)} \\
f_*f^*A & \longleftarrow & A
\end{array}
\qquad (3.2.1.3)
$$

To see then that the first row in (3.2.1.1) is the identity, i.e., that its composition with the canonical map $\zeta\colon f_* \to \mathbf{R}f_*$ is just ζ itself, consider the diagram (with obvious maps)

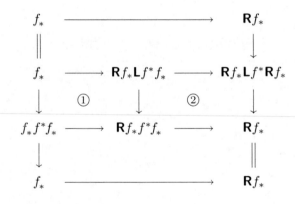

Subdiagrams ① and ② commute by the definitions of η and ϵ. The top and bottom rectangles clearly commute. Thus the whole diagram commutes, giving the desired conclusion.

A similar argument applies to the second row in (3.2.1.1).   Q.E.D.

Corollary 3.2.2. *The adjunction isomorphism (3.2.1) is the unique functorial map ρ making the following natural diagram commute for all $A \in \mathbf{K}(Y)$ and $B \in \mathbf{K}(X)$:*

$$\begin{array}{ccc}
\mathrm{Hom}_{\mathbf{K}(X)}(f^*A,\, B) \longrightarrow \mathrm{Hom}_{\mathbf{D}(X)}(f^*A,\, B) \longrightarrow \mathrm{Hom}_{\mathbf{D}(X)}(\mathbf{L}f^*A,\, B) \\
H^0(3.1.6)\Big\downarrow \simeq \qquad\qquad\qquad\qquad\qquad\qquad\qquad \Big\downarrow \rho \qquad\qquad (3.2.2.1)\\
\mathrm{Hom}_{\mathbf{K}(Y)}(A,\, f_*B) \xrightarrow{\;\nu\;} \mathrm{Hom}_{\mathbf{D}(Y)}(A,\, f_*B) \longrightarrow \mathrm{Hom}_{\mathbf{D}(Y)}(A,\, \mathbf{R}f_*B)
\end{array}$$

Moreover, ν is an isomorphism whenever A is left-f^*-acyclic (e.g., q-flat) and B is q-injective.

Proof. Suppose ρ is the adjunction isomorphism. To show (3.2.2) commutes, chase a $\mathbf{K}(X)$-map $\phi\colon f^*A \to B$ around it in both directions to reduce to showing that the following natural diagram commutes:

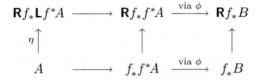

Here the left square commutes by the definition of η, and the right square commutes by functoriality of the natural map $f_* \to \mathbf{R}f_*$.

If, furthermore, A is left-f^*-acyclic (i.e., $\mathbf{L}f^*A \to f^*A$ is an isomorphism (2.2.6)) and B is q-injective, then all the maps in (3.2.2.1) other than ν are isomorphisms (see (2.3.8)(v)), so ν is an isomorphism too.

Finally, to prove the uniqueness of a functorial map $\rho(A, B)$ making diagram (3.2.2.1) commute, use the canonical maps $P_A \to A$ and $B \to I_B$ to map (3.2.2.1) to the corresponding diagram with P_A in place of A and I_B in place of B. As we have just seen, all the maps in this last diagram other than $\rho(P_A, I_B)$ are isomorphisms, so that $\rho(P_A, I_B)$ is uniquely determined by the commutativity condition; and since the sources and targets of $\rho(P_A, I_B)$ and $\rho(A, B)$ are isomorphic, it follows that $\rho(A, B)$ is uniquely determined.
$$\text{Q.E.D.}$$

Exercise. With $\psi_A\colon P_A \to A$ (resp. $\varphi_B\colon B \to I_B$) the canonical isomorphism in $\mathbf{D}(Y)$ (resp. $\mathbf{D}(X)$), see (3.1.1), $\eta(A)$ and $\epsilon(B)$ are the respective compositions

$$A \xrightarrow{\psi_A^{-1}} P_A \xrightarrow{\text{natural}} f_*(f^*P_A) \xrightarrow{f_*(\varphi_{f^*P_A})} f_*(I_{f^*P_A}) = \mathbf{R}f_*\mathbf{L}f^*A,$$

$$B \xrightarrow[\varphi_B^{-1}]{} I_B \xleftarrow{\text{natural}} f^*(f_*I_B) \xrightarrow{f^*(\psi_{f_*I_B})} f^*(P_{f_*I_B}) = \mathbf{L}f^*\mathbf{R}f_*B.$$

Recall from §2.4 the derived functors $\mathbf{R}\mathrm{Hom}^{\bullet}$ and $\mathbf{R}\mathcal{H}om^{\bullet}$. We write $\mathbf{R}\mathrm{Hom}_X^{\bullet}$ and $\mathbf{R}\mathcal{H}om_X^{\bullet}$ to specify that we are working on the ringed space X. For $E, F \in \mathbf{K}(X)$, and I_F as in (3.1.1), we have then, in $\mathbf{D}(X)$,

$$\mathbf{R}\mathrm{Hom}_X^{\bullet}(E, F) = \mathrm{Hom}^{\bullet}(E, I_F),$$
$$\mathbf{R}\mathcal{H}om_X^{\bullet}(E, F) = \mathcal{H}om^{\bullet}(E, I_F).$$

Proposition 3.2.3. (see [Sp, p. 147]). *Let $f\colon X \to Y$ be a ringed-space map.*

(i) *There is a unique Δ-functorial isomorphism*

$$\alpha\colon \mathbf{R}\mathrm{Hom}_X^{\bullet}(\mathbf{L}f^*A, B) \xrightarrow{\ \sim\ } \mathbf{R}\mathrm{Hom}_Y^{\bullet}(A, \mathbf{R}f_*B) \qquad (3.2.3.1)$$
$$\big(A \in \mathbf{K}(Y),\ B \in \mathbf{K}(X)\big)$$

such that the following natural diagram in $\mathbf{D}(X)^8$ commutes:

$$
\begin{array}{ccccc}
\mathrm{Hom}_X^{\bullet}(f^*A,\, B) & \longrightarrow & \mathbf{R}\mathrm{Hom}_X^{\bullet}(f^*A,\, B) & \longrightarrow & \mathbf{R}\mathrm{Hom}_X^{\bullet}(\mathbf{L}f^*A,\, B) \\
{\scriptstyle(3.1.6)}\big\downarrow{\simeq} & & & & \simeq\big\downarrow{\alpha} \\
\mathrm{Hom}_Y^{\bullet}(A,\, f_*B) & \longrightarrow & \mathbf{R}\mathrm{Hom}_Y^{\bullet}(A,\, f_*B) & \longrightarrow & \mathbf{R}\mathrm{Hom}_Y^{\bullet}(A,\, \mathbf{R}f_*B).
\end{array}
$$

Moreover, the induced homology map

$$H^0(\alpha)\colon \mathrm{Hom}_{\mathbf{D}(X)}(\mathbf{L}f^*A,\, B) \xrightarrow{\ \sim\ } \mathrm{Hom}_{\mathbf{D}(Y)}(A,\, \mathbf{R}f_*B)$$

(see (2.4.2)) is just the adjunction isomorphism in (3.2.1).

(ii) *There is a unique Δ-functorial isomorphism*

$$\beta\colon \mathbf{R}f_*\mathbf{R}\mathcal{H}om_X^{\bullet}(\mathbf{L}f^*A,\, B) \xrightarrow{\ \sim\ } \mathbf{R}\mathcal{H}om_Y^{\bullet}(A,\, \mathbf{R}f_*B) \qquad (3.2.3.2)$$
$$\big(A \in \mathbf{K}(Y),\ B \in \mathbf{K}(X)\big)$$

such that the following natural diagram commutes

$$
\begin{array}{ccccc}
f_*\mathcal{H}om_X^{\bullet}(f^*A, B) & \longrightarrow & \mathbf{R}f_*\mathbf{R}\mathcal{H}om_X^{\bullet}(f^*A, B) & \longrightarrow & \mathbf{R}f_*\mathbf{R}\mathcal{H}om_X^{\bullet}(\mathbf{L}f^*A, B) \\
{\scriptstyle(3.1.6)}\big\downarrow{\simeq} & & & & \simeq\big\downarrow{\beta} \\
\mathcal{H}om_Y^{\bullet}(A,\, f_*B) & \longrightarrow & \mathbf{R}\mathcal{H}om_Y^{\bullet}(A,\, f_*B) & \longrightarrow & \mathbf{R}\mathcal{H}om_Y^{\bullet}(A,\, \mathbf{R}f_*B)
\end{array}
$$

Proof. (i) For the first assertion it suffices, as in (2.6.5), that in the derived category of abelian groups the natural compositions

$$\mathrm{Hom}_X^{\bullet}(f^*A,\, B) \xrightarrow{a} \mathbf{R}\mathrm{Hom}_X^{\bullet}(f^*A,\, B) \xrightarrow{b} \mathbf{R}\mathrm{Hom}_X^{\bullet}(\mathbf{L}f^*A,\, B)$$
$$\mathrm{Hom}_Y^{\bullet}(A,\, f_*B) \xrightarrow{c} \mathbf{R}\mathrm{Hom}_Y^{\bullet}(A,\, f_*B) \xrightarrow{d} \mathbf{R}\mathrm{Hom}_Y^{\bullet}(A,\, \mathbf{R}f_*B)$$

be isomorphisms whenever A is q-flat and B is q-injective. But in this case we have $A = P_A$ and $B = I_B$, so that a, b, and d are identity maps. As for c, we need only note that by the last assertion of (3.2.2), the induced homology maps

$$H^i(c)\colon \mathrm{Hom}_{\mathbf{K}(Y)}(A[-i],\, f_*B) \to \mathrm{Hom}_{\mathbf{D}(Y)}(A[-i],\, f_*B)$$

are isomorphisms, see (1.2.2) and (2.4.2).

Now apply the functor H^0 to the diagram and conclude by the uniqueness of ρ in (3.2.2) that $H^0(\alpha)$ is as asserted.

---

8 With missing Q's left to the reader.

(ii) As above, it comes down to showing that the natural maps

$$f_* \mathcal{H}om_X^\bullet(f^*A, B) \xrightarrow{a'} \mathbf{R}f_* \mathcal{H}om_X^\bullet(f^*A, B)$$

$$\mathcal{H}om_Y^\bullet(A, f_*B) \xrightarrow{c'} \mathbf{R}\mathcal{H}om_Y^\bullet(A, f_*B) = \mathcal{H}om_Y^\bullet(A, I_{f_*B})$$

are isomorphisms (in $\mathbf{D}(X)$, $\mathbf{D}(Y)$ respectively) whenever A is q-flat and B is q-injective. The stalk $(f^*A)_x$ ($x \in X$) being isomorphic to $\mathcal{O}_{X,x} \otimes_{\mathcal{O}_{Y,f(x)}} A_{f(x)}$, (2.5.2) shows that f^*A is q-flat, and then (2.3.8)(iv) shows (via (2.6.2)) that $\mathcal{H} := \mathcal{H}om_X^\bullet(f^*A, B)$ is q-injective; so $\mathcal{H} = I_\mathcal{H}$ and $a' \colon f_*\mathcal{H} \to f_*I_\mathcal{H}$ is in fact an identity map.

For c', it is enough to check that we get an isomorphism after applying the functor Γ_U (sections over U) for arbitrary open $U \subset Y$, since then c' induces isomorphisms of the homology presheaves—and hence of the homology sheaves—of its source and target (see (1.2.2)). Let $i \colon U \to Y$, $j \colon f^{-1}U \to X$ be the inclusion maps, and let $g \colon f^{-1}U \to U$ be the map induced by f.

We have then by (2.3.1) a commutative diagram of quasi-isomorphisms

$$
\begin{array}{ccc}
i^*f_*B & \longrightarrow & i^*I_{f_*B} \\
\| & & \downarrow{\scriptstyle\gamma} \\
i^*f_*B & \longrightarrow & I_{i^*f_*B}
\end{array}
$$

Since $i^*I_{f_*B}$ is q-injective (2.4.5.2), γ is an *isomorphism* in $\mathbf{K}(U)$ (2.3.2.2). Keeping in mind that $i^*f_* = g_*j^*$, consider the commutative diagram

$$
\begin{array}{ccc}
\Gamma_U \mathcal{H}om_Y^\bullet(A, f_*B) & \xrightarrow{\ \Gamma_U(c')\ } & \Gamma_U \mathcal{H}om_Y^\bullet(A, I_{f_*B}) \\
\| & & \| \\
\mathrm{Hom}_U^\bullet(i^*A, i^*f_*B) & \longrightarrow & \mathrm{Hom}_U^\bullet(i^*A, i^*I_{f_*B}) \\
\| & & {\scriptstyle\simeq}\downarrow{\scriptstyle\text{via }\gamma} \\
\mathrm{Hom}_U^\bullet(i^*A, i^*f_*B) & \longrightarrow & \mathrm{Hom}_U^\bullet(i^*A, I_{i^*f_*B}) \\
\| & & \| \\
\mathrm{Hom}_U^\bullet(i^*A, g_*j^*B) & \xrightarrow{\quad c_U \quad} & \mathbf{R}\mathrm{Hom}_U^\bullet(i^*A, g_*j^*B)
\end{array}
$$

As in the proof of (i), since j^*B is q-injective and i^*A is q-flat (see above), therefore c_U is an isomorphism; and hence so is $\Gamma_U(c')$. Q.E.D.

Corollary 3.2.3.3. *Let $U \subset Y$ be open and let $\Gamma_U \colon \mathcal{A}_Y \to \mathfrak{Ab}$ be the abelian functor "sections over U." Then for any q-injective $B \in \mathbf{K}(X)$, f_*B is right-Γ_U-acyclic.*

Consequently, by (2.2.7) or (2.6.5), there is a unique Δ-functorial isomorphism $\mathbf{R}\Gamma_{f^{-1}U} \xrightarrow{\sim} \mathbf{R}\Gamma_U \mathbf{R}f_*$ making the following natural diagram commute for all $B \in \mathbf{K}(X)$:

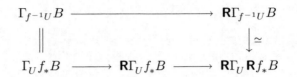

Proof. Let $\mathcal{O}'_U \in \mathcal{A}_Y$ be the "extension by zero" of $\mathcal{O}_U \in \mathcal{A}_U$, i.e., the sheaf associated to the presheaf taking an open $V \subset Y$ to $\mathcal{O}_U(V)$ if $V \subset U$, and to 0 otherwise. Then there is a natural functorial identification $\Gamma_U(-) = \mathrm{Hom}_Y(\mathcal{O}'_U, -)$. Since $\mathcal{O}'_U$ is flat, we have as in the proof of (3.2.3)(i) that the map $c\colon \mathrm{Hom}^{\bullet}(\mathcal{O}'_U, f_*B) \to \mathbf{R}\mathrm{Hom}^{\bullet}(\mathcal{O}'_U, f_*B)$ is an isomorphism, i.e., $\Gamma_U(f_*B) \to \mathbf{R}\Gamma_U(f_*B)$ is an isomorphism, whence the conclusion (see last assertion in (2.2.6)).   Q.E.D.

Proposition 3.2.4. (i) *For any ringed-space map $f\colon X \to Y$, there is a unique Δ-bifunctorial isomorphism*

$$\mathbf{L}f^*(A \otimes_Y B) \xrightarrow{\sim} \mathbf{L}f^*A \otimes_X \mathbf{L}f^*B \qquad (A, B \in \mathbf{D}(Y))$$

making the following natural diagram commute for all A, B:

$$
\begin{array}{ccc}
\mathbf{L}f^*(A \otimes_Y B) & \xrightarrow{\sim} & \mathbf{L}f^*A \otimes_X \mathbf{L}f^*B \\
\downarrow & & \downarrow \\
f^*(A \otimes_Y B) & \xrightarrow[\;(3.1.9)\;]{\sim} & f^*A \otimes_X f^*B
\end{array}
\tag{3.2.4.1}
$$

This isomorphism is conjugate (3.3.5) to the isomorphism β in (3.2.3.2).

(ii) *With $\eta'\colon E \to \mathbf{R}\mathcal{H}om^{\bullet}_X(F, E \otimes F)$ corresponding via $(2.6.1)^*$ to the identity map of $E \otimes F$, and $\epsilon\colon \mathbf{L}f^*\mathbf{R}f_* \to 1$ as in (3.2.1.0), the (Δ-functorial) map*

$$\gamma\colon \mathbf{R}f_*(E) \otimes \mathbf{R}f_*(F) \longrightarrow \mathbf{R}f_*(E \otimes F) \qquad (E, F \in \mathbf{D}(X)) \tag{3.2.4.2}$$

adjoint to the composed map

$$\mathbf{L}f^*\big(\mathbf{R}f_*E \otimes \mathbf{R}f_*F\big) \xrightarrow{\sim} \mathbf{L}f^*\mathbf{R}f_*E \otimes \mathbf{L}f^*\mathbf{R}f_*F \xrightarrow[\epsilon \otimes \epsilon]{} E \otimes F \tag{3.2.4.3}$$

corresponds via $(2.6.1)^*$ *to the composed map*

$$\mathbf{R}f_*E \xrightarrow{\mathbf{R}f_*\eta'} \mathbf{R}f_*\mathbf{R}\mathcal{H}om_X^{\bullet}(F,\, E \underset{\approx}{\otimes} F)$$

$$\xrightarrow{\text{via } \epsilon} \mathbf{R}f_*\mathbf{R}\mathcal{H}om_X^{\bullet}(\mathbf{L}f^*\mathbf{R}f_*F,\, E \underset{\approx}{\otimes} F)$$

$$\xrightarrow[\;(3.2.3.2)\;]{\beta} \mathbf{R}\mathcal{H}om_X^{\bullet}(\mathbf{R}f_*F,\, \mathbf{R}f_*(E \underset{\approx}{\otimes} F))\,.$$

(3.2.4.4)

Proof. (i) For $x \in X$, the stalk $(f^*A)_x$ is $\mathcal{O}_{X,x} \otimes_{\mathcal{O}_{Y,f(x)}} A_{f(x)}$, and so (2.5.2) shows that f^*A is q-flat whenever A is. Hence if A and B are both q-flat (whence so, clearly, is $A \otimes_Y B$), then the vertical arrows in (3.2.4.1) are isomorphisms, and the first assertion follows from (2.6.5) (dualized).

The second assertion amounts to commutativity, for any complexes $E, F, G \in \mathbf{D}(X)$, of the following diagram of natural isomorphisms:

$$
\begin{array}{ccc}
\mathrm{Hom}_{\mathbf{D}(X)}\big(\mathbf{L}f^*E,\, \mathbf{R}\mathcal{H}om_X^{\bullet}(\mathbf{L}f^*F, G)\big) & \xrightarrow{(2.6.1)^*} & \mathrm{Hom}_{\mathbf{D}(X)}\big(\mathbf{L}f^*E \underset{\approx}{\otimes} \mathbf{L}f^*F,\, G\big) \\[2mm]
{\scriptstyle(3.2.1)}\Big\downarrow & & \Big\downarrow{\scriptstyle\simeq} \\[2mm]
\mathrm{Hom}_{\mathbf{D}(Y)}\big(E,\, \mathbf{R}f_*\mathbf{R}\mathcal{H}om_X^{\bullet}(\mathbf{L}f^*F, G)\big) & & \mathrm{Hom}_{\mathbf{D}(X)}\big(\mathbf{L}f^*(E \underset{\approx}{\otimes} F),\, G\big) \\[2mm]
{\scriptstyle\text{via }\beta}\Big\downarrow & & \Big\downarrow{\scriptstyle(3.2.1)} \\[2mm]
\mathrm{Hom}_{\mathbf{D}(Y)}\big(E,\, \mathbf{R}\mathcal{H}om_Y^{\bullet}(F, \mathbf{R}f_*G)\big) & \xrightarrow[(2.6.1)^*]{} & \mathrm{Hom}_{\mathbf{D}(Y)}\big(E \underset{\approx}{\otimes} F,\, \mathbf{R}f_*G\big)
\end{array}
$$

(3.2.4.5)

in proving which, we may replace E by P_E, F by p_f, and G by I_G, i.e., we may assume E and F to be q-flat and G to be q-injective. Using the commutativity in $(2.6.1)^*$ (after applying homology H^0), (3.2.2.1), (3.2.3.2), and (3.2.4.1), we find that (3.2.4.5) is the target of a natural map, in the category of diagrams of abelian groups, coming from the diagram of isomorphisms (see (3.1.6), and recall that $H^0\mathrm{Hom}_X^{\bullet} = \mathrm{Hom}_{\mathbf{K}(X)}$):

$$
\begin{array}{ccc}
\mathrm{Hom}_{\mathbf{K}(X)}\big(f^*E,\, \mathcal{H}om_X^{\bullet}(f^*F, G)\big) & \longrightarrow & \mathrm{Hom}_{\mathbf{K}(X)}\big(f^*E \otimes f^*F,\, G\big) \\[2mm]
\Big\downarrow & & \Big\downarrow \\[2mm]
\mathrm{Hom}_{\mathbf{K}(Y)}\big(E,\, f_*\mathcal{H}om_X^{\bullet}(f^*F, G)\big) & & \mathrm{Hom}_{\mathbf{K}(X)}\big(f^*(E \otimes F),\, G\big) \\[2mm]
\Big\downarrow & & \Big\downarrow \\[2mm]
\mathrm{Hom}_{\mathbf{K}(Y)}\big(E,\, \mathcal{H}om_Y^{\bullet}(F, f_*G)\big) & \longrightarrow & \mathrm{Hom}_{\mathbf{K}(X)}\big(E \otimes F,\, f_*G\big)
\end{array}
$$

(3.2.4.6)

Also, E and F are q-flat (so that $\mathbf{L}f^*E \underset{\approx}{\otimes} \mathbf{L}f^*F \xrightarrow{\sim} f^*E \otimes f^*F$) and G is q-injective, so any $\mathbf{D}(X)$-map $\mathbf{L}f^*E \underset{\approx}{\otimes} \mathbf{L}f^*F \to G$ is represented by a map of complexes $f^*E \otimes f^*F \to G$, see (2.3.8)(v). Hence one need only show that (3.2.4.6) commutes. This is exercise (3.1.10), left to the reader.

(ii) With $\eta\colon 1 \to \mathbf{R}f_*\mathbf{L}f^*$ as in (3.2.1.0), the map (3.2.4.2) is the composition

$$\mathbf{R}f_*(E) \otimes \mathbf{R}f_*(F) \xrightarrow{\eta} \mathbf{R}f_*\mathbf{L}f^*(\mathbf{R}f_*(E) \otimes \mathbf{R}f_*(F)) \xrightarrow{\mathbf{R}f_* \; (3.2.4.3)} \mathbf{R}f_*(E \otimes F)$$

which is clearly Δ-functorial. The rest of the statement is best understood in the formal context of closed categories, see (3.5.4). In the present instance of that context—see (3.5.2)(d) and (3.4.4)(b)—the map (3.4.2.1) is just γ, and hence the adjoint (3.5.4.1) of (3.4.2.1) is the map in (i) above. Commutativity of (3.5.5.1) says that (3.4.5.1) is conjugate to the map (3.5.4.2), which must then, by (i), be β. Hence (ii) follows from the sentence preceding (3.5.4.2) and the description of (3.5.4.1) immediately following (3.5.4.2).   Q.E.D.

Remark. Commutativity of (3.2.4.5) yields another proof that β is an isomorphism, since the maps labeled (3.2.1) and (2.6.1)* are isomorphisms.

Exercises 3.2.5. $f\colon X \to Y$ is a ringed-space map, $A \in \mathbf{D}(A)$, $B \in \mathbf{D}(X)$.

(a) Show that the following two natural composed maps correspond under the adjunction isomorphism (3.2.1):

$$\mathbf{L}f^*\mathcal{O}_Y \to f^*\mathcal{O}_Y \to \mathcal{O}_X\,, \qquad \mathcal{O}_Y \to f_*\mathcal{O}_X \to \mathbf{R}f_*\mathcal{O}_X\,.$$

(b) Write τ_n for the truncation functor $\tau_{\geq n}$ of §1.10. Write f_* (resp. f^*) for $\mathbf{R}f_*$ (resp. $\mathbf{L}f^*$). Define the functorial map

$$\psi\colon f^*\tau_n \longrightarrow \tau_n f^*$$

to be the adjoint of the natural composed map

$$\tau_n \longrightarrow \tau_n f_* f^* \longrightarrow \tau_n f_* \tau_n f^* \xrightarrow{\sim} f_* \tau_n f^*.$$

(The isomorphism obtains because $f_*\mathbf{D}_{\geq \mathbf{n}}(X) \subset \mathbf{D}_{\geq \mathbf{n}}(Y)$, see (2.3.4).) Show that the following natural diagram commutes:

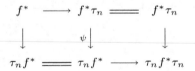

(One way is to check commutativity of the diagram whose columns are adjoint to those of the one in question. For this, (1.10.1.2) may be found useful.)

(c) The natural map $\mathrm{Hom}^{\bullet}_Y(A, f_*B) \to \mathbf{R}\mathrm{Hom}^{\bullet}_Y(A, \mathbf{R}f_*B)$ is an isomorphism for all q-injective $B \in \mathbf{K}(X)$ iff $\mathbf{L}f^*A \to f^*A$ is an isomorphism.

(d) Formulate and prove a statement to the effect that the map β in (3.2.3.2) is compatible with open immersions $U \hookrightarrow Y$.

(e) With Γ_Y as in (3.2.3.3), show that the natural map

$$\Gamma_Y f_* \mathcal{H}om^{\bullet}_X(f^*A, B) \to \mathbf{R}\Gamma_Y \mathbf{R}f_* \mathbf{R}\mathcal{H}om^{\bullet}_X(\mathbf{L}f^*A, B)$$

is an isomorphism if A is q-flat and B is q-injective.

(f) Show that there is a natural diagram of isomorphisms

$$
\begin{CD}
\mathbf{R}\Gamma_Y\,\mathbf{R}f_*\,\mathbf{R}\mathcal{H}om_X^\bullet(\mathbf{L}f^*A,\,B) @>\sim>{(3.2.3.2)}> \mathbf{R}\Gamma_Y\,\mathbf{R}\mathcal{H}om_Y^\bullet(A,\,\mathbf{R}f_*B) \\
@V\simeq VV @VV\simeq V \\
\mathbf{R}\mathrm{Hom}_X^\bullet(\mathbf{L}f^*A,\,B) @>\sim>{(3.2.3.1)}> \mathbf{R}\mathrm{Hom}_Y^\bullet(A,\,\mathbf{R}f_*B)
\end{CD}
$$

see (2.5.10)(b) and (3.2.3.3).

(First show the same with all $\mathbf{R}$'s and $\mathbf{L}$'s dropped; then apply (e) and (2.6.5).)

3.3 Δ-Adjoint Functors

We now run through the sorites related to adjointness of Δ-functors. Later, we will be constructing numerous functorial maps between multivariate Δ-functors by purely formal (category-theoretic) methods. The results in this section, together with the Proposition in §1.5, will guarantee that the so-constructed maps are in fact Δ-functorial.

Let $\mathbf{K_1}$ and $\mathbf{K_2}$ be Δ-categories with respective translation functors T_1 and T_2, and let $(f_*,\theta_*)\colon \mathbf{K_1} \to \mathbf{K_2}$ and $(f^*,\theta^*)\colon \mathbf{K_2} \to \mathbf{K_1}$ be Δ-functors such that f^* is left-adjoint to f_*. (Recall from §1.5 that $\theta_*\colon f_*T_1 \xrightarrow{\sim} T_2f_*$, and similarly $\theta^*\colon f^*T_2 \xrightarrow{\sim} T_1f^*$.) Let $\eta\colon 1 \to f_*f^*$, $\epsilon\colon f^*f_* \to 1$ be the functorial maps corresponding by adjunction to the identity maps of f^*, f_* respectively.

Lemma-Definition 3.3.1. *In the above circumstances, the following conditions are equivalent:*

(i) *η is Δ-functorial.*
(i)′ *ϵ is Δ-functorial.*
(ii) *For all $A \in \mathbf{K_2}$ and $B \in \mathbf{K_1}$, the following natural diagram commutes:*

$$
\begin{CD}
\mathrm{Hom}_{\mathbf{K_1}}(f^*A,\,B) @>>> \mathrm{Hom}_{\mathbf{K_1}}(T_1f^*A,\,T_1B) @>\theta^*>> \mathrm{Hom}_{\mathbf{K_1}}(f^*T_2A,\,T_1B) \\
@V\simeq VV @. @VV\simeq V \\
\mathrm{Hom}_{\mathbf{K_2}}(A,\,f_*B) @>>> \mathrm{Hom}_{\mathbf{K_2}}(T_2A,\,T_2f_*B) @>>\theta_*> \mathrm{Hom}_{\mathbf{K_2}}(T_2A,\,f_*T_1B)
\end{CD}
$$

When these conditions hold, we say that (f^*,θ^*) and (f_*,θ_*) are Δ-adjoint, or—leaving θ^* and θ_* to the reader—that (f^*,f_*) is a Δ-adjoint pair.

Proof. (i) $\Rightarrow$ (ii). Chase a map $\xi\colon f^*A \to B$ around the diagram in both directions to reduce to showing that the following diagram commutes:

$$T_2A \xrightarrow{\;T_2\eta(A)\;} T_2f_*f^*A \xrightarrow{\;T_2f_*\xi\;} T_2f_*B$$

$$\eta(T_2A)\Big\downarrow \qquad\qquad \Big\downarrow\theta_*^{-1}(f^*A) \qquad \Big\downarrow\theta_*^{-1}(B) \qquad (3.3.1.1)$$

$$f_*f^*T_2A \xrightarrow[\;f_*\theta^*(A)\;]{} f_*T_1f^*A \xrightarrow[\;f_*T_1\xi\;]{} f_*T_1B$$

The first square commutes by (i), and the second by functoriality of θ_*.

Conversely, (i) is just commutativity of (3.3.1.1) when $B := f^*A$ and ξ is the identity map.

Thus (i) $\Leftrightarrow$ (ii); and a similar proof (starting with a map $\xi': A \to f_*B$) yields (i)$'$ $\Leftrightarrow$ (ii).   Q.E.D.

Example 3.3.2. Quasi-inverse Δ-equivalences of categories (1.7.2) are Δ-adjoint pairs.

Example 3.3.3. The pair $(\mathbf{L}f^*, \mathbf{R}f_*)$ in (3.2.1) is Δ-adjoint. Indeed, in the proof of (3.2.1) the associated η and ϵ were defined to be certain Δ-functorial maps.

Example 3.3.4. With reference to (2.6.1)*, let $\mathbf{K_1} := \mathbf{D}(\mathcal{A}) =: \mathbf{K_2}$, fix $F \in \mathbf{D}(\mathcal{A})$, and for any $A, B \in \mathbf{D}(\mathcal{A})$ set

$$f^*A := A \underset{=}{\otimes} F, \qquad f_*B := \mathbf{R}\mathcal{H}om^\bullet(F, B).$$

Then this pair (f^*, f_*) is Δ-adjoint. To verify condition (ii) in (3.3.1), consider the following diagram of natural isomorphisms, where $\mathrm{H}^\bullet$ stands for $\mathbf{R}\mathrm{Hom}^\bullet$ and $\mathcal{H}^\bullet$ stands for $\mathbf{R}\mathcal{H}om^\bullet$:

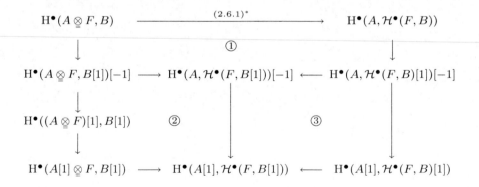

Subdiagram ① commutes because (2.6.1)* is Δ-functorial in the last variable; ② commutes because (2.6.1)* is Δ-functorial in the first variable; and ③ commutes for obvious reasons. One checks that application of the functor H^0 to this big commutative diagram gives (ii) in (3.3.1).   Q.E.D.

In particular, we have the canonical Δ-*functorial* maps

$$\eta' : A \to \mathbf{R}\mathcal{H}om^\bullet(F, A \underline{\underline{\otimes}} F),$$

$$\epsilon' : \mathbf{R}\mathcal{H}om^\bullet(F, B) \underline{\underline{\otimes}} F \to B. \qquad (3.3.4.1)$$

Lemma-Definition 3.3.5. *If* $f_* : \mathbf{X} \to \mathbf{Y}$, $g_* : \mathbf{X} \to \mathbf{Y}$ *are functors with respective left adjoints* $f^* : \mathbf{Y} \to \mathbf{X}$, $g^* : \mathbf{Y} \to \mathbf{X}$, *then with "Hom" denoting "functorial morphisms," the following natural compositions are inverse isomorphisms:*

$$\mathrm{Hom}(f_*, g_*) \longrightarrow \mathrm{Hom}(f_* f^*, g_* f^*) \longrightarrow \mathrm{Hom}(1, g_* f^*) \overset{\sim}{\longrightarrow} \mathrm{Hom}(g^*, f^*),$$

$$\mathrm{Hom}(f_*, g_*) \overset{\sim}{\longleftarrow} \mathrm{Hom}(g^* f_*, 1) \longleftarrow \mathrm{Hom}(g^* f_*, f^* f_*) \longleftarrow \mathrm{Hom}(g^*, f^*).$$

Functorial morphisms $f_* \to g_*$ *and* $g^* \to f^*$ *which correspond under these isomorphisms will be said to be* conjugate (*the first* right-conjugate *to the second, the second* left-conjugate *to the first*).

Proof. Exercise, or see [M, p. 100, Theorem 2].

Corollary 3.3.6. *Let* (f^*, f_*) *and* (g^*, g_*) *be* Δ-*adjoint pairs of* Δ-*functors between* $\mathbf{K_1}$ *and* $\mathbf{K_2}$. *Then a functorial morphism* $\alpha : f_* \to g_*$ *is* Δ-*functorial if and only if so is its conjugate* $\beta : g^* \to f^*$. *In particular,* f_* *and* g_* *are isomorphic* Δ-*functors* ⟺ *so are* f^* *and* g^*.

The first assertion follows from (3.3.1) since, for example, α is the composition

$$f_* \overset{\eta}{\to} g_* g^* f_* \xrightarrow{\text{via } \beta} g_* f^* f_* \overset{\epsilon}{\to} g_*.$$

That the conjugate of a functorial isomorphism is an isomorphism follows from Exercise (3.3.7)(c) below.

Exercises 3.3.7. (a) Maps $\alpha : f_* \to g_*$ and $\beta : g^* \to f^*$ are conjugate ⟺ (either of) the following diagrams commute:

(b) The conditions in (a) are equivalent to commutativity, for all $X \in \mathbf{X}$, $Y \in \mathbf{Y}$ of the diagram

$$
\begin{array}{ccc}
\mathrm{Hom}(Y, f_* X) & \xrightarrow{\text{via } \alpha} & \mathrm{Hom}(Y, g_* X) \\
\simeq \downarrow & & \downarrow \simeq \\
\mathrm{Hom}(f^* Y, X) & \xrightarrow[\text{via } \beta]{} & \mathrm{Hom}(g^* Y, X)
\end{array}
$$

(c) Denoting the conjugate of a functorial map α by α' we have (with the obvious interpretation) $1' = 1$ and $(\alpha_2 \alpha_1)' = \alpha_1' \alpha_2'$.

(d) The conditions in (3.3.1) are equivalent to either one of:

(iii) The functorial map $\theta^*: f^*T_2 \xrightarrow{\ \sim\ } T_1 f^*$ is left-conjugate to

$$f_*T_1^{-1} = T_2^{-1}T_2 f_* T_1^{-1} \xrightarrow{\ \theta_*^{-1}\ } T_2^{-1} f_* T_1 T_1^{-1} = T_2^{-1} f_* \,.$$

(iii)′ The functorial map $\theta_*: f_*T_1 \xrightarrow{\ \sim\ } T_2 f_*$ is right-conjugate to

$$f^*T_2^{-1} = T_1^{-1}T_1 f^* T_2^{-1} \xrightarrow{\ \theta^{*-1}\ } T_1^{-1} f^* T_2 T_2^{-1} = T_1^{-1} f^* .$$

The next Proposition, generalizing some of (1.7.2), says that a left adjoint of a Δ-functor can be made into a left Δ-adjoint, in a unique way.

Let $\mathbf{K_1}$, $\mathbf{K_2}$ be Δ-categories with respective translation functors T_1, T_2, and let $(f_*, \theta_*): \mathbf{K_1} \to \mathbf{K_2}$ be a Δ-functor such that f_* has a left adjoint $f^*: \mathbf{K_2} \to \mathbf{K_1}$ (automatically additive, see first footnote in §3.1).

Proposition 3.3.8. *There exists a unique functorial isomorphism*

$$\theta^*: f^*T_2 \xrightarrow{\ \sim\ } T_1 f^*$$

such that

 (i) (f^*, θ^*) *is a* Δ*-functor, and*
 (ii) *the* Δ*-functors* (f^*, θ^*) *and* (f_*, θ_*) *are* Δ*-adjoint.*

Proof. The functors f^*T_2 and $T_1 f^*$ are left-adjoint to $T_2^{-1} f_*$ and $f_* T_1^{-1}$ respectively; and since the latter two are isomorphic (in the obvious way via θ_*), so are the former two, and one checks that the conjugate isomorphism θ^* between them is adjoint to the composite map

$$T_2 \xrightarrow{\ \eta\ } T_2 f_* f^* \xrightarrow{\ \theta_*^{-1}\ } f_* T_1 f^*,\ ^9$$

i.e., θ^* is the *unique* map making the following diagram commute:

$$
\begin{array}{ccc}
T_2 & =\!=\!=\!=\!=\!=\!=\!= & T_2 \\[2pt]
\eta \downarrow & & \downarrow \eta \\[6pt]
f_* f^* T_2 \xrightarrow[\ f_*\theta^*\ ]{} f_* T_1 f^* \xrightarrow[\ \theta_*\ ]{} & & T_2 f_* f^*
\end{array}
\qquad (3.3.8.1)
$$

If (i) holds, then commutativity of (3.3.8.1) also expresses the condition that $\eta: 1 \to f_* f^*$ be Δ-functorial, i.e., that (ii) hold. Thus no other θ^* can satisfy (i) and (ii). (So far, the argument is just a variation on (3.3.7)(d).)

We still have to show that (i) holds for the θ^* we have specified. So let $A \xrightarrow{u} B \xrightarrow{v} C \xrightarrow{w} T_2 A$ be a triangle in $\mathbf{K_2}$. Apply (Δ3)′ in (1.4) to embed f^*u into a triangle $f^*A \xrightarrow{f^*u} f^*B \xrightarrow{p} C^* \xrightarrow{q} T_1 f^*A$. I claim that

(a) *there is a map* $\gamma: f^*C \to C^*$ *making the following diagram commute:*

$$
\begin{array}{ccccccc}
f^*A & \xrightarrow{\ f^*u\ } & f^*B & \xrightarrow{\ f^*v\ } & f^*C & \xrightarrow{\ \theta^* \circ f^*w\ } & T_1 f^*A \\[2pt]
\| & & \| & & \gamma \downarrow & & \| \\[6pt]
f^*A & \xrightarrow[\ f^*u\ ]{} & f^*B & \xrightarrow[\ p\ ]{} & C^* & \xrightarrow[\ q\ ]{} & T_1 f^*A
\end{array}
$$

and that (b) *any such* γ *must be an isomorphism.*

Given (a) and (b), condition (Δ1)″ in (1.4) ensures that the top row in the preceding diagram is a triangle, so that (f^*, θ^*) is indeed a Δ-functor.

---

[9] Whence, dually, θ_*^{-1} is adjoint to $T_1 \xleftarrow{\ \epsilon\ } T_1 f^* f_* \xleftarrow{\ \theta^*\ } f^* T_2 f_*$.

Assertion (a) results by adjunction from the map of triangles

$$
\begin{array}{ccccccc}
A & \xrightarrow{\ u\ } & B & \xrightarrow{\ v\ } & C & \xrightarrow{\ w\ } & T_2 A \\
\downarrow{\scriptstyle\eta} & & \downarrow{\scriptstyle\eta} & & \downarrow{\scriptstyle\gamma'} & & \downarrow{\scriptstyle T_2\eta} \\
f_* f^* A & \xrightarrow[f_* f^* u]{} & f_* f^* B & \xrightarrow[f_* p]{} & f_* C^* & \xrightarrow[\theta_* \circ f_* q]{} & T_2 f_* f^* A
\end{array}
$$

where γ' is given by $(\Delta 3)''$ in (1.4).

For (b), consider the commutative diagram (with $D \in \mathbf{K_1}$, and obvious maps):

$$
\begin{array}{ccccc}
\mathrm{Hom}(T_1 f^* B, D) & =\!=\!= & \mathrm{Hom}(T_1 f^* B, D) & \xrightarrow{\ \sim\ } & \mathrm{Hom}(T_2 B, f_* D) \\
\downarrow & & \downarrow & & \downarrow \\
\mathrm{Hom}(T_1 f^* A, D) & =\!=\!= & \mathrm{Hom}(T_1 f^* A, D) & \xrightarrow{\ \sim\ } & \mathrm{Hom}(T_2 A, f_* D) \\
\downarrow & & \downarrow & & \downarrow \\
\mathrm{Hom}(C^*, D) & \xrightarrow{\ \text{via } \gamma\ } & \mathrm{Hom}(f^* C, D) & \xrightarrow{\ \sim\ } & \mathrm{Hom}(C, f_* D) \\
\downarrow & & \downarrow & & \downarrow \\
\mathrm{Hom}(f^* B, D) & =\!=\!= & \mathrm{Hom}(f^* B, D) & \xrightarrow{\ \sim\ } & \mathrm{Hom}(B, f_* D) \\
\downarrow & & \downarrow & & \downarrow \\
\mathrm{Hom}(f^* A, D) & =\!=\!= & \mathrm{Hom}(f^* A, D) & \xrightarrow{\ \sim\ } & \mathrm{Hom}(A, f_* D)
\end{array}
$$

The left and right columns are *exact* [**H**, p. 23, Prop. 1.1, b], hence the map "via γ" is an isomorphism for all D, i.e., γ is an isomorphism. Q.E.D.

3.4 Adjoint Functors between Monoidal Categories

This section and the following one introduce some of the formalism arising from a pair of adjoint monoidal functors between closed categories. A simple example of such a pair occurs with respect to a map $R \to S$ of commutative rings, namely extension and restriction of scalars on the appropriate module categories. The module functors f^* and f_* associated with a map $f: X \to Y$ of ringed spaces form another such pair. The example which mosts interests us is that of the pair $(\mathbf{L}f^*, \mathbf{R}f_*)$ of §3.2. The point is to develop by purely categorical methods a host of relations, expressed by commutative functorial diagrams, among the four operations $\otimes$, $\mathbf{R}\mathcal{H}om^\bullet$, $\mathbf{L}f^*$ and $\mathbf{R}f_*$.

But even the purified categorical approach leads quickly to stultifying complexity—at which the exercises (3.5.6) merely hint. Ideally, we would like to have an implementable algorithm for deciding when a functorial diagram built up from the data given in the relevant categorical definitions (see (3.4.1), (3.4.2), (3.5.1)) commutes; or in other words, to prove a "constructive coherence theorem" for the generic context "monoidal functor between closed categories, together with left adjoint." (Lewis [**Lw**] does this, to some extent, without the left adjoint.) Though there exists a substantial body of results on "coherence in categories," see e.g., [**K′**], [**Sv**], and their references, it does not yet suffice; we will have to be content with subduing individual diagrams as needs dictate.

We treat symmetric monoidal categories in this section, leaving the additional "closed" structure to the next.

Definition 3.4.1. A *symmetric monoidal category*

$$\mathbf{M} = (\mathbf{M}_0, \otimes, \mathcal{O}_M, \alpha, \lambda, \rho, \gamma)$$

consists of a category $\mathbf{M}_0$, a "product" functor $\otimes \colon \mathbf{M}_0 \times \mathbf{M}_0 \to \mathbf{M}_0$, an object $\mathcal{O}_M$ of $\mathbf{M}_0$, and functorial isomorphisms

$$\alpha : (A \otimes B) \otimes C \xrightarrow{\sim} A \otimes (B \otimes C) \qquad \text{(associativity)}$$

$$\lambda : \mathcal{O}_M \otimes A \xrightarrow{\sim} A \qquad \rho \colon A \otimes \mathcal{O}_M \xrightarrow{\sim} A \qquad \text{(units)}$$

$$\gamma \colon A \otimes B \xrightarrow{\sim} B \otimes A \qquad \text{(symmetry)}$$

(where A, B, C are objects in $\mathbf{M}_0$) such that $\gamma \circ \gamma = 1$ and the following diagrams (3.4.1.1) commute.

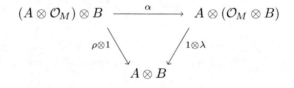

$$(A \otimes \mathcal{O}_M) \otimes B \xrightarrow{\ \alpha\ } A \otimes (\mathcal{O}_M \otimes B)$$

$$\rho \otimes 1 \searrow \qquad \swarrow 1 \otimes \lambda$$

$$A \otimes B$$

$$((A \otimes B) \otimes C) \otimes D \xrightarrow{\ \alpha\ } (A \otimes B) \otimes (C \otimes D) \xrightarrow{\ \alpha\ } A \otimes (B \otimes (C \otimes D))$$

$$\alpha \otimes 1 \downarrow \qquad\qquad\qquad\qquad\qquad\qquad\qquad\qquad \downarrow 1 \otimes \alpha$$

$$(A \otimes (B \otimes C)) \otimes D \xrightarrow{\hspace{6cm}\alpha\hspace{6cm}} A \otimes ((B \otimes C) \otimes D)$$

$$(A \otimes B) \otimes C \xrightarrow{\ \alpha\ } A \otimes (B \otimes C) \xrightarrow{\ \gamma\ } (B \otimes C) \otimes A$$

$$\gamma \otimes 1 \downarrow \qquad\qquad\qquad\qquad\qquad\qquad\qquad \downarrow \alpha$$

$$(B \otimes A) \otimes C \xrightarrow[\ \alpha\ ]{} B \otimes (A \otimes C) \xrightarrow[\ 1 \otimes \gamma\ ]{} B \otimes (C \otimes A)$$

$$\mathcal{O}_M \otimes A \xrightarrow{\ \gamma\ } A \otimes \mathcal{O}_M$$

$$\lambda \searrow \qquad \swarrow \rho$$

$$A$$

$$(3.4.1.1)$$

Definition 3.4.2. A symmetric monoidal functor $f_* \colon \mathbf{X} \to \mathbf{Y}$ between symmetric monoidal categories $\mathbf{X}, \mathbf{Y}$ is a functor $f_{*0} \colon \mathbf{X}_0 \to \mathbf{Y}_0$ together with two functorial maps

$$f_* A \otimes f_* B \longrightarrow f_*(A \otimes B)$$
$$\mathcal{O}_Y \longrightarrow f_* \mathcal{O}_X \tag{3.4.2.1}$$

(where we have abused notation, *as we will henceforth,* by omitting the subscript "$_0$" and by not distinguishing notationally between $\otimes$ in $\mathbf{X}$ and $\otimes$ in $\mathbf{Y}$), such that the following natural diagrams (3.4.2.2) commute.

$$
\begin{array}{ccc}
f_* \mathcal{O}_X \otimes f_* A & \longrightarrow & f_*(\mathcal{O}_X \otimes A) \\
\uparrow & & \downarrow{\scriptstyle f_*(\lambda_X)} \\
\mathcal{O}_Y \otimes f_* A & \xrightarrow[\lambda_Y]{} & f_* A
\end{array}
\qquad
\begin{array}{ccc}
f_* A \otimes f_* B & \longrightarrow & f_*(A \otimes B) \\
{\scriptstyle \gamma_Y}\downarrow & & \downarrow{\scriptstyle f_*(\gamma_X)} \\
f_* B \otimes f_* A & \longrightarrow & f_*(B \otimes A)
\end{array}
$$

$$
\begin{array}{ccccc}
(f_* A \otimes f_* B) \otimes f_* C & \longrightarrow & f_*(A \otimes B) \otimes f_* C & \longrightarrow & f_*((A \otimes B) \otimes C) \\
{\scriptstyle \alpha}\downarrow & & & & \downarrow{\scriptstyle f_*(\alpha)} \\
f_* A \otimes (f_* B \otimes f_* C) & \longrightarrow & f_* A \otimes f_*(B \otimes C) & \longrightarrow & f_*(A \otimes (B \otimes C))
\end{array}
$$

$$\tag{3.4.2.2}$$

(3.4.3). We assume further that the symmetric monoidal functor f_* has a *left adjoint* $f^* \colon \mathbf{Y} \to \mathbf{X}$. In other words we have functorial maps

$$\eta \colon 1 \to f_* f^* \qquad\qquad \epsilon \colon f^* f_* \to 1$$

such that the composites

$$f_* \xrightarrow{\ \text{via } \eta\ } f_* f^* f_* \xrightarrow{\ \text{via } \epsilon\ } f_* \qquad\qquad f^* \xrightarrow{\ \text{via } \eta\ } f^* f_* f^* \xrightarrow{\ \text{via } \epsilon\ } f^*$$

are identities, giving rise to a bifunctorial isomorphism

$$\operatorname{Hom}_{\mathbf{X}}(f^* F, G) \xrightarrow{\ \sim\ } \operatorname{Hom}_{\mathbf{Y}}(F, f_* G) \quad (F \in \mathbf{Y},\ G \in \mathbf{X}). \tag{3.4.3.1}$$

Example 3.4.4. (a) Let $f \colon X \to Y$ be a map of ringed spaces, $\mathbf{X}$ (resp. $\mathbf{Y}$) the category of $\mathcal{O}_X$- (resp. $\mathcal{O}_Y$-)modules with its standard structure of symmetric monoidal category ($\otimes$ having its usual meaning, etc. etc.), f_* and f^* the usual direct and inverse image functors, see (3.1.7).

(b) Let $f \colon X \to Y$ be a ringed-space map, $\mathbf{X} := \mathbf{D}(X)$, $\mathbf{Y} := \mathbf{D}(Y)$, $\otimes := \underset{=}{\otimes}$, $f_* := \mathbf{R}f_*$, $f^* := \mathbf{L}f^*$ (see (3.2.1)). To establish symmetric monoidality of, e.g., $\mathbf{D}(X)$, one need only work with q-flat complexes, ...

For (3.4.2.1), use the map γ from (3.2.4.2) and the natural composition $\mathcal{O}_Y \to f_*\mathcal{O}_X \to \mathbf{R}f_*\mathcal{O}_X$. One can then deduce via adjointness that $\mathbf{R}f_*$ is symmetric monoidal from the fact that $\mathbf{L}f^*$ is symmetric monoidal when considered as a functor from $\mathbf{Y}^{\mathrm{op}}$ to $\mathbf{X}^{\mathrm{op}}$, see (3.2.4). For this property of $\mathbf{L}f^*$, one can check the requisite commutativity in (3.4.2.2) after replacing each object A in $\mathbf{X}$ by an isomorphic q-flat complex, and recalling that if A is q-flat, then so is f^*A (see proof of (3.2.3)(ii)); in view of (3.1.3), the checking is thereby reduced to the context of (a) above, where one can use adjointness (see (3.1.9)) to deduce what needs to be known about f^* after showing directly that f_* is symmetric monoidal!

For example, to show commutativity of

$$
\begin{array}{ccc}
\mathbf{R}f_*(\mathcal{O}_X) \underline{\otimes} \mathbf{R}f_*(A) & \xrightarrow{\ \gamma\ } & \mathbf{R}f_*(\mathcal{O}_X \underline{\otimes} A) \\[1em]
\uparrow & & \downarrow \lambda_X \\[1em]
\mathcal{O}_Y \underline{\otimes} \mathbf{R}f_*(A) & \xrightarrow[\ \lambda_Y\ ]{} & \mathbf{R}f_*(A)
\end{array}
$$

consider the following natural diagram, in which we have written f^*, f_*, $\otimes$ for $\mathbf{L}f^*$, $\mathbf{R}f_*$, $\underline{\otimes}$ respectively:

$$
\begin{array}{ccccc}
 & & f^*f_*\mathcal{O}_X \otimes f^*f_*A & \longrightarrow & \mathcal{O}_X \otimes A \\[0.5em]
 & \nearrow & \uparrow \quad\quad\quad \textcircled{1} & & \uparrow \quad\quad \searrow \\[0.5em]
f^*(f_*\mathcal{O}_X \otimes f_*A) & & f^*\mathcal{O}_Y \otimes f^*f_*A \longrightarrow \mathcal{O}_X \otimes f^*f_*A & & A \\[0.5em]
 & \searrow & \uparrow \quad\quad\quad \textcircled{2} & & \downarrow \quad\quad \nearrow \\[0.5em]
 & & f^*(\mathcal{O}_Y \otimes f_*A) & \longrightarrow & f^*f_*A
\end{array}
$$

It will be enough to show that the outer border commutes, because it is "adjoint" to the preceding diagram, see (3.4.5.2). Subdiagram ① commutes by exercise (3.2.5)(a). For commutativity of ② replace f_*A by an isomorphic q-flat complex to reduce to showing commutativity of the corresponding diagram in context (a); then reduce via adjointness to checking (easily) that in that context the following natural diagram commutes:

$$
\begin{array}{ccc}
f_*(\mathcal{O}_X) \underline{\otimes} f_*(A) & \xrightarrow{\ (3.1.7)\ } & f_*(\mathcal{O}_X \underline{\otimes} A) \\[1em]
\uparrow & & \downarrow \\[1em]
\mathcal{O}_Y \underline{\otimes} f_*(A) & \longrightarrow & f_*(A)
\end{array}
$$

The rest is evident.

Exercise 3.4.4.1. Let R be a commutative ring, $Z := \mathrm{Spec}(R)$, T an indeterminate, $X := \mathrm{Spec}(R[T])$ with its obvious Z-scheme structure, $\delta\colon X \to Y := X \times_Z X$ the diagonal map, and $\sigma\colon Y \xrightarrow{\sim} Y$ the symmetry isomorphism, i.e., $\pi_1\sigma = \pi_2$ and $\pi_2\sigma = \pi_1$ where π_1 and π_2 are the canonical projections from Y to X.

Show that in the context of (3.4.4)(a) the natural composite $\mathcal{O}_X$-module map

$$\delta^*\delta_* F = (\sigma\delta)^*(\sigma\delta)_* F \xrightarrow{\sim} \delta^*\sigma^*\sigma_*\delta_* F \to \delta^*\delta_* F$$

is the identity map for all $\mathcal{O}_X$-modules F; but that in the context of (3.4.4)(b) the natural composite $\mathbf{D}(X)$-map

$$\mathbf{L}\delta^*\delta_*\mathcal{O}_X = \mathbf{L}(\sigma\delta)^*(\sigma\delta)_*\mathcal{O}_X \xrightarrow{\sim} \mathbf{L}\delta^*\sigma^*\sigma_*\delta_*\mathcal{O}_X \to \mathbf{L}\delta^*\delta_*\mathcal{O}_X$$

is not the identity map unless $2 = 0$ in R.

(More challenging.) Show: if $\iota\colon Z \to X$ is the closed immersion corresponding to the R-homomorphism $R[T] \twoheadrightarrow R$ taking T to 0, then the natural composite $\mathbf{D}(X)$-map

$$\mathbf{L}\delta^*\delta_*\iota_*\mathcal{O}_Z = \mathbf{L}(\sigma\delta)^*(\sigma\delta)_*\iota_*\mathcal{O}_Z \xrightarrow{\sim} \mathbf{L}\delta^*\sigma^*\sigma_*\delta_*\iota_*\mathcal{O}_Z \to \mathbf{L}\delta^*\delta_*\iota_*\mathcal{O}_Z$$

is an automorphism of order 2, inducing the identity map on homology.

(3.4.5) (Duality principle). From (3.4.2.1) we get, by adjunction, functorial maps

$$
\begin{aligned}
f^*C \otimes f^*D &\longleftarrow f^*(C \otimes D)\,, \\
\mathcal{O}_X &\longleftarrow f^*\mathcal{O}_Y\,.
\end{aligned}
\tag{3.4.5.1}
$$

Specifically, the second of these maps is defined to be adjoint to the map $\mathcal{O}_Y \to f_*\mathcal{O}_X$ in (3.4.2.1) (i.e., the two maps correspond under the isomorphism (3.4.3.1)); and the first adjoint to the composition

$$C \otimes D \xrightarrow{\;\eta\otimes\eta\;} f_*f^*C \otimes f_*f^*D \xrightarrow{\;(3.4.2.1)\;} f_*(f^*C \otimes f^*D)\,.$$

It follows that "dually,"

(3.4.5.2): $f_*A \otimes f_*B \xrightarrow{\;(3.4.2.1)\;} f_*(A \otimes B)$ *is adjoint to the composition*

$$A \otimes B \xleftarrow{\;\epsilon\otimes\epsilon\;} f^*f_*A \otimes f^*f_*B \xleftarrow{\;(3.4.5.1)\;} f^*(f_*A \otimes f_*B)\,.$$

To see this, it suffices to note that the following diagram, whose top row composes to the identity, commutes:

$$
\begin{array}{ccccc}
f_*A \otimes f_*B & \xleftarrow{\;\epsilon\otimes\epsilon\;} & f_*f^*f_*A \otimes f_*f^*f_*B & \xleftarrow{\;\eta\otimes\eta\;} & f_*A \otimes f_*B \\[2pt]
{\scriptstyle(3.4.2.1)}\big\downarrow & \textcircled{1} & \big\downarrow{\scriptstyle(3.4.2.1)} & \textcircled{2} & \big\downarrow{\scriptstyle\eta} \\[2pt]
f_*(A \otimes B) & \xleftarrow[\epsilon\otimes\epsilon]{} & f_*(f^*f_*A \otimes f^*f_*B) & \xleftarrow[(3.4.5.1)]{} & f_*f^*(f_*A \otimes f_*B)
\end{array}
$$

(Subdiagram ① commutes by functoriality of (3.4.2.1), and ② commutes by the above definition of (3.4.5.1).)

The maps (3.4.5.1) satisfy compatibility conditions with the associativity, unit, and symmetry isomorphisms in the symmetric monoidal categories $\mathbf{X}$

and **Y**, conditions dual to those expressed by the commutativity of the diagrams (3.4.2.2) (i.e., in (3.4.2.2) replace f_* by f^*, interchange $\mathcal{O}_X$ and $\mathcal{O}_Y$, and reverse all arrows). Proofs are left to the reader.

The maps (3.4.5.1) do not make f^* monoidal, since they point in the wrong direction (and we do *not* assume in general that they are isomorphisms, as happens to be the case in (3.4.4(a)) and (3.4.4(b)), so we cannot use their inverses).

However, to any symmetric monoidal category

$$\mathbf{M} = (\mathbf{M}_0, \otimes, \mathcal{O}_M, \alpha, \lambda, \rho, \gamma)$$

we can associate the *dual* symmetric monoidal category

$$\mathbf{M}^{\mathsf{op}} = (\mathbf{M}_0^{\mathsf{op}}, \otimes^{\mathsf{op}}, \mathcal{O}_M, \overline{\alpha}, \overline{\lambda}, \overline{\rho}, \overline{\gamma})$$

where $\mathbf{M}_0^{\mathsf{op}}$ is the dual category of $\mathbf{M}_0$ (same objects; arrows reversed), $\otimes^{\mathsf{op}}$ is the functor

$$\mathbf{M}_0^{\mathsf{op}} \times \mathbf{M}_0^{\mathsf{op}} = (\mathbf{M}_0 \times \mathbf{M}_0)^{\mathsf{op}} \xrightarrow{\ \otimes^{\mathsf{op}}\ } \mathbf{M}_0^{\mathsf{op}}$$

(so that $A \otimes^{\mathsf{op}} B = A \otimes B$ for all objects $A, B \in \mathbf{M}_0$),

$$\overline{\alpha} = (\alpha^{\mathsf{op}})^{-1} = (\alpha^{-1})^{\mathsf{op}} : (A \otimes B) \otimes C \xrightarrow{\ \sim\ } A \otimes (B \otimes C) \quad (\text{in } \mathbf{M}_0^{\mathsf{op}})$$

and similarly for $\overline{\lambda}, \overline{\rho}, \overline{\gamma}$.

Then one checks that the functor

$$(f^*)^{\mathsf{op}} \colon \mathbf{Y}^{\mathsf{op}} \to \mathbf{X}^{\mathsf{op}}$$

together with the maps (3.4.5.1) is indeed symmetric monoidal;[10] and it has a left adjoint

$$(f_*)^{\mathsf{op}} \colon \mathbf{X}^{\mathsf{op}} \to \mathbf{Y}^{\mathsf{op}}$$

(which need no longer be monoidal, because, for example, there may be no good map $\mathcal{O}_Y \to f_*\mathcal{O}_X$ in $\mathbf{Y}^{\mathsf{op}}$). Thus to every pair f_*, f^* as in (3.4.3), we can associate a "dual" such pair $(f^*)^{\mathsf{op}}, (f_*)^{\mathsf{op}}$.

This gives rise to a **duality principle**, which we now state rather imprecisely, but whose meaning should be clarified by the illustrations which follow (in connection with projection morphisms). We will be considering numerous diagrams whose vertices are functors build up from the constant functors $\mathcal{O}_X$ and $\mathcal{O}_Y$ (on **X**, **Y** respectively), identity functors, f_*, f^*, and $\otimes$, and whose arrows are morphisms of functors built up from those which express the "monoidality" of f_*, and from the adjunction isomorphism (3.4.3.1). (For

---

[10] f^* may then be said to be "op-monoidal" or "co-monoidal."

example the above-mentioned "compatibility conditions" state that certain such diagrams commute.) By interpreting any such diagram in the dual context, we get another such diagram: specifically, in the original diagram, *interchange*

- $\mathcal{O}_X$ and $\mathcal{O}_Y$
- the identity functors of $\mathbf{X}$ and $\mathbf{Y}$
- the adjunction maps η and ϵ
- the functors f^* and f_*
- the maps in (3.4.2.1) and (3.4.5.1).

If the original diagram commutes solely by virtue of the fact that f_* is a monoidal functor with left adjoint f^*, then the second diagram must also commute (because $(f^*)^{\mathsf{op}}$ is a monoidal functor with left adjoint $(f_*)^{\mathsf{op}}$).

Example 3.4.6 (Projection morphisms). With preceding notation, and $F \in \mathbf{X}$, $G \in \mathbf{Y}$, the bifunctorial *projection morphisms*

$$p_1 = p_1(F, G) \colon f_*F \otimes G \longrightarrow f_*(F \otimes f^*G)$$
$$p_2 = p_2(G, F) \colon G \otimes f_*F \longrightarrow f_*(f^*G \otimes F)$$

are the respective compositions

$$f_*F \otimes G \xrightarrow{\ 1 \otimes \eta\ } f_*F \otimes f_*f^*G \xrightarrow{\ (3.4.2.1)\ } f_*(F \otimes f^*G)$$
$$G \otimes f_*F \xrightarrow{\ \eta \otimes 1\ } f_*f^*G \otimes f_*F \xrightarrow{\ (3.4.2.1)\ } f_*(f^*G \otimes F).$$

Remarks 3.4.6.1. p_1 and p_2 determine each other via the following commutative diagram, in which γ_X, γ_Y are the respective symmetry isomorphisms in $\mathbf{X}$, $\mathbf{Y}$:

$$
\begin{array}{ccc}
f_*F \otimes G & \xrightarrow{\ p_1\ } & f_*(F \otimes f^*G) \\
\gamma_Y \downarrow & & \downarrow f_*(\gamma_X) \\
G \otimes f_*F & \xrightarrow{\ p_2\ } & f_*(f^*G \otimes F)
\end{array}
$$

The commutativity of this diagram follows from that of

$$
\begin{array}{ccc}
f_*F \otimes f_*f^*G & \xrightarrow{\ (3.4.2.1)\ } & f_*(F \otimes f^*G) \\
\gamma_Y \downarrow & & \downarrow f_*(\gamma_X) \\
f_*f^*G \otimes f_*F & \xrightarrow[\ (3.4.2.1)\ ]{} & f_*(f^*G \otimes F)
\end{array}
$$

which holds as part of the definition of "symmetric monoidal functor" (see (3.4.2.2)).

(3.4.6.2). The map $p_1(F, G)$ is adjoint to the composed map

$$f^*(f_*F \otimes G) \xrightarrow{(3.4.5.1)} f^*f_*F \otimes f^*G \xrightarrow{\epsilon \otimes 1} F \otimes f^*G$$

(a map which is *dual* (3.4.5) to $p_2(F, G)$): this follows from commutativity of the natural diagram

$$
\begin{array}{ccccc}
f^*(f_*F \otimes G) & \xrightarrow{\text{via } \eta} & f^*(f_*F \otimes f_*f^*G) & \xrightarrow{(3.4.2.1)} & f^*f_*(F \otimes f^*G) \\
{\scriptstyle(3.4.5.1)}\downarrow & \textcircled{1} & {\scriptstyle(3.4.5.1)}\downarrow & \textcircled{2} & \downarrow{\scriptstyle\epsilon} \\
f^*f_*F \otimes f^*G & \xrightarrow{\text{via } \eta} & f^*f_*F \otimes f^*f_*f^*G & \xrightarrow{\epsilon \otimes \epsilon} & F \otimes f^*G.
\end{array}
$$

(Here commutativity of ① is clear, and that of ② results from (3.4.5.2).) Similarly $p_2(G, F)$ is adjoint to the dual of $p_1(G, F)$.

Lemma 3.4.7. *The following diagrams commute:*

(i)
$$
\begin{array}{ccc}
A \otimes B & \xrightarrow{\eta} & f_*f^*(A \otimes B) \\
{\scriptstyle 1 \otimes \eta}\downarrow & & \downarrow{\scriptstyle(3.4.5.1)} \\
A \otimes f_*f^*B & \xrightarrow{p_2} & f_*(f^*A \otimes f^*B)
\end{array}
$$

(ii)
$$
\begin{array}{ccccc}
A \otimes \mathcal{O}_Y & \xrightarrow{(3.4.2.1)} & A \otimes f_*\mathcal{O}_X & \xrightarrow{p_2} & f_*(f^*A \otimes \mathcal{O}_X) \\
{\scriptstyle \rho}\downarrow & & & & \downarrow{\scriptstyle f_*(\rho)} \\
A & \xrightarrow{\hspace{5cm}\eta} & & & f_*f^*A
\end{array}
$$

(iii)
$$
\begin{array}{ccc}
f_*B \otimes \mathcal{O}_Y & \xrightarrow{p_1} & f_*(B \otimes f^*\mathcal{O}_Y) \\
{\scriptstyle \rho}\downarrow & & \downarrow{\scriptstyle(3.4.5.1)} \\
f_*B & \xleftarrow{f_*(\rho)} & f_*(B \otimes \mathcal{O}_X)
\end{array}
$$

(iv)
$$
\begin{array}{ccccc}
(A \otimes B) \otimes f_*C & \xrightarrow{\alpha} & A \otimes (B \otimes f_*C) & \xrightarrow{1 \otimes p_2} & A \otimes f_*(f^*B \otimes C) \\
{\scriptstyle p_2}\downarrow & & & & \downarrow{\scriptstyle p_2} \\
f_*(f^*(A \otimes B) \otimes C) & \xrightarrow{(3.4.5.1)} & f_*((f^*A \otimes f^*B) \otimes C) & \xrightarrow{\alpha} & f_*(f^*A \otimes (f^*B \otimes C))
\end{array}
$$

Proof. (i) The commutativity of this diagram simply states that the first map in (3.4.5.1) is adjoint to the composition

$$A \otimes B \xrightarrow{1 \otimes \eta} A \otimes f_*f^*B \xrightarrow{\eta \otimes 1} f_*f^*A \otimes f_*f^*B \xrightarrow{(3.4.2.1)} f_*(f^*A \otimes f^*B)$$

which is so by definition (see beginning of (3.4.5)).

(ii) We expand the diagram in question as follows:

$$
\begin{array}{ccccc}
A \otimes f_*\mathcal{O}_X & \xrightarrow{\;\eta\otimes1\;} & f_*f^*A \otimes f_*\mathcal{O}_X & \xrightarrow{(3.4.2.1)} & f_*(f^*A \otimes \mathcal{O}_X) \\
{\scriptstyle(3.4.2.1)}\uparrow & \textcircled{1} & \uparrow & \textcircled{2} & \downarrow{\scriptstyle f_*(\rho)} \\
A \otimes \mathcal{O}_Y & \xrightarrow[\eta\otimes1]{} & f_*f^*A \otimes \mathcal{O}_Y & \xrightarrow[\rho]{} & f_*f^*A \\
\rho\downarrow & & \textcircled{3} & & \| \\
A & & \xrightarrow[\hspace{5cm}\eta\hspace{5cm}]{} & & f_*f^*A
\end{array}
$$

Subdiagrams ① and ③ clearly commute; and so does ② because of the compatibility of (3.4.2.1) and ρ, which can be deduced from the two top diagrams in (3.4.2.2) (the first of which expresses the compatibility of (3.4.2.1) and λ) and the bottom diagram in (3.4.1.1).

(iii) The diagram expands as

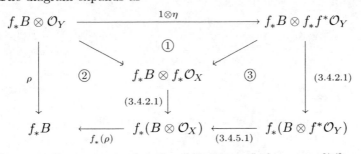

Subdiagram ① commutes by the definition of the map $f^*\mathcal{O}_Y \to \mathcal{O}_X$ in (3.4.5.1), ② by the compatibility of (3.4.2.1) and ρ (see preceding proof of (ii)), and ③ by functoriality of (3.4.2.1).

(iv) An expanded version of this diagram can be obtained by fitting together the following two diagrams (whose maps are the obvious ones):

$$
\begin{array}{ccccc}
(A \otimes B) \otimes f_*C & \longrightarrow & A \otimes (B \otimes f_*C) & \longrightarrow & A \otimes (f_*f^*B \otimes f_*C) \\
\| & & \textcircled{1} & & \downarrow{\scriptstyle a} \\
(A \otimes B) \otimes f_*C & \longrightarrow & (f_*f^*A \otimes f_*f^*B) \otimes f_*C & \xrightarrow[b]{} & f_*f^*A \otimes (f_*f^*B \otimes f_*C) \\
\downarrow & \textcircled{2} & \downarrow{\scriptstyle c} & & \\
f_*f^*(A \otimes B) \otimes f_*C & \longrightarrow & f_*(f^*A \otimes f^*B) \otimes f_*C & & \\
\downarrow & \textcircled{3} & \downarrow{\scriptstyle d} & & \\
f_*(f^*(A \otimes B) \otimes C) & \longrightarrow & f_*((f^*A \otimes f^*B) \otimes C) & &
\end{array}
$$

$$A \otimes (f_* f^* B \otimes f_* C) \longrightarrow A \otimes f_* (f^* B \otimes C)$$

$$a \downarrow \qquad\qquad ⑤ \qquad\qquad \downarrow$$

$$(f_* f^* A \otimes f_* f^* B) \otimes f_* C \xrightarrow{\ b\ } f_* f^* A \otimes (f_* f^* B \otimes f_* C) \longrightarrow f_* f^* A \otimes f_* (f^* B \otimes C)$$

$$c \downarrow$$

$$f_*(f^* A \otimes f^* B) \otimes f_* C \qquad\qquad ④$$

$$d \downarrow$$

$$f_*((f^* A \otimes f^* B) \otimes C) \xrightarrow{\hspace{8cm}} f_*(f^* A \otimes (f^* B \otimes C))$$

Subdiagram ① commutes by functoriality of a; ② by the definition of (3.4.5.1); ③ by functoriality of (3.4.2.1); ④ by commutativity of the bottom diagram in (3.4.2.2); and ⑤ for obvious reasons.   Q.E.D.

Remarks 3.4.7.1. By duality (3.4.5) we get four other commutative diagrams out of (3.4.7). For example, the dual of (ii) is

$$A \otimes \mathcal{O}_X \xleftarrow{\ (3.4.5.1)\ } A \otimes f^* \mathcal{O}_Y \xleftarrow{\ (3.4.6.2)\ } f^*(f_* A \otimes \mathcal{O}_Y)$$

$$\rho \downarrow \qquad\qquad\qquad\qquad\qquad\qquad \downarrow f^*(\rho)$$

$$A \xleftarrow{\hspace{6cm}}_{\epsilon} f^* f_* A$$

Using the symmetry isomorphism γ, Remark (3.4.6.1), the bottom diagram in (3.4.1.1), etc., we can also transform the commutative diagrams in (3.4.7) into similar ones with p_2 (resp. p_1) replaced by p_1 (resp. p_2), and with ρ replaced by λ.

3.5 Adjoint Functors between Closed Categories

The adjoint symmetric functors f_*, f^*, remain as in (3.4.3). Additional structure comes into play when the monoidal categories **X** and **Y** are *closed*, in the following sense.

Definition 3.5.1. A *symmetric monoidal closed category* (briefly, a *closed category*) is a symmetric monoidal category

$$\mathbf{M} = (\mathbf{M_0}, \otimes, \mathcal{O}_M, \alpha, \lambda, \rho, \gamma)$$

as in (3.4.1), together with a functor, called "internal hom,"

$$[-, -] \colon \mathbf{M_0^{op}} \times \mathbf{M_0} \to \mathbf{M_0} \tag{3.5.1.1}$$

(where $\mathbf{M_0^{op}}$ is the dual category of $\mathbf{M_0}$) and a functorial isomorphism

$$\pi \colon \operatorname{Hom}(A \otimes B, C) \; \xrightarrow{\sim} \; \operatorname{Hom}(A, [B, C]). \qquad (3.5.1.2)$$

The notion of closed category reduces myriad relations among, and maps involving, "tensor" and "hom" to the few basic ones appearing in the definition. (See, e.g., the following exercises (3.5.3).)[11] The original treatise on closed categories is [**EK**], in particular Chap. III, (p. 512 *ff*). Some more recent theory can be found starting with [**Sv**] and its references.

Example 3.5.2. (a) Let $\mathbf{M_0}$ be the category of modules over a given commutative ring R. Let $\otimes$ be the usual tensor product, $\mathcal{O}_M := R$, and $[B, C] := \operatorname{Hom}_R(B, C)$. Fill in the rest.

(b) $\mathbf{M_0}$ is the category of $\mathcal{O}_X$-modules on a ringed space X. Let $\otimes$ be the usual tensor product, $\mathcal{O}_M := \mathcal{O}_X$, and $[B, C] := \mathcal{H}om_X(B, C) \ldots$.

(c) Let $\mathbf{M_0'} := \mathbf{K}(X)$ be the homotopy category of complexes in the category $\mathbf{M_0}$ of (b). Let $\otimes$ be the tensor product in (1.5.4), set $\mathcal{O}_{M'} := \mathcal{O}_X$ (a complex vanishing in all nonzero degrees), and set $[B, C] := \mathcal{H}om_X^\bullet(B, C)$, see (2.4.5), (2.6.7), $\ldots$.

(d) $\mathbf{M_0''} := \mathbf{D}(X)$, the derived category of $\mathbf{M_0}$ in (b), $\otimes := \overset{\scriptscriptstyle\underset{\approx}{}}{\otimes}$ (2.5.7), $\mathcal{O}_{M''} := \mathcal{O}_X$, $[B, C] := \mathbf{R}\mathcal{H}om_X^\bullet(B, C)$, see (2.6.1)', (3.4.4)(b), $\ldots$.

Exercises 3.5.3. Let $(\mathbf{M}, [-, -], \pi)$ as above be a closed category. Write (A, B) for $\operatorname{Hom}_{\mathbf{M_0}}(A, B)$.

(a) Define the set-valued functor Γ on $\mathbf{M_0}$ to be the usual functor $(\mathcal{O}_M, -)$. Establish a bifunctorial isomorphism

$$\Gamma[A, B] \; \xrightarrow{\sim} \; (A, B).$$

(b) Let $t_{AB} \colon [A, B] \otimes A \to B$ correspond under π to the identity map of $[A, B]$. Use t_{AB} and π to obtain a natural map $[A, B] \to [A \otimes C, B \otimes C]$.

(c) Use π, t_{CA}, and t_{AB} (see (b)) to get a natural "internal composition" map

$$c \colon [A, B] \otimes [C, A] \to [C, B].$$

Prove associativity (up to canonical isomorphism) for this c.

(d) Show that the map

$$\ell = \ell_{A, B, C} \colon [A, B] \to [[C, A], [C, B]]$$

corresponding under π to internal composition (see (c)) is compatible with ordinary composition in $\mathbf{M_0}$ in that the following natural diagram (with Γ as in (a) and "Hom" meaning "set maps") commutes:

$\Gamma[A, B]$	$\xrightarrow{\;\sim\;}$	(A, B)	$\xrightarrow{\text{composition}}$	$\operatorname{Hom}\big((C, A), (C, B)\big)$
$\Gamma(\ell) \downarrow$		functoriality $\downarrow$ of $[C, -]$		$\downarrow \simeq$
$\Gamma[[C, A], [C, B]]$	$\xrightarrow{\;\sim\;}$	$([C, A], [C, B])$	$\xrightarrow[\text{of } \Gamma]{\text{functoriality}}$	$\operatorname{Hom}\big(\Gamma[C, A], \Gamma[C, B]\big)$

---

[11] When $\mathbf{M_0}$ has direct sums, π gives rise to a distributivity isomorphism

$$(A' \oplus A'') \otimes B \; \xrightarrow{\sim} \; (A' \otimes B) \oplus (A'' \otimes B)$$

whose consequences we will not follow up here—but see [**L**], [**L'**], [**K'**].

(e) From the sequence of functorial isomorphisms

$$(D, [A \otimes B, C]) \xrightarrow{\pi} (D \otimes (A \otimes B), C) \xrightarrow{\alpha} ((D \otimes A) \otimes B, C)$$
$$\xrightarrow{\pi} (D \otimes A, [B, C]) \xrightarrow{\pi} (D, [A, [B, C]])$$

deduce a functorial isomorphism

$$p = p_{A,B,C} \colon [A \otimes B, C] \xrightarrow{\sim} [A, [B, C]] .$$

(Take $D := [A \otimes B, C]$.) Referring to (a), show that $\Gamma(p) = \pi$. In example (3.5.2)(d), does this p coincide with the isomorphism in (2.6.1)*?

(f) Let $u_{AB} \colon A \to [B, A \otimes B]$ correspond under π to the identity map of $A \otimes B$. Show that the map $p_{A,B,C}$ in (e) factors as

$$[A \otimes B, C] \xrightarrow{\ell_{A \otimes B,C,B}} [[B, A \otimes B], [B, C]] \xrightarrow{\text{via } u_{AB}} [A, [B, C]].$$

with ℓ as in (d).

Let $t_{AB} \colon [A, B] \otimes A \to B$ correspond under π to the identity map of $[A, B]$. Show that $\ell_{A,B,C}$ factors as

$$[A, B] \xrightarrow{\text{via } t_{AC}} [[C, A] \otimes C, B] \xrightarrow{p_{[C,A],C,B}} [[C, A], [C, B]] .$$

(g) The preceding exercises make no use of the symmetry isomorphism γ, but this one does. Construct functorial maps

$$[A, B] \otimes [C, D] \to [[B, C], [A, D]] ,$$
$$[A, B] \otimes [C, D] \to [A \otimes C, B \otimes D] .$$

using π, c and γ for the first (see (c)), π, t and γ for the second (see (b)).

(h) Let $\alpha \colon B \to A$ be an $\mathbf{M}$-map. Show that the following diagrams—in which unlabeled maps correspond under π to identity maps—commute for any C:

Hint. For the first diagram, consider the adjoint (via π) diagram, with D arbitrary,

$$\begin{array}{ccc}
\mathrm{Hom}([A, C] \otimes B, D) & \longleftarrow & \mathrm{Hom}([A, C] \otimes A, D) \\
\uparrow & & \uparrow \\
\mathrm{Hom}([B, C] \otimes B, D) & \longleftarrow & \mathrm{Hom}(C, D)
\end{array}$$

Commutativity of the second diagram can be deduced from that of the first (and vice-versa), or proved independently.

(3.5.4). Now let us see how f_* and f^* interact with closed structures (assumed given) on $\mathbf{X}$ and $\mathbf{Y}$.

First we have a functorial map, with A, $B \in \mathbf{X}$,

$$f_*[A, B] \longrightarrow [f_*A, f_*B] \tag{3.5.4.1}$$

corresponding under π (3.5.1.2) to the composed map

$$f_*[A, B] \otimes f_*A \xrightarrow[(3.4.2.1)]{} f_*\big([A, B] \otimes A\big) \xrightarrow[(3.5.3)(b)]{f_* t_{AB}} f_*B\,.$$

Conversely (verify!), the functorial map

$$f_*(A \otimes B) \longleftarrow f_*A \otimes f_*B$$

in (3.4.2.1) corresponds to the composition

$$\big[f_*B,\, f_*(A \otimes B)\big] \xleftarrow[(3.5.4.1)]{} f_*[B,\, A \otimes B] \xleftarrow[(3.5.3)(f)]{f_* u_{AB}} f_*A\,.$$

There results a functorial composition

$$f_*[f^*A, B] \xrightarrow[(3.5.4.1)]{} [f_*f^*A,\, f_*B] \xrightarrow[(3.4.3)]{\text{via } \eta} [A,\, f_*B]\,,$$

(3.5.4.2)

from which (verify!) (3.5.4.1) can be recovered as the composition

$$f_*[A, B] \xrightarrow[(3.4.3)]{\text{via } \epsilon} f_*[f^*f_*A, B] \xrightarrow[(3.5.4.2)]{} [f_*A,\, f_*B]\,.$$

The functors $C \mapsto f^*(C \otimes A)$ and $C \mapsto f^*C \otimes f^*A$ (from **Y** to **X**) both have right adjoints, namely $B \mapsto [A, f_*B]$ and $B \mapsto f_*[f^*A, B]$. Hence there is a functorial map

$$[A, f_*B] \longleftarrow f_*[f^*A, B] \qquad\qquad (3.5.4.3)$$

right-conjugate (see (3.3.5)) to the functorial map $f^*(C \otimes A) \to f^*C \otimes f^*A$ in (3.4.5.1).

Similarly, there is a functorial map

$$f_*[B, A] \longrightarrow [f_*B, f_*A] \qquad\qquad (3.5.4.4)$$

right-conjugate to the adjoint $f^*C \otimes B \leftarrow f^*(C \otimes f_*B)$ of $p_2(C, B)$ (3.4.6).

If $f^*(C \otimes A) \to f^*C \otimes f^*A$—and hence its conjugate (3.5.4.3)—is a functorial *isomorphism*, then we have the functorial map

$$f^*[A, B] \longrightarrow [f^*A, f^*B] \qquad\qquad (3.5.4.5)$$

which is adjoint to the composition

$$[A, B] \xrightarrow{\eta} [A, f_*f^*B] \xrightarrow{(3.5.4.3)^{-1}} f_*[f^*A, f^*B]\,;$$

and (verify!) $(3.5.4.3)^{-1}$ is the map adjoint to the composition

$$f^*[A, f_*B] \xrightarrow{(3.5.4.5)} [f^*A, f^*f_*B] \xrightarrow{\text{via } \epsilon} [f^*A, B],$$

from which (3.5.4.5) can be recovered as the composition

$$f^*[A, B] \xrightarrow{\text{via } \eta} f^*[A, f_*f^*B] \longrightarrow [f^*A, f^*B].$$

This all holds in the most relevant (for us) cases, see e.g., (3.4.4)(a), (b), and (3.5.2).

Does the map in (3.5.4.3) (resp. (3.5.4.4)) coincide with that in (3.5.4.2) (resp. (3.5.4.1))? Of course, but it's not entirely obvious: it amounts to commutativity of the respective diagrams in (3.5.5) below. (Cf. (3.2.4)(i), but recall that in proving (3.2.4)(i), we used (3.1.10), for whose last assertion, given (3.1.8), (3.5.5) provides a formal proof.)[12]

Proposition 3.5.5. *The following functorial diagrams—where $A, B, G \in \mathbf{X}_0$, $E, F, C \in \mathbf{Y}_0$, H_X, H_Y stand for $\mathrm{Hom}_{\mathbf{X}_0}$, $\mathrm{Hom}_{\mathbf{Y}_0}$ respectively, and with maps arising naturally from those defined above—commute:*

$$
\begin{array}{ccc}
H_X\big(f^*E, [f^*F, G]\big) & \xrightarrow{\ \sim\ } & H_X\big(f^*E \otimes f^*F, G\big) \\[2mm]
\simeq \big\downarrow & & \big\downarrow {\scriptstyle (3.4.5.1)} \\[2mm]
H_Y\big(E, f_*[f^*F, G]\big) & & H_X\big(f^*(E \otimes F), G\big) \\[2mm]
{\scriptstyle (3.5.4.2)} \big\downarrow & & \big\downarrow \simeq \\[2mm]
H_Y\big(E, [F, f_*G]\big) & \xrightarrow{\ \sim\ } & H_Y\big(E \otimes F, f_*G\big)
\end{array}
\qquad (3.5.5.1)
$$

$$
\begin{array}{ccc}
H_Y\big(C \otimes f_*B, f_*A\big) & \xleftarrow{\ \sim\ } & H_Y\big(C, [f_*B, f_*A]\big) \\[2mm]
\simeq \big\uparrow & & \big\uparrow {\scriptstyle (3.5.4.1)} \\[2mm]
H_X\big(f^*(C \otimes f_*B), A\big) & & H_Y\big(C, f_*[B, A]\big) \\[2mm]
{\scriptstyle (3.4.6.2)} \big\uparrow & & \big\uparrow \simeq \\[2mm]
H_X\big(f^*C \otimes B, A\big) & \xleftarrow{\ \sim\ } & H_X\big(f^*C, [B, A]\big)
\end{array}
\qquad (3.5.5.2)
$$

---

[12] Diagram (3.2.4.6) is, in view of (3.1.8), an instance of (3.5.5.1). So is (3.2.4.5); but we don't know that *a priori*, because we don't know that the maps in (3.2.3.2) and (3.5.4.2) coincide until after proving either (3.2.4)(i) or the derived-category analog of (3.1.8), viz. (3.2.4)(ii)—in whose proof (3.5.5) was used.

The *proof* will be based on:

Lemma 3.5.5.3. *The following diagram (with preceding notation) commutes:*

$$
\begin{array}{ccccc}
H_X\left(A,\,[B,G]\right) & \xrightarrow{\text{ natural }} & H_Y\left(f_*A,\,f_*[B,G]\right) & \xrightarrow{(3.5.4.1)} & H_Y\left(f_*A,\,[f_*B,f_*G]\right) \\
\Big\downarrow\simeq & & & & \Big\downarrow\simeq \\
H_X\left(A\otimes B,\,G\right) & \xrightarrow[\text{natural}]{} & H_Y\left(f_*(A\otimes B),\,f_*G\right) & \xrightarrow[(3.4.2.1)]{} & H_Y\left(f_*A\otimes f_*B,\,f_*G\right)
\end{array}
$$

Proof. Chasing a map $\varphi\colon A\to[B,G]$ around the diagram both clockwise and counterclockwise from upper left to lower right, one comes down to showing commutativity of the following diagram (with t as in (3.5.3(b))):

$$
\begin{array}{ccccc}
f_*A\otimes f_*B & \xrightarrow{f_*\varphi\otimes 1_{f_*B}} & f_*[B,G]\otimes f_*B & \xrightarrow{(3.5.4.1)} & [f_*B,f_*G]\otimes f_*B \\
{\scriptstyle(3.4.2.1)}\Big\downarrow & & {\scriptstyle(3.4.2.1)}\Big\downarrow & & \Big\downarrow{\scriptstyle t_{f_*B,f_*G}} \\
f_*(A\otimes B) & \xrightarrow[f_*(\varphi\otimes 1_B)]{} & f_*\left([B,G]\otimes B\right) & \xrightarrow[f_*(t_{BG})]{} & f_*G
\end{array}
$$

The left square commutes by functoriality, and the right one by the definition of (3.5.4.1).   Q.E.D.

Proof of (3.5.5). Expand (3.5.5.1) to (3.5.5.1.)*, shown on the next page, where the map ξ is induced by the map $\xi'\colon E\otimes F\to f_*(f^*E\otimes f^*F)$ adjoint to $f^*(E\otimes F)\to f^*E\otimes f^*F$, see (3.4.5.1); and the other maps are the obvious ones. The outer border of (3.5.5.1)* commutes, by (3.5.5.3) (with $A:=f^*E$, $B:=f^*F$). Hence if all the subdiagrams other than (3.5.5.1) commute, then so does (3.5.5.1), as desired.

Commutativity of ① follows from adjointness of f_* and f^*.

Commutativity of ② follows from the definition (3.5.4.2) of the map $f_*[f^*F,G]\to[F,f_*G]$.

Commutativity of ③ follows from functoriality of π (3.5.1.2).

Commutativity of ④ and of ⑤ result respectively from the following two factorizations of the map ξ':

$$
\begin{array}{ccccc}
E\otimes F & \xrightarrow{\ \eta\ } & f_*f^*(E\otimes F) & \xrightarrow{(3.4.5.1)} & f_*(f^*E\otimes f^*F), \\
E\otimes F & \xrightarrow{\ \eta\otimes\eta\ } & f_*f^*E\otimes f_*f^*F & \xrightarrow{(3.4.2.1)} & f_*(f^*E\otimes f^*F).
\end{array}
$$

Thus (3.5.5.1) does commute.

$$H_X(f^*E, [f^*F, G]) \xrightarrow{\quad ① \quad} H_Y(f_*f^*E, [f_*f^*F, f_*G])$$

$$\big\| \qquad\qquad\qquad\qquad =$$

$$H_X(f^*E, [f^*F, G]) \xrightarrow{\quad\quad} H_Y(f_*f^*E, [f_*f^*F, f_*G])$$

$$\downarrow \simeq \qquad\qquad ② \qquad\qquad \downarrow$$

$$H_Y(E, f_*[f^*F, G]) \xrightarrow{\quad\quad} H_Y(E, [F, f_*G]) \qquad ③ \qquad \xrightarrow{\simeq} H_Y(f_*f^*E \otimes f_*f^*F, f_*G)$$

$$\text{(3.5.5.1)} \qquad\qquad\qquad \downarrow \simeq$$

$$H_X(f^*E \otimes f^*F, G) \xrightarrow{\quad\quad} H_Y(E \otimes F, f_*G) \qquad ④ \qquad = \ H_Y(f_*f^*E \otimes f_*f^*F, f_*G)$$

$$\big\| \qquad\qquad ⑤ \qquad\qquad \uparrow \xi$$

$$H_X(f^*E \otimes f^*F, G) \xrightarrow{\quad\quad} H_Y(f_*(f^*E \otimes f^*F), f_*G)$$

$$\text{(3.5.5.1)}^*$$

$$H_X(f^*C, [B, A]) \xrightarrow{\quad\quad} H_Y(C, f_*[B, A])$$

$$\big\| \qquad\qquad ② \qquad\qquad \big\|$$

$$H_X(f^*C, [B, A]) \xrightarrow{\quad ① \quad} H_Y(C, f_*[f^*f_*B, A])$$

$$\downarrow \simeq \qquad\qquad\qquad \downarrow$$

$$H_X(f^*C, [f^*f_*B, A]) \xrightarrow{\quad\quad} H_Y(C, [f_*B, f_*A])$$

$$\downarrow \simeq \qquad\qquad ③ \qquad\qquad ④ \qquad\qquad \downarrow \simeq$$

$$H_X(f^*C \otimes f^*f_*B, A) \xrightarrow{\quad\quad} H_X(f^*(C \otimes f_*B), A) \xrightarrow{\quad\quad} H_Y(C \otimes f_*B, f_*A)$$

$$H_X(f^*C \otimes B, A)$$

$$\text{(3.5.5.2)}^*$$

Now look at $(3.5.5.2)^*$, whose outer border is identical with $(3.5.5.2)$. Subdiagrams ① and ③ commute by functoriality. Commutativity of ② comes from the statement immediately following $(3.5.4.2)$. Subdiagram ④ is just $(3.5.5.1)$ with $E := C$, $F := f_* B$, $G := A$; so it commutes too. Thus $(3.5.5.2)^*$ commutes.   Q.E.D.

Exercises 3.5.6. (a) Show that if the natural map $f^*(C \otimes A) \to f^*C \otimes f^*A$ is an isomorphism for all C and A, then $(3.5.4.5)$ corresponds under π (see $(3.5.1.2)$) to the natural composite map $f^*[A, B] \otimes f^*A \xrightarrow{\sim} f^*([A, B] \otimes A) \longrightarrow f^*B$.

(b) Given a fixed map $e: B' \to B$, show that the functorial maps

$$f_*[B, A] \xrightarrow{\text{via } e} f_*[B', A] \qquad \text{and} \qquad f^*C \otimes B \xleftarrow{\text{via } e} f^*C \otimes B'$$

are conjugate; and then deduce the equality of the maps $(3.5.4.1)$ and $(3.5.4.4)$ from that of $(3.5.4.2)$ and $(3.5.4.3)$.

(c) In $(3.5.5.i)$ $(i = 1, 2, 3)$, replace $H_X(-, -)$ by $f_*[-, -]$, and $H_Y(-, -)$ by $[-, -]$. Show that the resulting diagrams commute. (For example, reduce to commutativity of $(3.5.5.i)$, by applying the functor $H_Y(D, -)$ to the diagram in question.)

Show that $(3.5.5.i)$ can be recovered from "the resulting diagram" by application of the functor $\Gamma_Y := H_Y(\mathcal{O}_Y, -)$ of $(3.5.3)(a)$.

(d) By Yoneda's principle, commutativity of $(3.5.5.1)$ can be proved by taking $E = f_*[f^*F, G]$ and chasing the identity map of $f_*[f^*F, G]$ around the diagram in both directions. Deduce that commutativity of $(3.5.5.1)$ is equivalent to that of the diagram

$$
\begin{array}{ccccc}
f^*\big(f_*[f^*F, G] \otimes F\big) & \xrightarrow[(3.5.4.2)]{} & f^*\big([F, f_*G] \otimes F\big) & \xrightarrow[(3.5.3)(b)]{t_{F, f_*G}} & f^*f_*G \\
\scriptstyle(3.4.5.1)\downarrow & & & & \downarrow\scriptstyle\epsilon \\
f^*f_*[f^*F, G] \otimes f^*F & \xrightarrow[\text{via } \epsilon]{(3.4.3)} & [f^*F, G] \otimes f^*F & \xrightarrow[t_{f^*F, G}]{} & G
\end{array}
$$

(e) In a closed category $\mathbf{X}$ the natural composite functorial map

$$\operatorname{Hom}(F, G) \xrightarrow{\sim} \operatorname{Hom}(F \otimes \mathcal{O}_X, G) \xrightarrow{\sim} \operatorname{Hom}(F, [\mathcal{O}_X, G]),$$

being an isomorphism, takes (when $F = G$) the identity map of G to an *isomorphism* $G \xrightarrow{\sim} [\mathcal{O}_X, G]$. Let $\mathbf{Y}$ be another closed category, and (f^*, f_*) be as in $(3.4.3)$. Show that for $G \in \mathbf{X}$ and $E \in \mathbf{Y}$ the following natural diagrams commute:

$$
\begin{array}{ccc}
f_*[\mathcal{O}_X, G] \longrightarrow [f_*\mathcal{O}_X, f_*G] & \qquad & f^*[\mathcal{O}_Y, E] \longrightarrow [f^*\mathcal{O}_Y, f^*E] \\
\scriptstyle\simeq\downarrow \qquad\qquad \downarrow & & \scriptstyle\simeq\downarrow \qquad\qquad \downarrow\scriptstyle\simeq \\
f_*G \xrightarrow{\quad\sim\quad} [\mathcal{O}_Y, f_*G] & & f^*E \xrightarrow{\quad\sim\quad} [\mathcal{O}_X, f^*E]
\end{array}
$$

Hint. The first diagram is right-conjugate to the *dual* $(3.4.5)$ of $(3.4.7)(iii)$. For the second diagram, use, e.g., (a) above.

(f) With notation as in (e), and $\pi_{\mathbf{X}}$, $\pi_{\mathbf{Y}}$ as in $(3.5.1.2)$, and assuming the functorial map $f^*(C \otimes D) \to f^*C \otimes f^*D$ in $(3.4.5.1)$ to be a functorial isomorphism, show that $\pi_{\mathbf{X}}$ takes the inverse of the isomorphism $f^*(G \otimes B) \to f^*G \otimes f^*B$ to the composite map

$$f^*G \xrightarrow{\text{natural}} f^*[B, G \otimes B] \xrightarrow{(3.5.4.5)} [f^*B, f^*(G \otimes B)],$$

or, equivalently, that the following diagram commutes.

$$[f^*B, f^*(G \otimes B)] \xleftarrow{\;(3.4.5.1)^{-1}\;} [f^*B, f^*G \otimes f^*B]$$

$$(3.5.4.5)\uparrow \qquad\qquad\qquad\qquad\qquad \uparrow \text{via } \pi_{\mathbf{X}}$$

$$f^*[B, G \otimes B] \xleftarrow[\text{via } \pi_{\mathbf{Y}}]{} \qquad f^*G$$

(g) With assumptions as in (f), and using the commutative diagram in (f)—or otherwise—show that for any **Y**-map $\alpha\colon C \otimes D \to E$, and α^f the composite map

$$f^*C \otimes f^*D \xrightarrow{\;(3.4.5.1)^{-1}\;} f^*(C \otimes D) \xrightarrow{\;f^*\alpha\;} f^*E,$$

it holds that $\pi_{\mathbf{X}}(\alpha^f)$ is the composite map

$$f^*C \xrightarrow{\;f^*(\pi_{\mathbf{Y}}\alpha)\;} f^*[D, E] \xrightarrow{\;(3.5.4.5)\;} [f^*D, f^*E].$$

3.6 Adjoint Monoidal Δ-Pseudofunctors

We review next the behavior of derived direct and inverse image functors vis-à-vis a pair of ringed-space maps $X \xrightarrow{f} Y \xrightarrow{g} Z$.

First, relative to the categories of $\mathcal{O}_X$- ($\mathcal{O}_Y$-, $\mathcal{O}_Z$-) modules we have the functorial *isomorphism* (in fact *equality*)

$$(gf)_* \xrightarrow{\sim} g_*f_* \tag{3.6.1}_*$$

and hence, since f^*g^* is left-adjoint to g_*f_* and $(fg)^*$ is left-adjoint to $(gf)_*$ there is a unique functorial isomorphism

$$f^*g^* \xrightarrow{\sim} (gf)^* \tag{3.6.1}^*$$

such that the following natural diagram of functors commutes:

$$
\begin{array}{ccccc}
1 & \longrightarrow & g_*g^* & \longrightarrow & g_*(f_*f^*g^*) \\
\downarrow & & & & \| \\
(gf)_*(gf)^* & \xrightarrow{\;\sim\;} & g_*f_*(gf)^* & \xleftarrow{\;\sim\;} & g_*f_*f^*g^*
\end{array}
\tag{3.6.2}
$$

or, equivalently, such that the "dual" diagram

$$
\begin{array}{ccccc}
1 & \longleftarrow & f^*f_* & \longleftarrow & f^*(g^*g_*f_*) \\
\uparrow & & & & \| \\
(gf)^*(gf)_* & \xleftarrow{\;\sim\;} & f^*g^*(gf)_* & \xrightarrow{\;\sim\;} & f^*g^*g_*f_*
\end{array}
\tag{3.6.2}^{\mathsf{op}}
$$

commutes. (This statement follows from (3.3.5), see also (3.3.7)(a)).

Given a third map $Z \xrightarrow{\ h\ } W$, we have the commutative diagram of functorial isomorphisms (actually equalities)

$$
\begin{array}{ccc}
(hgf)_* & \longrightarrow & (hg)_* f_* \\
\downarrow & & \downarrow \\
h_*(gf)_* & \longrightarrow & h_* g_* f_*
\end{array}
\qquad (3.6.3)_*
$$

from which we deduce formally, via adjunction, a commutative diagram of functorial isomorphisms

$$
\begin{array}{ccc}
(hgf)^* & \longleftarrow & f^*(hg)^* \\
\uparrow & & \uparrow \\
(gf)^* h^* & \longleftarrow & f^* g^* h^*
\end{array}
\qquad (3.6.3)^*
$$

From these observations we can derive similar ones involving the corresponding derived functors.

Indeed, taking $U := g^{-1}V$ (V open $\subset Z$) in (3.2.3.3), we find that f_*B is g_*-acyclic for any q-injective $B \in \mathbf{K}(X)$, whence, by (2.2.7), there is a unique Δ-functorial isomorphism

$$
\mathbf{R}(gf)_* \xrightarrow{\ \sim\ } \mathbf{R}g_* \mathbf{R}f_* \qquad (3.6.4)_*
$$

making the following natural diagram commute:

$$
\begin{array}{ccc}
(gf)_* & \xrightarrow{\ \sim\ } g_* f_* \longrightarrow & (\mathbf{R}g_*)f_* \\
\downarrow & & \downarrow \\
\mathbf{R}(gf)_* & \xrightarrow{\hspace{4cm}\sim\hspace{4cm}} & \mathbf{R}g_* \mathbf{R}f_*
\end{array}
\qquad (3.6.4.1)
$$

This allows us to build a diagram analogous to $(3.6.3)_*$, with $\mathbf{R}e_*$ in place of e_* for each map e involved. The resulting derived functor diagram still commutes, as can be seen by reduction (via suitable quasi-isomorphisms) to the case of q-injective complexes in $\mathbf{D}(X)$, for which the diagram in question is essentially $(3.6.3)_*$.

In a parallel fashion, using q-flat instead of q-injective complexes, and recalling that f^* transforms q-flat complexes into q-flat complexes (see proof of (3.2.4)(i)), etc., we get a natural Δ-functorial isomorphism

$$
\mathbf{L}f^* \mathbf{L}g^* \xrightarrow{\ \sim\ } \mathbf{L}(gf)^*, \qquad (3.6.4)^*
$$

and a commutative diagram analogous to $(3.6.3)^*$, with $\mathbf{L}e^*$ in place of e^*. By (3.3.5), we also have commutative diagrams like (3.6.2) and $(3.6.2)^{\mathrm{op}}$, with f_*, f^* etc. replaced by their respective derived functors.

It is helpful to conceptualize some of the foregoing, as follows, leading up to (3.6.10). We begin with some standard terminology.[13]

(3.6.5). Let $\mathbf{S}$ be a category. A *covariant pseudofunctor* # on $\mathbf{S}$ assigns to each object $X \in \mathbf{S}$ a category $\mathbf{X}_\#$, to each map $f\colon X \to Y$ in $\mathbf{S}$ a functor $f_\#\colon \mathbf{X}_\# \to \mathbf{Y}_\#$, with $f_\#$ the identity functor if $X = Y$ and $f = 1_X$, and to each pair of maps $X \xrightarrow{f} Y \xrightarrow{g} Z$ in $\mathbf{S}$ an isomorphism of functors

$$c_{f,g}\colon (gf)_\# \xrightarrow{\sim} g_\# f_\#$$

such that

1) $c_{1,g} = c_{f,1} = $ identity, and

2) for any triple of maps $X \xrightarrow{f} Y \xrightarrow{g} Z \xrightarrow{h} W$ the following diagram commutes:

$$
\begin{array}{ccc}
(hgf)_\# & \xrightarrow{\ c_{f,hg}\ } & (hg)_\# f_\# \\
{\scriptstyle c_{gf,h}}\downarrow & & \downarrow{\scriptstyle c_{g,h}} \\
h_\#(gf)_\# & \xrightarrow[\ c_{f,g}\ ]{} & h_\# g_\# f_\#
\end{array}
\qquad (3.6.5.1)
$$

Similarly, a *contravariant pseudofunctor* on $\mathbf{S}$ assigns to each $X \in \mathbf{S}$ a category $\mathbf{X}^\#$, to each map $f\colon X \to Y$ a functor $f^\#\colon \mathbf{Y}^\# \to \mathbf{X}^\#$ (with $1^\# = 1$), and to each map-pair $X \xrightarrow{f} Y \xrightarrow{g} Z$ a functorial isomorphism

$$d_{f,g}\colon f^\# g^\# \to (gf)^\#$$

satisfying $d_{1,g} = d_{g,1} = $ identity, and such that for each triple of maps $X \xrightarrow{f} Y \xrightarrow{g} Z \xrightarrow{h} W$ the following diagram commutes:

$$
\begin{array}{ccc}
(hgf)^\# & \xleftarrow{\ d_{f,hg}\ } & f^\#(hg)^\# \\
{\scriptstyle d_{gf,h}}\uparrow & & \uparrow{\scriptstyle d_{g,h}} \\
(gf)^\# h^\# & \xleftarrow[\ d_{f,g}\ ]{} & f^\# g^\# h^\#
\end{array}
\qquad (3.6.5.2)
$$

There is an obvious way of identifying contravariant pseudofunctors on $\mathbf{S}$ with pseudofunctors on the dual category $\mathbf{S}^{\mathrm{op}}$.

(3.6.6). Given covariant pseudofunctors $*$ and # with $\mathbf{X}_* = \mathbf{X}_\#$ for all $X \in \mathbf{S}$, a *morphism of pseudofunctors* $* \to $ # is a family of morphisms of functors

$$\alpha_f\colon f_* \to f_\#$$

(one for each map f in $\mathbf{S}$) such that for any pair of maps $X \xrightarrow{f} Y \xrightarrow{g} Z$,

---

[13] Pseudofunctors can also be interpreted as 2-functors.

the following diagram commutes:

$$
\begin{array}{ccc}
(gf)_* & \xrightarrow{\ \ \alpha_{gf}\ \ } & (gf)_\# \\
{\scriptstyle c_{f,g}}\Big\downarrow & & \Big\downarrow{\scriptstyle c_{f,g}} \\
g_* f_* \xrightarrow{\ g_*\alpha_f\ } g_* f_\# & \xrightarrow{\ \alpha_g\ } & g_\# f_\#
\end{array}
$$

and such that for all $X \in \mathbf{S}$, with identity map 1_X, $\alpha_{1_X} : (1_X)_* \to (1_X)_\#$ is the identity automorphism of $\mathbf{X}_* = \mathbf{X}_\#$. Morphisms of contravariant pseudofunctors are defined analogously.

Suppose we are given a pseudofunctor $*$, and functors $f_\# : \mathbf{X}_* \to \mathbf{Y}_*$, one for each $\mathbf{S}$-morphism $f : X \to Y$, such that $f_\#$ is an identity functor whenever f is an identity map, and a family of functorial isomorphisms $\alpha_f : f_* \to f_\#$. It is left as an exercise to show that then *there is a unique family of isomorphisms of functors* $c_{f,g} : (gf)_\# \xrightarrow{\sim} g_\# f_\#$ *which together with the family* $(f_\#)$ *constitute a pseudofunctor such that the family* (α_f) *is an isomorphism of pseudofunctors.*

(3.6.7). Various refinements of these notions can be made.

(a). Assume that each category $\mathbf{X}_\#$ is a Δ-category, that each $f_\#$ (resp. $f^\#$) is a Δ-functor, and that each $c_{f,g}$ (resp. $d_{f,g}$) is an isomorphism of Δ-functors. We say then that $\#$ is a covariant (resp. contravariant) Δ-pseudofunctor.

A morphism of Δ-pseudofunctors is then a family α_f as in (3.6.6), with each α_f a morphism of Δ-functors.

(b). Assume that each category $\mathbf{X}_\#$ is a symmetric monoidal category, see (3.4.1), that each $f_\#$ is a symmetric monoidal functor (3.4.2), and that each $c_{f,g}$ is a morphism of symmetric monoidal functors [**EK**, p. 474], i.e., that the following natural diagrams commute (where $\otimes$ denotes the appropriate product functor, and $\mathcal{O}$ the unit; and $A, B \in \mathbf{X}_\#$):

$$
\begin{array}{ccc}
\mathcal{O}_Z & \longrightarrow & (gf)_\# \mathcal{O}_X \\
\Big\downarrow & & \Big\downarrow \\
g_\# \mathcal{O}_Y & \longrightarrow & g_\# f_\# \mathcal{O}_X
\end{array}
\tag{3.6.7.1}
$$

$$
\begin{array}{ccc}
(gf)_\# A \otimes (gf)_\# B & \longrightarrow & (gf)_\#(A \otimes B) \\
\Big\downarrow & & \Big\downarrow \\
g_\# f_\# A \otimes g_\# f_\# B \longrightarrow g_\#(f_\# A \otimes f_\# B) & \longrightarrow & g_\# f_\#(A \otimes B)
\end{array}
\tag{3.6.7.2}
$$

We say then that $\#$ is a monoidal pseudofunctor.

We say that a contravariant pseudofunctor $\#$ is monoidal if for each map $f : X \to Y$ in $\mathbf{S}$, the *opposite* functor $(f^\#)^{\mathsf{op}} : (\mathbf{Y}^\#)^{\mathsf{op}} \to (\mathbf{X}^\#)^{\mathsf{op}}$ is monoidal. In other words, we have functorial maps

$$
f^\#(A \otimes B) \to f^\# A \otimes f^\# B
$$

and a map

$$f^\# \mathcal{O}_Y \to \mathcal{O}_X$$

satisfying the obvious conditions.

A morphism of monoidal pseudofunctors is a family α_f as in (3.6.6) such that each α_f is a morphism of symmetric monoidal functors (i.e., α_f is compatible, in an obvious sense, with the maps (3.4.2.1)).

(c). If every $\mathbf{X}_\#$ is both a Δ-category and a symmetric monoidal category, and if the multiplication $\mathbf{X}_\# \times \mathbf{X}_\# \to \mathbf{X}_\#$ is a Δ-functor (see (2.4.3)), then we say that $\mathbf{X}_\#$ is a *monoidal Δ-category*; and we speak correspondingly of monoidal Δ-pseudofunctors and their morphisms.

(d). A pair $({}^*, {}_*)$ with ${}_*$ a pseudofunctor and * a contravariant pseudofunctor on $\mathbf{S}$ are said to be *adjoint* if the following conditions hold:

(i) $\mathbf{X}_* = \mathbf{X}^*$ *for all objects X in* $\mathbf{S}$.

(ii)   *For every $f\colon X \to Y$ in $\mathbf{S}$ there are bifunctorial isomorphisms*

$$\mathrm{Hom}_{\mathbf{X}^*}(f^*C, D) \xrightarrow{\ \sim\ } \mathrm{Hom}_{\mathbf{Y}_*}(C, f_*D) \qquad (C \in \mathbf{Y}^*,\ D \in \mathbf{X}_*),$$

i.e., the functor $f_*\colon \mathbf{X}_* \to \mathbf{Y}_*$ is right adjoint *to $f^*\colon \mathbf{Y}^* \to \mathbf{X}^*$.*

(iii)   *The resulting functorial diagrams* (3.6.2) *(or* (3.6.2)^{op}*) commute.*
In the monoidal case, we also require:

(iv)   *The natural maps*

$$f_*(A) \otimes f_*(B) \to f_*(A \otimes B),$$

$$f^*(f_*A \otimes f_*B) \to f^*f_*A \otimes f^*f_*B \to A \otimes B$$

correspond under the adjunction isomorphism of (ii) *above, as do the natural maps $f^*\mathcal{O}_Y \to \mathcal{O}_X$, $\mathcal{O}_Y \to f_*\mathcal{O}_X$.*
In the Δ-case, we also require that f^* and f_* be Δ-adjoint (3.3.1), i.e.,

(v)   *The natural functorial morphisms*

$$1 \to f_*f^* \quad and \quad f^*f_* \to 1$$

are both morphisms of Δ-functors.

(3.6.8). We add some remarks on existence and uniqueness, some of which are relevant to the subsequent construction and understanding of specific adjoint pairs of pseudofunctors.

(3.6.8.1). If ${}_*$ is a monoidal pseudofunctor on $\mathbf{S}$, and if for each map $f\colon X \to Y$ in $\mathbf{S}$ the functor $f_*\colon \mathbf{X}_* \to \mathbf{Y}_*$ has a left adjoint f^*, then *there is a unique contravariant monoidal pseudofunctor * on $\mathbf{S}$ such that $\mathbf{X}^* = \mathbf{X}_*$ for all objects $X \in \mathbf{S}$, f^* is the said left adjoint for each $f\colon X \to Y$, and the pair $({}^*, {}_*)$ is adjoint.*

Indeed, condition (iii) in (d) above forces $d_{f,g}\colon f^*g^* \to (gf)^*$ to be the left conjugate of the given $c_{f,g}\colon (gf)_* \to g_*f_*$ (see beginning of this section, up to (3.6.3)*). Similarly, (iv) imposes a unique monoidal structure on $(f^*)^{\mathrm{op}}$: given (ii), we see as in (3.4.5) that (iv) is equivalent to the following dual statement:

(iv)′ *The natural maps*

$$f^*(A) \otimes f^*(B) \leftarrow f^*(A \otimes B),$$

$$f_*(f^*A \otimes f^*B) \leftarrow f_*f^*A \otimes f_*f^*B \leftarrow A \otimes B$$

correspond under the above adjunction isomorphism (ii), *as do the natural maps*

$$f_*\mathcal{O}_X \leftarrow \mathcal{O}_Y, \qquad \mathcal{O}_X \leftarrow f^*\mathcal{O}_Y.$$

The rest of the proof is left to the reader.

(3.6.8.2). If $*$ is a Δ-pseudofunctor on $\mathbf{S}$, and if for each map $f\colon X \to Y$ in $\mathbf{S}$ the functor $f_*\colon \mathbf{X}_* \to \mathbf{Y}_*$ has a left adjoint f^*, then *there is a unique contravariant Δ-pseudofunctor * on $\mathbf{S}$ such that $\mathbf{X}^* = \mathbf{X}_*$ for all objects $X \in \mathbf{S}$, f^* is the said left adjoint for each $f\colon X \to Y$, and the pair $(^*, {}_*)$ is adjoint.*

Indeed, by (3.3.8), each f^* carries a unique structure of Δ-functor such that (v) above holds; and for $X \xrightarrow{f} Y \xrightarrow{g} Z$ in $\mathbf{S}$, the isomorphism $d_{f,g}$—forced by (iii) to be the conjugate of the given Δ-functorial isomorphism $c_{f,g}$—is Δ-functorial, by (3.3.6).

(3.6.8.3). Here is another form of uniqueness:

If $(^*, {}_*)$ and $(^\#, {}_*)$ are adjoint pairs of monoidal (or Δ-)pseudofunctors, and if for each $f\colon X \to Y$ we define the morphism $\alpha_f\colon f^* \to f^\#$ to be adjoint to the natural morphism $1 \to f_*f^\#$, then the family α_f is an isomorphism of monoidal (or Δ-)pseudofunctors.

Remarks 3.6.9 (Duality principle II). To each adjoint pair of monoidal pseudofunctors $(^*, {}_*)$ on $\mathbf{S}$, (3.6.7)(d), associate a *dual pair* $(^\#, {}_\#)$ of monoidal pseudofunctors on the dual category $\mathbf{S}^{\mathrm{op}}$ as follows:

$$\mathbf{X}^\# := (\mathbf{X}_*)^{\mathrm{op}}, \qquad \mathbf{X}_\# := (\mathbf{X}^*)^{\mathrm{op}}$$

for objects $X \in \mathbf{S}^{\mathrm{op}}$, and

$$f^\# := (f_*)^{\mathrm{op}}\colon (\mathbf{X}_*)^{\mathrm{op}} \to (\mathbf{Y}_*)^{\mathrm{op}}, \qquad f_\# := (f^*)^{\mathrm{op}}\colon (\mathbf{Y}^*)^{\mathrm{op}} \to (\mathbf{X}^*)^{\mathrm{op}}$$

for each map $f\colon Y \to X$ in $\mathbf{S}^{\mathrm{op}}$ (i.e., for each map $f\colon X \to Y$ in $\mathbf{S}$), the isomorphisms $f^\#g^\# \xrightarrow{\sim} (gf)^\#$ and $(gf)_\# \xrightarrow{\sim} g_\#f_\#$ being the obvious ones. The monoidal structure on the category $\mathbf{X}_\# = \mathbf{X}^\#$ is defined to be dual to that on $\mathbf{X}^* = \mathbf{X}_*$ see (3.4.5), and then each functor $f_\#$ is monoidal, with left adjoint $f^\#$. It follows that:

Each diagram built up from the basic data defining adjoint monoidal pairs can be interpreted in the dual context, giving rise to a "dual" diagram, obtained by interchanging $_*$ and * and reversing arrows, etc., etc.

This somewhat imprecise statement will be illustrated in Ex. (3.7.1.1) and in the proof of Prop. (3.7.2) below.

(3.6.10). With the terminology of (3.6.7), and with (3.5.2)(d) in mind, we can formally summarize many preceding results as follows.

Scholium. *Let* $\mathbf{S}$ *be the category of ringed spaces. For each object* $X \in \mathbf{S}$, *set* $\mathbf{X}^* = \mathbf{X}_* := \mathbf{D}(X)$ *(the derived category of the category of* $\mathcal{O}_X$*-modules), a closed* Δ*-category with product* $\otimes$*, unit* $\mathcal{O}_X$*, and internal hom* $\mathbf{R}\mathcal{H}\mathrm{om}$. *For* $X \xrightarrow{f} Y \xrightarrow{g} Z$ *in* $\mathbf{S}$*, write*

$$f^* \text{ for } \mathbf{L}f^* \colon \mathbf{Y}^* \to \mathbf{X}^*, \qquad d_{f,g} \text{ for the map } (3.6.4)^*,$$
$$f_* \text{ for } \mathbf{R}f_* \colon \mathbf{X}_* \to \mathbf{Y}_*, \qquad c_{f,g} \text{ for the map } (3.6.4)_*.$$

This defines an adjoint pair $(^*, {}_*)$ *of monoidal* Δ*-pseudofunctors on* $\mathbf{S}$.

Proof. Essentially everything has already been proved, in (3.4.4)(b) and at the beginning of this §3.6, except for the commutativity of (3.6.7.1) and (3.6.7.2) (with $_*$ in place of $_\#$).

Commutativity of (3.6.7.1) is left to the reader.

To show that (3.6.7.2) commutes, first do it in the context of sheaves of modules—with the ordinary direct image functors see (3.1.7)—where it follows easily from definitions. A formal argument, using (iv) or (iv)′ above (details left to the reader), then yields the commutativity of the corresponding (dual) sheaf diagram with * in place of $_*$, and all arrows reversed. In this latter diagram, we can then replace f^* etc. by $\mathbf{L}f^*$, etc., and commutativity is preserved since the resulting derived functor diagram need only be checked when A and B are q-flat complexes, in which case it does not differ essentially from the original sheaf diagram.

Finally, the preceding formal (adjunction) argument, applied this time to derived functors, gives us commutativity in (3.6.7.2).

3.7 More Formal Consequences: Projection, Base Change

We give some additional consequences, to be used later, of the formalism in §3.6. Again, the introductory remarks in §3.4, suitably modified, are relevant.

We consider an adjoint monoidal pair $(^*, {}_*)$ as in (d) of (3.6.7).

Condition (ii) there means that for $f \colon X \to Y$ in $\mathbf{S}$, we have functorial maps

$$\eta \colon 1 \to f_* f^*, \qquad \epsilon \colon f^* f_* \to 1$$

such that the resulting compositions

$$f^* \xrightarrow{\eta} f^* f_* f^* \xrightarrow{\epsilon} f^*, \qquad f_* \xrightarrow{\eta} f_* f^* f_* \xrightarrow{\epsilon} f_*$$

are both identities.

For $X \in \mathbf{S}$, the product functor on the monoidal category $\mathbf{X}^* = \mathbf{X}_*$ will be denoted by $\otimes$.

For a map $f\colon X \to Y$ in $\mathbf{S}$, the functorial "projection" map

$$p_f\colon G \otimes f_*F \to f_*(f^*G \otimes F) \qquad (G \in \mathbf{Y}^*,\ F \in \mathbf{X}_*)$$

is defined as in (3.4.6). It is compatible with pseudofunctoriality, in the following sense.

Proposition 3.7.1 *For any* $X \xrightarrow{f} Y \xrightarrow{g} Z$ *in* $\mathbf{S}$, *the following diagram, with* $F \in \mathbf{X}_*$ *and* $G \in \mathbf{Z}^*$, *commutes.*

$$
\begin{array}{ccccc}
G \otimes g_* f_* F & \xrightarrow{\ p_g\ } & g_*(g^*G \otimes f_*F) & \xrightarrow{\ g_*(p_f)\ } & g_* f_*(f^*g^*G \otimes F) \\[2pt]
{\scriptstyle \text{via } c_{f,g}} \uparrow {\scriptstyle \simeq} & & & & \downarrow {\scriptstyle \text{via } d_{f,g}} \\[2pt]
G \otimes (gf)_* F & \xrightarrow[\ p_{gf}\ ]{} & (gf)_*((gf)^*G \otimes F) & \xrightarrow[\ c_{f,g}\ ]{\sim} & g_* f_*((gf)^*G \otimes F)
\end{array}
$$

Proof. An expanded form of the diagram is obtained by pasting the first of the following diagrams, along its right edge, to the second, along its left edge. (All the arrows have an obvious interpretation.)

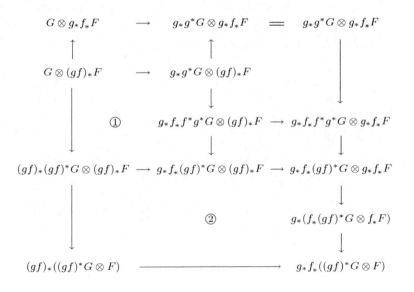

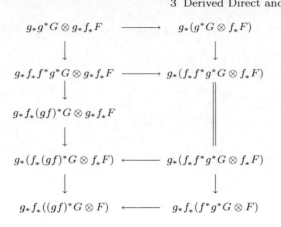

Subdiagram ① commutes because of commutativity of (3.6.2) (see condition (iii) in (3.6.7)(d)), Subdiagram ② commutes because of the commutativity of (3.6.7.2) (which is part of the definition of *monoidal* pseudofunctor); and commutativity of the remaining subdiagrams is clear. The conclusion follows.

Exercise 3.7.1.1 The preceding Proposition expresses the compatibility of the projection map with the structure "adjoint pair of monoidal pseudofunctors." One can ask about similar compatibilities for any of the maps introduced in §3.5. Here are some examples which will be needed later.

(*Challenge*: Establish metaresults of which such examples would be instances.)

With notation as in (3.7.1), ϱ as in (3.5.4.1), and $\beta_f : f_*[f^*-, -] \to [-, f_*-]$ as in (3.5.4.2) or (3.5.4.3), show that the following diagrams commute:

$$
\begin{array}{ccc}
(gf)_*[(gf)^*G, F] & \xrightarrow{\ \ \ \ \ \ \ \ \ \ \ \beta_{gf}\ \ \ \ \ \ \ \ \ \ \ } & [G, (gf)_*F] \\
{\scriptstyle \text{via } c_{f,g} }\downarrow{\scriptstyle \text{ and } d_{f,g}} & & \downarrow{\scriptstyle \text{via } c_{f,g}} \\
g_*f_*[f^*g^*G, F] & \xrightarrow[g_*\beta_f]{} g_*[g^*G, f_*F] \xrightarrow[\beta_g]{} & [G, g_*f_*F]
\end{array}
$$

$$
\begin{array}{ccc}
f_*g_*[F, F'] & \xrightarrow{f_*\varrho_g} f_*[g_*F, g_*F'] \xrightarrow{\varrho_f} & [f_*g_*F, f_*g_*F'] \\
{\scriptstyle c_{f,g}^{-1}}\downarrow & & \downarrow{\scriptstyle \text{via } c_{f,g}^{-1}} \\
(gf)_*[F, F'] & \xrightarrow[\varrho_{gf}]{} [(gf)_*F, (gf)_*F'] \xrightarrow[\text{via } c_{f,g}^{-1}]{} & [f_*g_*F, (gf)_*F']
\end{array}
$$

Deduce from the first diagram that with $\rho_f : f^*[-, -] \to [f^*-, f^*-]$ as in (3.5.4.5), the next diagram commutes:

$$f^*g^*[G, G'] \xrightarrow{\;f^*\rho_g\;} f^*[g^*G, g^*G'] \xrightarrow{\;\rho_f\;} [f^*g^*G, f^*g^*G']$$

$$d_{f,g} \Big\downarrow \simeq \qquad\qquad\qquad\qquad\qquad \simeq \Big\downarrow \text{via } d_{f,g}$$

$$(gf)^*[G, G'] \xrightarrow{\;\rho_{gf}\;} [(gf)^*G, (gf)^*G'] \xrightarrow[\text{via } d_{f,g}]{\;\sim\;} [f^*g^*G, (gf)^*G']$$

Hints. Apply (3.6.9) to the diagram in (3.6.7.2), resp. Prop. (3.7.1), and compare the result with the diagram left-conjugate to the first, resp. second, one above. The third diagram expands naturally as follows.

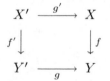

In this diagram, all but three subdiagrams clearly commute, and those three are taken care of by (3.6.2), (3.6.2)^{op}, and the first diagram above.

Next, we introduce an oft-to-be-encountered "base change" morphism.

Proposition 3.7.2 (i) *To each commutative square σ in* **S**:

$$
\begin{array}{ccc}
X' & \xrightarrow{\;g'\;} & X \\
f'\downarrow & & \downarrow f \\
Y' & \xrightarrow[g]{} & Y
\end{array}
$$

there is associated a natural map of functors

$$\theta = \theta_\sigma : g^*f_* \longrightarrow f'_*g'^*,$$

equal to each of the following four compositions (with $h = fg' = gf'$):

(a)
$$g^*f_* \xrightarrow{\;\eta\;} g^*f_*g'_*g'^* \xrightarrow{\;(c_{f',g})(c^{-1}_{g',f})\;} g^*g_*f'_*g'^* \xrightarrow{\;\epsilon\;} f'_*g'^*$$

(b)
$$g^*f_* \xrightarrow{\;\eta(\;)\eta\;} f'_*f'^*g^*f_*g'_*g'^* \xrightarrow{\;(d_{f',g})(c^{-1}_{g',f})\;} f'_*h^*h_*g'^* \xrightarrow{\;\epsilon\;} f'_*g'^*$$

(c) $\qquad g^*f_* \xrightarrow{\eta} f'_*f'^*g^*f_* \xrightarrow{(d_{g',f}^{-1})(d_{f',g})} f'_*g'^*f^*f_* \xrightarrow{\epsilon} f'_*g'^*$

(d) $\qquad g^*f_* \xrightarrow{\eta} g^*h_*h^*f_* \xrightarrow{(c_{f',g})(d_{g',f}^{-1})} g^*g_*f'_*g'^*f^*f_* \xrightarrow{\epsilon(\ )\epsilon} f'_*g'^*$

(ii) *Given a pair of commutative squares*

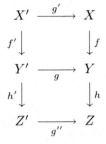

the following resulting diagram commutes:

$$
\begin{array}{ccc}
g''^*(hf)_* & \xrightarrow{\quad\theta\quad} & (h'f')_*g'^* \\
{\scriptstyle c_{f,h}}\downarrow & & \downarrow{\scriptstyle c_{f',h'}} \\
g''^*h_*f_* & \xrightarrow[\theta]{} h'_*g^*f_* \xrightarrow[\theta]{} & h'_*f'_*g'^*
\end{array}
$$

(iii) *Given a pair of commutative squares*

$$
\begin{array}{ccccc}
X'' & \xrightarrow{\ h\ } & X' & \xrightarrow{\ f\ } & X \\
{\scriptstyle g''}\downarrow & & \downarrow{\scriptstyle g} & & \downarrow{\scriptstyle g'} \\
Y'' & \xrightarrow[\ h'\ ]{} & Y' & \xrightarrow[\ f'\ ]{} & Y
\end{array}
$$

the following resulting diagram commutes:

$$
\begin{array}{ccc}
g''_*(fh)^* & \xleftarrow{\quad\theta\quad} & (f'h')^*g'_* \\
{\scriptstyle d_{h,f}}\uparrow & & \uparrow{\scriptstyle d_{h',f'}} \\
g''_*h^*f^* & \xleftarrow[\theta]{} h'^*g_*f^* \xleftarrow[\theta]{} & h'^*f'^*g'_*
\end{array}
$$

Proof. (i) To get convinced that (a), (b) and (c) are the same, contemplate the following commutative diagram, noting that $\epsilon \circ \eta$ on the right (resp. bottom) edge is the identity map, and recalling for subdiagrams ① and ② the condition (iii) in the definition (3.6.7)(d) of "adjoint pair."

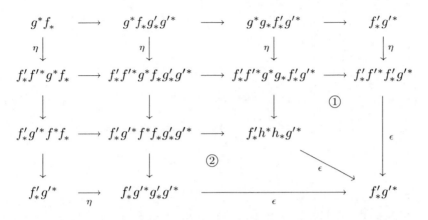

The equality (c) = (d) is obtained from (a) = (b) by *duality* (3.6.9).

(ii) Consider the expanded diagram (3.7.2.2) on the following page.

Recall that the composition $\epsilon \circ \eta$ of the adjacent arrows in the middle is the identity. Commutativity of subdiagram ① is an easy consequence of the commutativity of (3.6.5.1) (axiom for pseudofunctors). Commutativity of the other subdiagrams is straightforward, and the conclusion follows.

(iii) is simply the *dual* of (ii) (see (3.6.9)).   Q.E.D.

Proposition 3.7.3 (Base change and projection). *Let*

$$
\begin{array}{ccc}
X' & \xrightarrow{\ g'\ } & X \\
{\scriptstyle f'}\downarrow & & \downarrow{\scriptstyle f} \\
Y' & \xrightarrow[\ g\ ]{} & Y
\end{array}
$$

*be a commutative **S**-diagram, $P \in \mathbf{Y}_*$, $Q \in \mathbf{X}_*$. Then with θ as in (3.7.2), $h = fg' = gf'$, and p the projection map, the following diagram commutes:*

$$
\begin{array}{ccc}
g^*P \otimes g^*f_*Q & \xleftarrow{\ (3.4.5.1)\ } g^*(P \otimes f_*Q) \xrightarrow{\ g^*(p_f)\ } & g^*f_*(f^*P \otimes Q) \\
{\scriptstyle 1\otimes\theta}\downarrow & & \downarrow{\scriptstyle \theta} \\
g^*P \otimes f'_*g'^*Q & & f'_*g'^*(f^*P \otimes Q) \\
{\scriptstyle p_{f'}}\downarrow & & \downarrow{\scriptstyle (3.4.5.1)} \\
f'_*(f'^*g^*P \otimes g'^*Q) & \xrightarrow[\ d_{f',g}\ ]{} f'_*(h^*P \otimes g'^*Q) \xleftarrow[\ d_{g',f}\ ]{} & f'_*(g'^*f^*P \otimes g'^*Q)
\end{array}
$$

Proof. Consider the expanded diagram (3.7.3.1) on the following page (a diagram in which the arrows are self-explanatory). With a bit of patience, one checks that it suffices to show its commutativity.

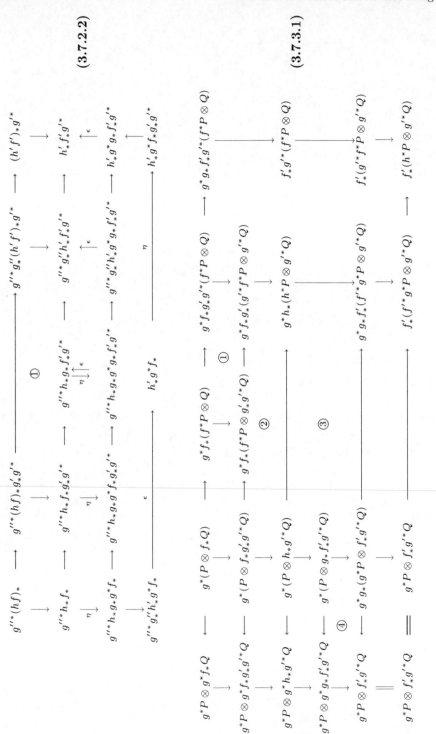

(3.7.2.2)

(3.7.3.1)

Subdiagram ① commutes by (3.4.7)(i), subdiagrams ② and ③ by (3.7.1), and ④ by the last sentence in (3.4.6.2). Commutativity of the other subdiagrams is straightforward to check.

Remarks 3.7.3.1 In the case of ringed spaces (3.6.10), the unlabeled arrows in the preceding diagram represent isomorphisms. So if θ is an isomorphism too, then the maps $g^*(p_f)$ and $p_{f'}$ are isomorphic. For such diagrams we can say then that "projection commutes with base change."

For example, when g is an open immersion, then θ is an isomorphism. That amounts to *compatibility of* $\mathbf{R}f_*$ *with open immersions,* which is also an immediate consequence of (2.4.5.2).

For other situations in which θ is an isomorphism, see (3.9.5) and its generalization (3.10.3).

3.8 Direct Sums

Proposition 3.8.1 *Let X be a ringed space. Arbitrary (small) direct sums exist in $\mathbf{K}(X)$ and in $\mathbf{D}(X)$; and the canonical functor $Q\colon \mathbf{K}(X) \to \mathbf{D}(X)$ preserves them. In both $\mathbf{K}(X)$ and $\mathbf{D}(X)$, natural maps of the type $\oplus_{\alpha \in A}(C_\alpha[1]) \;\to\; (\oplus_{\alpha \in A} C_\alpha)[1]$ are always isomorphisms—direct sums commute with translation; and any direct sum of triangles is a triangle.*

Proof. Let $(C_\alpha)_{\alpha \in A}$ (A small) be a family of complexes of $\mathcal{O}_X$-modules. The usual direct sum C of the family (C_α)—together with the homotopy classes of the canonical maps $C_\alpha \to C$—is also a direct sum in the category $\mathbf{K}(X)$. Since any complex in $\mathbf{D}(X)$ is isomorphic to a q-injective one, and since $\mathrm{Hom}_{\mathbf{D}(X)}(-, I) = \mathrm{Hom}_{\mathbf{K}(X)}(-, I)$ for any q-injective I, see (2.3.8(v)), it follows that C is also a direct sum in $\mathbf{D}(X)$.[14] The remaining assertions are easily checked for $\mathbf{K}(X)$, where we need only consider standard triangles, see (1.4.3); and they follow for $\mathbf{D}(X)$ upon application of Q, see (1.4.4). Q.E.D.

Proposition 3.8.2 *Let Y be a ringed space, and let $(C_\alpha)_{\alpha \in A}$ be a small family of complexes of $\mathcal{O}_Y$-modules. Then:*

(i) *For any $D \in \mathbf{D}(Y)$, the canonical map is an isomorphism*

$$\oplus_\alpha (C_\alpha \underline{\otimes} D) \xrightarrow{\sim} (\oplus_\alpha C_\alpha) \underline{\otimes} D.$$

(ii) *For any ringed-space map $f\colon X \to Y$, the canonical map is an isomorphism*

$$\oplus_\alpha \mathbf{L}f^* C_\alpha \xrightarrow{\sim} \mathbf{L}f^*(\oplus_\alpha C_\alpha).$$

---

[14] A more elementary proof, not using q-injective resolutions, is given in [**BN**, §1].

Proof. Each C_α is isomorphic to a q-flat complex; and any direct sum of q-flat complexes is still q-flat, see §2.5. Hence the assertions reduce to the corresponding ones for ordinary complexes, with $\otimes$ in place of $\underset{=}{\otimes}$ and f^* in place of $\mathbf{L}f^*$.

Alternatively, in view of $(2.6.1)^*$ and $(3.2.1)$ one can use the fact that any functor having a right adjoint respects direct sums.   Q.E.D.

Proposition 3.8.3 (See $[\mathbf{N}'$, p. 38, Remark 1.2.2]). *Let Y be a ringed space and*

$$C'_\alpha \longrightarrow C_\alpha \longrightarrow C''_\alpha \longrightarrow TC'_\alpha \qquad (\alpha \in A)$$

a small family of $\mathbf{D}(Y)$-triangles. Then the naturally resulting sequence

$$\oplus_\alpha C'_\alpha \longrightarrow \oplus_\alpha C_\alpha \longrightarrow \oplus_\alpha C''_\alpha \longrightarrow \oplus_\alpha TC'_\alpha \cong T(\oplus_\alpha C'_\alpha) \qquad (\alpha \in A)$$

is also a $\mathbf{D}(Y)$-triangle.

Exercise. Deduce $(3.8.2)(i)$ from $(2.5.10)(c)$. Using, e.g., $(2.5.5)$, prove an analogous generalization of $(3.8.2)(ii)$, i.e., show that if (C_α) is a (small, directed) inductive system of complexes of $\mathcal{O}_Y$-modules, then there are natural isomorphisms

$$\varinjlim_\alpha H^n \mathbf{L}f^* C_\alpha \xrightarrow{\ \sim\ } H^n \mathbf{L}f^* (\varinjlim_\alpha C_\alpha) \qquad (n \in \mathbb{Z}).$$

3.9 Concentrated Scheme-Maps

This section contains some refinements of preceding considerations as applied to a map $f\colon X \to Y$ of *schemes,* see $(3.4.4)(b)$. Except in $(3.9.1)$, which does not involve $\mathbf{R}f_*$, we need f to be *concentrated* ($=$ quasi-compact and quasi-separated). The main result $(3.9.4)$ asserts that under mild restrictions on f or on the $\mathcal{O}_X$-complex F, the projection map

$$p\colon \mathbf{R}f_* F \underset{=}{\otimes} G \to \mathbf{R}f_* (F \underset{=}{\otimes} \mathbf{L}f^* G) \qquad (\text{see } (3.4.6))$$

is an *isomorphism* for any $\mathcal{O}_Y$-complex G having quasi-coherent homology. The results of $(3.9.1)$ and $(3.9.2)$ on good behavior, vis-à-vis quasi-coherence, of the derived direct and inverse image functors of a concentrated map allow "way-out" reasoning to reduce $(3.9.4)$ essentially to the trivial case $G = \mathcal{O}_Y$, provided that F and G are bounded above; homological compatibility of $\mathbf{R}f_*$ and $\varinjlim$ (proved in $(3.9.3)$) then gets rid of the boundedness.

Another Proposition, $(3.9.5)$, says that for concentrated f the map θ associated as in $(3.7.2)$ to certain *flat* base changes is an isomorphism. A stronger result will be given in Theorem $(3.10.3)$, which contains $(3.9.4)$ as well. (But $(3.9.4)$ is used in the proof of $(3.10.3)$).

Proposition (3.9.6) takes note of, among other things, the fact that on a quasi-compact separated scheme, complexes with quasi-coherent homology are $\mathbf{D}$-isomorphic to quasi-coherent complexes.

We begin with some notation and terminology relative to any ringed space X, with $\mathbf{K}(X)$ and $\mathbf{D}(X)$ as in §3.1.

As in (1.6)–(1.8), we have various triangulated (i.e., Δ-)subcategories of $\mathbf{K}(X)$, denoted $\mathbf{K}^*(X)$, $\overline{\mathbf{K}}^*(X)$ (with "$*$" indicating a boundedness condition—below ($* = +$), above ($* = -$), or both above and below ($* = \mathrm{b}$)— and "$^-$" indicating application of the boundedness condition to the homology of a complex rather than to the complex itself); and we have the corresponding derived categories $\mathbf{D}^*(X)$, $\overline{\mathbf{D}}^*(X)$, which are Δ-subcategories of $\mathbf{D}(X)$. For example, $\mathbf{K}^+(X)$ is the full subcategory of $\mathbf{K}(X)$ whose objects are complexes $A^\bullet$ of $\mathcal{O}_X$-modules such that $A^n = 0$ for all $n \le n_0(A^\bullet)$ (where $n_0(A^\bullet)$ is some integer depending on $A^\bullet$); and $\overline{\mathbf{D}}^-(X)$ is the full subcategory of $\mathbf{D}(X)$ whose objects are complexes $A^\bullet$ such that $H^n(A^\bullet) = 0$ for all $n \ge n_1(A^\bullet)$.

The subscript "qc" indicates collections of $\mathcal{O}_X$-complexes whose homology sheaves are all *quasi-coherent* (see (1.9), with $\mathcal{A}^\#$ the category of quasi-coherent $\mathcal{O}_X$-modules, which is a plump subcategory of the category of all $\mathcal{O}_X$-modules [**GD**, p. 217, (2.2.2) (iii)]). For example $\overline{\mathbf{D}}_{\mathsf{qc}}^+(X)$ is the Δ-subcategory of $\mathbf{D}(X)$ whose objects are complexes $A^\bullet$ such that $H^n(A^\bullet)$ is quasi-coherent for all $n \in \mathbb{Z}$, and $H^n(A^\bullet) = 0$ for $n \le n_0(A^\bullet)$.

Proposition 3.9.1 *For any scheme-map $f\colon X \to Y$ we have*

$$\mathbf{L}f^*\big(\mathbf{D}_{\mathsf{qc}}(Y)\big) \subset \mathbf{D}_{\mathsf{qc}}(X).$$

Proof. For $C \in \mathbf{D}_{\mathsf{qc}}(Y)$ and $C_m := \tau_{\le m} C$ (1.10), there exists a q-flat resolution

$$\varinjlim Q_m = Q \to C = \varinjlim C_m \qquad (m \ge 0)$$

where for each i, Q_m is a bounded-above flat resolution of C_m, see (2.5.5). The resulting maps

$$\varinjlim f^*Q_m \to f^*Q \leftarrow \mathbf{L}f^*Q \to \mathbf{L}f^*C$$

are all *isomorphisms* in $\mathbf{D}(X)$ (recall that, as indicated just before (3.1.3), q-flat $\Rightarrow$ left-f^*-acyclic, and dualize the last assertion in (2.2.6)); and hence

$$H^n(\mathbf{L}f^*C) \cong \varinjlim H^n(f^*Q_m) \cong \varinjlim H^n(\mathbf{L}f^*C_m) \qquad (n \in \mathbb{Z}).$$

Since $\varinjlim$ preserves quasi-coherence, we need only deal with the case where $C = C_m \in \mathbf{D}_{\mathsf{qc}}^-(Y)$; and then way-out reasoning [**H**, p. 73, (ii) (dualized)] reduces us further to showing that *for any quasi-coherent $\mathcal{O}_Y$-module F and any $i \in \mathbb{Z}$, the $\mathcal{O}_X$-modules $L_i f^*(F) := H^{-i}\mathbf{L}f^*(F)(i \ge 0)$ are also quasi-coherent.*

For this, note that the restriction of a flat resolution of F to an open subset $U \subset Y$ is a flat resolution of the restriction $F|_U$, whence formation of $L_i f^*(F)$ "commutes" (in an obvious sense) with open immersions on Y; so we can assume X and Y to be affine, say $X = \mathrm{Spec}(B)$, $Y = \mathrm{Spec}(A)$, and $F = \widetilde{G}$, the quasi-coherent $\mathcal{O}_Y$-module associated to some A-module G; and then if $G_\bullet \to G$ is an A-free resolution of G, it is easily seen (since $M \mapsto \widetilde{M}$ is an exact functor of A-modules M [**GD**, p. 198, (1.3.5)], and since $f^* \widetilde{M} = (B \otimes_A M)^{\widetilde{}}$ [*ibid.*, p. 213, (1.7.7)]) that $L_i f^*(F)$ is the quasi-coherent $\mathcal{O}_X$-module $\widetilde{H_i}$, where H_i is the homology $H_i := H_i(B \otimes_A G_\bullet) = \mathrm{Tor}_i^A(B, G)$.

$$\text{Q.E.D.}$$

We will use the adjective *concentrated* as a less cumbersome synonym for *quasi-compact and quasi-separated*. Elementary properties of concentrated schemes and scheme-maps can be found in [**GD**, pp. 290 *ff*]. In particular, if $f\colon X \to Y$ is a scheme-map with Y concentrated, then X is concentrated iff f is a concentrated map [*ibid.*, p. 295, (6.1.10)].

Proposition 3.9.2 *Let* $f\colon X \to Y$ *be a concentrated map of schemes. Then*

$$\mathbf{R}f_*\big(\mathbf{D_{qc}}(X)\big) \subset \mathbf{D_{qc}}(Y). \tag{3.9.2.1}$$

Moreover, with notation as in §1.10, for all $n \in \mathbb{Z}$ it holds that

$$\mathbf{R}f_*\big(\mathbf{D_{qc}}(X)_{\geq \mathbf{n}}\big) \subset \mathbf{D_{qc}}(Y)_{\geq \mathbf{n}}; \tag{3.9.2.2}$$

and if Y is quasi-compact, then there exists an integer d such that for every $n \in \mathbb{Z}$,

$$\mathbf{R}f_*\big(\mathbf{D_{qc}}(X)_{\leq \mathbf{n}}\big) \subset \mathbf{D_{qc}}(Y)_{\leq \mathbf{n+d}}. \tag{3.9.2.3}$$

Proof. The fact that $\mathbf{R}f_*(\mathbf{D}(X)_{\geq \mathbf{n}}) \subset \mathbf{D}(Y)_{\geq \mathbf{n}}$ is, implicitly, in (2.7.3): any $F \in \mathbf{D}(X)_{\geq \mathbf{n}}$ admits the quasi-isomorphism $(1.8.1)^+\colon F \to \tau^+ F$, and there is a quasi-isomorphism $\tau^+ F \to I$ where I is a flasque complex with $I^m = 0$ for all $m < n$, so that $\mathbf{R}f_* F \cong f_* I \in \mathbf{D}(Y)_{\geq \mathbf{n}}$.

To finish proving (3.9.2.2), i.e., to show that if I has quasi-coherent homology then so does $f_* I$, use the standard spectral sequence

$$R^p f_*\big(H^q(I)\big) \Rightarrow H^\bullet\big(f_* I\big) \qquad (R^p f_* := H^p \mathbf{R}f_*)$$

and the fact (proved in [**AHK**, p. 33, Thm. (5.6)] or [**Kf**, p. 643, Cor. 11]) that $R^p f_*$ preserves quasi-coherence of sheaves. Or, reduce to this fact by "way-out" reasoning, see [**H**, p. 88, Prop. 2.1].

For the rest, we need:

Lemma 3.9.2.4 *If Y is quasi-compact then there is an integer d such that for any quasi-coherent $\mathcal{O}_X$-module $\mathcal{F}$ and any $i > d$, $R^i f_* \mathcal{F} = 0$.*

Proof. Since Y is covered by finitely many affine open subschemes Y_k and since for each k the restriction $R^i f_* \mathcal{F}|_{Y_k}$ is the quasi-coherent sheaf associated to the $\Gamma(Y_k, \mathcal{O}_Y)$-module $H^i(f^{-1}(Y_k), \mathcal{F})$ [**Kf**, p. 643, Cor. 11], we need only show that *if Y is affine then there is an integer d such that $H^i(X, \mathcal{F}) = 0$ for all $i > d$.*

Note that X is now a concentrated scheme. We proceed by induction on the unique integer $n = n(X)$ such that X can be covered by n quasi-compact separated open subschemes, but not by any $n - 1$ such subschemes. (This integer exists because X is quasi-compact and its affine open subschemes are quasi-compact and separated.)

If $n = 1$, i.e., X is separated, then $H^i(X, \mathcal{F})$ is the Čech cohomology with respect to a finite cover $X = \cup_{j=0}^d X_j$ by affine open subschemes, so it vanishes for $i > d$.

Suppose next that

$$ X = X_1 \cup X_2 \cup \cdots \cup X_n \qquad (n = n(X) > 1) $$

with each X_j a quasi-compact separated open subscheme of X. Since X is quasi-separated therefore $X_j \cap X_1$ is quasi-compact and separated,[15] so setting

$$ X_0 := X_2 \cup \cdots \cup X_n $$

we have $n(X_0) < n$ and $n(X_0 \cap X_1) < n$. The desired conclusion follows then from the inductive hypothesis and from the long exact sequence

$$ \cdots \to H^{i-1}(X_0 \cap X_1, \mathcal{F}) \to H^i(X, \mathcal{F}) \to H^i(X_0, \mathcal{F}) \oplus H^i(X_1, \mathcal{F}) \to \cdots $$

associated to the obvious short exact sequence of complexes

$$ 0 \to \Gamma(X, \mathcal{I}^\bullet) \to \Gamma(X_0, \mathcal{I}^\bullet) \oplus \Gamma(X_1, \mathcal{I}^\bullet) \to \Gamma(X_0 \cap X_1, \mathcal{I}^\bullet) \to 0 $$

where $\mathcal{I}^\bullet$ is a flasque resolution of $\mathcal{F}$.   Q.E.D.

Now let $F \in \mathbf{D}_{\mathsf{qc}}(X)$ and $N \in \mathbb{Z}$. Starting with an injective resolution $\tau_{\geq N} F \to I_N$, and using (3.9.2.5)(ii) below (with $\mathbf{J}$ the category of bounded-below injective complexes), we build inductively a commutative ladder

$$ \cdots \longrightarrow \tau_{\geq n} F \xrightarrow{\ \alpha_n\ } \tau_{\geq n+1} F \longrightarrow \cdots \longrightarrow \tau_{\geq N} F $$
$$ \beta_n \downarrow \qquad\qquad \downarrow \beta_{n+1} \qquad\qquad\qquad\qquad \downarrow $$
$$ \cdots \longrightarrow I_n \xrightarrow[\ \gamma_n\ ]{} I_{n+1} \longrightarrow \cdots \longrightarrow I_N $$

where for $-\infty < n < N$, α_n is the natural map, β_n is a quasi-isomorphism, I_{n+1} is a bounded-below injective (hence, by (2.3.4), q-injective) complex,

---

[15] Quasi-compactness holds by [**GD**, p. 296, (6.1.12)], where (U_α) should be a *base* of the topology.

and γ_n is split-surjective in each degree. Then $I := \varprojlim I_n$ is q-injective [**Sp**, p. 130, 2.5]; and the natural map $\varprojlim \tau_{\geq n} F = F \to I$ is a quasi-isomorphism [**Sp**, p. 134, 3.13]. So we have an isomorphism $\mathbf{R} f_* F \xrightarrow{\sim} f_* I$.

It follows from (2.4.5.2) that $\mathbf{R} f_*$ is compatible with open immersions on Y, and hence if (3.9.2.1) holds whenever Y is quasi-compact (indeed, affine) then it holds always. Assuming Y to be quasi-compact, we argue further as in *loc. cit.* Since γ_n is split surjective in each degree m, its kernel C_n is a bounded-below injective complex, and for any affine open $U \subset Y$, γ_n induces a surjection $\Gamma(f^{-1}U, I_n^m) \twoheadrightarrow \Gamma(f^{-1}U, I_{n+1}^m)$ with kernel $\Gamma(f^{-1}U, C_n^m)$. The five-lemma yields that β_n induces a quasi-isomorphism to C_n from the kernel A_n of the surjection α_n; and in $\mathbf{D}(X)$, $A_n \cong H^n(F)[-n]$. Thus $C_n[n]$ is an injective resolution of $H^n(F)$, and so if d is the integer in (3.9.2.4) then for any $m > n + d$,

$$H^m\big(\Gamma(f^{-1}U, C_n)\big) \cong H^{m-n}\big(f^{-1}U, H^n(F)\big) \cong \Gamma\big(U, R^{m-n}f_* H^n(F)\big) = 0,$$

so that the sequence

$$\Gamma(f^{-1}U, C_n^{m-1}) \to \Gamma(f^{-1}U, C_n^m) \to \Gamma(f^{-1}U, C_n^{m+1}) \to \Gamma(f^{-1}U, C_n^{m+2})$$

is *exact*. A Mittag-Leffler-like diagram chase ([**Sp**, p. 126, Lemma], applied to the inverse system of diagrams

$$\Gamma(f^{-1}U, I_n^{m-1}) \to \Gamma(f^{-1}U, I_n^m) \to \Gamma(f^{-1}U, I_n^{m+1}) \to \Gamma(f^{-1}U, I_n^{m+2})$$

where n runs through $\mathbb{Z}$ and $I_n := I_N$ for all $n > N$) shows then that if $m \geq N + d$ then the natural map

$$H^m\big(\Gamma(U, f_*I)\big) = H^m\big(\varprojlim \Gamma(f^{-1}U, I_n)\big)$$

$$\to H^m\big(\Gamma(f^{-1}U, I_N)\big) = H^m\big(\Gamma(U, f_*I_N)\big)$$

is an isomorphism. Sheafifying on Y, we get that *for any* $m \geq N + d$, *the natural composition*

$$R^m f_* F = H^m(\mathbf{R} f_* F) \xrightarrow{\sim} H^m(f_*I) \longrightarrow H^m(f_*I_N) \xrightarrow{\sim} R^m f_*(\tau_{\geq N} F)$$

is an isomorphism. From (3.9.2.2) we conclude then that $R^m f_* F$ is quasi-coherent, which gives (3.9.2.1) (since N is arbitrary); and furthermore if $\tau_{\geq N} F \cong 0$, then $\tau_{\geq N+d} \mathbf{R} f_* F \cong 0$, proving (3.9.2.3). Q.E.D.

Lemma 3.9.2.5 *Let $\mathcal{A}$ be an abelian category, and let* $\mathbf{J}$ *be a full subcategory of the category* $\mathbf{C}$ *of $\mathcal{A}$-complexes such that* (1): *a complex B is in* $\mathbf{J}$ *iff $B[1]$ is, and* (2): *for any map f in* $\mathbf{J}$, *the cone C_f (§1.3) is in* $\mathbf{J}$.

(i) *Let $u\colon P \to C$ be a map in $\mathbf{C}$ with $P \in \mathbf{J}$ and such that there exists a quasi-isomorphism $h\colon Q \to C_u$ with $Q \in \mathbf{J}$. Then u factors as $P \xrightarrow{v} P_1 \xrightarrow{u_1} C$ where $P_1 \in \mathbf{J}$, u_1 is a quasi-isomorphism, and in each degree m, $v^m\colon P^m \to P_1^m$ is a split monomorphism, i.e., has a left inverse.*

(ii) *Let $s\colon C \to I$ be a map in $\mathbf{C}$ with $I \in \mathbf{J}$ and such that there exists a quasi-isomorphism $C_s \to J$ with $J \in \mathbf{J}$. Then s factors as $C \xrightarrow{s_1} I_1 \xrightarrow{t} I$ where $I_1 \in \mathbf{J}$, s_1 is a quasi-isomorphism, and in each degree m, $t^m\colon I_1^m \to I^m$ is a split epimorphism, i.e., has a right inverse.*

Proof. (i) We have a diagram in $\mathbf{C}$

$$
\begin{array}{ccccccc}
P & \xrightarrow{\ v\ } & C_{wh}[-1] & \longrightarrow & Q & \xrightarrow{\ wh\ } & P[1] \\
\| & & \downarrow{g} & & \downarrow{h} & & \| \\
P & \longrightarrow & C_w[-1] & \underset{w}{\longrightarrow} & C_u & \longrightarrow & P[1] \\
\| & & \downarrow{\varphi} & \textcircled{1} & \| & & \| \\
P & \underset{u}{\longrightarrow} & C & \longrightarrow & C_u & \underset{w}{\longrightarrow} & P[1]
\end{array}
$$

where the bottom row is the standard triangle associated to u, the top two rows are made up of natural maps, φ is as in (1.4.3.1), and g is given in degree m by the map

$$
g^m = 1 \oplus h^m\colon C_{wh}[-1]^m = P^m \oplus Q^m \to P^m \oplus C_u^m = C_w[-1]^m .
$$

Here all the subdiagrams other than $\textcircled{1}$ commute, and $\textcircled{1}$ is homotopy-commutative (see (1.4.3.1)). By ($\Delta 2$) in §1.4, the rows of the diagram become triangles in $\mathbf{K}(\mathcal{A})$. Since h is a quasi-isomorphism, we see, using the exact homology sequences $(1.4.5)^{\mathrm{H}}$ of these triangles, that the composed map $\varphi \circ g$ is also a quasi-isomorphism. Since P and Q are in $\mathbf{J}$, so is $C_{wh}[-1]$. Thus we can take $P_1 := C_{wh}[-1]$ and $u_1 := \varphi \circ g$.

(ii) A proof resembling that of (i) (with arrows reversed) is left to the reader. See also the following exercise (a), or [**Sp**, p. 132, proof of 3.3].

Q.E.D.

Exercises 3.9.2.6 (a) Convince yourself that (i) and (ii) in (3.9.2.5) are *dual*, i.e., (ii) is essentially the statement about $\mathcal{A}$ obtained by replacing $\mathcal{A}$ in (i) by its opposite category $\mathcal{A}^{\mathrm{op}}$.

(b) (Cf. (1.11.2)(iv).) Let X be a scheme and let $\mathcal{A}_X$ (resp. $\mathcal{A}_X^{\mathrm{qc}}$) be the category of all $\mathcal{O}_X$-modules (resp. quasi-coherent $\mathcal{O}_X$-modules). Let $\phi\colon \mathcal{A}_X \to \mathfrak{Ab}$ be an additive functor satisfying $\phi(\varprojlim I_n) = \varprojlim \phi(I_n)$ for any inverse system $(I_n)_{n<0}$ of $\mathcal{A}_X$-injectives in which all the maps $I_n \to I_{n+1}$ are *split surjective*. Then

$$
\dim^+\!\big(\mathbf{R}\phi|_{\mathbf{D}_{\mathrm{qc}}(X)}\big) = \dim^+\!\big(\mathbf{R}\phi|_{\mathcal{A}_X^{\mathrm{qc}}}\big) .
$$

(c) Show: for any proper map $f\colon X \to Y$ of noetherian schemes, $\mathbf{R}f_*\mathbf{D}_{\mathrm{c}}(X) \subset \mathbf{D}_{\mathrm{c}}(Y)$.
<u>Hint.</u> (3.9.2), [**H**, p. 74, (iii)], [**EGA**, III, (3.2.1)].

(3.9.3). Henceforth, index sets A for inductive systems are assumed to be (small and) filtered: $\alpha, \beta \in A \Rightarrow \exists \gamma \in A$ with $\gamma \geq \alpha$ and $\gamma \geq \beta$. (More generally, the results will be valid for limits over filtered—or even pseudo-filtered—categories [**GV**, pp. 14–15], [**M**, p. 211].)

Lemma 3.9.3.1 *Let* $f \colon X \to Y$ *be a concentrated scheme-map. Fix* $n \in \mathbb{Z}$, *let* $(C_\alpha, \varphi_{\beta\alpha})_{\alpha,\beta \in A}$ *be an inductive system of* $\mathcal{O}_X$*-complexes all of whose homology vanishes in degree* $< n$, *and set* $C := \varinjlim_\alpha C_\alpha$. *Then we have natural isomorphisms*

$$\varinjlim_\alpha \mathbf{R}^i f_*(C_\alpha) \xrightarrow{\sim} \mathbf{R}^i f_*(C) \qquad (\mathbf{R}^i f_* := H^i \mathbf{R} f_*, \ i \in \mathbb{Z}).$$

Proof.[16]   In the category of bounded-below $\mathcal{O}_X$-complexes D, we can choose flasque resolutions $D \to F$ *functorially,* as follows: for each $q \in \mathbb{Z}$, let $0 \to D^q \to F^{0q} \to F^{1q} \to F^{2q} \to \dots$ be the (flasque) Godement resolution of D^q [**G**, p. 167, 4.3], set $F^{pq} := 0$ if $p < 0$, and let F be the complex coming from the double complex F^{pq}, i.e., $F^m := \oplus_{p+q=m} F^{pq}$, etc; then F^m is flasque, and diagram chasing, or a simple spectral sequence argument, shows that the family of natural maps $D^m \to F^{0m} \subset F^m$ gives a quasi-isomorphism $g_D \colon D \to F$. We will refer to this g_D (or simply F) as the *Godement resolution* of D.

With C_α and n as above, the truncation operator $\tau_{\geq n}$ as in §1.10, and F_α the Godement resolution of $\tau_{\geq n} C_\alpha$, we have an inductive system of quasi-isomorphisms $C_\alpha \to \tau_{\geq n} C_\alpha \to F_\alpha$, and hence a quasi-isomorphism $C \to F := \varinjlim F_\alpha$. Each F_α is flasque, hence f_*-acyclic (2.7.3). By [**Kf**, p. 641, Cor. 5 and 7], F is a complex of f_*-acyclic sheaves, and so, being bounded below, F itself is f_*-acyclic, see (2.7.4) (dualized). The last assertion in (2.2.6) shows then that the (obvious) map in (3.9.3.1) is isomorphic to the natural map

$$\varinjlim H^i(f_* F_\alpha) = H^i(\varinjlim f_* F_\alpha) \to H^i(f_* \varinjlim F_\alpha) = H^i(f_* F),$$

which is an isomorphism since f_* commutes with $\varinjlim$ [**Kf**, p. 641, Prop. 6].
                                                                          Q.E.D.

Corollary 3.9.3.2 *Let* $f \colon X \to Y$ *be a concentrated scheme-map. With notation as in* §1.9, *let* $\mathcal{A}^{\#}$ *be a plump subcategory of the category* $\mathcal{A}_X$ *of* $\mathcal{O}_X$*-modules, such that any* $\varinjlim$ *of objects in* $\mathcal{A}^{\#}$ *is itself in* $\mathcal{A}^{\#}$ *and such that the restriction of* $\mathbf{R}f_*$ *to* $\mathbf{D}_{\#}(X)$ *is bounded above* (§1.11). *Let* $(C_\alpha, \varphi_{\beta\alpha})_{\alpha,\beta \in A}$ *be an inductive system of complexes all of whose homology lies in* $\mathcal{A}^{\#}$, *and set* $C := \varinjlim_\alpha C_\alpha$. *Then we have natural isomorphisms*

$$\varinjlim_\alpha \mathbf{R}^i f_*(C_\alpha) \xrightarrow{\sim} \mathbf{R}^i f_*(C) \qquad (\mathbf{R}^i f_* := H^i \mathbf{R} f_*, \ i \in \mathbb{Z}).$$

---

[16] Cf. [**EGA**, III, Chap. 0, p. 36, (11.5.1)].

Remarks. (a) If the map f is *finite-dimensional* (2.7.6), (e.g., if X is noetherian, of finite Krull dimension (2.7.6.2)), then all the hypotheses in (3.9.3.2) are satisfied when $\mathcal{A}^{\#} = \mathcal{A}_X$.

(b) By (3.9.2.3), if Y is quasi-compact then all the hypotheses in (3.9.3.2) are satisfied when $\mathcal{A}^{\#} = \mathcal{A}^{qc}$, the category of quasi-coherent $\mathcal{O}_X$-modules. Even if Y is not quasi-compact, the conclusion of (3.9.3.2) still holds, because $\mathbf{R}f_*$ and $\varinjlim$ "commute" with open immersions on Y (see (2.4.5.2)), so it suffices to check over affine open subsets of Y.

Proof of (3.9.3.2). By (1.11.2)(ii) we have natural isomorphisms

$$\mathbf{R}^i f_*(D) \xrightarrow{\sim} \mathbf{R}^i f_*(\tau_{\geq i-d} D) \qquad \left(D \in \mathbf{D}_{\#}(X), \ d := \dim^+(\mathbf{R}f_*|_{\mathbf{D}_{\#}(X)}) \right).$$

Note that $C \in \mathbf{D}_{\#}(X)$ since homology commutes with $\varinjlim$; and clearly $\tau_{\geq i-d} C = \varinjlim \tau_{\geq i-d} C_\alpha$. Fixing i, we conclude by applying (3.9.3.1) to the inductive system $\tau_{\geq i-d} C_\alpha$.   Q.E.D.

Corollary 3.9.3.3 *Let $(C_\beta)_{\beta \in B}$ be a small family of complexes in $\mathbf{D}_{\geq \mathbf{n}}$ (n fixed, see (1.10)) or in $\mathbf{D}_{\#}(X)$ ($\mathcal{A}^{\#}$ as in (3.9.3.2)). Then the natural map $\oplus_\beta \mathbf{R}f_* C_\beta \to \mathbf{R}f_*(\oplus_\beta C_\beta)$ (see (3.8.1)) is an isomorphism.*

Proof. We need only check that the induced homology maps are isomorphisms, which follows from (3.9.3.1) or (3.9.3.2), a direct sum over B being a $\varinjlim$ of the family of direct sums over finite subsets of B.   Q.E.D.

Corollary 3.9.3.4 *Under the hypotheses of (3.9.3.1) or (3.9.3.2), if each C_α is f_*-acyclic then so is C.*

Proof. The assertion is that the natural map $f_* C \to \mathbf{R}f_* C$ is an isomorphism in $\mathbf{D}(Y)$, i.e., that the induced maps $H^i(f_* C) \to H^i(\mathbf{R}f_* C)$ are all isomorphisms. By assumption, this holds with C_α in place of C; and since H^i and f_* commute with $\varinjlim$ [**Kf**, p. 641, Prop. 6], it also holds, by (3.9.3.1) or (3.9.3.2), for C.   Q.E.D.

Corollary 3.9.3.5 *With $\mathcal{A}^{\#}$ as in (3.9.3.2), any complex C of f_*-acyclic $\mathcal{A}^{\#}$-objects is itself f_*-acyclic.*

Proof. The complexes $\cdots \to 0 \to 0 \to C^{-n} \to C^{-n+1} \to \cdots$ ($n \in \mathbb{Z}$) form an inductive system of f_*-acyclic complexes (see (2.7.2), dualized), whose $\varinjlim$ is C. Conclude by (3.9.3.4).   Q.E.D.

Proposition 3.9.4 *Let $f : X \to Y$ be a concentrated scheme-map, and let $F \in \mathbf{D}(X)$, $G \in \mathbf{D}_{qc}(Y)$. If f is finite-dimensional (2.7.6), or if $F \in \mathbf{D}_{qc}(X)$, then the projection maps*

$$p_1 : (\mathbf{R}f_* F) \underset{=}{\otimes} G \to \mathbf{R}f_*(F \underset{=}{\otimes} \mathbf{L}f^* G), \qquad p_2 : G \underset{=}{\otimes} \mathbf{R}f_* F \to \mathbf{R}f_*(\mathbf{L}f^* G \underset{=}{\otimes} F)$$

(see (3.4.6)) are isomorphisms.

Proof. We treat only p_1 (p_2 can be handled similarly; or (3.4.6.1) can be applied). The question is local on Y (check directly, or see (3.7.3.1)), so we may assume Y affine.

Suppose first that both F and G are bounded-above complexes. Then the source and target of p_1 are, for fixed F, bounded-above functors of G: this is clear when f is finite-dimensional, and if $F \in \mathbf{D}_{\mathsf{qc}}(X)$ then it follows from (3.9.2.3) since $F \underset{=}{\otimes} \mathbf{L}f^*G \in \mathbf{D}_{\mathsf{qc}}(X)$, see (3.9.1) and (2.5.8). By (1.11.3.1), with $\mathcal{A}^{\#}$ the category of quasi-coherent $\mathcal{O}_Y$-modules on the affine scheme Y, we reduce the question to where G is a single *free* $\mathcal{O}_Y$-module G^0, whence $\mathbf{L}f^*G$ is isomorphic to the free $\mathcal{O}_X$-module f^*G^0. After verifying via (3.8.2) and (3.9.3.3) that everything in sight commutes with direct sums, we have a further reduction to the case $G = \mathcal{O}_Y$. We check then, via (3.2.5)(a) and commutativity of the upper diagrams in (3.4.2.2), that p_1 is isomorphic to the identity map of $\mathbf{R}f_*F$.

Next, drop the assumption that F is bounded above. For any integer i and any triangle in $\mathbf{D}(X)$ based on the natural map $F \to \tau_{\geq i}F$, the vertex C_i (depending, up to isomorphism, only on F) lies in $\mathbf{D}_{<i}(X)$, see §§1.4, 1.10. We still assume that $G \in \mathbf{D}_{\leq e}(Y)$ for some e, so that $C_i \underset{=}{\otimes} \mathbf{L}f^*G \in \mathbf{D}_{<i+e}(X)$ (as one sees upon replacing C_i and G, via (1.8.1)$^-$, by quasi-isomorphic flat complexes vanishing in degrees above $i - 1$ and e respectively). As above, $C \in \mathbf{D}_{\mathsf{qc}}(X) \Rightarrow C \underset{=}{\otimes} \mathbf{L}f^*G \in \mathbf{D}_{\mathsf{qc}}(X)$. The finite dimensionality of $\mathbf{R}f_*|_{\mathbf{D}_{\mathsf{qc}}(X)}$ (3.9.2.3), or of $\mathbf{R}f_*$ itself when f is finite-dimensional, then gives $\mathbf{R}f_*(C_i \underset{=}{\otimes} \mathbf{L}f^*G) \in \mathbf{D}_{<i+e+d}(Y)$ for some integer d depending only on f, and so from the homology sequence (1.4.5)^{H} of the triangle

$$\mathbf{R}f_*(F \underset{=}{\otimes} \mathbf{L}f^*G) \to \mathbf{R}f_*(\tau_{\geq i}F \underset{=}{\otimes} \mathbf{L}f^*G) \to \mathbf{R}f_*(C_i \underset{=}{\otimes} \mathbf{L}f^*G) \to \mathbf{R}f_*(F \underset{=}{\otimes} \mathbf{L}f^*G)[1]$$

we get isomorphisms

$$H^j\big(\mathbf{R}f_*(F \underset{=}{\otimes} \mathbf{L}f^*G)\big) \xrightarrow{\sim} H^j\big(\mathbf{R}f_*(\tau_{\geq i}F \underset{=}{\otimes} \mathbf{L}f^*G)\big)$$

for all $j > i + e + d$. Similarly, we have natural isomorphisms

$$H^j\big(\mathbf{R}f_*F \underset{=}{\otimes} G\big) \xrightarrow{\sim} H^j\big(\mathbf{R}f_*\tau_{\geq i}F \underset{=}{\otimes} G\big).$$

Therefore, to show for any given j that the homology map $H^j(p_1)$ is an isomorphism—which suffices, by (1.2.2)—we can replace F by $\tau_{\geq j-1-e-d}F$. Thus we may assume that F is bounded below. Also, as above, we may assume that G is flat, whence so is $f^*G \cong \mathbf{L}f^*G$.

Let F_m be the Godement resolution of $\tau_{\leq m}F$ ($m \in \mathbb{Z}$), see proof of (3.9.3.1), so that the canonical map

$$F = \varinjlim_m \tau_{\leq m}F \to \varinjlim_m F_m$$

is the Godement resolution of F.

By the first part of this proof, there is a natural isomorphism

$$H^j\big(f_* F_m \otimes G\big) \cong H^j\big(\mathbf{R}f_* \tau_{\leq m} F \otimes G\big)$$

$$\xrightarrow{\sim}\; H^j\big(\mathbf{R}f_*(\tau_{\leq m} F \underset{=}{\otimes} \mathbf{L}f^*G)\big) \cong H^j\big(\mathbf{R}f_*(F_m \otimes f^*G)\big).$$

As before, if $F \in \mathbf{D}_{\mathsf{qc}}(X)$ then $(F_m \otimes f^*G) \cong (\tau_{\leq m} F \underset{=}{\otimes} \mathbf{L}f^*G) \in \mathbf{D}_{\mathsf{qc}}(X)$. Using (3.9.3.2) and—as in the proof of (3.9.3.1)—the commutativity of $\varinjlim$ with f_*, $\otimes$, and H^j, we find then that $H^j(p_1)$ factors as the composition of the natural isomorphisms

$$H^j\big(\mathbf{R}f_* F \underset{=}{\otimes} G\big) \xrightarrow{\sim} H^j\big(f_* \varinjlim F_m \otimes G\big)$$

$$\xrightarrow{\sim}\; \varinjlim H^j\big(f_* F_m \otimes G\big)$$

$$\xrightarrow{\sim}\; \varinjlim H^j\big(\mathbf{R}f_*(F_m \otimes f^*G)\big)$$

$$\xrightarrow{\sim}\; H^j\big(\mathbf{R}f_* \varinjlim (F_m \otimes f^*G)\big) \xrightarrow{\sim} H^j\big(\mathbf{R}f_*(F \underset{=}{\otimes} \mathbf{L}f^*G)\big),$$

proving (3.9.3) whenever G is bounded above.

Finally, to extend the assertion to any $G \in \mathbf{D}_{\mathsf{qc}}(Y)$, use a quasi-isomorphism $Q \to G$ where $Q = \varinjlim Q_m$ with $Q_m \in \mathbf{D}^-_{\mathsf{qc}}(Y)$ bounded-above and flat, so that $\mathbf{L}f^*G \cong f^*Q$, see proof of (3.9.1). As in (3.1.2), $\mathbf{R}f_*F = f_* I_F$; and, again, if $F \in \mathbf{D}_{\mathsf{qc}}(X)$ then $I_F \otimes f^*Q_m \in \mathbf{D}_{\mathsf{qc}}(X)$. Applying $\varinjlim_m$ to the system of natural maps

$$H^j\big(f_* I_F \otimes Q_m\big) \cong H^j\big(\mathbf{R}f_* F \underset{=}{\otimes} Q_m\big)$$

$$\longrightarrow H^j\big(\mathbf{R}f_*(F \underset{=}{\otimes} \mathbf{L}f^*Q_m)\big) \cong H^j\big(\mathbf{R}f_*(I_F \otimes f^*Q_m)\big),$$

maps which we have already seen to be isomorphisms, we find, via (3.9.3.2) and commutativity of $\varinjlim$ with H^j, with $\otimes$, and with f^*, that the maps

$$H^j(p_1)\colon H^j\big(\mathbf{R}f_* F \underset{=}{\otimes} Q\big) \longrightarrow H^j\big(\mathbf{R}f_*(F \underset{=}{\otimes} \mathbf{L}f^*Q)\big) \qquad (j \in \mathbb{Z})$$

are all isomorphisms, whence the conclusion.   Q.E.D.

Remarks 3.9.4.1 The projection map p_1 need not be an isomorphism for non-quasi-coherent $\mathcal{O}_Y$-modules G. For example, let R be a two-dimensional noetherian local ring with maximal ideal $\mathfrak{m}$, $Y = \mathrm{Spec}(R)$, $X = \mathrm{Spec}(R) - \{\mathfrak{m}\}$, $f\colon X \to Y$ the inclusion, $F = \mathcal{O}_Y$, and $G = \mathcal{O}_X$ extended by zero (so that G is a *flat* $\mathcal{O}_Y$-module). Then the stalk of $R^1 f_*(F) \otimes G$ at $\mathfrak{m}$ is 0, whereas the stalk of $R^1 f_*(F \otimes f^*G) = R^1 f_*(\mathcal{O}_X)$ is $H^1(X, \mathcal{O}_X) = H^2_{\mathfrak{m}}(R) \neq 0$ (where $H_{\mathfrak{m}}$ denotes local cohomology supported at $\mathfrak{m}$).

Exercises 3.9.4.2 Let X be a ringed space.

(a) Show that an $\mathcal{O}_X$-module F is flat iff $\mathcal{T}or_i(F,G):=H^{-i}(F\otimes G)=0$ for all $\mathcal{O}_X$-modules G and all $i\neq 0$. (One need only consider $i=1$, see proof of (2.7.6.4).)

(b) [**I**, p. 131]. A complex F of $\mathcal{O}_X$-modules has *finite flat amplitude* (or *finite tor-dimension*) if for some integers $d_1\leq d_2$, $\mathcal{T}or_i(F,G)=0$ for all $\mathcal{O}_X$-modules G and all i outside the interval $[d_1,d_2]$. Show that this condition is equivalent to there being a $\mathbf{D}(X)$-isomorphism $F\xrightarrow{\sim}P$ with P flat and $P^i=0$ for all $i\notin[-d_2,-d_1]$. (See (2.7.6), with f the identity map of X.)

(c) [**I**, p. 249]. Suppose further in (3.9.4) that f has finite tor-dimension (2.7.6) and that F has finite flat amplitude (b). Show that then $\mathbf{R}f_*F$ also has finite flat amplitude.

(d) Show: if X is an affine scheme and if $F\in\mathbf{D}_{qc}(X)$ has finite flat amplitude, then the complex P in (b) may be assumed to be quasi-coherent. (Use (3.9.6) below.)

(e) *Let $f\colon X\to Y$ be a concentrated scheme-map. Let $F\in\overline{\mathbf{D}}^+(X)$ and let $G\in\mathbf{D}_{qc}(Y)$ have finite flat amplitude. Then the projection map p_1 in (3.9.4) is an isomorphism.*

Hint. We may assume Y to be affine. Induction on the number of non-zero terms of a bounded flat quasi-coherent complex $P\cong G$ (see (d)) reduces the question to where G is a single flat quasi-coherent $\mathcal{O}_Y$-module. Then by a theorem of Lazard [**GD**, p. 163, Prop. (6.6.24)], G is a direct limit of finite-rank free $\mathcal{O}_Y$-modules, and so (3.9.3.1) gives a reduction to the trivial case $G=\mathcal{O}_Y$.

(f) Let Y be a ringed space. Show that the following conditions on a complex G of $\mathcal{O}_Y$-modules are equivalent:

(i) For some $d\in\mathbb{Z}$, $\mathcal{T}or_i(F,G)=0$ for all $\mathcal{O}_Y$-modules F and all $i>d$.

(ii) The functor $E\mapsto E\otimes G$ ($E\in\mathbf{D}(Y)$ is bounded below (1.11.1).

(iii) In $\mathbf{D}(Y)$, $G\cong P$ with P bounded-below and q-flat.

(iv) In $\mathbf{D}(Y)$, $G\cong P$ with P bounded-below, *flat,* and q-flat.

When these conditions hold we say that G has *bounded-below flat amplitude.*

(g) Do exercise (e) assuming only that G has bounded-below flat amplitude.

Hint. Assuming G to be bounded-below, flat, and q-flat, show that it suffices to apply (e) to each of the complexes $\ldots\to G^{n-1}\to G^n\to 0\to 0\to\ldots$ ($n\in\mathbb{Z}$).

The following result will be generalized in (3.10.3).

Proposition 3.9.5 *Given a commutative square σ of scheme-maps*

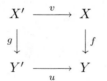

suppose that f is concentrated, that u is flat, and that σ is a fiber square (i.e., that the associated map $X'\to X\times_Y Y'$ is an isomorphism). Then for any $F\in\mathbf{D}_{qc}(X)$, the natural composed map (see (3.7.2)(a))

$$\theta_\sigma(F)\colon u^*\mathbf{R}f_*F\xrightarrow{\eta}u^*\mathbf{R}f_*\mathbf{R}v_*v^*F$$
$$\xrightarrow{\sim}u^*\mathbf{R}u_*\mathbf{R}g_*v^*F\xrightarrow{\epsilon}\mathbf{R}g_*v^*F$$

is an isomorphism.

Proof. It should be noted that since u, and hence v, is flat, we have functorial isomorphisms $\mathbf{L}u^* \xrightarrow{\sim} u^*$ and $\mathbf{L}v^* \xrightarrow{\sim} v^*$. (This follows from (2.2.6)(dualized), since the exactness of (e.g.) u^* implies at once that every $\mathcal{O}_X$-complex is u^*-acyclic.)

In view of (3.9.2.2) and (3.9.2.3), (1.11.3)(iv) allows us to assume that F is a single quasi-coherent $\mathcal{O}_X$-module. It will suffice then, by (1.2.2), to show that application of the homology functors H^n to $\theta_\sigma(F)$ produces (what else?) the "base change" isomorphisms $\alpha^n(F)$ of [**AHK**, p. 35, Theorem (6.7)].

For this purpose, we need to express θ_σ in terms of canonical flasque (Godement) resolutions—which we denote by $\mathcal{C}$. In [**AHK**, p. 28, §3] there is defined a map

$$\varphi\colon \mathcal{C}(F) \to v_* \mathcal{C}(v^* F)$$

(denoted there by $\theta_v^\bullet(F)$) which, as easily checked, makes the following natural diagram commute:

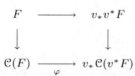

With the definitions of ϵ and η in §3.2, and the fact that the direct image of a flasque sheaf is still flasque, it is a straightforward exercise to verify that *the map $\theta_\sigma(F)$ is isomorphic to the derived category map given by the natural composition*

$$u^* f_* \mathcal{C}(F) \xrightarrow{\varphi} u^* f_* v_* \mathcal{C}(v^* F) \xrightarrow{\sim} u^* u_* g_* \mathcal{C}(v^* F) \longrightarrow g_* \mathcal{C}(v^* F).$$

Now applying H^n, and recalling that u is flat, we get a composed map

$$\alpha'^n\colon u^* H^n\big(f_* \mathcal{C}(F)\big) \xrightarrow{\varphi} u^* H^n\big(f_* v_* \mathcal{C}(v^* F)\big) \xrightarrow{\sim} u^* H^n\big(u_* g_* \mathcal{C}(v^* F)\big)$$
$$\xrightarrow{\gamma} H^n\big(g_* \mathcal{C}(v^* F)\big).$$

Let's look more closely at γ. Setting $g_* \mathcal{C}(v^* F) = E^\bullet$, let K^n be the kernel of the differential $E^n \to E^{n+1}$, and let $\delta\colon E^{n-1} \to K^n$ be the obvious map. Then γ can be identified with the map

$$\operatorname{coker}(u^* u_* \delta) = u^* \operatorname{coker}(u_* \delta) \to u^* u_* \operatorname{coker}(\delta) \to \operatorname{coker}(\delta)$$

which is adjoint to the natural map

$$\gamma'\colon H^n(u_* E^\bullet) = \operatorname{coker}(u_* \delta) \to u_* \operatorname{coker}(\delta) = u_* H^n(E^\bullet).$$

Note that $\operatorname{coker}(u_* \delta)$ is the sheaf associated to the presheaf

$$U \mapsto \operatorname{coker}\big(\delta(u^{-1} U)\big) = H^n\big(E^\bullet(u^{-1} U)\big) \qquad (U \text{ open in } Y)$$

and that γ' is the sheafification of the natural presheaf map

$$H^n\big(E^\bullet(u^{-1} U)\big) \to \Gamma\big(u^{-1} U, H^n(E^\bullet)\big).$$

It is then readily verified that the adjoint of α'^n, viz. the composed map

$$H^n\big(f_* \mathcal{C}(F)\big) \xrightarrow{\varphi} H^n\big(f_* v_* \mathcal{C}(v^* F)\big) \xrightarrow{\sim} H^n\big(u_* g_* \mathcal{C}(v^* F)\big)$$
$$\xrightarrow{\gamma'} u_* H^n\big(g_* \mathcal{C}(v^* F)\big),$$

is the map $\beta^n(f, g, u, v, F)$ near the top of p. 34 of [**AHK**]. But by definition the adjoint of this β^n is $\alpha^n(F)$; thus $\alpha'^n = \alpha^n(F)$, and we are done.   Q.E.D.

Here are two important results about quasi-coherence on quasi-compact *separated* schemes. Proofs can be found in the indicated references.

Proposition 3.9.6 *Let X be a quasi-compact separated scheme and $\mathcal{A}_X^{qc}$ the category of quasi-coherent $\mathcal{O}_X$-modules. Then:*

(a) [**BN**, p. 230, Corollary 5.5.] *The natural functor* $\mathbf{D}(\mathcal{A}_X^{qc}) \to \mathbf{D}_{qc}(X)$ *is an equivalence of categories.*

(b) [**AJL**, p. 10, Proposition 1.1.] *Every complex in* $\mathbf{D}_{qc}(X)$ *is* $\mathbf{D}(X)$*-isomorphic to a* quasi-coherent *q-flat complex.*

3.10 Independent Squares; Künneth Isomorphism

Throughout this section, $(^*, {}_*)$ will be the adjoint monoidal pair in (3.6.10), but with $\mathbf{S}$ restricted to be the category of quasi-separated schemes and concentrated (= quasi-compact and quasi-separated) maps between them [**GD**, p. 291, (6.1.5) and p. 294, (6.1.9)], and with the further restriction $\mathbf{X}_* = \mathbf{X}^* = \mathbf{D}_{qc}(X)$ for all $X \in \mathbf{S}$ (see (3.9.1), (3.9.2)). Note that any subscheme of a quasi-separated scheme is quasi-separated; and that the category $\mathbf{S}$ is closed under fiber product. Note also that if X and Y are quasi-separated then *any* scheme-map $f : X \to Y$ is quasi-separated, and further, quasi-compact if X is [**GD**, p. 295, (6.1.10)].

Accordingly (*except* in (3.10.1) and the proof of (3.10.2.2), where we need to distinguish between ordinary and derived functors), for any scheme-map α we write α_* for $\mathbf{R}\alpha_*$, and α^* for $\mathbf{L}\alpha^*$. We also write $\otimes$ for $\underset{=}{\otimes}$. These abbreviations should not be allowed to obscure the fact that we are working throughout with derived categories and derived functors.

After discussing some basic maps we define, in (3.10.2), various notions of *independence* of commutative $\mathbf{S}$-squares. The main result, (3.10.3), is that *all these independence conditions are equivalent.*[17] This implies, e.g., that the isomorphism in (3.9.5) holds for any tor-independent $\mathbf{S}$-square, as does a certain Künneth isomorphism, which subsumes the projection isomorphisms of (3.9.4).

Independent squares are important in Grothendieck duality theory, where they support base-change maps (Remark (3.10.2.1)(c)).

An *orientation* of a commutative $\mathbf{S}$-square σ

$$\begin{array}{ccc} X' & \xrightarrow{\ v\ } & X \\ {\scriptstyle g}\downarrow & \sigma & \downarrow{\scriptstyle f} \\ Y' & \xrightarrow[\ u\ ]{} & Y \end{array}$$

is an ordering of the pair (u, f).

---

[17] The knowledgeable reader might wish to place this result in the context of the *Künneth spectral sequences* of [**EGA**, III, (6.7.5)].

In this section, unless otherwise indicated, all commutative **S**-squares will be understood to be equipped with the orientation for which the bottom arrow precedes the right vertical one.

To such an oriented σ associate the functorial maps

$$\theta = \theta_\sigma : u^* f_* \to g_* v^* \quad \text{(see Proposition (3.7.2))}$$

and

$$\theta' = \theta'_\sigma := \theta_{\sigma'} : f^* u_* \to v_* g^*$$

where σ' is σ with its orientation reversed.

Setting $h := fv = ug$, define the functorial *Künneth map*

$$\eta = \eta_\sigma : u_* E \otimes f_* F \to h_*(g^* E \otimes v^* F) \qquad \left(E \in \mathbf{Y}'^*,\ F \in \mathbf{X}^*\right)$$

to be the natural composition

$$u_* E \otimes f_* F \to h_* h^* (u_* E \otimes f_* F)$$

$$\xrightarrow[\;(3.6.1)^*\;]{(3.4.5.1)} h_*(g^* u^* u_* E \otimes v^* f^* f_* F) \to h_*(g^* E \otimes v^* F).$$

The map η generalizes (3.4.2.1): let $X' = Y' = X$, let $v = g$ be the identity map, let $u = f$, so that $h = f$, and see (3.4.5.2) and 1) in (3.6.5).

The map η also generalizes the projection maps p_1 and p_2 in (3.4.6): for p_1, let f be the identity map of $X = Y$, let g be the identity map of $X' = Y'$, so that $h = v = u$, and see (3.4.6.2); and similarly for p_2 let u and v be identity maps, ...

Example 3.10.1 Let us see what the above θ_σ and η_σ look like in a concrete situation, when σ is a diagram of affine schemes. The results are hardly surprising, but do need proof.

(a) We deal first with θ. On **S** there is a second adjoint pair $(*, {}_\star)$ such that for each ringed space X, $\mathbf{X}^\star = \mathbf{X}_\star := \mathbf{K}(X)$, the homotopy category of $\mathcal{O}_X$-complexes, with monoidal structure given by the ordinary tensor product, and such that for each **S**-map $f : X \to Y$ the associated adjoint functors are the standard (sheaf-theoretic) inverse- and direct-image functors, $f^* := f^*$ and $f_\star := f_*$. So, as above, for each commutative **S**-square σ one gets functorial maps

$$\boldsymbol{\theta} = \boldsymbol{\theta}_\sigma : \mathsf{L} u^* \mathsf{R} f_* \to \mathsf{R} g_* \mathsf{L} v^*,$$

$$\theta = \theta_\sigma : u^* f_* \to g_* v^*,$$

(3.10.1.0)

related as follows.

Lemma 3.10.1.1 *With $Q\colon \mathbf{K} \to \mathbf{D}$ as usual, the following natural diagram of functors from $\mathbf{K}(X)$ to $\mathbf{D}(Y')$ commutes.*

$$
\begin{array}{ccccc}
\mathbf{L}u^*\mathbf{R}f_*Q & \xleftarrow{\ \alpha\ } & \mathbf{L}u^*Qf_* & \longrightarrow & Qu^*f_* \\[2pt]
\theta \downarrow & & & & \downarrow Q\theta \\[2pt]
\mathbf{R}g_*\mathbf{L}v^*Q & \xrightarrow[\ \beta\ ]{} & \mathbf{R}g_*Qv^* & \xleftarrow[\ \gamma\ ]{} & Qg_*v^*
\end{array}
$$

Proof. Expand the diagram (all maps being the obvious ones):

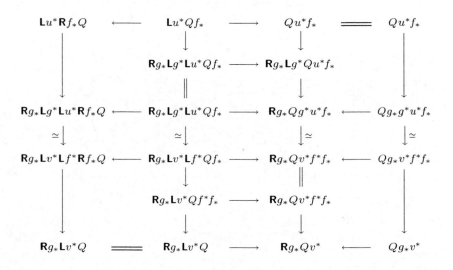

The upper right (resp. lower left) subdiagram commutes by (3.2.1.3) (resp. (3.2.1.2)). Commutativity of the rest is easy to verify.   Q.E.D.

Next, we make the map θ in (3.10.1.0) more explicit, at least locally.

Lemma 3.10.1.2 *Let*

be a commutative diagram of commutative-ring homomorphisms, let σ as above be the corresponding diagram of affine schemes ($Y := \mathrm{Spec}(R)$, etc.), and let $\theta = \theta_\sigma\colon u^*f_* \to g_*v^*$ be as in (3.10.1.0). For any S-complex E, let $\theta_0(E)$ be the natural composition $U \otimes_R E \to V \otimes_R E \to V \otimes_S E$, i.e., the U-homomorphism taking $1 \otimes_R e$ to $1 \otimes_S e$ for all $e \in E^n$ ($n \in \mathbb{Z}$).

Then there is a natural commutative diagram of $\mathcal{O}_{Y'}$-modules

$$
\begin{array}{ccc}
u^*f_*\widetilde{E} & \xrightarrow{\ \sim\ } & (U \otimes_R E)^{\sim} \\
{\scriptstyle \theta(\widetilde{E})}\Big\downarrow & & \Big\downarrow{\scriptstyle \widetilde{\theta_0(E)}} \\
g_*v^*\widetilde{E} & \xrightarrow{\ \sim\ } & (V \otimes_S E)^{\sim}
\end{array}
$$

where $^{\sim}$ denotes the usual functor from modules to quasi-coherent sheaves [**GD**, p. 197ff, §1.3], *and where the horizontal arrows are isomorphisms.*

Proof. The horizontal isomorphisms come from [**GD**, p. 213, (1.7.7)]. To check commutativity, expand the diagram as follows, where in the right hand column, the complexes to which $^{\sim}$ is applied are all regarded as U-complexes, and the maps are sheafifications of natural U-complex homomorphisms:

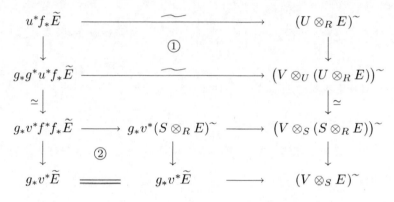

Commutativity of subdiagrams ① and ② is given by [**GD**, p. 214, (1.7.9)]. The rest is straightforward.   Q.E.D.

Under the hypotheses of (3.10.1.2), for any $G \in \mathbf{D}_{\mathsf{qc}}(X)$ the map $\boldsymbol{\theta}(G)\colon \mathbf{L}u^*\mathbf{R}f_*G \to \mathbf{R}g_*\mathbf{L}v^*G$ can now be described as follows.

By (3.9.6)(a), G is **D**-isomorphic to a quasi-coherent complex, which is $\widetilde{E}$ for some S-complex E. Arguing as in (2.5.5) (using that any S-module F is naturally a homomorphic image of the free S-module $P_0(F)$ with basis F) one sees that there exists a quasi-isomorphism $P \to E$ with P a $\varinjlim$ of bounded-above complexes of free S-modules. There results a quasi-isomorphism $\widetilde{P} \to \widetilde{E}$; and $\widetilde{P}$, being a $\varinjlim$ of bounded-above complexes of free $\mathcal{O}_X$-modules, is q-flat, as is $v^*\widetilde{P}$. One can replace $\widetilde{E}$ by $\widetilde{P}$, i.e., one may assume that there exists a **D**-isomorphism $\lambda\colon G \xrightarrow{\sim} \widetilde{E}$ such that both $\widetilde{E}$ and $v^*\widetilde{E}$ are q-flat as well as quasi-coherent.

Since f_* is an exact functor on the category of quasi-coherent $\mathcal{O}_X$-modules [**GD**, p. 214, (1.7.8)], therefore the natural map $f_*\widetilde{E} \to \mathbf{R}f_*\widetilde{E}$ is a $\mathbf{D}(Y)$-isomorphism. Also, the natural map $\mathbf{L}v^*\widetilde{E} \to v^*\widetilde{E}$ is a $\mathbf{D}(Y')$-isomorphism.

So the maps $\alpha(\widetilde{E})$ and $\beta(\widetilde{E})$ in (3.10.1.1) are isomorphisms. Moreover, the map $\theta(\widetilde{E})$ can be identified as in (3.10.1.2) with $\widetilde{\theta_0(E)}$. The map $\boldsymbol{\theta}(\widetilde{E})$ is thereby determined by (3.10.1.1) and (3.10.1.2); and via λ (a "quasi-coherent q-flat resolution"), so is the map $\boldsymbol{\theta}(G)$.

(b) We turn now to η. With σ, $(\overset{*}{\ },\underset{*}{\ })$ and $(\overset{\star}{\ },\underset{\star}{\ })$ as in (a), and $h = fv = gu$, one has for $\mathcal{O}_{Y'}$-complexes E and $\mathcal{O}_X$-complexes F the functorial maps

$$\boldsymbol{\eta} = \boldsymbol{\eta}_\sigma(E,F)\colon \mathbf{R}u_*E \underset{=}{\otimes} \mathbf{R}f_*F \to \mathbf{R}h_*(\mathbf{L}g^*E \underset{=}{\otimes} \mathbf{L}v^*F),$$

$$\eta = \eta_\sigma(E,F)\colon u_*E \otimes f_*F \to h_*(g^*E \otimes v^*F),$$

related as follows.

Lemma 3.10.1.3 *For all E and F as above, the following natural bifunctorial diagram—where appropriate insertions of "Q" are left to the reader—commutes.*

$$
\begin{array}{ccccc}
\mathbf{R}u_*E \underset{=}{\otimes} \mathbf{R}f_*F & \xleftarrow{\ \alpha'\ } & u_*E \underset{=}{\otimes} f_*F & \longrightarrow & u_*E \otimes f_*F \\
\Big\downarrow{\boldsymbol{\eta}} & & & & \Big\downarrow{\eta} \\
\mathbf{R}h_*(\mathbf{L}g^*E \underset{=}{\otimes} \mathbf{L}v^*F) & \xrightarrow{\ \beta'\ } & \mathbf{R}h_*(g^*E \otimes v^*F) & \longleftarrow & h_*(g^*E \otimes v^*F)
\end{array}
$$

Proof. Paste the following two diagrams along their common edge:

$$
\begin{array}{ccc}
\mathbf{R}u_*E \underset{=}{\otimes} \mathbf{R}f_*F & \xleftarrow{\ \alpha'\ } & u_*E \underset{=}{\otimes} f_*F \\
\Big\downarrow & & \Big\downarrow \\
\mathbf{R}h_*\mathbf{L}h^*(\mathbf{R}u_*E \underset{=}{\otimes} \mathbf{R}f_*F) & \longleftarrow & \mathbf{R}h_*\mathbf{L}h^*(u_*E \underset{=}{\otimes} f_*F) \\
\simeq\Big\downarrow & & \Big\downarrow\simeq \\
\mathbf{R}h_*(\mathbf{L}h^*\mathbf{R}u_*E \underset{=}{\otimes} \mathbf{L}h^*\mathbf{R}f_*F) & \longleftarrow & \mathbf{R}h_*(\mathbf{L}h^*u_*E \underset{=}{\otimes} \mathbf{L}h^*f_*F) \\
\simeq\Big\downarrow & & \Big\downarrow\simeq \\
\mathbf{R}h_*(\mathbf{L}(g^*u^*)\mathbf{R}u_*E \underset{=}{\otimes} \mathbf{L}(v^*f^*)\mathbf{R}f_*F) & \longleftarrow & \mathbf{R}h_*(\mathbf{L}(g^*u^*)u_*E \underset{=}{\otimes} \mathbf{L}(v^*f^*)f_*F) \\
\simeq\Big\downarrow & & \Big\downarrow\simeq \\
\mathbf{R}h_*(\mathbf{L}g^*\mathbf{L}u^*\mathbf{R}u_*E \underset{=}{\otimes} \mathbf{L}v^*\mathbf{L}f^*\mathbf{R}f_*F) & \longleftarrow & \mathbf{R}h_*(\mathbf{L}g^*\mathbf{L}u^*u_*E \underset{=}{\otimes} \mathbf{L}v^*\mathbf{L}f^*f_*F) \\
\Big\downarrow & \textcircled{3} & \Big\downarrow \\
& & \mathbf{R}h_*(\mathbf{L}g^*u^*u_*E \underset{=}{\otimes} \mathbf{L}v^*f^*f_*F) \\
& & \Big\downarrow \\
\mathbf{R}h_*(\mathbf{L}g^*E \underset{=}{\otimes} \mathbf{L}v^*F) & =\!=\!= & \mathbf{R}h_*(\mathbf{L}g^*E \underset{=}{\otimes} \mathbf{L}v^*F)
\end{array}
$$

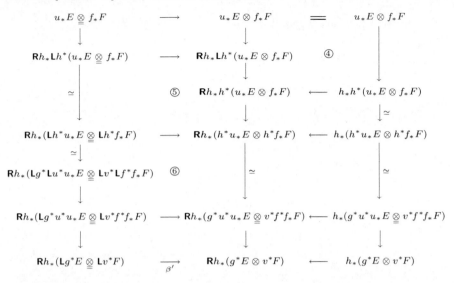

Commutativity of the unlabeled subdiagrams of the preceding diagrams is pretty clear.

Commutativity of subdiagram ③ follows from that of (3.2.1.2), of ④ from (3.2.1.3), of ⑤ from (3.2.4.1), and of ⑥ from the dual of the commutative diagram (3.6.4.1) (see the remarks surrounding (3.6.4)*).

Lemma (3.10.1.3) results. Q.E.D.

Lemma 3.10.1.4 *With notation as in (3.10.1.2), for any U-complex E and any S-complex F let $\eta = \eta_\sigma(\widetilde{E}, \widetilde{F})$ be as above, and let $\eta_0 = \eta_0(E, F)$ be the natural composition*

$$E \otimes_R F \to V \otimes_R (E \otimes_R F) \xrightarrow{\sim} (V \otimes_R E) \otimes_V (V \otimes_R F) \to (V \otimes_U E) \otimes_V (V \otimes_S F).$$

Then there is a natural commutative diagram of $\mathcal{O}_Y$-modules

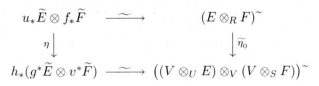

in which the horizontal arrows are isomorphisms.

Proof. The horizontal isomorphisms in the diagram are given by [**GD**, p. 213, (1.7.7) and p. 202, (1.3.12)(i)].

For commutativity, expand the diagram naturally as follows:

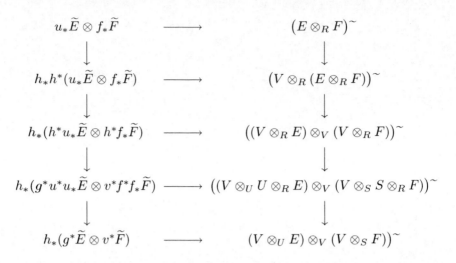

$$
\begin{array}{ccc}
u_*\widetilde{E} \otimes f_*\widetilde{F} & \longrightarrow & (E \otimes_R F)^\sim \\
\downarrow & & \downarrow \\
h_*h^*(u_*\widetilde{E} \otimes f_*\widetilde{F}) & \longrightarrow & \big(V \otimes_R (E \otimes_R F)\big)^\sim \\
\downarrow & & \downarrow \\
h_*(h^*u_*\widetilde{E} \otimes h^*f_*\widetilde{F}) & \longrightarrow & \big((V \otimes_R E) \otimes_V (V \otimes_R F)\big)^\sim \\
\downarrow & & \downarrow \\
h_*(g^*u^*u_*\widetilde{E} \otimes v^*f^*f_*\widetilde{F}) & \longrightarrow & \big((V \otimes_U U \otimes_R E) \otimes_V (V \otimes_S S \otimes_R F)\big)^\sim \\
\downarrow & & \downarrow \\
h_*(g^*\widetilde{E} \otimes v^*\widetilde{F}) & \longrightarrow & (V \otimes_U E) \otimes_V (V \otimes_S F)\big)^\sim
\end{array}
$$

Verification of commutativity of the subdiagrams is left as an exercise. (Suggestion: recall (3.1.9), and use [**GD**, p. 214, (1.7.9)(ii)].)   Q.E.D.

As in (a), Lemmas (3.10.1.3) and (3.10.1.4) determine (via quasi-coherent q-flat resolutions) the map $\boldsymbol{\eta}(G_1, G_2)$ for any $G_1 \in \mathbf{D_{qc}}(Y')$ and any $G_2 \in \mathbf{D_{qc}}(X)$, in terms of the concrete functorial map η_0.

Definition 3.10.2 A commutative oriented **S**-square

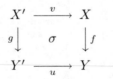

$$
\begin{array}{ccc}
X' & \xrightarrow{\ v\ } & X \\
{\scriptstyle g}\downarrow & \sigma & \downarrow{\scriptstyle f} \\
Y' & \xrightarrow[\ u\ ]{} & Y
\end{array}
$$

is said to be
  • *independent* if θ_σ is a functorial isomorphism;
  • *'-independent* if θ'_σ is a functorial isomorphism;
  • *Künneth-independent* if η_σ is a bifunctorial isomorphism;
  • *tor-independent* if σ is a fiber square (i.e., the map $X' \to X \times_Y Y'$ associated to σ is an isomorphism) and if the following equivalent conditions hold for all pairs of points $y' \in Y'$, $x \in X$ such that $y := u(y') = f(x)$:

(i) $\mathrm{Tor}_i^{\mathcal{O}_{Y,y}}(\mathcal{O}_{Y',y'}, \mathcal{O}_{X,x}) = 0$ for all $i > 0$.

(ii) There exist an affine open neighborhood $\mathrm{Spec}(A)$ of y and affine open sets $\mathrm{Spec}(A') \subset u^{-1}\mathrm{Spec}(A)$, $\mathrm{Spec}(B) \subset f^{-1}\mathrm{Spec}(A)$ such that

$$
\mathrm{Tor}_i^A(A', B) = 0 \quad \text{for all } i > 0.
$$

(ii)$'$ For any affine open neighborhood $\mathrm{Spec}(A)$ of y and affine open sets $\mathrm{Spec}(A') \subset u^{-1}\mathrm{Spec}(A)$, $\mathrm{Spec}(B) \subset f^{-1}\mathrm{Spec}(A)$,

$$\mathrm{Tor}_i^A(A', B) = 0 \quad \text{for all } i > 0.$$

Remarks 3.10.2.1 (a) The conditions of Künneth-independence and tor-independence do not depend on an orientation of σ.

(b) Condition (ii)$'$ in (3.10.2) implies condition (ii); and (ii) implies (i) because if $p \subset A$, $q \subset A'$, and $r \subset B$ are the prime ideals corresponding to y, y' and x respectively, then there are natural isomorphisms

$$\mathrm{Tor}_i^{A_p}(A'_q, B_r) \cong \mathrm{Tor}_i^A(A'_q, B_r) \cong A'_q \otimes_{A'} \mathrm{Tor}_i^A(A', B_r)$$
$$\cong A'_q \otimes_{A'} \mathrm{Tor}_i^A(A', B) \otimes_B B_r.$$

These isomorphisms also show that, conversely, (i) implies (ii)$'$: for if $m \subset A' \otimes_A B$ were a prime ideal in the support of $\mathrm{Tor}_i^A(A', B)$ and p, q, r were its inverse images in A, A' and B respectively, then $0 \neq \mathrm{Tor}_i^A(A', B)_m$ would be a localization of $\mathrm{Tor}_i^{A_p}(A'_q, B_r) = 0$.

(c) Let σ, as above, be an independent square; and suppose that the functors f_* and g_* have right adjoints $f^\times$ and $g^\times$ respectively. Then one can associate to σ a functorial *base-change map* (for $f^\times$ rather than f_*):

$$\beta_\sigma \colon v^* f^\times \to g^\times u^*,$$

adjoint to the natural composition $g_* v^* f^\times \xrightarrow{\theta^{-1}} u^* f_* f^\times \to u^*$.

This map plays a crucial role in Grothendieck duality theory on, say, the full subcategory of **S** whose objects are all the concentrated schemes, in which situation the right adjoints $f^\times$ and $g^\times$ exist, see (4.1.1) below.

(d) We call an **S**-map $f \colon X \to Y$ *isofaithful* if any **X***-map α such that $f_* \alpha$ is a **Y***-isomorphism is itself an isomorphism.

For example, if f is an open immersion then f is isofaithful because of the natural functorial isomorphism $G \xrightarrow{\sim} \mathbf{L}f^* \mathbf{R}f_* G$ $(G \in \mathbf{D}(Y))$.

Lemma 3.10.2.2 *If the* **S**-*map* $f \colon X \to Y$ *is affine* ([**GD**, p. 357, (9.1.10)]: *for each affine open* $U \subset Y$, $f^{-1}U$ *is affine) then* f *is isofaithful.*

Proof. In this proof only, $f_* \colon \mathbf{K}(X) \to \mathbf{K}(Y)$ will be the ordinary direct-image functor, and $\mathbf{R}f_* \colon \mathbf{D}(X) \to \mathbf{D}(Y)$ its derived functor.

From (2.4.5.2) it follows that $\mathbf{R}f_*$ "commutes" with open immersions, so the question is local, and we may assume that X and Y are affine, let us say $X = \mathrm{Spec}(B)$, $Y = \mathrm{Spec}(A)$.

By (3.9.6)(a), every complex in $\mathbf{D}_{\mathsf{qc}}(X)$ is **D**-isomorphic to a quasi-coherent complex. Therefore—and since a **D**-map α is an isomorphism iff the vertex of a triangle based on α is exact—we need only show: *if* C *is a quasi-coherent* $\mathcal{O}_X$-*complex such that* $\mathbf{R}f_*(C)$ *is exact then* C *is exact.*

Since the functor f_* of quasi-coherent $\mathcal{O}_X$-modules is exact, therefore, by (3.9.2.3) and the dual of (2.7.4), C is f_*-acyclic, so $f_*C \cong \mathbf{R}f_*C$ is exact, and for all i, $f_*H^iC \cong H^if_*C = 0$.

Finally, $C = \widetilde{E}$ for some B-complex E, so $H^iC = (H^iE)^\sim$, and when H^iE is regarded as an A-module, $f_*H^iC = (H^iE)^\sim$ (see [**GD**, p. 214, (1.7.7.2)]), whence $H^iE = 0$. The desired conclusion results.   Q.E.D.

The following assertions result at once from commutativity (to be shown) of diagram (3.10) below, for any $E \in \mathbf{Y}'^*$ and $F \in \mathbf{X}^*$.

• Independence or ′-independence of σ implies Künneth independence.

• If u (resp. f) is isofaithful then Künneth independence of σ implies independence (resp. ′-independence). (Take E (resp. F) to be $\mathcal{O}_{Y'}$ (resp. $\mathcal{O}_X$).) Thus:

• *If u and f are isofaithful then independence, ′-independence and Künneth independence are equivalent conditions on σ.*

This applies, for instance, if the schemes Y', Y and X are affine.

$$
\begin{array}{ccccc}
u_*(E \otimes u^*f_*F) & \xleftarrow[(3.9.4)]{\sim} & u_*E \otimes f_*F & \xrightarrow[(3.9.4)]{\sim} & f_*(f^*u_*E \otimes F) \\[4pt]
\text{\small via }\theta \downarrow & & \Big\downarrow & & \Big\downarrow \text{\small via }\theta' \\[4pt]
u_*(E \otimes g_*v^*F) & & \eta & & f_*(v_*g^*E \otimes F) \\[4pt]
\simeq \Big\downarrow {\scriptstyle (3.9.4)} & & & & {\scriptstyle (3.9.4)}\Big\downarrow \simeq \\[4pt]
u_*g_*(g^*E \otimes v^*F) & \xrightarrow[(3.6.4)_*]{\sim} & h_*(g^*E \otimes v^*F) & \xleftarrow[(3.6.4)_*]{\sim} & f_*v_*(g^*E \otimes v^*F)
\end{array}
$$
$$(3.10.2.3)$$

Proving commutativity of (3.10) is a formal exercise on adjoint monoidal pseudofunctors. For example, in view of the definition of $\theta_\sigma(F)$ in (3.7.2)(c), commutativity of the left half follows from that of the natural diagram

$$
\begin{array}{ccccc}
u_*E \otimes f_*F & \longrightarrow & u_*u^*(u_*E \otimes f_*F) & \longrightarrow & u_*g_*g^*u^*(u_*E \otimes f_*F) \\[4pt]
{\scriptstyle (3.9.4)}\Big\downarrow \simeq & \textcircled{1} & \Big\downarrow & & \Big\downarrow \\[4pt]
u_*(E \otimes u^*f_*F) & \longleftarrow & u_*(u^*u_*E \otimes u^*f_*F) & \longrightarrow & u_*g_*g^*(u^*u_*E \otimes u^*f_*F) \\[4pt]
\Big\downarrow & & \Big\downarrow & \textcircled{2} & \Big\downarrow \\[4pt]
u_*(E \otimes g_*g^*u^*f_*U) & \longleftarrow & u_*(u^*u_*E \otimes g_*g^*u^*f_*F) & \xrightarrow[(3.9.4)]{} & u_*g_*(g^*u^*u_*E \otimes g^*u^*f_*F) \\[4pt]
\Big\downarrow \simeq & & \Big\downarrow \simeq & & \simeq \Big\downarrow \\[4pt]
u_*(E \otimes g_*v^*f^*f_*U) & \longleftarrow & u_*(u^*u_*E \otimes g_*v^*f^*f_*F) & \xrightarrow[(3.9.4)]{} & u_*g_*(g^*u^*u_*E \otimes v^*f^*f_*F) \\[4pt]
\Big\downarrow & & \Big\downarrow & & \Big\downarrow \\[4pt]
u_*(E \otimes g_*v^*F) & = & u_*(E \otimes g_*v^*F) & \xrightarrow[(3.9.4)]{} & u_*g_*(g^*E \otimes v^*F)
\end{array}
$$

Commutativity of subsquare ① is given by 3.4.6.2, and of ② by (3.4.7)(i).
Commutativity of the other subsquares is straightforward to check.

Commutativity of the right half of (3.10.3.2) is shown similarly.

Theorem 3.10.3 *For any fiber square of concentrated maps of quasi-separated schemes*

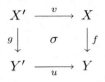

(σ *commutes and the associated map* $X' \to Y' \times_Y X$ *is an isomorphism*)*, the four independence conditions in Definition* (3.10.2) *are equivalent.*

   Proof. We first prove a special case.

Lemma 3.10.3.1 *Theorem* (3.10.3) *holds when all the schemes appearing in* σ *are affine.*

   Proof. We saw above (just before (3.10)) that the first three independence conditions are equivalent. From (3.10.2.2) and (3.10) with $F = \mathcal{O}_X$, it follows that if $\theta(\mathcal{O}_X)$ is an isomorphism then $\theta'(E)$ is an isomorphism for all E, i.e., σ is $'$-independent. Thus it will suffice to show that $\theta(\mathcal{O}_X)$ *is an isomorphism if and only if* σ *is tor-independent.*

From (3.10.1.2) with $E = S$, and the assumption that σ is a fiber square, one sees that when applied to $\mathcal{O}_X$ the right column in (3.10.1.1) becomes an isomorphism. As $\mathcal{O}_X$ is flat and quasi-coherent, the maps $\alpha(\mathcal{O}_X)$, $\beta(\mathcal{O}_X)$ and $\gamma(\mathcal{O}_X)$ in (3.10.1.1) are isomorphisms, and hence the left column—which is what we are now denoting by $\theta(\mathcal{O}_X)$—is an isomorphism iff so is the canonical map $\psi \colon \mathbf{L}u^* f_* \mathcal{O}_X \to u^* f_* \mathcal{O}_X$. Since sheafification is exact and preserves flatness (flatness of a sheaf being guaranteed by flatness of its stalks), using [**GD**, p. 214, (1.7.7.2)] one finds that ψ is $\mathbf{D}(Y')$-isomorphic to the sheafification $\tilde{\phi}$ of the natural U-homomorphism $\phi \colon U \otimes_R P^\bullet \to U \otimes_R S$, where U, R and S are as in (3.10.1.2) and $P^\bullet \to S$ is an R-flat resolution of S. Since ϕ is a quasi-isomorphism precisely when $\mathrm{Tor}_i^R(U, S) = 0$ for all $i > 0$, that is, when σ is tor-independent, the desired conclusion results.   Q.E.D.

   The *strategy* now is to show that:

   (A) *Independence is a local condition*, i.e., it holds for σ iff it holds for every induced fiber square

$$
\begin{array}{ccc}
X_0' & \xrightarrow{\ v\ } & X_0 \\
{\scriptstyle g}\big\downarrow & \sigma_0 & \big\downarrow{\scriptstyle f} \\
Y_0' & \xrightarrow[\ u\ ]{} & Y_0
\end{array}
$$

such that Y_0 is an affine open subscheme of Y, and Y_0', X_0 are affine open subschemes of $u^{-1}Y_0$, $f^{-1}Y_0$ respectively. (See first paragraph of §3.10.)

It follows then from (3.10.3.1) that tor-independence for σ in (3.10.3) implies independence and, by symmetry, $'$-independence.

It has already been noted (before (3.10)) that independence or $'$-independence implies Künneth independence. To finish proving (3.10.3) it will therefore suffice to show that:

(B) *Künneth independence for σ implies the same for any σ_0 as above.*

For then it will follow from (3.10.3.1) that Künneth-independence implies tor-independence.

Finally, (A) and (B) result at once from the first assertion in (3.10.3.3) and the last assertion in (3.10.3.4) below.

Lemma 3.10.3.2 (Independence and concatenation). *For each one of the following* **S**-*diagrams, assumed commutative,*

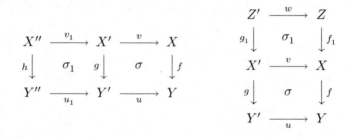

if σ and σ_1 are independent (resp. $'$-independent, Künneth-independent) then so is the rectangle $\sigma_0 := \sigma\sigma_1$ enclosed by the outer border.

Proof. As in (3.7.2)(iii), the following natural diagram commutes for any $G \in \mathbf{X}^*$:

$$
\begin{array}{ccc}
(uu_1)^* f_* G & \xrightarrow{\ \theta_{\sigma_0}(G)\ } & h_*(vv_1)^* G \\[4pt]
\simeq \downarrow & & \downarrow \simeq \\[4pt]
u_1^* u^* f_* G & \xrightarrow[u_1^*\theta_\sigma(G)]{} u_1^* g_* v^* G \xrightarrow[\theta_{\sigma_1}(v^*G)]{} & h_* v_1^* v^* G
\end{array}
\qquad (3.10.3.2.1)
$$

whence the independence assertion for the first of the diagrams in (3.10.3.2). The second is dealt with similarly via (3.7.2)(ii).

The assertion for $'$-independence follows by symmetry. (Reflection in the appropriate diagonal interchanges independence and $'$-independence.)

Künneth independence for the first diagram in (3.10.3.2)—and hence, since Künneth independence does not depend on orientation, for the second diagram too—is treated via commutativity of the following natural diagram (with $E \in \mathbf{Y}''^*$ and $F \in \mathbf{X}^*$):

$$
\begin{array}{ccc}
(uu_1)_* E \otimes f_* F & \xrightarrow{\ \eta_{\sigma_0}(E,F)\ } & (uu_1 h)_*(h^* E \otimes (vv_1)^* F) \\
\simeq \downarrow & & \downarrow \simeq \\
u_*(u_{1*} E) \otimes f_* F & & u_*(u_1 h)_*(h^* E \otimes v_1^* v^* F) \\
\eta_\sigma(u_{1*}E, F) \downarrow & & \uparrow u_* \eta_{\sigma_1}(E, v^* F) \\
(ug)_*(g^* u_{1*} E \otimes v^* F) \xrightarrow{\quad} u_* g_*(g^* u_{1*} E \otimes v^* F) \xrightarrow[(3.9.4)]{\ \sim\ } & & u_*(u_{1*} E \otimes g_* v^* F)
\end{array}
$$

$$(3.10.3.2.2)$$

Commutativity can be verified, e.g., by using the left half of the commutative diagram (3.10) to reduce the question to commutativity of the natural diagram:

$$
\begin{array}{ccccc}
(uu_1)_* E \otimes f_* F & \xrightarrow[(3.9.4)]{\sim} & (uu_1)_*(E \otimes (uu_1)^* f_* F) & \xrightarrow{\theta_{\sigma_0}} & (uu_1)_*(E \otimes h_*(vv_1)^* F) \\
\downarrow \simeq & & \downarrow \simeq & & \simeq \downarrow \ (3.9.4) \text{ etc.} \\
u_*(u_{1*} E) \otimes f_* F & \textcircled{1} & u_* u_{1*}(E \otimes (uu_1)^* f_* F) & & u_* u_{1*} h_*(h^* E \otimes v_1^* v^* F) \\
(3.9.4) \downarrow \simeq & & \downarrow \simeq & & \simeq \uparrow \ (3.9.4) \\
u_*(u_{1*} E \otimes u^* f_* F) & \xrightarrow[(3.9.4)]{\sim} & u_* u_{1*}(E \otimes u_1^* u^* f_* F) & \xrightarrow{\theta_{\sigma_v}} & u_* u_{1*}(E \otimes h_* v_1^* v^* F) \\
\theta_\sigma \downarrow & & \downarrow \theta_\sigma & \textcircled{2} & \| \\
u_*(u_{1*} E \otimes g_* v^* F) & \xrightarrow[(3.9.4)]{\sim} & u_* u_{1*}(E \otimes u_1^* g_* v^* F) & \xrightarrow{\theta_{\sigma_1}} & u_* u_{1*}(E \otimes h_* v_1^* v^* F)
\end{array}
$$

Commutativity of subdiagram ① follows from (3.7.1), and of subdiagram ② from (3.7.2)(iii). The rest is straightforward.   Q.E.D.

Corollary 3.10.3.3 *For σ as in* (3.10.3):

(i) *σ is independent if and only if for every diagram as in* (3.10.3.2) *with Y'' affine, $u_1 \colon Y'' \to Y'$ an open immersion and σ_1 a fiber square, $\sigma_0 := \sigma \circ \sigma_1$ is independent.*

(i)′ *σ is ′-independent if and only if for every diagram as in* (3.10.3.2) *with Z affine, $f_1 \colon Z \to X_1$ an open immersion and σ_1 a fiber square, $\sigma_0 := \sigma \circ \sigma_1$ is ′-independent.*

Proof. It follows from (1.2.2) that θ_σ is an isomorphism iff so is $u_1^* \theta_\sigma$ for all open immersions $u_1 \colon Y'' \to Y'$ with Y'' affine.

For such a u_1 the fiber square σ_1 is independent (as follows readily from (2.4.5.2)), so the commutative diagram (3.10.3.2.1) shows that $u_1^*\theta_\sigma$ is isomorphic to θ_{σ_0}, and (i) results.

Up to reversal of orientation, (i)′ is the same statement as (i).   Q.E.D.

Lemma 3.10.3.4 (Independence and base change). *Given σ as in (3.10.3) let $i\colon U \to Y$ be an open immersion, let $i^*\sigma$ be the fiber square*

$$
\begin{array}{ccc}
U \times_Y X' =: V' & \xrightarrow{\ v_1\ } & V:= \ U \times_Y X \\
{\scriptstyle g_1}\downarrow & & \downarrow{\scriptstyle f_1} \\
U \times_Y Y' =: U' & \xrightarrow{\ u_1\ } & U
\end{array}
$$

(with obvious maps) and let $j\colon V \to X$ and $i'\colon U' \to Y'$ be the projections. Then $i^*\sigma$ is an $\mathbf{S}$-square, and for any $G \in \mathbf{D}_{qc}(X)$ the map

$$
\theta_{i^*\sigma}(j^*G)\colon u_1^*f_{1*}j^*G \to g_{1*}v_1^*j^*G
$$

is isomorphic to the map

$$
i'^*\theta_\sigma(G)\colon i'^*u^*f_*G \to i'^*g_*v^*G\,.
$$

Moreover, for any $E \in \mathbf{D}_{qc}(U')$ and $F \in \mathbf{D}_{qc}(X)$ the map

$$
i_*\eta_{i^*\sigma}(E, j^*F)\colon i_*(u_{1*}E \otimes f_{1*}j^*F) \to i_*(u_1g_1)_*(g_1^*E \otimes v_1^*j^*F)
$$

is isomorphic to the map

$$
\eta_\sigma(i'_*E, F)\colon u_*(i'_*E) \otimes f_*F \to (ug)_*(g^*i'_*E \otimes v^*F)\,.
$$

Consequently, σ is independent if and only if $i^*\sigma$ is independent for every open immersion $i\colon U \hookrightarrow Y$ with U affine; and if σ is Künneth-independent then so is $i^*\sigma$ for all such i.

Proof. That U, U', V and V' are quasi-separated is given by [**GD**, p. 294, (6.1.9)(i) and (ii)]; and that u_1, f_1, g_1 and v_1 are quasi-compact by [**GD**, p. 291, (6.1.5)(iii)]. By (3.7.2)(iii), the diagrams

$$
\begin{array}{ccccc}
V' & \xrightarrow{\ v_1\ } & V & \xrightarrow{\ j\ } & X \\
{\scriptstyle g_1}\downarrow & i^*\sigma & \downarrow{\scriptstyle f_1}\ \ \sigma' & & \downarrow{\scriptstyle f} \\
U' & \xrightarrow[\ u_1\ ]{} & U & \xrightarrow[\ i\ ]{} & Y
\end{array}
\qquad
\begin{array}{ccccc}
V' & \xrightarrow{\ j'\ } & X' & \xrightarrow{\ v\ } & X \\
{\scriptstyle g_1}\downarrow & \sigma'' & \downarrow{\scriptstyle g}\ \ \sigma & & \downarrow{\scriptstyle f} \\
U' & \xrightarrow[\ i'\ ]{} & Y' & \xrightarrow[\ u\ ]{} & Y
\end{array}
$$

which are two decompositions of the same square—call it τ—give rise to a commutative diagram of functorial maps (cf. (3.10.3.2.1)):

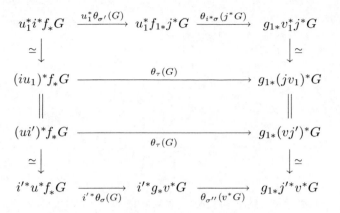

Since i and i' are open immersions, the maps $\theta_{\sigma'}$ and $\theta_{\sigma''}$ are isomorphisms (see proof of (3.10.3.3)), and the first isomorphism assertion in the Lemma results.

A similar argument using (3.10) proves the second isomorphism assertion.

The independence consequence for θ then follows from (1.2.2) and the fact that since j is an open immersion therefore $F \cong j^* j_* F$ for every $F \in \mathbf{D}(V)$.

The Künneth-independence consequence is proved similarly, with the additional observation that i is isofaithful (see (3.10.2.1)(d)). Q.E.D.

Exercises 3.10.4 (Conjugate base change). Let σ be a fiber square as in (3.10.3), and assume the schemes in σ are concentrated, so that by (4.1.1) below, f_* and g_* have right adjoints $f^\times$ and $g^\times$ respectively.

(a) Show that the map
$$\phi_\sigma : v_* g^\times \to f^\times u_*$$
(between functors from $\mathbf{D}_{qc}(Y')$ to $\mathbf{D}_{qc}(X)$) corresponding by adjunction to the natural composition $f_* v_* g^\times \overset{\sim}{\longrightarrow} u_* g_* g^\times \to u_*$ is right-conjugate to θ_σ.

Deduce that σ is independent iff ϕ_σ (or $\phi_{\sigma'}$) is an isomorphism.

Hint. The first assertion is that $\phi_\sigma(E)$ is the image of the identity map under the sequence of natural isomorphisms

$$\mathrm{Hom}(v_* g^\times E, v_* g^\times E) \overset{\sim}{\longrightarrow} \mathrm{Hom}(v^* v_* g^\times E, g^\times E) \overset{\sim}{\longrightarrow} \mathrm{Hom}(g_* v^* v_* g^\times E, E)$$
$$\overset{\sim}{\longrightarrow} \mathrm{Hom}(u^* f_* v_* g^\times E, E) \overset{\sim}{\longrightarrow} \mathrm{Hom}(f_* v_* g^\times E, u_* E)$$
$$\overset{\sim}{\longrightarrow} \mathrm{Hom}(v_* g^\times E, f^\times u_* E).$$

(b) Show that when σ is independent the map ϕ_σ^{-1}—right-conjugate to θ_σ^{-1}, see (a)—corresponds to the composition

$$v^* f^\times u_* \xrightarrow{\text{via } \beta_\sigma} g^\times u^* u_* \xrightarrow{\text{natural}} g^\times$$

with β_σ as in (3.10.2.1)(c).

(b)' Show that when σ is independent the map β_σ corresponds to the composition

$$f^\times \xrightarrow{\text{natural}} f^\times u_* u^* \xrightarrow{\text{via } \phi_\sigma^{-1}} v_* g^\times u^*.$$

Hint. To deduce (b)′ from (b), use the natural diagram (whose bottom row and right column both compose to the identity):

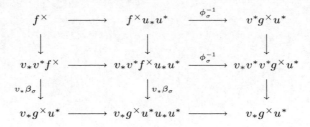

Similarly, (b)′ ⇒ (b).

(c) Show that ϕ_σ corresponds to the natural composition

$$g^\times \longrightarrow g^\times u^\times u_* \xrightarrow{\sim} v^\times f^\times u_*.$$

Chapter 4
Abstract Grothendieck Duality
for Schemes

In this chapter we review and elaborate on—with proofs and/or references—some basic abstract features of Grothendieck Duality for schemes with Zariski topology, a theory initially developed by Grothendieck [**Gr'**], [**H**], [**C**], Deligne [**De'**], and Verdier [**V'**].[1] The principal actor in this Chapter is the *twisted inverse image pseudofunctor*, described in the Introduction. The basic facts about this pseudofunctor—which may be seen as the main results in these Notes—are existence and flat base change, Theorems (4.8.1) and (4.8.3).

The abstract theory begins with Theorem (4.1) (Global Duality), asserting for any map $f\colon X \to Y$ of concentrated schemes the existence of a right adjoint $f^{\times}$ for the functor $\mathbf{R}f_{*}\colon \mathbf{D}_{\mathsf{qc}}(X) \to \mathbf{D}_{\mathsf{qc}}(Y)$. In order to sheafify this result, or, more generally, to prove tor-independent base change for $f^{\times}$—see (4.4.2) and (4.4.3), we need f to be *quasi-proper*, a condition which coincides with properness when the schemes involved are noetherian. This condition is discussed in section 4.3. The proofs of (4.4.2) and (4.4.3) are given in sections (4.5) and (4.6). That prepares the ground for the above main results.

Section (4.7) is concerned with *quasi-perfect* (= quasi-proper plus finite tor-dimension) maps of concentrated schemes. These maps have a number of especially nice properties with respect to $f^{\times}$.

Analogously, section (4.9) deals with *perfect* (= finite tor-dimension) finite-type separated maps of noetherian schemes. These maps behave nicely with respect to the twisted inverse image. For example, if $f\colon X \to Y$ is a finite-type separated map of noetherian schemes, and $f^{!}$ is the associated twisted inverse image functor, perfectness of f is characterized by boundedness of $f^{!}\mathcal{O}_{Y}$ plus the existence of a functorial isomorphism

$$f^{!}\mathcal{O}_{Y} \otimes \mathbf{L}f^{*}F \xrightarrow{\;\sim\;} f^{!}F \qquad \bigl(F \in \overline{\mathbf{D}}_{\mathsf{qc}}^{+}(Y)\bigr).$$

---

[1] As regards these Notes, see the Introduction for some comments on "abstract" vis-à-vis "concrete" duality. Exercise (4.8.12)(b) is an example of the latter.

J. Lipman, M. Hashimoto, *Foundations of Grothendieck Duality for Diagrams of Schemes*, Lecture Notes in Mathematics 1960,

159

This, and other characterizations, are in Theorem (4.9.4). Theorem (4.7.1) contains the corresponding result for the functor $f^\times$ associated to a quasi-perfect map f.

In an appendix, section (4.10), we say something about the role of dualizing complexes in duality theory. This is an important topic, but not a central one in these Notes.

Throughout, all schemes are assumed to be concentrated, i.e., quasi-separated and quasi-compact.

4.1 Global Duality

Fix once and for all a universe $\mathfrak{U}$ [**M**, p. 22]. Henceforth, any category is understood to have all its arrows and objects in $\mathfrak{U}$. Call a set *small* if it is a member of $\mathfrak{U}$. A *small category* is one whose arrows—and hence objects—form a small set. Every topological space X is understood to be small; and any sheaf E on X is understood to be such that for every open $U \subset X$, $\Gamma(U, E)$ is a small set.

For any scheme $(X, \mathcal{O}_X)$, $\mathcal{A}_X$ is, as before, the abelian category of $\mathcal{O}_X$-modules and their homomorphisms, and $\mathcal{A}_X^{\mathsf{qc}}$ is the full abelian subcategory whose objects are all the quasi-coherent $\mathcal{O}_X$-modules. Though these two categories are not small, they are *well-powered*, i.e., for each object E there is a small set J_E such that every subobject (or every quotient) of E is isomorphic to a member of J_E; and they have *small hom-sets*, i.e., for any objects E, F, the set $\mathrm{Hom}(E, F)$ is small.

"Global Duality" means:

Theorem 4.1 *Let X be a concentrated ($=$ quasi-compact, quasi-separated) scheme and $f \colon X \to Y$ a concentrated scheme-map. Then the Δ-functor $\mathbf{R}f_* \colon \mathbf{D}_{\mathsf{qc}}(X) \to \mathbf{D}(Y)$ has a bounded-below right Δ-adjoint.*

By (1.2.2), (2.4.2), and the description of θ^* in (3.3.8) (where it may be assumed that θ_* is the identity, see (2.7.3.2)), the following statement is equivalent to (4.1).

Theorem 4.1.1 *Let X be a concentrated (quasi-compact, quasi-separated) scheme and $f \colon X \to Y$ a concentrated scheme-map. Then there is a bounded-below Δ-functor $(f^\times, \text{identity}) \colon \mathbf{D}(Y) \to \mathbf{D}_{\mathsf{qc}}(X)$ and a map of Δ-functors $\tau \colon \mathbf{R}f_* f^\times \to \mathbf{1}$ such that for all $F \in \mathbf{D}_{\mathsf{qc}}(X)$ and $G \in \mathbf{D}(Y)$, the composite Δ-functorial map (in the derived category of abelian groups)*

$$\mathbf{R}\mathrm{Hom}_X^\bullet(F, f^\times G) \xrightarrow{\ (3.2.1.0)\ } \mathbf{R}\mathrm{Hom}_X^\bullet(\mathbf{L}f^* \mathbf{R}f_* F, f^\times G)$$

$$\xrightarrow{\ (3.2.3.1)\ } \mathbf{R}\mathrm{Hom}_Y^\bullet(\mathbf{R}f_* F, \mathbf{R}f_* f^\times G)$$

$$\xrightarrow{\ \text{via } \tau\ } \mathbf{R}\mathrm{Hom}_Y^\bullet(\mathbf{R}f_* F, G)$$

is a Δ-functorial isomorphism.

Corollary 4.1.2 *When restricted to concentrated schemes, the $\mathbf{D}_{\mathsf{qc}}$-valued pseudofunctor "derived direct image" (see (3.9.2)) has a pseudofunctorial right Δ-adjoint* $^\times$ *(see (3.6.7)(d)).*

Proofs. To get (4.1.2) from (4.1.1), recalling that a map $f\colon X \to Y$ of concentrated schemes is itself concentrated [**GD**, §6.1, pp. 290*ff*], choose for each such f a functor $f^\times$ right-Δ-adjoint to $\mathbf{R}f_*\colon \mathbf{D}_{\mathsf{qc}}(X) \to \mathbf{D}_{\mathsf{qc}}(Y)$, with $f^\times$ the identity functor whenever f is an identity map. For another such $g\colon Y \to Z$, define $d_{f,g}\colon f^\times g^\times \to (gf)^\times$ to be the functorial map adjoint to the natural composition

$$\mathbf{R}(gf)_* f^\times g^\times \xrightarrow{\;\sim\;} \mathbf{R}g_* \mathbf{R}f_* f^\times g^\times \to \mathbf{R}g_* g^\times \to \mathbf{1}.^2$$

This $d_{f,g}$ is an isomorphism, its inverse $(gf)^\times \to f^\times g^\times$ being the map adjoint to the natural composition

$$\mathbf{R}g_* \mathbf{R}f_* (gf)^\times \xrightarrow{\;\sim\;} \mathbf{R}(gf)_* (gf)^\times \to \mathbf{1}.$$

The verification of (4.1.2) is then straightforward (see (3.6.5)).

As for (4.1), the classical abstract method was introduced by Verdier in his treatment of duality for locally compact spaces, then adapted to schemes by Deligne [**De'**] to show that with $j\colon \mathbf{D}(\mathcal{A}_X^{\mathsf{qc}}) \to \mathbf{D}_{\mathsf{qc}}(X)$ the natural functor, $\mathbf{R}f_* \circ j$ has a right adjoint. This suffices only when f is *separated,* see (3.9.6). The proof given below (for historical reasons, because of the compactness of Deligne's original presentation) is just an elaboration of Deligne's arguments.

The reader may prefer to look up in [**N**] the more modern, lucidly exposed, approach of Neeman, who uses Brown Representability instead of, as below, the Special Adjoint Functor Theorem applied via injective resolutions. This is conceptually more elegant in that it gives a direct criterion for the existence of a right adjoint for a triangulated functor $\mathcal{F}$ on any compactly generated triangulated category, such as $\mathbf{D}_{\mathsf{qc}}(X)$. In analogy with the "cocontinuity" used in Deligne's method (see below), the condition on $\mathcal{F}$ is that it commute with small direct sums, a condition which follows for $\mathcal{F} = \mathbf{R}f_*$ from (3.9.3.3). The (nontrivial) proof in [**N**] that $\mathbf{D}_{\mathsf{qc}}(X)$ is compactly generated ostensibly requires X to be separated; but essentially the same proof shows that $\mathbf{D}_{\mathsf{qc}}(X)$ is compactly generated for any concentrated X, see [**BB**, §3], and this gives Theorem (4.1) in full generality.[3]

Proof of (4.1) (when X is separated, see above).

1. First, we review some terminology and basic results about abelian categories. Let $\mathcal{A}$ be an abelian category with small direct sums (i.e., every family of objects in $\mathcal{A}$ indexed by a small set has a direct sum). Any two

---

[2] This definition makes the property TRA 1 in [**H**, p. 207] tautologous.

[3] Arguments much like Deligne's or Neeman's apply also to *noetherian formal schemes,* see [**AJL'**, §4, pp. 42–46] resp. [**AJL'**, p. 41, 3.5.2] and [**AJS**, p. 245, Cor. 5.9].

arrows in $\mathcal{A}$ with the same source and target have a *coequalizer*, namely the cokernel of their difference [**M**, p. 70]. Hence $\mathcal{A}$ is *small-cocomplete*, i.e., any functor from a small category into $\mathcal{A}$ has a colimit, see [**M**, p. 113, Cor. 2] (dualized). An additive functor $\mathcal{F}$ from $\mathcal{A}$ to an abelian category $\mathcal{A}'$ is *cocontinuous* if $\mathcal{F}$ commutes with small colimits, in the sense that if $\mathcal{G}$ is any functor from a small category $\mathcal{C}$ into $\mathcal{A}$ and $\big(G, (g_c\colon \mathcal{G}c \to G)_{c \in \mathcal{C}}\big)$ is a colimit of $\mathcal{G}$ then $\big(\mathcal{F}G, (\mathcal{F}g_c)_{c \in \mathcal{C}}\big)$ is a colimit of $\mathcal{F}\mathcal{G}$. It follows from [**M**, p. 113, Thm. 2] that $\mathcal{F}$ *is cocontinuous iff it is right-exact and transforms small direct sums in $\mathcal{A}$ into small direct sums in $\mathcal{A}'$.*

We reserve the symbol $\varinjlim$ for denoting direct limits of small directed systems in $\mathcal{A}$, i.e., colimits of functors $\mathcal{G}\colon \mathcal{C} \to \mathcal{A}$ where $\mathcal{C}$ is the category associated to a small preordered set in which any two elements have an upper bound [**M**, p. 11, p. 211]. All such $\varinjlim$'s exist in an abelian category $\mathcal{A}$ iff $\mathcal{A}$ is small-cocomplete [**M**, p. 212, Theorem 1]. Similarly, an additive functor $\mathcal{F}\colon \mathcal{A} \to \mathcal{A}'$ is cocontinuous iff it is right-exact and commutes with all $\varinjlim$'s.

2. An essential ingredient of the proof of Theorem (4.1) is the following consequence of the Special Adjoint Functor Theorem [**M**, p. 130, Corollary]. (See also [**De′**, p. 408, Cor. 1]).

Proposition 4.1.3 *For a concentrated scheme X, an additive functor $\mathcal{F}$ from $\mathcal{A}_X^{\mathrm{qc}}$ to an abelian category $\mathcal{A}'$ with small hom-sets has a right adjoint if and (clearly) only if it is cocontinuous.*

(4.1.3.1). For the Special Adjoint Functor Theorem to be applicable here, the category $\mathcal{A}_X^{\mathrm{qc}}$—which, as above, is well-powered and has small hom-sets, and which is also small-cocomplete [**GD**, p. 217, (2.2.2)(iv)]—must have a small set of generators. Recall that an $\mathcal{O}_X$-module E on a ringed space X is *locally finitely presentable* (lfp for short) if X is covered by open subsets U such that for each U the restriction $E|_U$ is isomorphic to the cokernel of a map $\mathcal{O}_U^m \to \mathcal{O}_U^n$ with finite m and n. Since every quasi-coherent $\mathcal{O}_X$-module is the $\varinjlim$ of its lfp submodules [**GD**, p. 319, (6.9.9)], the small-generated property follows from the fact that *for any scheme X there exists a small set S of lfp $\mathcal{O}_X$-modules such that every lfp $\mathcal{O}_X$-module is isomorphic to a member of S.*

Proof. With U ranging over the small set of affine open subschemes of X, and $i_U\colon U \hookrightarrow X$ the inclusion, any $\mathcal{O}_X$-module E is isomorphic to a submodule of $\prod_U i_{U*}i_U^* E$. If E is lfp then so is the $\mathcal{O}_U$-module $i_U^* E$, so that $i_U^* E$ is a quotient of $\mathcal{O}_U^n$ for some finite n [**GD**, p. 207, (1.4.3)]. Thus every lfp E is isomorphic to a subsheaf of a sheaf of the form $\prod_U i_{U*}E_U$ where for each U, E_U ranges over a fixed small set of $\mathcal{O}_U$-modules, whence the conclusion.   Q.E.D.

(For another argument see [**Kn**, pp. 43–44, proof of Thm. 4.]).

3. The basic idea for proving (4.1) is to show that *there is a functorial exact $\mathcal{A}_X$-sequence (i.e., a finite resolution of the inclusion $\mathcal{A}_X^{\mathsf{qc}} \hookrightarrow \mathcal{A}_X$)*

$$0 \to M \xrightarrow{\;\delta(M)\;} \mathcal{D}^0(M) \xrightarrow{\;\delta^0(M)\;} \mathcal{D}^1(M) \xrightarrow{\;\delta^1(M)\;} \cdots \xrightarrow{\;\delta^{d-1}(M)\;} \mathcal{D}^d(M) \to 0 \quad (4.1.4)$$

$$\left(M \in \mathcal{A}_X^{\mathsf{qc}} \right)$$

such that the functors $\mathcal{D}^i \colon \mathcal{A}_X^{\mathsf{qc}} \to \mathcal{A}_X$ $(0 \le i \le d)$ *are additive and cocontinuous, such that for all M, $\mathcal{D}^i(M)$ is f_*-acyclic, and such that the functors $f_* \mathcal{D}^i$ are right-exact.*

Here is one way to do this. Recall the Godement resolution

$$0 \to M \to \mathcal{G}^0(M) \to \mathcal{G}^1(M) \to \cdots$$

where, with $\mathcal{G}^{-2}(M) := 0$, $\mathcal{G}^{-1}(M) := M$, and $\mathcal{K}^i(M)$ $(i \ge 0)$ the cokernel of $\mathcal{G}^{i-2}(M) \to \mathcal{G}^{i-1}(M)$, the sheaf $\mathcal{G}^i(M)$ is defined inductively by

$$\mathcal{G}^i(M)(U) := \prod_{x \in U} \mathcal{K}^i(M)_x \qquad (U \text{ open in } X).$$

One shows by induction on i that all the functors $\mathcal{G}^i$ and $\mathcal{K}^i$ (from $\mathcal{A}_X$ to itself) are *exact*. Moreover, for $i \ge 0$, $\mathcal{G}^i(M)$ is flasque, hence f_*-acyclic. With d as in (3.9.2.4), the dual version of (2.7.5)(iii) shows that $\mathcal{K}^d(M)$ is f_*-acyclic. So, setting

$$\mathcal{D}^i(M) := \begin{cases} \mathcal{G}^i(M) & (0 \le i < d) \\ \mathcal{K}^d(M) & (i = d) \\ 0 & (i > d) \end{cases}$$

we get a finite resolution (4.1.4) having all the desired properties except for commutativity of the $\mathcal{D}^i$ with $\varinjlim$.

To get commutativity with $\varinjlim$ we use the next Lemma, proved below.

Lemma 4.1.5 *Let $\mathcal{A}'$ be a small-cocomplete abelian category in which $\varinjlim$ preserves exactness of sequences. Then with $\mathbf{F}$ the category of additive functors from $\mathcal{A}_X^{\mathsf{qc}}$ to $\mathcal{A}'$, there is a functor $(-)_{\mathrm{cts}} \colon \mathbf{F} \to \mathbf{F}$ and a functorial map $i_{\mathcal{D}} \colon \mathcal{D}_{\mathrm{cts}} \to \mathcal{D}$ $(\mathcal{D} \in \mathbf{F})$ such that:*

(i) *For all lfp $M \in \mathcal{A}_X^{\mathsf{qc}}$, $i_{\mathcal{D}}(M)$ is an isomorphism $\mathcal{D}_{\mathrm{cts}}(M) \xrightarrow{\sim} \mathcal{D}(M)$.*

(ii) *For any $\mathcal{D} \in \mathbf{F}$, $\mathcal{D}_{\mathrm{cts}}$ commutes with $\varinjlim$.*

(iii) *If $\mathcal{D}$ commutes with $\varinjlim$ then $i_{\mathcal{D}}$ is a functorial isomorphism.*

(iv) *If $\mathcal{D}$ is right-exact then so is $\mathcal{D}_{\mathrm{cts}}$.*

(v) *For any exact sequence $\mathcal{D}' \to \mathcal{D} \to \mathcal{D}''$ in $\mathbf{F}$ (i.e., the $\mathcal{A}'$-sequence $\mathcal{D}'(M) \to \mathcal{D}(M) \to \mathcal{D}''(M)$ is exact for all $M \in \mathcal{A}_X^{\mathsf{qc}}$), the corresponding sequence $\mathcal{D}'_{\mathrm{cts}} \to \mathcal{D}_{\mathrm{cts}} \to \mathcal{D}''_{\mathrm{cts}}$ is exact.*

(vi) *When $\mathcal{A}' = \mathcal{A}_X$, if $\mathcal{D}(M)$ is f_*-acyclic for all $M \in \mathcal{A}_X^{\mathsf{qc}}$ then $\mathcal{D}_{\mathrm{cts}}(M)$ is f_*-acyclic for all $M \in \mathcal{A}_X^{\mathsf{qc}}$; and if, further, $\mathcal{D}$ is exact, then the functor $f_* \mathcal{D}_{\mathrm{cts}} \colon \mathcal{A}_X^{\mathsf{qc}} \to \mathcal{A}_Y$ is right-exact.*

Indeed, one can apply any such $(-)_{\mathrm{cts}}$ for $\mathcal{A}' = \mathcal{A}_X$ to the just-constructed truncated Godement resolution, to produce a resolution with all the desired properties. (For this, condition (4.1.5)(iii) is needed only when $\mathcal{D} = $ identity functor.)

From (4.1.4) there results a Δ-functor

$$(\mathcal{D}, \text{Identity}) \colon \mathbf{K}(\mathcal{A}_X^{\mathrm{qc}}) \to \mathbf{K}(\mathcal{A}_X) =: \mathbf{K}(X)$$

taking each $\mathcal{A}_X^{\mathrm{qc}}$-complex (M, d) to the f_*-acyclic $\mathcal{A}_X$-complex $\mathcal{D}(M)$ with

$$\mathcal{D}(M)^m := \oplus_{p+q=m} \mathcal{D}^q(M^p) \qquad (m \in \mathbb{Z}, \ 0 \le q \le d)$$

and with differential $\mathcal{D}(M)^m \to \mathcal{D}(M)^{m+1}$ defined on $\mathcal{D}^q(M^p)$ $(p+q=m)$ to be $\mathcal{D}^q(d^p) + (-1)^p \delta^q(M^p)$. One checks by elementary diagram chasing—or spectral sequences—that *the natural* $\mathbf{K}(X)$-*map* $\delta(M) \colon M \to \mathcal{D}(M)$ *is a quasi-isomorphism.*

It follows that the natural maps are $\mathbf{D}(Y)$-isomorphisms

$$f_*\mathcal{D}(M) \xrightarrow{\sim} \mathbf{R}f_*\mathcal{D}(M) \underset{\mathbf{R}f_* \delta(M)}{\overset{\sim}{\longleftarrow}} \mathbf{R}f_* j M, \qquad \left(M \in \mathbf{K}(\mathcal{A}_X^{\mathrm{qc}}) \right) \quad (4.1.6)$$

the first, in view of (3.9.2.4), by the dual version of (2.7.5)(a). Thus we have realized $\mathbf{R}f_* \circ j$ (up to isomorphism) at the homotopy level, as the functor $\mathcal{C}^\bullet := f_*\mathcal{D}$. Let us find a right adjoint at this level.

4.   Each functor $\mathcal{C}^q := f_*\mathcal{D}^q \colon \mathcal{A}_X^{\mathrm{qc}} \to \mathcal{A}_Y$ $(0 \le q \le d)$ is right-exact. Also, $\mathcal{C}^q$ commutes with $\varinjlim$ since both $\mathcal{D}^q$ and f_* do. (For f_* see [**Kf**, p. 641, Prop. 6], or imitate the proof on p. 163 of [**G**]). Thus $\mathcal{C}^q$ is cocontinuous, and so by (4.1.3), $\mathcal{C}^q$ *has a right adjoint* $\mathcal{C}_q \colon \mathcal{A}_Y \to \mathcal{A}_X^{\mathrm{qc}}$.

There are then functorial maps $\delta_s \colon \mathcal{C}_{s+1} \to \mathcal{C}_s$ right-conjugate to $f_*(\delta^s) \colon \mathcal{C}^s \to \mathcal{C}^{s+1}$, see (3.3.5).

For each $\mathcal{A}_Y$-complex $(F, d_{\prime})$, let $\mathcal{C}_\bullet F$ be the $\mathcal{A}_X^{\mathrm{qc}}$-complex with

$$(\mathcal{C}_\bullet F)^m := \prod_{p-q=m} \mathcal{C}_q F^p \qquad (m \in \mathbb{Z}, \ 0 \le q \le d),$$

and whose differential $(\mathcal{C}_\bullet F)^m \to (\mathcal{C}_\bullet F)^{m+1}$ is the unique map making the following diagram (with vertical arrows coming from projections) commute for all r, s with $r - s = m + 1$:

$$
\begin{array}{ccc}
\displaystyle\prod_{p-q=m} \mathcal{C}_q F^p = (\mathcal{C}_\bullet F)^m & \longrightarrow & (\mathcal{C}_\bullet F)^{m+1} = \displaystyle\prod_{p-q=m+1} \mathcal{C}_q F^p \\
\downarrow & & \downarrow \\
\mathcal{C}_s F^{r-1} \oplus \mathcal{C}_{s+1} F^r & \xrightarrow[\ \mathcal{C}_s d_{\prime}^{r-1} + (-1)^{r+s}\delta_s(F^r)\ ]{} & \mathcal{C}_s F^r
\end{array}
$$

There results naturally a Δ-functor $(\mathcal{C}_\bullet, \text{Identity}) \colon \mathbf{K}(Y) \to \mathbf{K}(\mathcal{A}_X^{\mathrm{qc}})$.

One checks that, applied componentwise, the adjunction isomorphism

$$\operatorname{Hom}_{\mathcal{A}_X^{\mathrm{qc}}}(M, \mathcal{C}_p N) \xrightarrow{\ \sim\ } \operatorname{Hom}_{\mathcal{A}_Y}(\mathcal{C}^p M, N) \qquad (M \in \mathcal{A}_X^{\mathrm{qc}},\ N \in \mathcal{A}_Y)$$

produces an isomorphism of complexes of abelian groups

$$\operatorname{Hom}_{\mathcal{A}_X^{\mathrm{qc}}}^{\bullet}(G, \mathcal{C}_{\bullet} F) \xrightarrow{\ \sim\ } \operatorname{Hom}_{\mathcal{A}_Y}^{\bullet}(\mathcal{C}^{\bullet} G, F) \qquad\qquad (4.1.7)$$

for all $\mathcal{A}_X^{\mathrm{qc}}$-complexes G and $\mathcal{A}_Y$-complexes F.

5. The isomorphism (4.1.7) suggests using $\mathcal{C}_{\bullet}$ to construct $f^{\times}$, as follows. Recall that a complex $J \in \mathbf{K}(\mathcal{A}_X^{\mathrm{qc}})$ is K-injective iff for each exact $G \in \mathbf{K}(\mathcal{A}_X^{\mathrm{qc}})$, the complex $\operatorname{Hom}_{\mathcal{A}_X^{\mathrm{qc}}}^{\bullet}(G, J)$ is exact too. The isomorphisms (4.1.6) show that $\mathcal{C}^{\bullet} G$ is exact if G is; so it follows from (4.1.7) that *if F is K-injective in $\mathbf{K}(Y)$ then $\mathcal{C}_{\bullet} F$ is K-injective in $\mathbf{K}(\mathcal{A}_X^{\mathrm{qc}})$*. Thus if $\mathbf{K}_{\mathbf{I}}(-) \subset \mathbf{K}(-)$ is the full subcategory whose objects are all the K-injective complexes, then we have a Δ-functor $(\mathcal{C}_{\bullet}, \operatorname{Id}) \colon \mathbf{K}_{\mathbf{I}}(Y) \to \mathbf{K}_{\mathbf{I}}(\mathcal{A}_X^{\mathrm{qc}})$.

Associating a K-injective resolution to each complex in $\mathcal{A}_Y$ leads to a Δ-functor $(\rho, \theta) \colon \mathbf{D}(Y) \to \mathbf{K}_{\mathbf{I}}(Y)$. In fact (ρ, θ) is an equivalence of Δ-categories, see §1.7. This ρ is *bounded below*: an $\mathcal{A}_Y$-complex E such that $H^i(E) = 0$ for all $i < n$ is quasi-isomorphic to its truncation $\tau_{\geq n} E$, which is quasi-isomorphic to an injective complex F vanishing in all degrees below n; and such an F is K-injective.

Finally, one defines $f^{\times}$ to be the composition of the functors

$$\mathbf{D}(Y) \xrightarrow{\ \rho\ } \mathbf{K}_{\mathbf{I}}(Y) \xrightarrow{\ \mathcal{C}_{\bullet}\ } \mathbf{K}_{\mathbf{I}}(\mathcal{A}_X^{\mathrm{qc}}) \xrightarrow{\ \text{natural}\ } \mathbf{D}(\mathcal{A}_X^{\mathrm{qc}}),$$

and checks, via (4.1.6), (4.1.7), (2.3.8.1) and (2.3.8)(v), that $(f^{\times}, \text{identity})$ is indeed a bounded-below right Δ-adjoint of $\mathbf{R}f_* \circ j$. (Checking the Δ-details can be tedious. Note that by (2.7.3.2) and (3.3.8), we can at least assume that $f^{\times}$ commutes with translation of complexes.)

That $f^{\times}$ is bounded below results from (3.9.2.3) and the following general fact.

Lemma 4.1.8 *Let $\mathfrak{A}^{\#}$, $\mathfrak{B}^{\#}$ be plump subcategories of the abelian categories $\mathfrak{A}$, $\mathfrak{B}$ respectively, let $\mathbf{E} = \mathbf{D}_{\#}(\mathfrak{A})$, $\mathbf{D}_{\#}^{*}(\mathfrak{A})$, or $\overline{\mathbf{D}}_{\#}^{*}(\mathfrak{A})$, see (1.9), and let $\mathbf{E}' = \mathbf{D}_{\#}(\mathfrak{B})$, $\mathbf{D}_{\#}^{*}(\mathfrak{B})$, or $\overline{\mathbf{D}}_{\#}^{*}(\mathfrak{B})$. If the functor $F \colon \mathbf{E} \to \mathbf{E}'$ has a right adjoint G, then for any $n, d \in \mathbb{Z}$:*

$$F(\mathbf{E}_{\leq \mathbf{n}}) \subset \mathbf{E}'_{\leq \mathbf{n+d}} \iff G(\mathbf{E}'_{\geq \mathbf{n}}) \subset \mathbf{E}_{\geq \mathbf{n-d}}.$$

Proof. Let $B \in \mathbf{E}'_{\geq \mathbf{n}}$. For $A = \tau_{\leq n-d-1} G(B)$, the natural $\alpha \colon A \to G(B)$ induces homology isomorphisms in all degrees $< n - d$, see (1.10). But since $F(A) \in \mathbf{E}'_{\leq \mathbf{n-1}}$ and $\tau_{\leq n-1} B \cong 0$, we have by adjointness and by (1.10.1.1):

$$\alpha \in \operatorname{Hom}_{\mathbf{E}}\big(A, G(B)\big) \cong \operatorname{Hom}_{\mathbf{E}'}\big(F(A), B\big) \cong \operatorname{Hom}_{\mathbf{E}'}\big(F(A), \tau_{\leq n-1} B\big) = 0.$$

Hence $H^j G(B) = 0$ for all $j < n - d$, i.e., $G(B) \in \mathbf{E}_{\geq \mathbf{n-d}}$.

A dual argument gives the opposite implication.   Q.E.D.

This completes the proof of Theorem (4.1), except for Lemma (4.1.5).

Proof of (4.1.5). For constructing $(-)_{\text{cts}}$ let S be a small set of lfp $\mathcal{O}_X$-modules such that every lfp $\mathcal{O}_X$-module is isomorphic to a member of S, see (4.1.3.1). For any $M \in \mathcal{A}_X^{\text{qc}}$ let $S{\downarrow}M$ be the *small category* whose objects are all the maps $s \to M$ ($s \in S$), a morphism from $\alpha\colon s \to M$ to $\beta\colon s' \to M$ being an $\mathcal{A}_X^{\text{qc}}$-map $\mu\colon s \to s'$ with $\beta\mu = \alpha$. Sending each $\alpha\colon s \to M$ in $S{\downarrow}M$ to its source $s_\alpha := s$, we get a functor $\mathbf{s}_M\colon S{\downarrow}M \to \mathcal{A}_X^{\text{qc}}$.

For any $\mathcal{D} \in \mathbf{F}$, the additive functor $\mathcal{D}_{\text{cts}} \in \mathbf{F}$ is defined as follows:

$$\mathcal{D}_{\text{cts}}(M) := \operatorname*{colim}_{S{\downarrow}M} \mathcal{D} \circ \mathbf{s}_M \qquad (M \in \mathcal{A}_X^{\text{qc}});$$

and for any $\mathcal{A}_X^{\text{qc}}$-map $\phi\colon M \to M'$, $\mathcal{D}_{\text{cts}}(\phi)$ is the $\mathcal{A}'$-map induced by the map $\mathbf{s}_M \to \mathbf{s}_{M'}$ given by composition with ϕ.[4] The map $i_{\mathcal{D}}\colon \mathcal{D}_{\text{cts}}(M) \to \mathcal{D}(M)$ is the one whose composition with the canonical map $\mathcal{D}(s_\alpha) = \mathcal{D}\mathbf{s}_M(\alpha) \to \mathcal{D}_{\text{cts}}(M)$ is $\mathcal{D}(\alpha)\colon \mathcal{D}(s_\alpha) \to \mathcal{D}(M)$ for each object $\alpha\colon s_\alpha \to M$ in $S{\downarrow}M$.

Condition (4.1.5)(i) follows from the observation that when M is lfp, the identity map of M is a final object in the category $S{\downarrow}M$.

To prove (ii) we need:

$(*)$: *For any lfp E and directed system N_σ of quasi-coherent $\mathcal{O}_X$-modules the natural map is an isomorphism*

$$\varinjlim_\sigma \operatorname{Hom}_{\mathcal{O}_X}(E, N_\sigma) \xrightarrow{\;\sim\;} \operatorname{Hom}_{\mathcal{O}_X}(E, \varinjlim_\sigma N_\sigma).$$

(*Proof*: Since X is concentrated, therefore $\Gamma(X, -)$ commutes with $\varinjlim$ [**Kf**, p. 641, Prop. 6], so it suffices to prove the statement with $\mathcal{H}om$ in place of Hom. Thus the statement is local, and so equivalent to the analogous well-known—and easily verifiable—one for modules over rings.)

Given a small directed system $\left(M_\gamma, (\phi_{\delta\gamma}\colon M_\gamma \to M_\delta)_{\delta \geq \gamma}\right)$ in $\mathcal{A}_X^{\text{qc}}$, $(*)$ shows that each map $s \to M := \varinjlim M_\gamma$ with $s \in S$ is determined by a unique equivalence class of maps $s \to M_\gamma$ (s fixed, γ variable), where $[s \to M_{\gamma'}] \equiv [s \to M_{\gamma''}]$ if and only if there exists a commutative diagram

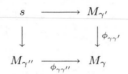

This is the least equivalence relation such that $[s \to M_\gamma] \equiv [s \to M_\gamma \xrightarrow{\phi_{\delta\gamma}} M_\delta]$ for all $\delta \geq \gamma$. Moreover, $\mathcal{A}'$-maps $f\colon \mathcal{D}_{\text{cts}}(M) \to A$ correspond naturally to families of maps $\left(f_\alpha\colon \mathcal{D}(s_\alpha) \to A\right)_{\alpha \in S{\downarrow}M}$ such that for any $\mathcal{O}_X$-homomorphism $\mu\colon s' \to s_\alpha$ ($s' \in S$), $f_{\alpha\circ\mu} = f_\alpha \circ \mathcal{D}(\mu)$. Hence an $\mathcal{A}'$-map $g\colon \mathcal{D}_{\text{cts}}(M) \to A$ corresponds to a family of maps $g_\alpha\colon \mathcal{D}(s_\alpha) \to A$ indexed by $\mathcal{O}_X$-homomorphisms $\alpha\colon s \to M_\gamma$ with variable $s \in S$ and γ, such that for any $\phi = \phi_{\delta\gamma}$ ($\delta \geq \gamma$),

$$g_{s \to M_\gamma \xrightarrow{\phi} M_\delta} = g_{s \to M_\gamma}$$

---

[4] For example, if X is noetherian then $\mathcal{D}_{\text{cts}}(M) \cong \varinjlim \mathcal{D}(N)$ where N runs through all finite-type $\mathcal{O}_X$-submodules of M.

and such that for any $\mathcal{O}_X$-homomorphism $\mu\colon s' \to s_\alpha$ with $s' \in S$,

$$g_{\alpha \circ \mu} = g_\alpha \circ \mathcal{D}(\mu).$$

One checks that an $\mathcal{A}'$-map $\varinjlim \mathcal{D}_{\mathrm{cts}}(M_\gamma) \to A$ is specified by a family g_α subject to exactly the same conditions, whence the natural map is an isomorphism

$$\varinjlim \mathcal{D}_{\mathrm{cts}}(M_\gamma) \xrightarrow{\sim} \mathcal{D}_{\mathrm{cts}}(M) = \mathcal{D}_{\mathrm{cts}}(\varinjlim M_\gamma),$$

proving (ii).

Then (iii) results by application of $\varinjlim$ to (i), since by [**GD**, p. 320, (6.9.12)] every $M \in \mathcal{A}_X^{\mathrm{qc}}$ is a $\varinjlim$ of lfp $\mathcal{O}_X$-modules.

Again, [**GD**, p. 320, (6.9.12)] allows each $M \in \mathcal{A}_X^{\mathrm{qc}}$ to be represented in the form $M = \varinjlim (M_\lambda)$ with each M_λ lfp. From $(*)$ above we get a natural isomorphism

$$\mathcal{D}_{\mathrm{cts}}(M) \cong \varinjlim \mathcal{D}(M_\lambda).$$

Since $\varinjlim$ preserves both exactness and f_*-acyclicity in $\mathcal{A}_X^{\mathrm{qc}}$ (see [**Kf**, p. 641, Thm. 8] for acyclicity), assertion (v) and the first part of (vi) follow.

As for (iv), for any exact $\mathcal{A}_X^{\mathrm{qc}}$-sequence $(\sharp)\colon 0 \to M' \to M \xrightarrow{\rho} M'' \to 0$ we must show exactness of the resulting sequence $\mathcal{D}_{\mathrm{cts}}(M') \to \mathcal{D}_{\mathrm{cts}}(M) \to \mathcal{D}_{\mathrm{cts}}(M'') \to 0$. As in the preceding paragraph, write $M = \varinjlim (M_\lambda)$ with each M_λ lfp, and let $\phi_\lambda\colon M_\lambda \to M$ be the natural maps. Then $(\sharp)$ is the $\varinjlim$ of the exact $\mathcal{A}_X^{\mathrm{qc}}$-sequences

$$(\sharp)_\lambda\colon 0 \to \ker(\rho\phi_\lambda) \to M_\lambda \to \mathrm{im}(\rho\phi_\lambda) \to 0.$$

Since $\mathcal{D}_{\mathrm{cts}}$ commutes with $\varinjlim$ and $\varinjlim$ preserves exactness, we can replace $(\sharp)$ by $(\sharp)_\lambda$, i.e., *we may assume that M is lfp.*

Now write $M' = \varinjlim (M'_\mu)$ with lfp M'_μ, so that as above, $\mathcal{D}_{\mathrm{cts}}(M') \cong \varinjlim \mathcal{D}(M'_\mu)$. If M''_μ is the cokernel of the natural composition $M'_\mu \to M' \to M$ then M''_μ is lfp; and since $\varinjlim$ preserves exactness, $M'' \cong \varinjlim M''_\mu$ and $\mathcal{D}_{\mathrm{cts}}(M'') \cong \varinjlim \mathcal{D}(M''_\mu)$. Applying $\varinjlim$ to the exact sequences $\mathcal{D}(M'_\mu) \to \mathcal{D}(M) \to \mathcal{D}(M''_\mu) \to 0$, we conclude that $\mathcal{D}_{\mathrm{cts}}$ is right-exact.

Finally, for the last part of (vi), note that if $\mathcal{D}$ is exact then since $R^1 f_* \mathcal{D}(M) = 0$ for all $M \in \mathcal{A}_X^{\mathrm{qc}}$ (because $\mathcal{D}(M)$ is f_*-acyclic), therefore $f_* \mathcal{D}$ is exact, and hence by (iv), $(f_* \mathcal{D})_{\mathrm{cts}}$ is right-exact. But since, as above, f_* commutes with $\varinjlim$, there are functorial isomorphisms

$$(f_* \mathcal{D})_{\mathrm{cts}}(M) \cong \varinjlim f_* \mathcal{D}(M_\lambda) \cong f_* \varinjlim \mathcal{D}(M_\lambda) \cong f_* \mathcal{D}_{\mathrm{cts}}(M),$$

and so $f_* \mathcal{D}_{\mathrm{cts}}$ is right-exact, as asserted.   Q.E.D.

Exercise 4.1.9 (a) In (4.1.1), suppose only that X is noetherian as a topological space (resp. that both X and Y are concentrated). Then the conclusion is valid for *any* scheme-map $f\colon X \to Y$.

Hint. See the remarks just before the proof of (4.1), resp. [**GD**, p. 295, (6.1.10(i) and (iii))]).

(b) If $f\colon X \to Y$ is a concentrated scheme-map and Y is a finite union of open subschemes Y_i with $f^{-1}Y_i$ concentrated, then the conclusion of Theorem (4.1.1) holds.

Hint. Arguing as in [**AJL′**, p. 60, 6.1.1], by induction on the least possible number of Y_i, one reduces via [**GD**, p. 296, (6.1.12), a)$\Rightarrow$c)] to where X itself is concentrated; and then the remarks just before the proof of (4.1) apply.

(c) Let $f\colon X \hookrightarrow Y$ be an open-and-closed immersion of concentrated schemes (i.e., an isomorphism of X onto a union of connected components of Y). Then the sheaf-functors f_* and f^* are exact, so may also be regarded as derived functors.

Establish, for $E \in \mathbf{D}(Y)$, $F \in \mathbf{D}(X)$, natural *bifunctorial isomorphisms*

$$\mathrm{Hom}_{\mathbf{D}(X)}(f_*E, F) \xrightarrow{\ \sim\ } \mathrm{Hom}_{\mathbf{D}(X)}(f^*f_*E, f^*f) \xleftarrow{\ \sim\ } \mathrm{Hom}_{\mathbf{D}(Y)}(E, f^*F),$$

whence, with $f^\times$ as in (b), for $F \in \mathbf{D}_{\mathrm{qc}}(Y)$ there is a functorial isomorphism

$$\xi(F)\colon f^\times F \xrightarrow{\ \sim\ } f^*F,$$

corresponding under the preceding isomorphism (with $E = f^\times F$) to the natural map $f_*f^\times F \to F$, and with inverse adjoint to the natural map $f_*f^*F \to F = f_*f^*F \oplus g_*g^*F$ where g is the inclusion $(Y \setminus X) \hookrightarrow Y$.

Verify that for the independent square

the associated map $\theta_\tau\colon f^*f_* \to 1_*1^* = \mathbf{1}$ is the *identity*, and hence the functorial base-change map from (3.10.2.1)(c)

$$\beta_\tau\colon 1^*f^\times = f^\times \to f^* = 1^\times f^*$$

is just the above isomorphism ξ.

Deduce (or prove directly) that ξ is a *pseudofunctorial* isomorphism. (Cf. (4.6.8), (4.8.1) and (4.8.7) below.)

(d) (Cf. [**Kn**, p. 43, Thm. 4].) Let $f\colon X \to Y$ be as in Theorem (4.1.1), with Y quasi-compact, and let d be an integer as in (3.9.2.3). Deduce from (4.1.1) a natural bifunctorial isomorphism

$$\mathrm{Hom}_X\big(A,\, H^{-d}f^\times(B)\big) \xrightarrow{\ \sim\ } \mathrm{Hom}_Y\big(R^d f_*(A),\, B\big)$$

for all quasi-coherent $\mathcal{O}_X$-modules A and all $\mathcal{O}_Y$-modules B.

For the smallest such d, i.e., $\dim^+ \mathbf{R}f_*|_{\mathbf{D}_{\mathrm{qc}(X)}}$, the quasi-coherent $\mathcal{O}_X$-module $D_f := H^{-d}f^\times \mathcal{O}_Y$ is the lowest-degree nonvanishing homology of $f^\times \mathcal{O}_Y$. When f is proper, D_f is often called a *relative dualizing sheaf for f*. (But certain features of the duality theory for sheaves do not just come out of the abstract theory—see [**Kn**], [**S**].)

(e) Show that the inclusion $\mathcal{A}_X^{\mathrm{qc}} \hookrightarrow \mathcal{A}_X$ has a right inverse. Deduce that every $M \in \mathcal{A}_X^{\mathrm{qc}}$ admits a monomorphism into an $\mathcal{A}_X^{\mathrm{qc}}$-injective $\mathcal{O}_X$-module.

(f) Show that the functor $(-)_{\mathrm{cts}}\colon \mathbf{F} \to \mathbf{F}$ constructed in the proof of (4.1.5) is right-adjoint to the inclusion into $\mathbf{F}$ of the full subcategory of functors that commute with *filtered* colimits (see [**M**, p. 212]). Also, the restriction of $(-)_{\mathrm{cts}}$ to the full subcategory of right-exact functors is right adjoint to the inclusion of the full subcategory of cocontinuous functors.

4.2 Sheafified Duality—Preliminary Form

Theorem 4.2 *Let* $f: X \to Y$, $f^\times$ *and* τ *be as in Theorem* (4.1.1). *Then with* $\mathrm{Hom} := \mathrm{Hom}_{\mathbf{D}(Y)}$, *for any* $E \in \mathbf{D}_{\mathsf{qc}}(Y)$, $F \in \mathbf{D}_{\mathsf{qc}}(X)$ *and* $G \in \mathbf{D}(Y)$, *the composite map*

$$\mathrm{Hom}\big(E, \mathbf{R}f_* \mathbf{R}\mathcal{H}om_X^\bullet(F, f^\times G)\big)$$

$$\xrightarrow{\;(3.2.1.0)\;} \mathrm{Hom}\big(E, \mathbf{R}f_* \mathbf{R}\mathcal{H}om_X^\bullet(\mathbf{L}f^* \mathbf{R}f_* F, f^\times G)\big)$$

$$\xrightarrow{\;(3.2.3.2)\;} \mathrm{Hom}\big(E, \mathbf{R}\mathcal{H}om_Y^\bullet(\mathbf{R}f_* F, \mathbf{R}f_* f^\times G)\big)$$

$$\xrightarrow{\;\text{via } \tau\;} \mathrm{Hom}\big(E, \mathbf{R}\mathcal{H}om_Y^\bullet(\mathbf{R}f_* F, G)\big)$$

is an isomorphism.

Proof.[5] Using $(2.6.2)^*$ and $(3.2.3)$, and checking all the requisite commutativities, one shows for fixed $F \in \mathbf{D}_{\mathsf{qc}}(Y)$ that the composite *duality map*

$$\mathbf{R}f_* \mathbf{R}\mathcal{H}om_X^\bullet(F, f^\times G) \xrightarrow{\;(3.2.1.0)\;} \mathbf{R}f_* \mathbf{R}\mathcal{H}om_X^\bullet(\mathbf{L}f^* \mathbf{R}f_* F, f^\times G)$$

$$\xrightarrow{\;(3.2.3.2)\;} \mathbf{R}\mathcal{H}om_Y^\bullet(\mathbf{R}f_* F, \mathbf{R}f_* f^\times G) \qquad (4.2.1)$$

$$\xrightarrow{\;\text{via } \tau\;} \mathbf{R}\mathcal{H}om_Y^\bullet(\mathbf{R}f_* F, G)$$

(functorial in G) is right-conjugate (see $(3.3.5)$) to the functorial (in E) projection map $p_2: E \otimes \mathbf{R}f_* F \to \mathbf{R}f_*(\mathbf{L}f^* E \otimes F)$, which, by $(3.9.4)$, is an isomorphism when $E \in \mathbf{D}_{\mathsf{qc}}(Y)$. Now apply Exercise $(3.3.7)(b)$ (with $Y = E$ and $X = G$). Q.E.D.

For *proper* maps $f: X \to Y$ one writes $f^!$ instead of $f^\times$. When Y is noetherian and f is proper, it holds that $\mathbf{R}f_* \overline{\mathbf{D}}_{\mathsf{c}}^-(X) \subset \overline{\mathbf{D}}_{\mathsf{c}}^-(Y)$ (where the subscript c indicates "coherent homology")—see [**H**, p. 89, Prop. 2.2] in which, owing to $(3.9.2.3)$ above, it is not necessary to assume that X has finite Krull dimension. So if $F \in \overline{\mathbf{D}}_{\mathsf{c}}^-(X)$ and $G \in \overline{\mathbf{D}}_{\mathsf{qc}}^+(Y)$, then $\mathbf{R}f_* F \in \overline{\mathbf{D}}_{\mathsf{c}}^-(Y)$ and $f^! G \in \overline{\mathbf{D}}_{\mathsf{qc}}^+(X)$, whence both $\mathbf{R}f_* \mathbf{R}\mathcal{H}om_X^\bullet(F, f^! G)$ and $\mathbf{R}\mathcal{H}om_Y^\bullet(\mathbf{R}f_* F, G)$ are in $\overline{\mathbf{D}}_{\mathsf{qc}}^+(X)$, see [**H**, p. 92, 3.3] or [**AJL′**, p. 35, 3.2.4]. One concludes that:

Corollary 4.2.2 *If* $f: X \to Y$ *is a proper map of noetherian schemes then for all* $F \in \overline{\mathbf{D}}_{\mathsf{c}}^-(X)$ *and* $G \in \overline{\mathbf{D}}_{\mathsf{qc}}^+(Y)$, *the duality map* (4.2.1) *is an isomorphism*

$$\mathbf{R}f_* \mathbf{R}\mathcal{H}om_X^\bullet(F, f^! G) \xrightarrow{\;\sim\;} \mathbf{R}\mathcal{H}om_Y^\bullet(\mathbf{R}f_* F, G).$$

One of our goals is to prove this Corollary under considerably weaker hypotheses—see $(4.4.2)$ below. For this purpose we need some facts about *pseudo-coherence*, reviewed in the following section.

---

[5] Cf. [**V**, p. 404, Proof of Prop. 3].

Exercises 4.2.3 Let X be a concentrated scheme. Ex. (4.1.9)(e) says that the inclusion $\mathcal{A}_X^{\mathrm{qc}} \hookrightarrow \mathcal{A}_X$ has a right adjoint Q_X, the "quasi-coherator." (Cf. [**I**, p. 186, §3].)

(a) Show that $\mathbf{R}Q_X$ is right-adjoint to the natural functor $\boldsymbol{j} \colon \mathbf{D}(\mathcal{A}_X^{\mathrm{qc}}) \to \mathbf{D}(\mathcal{A}_X)$; in other words, $\mathbf{R}Q_X = (1_X)^\times$. (Cf. [**AJL′**, p. 49, 5.2.2], where "let" in the second line should be "let $\boldsymbol{j}$ be the".)

In the rest of these exercises, assume all schemes to be quasi-compact and *separated*, so that by (3.9.6), $\boldsymbol{j}$ induces an equivalence $\boldsymbol{j}_{\mathrm{qc}} \colon \mathbf{D}(\mathcal{A}^{\mathrm{qc}}) \overset{\approx}{\longrightarrow} \mathbf{D}_{\mathrm{qc}}$. Also, $\mathbf{Q}$ denotes the functor $\boldsymbol{j}_{\mathrm{qc}} \circ \mathbf{R}Q$, right-adjoint (from (a)) to the inclusion $\mathbf{D}_{\mathrm{qc}} \hookrightarrow \mathbf{D}$; and $[-,-]$ denotes the functor $\mathbf{Q} \circ \mathbf{R}\mathcal{H}om^\bullet(-,-) \colon \mathbf{D} \times \mathbf{D} \to \mathbf{D}_{\mathrm{qc}}$.

(b) Redo 3.6.10 with $\mathbf{S}$ the category of quasi-compact separated schemes and with $\mathbf{X}^* = \mathbf{X}_* := \mathbf{D}_{\mathrm{qc}}(X)$. (Recall (2.5.8.1), (3.9.1), (3.9.2); and use the preceding $[-,-]$.)

(c) For any scheme-map $f \colon X \to Y$ there are natural functorial isomorphisms

$$\mathbf{R}\Gamma(X, \mathbf{Q}_X -) \overset{\sim}{\longrightarrow} \mathbf{R}\Gamma(X, -), \qquad \mathbf{R}f_* \mathbf{Q}_X \overset{\sim}{\longrightarrow} \mathbf{Q}_Y \mathbf{R}f_*, \qquad f^\times \mathbf{Q}_Y \overset{\sim}{\longrightarrow} f^\times.$$

(d) Deduce from Theorem (4.2) a functorial isomorphism

$$\mathbf{R}f_* [F, f^\times G]_X \overset{\sim}{\longrightarrow} [\mathbf{R}f_* F, G]_Y$$

to which application of the functor $\mathrm{H}^0 \mathbf{R}\Gamma(Y, -)$ produces the adjunction isomorphism $\mathrm{Hom}_{\mathbf{D}_{\mathrm{qc}}(X)}(F, f^\times G) \overset{\sim}{\longrightarrow} \mathrm{Hom}_{\mathbf{D}(Y)}(\mathbf{R}f_* F, G)$.

In particular, if f is an open immersion then there is a functorial isomorphism

$$f^\times G \overset{\sim}{\longrightarrow} f^* [\mathbf{R}f_* \mathcal{O}_X, G]_Y \qquad (G \in \mathbf{D}(Y)).$$

(e) Under the conditions of Theorem (4.1.1), show that the map right-conjugate to $p_1 \colon \mathbf{R}f_* E \otimes F \to \mathbf{R}f_*(E \otimes \mathbf{L}f^* F)$ (where $F \in \mathbf{D}_{\mathrm{qc}}(Y)$ is fixed, and both functors of $E \in \mathbf{D}_{\mathrm{qc}}(X)$ take values in $\mathbf{D}(Y)$) is a functorial isomorphism

$$[\mathbf{L}f^* F, f^\times G]_X \overset{\sim}{\longrightarrow} f^\times [F, G]_Y \qquad (G \in \mathbf{D}(Y)),$$

adjoint to the natural composition $\mathbf{R}f_* [\mathbf{L}f^* F, f^\times G]_X \overset{(\mathrm{d})}{\longrightarrow} [\mathbf{R}f_* \mathbf{L}f^* F, G]_Y \to [F, G]_Y$.

(f) Establish a natural commutative diagram, for $F \in \mathbf{D}_{\mathrm{qc}}(Y)$, $G \in \mathbf{D}(Y)$:

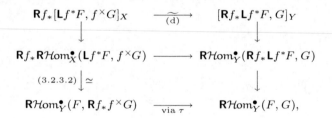

and show that the isomorphism in (e) is adjoint to the map obtained by going from the upper left to the lower right corner of this diagram.

(g) Show, via the lower square in (f), or via (3.5.6)(e), or otherwise, that the following natural diagram commutes:

$$
\begin{array}{ccc}
\mathbf{R}f_* f^\times G & \overset{(4.2.1)}{\longrightarrow} & \mathbf{R}\mathcal{H}om_Y^\bullet(\mathbf{R}f_* \mathcal{O}_X, G) \\
{\scriptstyle \tau} \downarrow & & \downarrow \\
G & \overset{\sim}{\longrightarrow} & \mathbf{R}\mathcal{H}om_Y^\bullet(\mathcal{O}_Y, G)
\end{array}
$$

In the next three exercises, for a scheme-map h we use the abbreviations $h_* := \mathbf{R}h_*$ and $h^* := \mathbf{L}h^*$.

(h) Let $X \xrightarrow{f} Y \xrightarrow{g} Z$ be maps of concentrated schemes. Referring to (e), show that for any $E, F \in \mathbf{D}_{\mathrm{qc}}(Z)$, the following diagram of natural isomorphisms commutes.

$$
\begin{array}{ccc}
[(gf)^*E, (gf)^\times F]_X \longrightarrow [f^*g^*E, g^\times f^\times F]_X \longrightarrow f^\times[g^*E, g^\times F]_Y \\
\downarrow \qquad\qquad\qquad\qquad\qquad\qquad\qquad\qquad\qquad\qquad \downarrow \\
(gf)^\times[E, F]_Z \xrightarrow{\hspace{6cm}} f^\times g^\times[E, F]_Z
\end{array}
$$

(i) Let $\beta_\sigma : v_*g^\times \to f^\times u_*$ be as in (3.10.2.1)(c). Taking into account (3.9.1), show that for any $E, F \in \mathbf{D}_{\mathrm{qc}}(Z)$ the following diagram commutes.

$$
\begin{array}{ccccc}
v^*f^\times[E, F]_Y & \xleftarrow{\;(e)\;} & v^*[f^*E, f^\times F]_X & \xrightarrow{(3.2.4)} & [v^*f^*E, v^*f^\times F]_{X'} \\
\beta_\sigma \downarrow & & & & \text{via } (3.6.4)^* \downarrow \text{ and } \beta_\sigma \\
g^\times u_*[E, F]_Y & \xrightarrow[(3.2.4)]{} & g^\times[u^*E, u^*F]_{Y'} & \xleftarrow[(e)]{} & [g^*u^*E, g^\times u^*F]_{X'}
\end{array}
$$

(j) Let $\phi_\sigma : v_*g^\times \to f^\times u_*$ be as in (3.10.4). Taking into account (3.9.2.1), show that for any $E, F \in \mathbf{D}_{\mathrm{qc}}(Z)$ the following diagram, with θ' as near the beginning of §3.10, commutes.

$$
\begin{array}{ccccc}
v_*g^\times[E, F]_{Y'} & \xleftarrow{\;(e)\;} & v_*[g^*E, g^\times F]_{X'} & \xrightarrow{(3.5.4.1)} & [v_*g^*E, v_*g^\times F]_X \\
\phi_\sigma \downarrow & & & & \text{via } \theta'_\sigma \downarrow \text{ and } \phi_\sigma \\
f^\times u_*[E, F]_{Y'} & \xrightarrow[(3.5.4.1)]{} & f^\times[u_*E, u_*F]_Y & \xleftarrow[(e)]{} & [f^*u_*E, f^\times u_*F]_X
\end{array}
$$

4.3 Pseudo-Coherence and Quasi-Properness

(4.3.1). Let us recall briefly some relevant definitions and results concerning *pseudo-coherence*. Details can be found in [**I**], as indicated, or, perhaps more accessibly, in [**TT**, pp. 283*ff*, §2].[6]

Let X be a scheme. A complex $F \in \overline{\mathbf{D}}^{\mathrm{b}}(X)$ is pseudo-coherent if each $x \in X$ has a neighborhood in which F is **D**-isomorphic to a bounded-above complex of finite-rank free $\mathcal{O}_X$-modules [**I**, p. 175, 2.2.10]. If X is divisorial, and either separated or noetherian, such an F is (globally) $\mathbf{D}(X)$-isomorphic to a bounded-above complex of finite-rank locally free $\mathcal{O}_X$-modules [*ibid.*, p. 174, Cor. 2.2.9]. If $\mathcal{O}_X$ is coherent, pseudo-coherence of F means simply that F has coherent homology [*ibid.*, p. 115, Cor. 3.5 b)]. If X is noetherian,

---

[6] Though [**I**] is written in the language of ringed topoi, the reader who, like me, is uncomfortable with that level of generality, ought with sufficient patience to be able to translate whatever's needed into the language of ringed spaces. A good starting point is 2.2.1 on p. 167 of *loc. cit.*, with examples b) on p. 88 and 2.15 on p. 108 kept in mind.

pseudo-coherence means that F is $\mathbf{D}(X)$-isomorphic to a bounded complex of coherent $\mathcal{O}_X$-modules [*ibid.*, p. 168, Cor. 2.2.2.1].

A scheme-map $f\colon X \to Y$ is pseudo-coherent if it factors locally as $f = p \circ i$ where $i\colon U \to Z$ (U open in X) is a closed immersion such that $i_*\mathcal{O}_U$ is pseudo-coherent on Z, and $p\colon Z \to Y$ is smooth [*ibid.*, p. 228, Déf. 1.2]. Pseudo-coherent maps are locally finitely-presentable (smooth maps being so by definition).

For example, any smooth map is pseudo-coherent, any regular immersion (= closed immersion corresponding to a quasi-coherent ideal generated locally by a regular sequence) is pseudo-coherent, and any composition of pseudo-coherent maps is still pseudo-coherent [*ibid.*, p. 236, Cor. 1.14].[7]

If $f\colon X \to Y$ is a proper map, and $\mathcal{L}$ is an f-ample invertible sheaf, then f is pseudo-coherent if and only if the $\mathcal{O}_Y$-complex $\mathbf{R}f_*(\mathcal{L}^{\otimes -n})$ is pseudo-coherent for all $n \gg 0$. (The proof is indicated below, in (4.3.8)). In particular, a finite map $f\colon X \to Y$ is pseudo-coherent if and only if $f_*\mathcal{O}_X$ is a pseudo-coherent $\mathcal{O}_Y$-module.

For noetherian Y, any finite-type map $f\colon X \to Y$ is pseudo-coherent. Pseudo-coherence persists under tor-independent base change [**I**, p. 233, Cor. 1.10]. Hence, by descent to the noetherian case [**EGA**, IV, (11.2.7) and its proof], *any flat finitely-presentable scheme-map is pseudo-coherent.*

Kiehl's Finiteness Theorem [**Kl**, p. 315, Thm. 2.9′] (due to Illusie for projective maps [**I**, p. 236, Thm. 2.2]) generalizes preservation of coherence by higher direct images under proper maps of noetherian schemes:

If $f\colon X \to Y$ is a proper pseudo-coherent map of quasi-compact schemes, and if $F \in \overline{\mathbf{D}}^{\mathbf{b}}(X)$ is pseudo-coherent, then so is $\mathbf{R}f_*F \in \overline{\mathbf{D}}^{\mathbf{b}}(Y)$.[8]

(**4.3.2**). For simplicity, we introduced pseudo-coherence only for complexes in $\overline{\mathbf{D}}^{\mathbf{b}}$, but that won't be enough. So let us recall [**I**, p. 98, Déf. 2.3]:

Let X be a ringed space, and let $n \in \mathbb{Z}$. A complex $F \in \mathbf{D}(X)$ is said to be *n-pseudo-coherent* if locally it is $\mathbf{D}$-isomorphic to a bounded-above complex E such that E^i is free of finite rank for all $i \geq n$. It is equivalent to say that each $x \in X$ has a neighborhood U over which there exists such an $E = E_U$ together with a quasi-isomorphism $E_U \to F|U$.

If $\mathcal{O}_X$ is coherent, then $F \in \overline{\mathbf{D}}^{-}(X)$ is n-pseudo-coherent $\Leftrightarrow$ $H^i(F)$ is coherent for all $i > n$ and $H^n(F)$ is of finite type [**I**, p. 115, Cor. 3.5 b)].

F is called pseudo-coherent if F is n-pseudo-coherent for all $n \in \mathbb{Z}$. For $F \in \overline{\mathbf{D}}^{\mathbf{b}}(X)$, this defining condition is equivalent to the one given in (4.3.1). Moreover, when X is a *quasi-compact separated scheme,* then by (3.9.6)(a), [**I**, p. 173, 2.2.8] shows the same for *any* $F \in \mathbf{D}(X)$.

---

[7] In the triangle at the top of [*ibid.*, p. 234], the map $X \to Z$ should be labeled h.

[8] The theorem actually involves a notion of pseudo-coherence of a complex *relative to a map* f; but when f itself is pseudo-coherent, relative pseudo-coherence coincides with pseudo-coherence [**I**, p. 236, Cor. 1.12].

(4.3.3). Now the above Finiteness Theorem can be put more precisely (as can be seen from the statement of [**Kl**, p. 308, Satz 2.8] and the proof of [*ibid.*, p. 310, Thm. 2.9]):

For any proper pseudo-coherent map $f\colon X \to Y$ of quasi-compact schemes, there is an integer k such that for any $n \in \mathbb{Z}$ and any n-pseudo-coherent complex $F \in \overline{\mathbf{D}}^{\mathbf{b}}(X)$, the complex $\mathbf{R}f_*F$ is $(n+k)$-pseudo-coherent.

Definition 4.3.3.1 A map $f\colon X \to Y$ is *quasi-proper* if $\mathbf{R}f_*$ takes pseudo-coherent $\mathcal{O}_X$-complexes to pseudo-coherent $\mathcal{O}_Y$-complexes.

Corollary 4.3.3.2 *Proper pseudo-coherent maps are quasi-proper. In particular, flat finitely-presentable proper maps are quasi-proper.*

Proof. The question is easily seen to be local on Y, so we may assume that both X and Y are quasi-compact. Let F be a pseudo-coherent $\mathcal{O}_X$-complex. It follows from [**I**, p. 96, Prop. 2.2, b)(ii′)] that for each n, the truncation $\tau_{\geq n}F \in \overline{\mathbf{D}}^{\mathbf{b}}(X)$ (see §1.10) is n-pseudo-coherent, and so there exists an integer k depending only on f such that $\mathbf{R}f_*\tau_{\geq n}F$ is $(n+k)$-pseudo-coherent.

Let $C \in (\mathbf{D}_{\mathsf{qc}})_{\leq n-1}$ be the summit of a triangle whose base is the natural map $F \to \tau_{\geq n}F$. With d be as in (3.9.2), application of $\mathbf{R}f_*$ to this triangle shows that $\mathbf{R}f_*(C)$ is exact in all degrees $\geq n+d-1$, so the natural map is an isomorphism $\tau_{\geq n+d}\mathbf{R}f_*F \xrightarrow{\sim} \tau_{\geq n+d}\mathbf{R}f_*\tau_{\geq n}F$ (see (1.4.5), (1.2.2)). Hence by [**I**, p. 96, Prop. 2.2, b)(ii′)], $\tau_{\geq n+d}\mathbf{R}f_*F$ is $(n+d+k)$-pseudo-coherent for all n, whence $\mathbf{R}f_*F$ is pseudo-coherent. Q.E.D.

Remark. A *projective* map is quasi-proper iff it is pseudo-coherent, see the Remark following (4.7.3.3) below. See also Example (4.3.8).

As noted above, finite-type maps of noetherian schemes are pseudo-coherent. Using Exercise (4.3.9) below, one concludes that:

Corollary 4.3.3.3 *If Y is noetherian then a map $f\colon X \to Y$ is proper iff it is finite-type, separated and quasi-proper.*

The next two Lemmas are elementary.

Lemma 4.3.4 *For any scheme-map $f\colon X \to Y$, if $G \in \mathbf{D}(Y)$ is n-pseudo-coherent then so is $\mathbf{L}f^*G$.*

This is proved by reduction to the simple case where G is a bounded-above complex of finite-rank free $\mathcal{O}_Y$-modules, vanishing in all degrees $< n$, cf. [**I**, p. 106, proof of 2.13 and p. 130, 4.19.2].

Lemma 4.3.5 *If $F \in \mathbf{D}(X)$ is n-pseudo-coherent and if the complex $G \in \mathbf{D}_{\mathsf{qc}}(X)$ is such that $H^m(G) = 0$ for all $m < r$ then $H^j\mathbf{R}\mathcal{H}\mathrm{om}_X^{\bullet}(F, G)$ is quasi-coherent for all $j < r - n$.*

Thus if F is pseudo-coherent then $\mathbf{R}\mathcal{H}\mathrm{om}_X^{\bullet}(F, G) \in \mathbf{D}_{\mathsf{qc}}(X)$.

Proof. Replacing G by $\tau^+ G$ (1.8.1), we may assume that $G^m = 0$ for $m < r$. Also, the question being local, we may assume that F is bounded above and that F^i is free of finite rank for $i \geq n$. If $F' \subset F$ is the bounded free complex which vanishes in degree $< n$ and agrees with F in degree $\geq n$, then by (1.4.4) and (1.5.3) we have a triangle (with $\mathcal{H}_X = \mathbf{R}\mathcal{H}om_X^\bullet$):

$$\mathcal{H}_X(F/F', G) \to \mathcal{H}_X(F, G) \to \mathcal{H}_X(F', G) \to \mathcal{H}_X(F/F', G)[1].$$

The complex $\mathcal{H}_X(F/F', G)$ vanishes in degree $\leq r - n$; and so from the exact homology sequence associated (as in (1.4.5)) to the triangle, we get isomorphisms

$$H^j \mathcal{H}_X(F, G) \xrightarrow{\;\sim\;} H^j \mathcal{H}_X(F', G) \qquad (j < r - n).$$

A simple induction on the number of degrees in which F' doesn't vanish (using [**H**, p. 70, (1)] to pass from n to $n + 1$) yields $\mathcal{H}_X(F', G) \in \mathbf{D}_{qc}(X)$, whence the assertion.   Q.E.D.

There results a generalization of (4.2.2), with a similar proof (given (4.3.3.2) and (4.3.5)):

Corollary 4.3.6 *If* $f\colon X \to Y$ *is a quasi-proper concentrated scheme-map, with* X *concentrated, then for all pseudo-coherent* $F \in \mathbf{D}(X)$ *and all* $G \in \overline{\mathbf{D}}_{qc}^+(Y)$, *the duality map* (4.2.1) *is an isomorphism*

$$\mathbf{R}f_* \mathbf{R}\mathcal{H}om_X^\bullet(F, f^\times G) \xrightarrow{\;\sim\;} \mathbf{R}\mathcal{H}om_Y^\bullet(\mathbf{R}f_* F, G).$$

Here is a fact needed in the proof of Theorem (4.4.1), and elsewhere.

Lemma 4.3.7 *Let* $f\colon X \to Y$ *be a finitely-presentable scheme-map, and let* $\varphi\colon A_1 \to A_2$ *be a map in* $\overline{\mathbf{D}}_{qc}^+(X)$. *Suppose that for every pseudo-coherent* $F \in \mathbf{D}(X)$, *the resulting map*

$$\mathbf{R}f_* \mathbf{R}\mathcal{H}om_X^\bullet(F, A_1) \to \mathbf{R}f_* \mathbf{R}\mathcal{H}om_X^\bullet(F, A_2) \qquad (4.3.7.1)$$

is an isomorphism. Then φ *is an isomorphism.*

Proof. There are functorial isomorphisms (see (3.2.3.3), (2.5.10)(b)):

$$\mathbf{R}\Gamma_Y \mathbf{R}f_* \mathbf{R}\mathcal{H}om_X^\bullet \xrightarrow{\;\sim\;} \mathbf{R}\Gamma_X \mathbf{R}\mathcal{H}om_X^\bullet \xrightarrow{\;\sim\;} \mathbf{R}\mathrm{Hom}_X^\bullet.$$

Application of the functor $H^0 \mathbf{R}\Gamma_Y$ to (4.3.7.1) gives then, via (2.4.2), an isomorphism

$$\mathrm{Hom}_{\mathbf{D}(X)}(F, A_1) \xrightarrow{\;\sim\;} \mathrm{Hom}_{\mathbf{D}(X)}(F, A_2). \qquad (4.3.7.2)$$

Let $C \in \overline{\mathbf{D}}_{qc}^+(X)$ be the summit of a triangle with base φ. The exact homology sequence $(1.4.5)^H$ of this triangle shows, in view of (1.2.2), that φ is an isomorphism iff $H^n(C) = 0$ for all $n \in \mathbb{Z}$.

Let us suppose that $H^n(C) = 0$ for all $n < m$ while $H^m(C) \neq 0$, and derive a contradiction. The whole question being local on Y, we may assume that Y is affine. Since $H^m(C)$ is quasi-coherent, there exists then a finitely-presentable $\mathcal{O}_X$-module E together with a *non-zero* map $E \to H^m(C)$ [**GD**, p. 320, (6.9.12)].[9] By [**EGA**, IV, (8.9.1)], there exists a noetherian ring R, a map $Y \to \mathrm{Spec}(R)$, a finite-type map $X_0 \to \mathrm{Spec}(R)$, and a coherent $\mathcal{O}_{X_0}$-module E_0, such that, up to isomorphism, $X = X_0 \otimes_R Y$ and, with $w \colon X \to X_0$ the resulting map, $E = w^* E_0 = H^0(\mathbf{L}w^* E_0)$. It will be convenient to set $F := \mathbf{L}w^* E_0[-m]$, so that $\tau_{\geq m} F \cong E[-m]$ (see §1.10). Since X_0 is noetherian, therefore E_0 is pseudo-coherent, and hence, by (4.3.4), so is F.

Now by (1.4.2.1) there is an exact sequence (with $\mathrm{Hom} := \mathrm{Hom}_{\mathbf{D}(X)}$):

$$\mathrm{Hom}(F, A_1) \xrightarrow{\varphi} \mathrm{Hom}(F, A_2) \longrightarrow \mathrm{Hom}(F, C) \longrightarrow \mathrm{Hom}(F, A_1[1]) \longrightarrow \mathrm{Hom}(F, A_2[1])$$

$$\| \qquad\qquad\qquad \|$$

$$\mathrm{Hom}(F[-1], A_1) \xrightarrow[\varphi]{} \mathrm{Hom}(F[-1], A_2)$$

where, F and $F[-1]$ being pseudo-coherent, the maps labeled φ are isomorphisms, see (4.3.7.2). Thus,

$$
\begin{aligned}
0 = \mathrm{Hom}\big(F, C\big) & \\
\cong \mathrm{Hom}\big(\tau_{\geq m} F, C\big) &\qquad \text{see } (1.10.1.2) \\
\cong \mathrm{Hom}\big(E[-m], C\big) & \\
\cong \mathrm{Hom}\big(E[-m], \tau_{\leq m} C\big) &\qquad \text{see } (1.10.1.1) \\
\cong \mathrm{Hom}\big(E[-m], (H^m(C))[-m]\big) &\qquad \text{see } (1.2.3) \\
\neq 0, &
\end{aligned}
$$

a contradiction.   Q.E.D.

Example 4.3.8 Let $f \colon X \to Y$ be a proper map of schemes, and let $\mathcal{L}$ be an f-ample invertible sheaf [**EGA**, II, p. 89, Déf. (4.6.1)]. Then f is pseudo-coherent if and only if the $\mathcal{O}_Y$-complex $\mathbf{R}f_*(\mathcal{L}^{\otimes -n})$ is pseudo-coherent for all $n \gg 0$.

Proof. If f is pseudo-coherent then $\mathbf{R}f_*(\mathcal{L}^{\otimes -n})$ is pseudo-coherent, by the Finiteness Theorem (4.3.3) (in fact—since f is projective locally on Y [**EGA**, II, p. 104, Thm. (5.5.3)]—by [**I**, p. 236, Thm. 2.2 and Cor. 1.12]).

We first illustrate the converse by treating the special case where f is finite and $f_* \mathcal{O}_X$ is a pseudo-coherent $\mathcal{O}_Y$-module. To check that f is pseudo-coherent, we may assume that Y—and hence X—is affine, so that for some $r > 0$, f factors as $f = pi$ with $p \colon \mathbb{A}_Y^r \to Y$ the (smooth) projection and $i \colon X \hookrightarrow \mathbb{A}_Y^r$ a closed immersion; and we need to show that $i_* \mathcal{O}_X$ is pseudo-coherent.

In algebraic terms, we have a finite ring-homomorphism $A \to B = A[t_1, \ldots, t_r]$, such that the A-module B is resolvable by a complex $E_\bullet$ of finite-type free A-modules [**I**, p. 160, Prop. 1.1]. Let $T := (T_1, \ldots, T_r)$ be a sequence of indeterminates, and

---

[9] Recall that finitely-presentable maps are quasi-compact and quasi-separated, by definition [**GD**, p. 305, (6.3.7)], so that X is quasi-compact and quasi-separated.

let $\varphi\colon B[T] = B[T_1,\ldots,T_r] \to B$ be the unique B-homomorphism such that $\varphi(T_k) = t_k$ ($1 \le k \le r$). Then B is resolved as a $B[T]$-module by the Koszul complex $K_\bullet$ on $(T_1 - t_1,\ldots,T_r - t_r)$. Since the $A[T]$-module $B[T]$ is resolved by $E_\bullet \otimes_A A[T]$, therefore the free $B[T]$-modules K_j can be resolved by finite-type free $A[T]$-modules, whence so can B, giving the desired pseudo-coherence of $i_*\mathcal{O}_X$.

Now let us treat (sketchily) the general case. Assuming, as we may, that Y is affine, we have for some $r > 0$, a factorization $f = pi$ where $p\colon \mathbb{P}_Y^r \to Y$ is the (smooth) projection and $i\colon X \hookrightarrow \mathbb{P}_Y^r$ is a closed immersion [**EGA**, II, p. 104, (5.5.4)(ii)]. With $\gamma\colon X \to X \times_Y \mathbb{P}_Y^r = \mathbb{P}_X^r$ the graph of i, there is a natural diagram

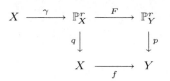

and it needs to be shown that $i_*\mathcal{O}_X = \mathbf{R}F_*(\gamma_*\mathcal{O}_X)$ is pseudo-coherent. Note that since γ is a regular immersion [**Bt**, p. 429, Prop. 1.10], therefore $\gamma_*\mathcal{O}_X$ is pseudo-coherent. So it's enough to show that F is quasi-proper.

By [**EGA**, II, p. 91, (4.6.13)(iii)], $L := q^*\mathcal{L}$ is F-ample; and for $n \gg 0$, say $n \ge m$,

$$\mathbf{R}F_*(L^{\otimes -n}) = \mathbf{R}F_*\big(q^*(\mathcal{L}^{\otimes -n})\big) \underset{(3.9.5)}{\cong} p^*\mathbf{R}f_*(\mathcal{L}^{\otimes -n})$$

is pseudo-coherent (4.3.4).

Imitating the proof of [**I**, p. 238, Thm. 2.2.2], we can then reduce the problem to showing that $\mathbf{R}F_*(E')$ is pseudo-coherent for any bounded $\mathcal{O}_X$-complex E' whose component in each degree is a finite direct sum of sheaves of the form $L^{\otimes -n}$; and this is easily done by induction on the number of nonzero components of E'. Q.E.D.

Exercises 4.3.9 (a) (Curve selection.) Let $\bar Z$ be a noetherian scheme, $Z \subset \bar Z$ a dense open subset, and $W := \bar Z \setminus Z$. Show that for each closed point $w \in W$ there is an integral one-dimensional subscheme $C \subset \bar Z$ such that w is an isolated point of $C \cap W$.

Hint. Use the *local nullstellensatz*: in any noetherian local ring A with $\dim A \ge 1$, the intersection of all those prime ideals p such that $\dim A/p = 1$ is the nilradical of A. (For this, note that the maximal ideal is contained in the union of all the height one primes, so that when $\dim A > 1$ there must be infinitely many height one primes; and deduce that if $q \subset A$ is a prime ideal with $\dim A/q > 1$ and $a \notin q$ then there exists a prime ideal $q' \ne m$ such that $q' \supsetneq q$ and $a \notin q'$.)

(b) Prove that if $f\colon X \to Y$ is a finite-type separated map of noetherian schemes such that $f_*(\mathcal{O}_X/\mathcal{I})$ is coherent for every coherent $\mathcal{O}_X$-ideal $\mathcal{I}$, then f is proper. In particular, if f is quasi-proper then f is proper.

Outline. If not, let $Z \subset X$ be a closed subscheme of Z minimal among those for which the restriction of f is not proper. Then Z is integral [**EGA**, II, p. 101, 5.4.5]. Let $\bar f\colon \bar Z \to Y$ be a *compactification* of $f|_Z$, see [**C'**], [**Lt**], [**Vj**], that is, $f = \bar f v$ with $\bar f$ proper and $v\colon Z \hookrightarrow \bar Z$ an open immersion. If $\dim Z > 1$ then by (a) there is a curve on Z for which the restriction of f is not proper, contradiction. So the problem is reduced to where X is integral, of dimension 1. Then if $\dim Y = 0$, and f is not proper, we may assume that $Y = \mathrm{Spec}(k)$, k a field, whence X is affine, and $f_*\mathcal{O}_X$ is not coherent. If $\dim(Y) = 1$ and $\bar f\colon \bar X \to Y$ is a compactification of f, then the map $\bar f$ is finite; and if $u\colon X \hookrightarrow \bar X$ is the inclusion, $u_*\mathcal{O}_X$ is coherent, whence, by [**EGA**, IV, p. 117, (5.10.10)(ii)], $X = \bar X$.

4.4 Sheafified Duality, Base Change

Unless otherwise indicated, *all schemes—and hence all scheme-maps—are assumed henceforth to be concentrated. All proper and quasi-proper maps are assumed to be finitely presentable.*

As in §4.3, a scheme-map $f: X \to Y$ is called quasi-proper if $\mathbf{R}f_*$ takes pseudo-coherent $\mathcal{O}_X$-complexes to pseudo-coherent $\mathcal{O}_Y$-complexes. For example, when Y is noetherian and f is of finite type and separated then f is quasi-proper iff it is proper, see (4.3.3.3). We will need the nontrivial fact that *quasi-properness of maps is preserved under tor-independent base change* [**LN**, Prop. 4.4].

The following abbreviations will be used, for a scheme-map h or a scheme Z:

$$h_* := \mathbf{R}h_*, \qquad h^* := \mathbf{L}h^*,$$
$$\mathcal{H}_Z := \mathbf{R}\mathcal{H}om_Z^{\bullet}, \qquad \mathbf{H}_Z := \mathbf{R}Hom_Z^{\bullet},$$
$$\otimes_Z := \underset{=}{\otimes}_Z, \qquad \mathbf{\Gamma}_Z(-) := \mathbf{R}\Gamma(Z, -).$$

Recall the characterizations of *independent fiber square* (3.10.3), of *finite tor-dimension map* (2.7.6), and of the "dualizing pair" $(f^{\times}, \tau)$ in (4.1.1). We write $f^!$ for $f^{\times}$ when f is quasi-proper.

Recall also the natural map (3.5.4.1) = (3.5.4.4) (see (3.5.2)(d)) associated to any ringed-space map $f: X \to Y$,

$$\nu: f_*\mathcal{H}_X(F, H) \to \mathcal{H}_Y(f_*F, f_*H) \qquad \big(F, H \in \mathbf{D}(X)\big). \tag{4.4.0}$$

The composition $(3.2.3.2) \circ (3.2.1.0)$ in (4.2.1) is an instance of this map. (See the line immediately following (3.5.4.2).)

Theorem 4.4.1 *Suppose one has an independent fiber square*

$$
\begin{array}{ccc}
X' & \xrightarrow{\;\;v\;\;} & X \\
{\scriptstyle g}\downarrow & \sigma & \downarrow{\scriptstyle f} \\
Y' & \xrightarrow[\;\;u\;\;]{} & Y
\end{array}
$$

with f (hence g) quasi-proper and u of finite tor-dimension.

Then for any $F' \in \mathbf{D}_{\mathsf{qc}}(X')$ and $G \in \overline{\mathbf{D}}_{\mathsf{qc}}^+(Y)$, the composition

$$g_*\mathcal{H}_{X'}(F', v^*f^!G) \xrightarrow{\;\;\nu\;\;} \mathcal{H}_{Y'}(g_*F', g_*v^*f^!G)$$

$$\xrightarrow[(3.10.3)]{\;\sim\;} \mathcal{H}_{Y'}(g_*F', u^*f_*f^!G) \xrightarrow[\;\tau\;]{} \mathcal{H}_{Y'}(g_*F', u^*G)$$

is an isomorphism.

If u and v are identity maps then so is the map labeled (3.10.3), and the resulting composition (with $F := F'$)

$$\delta(F, G): f_*\mathcal{H}_X(F, f^!G) \xrightarrow{\;\nu\;} \mathcal{H}_Y(f_*F, f_*f^!G) \xrightarrow{\;\tau\;} \mathcal{H}_Y(f_*F, G)$$

is just the duality map (4.2.1), whence the following generalization of (4.3.6):

Corollary 4.4.2 (Duality). *Let* $f: X \to Y$ *be quasi-proper. Then for any* $F \in \mathbf{D}_{\mathsf{qc}}(X)$ *and* $G \in \overline{\mathbf{D}}_{\mathsf{qc}}^{+}(Y)$, *the duality map* $\delta(F, G)$ *is an isomorphism.*

Moreover:

Corollary 4.4.3 (Base Change). *In* (4.4.1), *the functorial map adjoint to the composition*

$$g_* v^* f^! G \xrightarrow[\quad(3.10.3)\quad]{\sim} u^* f_* f^! G \xrightarrow[\quad u^*\tau\quad]{} u^* G,$$

is an isomorphism

$$\beta(G) = \beta_\sigma(G) \colon v^* f^! G \xrightarrow{\sim} g^! u^* G \qquad \left(G \in \overline{\mathbf{D}}_{\mathsf{qc}}^{+}(Y)\right).$$

To deduce (4.4.3) from (4.4.1), let $F' \in \mathbf{D}_{\mathsf{qc}}(X')$ and consider the next diagram, whose commutativity follows from the definition of $\beta = \beta(G)$:

$$
\begin{array}{ccc}
g_* \mathcal{H}_{X'}(F', v^* f^! G) & \xrightarrow{\ \beta\ } & g_* \mathcal{H}_{X'}(F', g^! u^* G) \\
{\scriptstyle \nu}\downarrow & & \downarrow{\scriptstyle \nu} \\
\mathcal{H}_{Y'}(g_* F', g_* v^* f^! G) & \xrightarrow{\ \beta\ } & \mathcal{H}_{Y'}(g_* F', g_* g^! u^* G) \\
{\scriptstyle (3.10.3)}\downarrow{\scriptstyle \simeq} & & \downarrow{\scriptstyle \tau} \\
\mathcal{H}_{Y'}(g_* F', u^* f_* f^! G) & \xrightarrow{\ \tau\ } & \mathcal{H}_{Y'}(g_* F', u^* G)
\end{array}
\qquad (4.4.3.1)
$$

By (4.4.1), $\tau \circ (3.10.3) \circ \nu$ is an isomorphism; and by (4.4.2) (a special case of (4.4.1)), the right column is an isomorphism too. (Note that by (2.7.5)(d) and (3.9.1), $u^* G \in \overline{\mathbf{D}}_{\mathsf{qc}}^{+}(Y')$.) It follows that the top row is an isomorphism, and applying the functor $H^0 \mathbf{\Gamma}_{Y'}$ we get as in (4.3.7.2) an isomorphism

$$\mathrm{Hom}_{\mathbf{D}(X')}(F', v^* f^! G) \xrightarrow{\ \text{via } \beta\ } \mathrm{Hom}_{\mathbf{D}(X')}(F', g^! u^* G);$$

and since this holds for any $F' \in \mathbf{D}_{\mathsf{qc}}(X')$, in particular for $F' = v^* f^! G$ and $F' = g^! u^* G$, it follows that β itself is an isomorphism.   Q.E.D.

Remarks 4.4.4 (a) Conversely, the commutativity of (4.4.3.1) shows that (4.4.2) and (4.4.3) together imply (4.4.1).

(b) An example of Neeman [**N**, p. 233, 6.5], with f the unique map $\mathrm{Spec}(\mathbb{Z}[T]/(T^2)) \to \mathrm{Spec}(\mathbb{Z})$ (T an indeterminate), shows that (4.4.2) and (4.4.3) can fail when G is not bounded below.

(c) In (4.4.1), tordim $v \leq$ tordim $u < \infty$.

To see this, let $x' \in X'$, $x = v(x')$, $y' = g(x')$, $y = u(y') = f(x)$, $A = \mathcal{O}_{Y,y}$, $A' = \mathcal{O}_{Y',y'}$, $B = \mathcal{O}_{X,x}$, and $B' = \mathcal{O}_{X',x'}$. By (2.7.6.4), the A-module A' has a flat resolution $P_\bullet$ of length $d := $ tordim $u < \infty$; and so by (i) in (3.10.2), $P_\bullet \otimes_A B$ is a flat resolution of the B-module $B^* = A' \otimes_A B$. Since B' is a localization of B^*, it holds for any B-module M that

$$\mathrm{Tor}_j^B(B', M) = B' \otimes_{B^*} \mathrm{Tor}_j^B(B^*, M) = 0 \qquad (j > d);$$

and it follows then from (2.7.6.4) that tordim $v \le d$.

(d) By definition, β is the unique functorial map making the following diagram commute:

$$
\begin{array}{ccc}
g_* v^* f^! & \xrightarrow{\ g_* \beta\ } & g_* g^! u^* \\[2pt]
{\scriptstyle (3.10.3)}\Big\downarrow{\scriptstyle \simeq} & & \Big\downarrow{\scriptstyle \tau_g} \\[2pt]
u^* f_* f^! & \xrightarrow[\ u^* \tau_f\ ]{} & u^*
\end{array}
$$

This diagram generalizes [**H**, p. 207, TRA 4.]

4.5 Proof of Duality and Base Change: Outline

In describing the organization of the proof of (4.4.1), we will attach symbols to labels of the form (4.4.x) to refer to special cases of (4.4.x):

$(4.4.1)^*_{\mathrm{pc}} := (4.4.1)$ with $F' = v^* F$, where $F \in \mathbf{D}(X)$ is pseudo-coherent.

$(4.4.2)_{\mathrm{pc}} := $ Corollary $(4.3.6) := (4.4.1)^*_{\mathrm{pc}}$ with $u = v = $ identity.

$(4.4.3)^{\circ} := (4.4.3)$ with the map u an open immersion.

$(4.4.3)^{\mathrm{af}} := (4.4.3)$ with the map u affine.

Having already proved $(4.4.2)_{\mathrm{pc}}$, our strategy is to prove the chain of implications

$$(4.4.2)_{\mathrm{pc}} \Leftrightarrow (4.4.1)^*_{\mathrm{pc}} \Rightarrow \big((4.4.3)^{\circ} + (4.4.3)^{\mathrm{af}}\big) \Rightarrow (4.4.3) \Rightarrow (4.4.3)^{\circ} \Leftrightarrow (4.4.2).$$

By (4.4.4)(a), then, (4.4.1) results.

Remark 4.5.1 For arbitrary finitely-presentable f, assertions (4.4.1)–(4.4.3) are meaningful—though not necessarily true—with $(f^\times, g^\times)$ in place of $(f^!, g^!)$. As will be apparent from the following proofs, the equivalence (4.4.1) $\Leftrightarrow$ (4.4.2) + (4.4.3) holds in this generality, as do the preceding implications except for $(4.4.2)_{\mathrm{pc}} \Rightarrow (4.4.1)^*_{\mathrm{pc}}$.

4.6 Steps in the Proof

I. Proof of $(4.4.2)_{\mathrm{pc}}$

This has already been done (Corollary (4.3.6)).

II. $(4.4.2)_{\mathrm{pc}} \Leftrightarrow (4.4.1)^*_{\mathrm{pc}}$

The implication $\Leftarrow$ is trivial.

The implication $\Rightarrow$ follows at once from:

Lemma 4.6.4 *With the assumptions of* $(4.4.1)^*_{\mathrm{pc}}$, *and* δ *the duality map in* (4.4.2), *there is a natural commutative* $\mathbf{D}(Y')$-*diagram*

$$
\begin{array}{ccc}
u^*f_*\mathcal{H}_X(F,\,f^!G) & \xrightarrow{\ u^*\delta\ } & u^*\mathcal{H}_Y(f_*F,\,G) \\
\simeq\Big\downarrow & & \Big\downarrow\simeq \\
g_*\mathcal{H}_{X'}(v^*F,\,v^*f^!G) & \xrightarrow[(4.4.1)^*_{\mathrm{pc}}]{} & \mathcal{H}_{Y'}(g_*v^*F,\,u^*G)
\end{array}
$$

in which the vertical arrows are isomorphisms.

Commutativity in (4.6.4) is derived from the following relation—to be proved below—among the canonical maps ν, θ (3.7.2), and ρ (3.5.4.5):

Lemma 4.6.5 *For any commutative diagram of ringed-space maps*

$$
\begin{array}{ccc}
X' & \xrightarrow{\ v\ } & X \\
g\Big\downarrow & & \Big\downarrow f \\
Y' & \xrightarrow[u]{} & Y
\end{array}
\qquad (4.6.5.1)
$$

and $F \in \mathbf{D}_{\mathrm{qc}}(X)$, $H \in \mathbf{D}(X)$, *the following diagram commutes:*

$$
\begin{array}{ccc}
u^*f_*\mathcal{H}_X(F,\,H) & \xrightarrow{\qquad\qquad\qquad \nu \qquad\qquad\qquad} & u^*\mathcal{H}_Y(f_*F,\,f_*H) \\
\theta\Big\downarrow & & \Big\downarrow\rho \\
g_*v^*\mathcal{H}_X(F,\,H) & & \mathcal{H}_{Y'}(u^*f_*F,\,u^*f_*H) \\
\rho\Big\downarrow & & \Big\downarrow(1,\theta) \\
g_*\mathcal{H}_{X'}(v^*F,\,v^*H) & \xrightarrow[\nu]{} \mathcal{H}_{Y'}(g_*v^*F,\,g_*v^*H) \xrightarrow[(\theta,1)]{} & \mathcal{H}_{Y'}(u^*f_*F,\,g_*v^*H)
\end{array}
$$

Indeed, if (4.6.5.1) is an independent fiber square of scheme-maps, so that by (3.10.3), $\theta(F)\colon u^*f_*F \to g_*v^*F$ is an isomorphism. If $G \in \mathbf{D}(Y)$ and $H := f^\times G$, so that there is a natural map $f_*H \to G$ (see (4.1.1)), then we get (a generalization of) commutativity in (4.6.4) by gluing the $\mathbf{D}(X')$-diagram in (4.6.5) and the following natural commutative diagram along the common column:

$$
\begin{array}{ccc}
u^*\mathcal{H}_Y(f_*F,\,f_*H) & \xrightarrow{\qquad\qquad\qquad\qquad\qquad} & u^*\mathcal{H}_Y(f_*F,\,G) \\
\rho\Big\downarrow & & \Big\downarrow\rho \\
\mathcal{H}_{Y'}(u^*f_*F,\,u^*f_*H) \ =\!=\ \mathcal{H}_{Y'}(u^*f_*F,\,u^*f_*H) & \longrightarrow & \mathcal{H}_{Y'}(u^*f_*F,\,u^*G) \\
(1,\theta)\Big\downarrow \qquad\qquad (\theta^{-1},1)\Big\downarrow\simeq & & \simeq\Big\downarrow(\theta^{-1},1) \\
\mathcal{H}_{Y'}(u^*f_*F,\,g_*v^*H) \xrightarrow[(\theta^{-1},\theta^{-1})]{\sim} \mathcal{H}_{Y'}(g_*v^*F,\,u^*f_*H) & \longrightarrow & \mathcal{H}_{Y'}(g_*v^*F,\,u^*G)
\end{array}
$$

Here is where we need f to be quasi-proper: since F is, by assumption, pseudo-coherent, therefore f_*F is pseudo-coherent. In view of (4.4.4)(c), the following Proposition gives then the isomorphism assertion in (4.6.4).

Proposition 4.6.6 *Let $u\colon Y' \to Y$ be any scheme-map of finite tor-dimension, and let $H \in \overline{\mathbf{D}}^+(Y)$. Then there is an integer e such that for all $m \in \mathbb{Z}$ and all m-pseudo-coherent $C \in \mathbf{D}(Y)$, the map*

$$\rho_u\colon u^*\mathcal{H}_Y(C,\,H) \to \mathcal{H}_{Y'}(u^*C,\,u^*H)$$

induces homology isomorphisms in all degrees $\leq e - m$. In particular, if C is pseudo-coherent then ρ_u is an isomorphism.

Proof. The question is local on Y, because if $i\colon U \to Y$ is an open immersion, $U' := U \times_Y Y'$, and $w\colon U' \to U$, $j\colon U' \to Y'$ are the projections (so that j is an open immersion), then $j^*\rho_u \cong \rho_w$—more precisely, the following natural diagram commutes for any $F, G \in \mathbf{D}(Y)$:

$$
\begin{array}{ccc}
j^*u^*\mathcal{H}_Y(F,G) & \xrightarrow{\;j^*\rho_u\;} & j^*\mathcal{H}_{Y'}(u^*f, u^*G) \\[2pt]
\Big\downarrow{\scriptstyle\simeq} & & {\scriptstyle\simeq}\Big\downarrow{\scriptstyle\rho_j} \\[2pt]
w^*i^*\mathcal{H}_Y(F,G) & & \mathcal{H}_{U'}(j^*u^*f, j^*u^*G) \\[2pt]
{\scriptstyle w^*\rho_i}\Big\downarrow{\scriptstyle\simeq} & & {\scriptstyle\simeq}\Big\downarrow \\[2pt]
w^*\mathcal{H}_U(i^*F, i^*G) & \xrightarrow{\;\;\rho_w\;\;} & \mathcal{H}_{U'}(w^*i^*F, w^*i^*G)
\end{array}
$$

Here ρ_i and ρ_j are isomorphisms by the last assertion in (4.6.7) (whose proof does not depend on (4.6.6)); and commutativity follows from (3.7.1.1).

So by [**I**, p. 98, 2.3] we may assume there is a $\mathbf{D}(Y)$-map $E \to C$ with E *strictly perfect* (i.e., E is a bounded complex of finite-rank locally free $\mathcal{O}_Y$-modules), such that the induced map is an isomorphism $\tau_{\geq m+1}E \xrightarrow{\sim} \tau_{\geq m+1}C$. The contravariant Δ-functors

$$\Phi_1(C) := u^*\mathcal{H}_Y(C,\,H), \qquad \Phi_2(C) := \mathcal{H}_{Y'}(u^*C,\,u^*H)$$

are both bounded below (1.11.1), and so arguing as in the proof of (4.3.3.2) we find that there is an integer e such that for $i = 1, 2$, the natural maps

$$\tau_{\leq e-m}\Phi_i(E) \leftarrow \tau_{\leq e-m}\Phi_i(\tau_{m+1}E) \xrightarrow{\sim} \tau_{\leq e-m}\Phi_i(\tau_{m+1}C) \to \tau_{\leq e-m}\Phi_i(C)$$

are isomorphisms.

Thus it will be more than enough to prove:

Proposition 4.6.7 *Let $u\colon Y' \to Y$ be a scheme-map, let E be a bounded-above complex of finite-rank locally free $\mathcal{O}_Y$-modules, and let $H \in \overline{\mathbf{D}}^+(Y)$.*

If E is strictly perfect or if u has finite tor-dimension then the map

$$\rho\colon u^*\mathcal{H}_Y(E, H) \to \mathcal{H}_{Y'}(u^*E, u^*H)$$

is an isomorphism.

The same holds for any $E, H \in \mathbf{D}(Y)$ if u is an open immersion.

Except for the proofs of (4.6.5) and (4.6.7), which are postponed to the end of this section 4.6, the proof of (4.6.4)—and hence of the implication $(4.4.2)_{\mathrm{pc}} \Rightarrow (4.4.1)^*_{\mathrm{pc}}$—is now complete.

III. $(4.4.1)^*_{\mathrm{pc}} \Rightarrow \big((4.4.3)^\circ + (4.4.3)^{\mathrm{af}}\big)$

Let $\beta = \beta(G)$ be as in (4.4.3). When u, hence v, is an open immersion or affine, then v is isofaithful ((3.10.2.1)(d) or (3.10.2.2)), so that for β to be an isomorphism it suffices that $v_*\beta$ be an isomorphism.

Let $F \in \mathbf{D}(X)$ be pseudo-coherent. From (4.4.3.1) with $F' = v^*F$ and with $^!$ replaced by $^\times$, one derives the following commutative diagram:

$$
\begin{array}{ccc}
f_*\mathcal{H}_X(F,\, v_*v^*f^\times G) & \xrightarrow{\ \text{via } v_*\beta\ } & f_*\mathcal{H}_X(F,\, v_*g^\times u^*G) \\[2pt]
{\scriptstyle (3.2.3.2)^{-1}}\Big\downarrow{\scriptstyle \simeq} & & {\scriptstyle \simeq}\Big\downarrow{\scriptstyle (3.2.3.2)^{-1}} \\[2pt]
f_*v_*\mathcal{H}_{X'}(v^*F,\, v^*f^\times G) & \xrightarrow{\ \text{via } \beta\ } & f_*v_*\mathcal{H}_{X'}(v^*F,\, g^\times u^*G) \\[2pt]
\Big\downarrow{\scriptstyle \simeq} & & {\scriptstyle \simeq}\Big\downarrow \\[2pt]
u_*g_*\mathcal{H}_{X'}(v^*F,\, v^*f^\times G) & \xrightarrow[\ \text{via } \beta\ ]{} & u_*g_*\mathcal{H}_{X'}(v^*F,\, g^\times u^*G) \\[2pt]
\Big\| & & {\scriptstyle \simeq}\Big\downarrow{\scriptstyle u_*\delta} \\[2pt]
u_*g_*\mathcal{H}_{X'}(v^*F,\, v^*f^\times G) & \xrightarrow[\ u_*(4.4.1)^*_{\mathrm{pc}}\ ]{\sim} & u_*\mathcal{H}_{Y'}(g_*v^*F,\, u^*G)
\end{array}
$$

The bottom row is an isomorphism by assumption, as is the right column, by the special case $(4.4.2)_{\mathrm{pc}}$ of $(4.4.1)^*_{\mathrm{pc}}$. Thus the top row is an isomorphism, and hence, by (4.3.7), so is $v_*\beta$.

IV. $\big((4.4.3)^\circ + (4.4.3)^{\mathrm{af}}\big) \Rightarrow (4.4.3)$

The essence of what follows is contained in the four lines preceding "CASE 1" on p. 401 of [**V**].

Denote the independent square in (4.4.1) by σ, and the corresponding functorial map $v^*f^\times \to g^\times u^*$ by β_σ (cf. (4.4.3), without assuming f and g to be quasi-proper). Let us first record the following elementary *transitivity* properties of β_σ.

Proposition 4.6.8 *For any commutative diagram*

$$
\begin{array}{ccc}
X'' & \xrightarrow{v_1} X' & \xrightarrow{v} X \\
h\downarrow \quad \sigma_1 \quad g\downarrow & \sigma & \downarrow f \\
Y'' & \xrightarrow{u_1} Y' & \xrightarrow{u} Y
\end{array}
\qquad or \qquad
\begin{array}{ccc}
Z' & \xrightarrow{w} & Z \\
g_1\downarrow & \sigma_1 & \downarrow f_1 \\
X' & \xrightarrow{v} & X \\
g\downarrow & \sigma & \downarrow f \\
Y' & \xrightarrow{u} & Y
\end{array}
$$

where both σ and σ_1 are independent squares—whence so is the composed square $\sigma_0 := \sigma\sigma_1$ see (3.10.3.2)—the following resulting diagrams of functorial maps commute:

$$
\begin{array}{ccc}
(vv_1)^* f^\times & \xrightarrow{\ \beta_{\sigma_0}\ } & h^\times (uu_1)^* \\
\simeq\downarrow & & \downarrow\simeq \\
v_1^* v^* f^\times & \xrightarrow[v_1^*\beta_\sigma]{} v_1^* g^\times u^* \xrightarrow[\beta_{\sigma_1}]{} & h^\times u_1^* u^*
\end{array}
$$

$$
\begin{array}{ccc}
w^*(ff_1)^\times & \xrightarrow{\ \beta_{\sigma_0}\ } & (gg_1)^\times u^* \\
\simeq\downarrow & & \downarrow\simeq \\
w^* f_1^\times f^\times & \xrightarrow[\beta_{\sigma_1}]{} g_1^\times v^* f^\times \xrightarrow[g_1^\times \beta_\sigma]{} & g_1^\times g^\times u^*
\end{array}
$$

Proof. (Sketch.) Using the definition of β, one reduces mechanically to proving the transitivity properties for θ in (3.7.2), (ii) and (iii).   Q.E.D.

Assuming $(4.4.3)^\circ$, we first reduce (4.4.3) to the case where Y is affine. Let $(\mu_i \colon Y_i \to Y)_{i\in I}$ be an open covering of Y with each Y_i affine. Consider the diagrams, with σ as in (4.4.1),

$$
\begin{array}{ccc}
X_i' & \xrightarrow{v_i} X_i & \xrightarrow{\nu_i} X \\
g_i\downarrow \quad \sigma_i \quad f_i\downarrow & \tau_i & \downarrow f \\
Y_i' & \xrightarrow{u_i} Y_i & \xrightarrow{\mu_i} Y
\end{array}
$$

$$
\begin{array}{ccc}
X_i' & \xrightarrow{\nu_i'} X' & \xrightarrow{v} X \\
g_i\downarrow \quad \tau_i' \quad g\downarrow & \sigma & \downarrow f \\
Y_i' & \xrightarrow{\mu_i'} Y' & \xrightarrow{u} Y
\end{array}
$$

where $Y_i' := Y' \times_Y Y_i$, u_i and μ_i' are the projections, and all the squares are fiber squares. The composed squares $\tau_i \sigma_i$ and $\sigma \tau_i'$ are identical. The squares τ_i and τ_i' are independent because μ_i and μ_i' are open immersions; and by $(4.4.3)^\circ$, β_{τ_i} and $\beta_{\tau_i'}$ are isomorphisms.

Furthermore, since f is quasi-proper therefore so are the maps f_i. The map u_i, which agrees over Y_i with u, has finite tor-dimension. By $(3.10.3.4)$, the square $\sigma_i \cong \mu_i^* \sigma$ is independent. Thus if $(4.4.3)$ holds whenever Y is affine, then β_{σ_i} is an isomorphism, and $(4.6.8)$ shows that so are $\beta_{\sigma\tau_i'}$ $(= \beta_{\tau_i\sigma_i})$ and $\nu_i'^* \beta_\sigma$. Since $(\nu_i' \colon X_i' \to X')_{i \in I}$ is an open covering of X', and since isomorphism can be checked locally (see $(1.2.2)$), it follows that β_σ is an isomorphism, whence the asserted reduction.

Next, again assuming $(4.4.3)^\circ$, we reduce $(4.4.3)$ with affine Y to where Y' too is affine. That will complete the proof, since when both Y and Y' are affine then so is u, and $(4.4.3)^{\text{af}}$ applies.

Let $(\nu_j \colon Y_j' \to Y')_{j \in J}$ be an open covering of Y' with each Y_j' affine. Consider the diagram, with affine Y and σ as in $(4.4.1)$,

$$
\begin{array}{ccccc}
X_j' & \xrightarrow{\;v_j\;} & X' & \xrightarrow{\;v\;} & X \\
{\scriptstyle g_j}\downarrow & {\scriptstyle \sigma_j} & {\scriptstyle g}\downarrow & {\scriptstyle \sigma} & \downarrow{\scriptstyle f} \\
Y_j' & \xrightarrow{\;\nu_j\;} & Y' & \xrightarrow{\;u\;} & Y
\end{array}
$$

where σ_j is a fiber square, hence independent. By $(4.4.3)^\circ$, β_{σ_j} is an isomorphism. If $(4.4.3)$ holds for independent squares whose bottom corners are affine, then $\beta_{\sigma\sigma_j}$ is an isomorphism; and so by $(4.6.8)$, $v_j^* \beta_\sigma$ is also an isomorphism. As before, then, β_σ is an isomorphism, and we have the desired reduction.   Q.E.D.

V. $(4.4.3) \Rightarrow (4.4.3)^\circ \Leftrightarrow (4.4.2)$

The first implication is trivial. The implication $(4.4.2) \Rightarrow (4.4.3)^\circ$ is contained in what we have already done, but it's more direct than that, as we'll see. Incidentally, the following argument does not need f to be quasi-proper.

Let us first deduce $(4.4.2)$ from $(4.4.3)^\circ$. As in $(4.6.4)$, via $(4.6.5)$, there is for any $F \in \mathbf{D}(X)$, $G \in \mathbf{D}(Y)$ a commutative diagram

$$
\begin{array}{ccc}
u^* f_* \mathcal{H}_X(F,\, f^\times G) & \xrightarrow{\;u^* \delta\;} & u^* \mathcal{H}_Y(f_* F,\, G) \\
\downarrow & & \downarrow \\
g_* \mathcal{H}_{X'}(v^* F,\, v^* f^\times G) & \xrightarrow[\;(4.4.1)\;]{} & \mathcal{H}_{Y'}(g_* v^* F,\, u^* G)
\end{array}
\qquad (4.6.9)
$$

When u (hence v) is an open immersion, then the vertical arrows in this diagram are *isomorphisms*. Indeed, these arrows are combinations of ρ and θ, ρ being an isomorphism by $(4.6.7)$, and $\theta(L) \colon u^* f_* L \to g_* v^* L$ being an isomorphism for *any* $L \in \mathbf{D}(X)$, as follows easily from $(2.4.5.2)$ after L is replaced by a q-injective resolution. Furthermore, the functor $\mathbf{\Gamma}_{Y'} := \mathbf{R}\Gamma(Y', -)$ transforms the bottom row of (4.6.9) into an isomorphism. This follows from commutativity of the next diagram, obtained via Exercise (3.2.5)(f) by application of $\mathbf{\Gamma}_{Y'}$ to the commutative diagram (4.4.3.1), and where, under the present assumption of $(4.4.3)^\circ$, β is an isomorphism:

$$
\mathbf{H}_{X'}(F', v^* f^\times G)
$$

$$
\mathbf{\Gamma}_{Y'}(4.4.1) \swarrow \qquad \searrow \text{ via } \beta \qquad\qquad (4.6.10)
$$

$$
\mathbf{H}_{Y'}(g_* F', u^* G) \xleftarrow[\;(4.1.1)\;]{\sim} \mathbf{H}_{X'}(F', g^\times u^* G)
$$

We conclude that $\mathbf{\Gamma}_{Y'} u^* \delta$ is an isomorphism whenever $u \colon Y' \to Y$ is an open immersion; and then (4.4.2) results from:

Lemma 4.6.11 *Let $\phi \colon G_1 \to G_2$ be a map in $\mathbf{D}(Y)$. Then ϕ is an isomorphism iff for every open immersion $u \colon Y' \hookrightarrow Y$ with Y' affine, the map*

$$
\mathbf{\Gamma}_{Y'} u^*(\phi) \colon \mathbf{\Gamma}_{Y'} u^*(G_1) \to \mathbf{\Gamma}_{Y'} u^*(G_2)
$$

is an isomorphism.

Proof. Write $\Gamma_{Y'}$ for the sheaf-functor $\Gamma(Y', -)$. We may assume that G_1 and G_2 are q-injective and that ϕ is actually a map of complexes, see (2.3.8)(v), so that $\mathbf{\Gamma}_{Y'} u^*(\phi)$ is the map $\Gamma_{Y'}(\phi) \colon \Gamma_{Y'}(G_1) \to \Gamma_{Y'}(G_2)$. If $\mathbf{\Gamma}_{Y'} u^*(\phi)$ is an isomorphism, then the homology maps

$$
H^p \Gamma_{Y'}(\phi) \colon H^p \Gamma_{Y'}(G_1) \to H^p \Gamma_{Y'}(G_2) \qquad (p \in \mathbb{Z})
$$

are all isomorphisms; and since $H^p(G_i)$ is the sheaf associated to the presheaf $Y' \mapsto H^p \Gamma_{Y'}(G_i)$ ($i = 1, 2$), it follows for every $p \in \mathbb{Z}$ that the map $H^p(\phi) \colon H^p(G_1) \to H^p(G_2)$ is an isomorphism, so that by (1.2.2), ϕ is an isomorphism. The converse is obvious.   Q.E.D.

Conversely, if (4.4.2) holds, then the top row—and hence the bottom row—in (4.6.9) is an isomorphism. We deduce from (4.6.10) that

$$
\mathbf{H}_{X'}(F', v^* f^\times G) \xrightarrow{\text{ via } \beta} \mathbf{H}_{X'}(F', g^\times u^* G)
$$

is an isomorphism for all F', whence (taking homology, see (2.4.2)) that

$$\mathrm{Hom}_{\mathbf{D}(X')}(F',\, v^*f^\times G) \xrightarrow{\;\;\text{via }\beta\;\;} \mathrm{Hom}_{\mathbf{D}(X')}(F',\, g^\times u^*G)$$

is an isomorphism for all F', so that β itself is an isomorphism.   Q.E.D.

It remains to prove (4.6.5) and (4.6.7).

Proof of (4.6.5). One verifies, using the definitions of ν, of θ (via (3.7.2)(a)) and of ρ, and the line following (3.5.4.2), that in the big diagram on the following page—with natural maps, and in which α denotes the map (3.5.4.2) = (3.5.4.3) (of which the isomorphism (3.2.3.2) is an instance, see (3.2.4)(i))—the outer border is adjoint to the diagram in (4.6.5). Therefore it will suffice to show that all the subdiagrams in the big diagram commute.

For the unnumbered subdiagrams commutativity is clear. Commutativity of ① follows from the definition of ρ; of ② from the definition of θ via (3.7.2)(a); of ③ from (3.7.1.1) (with β replaced by α, etc.); and of ④ from the definition of θ via (3.7.2)(c).   Q.E.D.

Proof of (4.6.7). For this proof, we drop the abbreviations introduced at the beginning of §4.4. Thus u_* and u^* will now denote the usual sheaf-functors, and $\mathbf{R}u_*$, $\mathbf{L}u^*$ their respective derived functors. Similarly, $\mathcal{H}$ will denote the functor $\mathcal{H}om^\bullet$ of complexes, and $\mathbf{R}\mathcal{H}om^\bullet$ its derived functor.

We need to understand ρ more concretely, and to that end we will establish commutativity of the following diagram of natural maps, for any complexes E, H of $\mathcal{O}_Y$-modules:

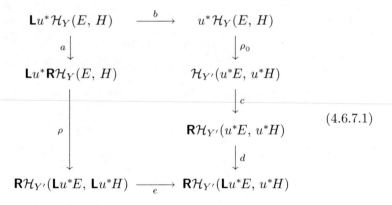

$$(4.6.7.1)$$

Here ρ_0 is adjoint to the natural composite map of complexes

$$\xi\colon \mathcal{H}_Y(E,\,H) \to \mathcal{H}_Y(E,\,u_*u^*H) \xrightarrow[\;(3.1.6)\;]{\sim} u_*\mathcal{H}_{Y'}(u^*E,\,u^*H).$$

This ξ is such that for any open $U \subset Y$, $\Gamma(U, \xi)$ is the map

$$\prod_{i\in\mathbb{Z}} \mathrm{Hom}_U(E^i,\,H^{i+n}) \to \prod_{i\in\mathbb{Z}} \mathrm{Hom}_{f^{-1}U}(u^*E^i,\,u^*H^{i+n})$$

arising from the functoriality of u^*.

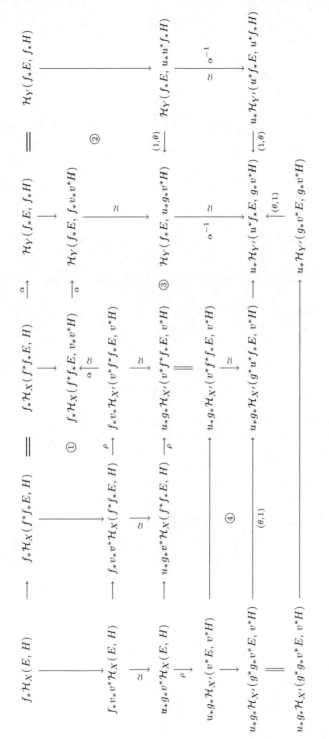

Commutativity of (4.6.7.1) is equivalent to commutativity of the following "adjoint" diagram:[10]

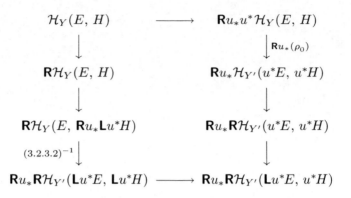

But in this diagram the two maps obtained by going around from the top left to the bottom right clockwise and counterclockwise respectively, are both equal to the natural composition

$$\mathcal{H}_Y(E, H) \longrightarrow \mathcal{H}_Y(E, u_*u^*H) \xrightarrow{(3.1.5)^{-1}} u_*\mathcal{H}_{Y'}(u^*E, u^*H)$$
$$\longrightarrow \mathbf{R}u_*\mathcal{H}_{Y'}(u^*E, u^*H) \longrightarrow \mathbf{R}u_*\mathbf{R}\mathcal{H}_{Y'}(u^*E, u^*H)$$
$$\longrightarrow \mathbf{R}u_*\mathbf{R}\mathcal{H}_{Y'}(\mathbf{L}u^*E, u^*H),$$

as shown by the commutativity of the following two diagrams. (In the first, the top three horizontal arrows come from the natural functorial composition $1 \to u_*u^* \to \mathbf{R}u_*u^*$; and the right column is $\mathbf{R}u_*(\rho_0)$.)

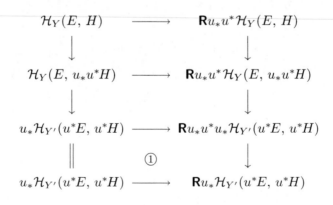

---

[10] Recall that by (3.2.4)(i), the map (3.2.3.2) is an instance of the map (3.5.4.3).

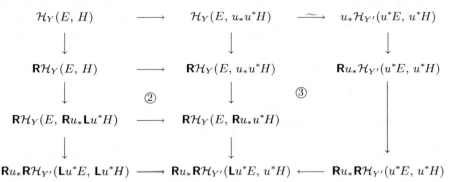

Commutativity of subdiagram ① follows from the natural functorial composition $u_* \to u_* u^* u_* \to u_*$ being the identity. Commutativity of ② follows from that of (3.2.1.3). Commutativity of ③ follows from that of the diagram immediately following (3.2.3.2).

Thus (4.6.7.1) does indeed commute.

Proceeding now with the proof of (4.6.7), suppose that E is a bounded-above complex of finite-rank locally free $\mathcal{O}_Y$-modules, and that $H \in \overline{\mathbf{D}}^+(Y)$. To show that ρ is an isomorphism, we may assume that H is a complex of u^*-acyclic $\mathcal{O}_Y$-modules, bounded below if u has finite tor-dimension, see (2.7.5)(vi). Then in (4.6.7.1), d and e are isomorphisms; and $\mathcal{H}_Y(E, H)$ is also a complex of u^*-acyclic $\mathcal{O}_Y$-modules (the question being local on Y), so that b too is an isomorphism, see (2.7.5)(a). That ρ_0 is an isomorphism follows from the fact that (exercise) its stalk at $y' \in Y'$ is—with $y := u(y')$, $R' := \mathcal{O}_{Y', y'}$ and $R := \mathcal{O}_{Y, y}$—the natural map

$$R' \otimes_R \operatorname{Hom}_R(E_y, H_y) \to \operatorname{Hom}_{R'}(R' \otimes_R E_y, R' \otimes_R H_y).$$

It remains to be shown that a and c are isomorphisms. For a, it suffices that if $H \to I$ is a quasi-isomorphism with I injective and bounded-below, then the resulting map $\mathcal{H}_Y(E, H) \to \mathcal{H}_Y(E, I)$ be an isomorphism. Since $\mathcal{H}_Y$ is a Δ-functor, and by the footnote under (1.5.1), it is equivalent to show that if C is the summit of a triangle whose base is $H \to I$ (so that C is exact), then $\mathcal{H}_Y(E, C)$ is exact. For any $n \in \mathbb{Z}$, to show that $H^n \mathcal{H}_Y(E, C) = 0$ we may assume that $E \neq 0$, let $m_0 = m_0(E)$ be the least integer such that $E^m = 0$ for all $m > m_0$, and argue by induction on m_0, as follows.

If $m_0 \ll 0$, then $\mathcal{H}_Y(E, C)$ vanishes in degree n, so the assertion is obvious. Proceeding inductively, set $i = m_0(E)$, and let $E_{<i}$ be the complex which agrees with E in all degrees $< i$, and vanishes in all degrees $\geq i$, so that we have a natural semi-split exact sequence

$$0 \to E^i[-i] \to E \to E_{<i} \to 0,$$

and a corresponding triangle, cf. (1.4.3.3). There results an exact homology sequence, see $(1.4.5)^H$:

$$H^n \mathcal{H}_Y(E_{<i}, C) \to H^n \mathcal{H}_Y(E, C) \to H^{n+i} \mathcal{H}_Y(E^i, C)$$

in which the first term vanishes by the inductive hypothesis, and the last term vanishes because E^i is locally free of finite rank and C is exact. Hence $H^n\mathcal{H}_Y(E, C)$ also vanishes, as desired. Thus a is indeed an isomorphism. Similarly c is an isomorphism. Hence, finally, so is ρ.

For the last assertion in (4.6.7), suppose u is an open immersion. It is left as an exercise to show that now ρ_0 is just the obvious restriction map. To show that ρ is an isomorphism we may assume that H—and hence u^*H—is q-injective, see (2.4.5.2). Clearly, then, all the maps in (4.6.7.1) other than ρ are isomorphisms, whence so is ρ.   Q.E.D.

4.7 Quasi-Perfect Maps

Again, all schemes are assumed to be concentrated.

In this section, for a scheme-map $f\colon X \to Y$ the functor $f^\times$ will be as in (4.1.1), *but* restricted to $\mathbf{D}_{\mathsf{qc}}(Y)$; in other words, $f^\times$ is always to be regarded as a functor from $\mathbf{D}_{\mathsf{qc}}(Y)$ to $\mathbf{D}_{\mathsf{qc}}(X)$.

Quasi-perfect maps are scheme-maps $f\colon X \to Y$ characterized by any one of several nice properties preserved by tor-independent base change (see (4.7.3.1)). Among those properties are the following, the first two by (4.7.1), and the next two by (4.7.4) and (4.7.6)(d):

• $f^\times$ commutes with small direct sum in $\mathbf{D}_{\mathsf{qc}}$ (i.e., direct sum of any family indexed by a small set, see §4.1).

• For all $F \in \mathbf{D}_{\mathsf{qc}}(Y)$ the natural map is an isomorphism

$$\chi_F\colon f^\times\mathcal{O}_Y \otimes \mathbf{L}f^*F \xrightarrow{\;\sim\;} f^\times F.$$

• $f^\times$ is a bounded functor, and it satisfies *universal* tor-independent base change, that is, for *any* independent square as in (4.4.1), and for any $G \in \mathbf{D}_{\mathsf{qc}}(Y)$—not necessarily in $\overline{\mathbf{D}}_{\mathsf{qc}}^+(Y)$—the base-change map $\beta(G)$ in (4.4.3) is an isomorphism.

• $f^\times$ is a bounded functor, and these two conditions hold:

(i) For all $F \in \mathbf{D}_{\mathsf{qc}}(X)$ the duality map (4.2.1) is an isomorphism

$$\mathbf{R}f_*\mathbf{R}\mathcal{H}om^\bullet(F, f^\times\mathcal{O}_Y) \xrightarrow{\;\sim\;} \mathbf{R}\mathcal{H}om_Y^\bullet(\mathbf{R}f_*F, \mathcal{O}_Y).$$

(ii) If (F_α) is a small directed system of flat quasi-coherent $\mathcal{O}_Y$-modules then for any $n \in \mathbb{Z}$ the natural map is an isomorphism

$$\varinjlim_\alpha H^n(f^\times F_\alpha) \xrightarrow{\;\sim\;} H^n(f^\times \varinjlim_\alpha F_\alpha).$$

It follows that quasi-perfection of f implies the following; and in fact when Y is separated the converse is true, see (4.7.4):

• $f^\times$ is a bounded functor, and the above natural map χ_F is an isomorphism whenever F is a flat quasi-coherent $\mathcal{O}_Y$-module.

Further, though we won't prove it here, the main result Theorem 1.2 in [**LN**] is the equivalence of the following conditions:

(i) f is quasi-perfect.

(ii) f is quasi-proper (4.3.3.1) and has finite tor-dimension.

(iii) f is quasi-proper and the functor $f^\times$ is bounded.

We call a scheme-map f perfect if f is pseudo-coherent and of finite tor-dimension. (For pseudo-coherent f, being of finite tor-dimension is equivalent to boundedness of $f^\times$, see [**LN**, Thm. 1.2]).

For example, since finite-type maps of noetherian schemes are always pseudo-coherent, the foregoing and (4.3.9) show that a separated such map is quasi-perfect if and only if it is proper and perfect.

Perfect maps of noetherian schemes will be treated in §4.9.

Before proceeding, we review a few basic facts about *perfect complexes*. A complex in $E \in \mathbf{D}(X)$ (X a scheme) is said to be perfect if it is locally **D**-isomorphic to a strictly perfect complex, i.e., a bounded complex of finite-rank free $\mathcal{O}_X$-modules. More precisely, E is said to have *perfect amplitude in* $[a, b]$ ($a \le b \in \mathbb{Z}$) if locally on X, E is **D**-isomorphic to a strictly perfect complex vanishing in all degrees which are $< a$ or $> b$. Thus E is perfect iff it has perfect amplitude in some interval $[a, b]$. By [**I**, p. 134, 5.8], this condition is equivalent to E being pseudo-coherent and also having flat amplitude in $[a, b]$ (i.e., being globally **D**-isomorphic to a flat complex vanishing in all degrees $< a$ and $> b$). So E is perfect iff it is pseudo-coherent and of finite tor-dimension (that is, **D**-isomorphic to a bounded flat complex, see (3.9.4.2)(b)).

Proposition 4.7.1 (Neeman). *For any scheme-map $f \colon X \to Y$, the following conditions, with $f^\times$ as in (4.1.1), are equivalent:*

(i) *$f^\times$ respects direct sums (see (3.8.1)) in $\mathbf{D}_{\mathsf{qc}}$, i.e., for any small $\mathbf{D}_{\mathsf{qc}}(Y)$-family (F_α) the natural map is an isomorphism*

$$\underset{\alpha}{\oplus} f^\times F_\alpha \xrightarrow{\ \sim\ } f^\times(\underset{\alpha}{\oplus} F_\alpha).$$

(ii) *The functor $\mathbf{R}f_*$ takes perfect complexes to perfect complexes.*

(iii) *The functor $f^\times$ has a right adjoint.*

(iv) *For all $F \in \mathbf{D}_{\mathsf{qc}}(Y)$, the map adjoint to*

$$\mathbf{R}f_*(f^\times \mathcal{O}_Y \underset{=}{\otimes} \mathbf{L}f^*F) \underset{(3.9.4)}{\xrightarrow{\ \sim\ }} \mathbf{R}f_* f^\times \mathcal{O}_Y \underset{=}{\otimes} F \underset{\text{via } \tau}{\xrightarrow{\hspace{1cm}}} F$$

is an isomorphism

$$f^\times \mathcal{O}_Y \underset{=}{\otimes} \mathbf{L}f^*F \xrightarrow{\ \sim\ } f^\times F.$$

Proof. (i) $\Leftrightarrow$ (ii): [**N**, p. 215, Prop. 2.5 and Cor. 2.3; and p. 224, Thm. 5.1 (where every $s \in S$ is implicitly assumed to be compact)].

(i) $\Rightarrow$ (iii): [**N**, p. 215, Prop. 2.5; p. 207, lines 12–13; and p. 223, Thm. 4.1].

(iii) $\Rightarrow$ (i): simple.

(i) $\Rightarrow$ (iv) $\Rightarrow$ (i): For the first $\Rightarrow$ see [**N**, p. 226, Thm. 5.4]. The second implication follows from (3.8.2).

Strictly speaking, the referenced results in [**N**] are proved for *separated* schemes; but in view of [**BB**, p. 9, Thm. 3.1.1] one readily verifies that the proofs are valid for any concentrated scheme.   Q.E.D.

Definition 4.7.2 A map $f: X \to Y$ is quasi-perfect if it satisfies the conditions in (4.7.1).

Remark. The fact, mentioned above, that quasi-perfect maps are quasi-proper results from (4.7.1)(ii) and [**LN**, Cor. 4.3.2], which says that f is quasi-proper if and (clearly) only if $\mathbf{R}f_*$ takes perfect complexes to pseudo-coherent complexes.

Example 4.7.3 (a)   *Any   quasi-proper   scheme-map f of   finite   tor-dimension—so by (4.3.3.2), any proper perfect map, in particular, any flat finitely-presentable proper map—is quasi-perfect.*

Indeed $\mathbf{R}f_*$ preserves both pseudo-coherence of complexes and—by [**I**, p. 250, 3.7.2] (a consequence of (3.9.4) above)—finite tor-dimensionality of complexes; so (4.7.1)(ii) holds.

(b) Let $f: X \to Y$ be a scheme-map with X *divisorial,* i.e., X has an ample family $(\mathcal{L}_i)_{i \in I}$ of invertible $\mathcal{O}_X$-modules [**I**, p. 171, Défn. 2.2.5].

Then [**N**, p. 211, Example 1.11 and p. 224, Theorem 5.1] imply that *f is quasi-perfect $\Leftrightarrow$ for each $i \in I$, there is an integer n_i such that the $\mathcal{O}_Y$-complex $\mathbf{R}f_*(\mathcal{L}_i^{\otimes -n})$ is perfect for all $n \geq n_i$.*

(c) (Cf. (4.3.8).) Let f be quasi-projective and let $\mathcal{L}$ be an f-ample invertible $\mathcal{O}_X$-module. Then:
f is quasi-perfect $\Leftrightarrow$ the $\mathcal{O}_Y$-complex $\mathbf{R}f_*(\mathcal{L}^{\otimes -n})$ is perfect for all $n \gg 0$
$$\Rightarrow f \text{ is perfect.}$$
Indeed, condition (4.7.1)(ii), together with the compatibility of $\mathbf{R}f_*$ and open base change, implies that quasi-perfection is a property of f which is local on Y, and the same holds for perfection of $\mathbf{R}f_*(\mathcal{L}^{\otimes -n})$; so for the $\Leftrightarrow$ we may assume Y affine, and apply (b). The $\Rightarrow$ is given by (4.7.3.3) below.

(d) For a *finite* map $f: X \to Y$ the following are equivalent:
   (i) *f is quasi-perfect.*
   (ii) *f is perfect.*
   (iii) *The complex $f_*\mathcal{O}_X \cong \mathbf{R}f_*\mathcal{O}_X$ is perfect.*
Indeed, the implication (i) $\Rightarrow$ (iii) is given by (4.7.1)(ii). If (iii) holds then f has finite tor-dimension (see (2.7.6.4)), and as in the first part of the proof of (4.3.8), f is pseudo-coherent; thus f is perfect. The implication (ii) $\Rightarrow$ (i) is given by (a).

Proposition 4.7.3.1 *For any independent square of scheme-maps,*

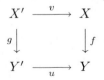

(i) *if f is quasi-perfect then so is g; and*

(ii) *if the (bounded-below) functor $f^\times\colon \mathbf{D}_{\mathsf{qc}}(Y) \to \mathbf{D}_{\mathsf{qc}}(X)$ is bounded above, then so is $g^\times\colon \mathbf{D}_{\mathsf{qc}}(Y') \to \mathbf{D}_{\mathsf{qc}}(X')$.*

Hence, if $(Y_i)_{i\in I}$ is an open cover of Y then

(iii) *f is quasi-perfect $\Leftrightarrow$ for all i, the same is true of the induced map $f^{-1}Y_i \to Y_i$; and*

(iv) *if f is quasi-proper then $f^\times$ is bounded above $\Leftrightarrow$ for all i, the same is true of the induced map $f^{-1}Y_i \to Y_i$.*

Proof. To begin with, (iii) follows easily from (i) and (4.7.1)(ii); and (iv) follows from (ii) and (4.4.3).

In the rest of this proof, quasi-perfection is characterized by (4.7.1)(i).

Suppose first that Y' is separated. We induct on $q = q(Y')$, the least number of affine open subschemes needed to cover Y'.

If $q = 1$ then the map u is affine, whence so is v [**GD**, p. 358, (9.1.16), (v) and (iii)]; so to prove (i) (resp. (ii)) it suffices, by (3.10.2.2), to show that for any small $\mathbf{D}_{\mathsf{qc}}(Y')$-family (F_α) the natural map is an isomorphism

$$\underset{\alpha}{\oplus}\mathbf{R}v_*g^\times F_\alpha \overset{(3.9.3.3)}{\cong} \mathbf{R}v_*\big(\underset{\alpha}{\oplus}g^\times F_\alpha\big) \overset{\sim}{\longrightarrow} \mathbf{R}v_*g^\times\big(\underset{\alpha}{\oplus}F_\alpha\big)$$

(resp.—since every $G \in \mathbf{D}_{\mathsf{qc}}(X')$ is isomorphic to a quasi-coherent, hence v_*-acyclic, $\mathcal{O}_{X'}$-complex G', see (2.7.5)(a), so that

$$H^n(\mathbf{R}v_*G) \cong H^n(v_*G') \cong v_*H^n(G') = 0 \implies H^n(G) \cong H^n(G') = 0$$

—that $\mathbf{R}v_*g^\times\colon \mathbf{D}_{\mathsf{qc}}(Y') \to \mathbf{D}_{\mathsf{qc}}(X)$ is bounded). Since $\mathbf{R}u_*$ is bounded (see (3.9.2.3)), the second of these facts results from the natural isomorphism $\mathbf{R}v_*g^\times \overset{\sim}{\longrightarrow} f^\times\mathbf{R}u_*$ of (3.10.4). The first results from the (easily-checked) commutativity of

$$
\begin{array}{ccccc}
\underset{\alpha}{\oplus}\mathbf{R}v_*g^\times F_\alpha & \xrightarrow[(3.9.3.3)]{\sim} & \mathbf{R}v_*\big(\underset{\alpha}{\oplus}g^\times F_\alpha\big) & \longrightarrow & \mathbf{R}v_*g^\times\big(\underset{\alpha}{\oplus}F_\alpha\big) \\
\simeq \big\downarrow {\scriptstyle(3.10.4)} & & & & {\scriptstyle(3.10.4)}\big\downarrow \simeq \\
\underset{\alpha}{\oplus}f^\times\mathbf{R}u_*F_\alpha & \longrightarrow & f^\times\big(\underset{\alpha}{\oplus}\mathbf{R}u_*F_\alpha\big) & \xrightarrow[(3.9.3.3)]{\sim} & f^\times\mathbf{R}u_*\big(\underset{\alpha}{\oplus}F_\alpha\big)
\end{array}
$$

Suppose $q > 1$, so $Y' = Y_1' \cup Y_2'$ with Y_i' open in Y', $q(Y_1') = q - 1$, and $q(Y_2') = 1$. Set $Y_{12}' := Y_1' \cap Y_2'$, so that $q(Y_{12}') \le q-1$. (Y' being separated, the intersection of affine subschemes of Y' is affine). We have the commutative diagram of immersions

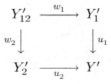

With $u_{12} := u_1 w_1 = u_2 w_2$ there is, for any $F \in \mathbf{D}(Y')$, a natural triangle

$$F \to \mathbf{R}u_{1*}u_1^* F \oplus \mathbf{R}u_{2*}u_2^* F \to \mathbf{R}u_{12*}u_{12}^* F \to F[1] \qquad (4.7.3.2)$$

obtained by applying the standard exact sequence—holding for any injective (or even flasque) $\mathcal{O}_{Y'}$-module G—

$$0 \to G \to u_{1*}u_1^* G \oplus u_{2*}u_2^* G \to u_{12*}u_{12}^* G \to 0$$

to an injective q-injective resolution of F (see paragraph around (1.4.4.2)).

The inductive hypothesis applied to the natural composite independent square (see (3.10.3.2)), with $i = 1, 2, 12$,

$$
\begin{array}{ccccc}
X_i' & \xrightarrow{\;v_i\;} & X' & \xrightarrow{\;v\;} & X \\
\Big\downarrow{\scriptstyle g_i} & & \Big\downarrow{\scriptstyle g} & & \Big\downarrow{\scriptstyle f} \\
Y_i' & \xrightarrow{\;u_i\;} & Y' & \xrightarrow{\;u\;} & Y
\end{array}
$$

gives that $g_i^\times$ is bounded. Since $\mathbf{R}v_{i*}$ is bounded (3.9.2.3), therefore so is

$$g^\times \mathbf{R}u_{i*}u_i^* \underset{(3.10.4)}{\cong} \mathbf{R}v_{i*}g_i^\times u_i^*.$$

Hence, application of the Δ-functor $g^\times$ to the triangle (4.7.3.2) shows that $g^\times$ is bounded above, proving (ii).

As for (i), in view of $(\Delta 3)^*$ of §1.4 it similarly suffices to show (left as an exercise) that the following natural diagram—whose columns are triangles (see (3.8.3)), and where the two middle arrows are isomorphisms by (3.9.3.3), by the inductive hypothesis, and by (3.8.2)(ii) (for the trivial case of an open immersion)—commutes:

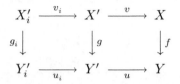

$$
\begin{array}{ccc}
\bigoplus\limits_{\alpha} g^\times F_\alpha & \longrightarrow & g^\times\!\left(\bigoplus\limits_{\alpha} F_\alpha\right) \\
\Big\downarrow & & \Big\downarrow \\
\bigoplus\limits_{\alpha}\!\left(\mathbf{R}v_{1*}g_1^\times u_1^* F_\alpha \oplus \mathbf{R}v_{2*}g_2^\times u_2^* F_\alpha\right) & \xrightarrow{\;\sim\;} & \left(\mathbf{R}v_{1*}g_1^\times u_1^* \bigoplus\limits_{\alpha} F_\alpha\right)\!\oplus\!\left(\mathbf{R}v_{2*}g_2^\times u_2^* \bigoplus\limits_{\alpha} F_\alpha\right) \\
\Big\downarrow & & \Big\downarrow \\
\bigoplus\limits_{\alpha} \mathbf{R}v_{12*}g_{12}^\times u_{12}^* F_\alpha & \xrightarrow{\;\sim\;} & \mathbf{R}v_{12*}g_{12}^\times u_{12}^* \bigoplus\limits_{\alpha} F_\alpha \\
\Big\downarrow & & \Big\downarrow \\
\bigoplus\limits_{\alpha} g^\times F_\alpha[1] & \longrightarrow & g^\times\!\left(\bigoplus\limits_{\alpha} F_\alpha\right)[1]
\end{array}
$$

Having thus settled the separated case, we can proceed similarly for arbitrary concentrated Y', with $q(Y')$ the least number of separated open subschemes needed to cover Y'.   Q.E.D.

Proposition 4.7.3.3 *Let* $f\colon X \to Y$ *be a* locally embeddable *scheme-map, i.e., every* $y \in Y$ *has an open neighborhood* V *over which the induced map* $f^{-1}V \to V$ *factors as* $f^{-1}V \xrightarrow{i} Z \xrightarrow{p} V$ *where* i *is a closed immersion and* p *is smooth. (For instance, any quasi-projective* f *satisfies this condition* [**EGA**, II, (5.3.3)].*) If* f *is quasi-perfect then* f *is perfect.*

Proof. (i) By (4.7.3.1)(iii), quasi-perfection is local over Y, and the same clearly holds for perfection; so we may as well assume that $X = f^{-1}V$. Then by [**I**, p. 252, Prop. 4.4] it suffices to show that the complex $i_*\mathcal{O}_X$ is perfect, or, more generally, that the map i is quasi-perfect. But i factors as $X \xrightarrow{\gamma} X \times_Y Z \xrightarrow{g} Z$ where γ is the graph of i and g is the projection. The map γ is a local complete intersection [**EGA**, IV, (17.12.3)], so the complex $\gamma_*\mathcal{O}_X$ is perfect, and by Example (4.7.3)(d) (or otherwise) γ is quasi-perfect. Also, g arises from f by flat base change, so by (4.7.3.1)(i), g is quasi-perfect. Hence $i = g\gamma$ is quasi-perfect, as desired.   Q.E.D.

Remark. Using the analog of (4.7.3.1)(i) with "quasi-proper" in place of "quasi-perfect" [**LN**, Prop. 4.4], one shows similarly for locally embeddable f that f *quasi-proper* $\Rightarrow$ f *pseudo-coherent*. The converse holds when f is also proper, see (4.3.3.2). Thus, e.g., a projective map is quasi-proper if and only if it is pseudo-coherent.

Exercises 4.7.3.4 For a scheme-map $f\colon X \to Y$ and for $E,\, F \in \mathbf{D}_{\mathrm{qc}}(Y)$, let

$$\chi_{E,F}\colon f^\times E \otimes \mathsf{L}f^*F \longrightarrow f^\times(E \otimes F).$$

be the map adjoint to

$$\mathsf{R}f_*(f^\times E \otimes \mathsf{L}f^*F) \xrightarrow[\ (3.9.4)\ ]{\sim} \mathsf{R}f_* f^\times E \otimes F \xrightarrow{\ \text{via } \tau\ } E \otimes F.$$

In particular, $\chi_{\mathcal{O}_Y, F}$ is the map in (4.7.1)(iv).

(a) Show that for any $E,\, F,\, G \in \mathbf{D}_{\mathrm{qc}}(Y)$, the following diagram commutes.

$$
\begin{array}{ccccc}
f^\times E \otimes (\mathsf{L}f^*F \otimes \mathsf{L}f^*G) & \xrightarrow{\ \text{via } (3.2.4)\ } & f^\times E \otimes \mathsf{L}f^*(F \otimes G) & \xrightarrow{\chi_{E,F\otimes G}} & f^\times(E \otimes (F \otimes G)) \\
\simeq \downarrow & & & & \downarrow \simeq \\
(f^\times E \otimes \mathsf{L}f^*F) \otimes \mathsf{L}f^*G & \xrightarrow{\ \chi_{E,F}\otimes 1\ } & f^\times(E \otimes F) \otimes \mathsf{L}f^*G & \xrightarrow{\chi_{E\otimes F,G}} & f^\times((E \otimes F) \otimes G)
\end{array}
$$

Taking $E = \mathcal{O}_Y$, deduce that f *is quasi-perfect if and only if* $\chi_{F,G}$ *is an isomorphism for all F and G.* (For this one needs that for any f the map defined in (4.7.1)(iv) is an isomorphism

$$f^\times \mathcal{O}_Y \otimes \mathsf{L}f^*\mathcal{O}_Y \xrightarrow{\sim} f^\times \mathcal{O}_Y, \tag{\#}$$

since, e.g., it factors naturally as $f^\times \mathcal{O}_Y \otimes \mathsf{L}f^*\mathcal{O}_Y \xrightarrow{\sim} f^\times \mathcal{O}_Y \otimes \mathcal{O}_X \xrightarrow{\sim} f^\times \mathcal{O}_Y$. In fact (#) obtains with any perfect complex in place of $\mathcal{O}_Y$: see [**N**, pp. 227–228 and p. 213]. Cf. also (4.7.5) below.)

Hint. Using 3.4.7(iv), show that the adjoint of the preceding diagram commutes.

(b) Show that, with 1 the identity map of Y, the map

$$\chi_{E,F} : E \underline{\otimes} F = \mathbf{1}^{\times} E \otimes \mathbf{1}^* F \to E \underline{\otimes} F$$

is the identity map.

(c) (Compatibility of χ and base change.) In this exercise, v^* is an abbreviation for $\mathbf{L}v^*$, and u^*, f^* and g^* are analogously understood. Also, $\otimes$ stands for $\underline{\otimes}$.

For any independent square

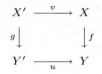

show that the following diagram, in which β comes from (4.4.3), and the unlabeled isomorphisms are the natural ones, commutes:

$$
\begin{array}{ccc}
v^* f^{\times} E \otimes v^* f^* F & \xrightarrow{\ \beta(E) \otimes 1\ } & g^{\times} u^* E \otimes v^* f^* F \\
\downarrow{\scriptstyle\simeq} & & {\scriptstyle\simeq}\downarrow \\
v^* (f^{\times} E \otimes f^* F) & & g^{\times} u^* E \otimes g^* u^* F \\
& & \downarrow{\scriptstyle\chi_{u^* E, u^* F}} \\
v^*\chi_{E,F} \Big\downarrow & & g^{\times} (u^* E \otimes u^* F) \\
& & \downarrow{\scriptstyle\simeq} \\
v^* f^{\times} (E \otimes F) & \xrightarrow[\ \beta(E \otimes F)\ ]{} & g^{\times} u^* (E \otimes F)
\end{array}
$$

Hint. It suffices to check commutativity of the following natural diagram, whose outer border is adjoint to that of the one in question.

$$
\begin{array}{ccc}
g_*(v^* f^{\times} E \otimes v^* f^* F) & \xrightarrow{\qquad\beta(E)\qquad} & g_*(g^{\times} u^* E \otimes v^* f^* F) \\
\| & & \downarrow{\scriptstyle\simeq}
\end{array}
$$

$$
\begin{array}{ccccc}
g_*(v^* f^{\times} E \otimes v^* f^* F) & \xrightarrow{\ \sim\ } & g_*(v^* f^{\times} E \otimes g^* u^* F) & \xrightarrow{\ \beta(E)\ } & g_*(g^{\times} u^* E \otimes g^* u^* F) \\
& & {\scriptstyle p}\downarrow & & \downarrow{\scriptstyle p} \\
& & g_* v^* f^{\times} E \otimes u^* F & \xrightarrow{\ \beta(E)\ } & g_* g^{\times} u^* E \otimes u^* F \\
& & \downarrow & & \downarrow \\
{\scriptstyle\simeq}\Big\downarrow \quad \text{cf. (3.7.3)} & & u^* f_* f^{\times} E \otimes u^* F & \longrightarrow & u^* E \otimes u^* F \\
& & \downarrow & & \downarrow{\scriptstyle\simeq} \\
& & u^*(f_* f^{\times} E \otimes F) & \longrightarrow & u^*(E \otimes F) \\
& & {\scriptstyle p}\downarrow & & \| \\
g_* v^*(f^{\times} E \otimes f^* F) & \longrightarrow & u^* f_*(f^{\times} E \otimes f^* F) & & \| \\
\downarrow & & \downarrow & & \| \\
g_* v^* f^{\times}(E \otimes F) & \xrightarrow{\ \sim\ } & u^* f_* f^{\times}(E \otimes F) & \longrightarrow & u^*(E \otimes F)
\end{array}
$$

(d) (Transitivity of χ). If $g\colon Y \to Z$ is a second scheme-map then the following natural diagram is commutative:

$$
\begin{array}{ccccc}
f^\times g^\times E \otimes \mathsf{L}f^*\mathsf{L}g^*F & \longrightarrow & f^\times(g^\times E \otimes \mathsf{L}g^*F) & \longrightarrow & f^\times(g^\times(E \otimes F)) \\
\simeq\downarrow & & & & \| \\
(gf)^\times E \otimes \mathsf{L}(gf)^*F & \longrightarrow & (gf)^\times(E \otimes F) & \xrightarrow{\;\sim\;} & f^\times g^\times(E \otimes F)
\end{array}
$$

Hint. Using (3.7.1), show that the adjoint diagram commutes.

(e) Show that $\chi_{E,F}$ corresponds via (2.6.1)$'$ to the composite map

$$
f^\times E \xrightarrow[\text{natural}]{} f^\times \mathsf{R}\mathcal{H}om^\bullet(F, E \otimes F) \xrightarrow[(4.2.3)(c)]{\;\sim\;} f^\times[F, E \otimes F]_Y
$$
$$
\xrightarrow[(4.2.3)(e)]{\;\sim\;} [\mathsf{L}f^*F, f^\times(E \otimes F)]_X
$$
$$
\xrightarrow[\text{natural}]{} \mathsf{R}\mathcal{H}om^\bullet(\mathsf{L}f^*F, f^\times(E \otimes F)).
$$

(f) With notation as in (4.2.3)(e), and $E, F, G \in \mathbf{D}_{\mathsf{qc}}(Y)$, establish a natural commutative functorial diagram

$$
\begin{array}{ccccc}
f^\times F \otimes \mathsf{L}f^*[E,G]_Y & \xrightarrow{\;\chi\;} & f^\times\big(F \otimes [E,G]_Y\big) & \longrightarrow & f^\times[E, F \otimes G]_Y \\
\downarrow & & & & \uparrow\simeq \\
f^\times F \otimes [\mathsf{L}f^*E, \mathsf{L}f^*G]_X & \longrightarrow & [\mathsf{L}f^*E, f^\times F \otimes \mathsf{L}f^*G]_X & \xrightarrow[\text{via }\chi]{} & [\mathsf{L}f^*E, f^\times(F \otimes G)]_X
\end{array}
$$

We adopt again the notations introduced at the beginning of §4.4.

Apropos of the next theorem, recall from the beginning of §4.7 that f quasi-perfect $\Longrightarrow f^\times$ bounded.

Theorem 4.7.4 *Let*

$$
\begin{array}{ccc}
X' & \xrightarrow{\;v\;} & X \\
g\downarrow & & \downarrow f \\
Y' & \xrightarrow{\;u\;} & Y
\end{array}
$$

be an independent square of scheme-maps, with f quasi-perfect. Then for all $E \in \mathbf{D}_{\mathsf{qc}}(Y)$ the base-change map of (4.4.3)—with $^\times$ in place of $^!$—is an isomorphism

$$
\beta(E)\colon v^*f^\times E \xrightarrow{\;\sim\;} g^\times u^*E.
$$

The same holds, with no assumption on f, whenever u is finite and perfect.

Conversely, the following conditions on a scheme-map $f\colon X \to Y$ are equivalent; and if Y is separated and $f^\times$ bounded above, they imply that f is quasi-perfect:

(i) *For any flat affine universally bicontinuous map $u\colon Y' \to Y$, (i.e., for any $Y'' \to Y$ the resulting projection $Y' \times_Y Y'' \to Y''$ is a homeomorphism onto its image [**GD**, p. 249, Défn. (3.8.1)]) the base-change map associated to* *the independent fiber square*

$$Y' \times_Y X = X' \xrightarrow{\ v\ } X$$

$$g \downarrow \qquad\qquad \downarrow f$$

$$Y' \xrightarrow{\ u\ } Y$$

is an isomorphism $\beta(\mathcal{O}_Y) \colon v^* f^\times \mathcal{O}_Y \xrightarrow{\sim} g^\times u^* \mathcal{O}_Y$.

(ii) *The map in* (4.7.1)(iv) *is an isomorphism*

$$\chi_F \colon f^\times \mathcal{O}_Y \otimes \mathsf{L} f^* F \xrightarrow{\ \sim\ } f^\times F$$

whenever F *is a flat quasi-coherent* $\mathcal{O}_Y$-*module.*

Proof. For the first assertion, using (4.7.3.1)(i) we reduce as in IV of §4.6 to where u, hence v, is an open immersion or affine, so that v is isofaithful ((3.10.2.1)(d) or (3.10.2.2)), and for β to be an isomorphism it suffices that $v_* \beta$ be an isomorphism.

For this purpose it will clearly suffice that the following diagram—in which $\mathcal{O}' := \mathcal{O}_{Y'}$, ϕ is the isomorphism in (3.10.4), θ' is as in (3.10.2) (see (3.10.3)), $\chi := \chi_{E, u_* \mathcal{O}'}$ is as in (4.7.3.4)(a), q is the natural composite isomorphism

$$f^\times E \otimes v_* g^* \mathcal{O}' \xrightarrow[(3.9.4)]{\sim} v_*(v^* f^\times E \otimes g^* \mathcal{O}') \xrightarrow{\sim} v_* v^* f^\times E$$

and r is the natural composite isomorphism

$$E \otimes u_* \mathcal{O}' \xrightarrow[(3.9.4)]{\sim} u_*(u^* E \otimes \mathcal{O}') \xrightarrow{\sim} u_* u^* E,$$

—is commutative:

$$
\begin{array}{ccccc}
f^\times E \otimes v_* g^* \mathcal{O}' & \xrightarrow[q]{\sim} & v_* v^* f^\times E & \xrightarrow{v_* \beta(E)} & v_* g^\times u^* E \\
{\scriptstyle 1 \otimes \theta'} \uparrow {\scriptstyle \simeq} & & & & {\scriptstyle \simeq} \downarrow {\scriptstyle \phi} \qquad (4.7.4.1) \\
f^\times E \otimes f^* u_* \mathcal{O}' & \xrightarrow[\chi]{\sim} & f^\times(E \otimes u_* \mathcal{O}') & \xrightarrow[f^\times r]{\sim} & f^\times u_* u^* E
\end{array}
$$

Since χ is an isomorphism whenever $u_* \mathcal{O}'$ is perfect (see the end of (4.7.3.4)(a)), and since finite maps are isofaithful (3.10.2.2), commutativity of (4.1) also implies the theorem's assertion about finite perfect u.

Now, commutativity of (4.1) results from commutativity of the following diagram (4.1)*, where q' is the composite isomorphism

$$f_* f^\times E \otimes u_* \mathcal{O}' \xrightarrow[(3.9.4)]{\sim} u_*(u^* f_* f^\times E \otimes \mathcal{O}') \xrightarrow{\sim} u_* u^* f_* f^\times E$$

and t and t' are the natural maps, a diagram whose outer border, with the isomorphism (3.4.9) replaced by its inverse, is adjoint to (4.1):

$$
\begin{array}{ccccc}
f_*(f^\times E \otimes v_* g^* \mathcal{O}') & \xrightarrow{f_* q} & f_* v_* v^* f^\times E & \xrightarrow{f_* v_* \beta} & f_* v_* g^\times u^* E \\[2pt]
{\scriptstyle f_*(1 \otimes \theta')} \Big\uparrow {\scriptstyle \simeq} & & \Big\| & & \Big\| \\[2pt]
f_*(f^\times E \otimes f^* u_* \mathcal{O}') & \textcircled{1} & u_* g_* v^* f^\times E & \xrightarrow{u_* g_* \beta} & u_* g_* g^\times u^* E \\[2pt]
{\scriptstyle (3.9.4)} \Big\uparrow {\scriptstyle \simeq} & & {\scriptstyle u_* \theta} \Big\uparrow {\scriptstyle \simeq} & \textcircled{2} & \Big\downarrow {\scriptstyle u_* t'} \\[2pt]
f_* f^\times E \otimes u_* \mathcal{O}' & \xrightarrow[q']{\sim} & u_* u^* f_* f^\times E & \xrightarrow[u^* u_* t]{} & u_* u^* E
\end{array}
$$

$$(4.7.4.1)^*$$

Subdiagram ② commutes by the very definition of β.

Expand subdiagram ① as follows, with an arbitrary $F \in \mathbf{D}(X)$ in place of $f^\times E$, with unlabeled maps being the natural ones, and with p denoting projection maps from (3.4.6) or (3.9.4):

$$
\begin{array}{ccccccc}
f_*(F \otimes v_* g^* \mathcal{O}') & \longrightarrow & f_*(v_* v^* F \otimes v_* g^* \mathcal{O}') & \xrightarrow{(3.4.2.1)} & f_* v_*(v^* F \otimes g^* \mathcal{O}') & \longrightarrow & f_* v_* v^* F \\[2pt]
{\scriptstyle \theta'} \uparrow & & {\scriptstyle \theta'} \uparrow & & \Big\| & & \Big\| \\[2pt]
f_*(F \otimes f^* u_* \mathcal{O}') & \longrightarrow & f_*(v_* v^* F \otimes f^* u_* \mathcal{O}') & & u_* g_*(v^* F \otimes g^* \mathcal{O}') & \longrightarrow & u_* g_* v^* F \\[2pt]
{\scriptstyle p} \uparrow & & {\scriptstyle p} \uparrow & \textcircled{3} & & & \\[2pt]
f_* F \otimes u_* \mathcal{O}' & \longrightarrow & f_* v_* v^* F \otimes u_* \mathcal{O}' & & {\scriptstyle u_* p} \uparrow & \textcircled{4} & \Big\| \\[2pt]
\Big\| & & \Big\| & & & & \\[2pt]
& \textcircled{5} & u_* g_* v^* F \otimes u_* \mathcal{O}' & \xrightarrow{(3.4.2.1)} & u_*(g_* v^* F \otimes \mathcal{O}') & \longrightarrow & u_* g_* v^* F \\[2pt]
\Big\| & & {\scriptstyle \theta} \uparrow & & {\scriptstyle \theta} \uparrow & & {\scriptstyle \theta} \uparrow \\[2pt]
f_* F \otimes u_* \mathcal{O}' & \longrightarrow & u_* u^* f_* F \otimes u_* \mathcal{O}' & \xrightarrow[(3.4.2.1)]{} & u_*(u^* f_* F \otimes \mathcal{O}') & \xrightarrow{\sim} & u_* u^* f_* F
\end{array}
$$

Commutativity of the unlabeled subdiagrams is clear. That of ⑤ follows from the definition (3.7.2)(a) of θ; and that of ④ follows from (3.4.7)(iii). Subdiagram ③ expands as follows:

$$
\begin{array}{ccccc}
f_*(v_* v^* F \otimes v_* g^* \mathcal{O}') & = & f_*(v_* v^* F \otimes v_* g^* \mathcal{O}') & \xrightarrow{(3.4.2.1)} & f_* v_*(v^* F \otimes g^* \mathcal{O}') \\[2pt]
\uparrow & & \uparrow {\scriptstyle (3.4.2.1)} & & \Big\| \\[2pt]
{\scriptstyle \theta'} \Big| & \textcircled{6} & f_* v_* v^* F \otimes f_* v_* g^* \mathcal{O}' & \textcircled{7} & \Big\| \\[2pt]
\Big| & & \Big\| & & \Big\| \\[2pt]
f_*(v_* v^* F \otimes f^* u_* \mathcal{O}') & & u_* g_* v^* F \otimes u_* g_* g^* \mathcal{O}' & \xrightarrow{(3.4.2.1)} & u_* g_*(v^* F \otimes g^* \mathcal{O}') \\[2pt]
{\scriptstyle p} \uparrow & & \uparrow & \textcircled{8} & {\scriptstyle u_* p} \uparrow \\[2pt]
f_* v_* v^* F \otimes u_* \mathcal{O}' & = & u_* g_* v^* F \otimes u_* \mathcal{O}' & \xrightarrow[(3.4.2.1)]{} & u_*(g_* v^* F \otimes \mathcal{O}')
\end{array}
$$

For commutativity of subdiagram ⑧, replace p by its definition (3.4.6), and apply commutativity of (3.6.7.2). Commutativity of ⑦ also follows from that of (3.6.7.2). Finally, subdiagram ⑥ expands as follows:

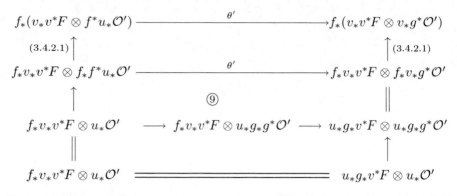

Commutativity of ⑨ is an easy consequence of the definition (3.7.2)(a) of θ'; and that of the other two subdiagrams is clear.

It is thus established that (4.1)* commutes.

We show next that (i) ⇔ (ii).

Assume (i). Let F be a flat quasi-coherent $\mathcal{O}_Y$-module. Let $\mathcal{F}$ be the $\mathcal{O}_Y$-algebra $\mathcal{O}_Y \oplus F$ with $F^2 = 0$ (i.e., the symmetric algebra on F, modulo everything of degree ≥ 2), and let $u \colon Y' \to Y$ be an affine scheme-map such that $u_*\mathcal{O}_{Y'} = \mathcal{F}$ (see [**GD**, p. 355, (9.1.4) and p. 370, (9.4.4)]). This u is a flat affine universally bicontinuous map. With $E = \mathcal{O}_Y$, all the maps in the commutative diagram (4.7.4.1) other than $\chi = \chi_{\mathcal{O}_Y} \oplus \chi_F$ are isomorphisms, and so χ must be an isomorphism too. But $\chi_{\mathcal{O}_Y}$ is an isomorphism (exercise), so χ_F is an isomorphism, i.e., (ii) holds.

Conversely, if u is any flat affine map and (ii) holds for the flat quasi-coherent $\mathcal{O}_Y$-module $F = u_*\mathcal{O}_{Y'}$ then (4.7.4.1) with $E = \mathcal{O}_Y$ shows that $v_*\beta(\mathcal{O}_Y)$ is an isomorphism, whence, v being affine, so is $\beta(\mathcal{O}_Y)$, see (3.10.2.2).

Finally, assuming (ii) and that Y is separated and $f^\times$ bounded-above, let us deduce that the map $\chi_E \colon f^\times \mathcal{O}_Y \otimes \mathbf{L}f^*E \to f^\times E$ is an isomorphism for all $E \in \mathbf{D}_{\mathrm{qc}}(Y)$, so that f is quasi-perfect (see 4.7.1(iv)).

Since Y is separated, we can replace E by a $\mathbf{D}$-isomorphic q-flat quasi-coherent complex, which is a $\varinjlim$ of bounded-above flat complexes, see [**AJL**, p. 10, (1.1)] and its proof. Since the functors $f^\times \mathcal{O}_Y \otimes \mathbf{L}f^*(-)$ and $f^\times(-)$ are both bounded-above, we may assume that E is bounded-below: for each $n \in \mathbb{Z}$, if E' is obtained by replacing all sufficiently-negative-degree components of E by (0) then χ_E and $\chi_{E'}$ induce identical homology maps in degree n, and (1.2.2) can be applied. Similarly, since $f^\times$ is bounded below, and $\mathbf{L}f^*E = f^*E$ when E is a $\varinjlim$ of bounded-above flat complexes, we can reduce further to where E is bounded, flat, and quasi-coherent.

Now an induction on the number of nonvanishing components of E (using the triangle [**H**, p. 70, (1)]) gives the desired conclusion.   Q.E.D.

For more along these lines see exercise 4.7.6(f) below.

Proposition 4.7.5 *If* $f\colon X \to Y$ *is quasi-proper and* $F \in \mathbf{D}_{qc}(Y)$ *has finite tor-dimension then for all* $E \in \mathbf{D}_{qc}(Y)$ *the map* $\chi_{E,F}$ *of* (4.7.3.4) *is an isomorphism*

$$f^{\times}E \underset{\cong}{\otimes} \mathbf{L}f^{*}F \xrightarrow{\;\sim\;} f^{\times}(E \underset{\cong}{\otimes} F).$$

Proof. If $U \hookrightarrow Y$ is an open immersion, then by [**LN**, Prop. 4.4], the projection $X \times_{Y} U \to U$ is quasi-proper. Together with (4.4.3) and (4.7.3.4)(c), this implies that the assertion in (4.7.5) is local on Y, so we may assume that Y is affine.

We can then replace F by a **D**-isomorphic bounded-above quasi-coherent complex—see (3.9.6)(a)—which by [**H**, p. 42, 4.6.1)] (dualized) may be assumed flat. Since F has finite tor-dimension, an application of [**I**, p. 131, 5.1.1] to a suitable **D**-isomorphic truncation of F allows one to assume further that F is bounded. Then an induction on the number of nonvanishing components of F (using the triangle [**H**, p. 70, (1)]) reduces the problem to where F is a single flat quasi-coherent $\mathcal{O}_{Y}$-module.

As in the proof of (4.7.4) ((i) $\Leftrightarrow$ (ii)), let $u\colon Y' \to Y$ be an affine scheme-map such that $u_{*}\mathcal{O}_{Y'} = \mathcal{O}_{Y} \oplus F$. The map u is flat, so u and f are two sides of an independent square, and by (4.4.3) the corresponding base-change map $\beta(E)$ in the commutative diagram (4.7.4.1) is an isomorphism. One concludes as before that $\chi_{E,F}$ is an isomorphism.   Q.E.D.

Exercises 4.7.6 (a). Let $f\colon X \to Y$ be a quasi-perfect scheme-map. Assume that X is *divisorial*—i.e., X has an ample family of invertible $\mathcal{O}_{X}$-modules—so that by [**I**, p. 173, 2.2.8 b)] every pseudo-coherent $\mathcal{O}_{X}$-complex is **D**-isomorphic to a bounded above complex of finite-rank locally free $\mathcal{O}_{X}$-modules. Show that an $\mathcal{O}_{X}$-complex F is pseudo-coherent iff for every $n \in \mathbb{Z}$ there is a triangle $P \to F \to R \to P[1]$ with P perfect and $R \in (\mathbf{D}_{qc})_{<\mathbf{n}}$; and using (3.9.2.3) above, deduce that f is quasi-proper.

(A similar result without the divisoriality assumption is [**LN**, Thm. 4.1].)

(b). Let $f\colon X \to Y$ be a quasi-proper scheme-map. Let $r \in \mathbb{Z}$ and let $(G_{\alpha})_{\alpha \in A}$ be a family of complexes in $\mathbf{D}_{qc}(X)_{\geq \mathbf{r}}$, i.e., for every α, $H^{m}(G_{\alpha}) = 0$ whenever $m < r$. Show that the natural map is an isomorphism[11]

$$\underset{\alpha}{\oplus} f^{\times}G_{\alpha} \xrightarrow{\;\sim\;} f^{\times}(\underset{\alpha}{\oplus}G_{\alpha}).$$

Hint. Write f_{*} for $\mathbf{R}f_{*}$, $\mathcal{H}_{X}$ for $\mathbf{R}\mathcal{H}om_{X}^{\bullet}$, etc. The triangulated category $\mathbf{D}_{qc}(X) \equiv \mathbf{D}(\mathcal{A}_{qc}(X))$ is generated by perfect complexes (see [**N**, pp. 215–216], or [**LN**, Thm. 4.2]), so a $\mathbf{D}_{qc}$-map $\varphi\colon A_{1} \to A_{2}$ is an isomorphism if and only if the induced map $\mathrm{Hom}(E, A_{1}) \to \mathrm{Hom}(E, A_{2})$ is an isomorphism for all perfect $E \in \mathbf{D}(X)$. In the following natural diagram, easily seen to commute,

$$
\begin{array}{ccccc}
f_{*}\mathcal{H}_{X}(E, \oplus f^{\times}G_{\alpha}) & \longrightarrow & f_{*}\mathcal{H}_{X}(E, f^{\times}(\oplus G_{\alpha})) & \xrightarrow[(4.3.6)]{\sim} & \mathcal{H}_{Y}(f_{*}E, \oplus G_{\alpha}) \\
\uparrow & & \uparrow & & \uparrow \\
f_{*}(\oplus \mathcal{H}_{X}(E, f^{\times}G_{\alpha})) & \xleftarrow[(3.9.3.1)]{\sim} & \oplus f_{*}\mathcal{H}_{X}(E, f^{\times}G_{\alpha}) & \xrightarrow[(4.3.6)]{\sim} & \oplus \mathcal{H}_{Y}(f_{*}E, G_{\alpha})
\end{array}
$$

the left and right vertical arrows are isomorphisms whenever E is pseudo-coherent.

---

[11] Cf. [**V′**, p. 396, Lemma 1], where the necessary *uniform* lower bound on the G_{α} is omitted.

(The question being local on X, one can, as in the proof of (4.3.5), replace E by a bounded finite-rank free complex E' and then, using the triangle [**H**, p. 70, (1)], proceed by induction on the number of degrees in which E' doesn't vanish.) Finally, apply the functor $H^0 \mathbf{R}\Gamma(Y, -)$.

(c) Deduce from (b) that *a quasi-proper scheme-map f with $f^\times$ bounded above is quasi-perfect.* (This is part of [**LN**, Thm. 1.2.])

(d) Let $f: X \to Y$ be a scheme-map. Show that if f is quasi-perfect then the following two conditions hold, and that the converse is true when $f^\times$ is bounded. (Apropos, recall again from the beginning of this section that if f is quasi-perfect then $f^\times$ is bounded.)

(i) If $u: Y' \to Y$ is an open immersion, and $v: f^{-1}U \to X$, $g: f^{-1}U \to U$ are the obvious induced maps, then the base-change map is an isomorphism

$$\beta(\mathcal{O}_Y): v^* f^\times \mathcal{O}_Y \xrightarrow{\sim} g^\times u^* \mathcal{O}_Y.$$

Equivalently (see subsection V in §4.6), for all $F \in \mathbf{D}_{qc}(X)$ the duality map $\delta(F, \mathcal{O}_Y)$ defined as in (4.4.2) is an isomorphism

$$\mathbf{R}f_* \mathbf{R}\mathcal{H}om_X^\bullet(F, f^\times \mathcal{O}_Y) \xrightarrow{\sim} \mathbf{R}\mathcal{H}om_Y^\bullet(\mathbf{R}f_* F, \mathcal{O}_Y)$$

(ii) If (F_α) is a small filtered direct system of flat quasi-coherent $\mathcal{O}_Y$-modules then for all $n \in \mathbb{Z}$ the natural map is an isomorphism

$$\varinjlim_\alpha H^n(f^\times F_\alpha) \xrightarrow{\sim} H^n(f^\times \varinjlim_\alpha F_\alpha).$$

Hint. Use (4.7.3.4)(c) and Lazard's theorem that over a commutative ring A any flat module is a $\varinjlim$ of finite-rank free A-modules [**GD**, p. 163, (6.6.24)] to show that (i) and (ii) imply condition (ii) in (4.7.4).

(e) (i) (Neeman). Using, e.g., (i) in (d) (with $F = \mathcal{O}_X$), show that if $f: X \to Y$ is quasi-perfect then the $\mathcal{O}_Y$-complex $\mathbf{R}f_* f^\times \mathcal{O}_Y$ is perfect; and deduce that for any perfect $\mathcal{O}_Y$-complex E, $\mathbf{R}f_* f^\times E$ is perfect.

(ii) (cf. [**I**, p. 257, 4.8]). Let $f: X \to Y$ be a concentrated quasi-proper map of quasi-compact schemes. Then for any f-perfect $\mathcal{O}_X$-complex E, $\mathbf{R}f_* E$ is a perfect $\mathcal{O}_Y$-complex.

(f) Let $U \xrightarrow{u} X \xrightarrow{f} Y$ be scheme-maps, with f quasi-proper, and let $E \in \mathbf{D}_{qc}(Y)$. Show that the following are equivalent.

(i) The functor $\mathbf{L}u^* f^\times(E \otimes F)$ $(F \in \mathbf{D}_{qc}(Y))$ is bounded above.

(ii) $\mathbf{L}u^* f^\times E \in \bar{\mathbf{D}}^-(X))$, and the map (see exercise (4.7.3.4) above)

$$\mathbf{L}u^* \chi_{E,F}: \mathbf{L}u^* f^\times E \otimes \mathbf{L}(fu)^* F \to \mathbf{L}u^* f^\times(E \otimes F),$$

is an isomorphism for all $F \in \mathbf{D}_{qc}(Y)$.

(iii) $\mathbf{L}u^* f^\times E \in \bar{\mathbf{D}}^-(X))$, and the functor $\mathbf{L}u^* f^\times(E \otimes F)$ $(F \in \mathbf{D}_{qc}(Y))$ respects direct sums (cf. (4.7.1)(i)).

Moreover, if u has finite tor-dimension, then the following are equivalent.

(i)′ The functor $\mathbf{L}u^* f^\times(E \otimes F)$ $(F \in \mathbf{D}_{qc}(Y))$ is bounded.

(ii)′ The complex $\mathbf{L}u^* f^\times E$ has finite flat fu-amplitude (2.7.6), and $\mathbf{L}u^* \chi_{E,F}$ is an isomorphism for all $F \in \mathbf{D}_{qc}(Y)$.

(iii)′ $\mathbf{L}u^*f^\times E$ has finite flat fu-amplitude, and the functor $\mathbf{L}u^*f^\times(E \otimes F)$ $(F \in \mathbf{D}_{qc}(Y))$ respects direct sums.

Hint. Given (i), one sees as in exercise (c) above that the functor $\mathbf{L}u^*f^\times(E \otimes F)$ respects direct sums; and then arguing as in [**N**, p. 226, Thm. 5.4], one see that $\mathbf{L}u^*\chi_{E,F}$ in (ii) is an isomorphism. It follows then from [**I**, p. 242, 3.3(iv)], and the fact that if $V \subset Y$ is open then any quasi-coherent $\mathcal{O}_V$-module M is the restriction of a quasi-coherent $\mathcal{O}_Y$-module, that if (i)′ holds then $\mathbf{L}u^*f^\times E$ has finite flat fu-amplitude.

4.8 Two Fundamental Theorems

Up to now we have dealt with the pseudofunctor $^\times$ (see (4.1.1)) for quite general maps—it cost nothing to do so. But for non-proper maps this pseudofunctor may still be of limited interest (see [**De′**, p. 416, line 3]).

As indicated in the Introduction to these notes, Grothendieck Duality is fundamentally concerned with a $\overline{\mathbf{D}}_{qc}^+$-valued pseudofunctor $^!$ over the category of say, separated finite-type maps of noetherian schemes, agreeing with $^\times$ on proper maps, but, unlike $^\times$ (see (4.2.3)(d)), agreeing with the usual pseudofunctor * on open immersions (more generally, on separated étale maps see [**EGA**, IV, §§17.3, 17.6]), and compatible in a suitable sense with flat base change. The existence and uniqueness, up to isomorphism, of this remarkable pseudofunctor is given by Theorem (4.8.1), and its behavior vis-à-vis flat base change is described in Theorem (4.8.3).

The proof of (4.8.1) presented here is based on a formal method of Deligne for pasting pseudofunctors (see Proposition (4.8.4)), and on the compactification theorem of Nagata, that any finite-type separable map of noetherian schemes factors as an open immersion followed by a proper map (see [**Lt**], [**C′**], [**Vj**]). The proof of (4.8.3) is based on a formal pasting procedure for *base-change setups* (see (4.8.2), (4.8.5)).

There are other pasting techniques, due to Nayak [**Nk**], to establish the two basic theorems, (4.8.1) and (4.8.3).[12] As mentioned in the Introduction, Nayak's methods avoid using Nagata's theorem, and so apply in contexts where Nagata's theorem may not hold. For example, the results in [**Nk**, §7.1] are generalizations of (4.8.1) and (4.8.3) to the case of noetherian formal schemes (except for "thickening" as in (4.8.11) below, which allows flat base-change isomorphisms for *admissible squares* (4.8.3.0) rather than just fiber squares, see Exercise (4.8.12)(d).)

All commutative squares will be considered to be *oriented,* as in §3.10.

The first main result defines (up to isomorphism) the *twisted inverse image pseudofunctor.*

---

[12] [**Nk**, §7.5] discusses the relation between Nayak's methods and Deligne's. On the other hand, in [**Nk′**] Nayak extends Nagata compactification—and hence Theorems (4.8.1) and (4.8.3)—to separated maps which are *essentially* of finite type.

Theorem 4.8.1 *On the category* $\mathbf{S_f}$ *of finite-type separated maps of noetherian schemes, there is a* $\overline{\mathbf{D}}^+_{\mathsf{qc}}$-*valued pseudofunctor* $^!$ *that is uniquely determined up to isomorphism by the following three properties:*

(i) *The pseudofunctor* $^!$ *restricts on the subcategory of proper maps to a right adjoint of the derived direct-image pseudofunctor, see* (3.6.7)(d).

(ii) *The pseudofunctor* $^!$ *restricts on the subcategory of étale maps to the usual inverse-image pseudofunctor* *.

(iii) *For any fiber square in* $\mathbf{S_f}$:

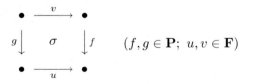

$$(f, g \text{ proper}; u, v \text{ étale}),$$

the base-change map β_σ *of* (4.4.3) *is the natural composite isomorphism*

$$v^* f^! = v^! f^! \xrightarrow{\sim} (fv)^! = (ug)^! \xrightarrow{\sim} g^! u^! = g^! u^*.$$

Remark 4.8.1.1 It follows that when f is both étale and proper (hence by [**EGA**, III, 4.4.11], finite), then the natural map $f_* f^* = f_* f^! \to 1$ is precisely—not just up to isomorphism—the standard trace map, see Exercise (4.8.12)(b)(vii).

For subsequent considerations, involving base-change isomorphisms and their properties, the following definition will be convenient to have.

Definition 4.8.2 A *base-change setup* $\mathcal{B}\big(\mathbf{S}, \mathbf{P}, \mathbf{F}, {}^!, {}^*, (\beta_\sigma)_{\sigma \in \square}\big)$ consists of the following data (a)–(d), subject to conditions (1)–(3):

(a) Subcategories $\mathbf{P}$ and $\mathbf{F}$ of a category $\mathbf{S}$, each containing every object of $\mathbf{S}$.

(b) Contravariant pseudofunctors $^!$ on $\mathbf{P}$ and * on $\mathbf{F}$ such that for all objects $X \in \mathbf{S}$, the categories $\mathbf{X}^!$ and $\mathbf{X}^*$ coincide (see §3.6.5).

(c) A class $\square$ of (oriented) commutative $\mathbf{S}$-squares, the *distinguished squares,* each member of which has the form

$$(f, g \in \mathbf{P}; \ u, v \in \mathbf{F})$$

(where u precedes f in the orientation of σ, see §3.10).

(d) For each distinguished σ as in (c), an isomorphism of functors

$$\beta_\sigma \colon v^* f^! \xrightarrow{\sim} g^! u^*.$$

(1) If two commutative **S**-squares

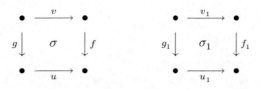

are *isomorphic,* i.e., there exists a commutative cube with front and rear faces σ and σ_1 respectively, and i, i_1, j, j_1 isomorphisms:

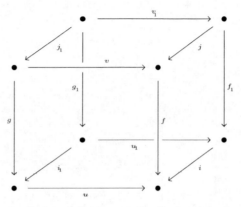

then σ is distinguished $\Leftrightarrow \sigma_1$ is distinguished.

(2) For every **P**-map f, the square

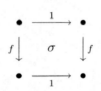

is distinguished, and $\beta_\sigma : f^! \to f^!$ is the identity map.

(2)′ For every **F**-map u, the square

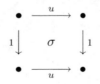

is distinguished, and $\beta_\sigma : u^* \to u^*$ is the identity map.

(3) (Horizontal and vertical transitivity.) If the square $\sigma_0 = \sigma_2 \circ \sigma_1$ (with g resp. v deleted)

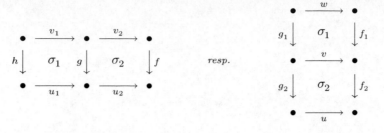

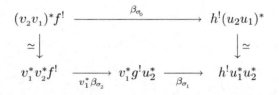

as well as its constituents σ_2 and σ_1 are all distinguished, then the corresponding natural diagram of functorial maps commutes:

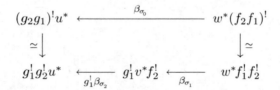

resp.

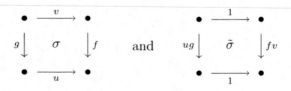

Remarks 4.8.2.1 (a) Let u and v be **S**-isomorphisms. If f and g are **S**-maps such that $fv = ug$ is in **P**, then the squares

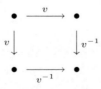

are isomorphic, so that by (1) and (2), σ is distinguished—which entails that u and v are in **F** and that f and g are in **P**. In particular,

$$\begin{array}{ccc} \bullet & \xrightarrow{v} & \bullet \\ v\downarrow & & \downarrow v^{-1} \\ \bullet & \xrightarrow{v^{-1}} & \bullet \end{array}$$

is distinguished, and consequently every **S**-isomorphism lies in $\mathbf{P} \cap \mathbf{F}$ (whence $fv \in \mathbf{P} \iff f \in \mathbf{P}$, and $ug \in \mathbf{P} \iff g \in \mathbf{P}$).

Similarly, if f and g are **S**-isomorphisms, and u and v are any **F**-maps such that $fv = ug$, then σ is distinguished.

(b) That the isomorphism β_σ in (2) is idempotent, hence the identity, actually follows from (3), with $u_i = v_i = 1$ (resp. $f_i = g_i = 1$).

(c) To each base-change setup $\mathcal{B} = \mathcal{B}(\mathbf{S}, \mathbf{P}, \mathbf{F}, {}^!, {}^*, (\beta_\sigma)_{\sigma \in \square})$ is associated a *dual setup* $\mathcal{B}^{\mathrm{op}} := \mathcal{B}(\mathbf{S}, \mathbf{F}, \mathbf{P}, {}^*, {}^!, (\beta_{\sigma'} := \beta_\sigma^{-1})_{\sigma' \in \square'})$, where σ' is the transpose of σ (i.e., σ with its orientation reversed, or, visually, the reflection of σ in its upper-left to lower-right diagonal), and $\square'$ consists of all transposes of squares in $\square$.

Example 4.8.2.2 Let **S** be a category, take $\mathbf{P} = \mathbf{F} = \mathbf{S}$, let $^! = {}^*$ be a contravariant pseudofunctor on **S**, let all commutative squares in **S** be distinguished, and for any such square σ, let

$$\beta_\sigma : v^* f^* \xrightarrow{\ \sim\ } (fv)^* = (ug)^* \xrightarrow{\ \sim\ } g^* u^*$$

be the isomorphism naturally associated with the pseudofunctor *.

Then (4.8.2)(1) holds trivially, and (2), (2)$'$, (3) follow readily from the definition of "pseudofunctor."

We will denote such a base-change setup by $\mathcal{B}(\mathbf{S}, {}^*)$.

Example 4.8.2.3 Let **S** be a subcategory of the category of quasi-compact separated schemes, $\mathbf{P} \subset \mathbf{S}$ the subcategory of *quasi-proper* maps, and $\mathbf{F} \subset \mathbf{S}$ the subcategory of *finite-tor-dimension* maps. On **P** there is the $\overline{\mathbf{D}}_{\mathsf{qc}}^+$-valued pseudofunctor $^\times$ (see (4.1.2)); and on **F** there is the $\overline{\mathbf{D}}_{\mathsf{qc}}^+$-valued pseudofunctor * with $u^* := \mathsf{L}u^*$ for any **F**-map u. Let $\square$ be the class of *independent fiber squares* of the form specified in 4.8.2(c). For $\sigma \in \square$, let $\beta_\sigma : v^* f^\times \to g^\times u^*$ be the corresponding base-change isomorphism from (4.4.3).

Conditions (1), (2) and (2)$'$ in (4.8.2) are then easily verified; and as in (4.6.8), (3) follows formally from (3.7.2), (ii) and (iii). So we have a base-change setup $\mathcal{B}(\mathbf{S}, \mathbf{P}, \mathbf{F}, {}^\times, {}^*, (\beta_\sigma)_{\sigma \in \square})$.

Example 4.8.2.4 As a special case, we have the base-change setup $\mathcal{B}(\mathbf{S_f}, \mathbf{P}, \mathbf{E}, {}^\times, {}^*, (\beta_\sigma)_{\sigma \in \square})$ with $\mathbf{S_f}$ as in (4.8.1), $\mathbf{P} \subset \mathbf{S_f}$ the subcategory of proper maps, $\mathbf{E} \subset \mathbf{S_f}$ the subcategory of étale maps, and $^\times, {}^*, \square, \beta_\sigma$ as in the preceding example (4.8.2.3) (with **F** replaced by **E**).

To prove (4.8.1), we will need to show that there is a unique way to enlarge the preceding setup to a setup $\mathcal{B}(\mathbf{S_f}, \mathbf{P}, \mathbf{E}, {}^\times, {}^*, (\beta_\sigma')_{\sigma \in \square'})$ where $\square'$ consists of *all* commutative $\mathbf{S_f}$-squares

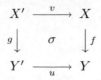

with f, g proper and u, v étale.

This, and more, will be done in (4.8.11). Meanwhile, we'll refer to this *unique* enlarged setup as Example (4.8.2.4)′.

Notation-Definition (4.8.3.0). A category **S** having been given, for **S**-maps v, f, g, u with $fv = ug$, $\sigma_{v,f,g,u}$ is the commutative square

In the category of schemes, such a $\sigma_{v,f,g,u}$:

$$
\begin{array}{ccc}
X' & \xrightarrow{\ v\ } & X \\
{\scriptstyle g}\downarrow & & \downarrow{\scriptstyle f} \\
Y' & \xrightarrow[\ u\ ]{} & Y
\end{array}
$$

is an *admissible square* if u is flat, f is finitely presentable, and in the associated diagram

$$
\begin{array}{ccccc}
X' & \xrightarrow{\ i\ } & X \times_Y Y' & \xrightarrow{\ q_1\ } & X \\
& & {\scriptstyle q_2}\downarrow & & \downarrow{\scriptstyle f} \\
& & Y' & \xrightarrow[\ u\ ]{} & Y
\end{array}
$$

where q_1, q_2 are the projections, $q_1 i = v$ and $q_2 i = g$, the map i is *étale*. (Note that then $g = q_2 i$ is finitely presentable, and $v = q_1 i$ is flat, so that $\mathbf{L}v^* = v^*$.)

Theorem 4.8.3 *Let* **S** *be the category of separated maps of noetherian schemes, let* $\mathbf{S_f} \subset \mathbf{S}$ *and* ! *be as in* (4.8.1), *let* $\mathbf{F} \subset \mathbf{S}$ *be the subcategory of flat maps, and let* * *be the usual* $\overline{\mathbf{D}}_{\mathsf{qc}}^{+}$*-valued inverse-image pseudofunctor on* **F**. *Then there is a unique base-change setup* $\mathcal{B}(\mathbf{S}, \mathbf{S_f}, \mathbf{F}, !, *, (\beta_\sigma)_{\sigma \in \square})$ *with* $\square$ *the class of admissible* **S**-*squares, such that the following conditions hold for any admissible* **S**-*square* $\sigma = \sigma_{v,f,g,u}$:

(i) *If σ is a fiber square with f proper then β_σ is the base-change isomorphism in* (4.4.3).

(ii) *If f—and hence g—is étale, so that $f^! = f^*$ and $g^! = g^*$, then β_σ is the natural isomorphism $v^* f^* \xrightarrow{\sim} g^* u^*$.*

(iii) *If u—and hence v—is étale, so that $u^* = u^!$ and $v^* = v^!$, then β_σ is the natural isomorphism $v^! f^! \xrightarrow{\sim} g^! u^!$.*

Remarks 4.8.3.1 (a) Since étale maps are unramified [**EGA**, IV, (17.6.2)], therefore by [**EGA**, IV, (17.3.3)(iii) and (17.3.4)], every commutative $\mathbf{S_f}$-square $\sigma_{v,f,g,u}$ with u and v flat and such that either f and g or u and v are étale is admissible.

(b) *Uniqueness* in (4.8.3) is implied by (i), (ii) and vertical transitivity as in (4.8.2)(3), because if $\sigma_{v,f,g,u}$ is admissible, then, by Nagata's theorem, $f = f_2 f_1$ with f_2 proper and f_1 an open immersion, whence σ decomposes as in the second diagram in (4.8.2)(3), with σ_1 having v, w flat and f_1, g_1 étale, and with σ_2 an admissible fiber square.

(c) As for *existence*, the preceding suggests defining β_σ via a choice of such factorizations, one for each f, then showing that the definition does not depend on the choice, and that (i)–(iii) in (4.8.3) are satisfied.

This purely formal procedure is straightforward in principle but, as will emerge, lengthy in practice.

In view of Nagata's compactification theorem, it is readily verified that the *existence* of the pseudofunctor $^!$ in Theorem (4.8.1) results from the next Proposition (4.8.4) on the pasting of pseudofunctors, as applied to the base-change setup $(4.8.2.4)'$.

Proposition 4.8.4 ([De, p. 318, Prop. 3.3.4]) *Let there be given a base-change setup $\mathcal{B} = \mathcal{B}(\mathbf{S}, \mathbf{P}, \mathbf{E}, {}^\times, {}^*, (\beta_\sigma)_{\sigma \in \square})$ such that:*

(a) *the fiber product in $\mathbf{P}$ of any two $\mathbf{P}$-maps with the same target exists, and is a fiber product in $\mathbf{S}$ of the same two maps;*

(b) *every map $f \in \mathbf{S}$ has a "compactification," i.e., a factorization $f = \bar{f} i$ with $\bar{f} \in \mathbf{P}$ and $i \in \mathbf{E}$; and*

(c) $\square$ *consists of all of the commutative $\mathbf{S}$-squares $\sigma_{v,f,g,u}$ for which $f, g \in \mathbf{P}$ and $u, v \in \mathbf{E}$.*

Then there exists a contravariant pseudofunctor $^!$ on $\mathbf{S}$, uniquely determined up to isomorphism by the properties that $\mathbf{X}^! = \mathbf{X}^\times = \mathbf{X}^*$ for all $X \in \mathbf{S}$ and that there exist isomorphisms of pseudofunctors (see (3.6.6)) $\alpha_\mathbf{P} : {}^!|_\mathbf{P} \xrightarrow{\sim} {}^\times$ and $\alpha_\mathbf{E} : {}^!|_\mathbf{E} \xrightarrow{\sim} {}^*$ such that for any $\sigma = \sigma_{v,f,g,u} \in \square$, β_σ is the natural composition (with first and last isomorphisms coming from $\alpha_\mathbf{P}$ and $\alpha_\mathbf{E}$):

$$v^* f^\times \xrightarrow{\sim} v^! f^! \xrightarrow{\quad\sim\quad} (fv)^! = (ug)^! \xrightarrow{\quad\sim\quad} g^! u^! \xrightarrow{\sim} g^\times u^*.$$

In other words, $\mathcal{B}(\mathbf{S}, {}^!)$ (see (4.8.2.2)) extends $\mathcal{B}$, via $\alpha_\mathbf{P}$ and $\alpha_\mathbf{E}$.

In fact there is a $^!$ such that, furthermore, $^!|_\mathbf{E} = {}^*$ and $\alpha_\mathbf{E}$ is the identity isomorphism.

Remark. Uniqueness (up to isomorphism) in (4.8.1) also results from (4.8.4), as follows. Let $\mathbf{P} \subset \mathbf{S_f}$, $\mathbf{E} \subset \mathbf{S_f}$ and $\square'$ be as in (4.8.2.4). If the pseudofunctor $^!$ satisfies the conditions in (4.8.1) then there is a natural pseudofunctorial isomorphism $\alpha_{\mathbf{P}} \colon {^!}|_{\mathbf{P}} \xrightarrow{\sim} {^\times}|_{\mathbf{P}}$ (since both $^!|_{\mathbf{P}}$ and $^\times|_{\mathbf{P}}$ have the same pseudofunctorial left adjoint). For any $\sigma_{v,f,g,u} \in \square'$ let β''_σ be the natural composite isomorphism

$$v^* f^\times \xrightarrow[v^* \alpha_{\mathbf{P}}^{-1}]{\sim} v^* f^! = v^! f^! \xrightarrow{\sim} g^! u^! = g^! u^* \xrightarrow[\alpha_{\mathbf{P}}]{\sim} g^\times u^*.$$

This gives a setup $\mathcal{B}'' = \mathcal{B}\big(\mathbf{S_f}, \mathbf{P}, \mathbf{E}, {^\times}, {^*}, (\beta''_\sigma)_{\sigma \in \square'}\big)$. (Check directly, or see Exercise (4.8.12)(a).) When σ is a fiber square then, one checks, β''_σ is the base change map of (4.4.3). Thus $\mathcal{B}''$ is the unique enlargement (4.8.2.4)$'$ of the setup (4.8.2.4), so that the uniqueness assertion in (4.8.4) gives the uniqueness in (4.8.1).

Proof of (4.8.4). (Outline: more details are in [**De**, pp. 304–318].[13])

If the pseudofunctor $^!$ exists then to each compactification $f = \bar{f}i$ there is naturally associated an isomorphism $f^! \xrightarrow{\sim} i^* \bar{f}^\times$; and for a composed map $f_1 f_2$ and compactifications $f_1 = \bar{f}_1 i_1$, $f_2 = \bar{f}_2 i_2$, $i_1 \bar{f}_2 = \bar{g}j$, with $\sigma := \sigma_{j, \bar{g}, \bar{f}_2, i_1}$, the canonical isomorphism $f_2^! f_1^! \xrightarrow{\sim} (f_1 f_2)^!$ factors naturally as

$$(\bar{f}_2 i_2)^! (\bar{f}_1 i_1)^! \xrightarrow{\sim} i_2^* \bar{f}_2^\times i_1^* \bar{f}_1^\times \xrightarrow[\beta_\sigma^{-1}]{\sim} i_2^* j^* \bar{g}^\times \bar{f}_1^\times$$

$$\xrightarrow{\sim} (ji_2)^* (\bar{f}_1 \bar{g})^\times \xrightarrow{\sim} (\bar{f}_1 \bar{g} j i_2)^! = (\bar{f}_1 i_1 \bar{f}_2 i_2)^!. \qquad (4.8.4.1)$$

If $^{!!}$ is another pseudofunctor with the same property as $^!$ then for each compactification $f = \bar{f}i$ we have a natural composite functorial isomorphism

$$f^! = (\bar{f}i)^! \xrightarrow{\sim} i^! \bar{f}^! \xrightarrow{\sim} i^* \bar{f}^\times \xrightarrow{\sim} i^{!!} \bar{f}^{!!} \xrightarrow{\sim} (\bar{f}i)^{!!} = f^{!!}. \qquad (4.8.4.2)$$

One must show that (4.8.4.2) depends only on the **S**-map $f \colon X \to Y$, not on any particular compactification. Then it is a simple exercise to check via (4.8.4.1) that these isomorphisms, for variable f, constitute an isomorphism of pseudofunctors, giving *uniqueness* of $^!$ (up to a pseudofunctorial isomorphism—itself unique if we require compatibility with $\alpha_{\mathbf{P}}$ and $\alpha_{\mathbf{E}}$).

For comparing (4.8.4.2) relative to various compactifications of f,

$$(i_s, \bar{f}_s) := \big(X \xrightarrow{i_s} X_s \xrightarrow{\bar{f}_s} Y \big),$$

let $[(i_1, \bar{f}_1), (i_2, \bar{f}_2)]$ be the natural composite isomorphism

$$i_2^* \bar{f}_2^\times \xrightarrow{\sim} i_2^! \bar{f}_2^! \xrightarrow{\sim} f^! \xrightarrow{\sim} i_1^! \bar{f}_1^! \xrightarrow{\sim} i_1^* \bar{f}_1^\times.$$

---

[13] Where there are a few minor misprints (for example, (3.2.4.∗) should be (3.2.5.∗)), and omissions of symbols.

Noting that the compactifications of f are the objects of a category $\mathcal{C}$ in which a morphism $(i_1, \bar{f}_1) \to (i_2, \bar{f}_2)$ is a **P**-map $g\colon X_1 \to X_2$ such that $g i_1 = i_2$ and $\bar{f}_2 g = \bar{f}_1$, one shows the following identity, transitivity and normalization properties (sketch the diagrams!):

(i) $[(i_1, \bar{f}_1), (i_1, \bar{f}_1)] = \text{identity}$.

(ii) $[(i_1, \bar{f}_1), (i_2, \bar{f}_2)] \circ [(i_2, \bar{f}_2), (i_3, \bar{f}_3)] = [(i_1, \bar{f}_1), (i_3, \bar{f}_3)]$.

(iii) For any $g\colon (i_1, \bar{f}_1) \to (i_2, \bar{f}_2)$, and $\sigma := \sigma_{i_1, g, 1, i_2}$, the isomorphism $[(i_2, \bar{f}_2), (i_1, \bar{f}_1)]$ factors naturally as $i_1^* \bar{f}_1^{\times} \xrightarrow{\;\sim\;} i_1^* g^{\times} \bar{f}_2^{\times} \xrightarrow[\beta_\sigma]{\;\sim\;} i_2^* \bar{f}_2^{\times}$.

Making use of condition (4.8.4)(a), Deligne shows in [**De**, p. 308, 3.2.6(ii)] that the opposite category $\mathcal{C}^{\mathrm{op}}$ is *filtered* (see [**M**, p. 211]).[14] It follows that the independence verification for (4.8.4.2) need only be done for a pair of compactifications of which one maps to the other. This is now a straightforward exercise, using isomorphisms of the form $[(i_1, \bar{f}_1), (i_2, \bar{f}_2)]$.

To prove *existence* of $^!$ Deligne constructs, for each map f, a family of functorial isomorphisms $[(i_1, \bar{f}_1), (i_2, \bar{f}_2)]\colon i_2^* \bar{f}_2^{\times} \xrightarrow{\;\sim\;} i_1^* \bar{f}_1^{\times}$, indexed by pairs of compactifications of f, and satisfying (i)–(iii) [**De**, p. 313, 3.3.2.1]. (There is a pretty obvious such isomorphism when $(i_1, \bar{f}_1)$ maps to $(i_2, \bar{f}_2)$; and the rest follows from the fact that $\mathcal{C}^{\mathrm{op}}$ is filtered.) He then makes an arbitrary choice of a compactification $f = \bar{f} i$, and sets $f^! := i^* \bar{f}^{\times}$. Thus for any compactification $f = \bar{f}_{\bullet} i_{\bullet}$ one has an isomorphism

$$[(i_{\bullet}, \bar{f}_{\bullet}), (i, \bar{f})]\colon f^! = i^* \bar{f}^{\times} \xrightarrow{\;\sim\;} i_{\bullet}^* \bar{f}_{\bullet}^{\times}. \qquad (4.8.4.3)$$

For $f \in \mathbf{E}$, taking $\bar{f}_{\bullet} = 1$, $i_{\bullet} = f$, one gets $f^! \xrightarrow{\;\sim\;} f^*$, giving $\alpha_{\mathbf{E}}$ at the functorial—but not yet the pseudofunctorial—level. Analogous remarks lead to $\alpha_{\mathbf{P}}$.

Substituting isomorphisms as in (4.8.4.3) at each of the three appropriate places in (4.8.4.1), one gets a *definition* of $d_{f_1, f_2}\colon f_2^! f_1^! \xrightarrow{\;\sim\;} (f_1 f_2)^!$, provided it is first shown that the result of this substitution does not depend on the choice of $\bar{g}$ and j. As before, since $\mathcal{C}^{\mathrm{op}}$ is filtered it suffices to show that (4.8.4.1) (as here modified) is unaltered by the substitution for $(j, \bar{g})$ of a compactification $(j_1, \bar{g}_1)$ of $i_1 \bar{f}_2$ such that there exists a **P**-map $\bar{h}$ with $j = j_1 \bar{h}$ and $\bar{g} \bar{h} = \bar{g}_1$. This is done in [**De**, pp. 314–316].

Finally, a brief check [**De**, p. 317, 3.3.2.4] ensures that this d endows $^!$, $\alpha_{\mathbf{P}}$ and $\alpha_{\mathbf{E}}$ with all the desired pseudofunctorial properties. The last assertion in (4.8.4) simply reflects the possibility in the above definition of $^!$ of making the obvious choice $\bar{f} = 1$, $i = f$ whenever $f \in \mathbf{E}$.   Q.E.D.

The proof of (4.8.3) will be based on the following pasting result for base-change setups.[15]

---

[14] In that proof take K to be the inverse image of the diagonal under the map $(r, s)\colon \overline{Y}_1 \to \overline{Y}_2 \times_X \overline{Y}_2$.

[15] This result should be compared with [**Nk**, p. 205, Thm. 2.3.2].

Proposition 4.8.5 *With notation and assumptions as in (4.8.4), let $\bar{\mathbf{S}}$ be a category containing $\mathbf{S}$ as a subcategory. Let*

$$\mathcal{B}' := \mathcal{B}\big(\bar{\mathbf{S}}, \mathbf{E}, \mathbf{F}, {}^*, {}^\#, (\beta'_\sigma)_{\sigma \in \square'}\big), \qquad \mathcal{B}'' := \mathcal{B}\big(\bar{\mathbf{S}}, \mathbf{P}, \mathbf{F}, {}^\times, {}^\#, (\beta''_\sigma)_{\sigma \in \square''}\big),$$

be base-change setups with $\square'$ (resp. $\square''$) the class of $\bar{\mathbf{S}}$-fiber squares $\sigma_{v,f,g,u}$ such that $f, g \in \mathbf{E}$ (resp. $\mathbf{P}$) and $u, v \in \mathbf{F}$. Assume that for any $f \in \mathbf{E}$ (resp. $\mathbf{P}$) and $u \in \mathbf{F}$, such a $\sigma_{v,f,g,u}$ exists.

Then there is at most one base-change setup

$$\bar{\mathcal{B}} := \mathcal{B}\big(\bar{\mathbf{S}}, \mathbf{S}, \mathbf{F}, {}^!, {}^\#, (\bar{\beta}_\sigma)_{\sigma \in \bar{\square}}\big)$$

which extends—in the obvious sense, via $\alpha_{\mathbf{P}}$ and $\alpha_{\mathbf{E}}$—both $\mathcal{B}'$ and $\mathcal{B}''$, and with $\bar{\square}$ the class of $\bar{\mathbf{S}}$-fiber squares $\sigma_{v,f,g,u}$ such that $f, g \in \mathbf{S}$ and $u, v \in \mathbf{F}$. Such a $\bar{\mathcal{B}}$ exists if and only if for any $\bar{\mathbf{S}}$-cube with $i, i_1, j, j_1 \in \mathbf{E}$, $f, f_1, g, g_1 \in \mathbf{P}$, and $u, u_1, v, v_1 \in \mathbf{F}$, and in which all the faces are distinguished (for the appropriate one of $\mathcal{B}$, $\mathcal{B}'$, or $\mathcal{B}''$):

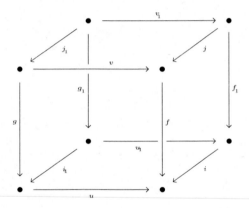

the following diagram commutes:

$$
\begin{array}{ccccc}
v_1^\# j^* f^\times & \xrightarrow{\ \beta'\ } & j_1^* v^\# f^\times & \xrightarrow{\ \beta''\ } & j_1^* g^\times u^\# \\
\ \downarrow{\scriptstyle\beta} & & & & \ \downarrow{\scriptstyle\beta} \\
v_1^\# f_1^\times i^* & \xrightarrow[\ \beta''\ ]{} & g_1^\times u_1^\# i^* & \xrightarrow[\ \beta'\ ]{} & g_1^\times i_1^* u^\#
\end{array}
\qquad (4.8.5.1)
$$

Remark 4.8.5.2 The existence part of Theorem (4.8.3), weakened by substituting for $\square$ the class of *fiber squares* $\sigma_{v,f,g,u}$ with u, v flat and f, g finitely presentable, and by leaving aside conditions (4.8.3)(iii), results from an application of (4.8.5) to the following base-change setups $\mathcal{B}'$ and $\mathcal{B}''$.

For $\mathcal{B}'$, let $\bar{\mathbf{S}}$ be the category of separated maps of noetherian schemes; $\mathbf{F}$ the subcategory of flat maps, with ${}^\# = {}^*$, the usual $\bar{\mathbf{D}}_{\mathsf{qc}}^+$-valued inverse-image pseudofunctor; $\mathbf{E} \subset \mathbf{F}$ the subcategory of étale maps, with the same inverse-image pseudofunctor $*$; $\square'$ the class of all $\overline{\mathbf{S}}$-fiber squares $\sigma_{v,f,g,u}$ with f, g étale and u, v flat; and $\beta_\sigma\colon v^*f^* \xrightarrow{\ \sim\ } g^*u^*$ the natural isomorphism. (This is just a "subsetup" of $\mathcal{B}(\overline{\mathbf{S}}, \mathbf{L}{-}^*)$, see (4.8.2.2).)

For $\mathcal{B}''$, let $\overline{\mathbf{S}}$ and $(\mathbf{F},^{\#})$ be the same as for $\mathcal{B}'$; let $\mathbf{P}$ be the subcategory of proper maps, with the $\overline{\mathbf{D}}^{+}_{\mathsf{qc}}$-valued pseudofunctor $^\times$ (see (4.1.2)); $\square''$ the class of those $\overline{\mathbf{S}}$-fiber squares $\sigma_{v,f,g,u}$ with f and g proper, u and v flat; and β_σ ($\sigma \in \square'$) the base-change isomorphism from (4.4.3).

In this situation, commutativity of (4.8.5.1) is easily checked, via "horizontal transitivity" in Example (4.8.2.3).

In (4.8.6)–(4.8.11), the resulting base-change setup $\overline{\mathcal{B}}$ will be extended to where $\overline{\square}$ consists of all admissible $\overline{\mathbf{S}}$-squares.

Proof of (4.8.5). Fiber products being unique up to isomorphism, it follows from (4.8.2.1)(a) and the assumption in (4.8.5) that *any* $\overline{\mathbf{S}}$-fiber square $\sigma_{v,f,g,u}$ with $f \in \mathbf{E}$ (resp. $\mathbf{P}$) and $u \in \mathbf{F}$ is in $\square'$ (resp. $\square''$). It is then straightforward to see via (4.8.4)(b) that any $\sigma \in \overline{\square}$ is a vertical composite $\sigma_2 \circ \sigma_1$ with $\sigma_1 \in \square'$ and $\sigma_2 \in \square''$:

$$
\sigma \;=\;
\begin{array}{ccc}
\bullet & \xrightarrow{\ \ v\ \ } & \bullet \\
{\scriptstyle j}\big\downarrow & {\scriptstyle \sigma_1} & \big\downarrow{\scriptstyle i} \\
\bullet & \xrightarrow{\ \ w\ \ } & \bullet \\
{\scriptstyle \bar{g}}\big\downarrow & {\scriptstyle \sigma_2} & \big\downarrow{\scriptstyle \bar{f}} \\
\bullet & \xrightarrow{\ \ u\ \ } & \bullet,
\end{array}
\tag{4.8.5.3}
$$

and to check that if $\overline{\mathcal{B}}$ exists then $\bar{\beta}_\sigma$ has to be the natural composition

$$
v^{\#}(\bar{f}i)^! \xrightarrow{\ \sim\ } v^{\#}i^!\bar{f}^! \xrightarrow[\alpha_{\mathbf{P}}]{\ \sim\ } v^{\#}i^!\bar{f}^\times \xrightarrow[\alpha_{\mathbf{E}}]{\ \sim\ } v^{\#}i^*\bar{f}^\times \xrightarrow[\beta']{\ \sim\ } j^*w^{\#}\bar{f}^\times
$$
$$
\xrightarrow[\beta'']{\ \sim\ } j^*\bar{g}^\times u^{\#} \xrightarrow[\alpha_{\mathbf{E}}^{-1}]{\ \sim\ } j^!\bar{g}^\times u^{\#} \xrightarrow[\alpha_{\mathbf{P}}^{-1}]{\ \sim\ } j^!\bar{g}^!u^{\#} \xrightarrow{\ \sim\ } (\bar{g}j)^!u^{\#},
$$

whence the uniqueness of $\overline{\mathcal{B}}$ (if it exists). Expanding the two instances of β in (4.8.5.1) according to the description of β_σ in (4.8.4), one finds then that (4.8.5.1) commutes. (The commutativity amounts to two ways of expanding $\bar{\beta}\colon v_1^{\#}(fj)^! = v_1^{\#}(if_1)^! \xrightarrow{\ \sim\ } (gj_1)^!u^{\#} = (i_1g_1)^!u^{\#}$ according to vertical transitivity (4.8.2)(3).)

To prove the existence of $\overline{\mathcal{B}}$, we first show that the above expression for $\bar{\beta}_\sigma$ depends only on σ.

For this purpose, consider the category $\widetilde{\mathbf{S}}$ whose objects are $\mathbf{F}$-maps, the morphisms from an $\mathbf{F}$-map $v\colon X' \to X$ to an $\mathbf{F}$-map $u\colon Y' \to Y$ being the fibre squares $\sigma_{v,f,g,u} \in \overline{\square}$, with the obvious definition of composition. Define the subcategory $\widetilde{\mathbf{E}} \subset \widetilde{\mathbf{S}}$ (resp. $\widetilde{\mathbf{P}} \subset \widetilde{\mathbf{S}}$) to be the one having the same objects as $\widetilde{\mathbf{S}}$, but with morphisms $\sigma_{v,f,g,u} \in \overline{\square}$ such that $f, g \in \mathbf{E}$ (resp. $\mathbf{P}$).

The above decomposition $\sigma = \sigma_2 \circ \sigma_1$ signifies that every $\widetilde{\mathbf{S}}$-morphism has an $(\widetilde{\mathbf{E}}, \widetilde{\mathbf{P}})$-*compactification,* i.e., it factors as an $\widetilde{\mathbf{E}}$-morphism followed by a $\widetilde{\mathbf{P}}$-morphism.

It is left as an exercise to deduce from (4.8.4)(a) its analogue for $\widetilde{\mathbf{P}} \subset \widetilde{\mathbf{S}}$.

It follows then, as in the proof of (4.8.4), that it will be enough to show that two different compactifications of $\sigma \in \Box$ give the same $\bar{\beta}_\sigma$ when one of them maps to the other, via $\widetilde{\mathbf{P}}$—cf. the definition of morphisms of compactifications which appears in the proof of (4.8.4). Let the target compactification be given by factorizations $f = \bar{f}i$, $g = \bar{g}j$ (see (4.8.5.3)); let the source compactification be given similarly by factorizations $f = \bar{f}_1 i_1$, $g = \bar{g}_1 j_1$. Then the map of compactifications is given by $\mathbf{P}$-maps p and q fitting into commutative cubes (with a common face), whose horizontal arrows are $\mathbf{F}$-maps:

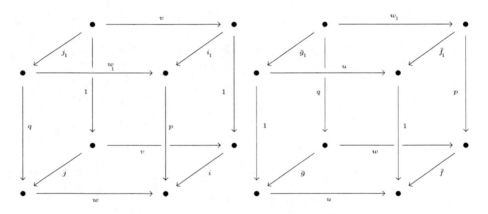

The first cube entails, via (4.8.5.1), a commutative diagram

$$
\begin{array}{ccccc}
v^{\#} i_1^{*} p^{\times} & \xrightarrow{\ \beta'\ } & j_1^{*} w_1^{\#} p^{\times} & \xrightarrow{\ \beta''\ } & j_1^{*} q^{\times} w^{\#} \\[2pt]
{\scriptstyle\beta}\big\downarrow & & & & \big\downarrow{\scriptstyle\beta} \\[2pt]
v^{\#} i^{*} & =\!=\!= & v^{\#} i^{*} & \xrightarrow[\ \beta'\ ]{} & j^{*} w^{\#}
\end{array}
\qquad (4.8.5.4)
$$

Vertical transitivity (4.8.2)(3) for the setup $\mathcal{B}\big(\bar{\mathbf{S}}, \mathbf{P}, \mathbf{F}, {}^{\times}, {}^{\#}, (\beta''_\sigma)_{\sigma \in \Box''}\big)$, applied to the composite diagram consisting of the rear and bottom faces of the second cube, yields a commutative diagram

$$
\begin{array}{ccc}
\bar{g}_1^{\times} u^{\#} & \xrightarrow{\hspace{4cm}} & w_1^{\#} \bar{f}_1^{\times} \\[2pt]
\big\uparrow & & \big\uparrow \\[2pt]
q^{\times} \bar{g}^{\times} u^{\#} & \xleftarrow{\ } q^{\times} w^{\#} \bar{f}^{\times} \xleftarrow{\ } & w_1^{\#} p^{\times} \bar{f}^{\times}
\end{array}
\qquad (4.8.5.5)
$$

Now, by the definition of $\bar{\beta}_\sigma$ with respect to a given compactification, the present problem is to show commutativity of the outer border of the following diagram, in which the maps are the obvious isomorphisms. (Recall that $i \circ 1 = i = pi_1$, $\bar{f}_1 = \bar{f}p$, $j \circ 1 = j = qj_1$, $wq = pw_1$ and $\bar{g}_1 = \bar{g}q$.)

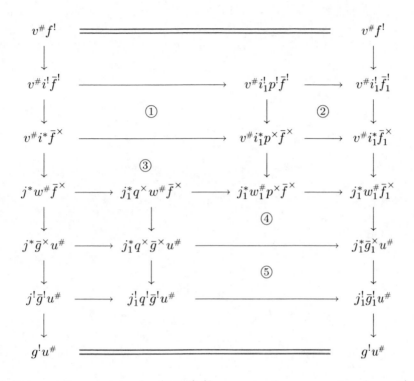

Subdiagram ① commutes by (4.8.4) (for $v := i_1$, $f := p$, $u := i$ and $g := 1$), ③ by (4.8.5.4), and ④ by (4.8.5.5). Subdiagrams ② and ⑤ commute because the isomorphism $\alpha_\mathbf{P}$ is pseudofunctorial. Commutativity of the remaining subdiagrams is clear. Thus the entire diagram does commute, and so $\bar{\beta}_\sigma$ depends only on σ.

It remains to check conditions (1)–(3) in (4.8.2), of which only "vertical transitivity for $\bar{\beta}_\sigma$" is not straightforward enough to be left to the reader.

So we need to consider a commutative diagram, with $\bar{f}_t, \bar{g}_t \in \mathbf{P}$ and $i_t, j_t \in \mathbf{E}$ ($t = 1, 2$), $w, x, y, z, u \in \mathbf{F}$, and in which all the squares are fiber squares:

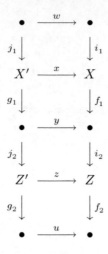

Let $i_2 f_1 = fi$ with $f\colon Y \to Z \in \mathbf{P}$ and $i\colon X \to Y \in \mathbf{E}$.

Let $g\colon Z' \times_Z Y \to Z'$ and $v\colon Z' \times_Z Y \to Y$ be the projections, so that $g \in \mathbf{P}$ and $v \in \mathbf{F}$.

Then there is a unique $\mathbf{E}$-map $j\colon X' \to Z' \times_Z Y$ such that $gj = j_2 g_1$ and $vj = ix$. One sees then that in the cube

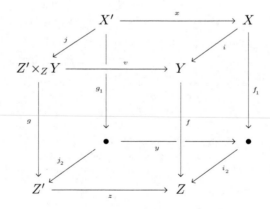

the top and bottom faces are $\mathcal{B}'$-distinguished, the front and back faces are $\mathcal{B}''$-distinguished, and the other two faces are $\mathcal{B}$-distinguished.

Now vertical transitivity amounts to commutativity of the diagram

$$
\begin{array}{ccccc}
w^{\#}(ii_1)^*(f_2 f)^{\times} & \longrightarrow & w^{\#}i_1^* i^* f^{\times} f_2^{\times} & \longrightarrow & w^{\#}i_1^* f_1^{\times} i_2^* f_2^{\times} \\[2ex]
\Big\downarrow & \textcircled{1} & \Big\downarrow & & \Big\downarrow \\[2ex]
 & & j_1^* x^{\#} i^* f^{\times} f_2^{\times} & \longrightarrow & j_1^* x^{\#} f_1^{\times} i_2^* f_2^{\times} \\[2ex]
 & & \Big\downarrow & & \Big\downarrow \\[2ex]
(jj_1)^* v^{\#}(f_2 f)^{\times} & \longrightarrow & j_1^* j^* v^{\#} f^{\times} f_2^{\times} & \textcircled{3} & j_1^* g_1^{\times} y^{\#} i_2^* f_2^{\times} \\[2ex]
\Big\downarrow & & \Big\downarrow & & \Big\downarrow \\[2ex]
 & \textcircled{2} & j_1^* j^* g^{\times} z^{\#} f_2^{\times} & \longrightarrow & j_1^* g_1^{\times} j_2^* z^{\#} f_2^{\times} \\[2ex]
 & & \Big\downarrow & & \Big\downarrow \\[2ex]
(jj_1)^*(g_2 g)^{\times} u^{\#} & \longrightarrow & j_1^* j^* g^{\times} g_2^{\times} u^{\#} & \longrightarrow & j_1^* g_1^{\times} j_2^* g_2^{\times} u^{\#}
\end{array}
$$

Subsquares ① and ② commute by vertical transitivity for $\mathcal{B}''$. Commutativity of ③ is the instance of (4.8.5.1) corresponding to the preceding cube. Commutativity of the remaining two subsquares is obvious.

This completes the proof of Proposition (4.8.5). Q.E.D.

As previously noted, to finish the proof of (4.8.1) we need to enlarge the setup (4.8.2.4) to (4.8.2.4)'. Similarly, to finish the proof of (4.8.3) we need to show that there exists a unique enlargement $\widetilde{\mathcal{B}}$ of the setup $\overline{\mathcal{B}}$ at the end of (4.8.5.2) such that all admissible **S**-squares are $\widetilde{\mathcal{B}}$-distinguished. In addition, we need to check that (4.8.3)(ii) and (iii) hold for this $\widetilde{\mathcal{B}}$.

All this will be done in (4.8.11), after the supporting formal details are developed in (4.8.6)–(4.8.10).

Definition 4.8.6 For a base-change setup $\mathcal{B}(\mathbf{S}, \mathbf{P}, \mathbf{F}, {}^{!}, {}^{*}, (\beta_\sigma)_{\sigma \in \square})$ a subcategory $\mathbf{E} \subset \mathbf{S}$ is *special* if for any maps $i\colon X \to Y$ in $\mathbf{E}$, $g\colon X' \to X$ in $\mathbf{P}$, and $v\colon X' \to X$ in $\mathbf{F}$, the squares

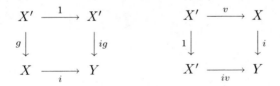

are *distinguished*.

Remarks 4.8.6.1 (a) If **E** is special then $\mathbf{E} \subset \mathbf{P} \cap \mathbf{F}$.

(b) If **E** is special for $\mathcal{B}$, then **E** is also special for the dual of $\mathcal{B}$ (see (4.8.2.1)(c)).

Example 4.8.6.2 For (4.8.2.4), or for $\mathcal{B}'$, $\mathcal{B}''$ or $\overline{\mathcal{B}}$ in (4.8.5.2), the category $\mathbf{E}$ whose maps are all the open-and-closed immersions of noetherian schemes is special. Indeed, since i is a monomorphism, both squares in (4.8.6) are fiber squares.

After fixing a special subcategory $\mathbf{E}$, we will call its maps special. For any special map $i \colon X \to Y$,

$$\beta_i \colon i^! \xrightarrow{\ \sim\ } i^* \tag{4.8.7.0}$$

is defined to be the isomorphism β_τ associated to the distinguished square

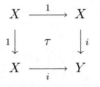

Proposition 4.8.7 *Let* $\mathcal{B}\big(\mathbf{S}, \mathbf{P}, \mathbf{F}, \,^!, \,^*, (\beta_\sigma)_{\sigma \in \square}\big)$ *be a base-change setup and* $\mathbf{E}$ *a special subcategory. Then the restrictions of the pseudofunctors* $^!$ *and* * *to* $\mathbf{E}$ *are naturally isomorphic.*

Proof. The family of isomorphisms β_i $(i \in \mathbf{E})$ of (4.8.7.0) is pseudo-functorial (see (3.6.6)): if $i \colon X \to Y$ and $j \colon Y \to Z$ are in $\mathbf{E}$, apply (3) and (2) of (4.8.2) to

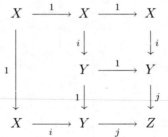

to see that the left and right halves of the following diagram commute:

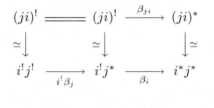

Q.E.D.

Proposition 4.8.8 *Let* $\mathcal{B}\big(\mathbf{S}, \mathbf{P}, \mathbf{F}, \,^!, \,^*, (\beta_\sigma)_{\sigma \in \square}\big)$ *be a base-change setup,* $\mathbf{E}$ *a special subcategory, and* β_i $(i \in \mathbf{E})$ *as in* (4.8.7.0)*. Then:*

(i) *For each distinguished square*

with f and g in **E**, *the following diagram commutes*:

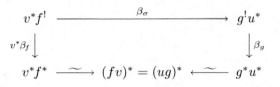

(ii) *For each distinguished square*

with u and v in **E**, *the following diagram commutes*:

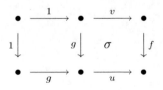

Proof. Definition (4.8.6) shows that the following composite square ρ is distinguished, as are its constituents:

$$
\begin{array}{ccccc}
\bullet & \xrightarrow{\ 1\ } & \bullet & \xrightarrow{\ v\ } & \bullet \\
\downarrow{\scriptstyle 1} & & \downarrow{\scriptstyle g} & \sigma & \downarrow{\scriptstyle f} \\
\bullet & \xrightarrow[\ g\ ]{} & \bullet & \xrightarrow[\ u\ ]{} & \bullet
\end{array}
$$

so horizontal transitivity (4.8.2)(3) gives a commutative diagram

$$
\begin{array}{ccc}
v^*f^! & \xrightarrow{\ \beta_\sigma\ } & g^!u^* \\
\downarrow{\scriptstyle \beta_\rho} & & \downarrow{\scriptstyle \beta_g} \\
(ug)^* & \xleftarrow[\ \sim\ ]{} & g^*u^*
\end{array}
\qquad\qquad (4.8.8.1)
$$

Also, the following decomposition of ρ

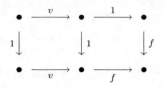

yields—via (2)$'$ and (3) of (4.8.2)—the commutative diagram

$$\begin{array}{ccc} v^*f^! & =\!=\!=\!=\!=\!=\!=\!=\!= & v^*f^! \\ {\scriptstyle v^*\beta_f}\downarrow & & \downarrow{\scriptstyle \beta_\rho} \\ v^*f^* & \xrightarrow{\ \sim\ } (fv)^* =\!=\!= (ug)^* \end{array} \qquad (4.8.8.2)$$

Pasting (4.8.8.1) and (4.8.8.2) along their common edge, we get (i).
  Assertion (ii) is just (i) for the dual setup (see (4.8.2.1)(c)).

Q.E.D.

  (4.8.9) We will now see how to enlarge certain base-change setups.
  Consider a category $\mathbf{S}$ in which for any maps $X \to Y$ and $Y' \to Y$ a fiber product $X \times_Y Y'$ exists. A square $\sigma_{v,f,g,u}$ in $\mathbf{S}$:

$$\begin{array}{ccc} X' & \xrightarrow{\ v\ } & X \\ {\scriptstyle g}\downarrow & & \downarrow{\scriptstyle f} \\ Y' & \xrightarrow{\ u\ } & Y \end{array} \qquad (4.8.9.1)$$

is, as usual, called a *fiber square* if the corresponding map $X' \to X \times_Y Y'$ is an isomorphism.
  Let $\mathcal{B} := \mathcal{B}(\mathbf{S}, \mathbf{P}, \mathbf{F}, {}^!, {}^*, (\beta_\sigma)_{\sigma \in \square})$ be a base-change setup, and $\mathbf{E}$ a special subcategory, see (4.8.6).
  We make the following assumptions, in addition to those in (4.8.2).
  (4) In the following $\mathbf{S}$-diagrams, suppose that $u_1 \in \mathbf{F}$ (resp. $f_1 \in \mathbf{P}$).

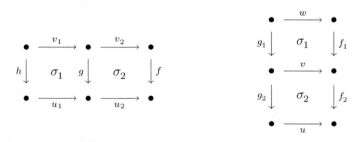

In either diagram, if σ_2 is a fiber square and the composed square $\sigma_2\sigma_1$ is in $\square$, then $\sigma_1 \in \square$.

(5) For any fiber square (4.8.9.1) in $\square$, if u (resp. f) is special (i.e., lies in $\mathbf{E}$) then so is v (resp. g).

(6) If the square (4.8.9.1) is in $\square$ then so is any fiber square with the same u and f,

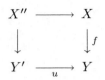

and furthermore, the resulting map $X' \to X''$ is special.

Example 4.8.9.2 Conditions (4)–(6) are easily seen to be satisfied in any of the situations in Example (4.8.6.2), where all distinguished squares are fiber squares.

Remark 4.8.9.3 Let $\mu \colon X' \to X''$ be an isomorphism and consider the following fiber squares, the first of which is, by (4.8.2)(2), distinguished:

From (6) it follows that μ is special. Thus *every isomorphism is special.*

Proposition 4.8.10 *Under the preceding assumptions, there is a unique base-change setup* $\mathcal{B}' = \mathcal{B}'_{\mathbf{E}} = \mathcal{B}(\mathbf{S}, \mathbf{P}, \mathbf{F}, {}^!, {}^*, (\beta'_\sigma)_{\sigma \in \square'})$ *such that:*

(i) *A commutative square*

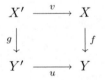

is in $\square'$ *if and only if there is a fiber square in* $\square$

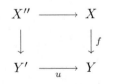

such that the resulting map $X' \to X''$ *is special.*

So by (4.8.9)(6) and (4.8.9.3), $\square \subseteq \square'$; and by (4.8.2)(1), every fiber square in $\square'$ is in $\square$.

(ii) *For every* $\sigma \in \square \subseteq \square'$ *it holds that* $\beta_\sigma = \beta'_\sigma$.

Proof. For uniqueness, suppose that $\mathcal{B}'$ satisfies (i) (which determines $\square'$) and (ii). We note first that if $i\colon X \to Y$ is a special map, then by (i), the square τ' in the following diagram is in $\square'$, as are the squares τ (by (4.8.6)) and $\tau'\tau$ (by (4.8.2)(2)'):

$$
\begin{array}{ccccc}
X & \xrightarrow{\ 1\ } & X & \xrightarrow{\ i\ } & Y \\
{\scriptstyle 1}\downarrow & \tau & {\scriptstyle i}\downarrow & \tau' & \downarrow{\scriptstyle 1} \\
X & \xrightarrow{\ i\ } & Y & \xrightarrow{\ 1\ } & Y
\end{array}
$$

It follows then from (4.8.2)(3) and (4.8.2)(2)' that

$$
\beta'_{\tau'} = (\beta'_{\tau})^{-1} \overset{\text{(ii)}}{=} (\beta_{\tau})^{-1} \overset{(4.8.7.0)}{=} (\beta_i)^{-1}.
$$

Now, any $\sigma \in \square'$:

$$
\begin{array}{ccc}
X' & \xrightarrow{\ v\ } & X \\
{\scriptstyle g}\downarrow & \sigma & \downarrow{\scriptstyle f} \\
Y' & \xrightarrow{\ u\ } & Y
\end{array}
$$

can, according to (i), be decomposed as

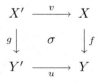

(4.8.10.1)

with $\sigma_3 \in \square$ a fiber square (so that $h \in \mathbf{P}$), and i special. The fiber square σ_2 is in $\square$, by (4.8.2)(2); and by (i), σ_1 and $\sigma_2\sigma_1 \in \square'$. We saw above that $\beta'_{\sigma_1} = (\beta_i)^{-1}$; and the maps β'_{σ_k} $(k = 2, 3)$ are determined by (ii). Hence $\beta'_{\sigma_2\sigma_1}$ is determined, and then so is β'_{σ} (see (4.8.2)(3)). Thus $\mathcal{B}'$ is unique.

For the existence, let $\square'$ be the class of all squares

$$
\begin{array}{ccc}
X' & \xrightarrow{\ v\ } & X \\
{\scriptstyle g}\downarrow & \sigma & \downarrow{\scriptstyle f} \\
Y' & \xrightarrow{\ u\ } & Y
\end{array}
$$

satisfying (i), that is, decomposing as in (4.8.10.1)—where $i \in \mathbf{P} \cap \mathbf{F}$ (see (4.8.6.1)), $h \in \mathbf{P}$ and $w \in \mathbf{F}$, so that $f, g \in \mathbf{P}$ and $u, v \in \mathbf{F}$, as required of distinguished squares.

To such a decomposition we associate the natural composite map

$$v^*f^! \xrightarrow{\sim} i^*w^*f^! \xrightarrow[i^*\beta_{\sigma_3}]{} i^*h^!u^* \xrightarrow[\beta_i^{-1}]{\sim} i^!h^!u^* \xrightarrow{\sim} g^!u^*. \qquad (4.8.10.2)$$

We will define β'_σ for $\mathcal{B}'$ to be (4.8), but first we need to show it independent of the chosen decomposition.

Suppose then that we have another decomposition with (X'', i, h, w) replaced by (X''_1, i_1, h_1, w_1), i.e., there is an isomorphism $\mu \colon X'' \xrightarrow{\sim} X''_1$ such that

$$i_1 = \mu i, \qquad h_1 = h\mu^{-1}, \qquad w_1 = w\mu^{-1}.$$

For the special map μ (see (4.8.9.3)), we have the isomorphism β_μ of (4.8.7.0). We have also the isomorphism β_ρ associated to the square

$$
\begin{array}{ccc}
X' & \xrightarrow{\ 1\ } & X' \\
{\scriptstyle i}\downarrow & {\scriptstyle \rho} & \downarrow{\scriptstyle i_1} \\
X'' & \xrightarrow[\ \mu\ ]{} & X''_1
\end{array}
$$

which is in $\square$ by (4.8.2.1)(a).

We want to show that the following diagram of natural maps (with outside columns as in (4.8)) commutes:

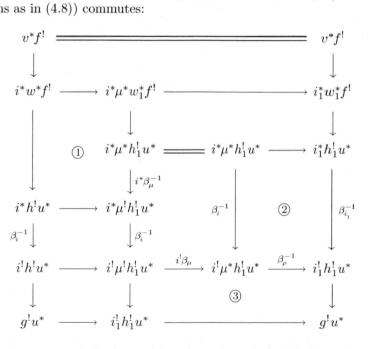

Commutativity of ② (resp. ③) follows from (4.8.8)(i) (resp. (4.8.8)(ii)) applied to ρ.

Commutativity of ① follows from (4.8.2)(3) and (4.8.8)(ii), applied respectively to the following fiber squares $\sigma_3 = \sigma_3'\sigma'$ and σ' (σ' being distinguished, by (4.8.2.1)(a)):

$$
\begin{array}{ccccc}
X'' & \xrightarrow{\ \mu\ } & X_1'' & \xrightarrow{\ w_1\ } & X \\
{\scriptstyle h}\downarrow & {\scriptstyle \sigma'} & \downarrow{\scriptstyle h_1}\ \ {\scriptstyle \sigma_3'} & & \downarrow{\scriptstyle f} \\
Y' & \xrightarrow[\ 1\ ]{} & Y' & \xrightarrow[\ u\ ]{} & Y
\end{array}
$$

Commutativity of the remaining subdiagrams is clear.

So we can indeed define β_σ' as indicated above.

Condition (i) in (4.8.10) is then obvious.

As for (ii), referring to a decomposition (4.8.10.1) of $\sigma \in \square$ (where $wi = v$ and $hi = g$), note that by (4.8.9)(4) the square $\sigma_2\sigma_1$ is in $\square$, so by (4.8.2)(3) the diagram

$$
\begin{array}{ccc}
v^*f^! & \xrightarrow{\quad\quad\beta_\sigma\quad\quad} & g^!u^* \\
{\scriptstyle \simeq}\downarrow & & \| \\
i^*w^*f^! & \xrightarrow[i^*\beta_{\sigma_3}]{} i^*h^!u^* \xrightarrow[\beta_{\sigma_2\sigma_1}]{} & g^!1^*u^*
\end{array}
$$

commutes. Also, (4.8.8)(ii) applied to $\sigma_2\sigma_1$ shows that $\beta_{\sigma_2\sigma_1}$ factors as

$$
i^*h^!u^* \xrightarrow[\beta_i^{-1}]{} i^!h^!u^* \xrightarrow{\ \sim\ } g^!u^*.
$$

Hence the composite map (4.8) is equal to β_σ, proving (ii).

Having thus defined $\mathcal{B}'$, we are left with proving (1)–(3) in (4.8.2).

For (1), assume, with notation as in (4.8.2), that $\sigma_1 \in \square'$. Consider a commutative decomposition of σ

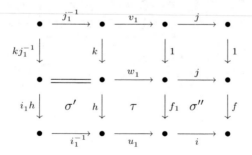

in which the middle third of the diagram is a decomposition of σ_1 with $\tau \in \square$ a fiber square and k special, and $v_1 := w_1k$; and the right third exists by assumption, σ'' being a fiber square because i and j are isomorphisms. (Note: $i_1hkj_1^{-1} = i_1g_1j_1^{-1} = g$.) The composed fiber square $\sigma''\tau\sigma'$, being isomorphic to τ, is in $\square$; and thus, since kj_1^{-1} is special (see (4.8.6.1)(a)), therefore $\sigma \in \square'$, proving (1).

Conditions (2) and (2)' for $\mathcal{B}'$ follow from the same for $\mathcal{B}$, because of (4.8.10)(ii).

As for (3), consider a composite diagram $\sigma_0 = \sigma_2 \sigma_1$:

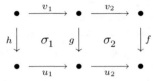

with σ_2, σ_1 and σ_0 in $\square'$. Using all the assumptions in (4.8.9), we find that this decomposes further as

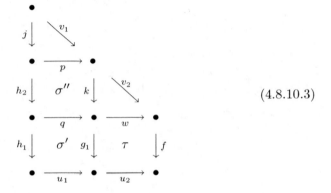

$$(4.8.10.3)$$

where σ'', σ' and τ are fiber squares in $\square$; the maps g_1, w, h_1, q, h_2, p are the natural projections; the maps j and k are special—whence so are h_2 and $h_2 j$ (see (4.8.9)(5)); the triangles commute; $g_1 k = g$ and $h_1 h_2 j = h$.

What (3) asserts is, first, that the following natural diagram commutes:

$$(4.8.10.4)$$

$$
\begin{array}{ccc}
(v_2 v_1)^* f^! & \xrightarrow{\;\beta_{\sigma_0}\;} & h^!(u_2 u_1)^* \\
\downarrow & & \downarrow \\
v_1^* v_2^* f^! & \quad v_1^* g^! u_2^* \xrightarrow{\;\beta_{\sigma_1}\;} h^! u_1^* u_2^* & \\
\simeq\downarrow & \uparrow\simeq & \\
j^* p^* v_2^* f^! & \xrightarrow[\text{via } \beta_{\sigma_2}]{} j^* p^* (g_1 k)^! u_2^* &
\end{array}
$$

Expanding β_{σ_2}, β_{σ_1}, and $\beta_{\sigma_2 \sigma_1}$, as in (4.8), one sees that for this it is enough to show commutativity of the outer border of the natural diagram on the following page, or just to show that each of its twelve undecomposed subdiagrams commutes.

But for the eight unlabeled subdiagrams, commutativity holds by elementary (pseudo)functorial considerations; for subdiagram ①, one can use (4.8.7); for ② and ④, (4.8.2)(3); and for ③, (4.8.8)(i).

This completes the proof of the "horizontal" part of (3).

The proof of the "vertical" part of (3) is similar. Alternatively, one can just dualize everything in sight, as indicated in (4.8.2.1)(c). The conditions in (4.8.6) defining a special subcategory are self-dual, so that if $\mathbf{E}$ is special for a setup $\mathcal{B}$, then $\mathbf{E}$ is also special for the dual setup $\mathcal{B}^{\mathrm{op}}$. Likewise, conditions (4)–(6) in (4.8.9) hold for $\mathcal{B}$ iff they hold for $\mathcal{B}^{\mathrm{op}}$. Then, one checks, vertical transitivity for $(\mathcal{B}^{\mathrm{op}})'$ (constructed as above) is identical with the just-proved horizontal transitivity for $\mathcal{B}'$.

This completes the proof of Proposition (4.8.10).   Q.E.D.

Corollary 4.8.10.5 *With notation and assumptions as in* (4.8.10), *let* $\mathbf{E}'$ *be a subcategory of* $\mathbf{S}$ *such that for every map* $i\colon X \to Y \in \mathbf{E}'$ *the diagonal map* $\delta_i\colon X \to X \times_Y X$ *is in* $\mathbf{E}$. *Assume further that for any fiber square* $\sigma_{v,f,g,u}$ *in* $\mathbf{S}$, *if* u (*resp.* f) *is in* $\mathbf{E}'$ *then so is* v (*resp.* g). *Then:*

(i) $\mathbf{E}'$ *is* $\mathcal{B}'$-*special; and conditions* (4)-(6) *in* (4.8.9) *hold for* $(\mathcal{B}', \mathbf{E}')$. *Thus it is meaningful to set* $\mathcal{B}'' := (\mathcal{B}')'_{\mathbf{E}'}$.

(ii) *If a fiber square* $\sigma = \sigma_{v,f,g,u}$ *with* $u \in \mathbf{E}'$ *is in* $\square$, *then any commutative* $\sigma_{v',f,g',u}$ *with* $v' \in \mathbf{E}'$ *and* $g' \in \mathbf{P}$ *is* $\mathcal{B}''$-*distinguished.*

Proof. (i) The second diagram in (4.8.6)—call it σ—expands as

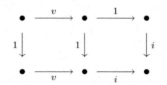

which when $i \in \mathbf{E}'$ can be further expanded in the form (4.8.10.3), with $j = 1$ and $k \in \mathbf{E}$, whence (since $\sigma'' \in \square$) $h_2 \in \mathbf{E}$, whence by (4.8.10)(i), $\sigma \in \square'$. In a similar way, or by dualizing (see (4.8.6.1(b))), one finds that the first diagram in (4.8.6) is in $\square'$.

For (4.8.9)(4), we can decompose, say, the horizontal $\sigma_2\sigma_1$ of that condition as

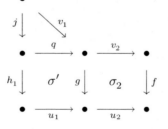

with $j \in \mathbf{E}$, $qj = v_1$, $h_1 j = h$, and σ_2, σ' fiber squares such that the fiber square $\sigma_2\sigma'$ is in $\square$.

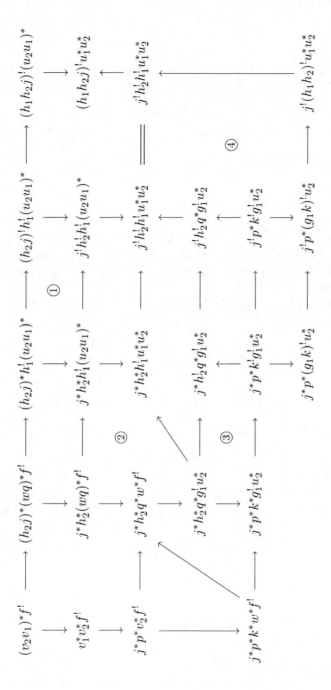

It follows from $(4.8.9)(4)$ for $\mathcal{B}$ that $\sigma' \in \square$, whence $\sigma_1 \in \square'$, proving the horizontal part of $(4.8.9)(4)$ for $\mathcal{B}'$. The vertical part is similar (or dual).

Since any fiber square in $\square'$ is in $\square$, $(4.8.9)(5)$ is essentially the "further" assumption on $\mathbf{E}'$.

Finally, $(4.8.9)(6)$ for $\mathcal{B}'$ follows from $(4.8.10)(i)$, $(4.8.9)(6)$ for $\mathcal{B}$, and $(4.8.9.3)$.

(ii) Consider a decomposition of $\sigma_{v',f,g',u}$

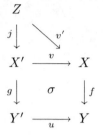

with $v' = vj$. We need only show that $j \in \mathbf{E}'$.

With Γ_j the graph map of j and $\pi_2\colon Z \times_X X' \to X'$ the projection, the map j factors as

$$Z \xrightarrow{\Gamma_j} Z \times_X X' \xrightarrow{\pi_2} X'.$$

The fiber square

$$\begin{array}{ccc}
Z \times_X X' & \xrightarrow{\pi_2} & X' \\
{\scriptstyle \pi_1}\downarrow & & \downarrow{\scriptstyle v} \\
Z & \xrightarrow[v']{} & X
\end{array}$$

shows that $\pi_2 \in \mathbf{E}'$; and the fiber square

$$\begin{array}{ccc}
Z & \xrightarrow{\Gamma_j} & Z \times_X X' \\
{\scriptstyle j}\downarrow & & \downarrow{\scriptstyle j \times 1_{X'}} \\
X' & \xrightarrow[\delta_v]{} & X' \times_X X'
\end{array}$$

shows that $\Gamma_j \in \mathbf{E}'$, whence the conclusion.   Q.E.D.

(4.8.11). Let us now complete the proof of $(4.8.1)$ and $(4.8.3)$ by doing what was indicated just before Definition $(4.8.6)$.

For either the setup $\mathcal{B}$ of $(4.8.2.4)$ or the larger setup $\overline{\mathcal{B}}$ of $(4.8.5.2)$, the category $\mathbf{E}$ of open-and-closed immersions is special, see $(4.8.6.2)$.

The diagonal of a separated étale map is an open-and-closed immersion [**EGA IV**, $(17.4.2)(b)$]; and maps which are étale (resp. separated, resp. proper) remain so after arbitrary base change [**EGA IV**, $(17.3.3)(iii)$]. Therefore the category $\mathbf{E}'$ of separated étale maps (resp. proper étale maps) satisfies the hypotheses of $(4.8.10.5)$ with respect to $(\overline{\mathcal{B}}, \mathbf{E})$ (resp. $(\mathcal{B}, \mathbf{E})$). Keeping in mind the uniqueness part of $(4.8.10)$, one see that the resulting base-change setup $\widetilde{\mathcal{B}} := \overline{\mathcal{B}}''$ is the sought-after unique enlargement of $\overline{\mathcal{B}}$, and that $\mathcal{B}''$ is the unique enlargement $(4.8.2.4)'$ of $\mathcal{B}$.

It remains to show that conditions $(4.8.3)$(ii) and (iii) hold for $\widetilde{\mathcal{B}}$.

Using the definition (4.8) of β_σ, one readily reduces the question to where σ is a fiber square. In that case, (ii) follows from the description of $\mathcal{B}'$ in $(4.8.5.2)$.

As for (iii), let $f = \bar{f}i$ be a compactification, and apply vertical transitivity $(4.8.2)$(3), to reduce to where *either* $f = i$ is an open immersion, a case covered by (ii), *or* $f = \bar{f}$ is proper, a case covered by $(4.8.1)$(iii).   Q.E.D.

Exercises 4.8.12 (a) Let $\mathcal{B}(\mathbf{S}, \mathbf{P}, \mathbf{F}, {}^!, {}^*, (\beta_\sigma)_{\sigma \in \square})$ be a base-change setup, and let there be given pseudofunctorial isomorphisms ${}^! \xrightarrow{\sim} {}^\times$, ${}^* \xrightarrow{\sim} {}^\#$. For any $\sigma_{v, f, g, u} \in \square$ let $\bar{\beta}_\sigma$ be the natural composite isomorphism

$$v^\# f^\times \xrightarrow{\;\sim\;} v^* f^! \xrightarrow[\beta_\sigma]{\;\sim\;} g^! u^* \xrightarrow{\;\sim\;} g^\times u^\#.$$

Show that $\mathcal{B}(\mathbf{S}, \mathbf{P}, \mathbf{F}, {}^\times, {}^\#, (\bar{\beta}_\sigma)_{\sigma \in \square})$ is a base-change setup.

(b) (generalizing $(4.1.9)$(c)). Notation is as in $(4.8.2.4)$. For a finite étale scheme-map $f \colon X \to Y$, the natural map is an isomorphism $f_* \xrightarrow{\sim} \mathbf{R}f_*$ of functors from $\mathbf{D}_{\mathrm{qc}}(X)$ to $\mathbf{D}_{\mathrm{qc}}(Y)$, see proof of $(3.10.2.2)$. Define the functorial "trace" map

$$f_* f^* E \underset{(3.9.4)}{\cong} f_* \mathcal{O}_X \otimes E \to \mathcal{O}_Y \otimes E \cong E \qquad (E \in \mathbf{D}_{\mathrm{qc}}(Y))$$

to be $\mathrm{tr}_f \otimes 1$ where tr_f is the natural composition

$$f_* \mathcal{O}_X \longrightarrow \mathcal{H}om^\bullet(f_* \mathcal{O}_X, f_* \mathcal{O}_X) \cong \mathcal{H}om^\bullet(f_* \mathcal{O}_X, \mathcal{O}_Y) \otimes f_* \mathcal{O}_X \longrightarrow \mathcal{O}_Y,$$

given locally by the usual linear-algebra trace map. (Note that, f being flat and finitely presented, $f_* f^* \mathcal{O}_Y$ is a locally free $\mathcal{O}_Y$-module.) There corresponds a functorial map $t_f \colon f^* \to f^\times$.

(i) Show that on finite étale maps, the map $t_{(-)} \colon (-)^* \to (-)^\times$ is *pseudofunctorial*, see $(3.6.6)$. (Reduction to the affine case may help.) Also, $t_{\mathrm{identity}} = \mathrm{identity}$.

(ii) (Compatibility of trace with base change.) Given a fiber square $\sigma = \sigma_{v, f, g, u}$ with f and g finite étale, u and v flat, show that the following diagram commutes:

$$
\begin{array}{ccc}
u^* f_* f^* & \xrightarrow{\;u^* \mathrm{tr}_f\;} & u^* \\[4pt]
{\scriptstyle (3.7.2)} \downarrow {\scriptstyle \simeq} & & \uparrow {\scriptstyle \mathrm{tr}_g} \\[4pt]
g_* v^* f^* & \xrightarrow[\mathrm{natural}]{\;\sim\;} & g_* g^* u^*
\end{array}
$$

(iii) For σ as in (ii), show that the following diagram commutes:

$$
\begin{array}{ccc}
v^* f^* & \xrightarrow{\;\mathrm{natural}\;} & g^* u^* \\[4pt]
{\scriptstyle v^* t_f} \downarrow & & \downarrow {\scriptstyle t_g} \\[4pt]
v^* f^\times & \xrightarrow[\beta_\sigma]{} & g^\times u^*
\end{array}
$$

(Commutativity of the adjoint diagram is a consequence of (ii).)

(iv) For any finite étale f show, using, e.g., (i), (iii), and (4.8), that with $\beta_f \colon f^\times \xrightarrow{\sim} f^*$ (see (4.8.7.0)) as in the base-change setup (4.8.2.4)′, $\beta_f t_f$ *is the identity* (whence t_f *is an isomorphism*—which can also be proved more directly).

(v) Deduce from (iv) that when ! is constructed as in the proof of (4.8.1), via application of (4.8.4) to (4.8.2.4)′, then the canonical map $f_* f^* = f_* f^! \to \mathbf{1}$ (arising from right-adjointness of $f^!$ to f_*) is just the trace map.

(vi) For any finite étale $f \colon X \to Y$, and $E, F \in \mathbf{D}_{qc}(X)$, show, via (v) or otherwise, that the map $\chi_{E,F}$ of (4.7.3.4) is just the isomorphism $f^* E \otimes f^* F \xrightarrow{\sim} f^*(E \otimes F)$ of (3.2.4).

(vii) Suppose that on the category $\mathbf{E}$ of finite étale maps of noetherian schemes there is associated to each $f \colon X \to Y$ a functorial map $\tau_f \colon f_* f^* \to \mathbf{1}$ in such a way that the pairs (f^*, τ_f) $(f \in \mathbf{E})$ form a pseudofunctorial right adjoint to the $\mathbf{D}_{qc}$-valued direct image pseudofunctor, and such that furthermore, the diagram in (ii) above still commutes when tr_f is replaced by τ_f. Prove that $\tau_f = \mathrm{tr}_f$ for all f.

Deduce that (v) holds for *any* $f^!$ satisfying (4.8.1).

Hint. Show that $\tau_f = \mathrm{tr}_f \circ \theta_f$ for some automorphism θ_f of the functor f^*, i.e., $\theta_f = $ multiplication by e_f for some unit $e_f \in \mathrm{H}^0(X, \mathcal{O}_X)$. Then check that pseudofunctoriality implies, for any composition $X \xrightarrow{f} Y \xrightarrow{g} Z$, that $e_{gf} = e_g(g^* e_f)$; and check that for any σ as in (ii), $e_g = v^* e_f$. Then deduce from (iii), *mutatis mutandis,* that for any open-and closed immersion δ, $e_\delta = 1$; and finally, from the diagram

$$X \xrightarrow{\delta} X \times_Y X \xrightarrow{\pi_2} X$$

$$\pi_1 \downarrow \qquad\qquad \downarrow f$$

$$X \xrightarrow{\quad f \quad} Y$$

($\delta := $ diagonal), that $e_f = 1$ for all f.

(c) Show that a horizontal or vertical composite of admissible squares is admissible.

(d) Adapt the arguments in §4.11 to extend [**Nk**, p. 268, Thm. 7.3.2]—which avoids noetherian hypotheses—to where $\mathfrak{s}$ can be any admissible square $\sigma_{v,f,g,u}$ with f and g composites of finitely-presentable proper flat maps and étale maps. (Recall that finitely-presentable flat maps are pseudo-coherent (4.3.1).)

4.9 Perfect Maps of Noetherian Schemes

In this section *all schemes are assumed noetherian and all scheme-maps finite-type and separated.* The abbreviations introduced at the beginning of §4.4 will be used throughout.

We will associate to any such scheme-map $f \colon X \to Y$ a canonical bifunctorial map, with $f^!$ as in (4.8.1), and both E and $E \otimes F$ in $\overline{\mathbf{D}}_{qc}^+(Y)$,

$$\chi_{E,F}^f \colon f^! E \otimes f^* F \to f^!(E \otimes F),$$

agreeing with the map $\chi_{E,F}$ in (4.7.3.4) when f is proper,. and with the inverse of the isomorphism in (3.2.4) when f is étale.

Any functorial relation involving $(-)^!$ ought to be examined with regard to pseudofunctoriality and base change (cf., e.g., (4.2.3)(h)–(j)). For χ, this is done in Corollary (4.9.5) and Exercise (4.9.3)(c).

The main result, Theorem (4.9.4), inspired by [**V**$'$, p. 396, Lemma 1 and Corollary 2], gives several criteria for f to be *perfect* (i.e., since f is pseudo-coherent, to have *finite tor-dimension*). Included there is the implication f *perfect* $\implies \chi_{E,F}^{f}$ *an isomorphism*.

In [**Nk**$'$, Theorem 5.9] Nayak extends these results to separated maps that are only *essentially* of finite type.

(4.9.1). For scheme-maps $X \xrightarrow{u} \overline{X} \xrightarrow{\overline{f}} Y$, u an open immersion, $\overline{f}$ proper, we define the bifunctorial map

$$\chi_{E,F}^{\overline{f}} : \overline{f}^{!}E \otimes \overline{f}^{*}F \longrightarrow \overline{f}^{!}(E \otimes F) \qquad (E, F \in \mathbf{D_{qc}}(Y))$$

to be the map adjoint to the natural composite map

$$\overline{f}_{*}(\overline{f}^{!}E \otimes \overline{f}^{*}F) \xrightarrow[\;(3.9.4)\;]{\sim} \overline{f}_{*}\overline{f}^{!}E \otimes F \longrightarrow E \otimes F,$$

and we define the bifunctorial map

$$\chi_{E,F}^{\overline{f},u} : u^{*}\overline{f}^{!}E \otimes f^{*}F \longrightarrow u^{*}\overline{f}^{!}(E \otimes F) \qquad (E, F \in \mathbf{D_{qc}}(Y))$$

to be the natural composite map

$$u^{*}\overline{f}^{!}E \otimes f^{*}F \xrightarrow{\sim} u^{*}\overline{f}^{!}E \otimes u^{*}\overline{f}^{*}F \xrightarrow{\sim} u^{*}(\overline{f}^{!}E \otimes \overline{f}^{*}F) \xrightarrow{u^{*}\chi_{E,F}^{\overline{f}}} u^{*}\overline{f}^{!}(E \otimes F).$$

When E and $E \otimes F$ are in $\overline{\mathbf{D}}_{\mathsf{qc}}^{+}(Y)$, setting $f := \overline{f}u$ we can write $f^{!}$ for $u^{*}\overline{f}^{!}$. In that case, we'll see below, in (4.9.2.2), that $\chi_{E,F}^{\overline{f},u}$ depends only on f, not on the factorization $f = \overline{f}u$, so we can denote the map $\chi_{E,F}^{\overline{f},u}$ by

$$\chi_{E,F}^{f} : f^{!}E \otimes f^{*}F \to f^{!}(E \otimes F). \tag{4.9.1.1}$$

In this connection, recall that by Nagata's compactification theorem, *any* (finite-type separated) scheme-map f factors as $f = \overline{f}u$.

Lemma 4.9.2 *Let there be given a commutative diagram*

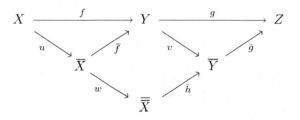

with u, v and w open immersions, $\overline{f}$, $\overline{g}$ and $\overline{h}$ proper.

Then for all $E, F \in \mathbf{D}(Z)$ such that E and $E \otimes F$ are in $\overline{\mathbf{D}}_{\mathsf{qc}}^{+}(Z)$, the following natural diagram commutes.

$$
\begin{array}{ccc}
(gf)^! E \otimes (gf)^* F & \xrightarrow{\;\;\chi_{E,F}^{\bar{g}\bar{h},wu}\;\;} & (gf)^! (E \otimes F) \\
\simeq \downarrow & & \downarrow \simeq \\
f^! g^! E \otimes f^* g^* F & \xrightarrow[\chi_{g^!E,g^*F}^{\bar{f},u}]{} \; u^* \bar{f}^! (g^! E \otimes g^* F) \xrightarrow[u^* \bar{f}^! \chi_{E,F}^{\bar{g},v}]{} \; f^! g^! (E \otimes F)
\end{array}
$$

Proof (Sketch). Set $\bar{E} := \bar{g}^! E$, $\bar{F} := \bar{g}^* F$ (so that $v^* \bar{E} \cong g^! E$ and $v^* \bar{F} \cong g^* F$). Let β be the natural composite functorial isomorphism

$$
w^* \bar{h}^! \xrightarrow{\;\sim\;} (\bar{h}w)^! = (v\bar{f})^! \xrightarrow{\;\sim\;} \bar{f}^! v^*. \tag{4.9.2.1}
$$

Straightforward—if a bit tedious—considerations, using the definitions of the maps involved (see, e.g., (4.8.4)), translate Lemma (4.9.2) into commutativity of the natural diagram

$$
\begin{array}{ccc}
(wu)^* \big((\bar{g}\bar{h})^! E \otimes (\bar{g}\bar{h})^* F\big) & \xrightarrow{\;\;(wu)^* \chi_{E,F}^{\bar{g}\bar{h}}\;\;} & (wu)^* (\bar{g}\bar{h})^! (E \otimes F) \\
\simeq \downarrow & \textcircled{1} & \downarrow \simeq \\
u^* w^* \bar{h}^! \bar{E} \otimes u^* w^* \bar{h}^* \bar{F} & \xrightarrow{u^* w^* \chi_{\bar{E},\bar{F}}^{\bar{h}}} u^* w^* \bar{h}^! (\bar{E} \otimes \bar{F}) \xrightarrow{u^* w^* \bar{h}^! \chi_{E,F}^{\bar{g}}} u^* w^* \bar{h}^! \bar{g}^! (E \otimes F) \\
\text{via } \beta \downarrow \simeq & \textcircled{2} \quad \text{via } \beta \downarrow \simeq \quad \textcircled{3} & \simeq \downarrow \text{via } \beta \\
u^* \bar{f}^! v^* \bar{E} \otimes u^* \bar{f}^* v^* \bar{F} & \xrightarrow[u^* \chi_{v^*\bar{E},v^*\bar{F}}^{\bar{f}}]{} u^* \bar{f}^! v^* (\bar{E} \otimes \bar{F}) \xrightarrow[u^* \bar{f}^! v^* \chi_{E,F}^{\bar{g}}]{} u^* \bar{f}^! v^* \bar{g}^! (E \otimes F),
\end{array}
$$

in which, commutativity of subdiagram ③ is obvious.

Commutativity of subdiagram ① follows from "transitivity" of χ with respect to proper maps (Exercise (4.7.3.4)(d)).

As for the remaining subdiagram ②, decomposing $\sigma_{w,\bar{h},\bar{f},v}$ as

$$
\begin{array}{ccc}
\overline{X} & \xrightarrow{\;i\;} Y \times_{\overline{Y}} \overline{\overline{X}} \xrightarrow{\;w_1\;} & \overline{\overline{X}} \\
f_1 \downarrow & \sigma & \downarrow \bar{h} \\
Y & \xrightarrow{\qquad v \qquad} & \overline{Y}
\end{array}
$$

with $w_1 i = w$, $f_1 i = \bar{f}$, and σ an independent fiber square (since v is flat), we see from (4.8) that β factors naturally as

$$
w^* \bar{h}^! \xrightarrow{\;\sim\;} i^* w_1^* \bar{h}^! \xrightarrow[\beta_\sigma]{} i^* f_1^! v^* \xrightarrow[\beta_i^{-1}]{} i^! f_1^! v^* \xrightarrow{\;\sim\;} \bar{f}^! v^*.
$$

Here i is an open and closed immersion, so that by (4.8.4), $i^! = i^*$ and the map β_i (see (4.8.7.0)) is the identity. Indeed, since if_1 and f_1 are both proper, therefore so is i [**EGA**, II, (5.4.3)(i)]; and since iw_1 and w_1 are both open immersions, therefore so is i (cf. (4.8.3.1)(a)).

It is left now to the reader to expand β as above and then to verify, with the aid of (4.7.3.4)(c) and (d), and of Exercise (4.8.12)(b)(vi) for open-and-closed immersions, that ② does commute.   Q.E.D.

Corollary 4.9.2.2 *If a map* $f\colon X \to Z$ *factors in two ways as*

$$ X \xrightarrow{\ u\ } Y \xrightarrow{\ \bar{f}\ } Z, \qquad X \xrightarrow{\ v\ } \overline{Y} \xrightarrow{\ \bar{g}\ } Z $$

($\bar{f}$ and $\bar{g}$ proper, u and v open immersions) then for all E, F as in (4.9.2), it holds that $\chi_{E,F}^{\bar{f},u} = \chi_{E,F}^{\bar{g},v}$.

Proof. The given data determine uniquely a map $\bar{w}\colon X \to Y \times_Z \overline{Y}$, whose schematic image we denote by $\overline{\overline{X}}$, see [**GD**, p. 324, (6.10.1) and p. 325, (6.10.5)]. The map $\bar{w}$ factors as $X \to X \times_Z X \to Y \times_Z \overline{Y}$, where the first map is the diagonal, a closed immersion, and the second is an open immersion. So $\bar{w}$ is an immersion, and hence induces an open immersion $w\colon X \to \overline{\overline{X}}$. Furthermore, the projections to Y and $\overline{Y}$ induce proper maps $h\colon \overline{\overline{X}} \to Y$ and $\bar{h}\colon \overline{\overline{X}} \to \overline{Y}$. It suffices then for (i) to prove the Corollary for each of the pairs of factorizations $f = \bar{g}v = (\bar{g}\bar{h})w$ and $f = \bar{f}u = (\bar{f}h)w$.

For the first pair, one need only look at the case $u = f = \bar{f} = 1$ of Lemma (4.9.2). The second pair, being of the same form as the first, is handled similarly.   Q.E.D.

Corollary 4.9.2.3 *For any étale* $g\colon Y \to Z$ *and E, F as in (4.9.2), the map* $\chi_{E,F}^{g}$ *(4.9.1.1) is the isomorphism* $f^*E \otimes f^*F \xrightarrow{\ \sim\ } f^*(E \otimes F)$ *coming from (3.2.4).*

Proof (Sketch). The idea is to redo everything in this section 4.9, up to this point, with "étale" in place of "open immersion." The first difficulty which arises is that in the last paragraph of the proof of Lemma (4.9.2), the map i is now finite étale, making it necessary to know (4.9.2.3) for finite étale f, a fact given by Exercise (4.8.12)(b)(vi). The only other nontrivial modification is in the proof of (4.9.2.2), where the map $X \times_Z X \to Y \times_Z \overline{Y}$ should now be factored as $X \times_Z X \hookrightarrow W \to Y \times_Z \overline{Y}$ with the first map an open immersion and the second proper, and then $\overline{\overline{X}}$ should be defined to be the schematic image of $X \to X \times_Z X \hookrightarrow W \dots$ Q.E.D.

Exercises 4.9.3 (a) In Ex. (4.7.3.4)(e) replace $f^\times$ by $\bar{f}^!$ and apply the functor u^* to get a natural map $u^*\bar{f}^! E \to \mathcal{H}_X(f^*F, u^*\bar{f}^!(E \otimes F))$. Then show that this map corresponds via (2.6.1)' to $\chi_{E,F}^{\bar{f},u}$.

(b) Let $f = \bar{f}u$ be as in (4.9.1). Show, for $E, F \in \mathbf{D}_{qc}(Y)$, that the composite map

$$u^*\bar{f}^!\mathcal{H}_Y(E, F) \otimes u^*\bar{f}^*E \xrightarrow{\ \chi^{\bar{f},u}\ } u^*\bar{f}^!(\mathcal{H}_Y(E, F) \otimes E) \xrightarrow{\ \text{natural}\ } u^*\bar{f}^!F$$

depends only on f, not on its factorization.

Deduce the existence, for any $E \in \mathbf{D}_c^-(Y)$ and $F \in \overline{\mathbf{D}}_{qc}^+(Y)$, of a canonical isomorphism

$$\bar{f}^!\mathcal{H}_Y(E, F) \xrightarrow{\ \sim\ } \mathcal{H}_X(f^*E, f^!F),$$

inverse to $u^*\zeta$ where ζ comes from (4.2.3)(e) applied to $\bar{f}$. (This can also be done without recourse to χ.)

(c) (Compatibility of χ with base change.) After replacing $(-)^\times$ by $(-)^!$, do exercise (4.7.3.4)(c), assuming that the square is an *admissible* square, and interpreting β as in (4.8.3). Do something similar with the map ϕ of (3.10.4) in place of β.

(d) Proceeding as in (a), work out exercises (4.7.3.4)(a), (d), and (f), with $(-)^\times$ replaced by $(-)^!$. This will likely involve verifications of compatibility with restriction to open subschemes for a number of functorial maps. Do similarly for (4.2.3)(h)–(j).

(e) Show that if $f: X \to Y$ is ètale then the map in (b) is the same as the map coming from (3.5.4.5).

(f) Explain the formal tensor-hom symmetry in the pair of natural isomorphisms

$$
\begin{aligned}
f^*E \otimes f^!F &\xrightarrow{\ \sim\ } f^!(E \otimes F) & (E, F \in \mathbf{D}_{qc(Y)}), \\
\mathcal{H}_X(f^*E, f^!F) &\xrightarrow{\ \sim\ } f^!\mathcal{H}_Y(E, F) & (E \in \overline{\mathbf{D}}_c^-(Y),\, F \in \overline{\mathbf{D}}_{qc}^+(Y)).
\end{aligned}
$$

Another such pair, coming from (3.9.4) and (3.2.3.2), is

$$
\begin{aligned}
E \otimes f_*F &\xrightarrow{\ \sim\ } f_*(f^*E \otimes F) & (E, F \in \mathbf{D}_{qc}(Y)), \\
\mathcal{H}_Y(E, f_*F) &\xrightarrow{\ \sim\ } f_*\mathcal{H}_X(f^*E, F) & (E, F \in \mathbf{D}(Y)).
\end{aligned}
$$

(I don't have an answer.)

With respect to a scheme-map $f: X \to Y$, an $\mathcal{O}_X$-complex E is *f-perfect* if E has coherent homology and finite flat f-amplitude. As noted in (2.7.6), f is perfect (i.e., of finite tor-dimension) $\iff \mathcal{O}_X$ is f-perfect.

When f is perfect, the natural map, taking $1 \in \mathrm{H}^0(X, \mathcal{O}_X)$ to the identity map of the *relative dualizing complex* $f^!\mathcal{O}_Y$ is an isomorphism

$$\xi: \mathcal{O}_X \xrightarrow{\ \sim\ } \mathcal{H}_X(f^!\mathcal{O}_Y, f^!\mathcal{O}_Y).$$

In fact, the functor $\mathcal{H}_X(-, f^!\mathcal{O}_Y)$ induces an antiequivalence of the full subcategory of f-perfect complexes in $\mathbf{D}(X)$ to itself [**I**, p. 259, 4.9.2].

Theorem 4.9.4 *For any finite-type separated map $f: X \to Y$ of noetherian schemes, the following conditions are equivalent.*

(i) *The map f is perfect, i.e., the complex $\mathcal{O}_X$ is f-perfect.*

(ii) *The complex $f^!\mathcal{O}_Y$ is f-perfect.*

(iii) *$f^!\mathcal{O}_Y \in \overline{\mathbf{D}}_{qc}^-(X)$, and for every $F \in \overline{\mathbf{D}}_{qc}^+(Y)$, the $\mathbf{D}_{qc}(X)$-map*

$$\chi_{\mathcal{O}_Y, F}^f: f^!\mathcal{O}_Y \otimes f^*F \longrightarrow f^!F$$

is an isomorphism.

(iii)′ *For every perfect $\mathcal{O}_Y$-complex E, $f^!E$ is f-perfect; and for all $E, F \in \mathbf{D}(Y)$ such that E and $E \otimes F$ are in $\overline{\mathbf{D}}_{\mathrm{qc}}^+(Y)$, the $\mathbf{D}_{\mathrm{qc}}(X)$-map*

$$\chi_{E,F}^f \colon f^!E \otimes f^*F \longrightarrow f^!(E \otimes F).$$

is an isomorphism.

(iv) *The functor $f^! \colon \overline{\mathbf{D}}_{\mathrm{qc}}^+(Y) \to \overline{\mathbf{D}}_{\mathrm{qc}}^+(X)$ is bounded.*

Proof. (i)⇔(ii). The question is local on X, and so we may assume that f factors as $X \xrightarrow{i} Z \xrightarrow{p} Y$ where Z is an affine open subscheme of $Y \otimes_{\mathbb{Z}} \mathbb{Z}[T_1, \ldots, T_n]$ (with independent indeterminates T_i), i is a closed immersion, and p is the obvious map.

By (4.4.2) (with $F = \mathcal{O}_X$), we have a functorial isomorphism

$$i_* i^! G \xrightarrow{\sim} \mathcal{H}_Z(i_* \mathcal{O}_X, G) \qquad \left(G \in \overline{\mathbf{D}}_{\mathrm{qc}}^+(Z) \right). \tag{4.9.4.1}$$

Also, with Ω_p^n the invertible $\mathcal{O}_Z$-module of relative Kähler n-forms, there is a natural isomorphism

$$p^!E \cong \Omega_p^n[n] \otimes p^*E \qquad (E \in \overline{\mathbf{D}}_{\mathrm{qc}}^+(Y)), \tag{4.9.4.2}$$

see [**V′**, p. 397, Thm. 3].[16]

Now, by [**I**, p. 250, 4.1, and p. 252, 4.4], (i) holds if and only if the $\mathcal{O}_Z$-complex $i_* \mathcal{O}_X$ is perfect; and (ii) holds if and only if the $\mathcal{O}_Z$-complex

$$i_* f^! \mathcal{O}_Y \cong i_* i^! p^! \mathcal{O}_Y \cong \mathcal{H}_Z(i_* \mathcal{O}_X, p^! \mathcal{O}_Y) \cong \mathcal{H}_Z(i_* \mathcal{O}_X, \Omega_p^n[n])$$

is perfect. Hence the equivalence of (i) and (ii) results from the following fact, in the case $F = i_* \mathcal{O}_X$.

Lemma 4.9.4.3 *On any noetherian scheme W, an $\mathcal{O}_W$-complex F is perfect $\iff F \in \overline{\mathbf{D}}_{\mathrm{c}}^{\mathrm{b}}(W)$ and $\mathcal{H}_W(F, \mathcal{O}_W)$ is perfect.*

Proof. The implication ⇒ results from [**I**, p. 148, 7.1].

For the converse, the question being local, we may assume that W is affine, say $W = \mathrm{Spec}(R)$, that F is a bounded-above complex of finite-rank locally free $\mathcal{O}_W$-modules (see 4.3.2), and that $\mathcal{H}_W(F, \mathcal{O}_W)$ is $\mathbf{D}(W)$-isomorphic to a strictly perfect $\mathcal{O}_W$-complex.

---

[16] The proof in *loc. cit.* can be imitated, without the assumption of finite Krull dimension, and with E in place of $\mathcal{O}_Y$; but instead of Corollary 2 one should use [**H**, p. 180, Cor. 7.3], noting that the graph map denoted by Δ is a local complete intersection map of codimension n [**EGA**, IV, (17.12.3)]. It might appear simpler to use [**V′**, p. 396, Lemma 1], whose proof, however, seems to need an isomorphism of the form (4.9.4.2) when Z is $\mathbf{P}_Y^1$. For this, see [**H**, p. 161, 5.1] (duality for $\mathbf{P}_Y^n$), except that the proof given there applies only to $F \in \overline{\mathbf{D}}_{\mathrm{qc}}^-(Y)$. That suffices, nevertheless, by (4.3.7) applied to the map $\phi \colon \Omega_p^1 \to p^! \mathcal{O}_Y$ corresponding by duality to the canonical isomorphism $R^1 p_*(\Omega_p^1) \xrightarrow{\sim} \mathcal{O}_Y$ [**H**, p. 155, 4.3].

Then $N := \Gamma(W, F)$ is a bounded-above complex of finite-rank projective R-modules, and with $\sim$ the usual sheafification functor, $F \cong N^{\sim}$.

Let $R \to I^{\bullet}$ be an R-injective resolution of R. By [**H**, p. 130, 7.14], the resulting map $\mathcal{O}_W = R^{\sim} \to I^{\bullet\sim}$ is an injective resolution of $\mathcal{O}_W$. So $\mathcal{H}om_W(N^{\sim}, I^{\bullet\sim}) \cong \mathcal{H}_W(F, \mathcal{O}_W)$ is $\mathbf{D}(W)$-isomorphic—and hence, by (3.9.6)(a), $\mathbf{D}(\mathcal{A}_W^{qc})$-isomorphic—to a strictly perfect $\mathcal{O}_W$-complex. Since $\Gamma(W, -)$ is exact on $\mathcal{A}_W^{qc}$, it follows that

$$\mathbf{R}\mathrm{Hom}_R(N, R) \cong \mathrm{Hom}_R(N, I^{\bullet}) \cong \Gamma(W, \mathcal{H}om_W(N^{\sim}, I^{\bullet\sim}))$$

is a perfect R-complex. So by [**AIL**, Prop. 4.1(ii)], N is perfect, whence so is $F \cong N^{\sim}$.   Q.E.D.

(i)$\Rightarrow$(iii). One may assume f factors as above: $X \xrightarrow{i} Z \xrightarrow{p} Y$.

By (4.9.4.2), for $f^! \mathcal{O}_Y = i^! p^! \mathcal{O}_Y$ to be in $\overline{\mathbf{D}}_{qc}^{-}(X)$ it suffices that the functor $i^!$ be bounded on $\overline{\mathbf{D}}_{qc}^{+}(Z)$, which it is, by (4.9.4.1), because $i_* \mathcal{O}_X$ is perfect. (For this boundedness, as in the proof of [**I**, p. 148, 7.1], after replacing $i_* \mathcal{O}_X$ by an arbitrary perfect $\mathcal{O}_X$-complex E and localizing, one may assume that E is a bounded complex of finite-rank free $\mathcal{O}_Z$-modules, and proceed by "dévissage," i.e., induction on the number of nonzero components of E, to reduce to noting that $\mathcal{H}_Z(E, G)$ is a bounded functor of $G \in \overline{\mathbf{D}}_{qc}^{+}(Z)$ when E is a finite-rank free $\mathcal{O}_Z$-module.)

Next, by (4.9.2), with $(f, g, u, \bar{f})$ replaced by $(i, p, 1, i)$, it suffices to show that $\chi_{p^! \mathcal{O}_Y, p^* F}^{i}$ and $\chi_{\mathcal{O}_Y, F}^{p}$ are isomorphisms.

By (4.9.3)(c), the question of whether $\chi_{p^! \mathcal{O}_Y, p^* F}^{i}$ is an isomorphism is local on Y, so we may assume Y affine, in which case every quasi-coherent $\mathcal{O}_Y$-module is a homomorphic image of a free one.

Since p is flat and, by (4.9.4.2), the complex $p^! \mathcal{O}_Y$ is perfect, therefore $p^! \mathcal{O}_Y \otimes p^* F$ is a bounded functor of F; and again by (4.9.4.2), so is $p^! F$. Hence, by (1.11.3.1), one need only note that by (4.7.5) applied to a compactification of p, $\chi_{p^! \mathcal{O}_Y, p^* F}^{i}$ is an isomorphism whenever F is a *free* $\mathcal{O}_Y$-module.

That $\chi_{p^! \mathcal{O}_Y, G}^{i}$ is an isomorphism for any $G \in \mathbf{D}_{qc}(Z)$ can be checked *after* application of the functor i_*. The source and target of $i_* \chi_{p^! \mathcal{O}_Y, G}^{i}$ are

$$i_*(i^! p^! \mathcal{O}_Y \otimes i^* G) \underset{(3.9.4)}{\cong} i_* i^! p^! \mathcal{O}_Y \otimes G \underset{(4.9.4.1)}{\cong} \mathcal{H}_Z(i_* \mathcal{O}_X, p^! \mathcal{O}_Y) \otimes G,$$

$$i_* i^!(p^! \mathcal{O}_Y \otimes G) \underset{(4.9.4.1)}{\cong} \mathcal{H}_Z(i_* \mathcal{O}_X, p^! \mathcal{O}_Y \otimes G).$$

Since $i_* \mathcal{O}_X$ is perfect, and, by (4.9.4.2), so is $p^! \mathcal{O}_Y$, therefore both the source and target are bounded functors of G, commuting with direct sums (3.8.2).

As before, one reduces to where Z is affine and G is a free $\mathcal{O}_Z$-module, in which case commutativity with direct sums gives a reduction to the trivial case $G = \mathcal{O}_Z$.

(Alternatively, it is a nontrivial exercise to show that (4.9.4.2) with $p^! \mathcal{O}_Y$ in place of $\Omega_p^n[n]$ is in fact $\chi_{\mathcal{O}_Y, E}^{p}$. One also shows, with $E := i_* \mathcal{O}_X$, $F := p^! \mathcal{O}_Y$, that $i_*\chi^i_{F,G}$ is isomorphic to the map

$$\zeta(E)\colon \mathcal{H}_Z(E, F) \otimes G \to \mathcal{H}_Z(E, F \otimes G)$$

associated by $(2.6.1)^*$ to the natural map $\mathcal{H}_Z(E, F) \otimes G \otimes E \to F \otimes G$, and then sees via dévissage to the trivial case $E = \mathcal{O}_Z$ that $\zeta(E)$ is an isomorphism for all perfect E. What is involved here is a *concrete* local interpretation of χ^f.)

(iii)$\Leftrightarrow$(iii)′ $\Rightarrow$(ii). The implications (iii)′ $\Rightarrow$ (ii) and (iii)′ $\Rightarrow$ (iii) are trivial. Assume, conversely, that (iii) holds.

To be shown first is that for a perfect $\mathcal{O}_Y$-complex E, $f^!E$ is f-perfect. Since $f^!$ commutes with open base change (4.8.3), one can replace Y by any open subset. Thus one may assume that E is a bounded complex of finite-rank free $\mathcal{O}_Y$-modules, and then proceed by dévissage to reduce to the case $E = \mathcal{O}_Y$, treated as follows.

Let $\mu\colon V \hookrightarrow Y$ be the inclusion of an open subscheme, $\nu\colon f^{-1}V \hookrightarrow X$ the inclusion, $g\colon f^{-1}V \to V$ the map induced by f, and M an $\mathcal{O}_V$-module. We have then the obvious isomorphisms

$$\nu^*f^!\mathcal{O}_Y \otimes g^*M \cong \nu^*(f^!\mathcal{O}_Y \otimes f^*\mu_*M) \underset{\text{(iii)}}{\cong} \nu^*f^!\mu_*M.$$

Since μ_*M is a bounded complex (3.9.2), and since $f^!$ is bounded below and, by (iii), bounded above, therefore there is an interval $[m,n]$ not depending on M such that

$$H^i(\nu^*f^!\mathcal{O}_Y \otimes g^*M) = 0 \quad \text{for all } i \notin [m,n].$$

So by [**I**, p. 242, 3.3(iv)], $f^!\mathcal{O}_Y$ has finite flat f-amplitude. Also, (4.9.4.1) and (4.9.4.2) imply that $f^!\mathcal{O}_Y \in \mathbf{D}_c(X)$. Thus $f^!\mathcal{O}_Y$ is f-perfect.

For the isomorphism in (iii)′, apply (4.7.3.4)(a) with $E = \mathcal{O}_Y$ to a compactification of f.

(i)$\Rightarrow$(iv). Theorem (4.1) gives that $f^!$ is bounded below. If (i) holds then by definition, the (derived) functor f^* is bounded above; and as shown above, (iii) holds, whence $f^!$ is bounded above. Thus $f^!$ is bounded.

(iv)$\Rightarrow$(i). With notation as in the proof of (i) $\Leftrightarrow$ (ii), we will show that if $f^!$ is bounded then so is $i^!$. By [**LN**, Thm. 1.2] (or (4.9.6(e)) below), this implies that i is perfect, whence so is $f = pi$.

Factor i as $X \xrightarrow{\gamma} X \times_Y Z \xrightarrow{g} Z$ where γ is the graph of i and g is the projection. The map γ, a local complete intersection [**EGA**, IV, (17.12.3)], is perfect, and so, as we've just seen, $\gamma^!$ is bounded.

Also, g arises from f by flat base change, so, as in (4.7.3.1)(ii) with $^\times$ replaced by $^!$, $g^!$ is bounded: to imitate the proof of (4.7.3.1)(ii) one just needs to associate a functorial isomorphism $v_*g^! \xrightarrow{\sim} f^!u_*$ to each composite fiber square

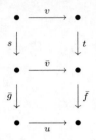

with u, $\bar{v}$ and v flat, $\bar{f}$ and $\bar{g}$ proper, t and s open immersions, $f = \bar{f}t$ and $g = \bar{g}s$. One such isomorphism is the natural composition

$$v_* g^! \xrightarrow{\;\sim\;} v_* s^* \bar{g}^! \xrightarrow{\;\sim\;} t^* \bar{v}_* \bar{g}^! \underset{(3.10.4)}{\xrightarrow{\;\sim\;}} t^* \bar{f}^! u_* \xrightarrow{\;\sim\;} f^! u_*.$$

Thus $i^! \cong \gamma^! g^!$ is bounded.   Q.E.D.

Corollary 4.9.5 *On the category of perfect maps there is a pseudo-functor* $(-)^{\#}$ *which associates to each such map* $f : X \to Y$ *the functor* $f^{\#} \colon \overline{\mathbf{D}}_{\mathsf{qc}}^{+}(Y) \to \overline{\mathbf{D}}_{\mathsf{qc}}^{+}(X)$ *given objectwise by*

$$f^{\#} F := f^! \mathcal{O}_Y \otimes F \qquad \left(F \in \overline{\mathbf{D}}_{\mathsf{qc}}^{+}(Y) \right).$$

For a composition $X \xrightarrow{f} Y \xrightarrow{g} Z$ of perfect maps, the resulting functorial isomorphism $f^{\#} g^{\#} G \xrightarrow{\;\sim\;} (gf)^{\#} G$ $(G \in \overline{\mathbf{D}}_{\mathsf{qc}}^{+}(Z))$ is the left column of the following diagram of natural isomorphisms, whose commutativity results from (4.7.3.4)(a) and (d), as treated in (4.9.3)(d), or from (4.9.2) with $E := \mathcal{O}_X$ and $F := G$.

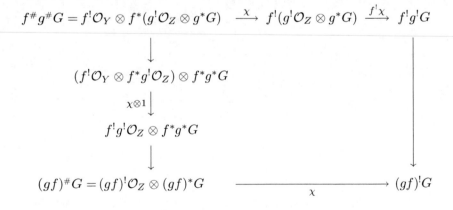

Exercises 4.9.6 (a) Show that $\chi_{E,F}^{f}$ is an isomorphism whenever $F \in \mathbf{D}_{\mathsf{qc}}(X)$ has finite tor-dimension. (Cf. (4.7.5).)

(b) Noting Ex. (3.5.3)(g), establish a natural commutative diagram

$$
\begin{array}{ccc}
f^!F \otimes \mathcal{H}_X(f^*E, f^*G) & \longrightarrow & \mathcal{H}_X(f^*E, f^!F \otimes f^*G) \longleftarrow \mathcal{H}_X(f^*E, f^!F) \otimes f^*G \\
\uparrow & & \downarrow \qquad\qquad\qquad\qquad \downarrow \\
f^!F \otimes f^*\mathcal{H}_Y(E, G) & & \mathcal{H}_X(f^*E, f^!(F \otimes G)) \qquad f^!\mathcal{H}_Y(E, F) \otimes f^*G \\
\downarrow & & \downarrow \qquad\qquad\qquad\qquad \downarrow \\
f^!(F \otimes \mathcal{H}_Y(E, G)) & \longrightarrow & f^!\mathcal{H}_Y(E, F \otimes G) \longleftarrow f^!(\mathcal{H}_Y(E, F) \otimes G)
\end{array}
$$

(c) (Neeman, van den Bergh). Show, for any perfect $f\colon X \to Y$ and $E \in \overline{\mathbf{D}}_{\mathrm{qc}}^+(Y)$, that the map $f^*E \to \mathcal{H}_X(f^!\mathcal{O}_Y, f^!E)$ induced via $(2.6.1)'$ by $\chi_{\mathcal{O}_Y, E}^f$ is an isomorphism.

Hint. Factor f locally as pi—see proof of (4.9.4), and apply i_*.

(d) Let X be a noetherian scheme, $E \in \mathbf{D}_c^b(X)$. Show that the functor $\mathcal{H}_X(E, -)$ from $\overline{\mathbf{D}}_{\mathrm{qc}}^+(X)$ to itself is bounded if and only if E is perfect.

Hint. Reduce to where $X = \mathrm{Spec}(A)$, and where E is the sheafficaton $\mathrm{E}^\sim$ of a bounded A-complex E of finitely generated A-modules. Use the fact that the sheafification of an A-injective module is $\mathcal{O}_X$-injective [**RD**, p. 130, 7.14], to show that for any $\mathrm{F} \in \overline{\mathbf{D}}^+(A)$, $\mathcal{H}_X(E, \mathrm{F}^\sim) = \mathbf{R}\mathrm{Hom}_A(\mathrm{E}, \mathrm{F})^\sim$, and hence to reduce further to the corresponding statement for A-modules.

(e) Using (d) and (4.4.2) with $F = \mathcal{O}_X$, show that a finite map $f\colon X \to Y$ of noetherian schemes is perfect iff the functor $f^!\colon \overline{\mathbf{D}}_{\mathrm{qc}}^+(Y) \to \overline{\mathbf{D}}_{\mathrm{qc}}^+(X)$ is bounded.

4.10 Appendix: Dualizing Complexes

Grothendieck's original strategy for proving duality—at least the version in Corollary (4.2.2)—for proper not-necessarily-projective maps, is based on pseudofunctorial properties of dualizing complexes. In this section, we sketch the idea. The principal result, Thm. (4.10.4), makes clear how the basic problem—not treated here—in this approach is the construction of a "coherent" family of dualizing complexes (in other words, a "Dualizing Complex," see below). What emerges is less than Thm. (4.8.1). But for formal schemes, this kind of approach yields results not otherwise obtainable (as of early 2009), see the remarks following Thm. (4.10.4).

Throughout this section, without further mention we restrict to schemes which are noetherian and to scheme-maps that are separated, of finite type. Also, we continue to use the notations introduced at the beginning of §4.4.

Let $\mathcal{A}^c(X) \subset \mathcal{A}(X)$ be the full subcategory whose objects are the coherent $\mathcal{O}_X$-modules; it is a plump subcategory [**GD**, 113, 5.3.5]. Additional notation will be as in §(1.9.1), with # = c.

For example, $\overline{\mathbf{D}}_c^+(X)$ is the Δ-subcategory of $\mathbf{D}(X)$ whose objects are the complexes whose homology modules vanish in all sufficiently negative degrees, and are coherent in all degrees.

A *dualizing complex* R on a noetherian scheme X is a complex in $\mathbf{D}_c(X)$ which is **D**-isomorphic to a bounded injective complex, and has the following equivalent properties [**H**, p. 258, 2.1]:

(i) For every $F \in \mathbf{D}_c(X)$, the map corresponding via $(2.6.1)'$ to the natural composition

$$F \otimes \mathbf{R}\mathcal{H}om(F, R) \xrightarrow{\sim} \mathbf{R}\mathcal{H}om(F, R) \otimes F \to R$$

is an isomorphism (called by some other authors the *Grothendieck Duality isomorphism*):

$$F \xrightarrow{\sim} \mathbf{R}\mathcal{H}om(\mathbf{R}\mathcal{H}om(F, R), R).$$

(ii) Condition (i) holds for $F = \mathcal{O}_X$, i.e., the map $\mathcal{O}_X \to \mathbf{R}\mathcal{H}om(R, R)$ which takes $1 \in \Gamma(X, \mathcal{O}_X)$ to the identity map of R is an isomorphism.

For connected X, dualizing $\mathcal{O}_X$-complexes, if they exist, are unique up to tensoring with a complex of the form $L[n]$ where L is an invertible $\mathcal{O}_X$-module and $n \in \mathbb{Z}$ [**H**, p. 266, 3.1].

The associated dualizing functor

$$\mathcal{D}_R := \mathbf{R}\mathcal{H}om_X(-, R)$$

satisfies $\mathcal{D}_R \circ \mathcal{D}_R \cong 1$, and it induces antiequivalences from $\mathbf{D}_c(X)$ to itself, and between $\overline{\mathbf{D}}_c^+(X)$ and $\overline{\mathbf{D}}_c^-(X)$ (in either direction).

The existence of a dualizing complex places restrictions on X—for instance, X must then be universally catenary and of finite Krull dimension [**H**, p. 300]. Sufficient conditions for the existence are given in [**H**, p. 299]. For example, any scheme of finite type over a regular (or even Gorenstein) scheme of finite Krull dimension has a dualizing complex.[17]

Henceforth we restrict schemes to those which, in addition to being noetherian, have dualizing complexes.

The relation between dualizing complexes and the pseudofunctor $^!$ of Thm. (4.8.1) is rooted in the following Proposition, see [**H**, Chapter V, §8], [**V'**, p. 396, Corollary 3], or [**N''**, Theorems 3.12 and 3.14].

Proposition 4.10.1 *Let $f \colon X \to Y$ be a scheme-map, and let R be a dualizing $\mathcal{O}_Y$-complex. Then with $R_f := f^! R$,*

(i) R_f *is a dualizing $\mathcal{O}_X$-complex.*

(ii) *There is a functorial isomorphism*

$$f^! \mathcal{D}_R F \xrightarrow{\sim} \mathcal{D}_{R_f} \mathbf{L}f^* F \qquad \bigl(F \in \overline{\mathbf{D}}_c^-(Y)\bigr)$$

or equivalently,

$$f^! E \xrightarrow{\sim} \mathcal{D}_{R_f} \mathbf{L}f^* \mathcal{D}_R E \qquad \bigl(E \in \overline{\mathbf{D}}_c^+(Y)\bigr).$$

---

[17] In [**N''**], Neeman studies a notion of dualizing complex which applies to infinite-dimensional schemes. Suresh Nayak observed, via [**C**, p. 121, Lemma 3.1.5], that Neeman's dualizing complexes are the same as pointwise dualizing complexes with bounded cohomology, cf. [**C**, p. 127, Lemma 3.2.1].

Proof. First, it follows from the construction of the functor $f^\times$ (see just before (4.1.8)) that it preserves finite injective dimension. So when f is proper, $f^! = f^\times$ preserves finite injective dimension. The same is clearly true for $f^! = f^*$ when f is an open immersion, and hence—via compactification— for any f.

The question of whether $f^! R \in \mathbf{D_c}(X)$ is local; hence an affirmative answer is provided by (4.9.4.1) and (4.9.4.2).

It remains to show that the natural map $\psi_f \colon \mathcal{O}_X \to \mathcal{D}_{R_f}\mathcal{D}_{R_f}\mathcal{O}_X$ is an isomorphism. Again, the question is local, so we reduce to the two cases (a) f is smooth, (b) f is a closed immersion.

(a) For smooth f, (4.9.4.2) and (4.6.7) provide natural isomorphisms

$$\mathbf{R}\mathcal{H}om_X(R_f, R_f) \overset{\sim}{\longrightarrow} \mathbf{R}\mathcal{H}om_X(p^*R, p^*R) \overset{\sim}{\longrightarrow} p^*\mathbf{R}\mathcal{H}om_Y(R, R).$$

One verifies then that ψ_f is isomorphic, via the preceding isomorphisms, to p^* applied to the isomorphism $\mathcal{O}_Y \overset{\sim}{\longrightarrow} \mathcal{D}_R\mathcal{D}_R\mathcal{O}_Y$.

(b) It suffices that $f_*\psi_f$ be an isomorphism, which it is, by (4.9.4.1) (with $i = f$), since $f_*\mathcal{O}_X \in \overline{\mathbf{D}}_{\mathbf{c}}^{\mathbf{b}}(Y)$ and therefore the canonical map $f_*\mathcal{O}_X \to \mathcal{D}_R\mathcal{D}_R f_*\mathcal{O}_X$ is an isomorphism.

Assertion (ii) follows immediately from Ex. (4.2.3)(e), as $\mathcal{D}_R$ and $\mathcal{D}_{R_f}$ are antiequivalences.   Q.E.D.

Definition 4.10.2 A *Dualizing Complex* on a scheme Y is a map which associates to each $f \colon X \to Y$ a dualizing complex R_f on X, to each open immersion $u \colon U \to X$ a $\mathbf{D}(X)$-isomorphism $\gamma_{f,u} \colon u^*R_f \overset{\sim}{\longrightarrow} R_{fu}$, and to each proper map $g \colon X' \to X$ a $\mathbf{D}(X)$-map $\tau_{f,g} \colon g_*R_{fg} \to R_f$, subject to the following conditions on each such f, u and g:

(a) If $v \colon V \to U$ is an open immersion, then the following diagram commutes:

$$
\begin{array}{ccc}
v^*u^*R_f & \xrightarrow{\;\;\underset{(3.6.4)^*}{\sim}\;\;} & (uv)^*R_f \\
{\scriptstyle v^*\gamma_{f,u}}\downarrow & & \downarrow{\scriptstyle \gamma_{f,uv}} \\
v^*R_{fu} & \xrightarrow[\;\gamma_{fu,v}\;]{} & R_{fuv}
\end{array}
$$

(b) The pair $(R_{fg}, \tau_{f,g})$ represents the functor

$$\mathrm{Hom}_{\mathbf{D}(X)}(g_*E, R_f) \colon \overline{\mathbf{D}}_{\mathbf{c}}^{+}(X') \to \overline{\mathbf{D}}_{\mathbf{c}}^{+}(X),$$

that is, the natural composite map

$$\mathrm{Hom}_{\mathbf{D}(X')}(E, R_{fg}) \longrightarrow \mathrm{Hom}_{\mathbf{D}(X)}(g_*E, g_*R_{fg}) \xrightarrow[\text{via } \tau]{} \mathrm{Hom}_{\mathbf{D}(X)}(g_*E, R_f)$$

is an isomorphism.

Further, if $h\colon X'' \to X'$ is proper then the following diagram commutes:

$$
\begin{array}{ccc}
g_* h_* R_{fgh} & \xrightarrow[\;(3.6.4)_*\;]{\;\sim\;} & (gh)_* R_{fgh} \\[2pt]
{\scriptstyle g_* \tau_{fg,h}}\big\downarrow & & \big\downarrow{\scriptstyle \tau_{f,gh}} \\[4pt]
g_* R_{fg} & \xrightarrow[\;\tau_{f,g}\;]{} & R_f
\end{array}
$$

(c) For any fiber square

$$
\begin{array}{ccc}
V & \xrightarrow{\;v\;} & Z \\[2pt]
{\scriptstyle h}\big\downarrow & & \big\downarrow{\scriptstyle g} \\[4pt]
U & \xrightarrow[\;u\;]{} & X
\end{array}
$$

with g (hence h) proper and u (hence v) an open immersion, the following natural diagram commutes:

$$
\begin{array}{ccc}
u^* g_* R_{fg} & \xrightarrow{\qquad\qquad\sim\qquad\qquad} & h_* v^* R_{fg} \\[2pt]
{\scriptstyle u^* \tau_{f,g}}\big\downarrow & & \simeq\big\downarrow{\scriptstyle h_* \gamma_{fg,v}} \\[4pt]
u^* R_f \;\xrightarrow[\;\gamma_{f,u}\;]{\sim}\; R_{fu} & \xleftarrow[\;\tau_{fu,h}\;]{} & h_* R_{fuh} = h_* R_{fgv}
\end{array}
$$

Remarks. In (4.10.2)(a) take $U = V = X$ and let u and v be identity maps, to get $\gamma_{f,u} \circ \gamma_{f,u} = \gamma_{f,u}$, whence the isomorphism $\gamma_{f,u}$ is the identity map $\mathbf{1}$ of R_f. Similarly, when g is the identity map of X, one deduces from (b) that $\tau_{f,g} \circ \tau_{f,g} = \tau_{f,g}$; but $(R_f, \tau_{f,g})$ and $(R_f, \mathbf{1})$ both represent the same functor, whence $\tau_{f,g}$ is an isomorphism, so $\tau_{f,g} = \mathbf{1}$. Also, when $Z = U = V$ and $g = u$ is an open and closed immersion, (c) shows that $\gamma_{f,g} \circ g^* \tau_{f,g}$ is the canonical isomorphism $g^* g_* R_{fg} \xrightarrow{\;\sim\;} R_{fg}$.

Example 4.10.2.1 (A) If R is a dualizing $\mathcal{O}_Y$-complex and $^!$ is as in (4.8.1), one can associate to each map $f\colon X \to Y$ the dualizing $\mathcal{O}_X$-complex $R_f :=$ $f^! R$, to each open immersion $u\colon U \to X$ the natural composition

$$
\gamma_{f,u}\colon u^* R_f = u^! f^! R \xrightarrow{\;\sim\;} (fu)^! R = R_{fu},
$$

and to each proper map $g\colon X' \to X$ the map $\tau = \tau_{f,g}\colon g_*(fg)^! R \to f^! R$ resulting from (4.1.1). Condition (a) is then clear, (b) follows from (4.1.2), and (c) from (4.4.4)(d).

(B) Let $\mathcal{R} = (R, \gamma, \tau)$ be a Dualizing Complex on Y. Then for any map $e\colon Y' \to Y$ we have a Dualizing Complex $\mathcal{R} \times_Y Y' := (R', \gamma', \tau')$ on Y', where for all $f\colon X \to Y'$ we set $R'_f := R_{ef}$, $\gamma'_{f,u} := \gamma_{ef,u}$ and $\tau'_{f,g} := \tau_{ef,g}$.

That $\mathcal{R} \times_Y Y'$ satisfies conditions (a), (b) and (c) is simple to check.

(C) Let $\mathcal{R} = (R, \gamma, \tau)$ be a Dualizing Complex on Y. Then for any invertible $\mathcal{O}_Y$-module $\mathcal{L}$ and any locally constant function $n\colon Y \to \mathbb{Z}$, we have a Dualizing Complex

$$\mathcal{R} \otimes \mathcal{L}[n] = (R \otimes \mathcal{L}[n], \gamma \otimes \mathcal{L}[n], \tau \otimes \mathcal{L}[n])$$

on Y, where for all $f \colon X \to Y$,

- $(R \otimes \mathcal{L}[n])_f := R_f \otimes f^*\mathcal{L}[n]$ (easily seen to be a dualizing $\mathcal{O}_X$-complex),
- $(\gamma \otimes \mathcal{L}[n])_{f,u}$ is the natural composition

$$u^*\big(R_f \otimes f^*\mathcal{L}[n]\big) \xrightarrow{\ \sim\ } u^*R_f \otimes u^*f^*\mathcal{L}[n] \xrightarrow{\ \sim\ } R_{fu} \otimes (fu)^*\mathcal{L}[n],$$

- $\big(\tau \otimes \mathcal{L}[n]\big)_{f,g}$ is the natural composition

$$g_*\big(R_{fg} \otimes (fg)^*\mathcal{L}[n]\big) \xrightarrow{\ \sim\ } g_*\big(R_{fg} \otimes g^*f^*\mathcal{L}[n]\big) \xrightarrow[(3.9.4)]{\ \sim\ } g_*R_{fg} \otimes f^*\mathcal{L}[n]$$

$$\longrightarrow R_f \otimes f^*\mathcal{L}[n].$$

Here, condition (a) is given by the (readily verified) commutativity of the natural diagram

$$
\begin{array}{ccc}
v^*u^*\big(R_f \otimes f^*\mathcal{L}[n]\big) & \xrightarrow{\hspace{6cm}} & (uv)^*\big(R_f \otimes f^*\mathcal{L}[n]\big) \\
\downarrow & & \downarrow \\
v^*\big(u^*R_f \otimes u^*f^*\mathcal{L}[n]\big) \longrightarrow v^*u^*R_f \otimes v^*u^*f^*\mathcal{L}[n] \longrightarrow & & (uv)^*R_f \otimes (uv)^*f^*\mathcal{L}[n] \\
\downarrow & \downarrow & \downarrow \\
v^*\big(R_{fu} \otimes (fu)^*\mathcal{L}[n]\big) \longrightarrow v^*R_{fu} \otimes v^*(fu)^*\mathcal{L}[n] \longrightarrow & & R_{fuv} \otimes (fuv)^*\mathcal{L}[n]
\end{array}
$$

Fix a $\mathbf{D}(X)$-isomorphism $\alpha \colon \mathcal{L}[n] \otimes \mathcal{L}^{-1}[-n] \xrightarrow{\ \sim\ } \mathcal{O}_Y$. The first part of condition (b) results from commutativity of the natural diagram

$$
\begin{array}{ccc}
\mathrm{Hom}_{\mathbf{D}(X)}\big(E, R_{fg} \otimes (fg)^*\mathcal{L}[n]\big) & \xrightarrow{\ \sim\ } & \mathrm{Hom}_{\mathbf{D}(X)}\big(E \otimes (fg)^*\mathcal{L}^{-1}[-n], R_{fg}\big) \\
\downarrow{\scriptstyle\simeq} & & {\scriptstyle\simeq}\downarrow \\
\mathrm{Hom}_{\mathbf{D}(X)}\big(E, R_{fg} \otimes g^*f^*\mathcal{L}[n]\big) & \longrightarrow & \mathrm{Hom}_{\mathbf{D}(X)}\big(E \otimes g^*f^*\mathcal{L}^{-1}[-n], R_{fg}\big) \\
\downarrow & & \downarrow \\
\mathrm{Hom}_{\mathbf{D}(Y)}\big(g_*E, g_*\big(R_{fg} \otimes g^*f^*\mathcal{L}[n]\big)\big) & & \mathrm{Hom}_{\mathbf{D}(Y)}\big(g_*\big(E \otimes g^*f^*\mathcal{L}^{-1}[-n]\big), g_*R_{fg}\big) \\
{\scriptstyle\text{via (3.9.4)}}\downarrow{\scriptstyle\simeq} & & {\scriptstyle\simeq}\downarrow{\scriptstyle\text{via (3.9.4)}} \\
\mathrm{Hom}_{\mathbf{D}(Y)}\big(g_*E, g_*R_{fg} \otimes f^*\mathcal{L}[n]\big) & \xleftarrow{\ \sim\ } & \mathrm{Hom}_{\mathbf{D}(Y)}\big(g_*E \otimes f^*\mathcal{L}^{-1}[-n], g_*R_{fg}\big) \\
\downarrow & & \downarrow \\
\mathrm{Hom}_{\mathbf{D}(Y)}\big(g_*E, R_f \otimes f^*\mathcal{L}[n]\big) & \xleftarrow{\ \sim\ } & \mathrm{Hom}_{\mathbf{D}(Y)}\big(g_*E \otimes f^*\mathcal{L}^{-1}[-n], R_f\big)
\end{array}
$$

where, with $\mathcal{L}_n := \mathcal{L}[n]$ and $\mathcal{L}_{-n}^{-1} := \mathcal{L}^{-1}[-n]$, the first row takes a map $\eta \colon E \to R_{fg} \otimes (fg)^*\mathcal{L}_n$ to the natural composition

$$E \otimes (fg)^* \mathcal{L}_{-n}^{-1} \xrightarrow{\text{via } \eta} \left(R_{fg} \otimes (fg)^* \mathcal{L}_n\right) \otimes (fg)^* \mathcal{L}_{-n}^{-1}$$

$$\xrightarrow{\sim} R_{fg} \otimes (fg)^* \left(\mathcal{L}_n \otimes \mathcal{L}_{-n}^{-1}\right) \xrightarrow{\text{via } \alpha} R_{fg} \otimes (fg)^* \mathcal{O}_Y \xrightarrow{\sim} R_{fg}$$

and the second row takes $\eta' \colon E \to R_{fg} \otimes g^* f^* \mathcal{L}_n$ to the natural composition

$$E \otimes g^* f^* \mathcal{L}_{-n}^{-1} \xrightarrow{\text{via } \eta'} \left(R_{fg} \otimes g^* f^* \mathcal{L}_n\right) \otimes g^* f^* \mathcal{L}_{-n}^{-1}$$

$$\xrightarrow{\sim} R_{fg} \otimes g^* f^* \left(\mathcal{L}_n \otimes \mathcal{L}_{-n}^{-1}\right) \xrightarrow{\text{via } \alpha} R_{fg} \otimes g^* f^* \mathcal{O}_Y \xrightarrow{\sim} R_{fg}.$$

The arrows in the last two rows are defined in a similar manner.

Commutativity of the bottom subrectangle is obvious. Checking commutativity of the other two subdiagrams is left as an exercise. (For the middle one, a variant of diagram (3.4.7)(iv) may prove useful.)

The second part of condition (b) follows from (3.7.1). (Details left as an exercise.)

Condition (c) is given by commutativity of the following natural diagram, where $\mathcal{L}[n]$ has been abbreviated to $\mathcal{L}$:

$$u^*(R_f \otimes f^*\mathcal{L}) \longleftarrow u^*(g_* R_{fg} \otimes f^*\mathcal{L}) \longleftarrow u^* g_*(R_{fg} \otimes g^* f^*\mathcal{L}) \longrightarrow h_* v^*(R_{fg} \otimes g^* f^*\mathcal{L})$$

$$u^* R_f \otimes u^* f^*\mathcal{L} \longleftarrow u^* g_* R_{fg} \otimes u^* f^*\mathcal{L} \qquad \textcircled{1}$$

$$h_* v^* R_{fg} \otimes u^* f^*\mathcal{L} \longleftarrow h_*(v^* R_{fg} \otimes h^* u^* f^*\mathcal{L}) \longleftarrow h_*(v^* R_{fg} \otimes v^* g^* f^*\mathcal{L})$$

$$h_*(R_{fgv} \otimes (fgv)^*\mathcal{L})$$

$$R_{fu} \otimes (fu)^*\mathcal{L} \longleftarrow h_* R_{fuh} \otimes (fu)^*\mathcal{L} \longleftarrow h_*(R_{fuh} \otimes h^*(fu)^*\mathcal{L}) \longleftarrow h_*(R_{fuh} \otimes (fuh)^*\mathcal{L})$$

Commutativity of subdiagram $\textcircled{1}$ is given by (3.7.3). Commutativity of the other subdiagrams is easy to check.

A *morphism of Dualizing Complexes on* Y, $\psi \colon (R, \gamma, \tau) \xrightarrow{\sim} (R', \gamma', \tau')$ is a map associating to each scheme-map $f \colon X \to Y$ a $\mathbf{D}(X)$-map $\psi_f \colon R_f \xrightarrow{\sim} R'_f$, such that for each open immersion $u \colon U \to X$ (resp. each proper map $g \colon X' \to X$) the following diagrams commute:

$$
\begin{array}{ccc}
u^* R_f & \xrightarrow{\gamma_{f,u}} & R_{fu} \\
{\scriptstyle u^*\psi_f}\downarrow & & \downarrow{\scriptstyle \psi_{fu}} \\
u^* R'_f & \xrightarrow{\gamma'_{f,u}} & R'_{fu}
\end{array}
\qquad
\begin{array}{ccc}
g_* R_{fg} & \xrightarrow{\tau_{f,g}} & R_f \\
{\scriptstyle g_*\psi_{fg}}\downarrow & & \downarrow{\scriptstyle \psi_f} \\
g_* R'_{fg} & \xrightarrow{\tau'_{f,g}} & R'_f
\end{array}
\qquad (4.10.2.2)
$$

In the next Proposition, 1 denotes the identity map of Y.

Proposition 4.10.3 *Let* (R, γ, τ) *and* (R', γ', τ') *be Dualizing Complexes on* Y, *and let* $\psi_0 \colon R_1 \to R'_1$ *be a* $\mathbf{D}(Y)$-*map. Then there exists a unique morphism* $\psi \colon (R, \gamma, \tau) \xrightarrow{\sim} (R', \gamma', \tau')$ *with* $\psi_1 = \psi_0$.

Corollary 4.10.3.1 (Uniqueness of Dualizing Complexes). *If* $\mathcal{R}$ *and* $\mathcal{R}'$ *are Dualizing Complexes on* Y *then there exists an invertible* $\mathcal{O}_Y$-*module* $\mathcal{L}$, *unique up to isomorphism, and a unique locally constant function* $n \colon Y \to \mathbb{Z}$ *such that* $\mathcal{R}' \cong \mathcal{R} \otimes \mathcal{L}[n]$. *Moreover, if* ψ *and* χ *are two isomorphisms from* $\mathcal{R}'$ *to* $\mathcal{R} \otimes \mathcal{L}[n]$ *then* $\psi^{-1}\chi$ *is multiplication by a unit in* $\mathrm{H}^0(Y, \mathcal{O}_Y)$.

Proof of (4.10.3.1). One reduces easily to where Y is connected. In view of (4.10.3), the first assertion follows then from the corresponding assertion for dualizing $\mathcal{O}_Y$-complexes [**H**, p. 266, Thm. 3.1]. The second assertion results from the sequence of natural ring isomorphisms and anti-isomorphisms—with R a dualizing $\mathcal{O}_Y$-complex and $\mathcal{D}_R(-) := \mathbf{R}\mathcal{H}om_X(-, R)$:

$$\mathrm{Hom}_{\mathbf{D}(Y)}(R, R) \cong \mathrm{Hom}_{\mathbf{D}(Y)}\big(\mathcal{D}_R(R), \mathcal{D}_R(R)\big) \cong \mathrm{Hom}_{\mathbf{D}(Y)}(\mathcal{O}_Y, \mathcal{O}_Y)$$
$$\cong \mathrm{H}^0\mathbf{R}\Gamma\mathbf{R}\mathcal{H}om(\mathcal{O}_Y, \mathcal{O}_Y) \cong \mathrm{H}^0\mathbf{R}\Gamma(\mathcal{O}_Y) \cong \mathrm{H}^0(Y, \mathcal{O}_Y).$$

Proof of (4.10.3). For any proper map $g \colon X \to Y$, since $(R'_g, \tau'_{1,g})$ represents the functor $\mathrm{Hom}_{\mathbf{D}(Y)}(g_* E, R'_1)$ (see (4.10.2)(b)), there exists a unique $\mathbf{D}(X)$-map $\psi_g \colon R_g \to R'_g$ making the following diagram commute:

$$
\begin{array}{ccc}
g_* R_g & \xrightarrow{\ g_* \psi_g\ } & g_* R'_g \\[4pt]
{\scriptstyle \tau_{1,g}} \downarrow & & \downarrow {\scriptstyle \tau'_{1,g}} \\[4pt]
R_1 & \xrightarrow[\ \psi_0\ ]{} & R'_1
\end{array}
$$

A general map $f \colon X \to Y$ factors as $X \xrightarrow{\ u\ } Z \xrightarrow{\ g\ } Y$ with g proper and u an open immersion. Let $\psi_{g,u} \colon R_f \to R'_f$ be the unique $\mathbf{D}(X)$-map making the following diagram commute:

$$
\begin{array}{ccc}
R_f & \xrightarrow{\ \psi_{g,u}\ } & R'_f \\[4pt]
{\scriptstyle \gamma_{g,u}} \uparrow {\scriptstyle \simeq} & & {\scriptstyle \simeq} \uparrow {\scriptstyle \gamma'_{g,u}} \\[4pt]
u^* R_g & \xrightarrow[\ u^* \psi_g\ ]{} & u^* R'_g
\end{array}
$$

Let us show that $\psi_{g,u}$ depends only on f, allowing us to write ψ_f instead of $\psi_{g,u}$. So let $X \xrightarrow{\ \tilde{u}\ } \tilde{Z} \xrightarrow{\ \tilde{g}\ } Y$ also be a factorization of f ($\tilde{u}$ an open immersion, $\tilde{g}$ proper). There results a natural diagram

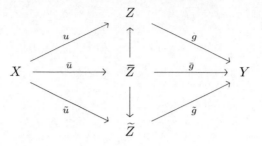

with $\bar{Z}$ the scheme-theoretic image [**GD**, p. 324, §6.10] of the composite immersion $X \xrightarrow{\text{diag}} X \times_Y X \xrightarrow{(u,\tilde{u})} Z \times_Y \tilde{Z}$, and $\bar{u}\colon X \to \bar{Z}$ the resulting open immersion; and where the vertical maps, induced by the canonical projections, are proper.

We need only show that $\psi_{u,g} = \psi_{\bar{u},\bar{g}} = \psi_{\tilde{u},\tilde{g}}$; so it's enough to treat the case $(\tilde{u}, \tilde{g}) = (\bar{u}, \bar{g})$, that is, we may assume that there is a proper map $p\colon \tilde{Z} \to Z$ such that $gp = \tilde{g}$ and $p\tilde{u} = u$, and that furthermore $\tilde{u}(X)$ is a dense open subset of $\tilde{Z}$:

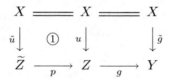

Here subdiagram ① is a fiber square, since the map $\tilde{u}_0\colon X \to p^{-1}(uX)$ induced by $\tilde{u}$ is both an open immersion (clearly) and a closed immersion (because $\tilde{u}_0$ has a left inverse, essentially $p|_{p^{-1}(uX)}$), so that $\tilde{u}X$ is open, closed and dense in $p^{-1}(uX)$, hence equal to $p^{-1}(uX)$. Consequently, there is a natural functorial isomorphism $\theta\colon u^*p_* \xrightarrow{\sim} \tilde{u}^*$.

It will be enough to show that the following diagram—whose top and bottom rows compose to $\psi_{g,u}$ and $\psi_{\tilde{g},\tilde{u}}$ respectively—commutes:

$$
\begin{array}{ccccccc}
R_{gu} & \xrightarrow[\gamma_{g,u}^{-1}]{\sim} & u^*R_g & \xrightarrow{u^*\psi_g} & u^*R_g' & \xrightarrow[\gamma_{g,u}']{\sim} & R_{gu}' \\
\Big\| & & \Big\uparrow{\scriptstyle u^*\tau_{g,p}} & \textcircled{3} & \Big\uparrow{\scriptstyle u^*\tau_{g,p}'} & & \Big\| \\
& \textcircled{2} & u^*p_*R_{gp} & \xrightarrow{u^*p_*\psi_{gp}} & u^*p_*R_{gp}' & \textcircled{5} & \\
& & {\scriptstyle\theta}\Big\downarrow{\scriptstyle\simeq} & \textcircled{4} & {\scriptstyle\simeq}\Big\downarrow{\scriptstyle\theta} & & \\
R_{\tilde{g}\tilde{u}} = R_{gp\tilde{u}} & \xrightarrow[\gamma_{gp,\tilde{u}}^{-1}]{\sim} & \tilde{u}^*R_{gp} & \xrightarrow{\tilde{u}^*\psi_{gp}} & \tilde{u}^*R_{gp}' & \xrightarrow[\gamma_{gp,\tilde{u}}']{\sim} & R_{gp\tilde{u}}' = R_{\tilde{g}\tilde{u}}'
\end{array}
$$

Commutativity of subdiagram ④ is clear. Subdiagrams ② and ⑤ commute by condition (c) in (4.10.2), applied to the above fiber square ①. Finally, the first part of (4.10.2)(b) guarantees the existence of a map $\hat{\psi}_{gp}\colon R_{gp} \to R'_{gp}$ such that the following diagram commutes:

$$
\begin{array}{ccc}
R_g & \xrightarrow{\psi_g} & R'_g \\
\tau_{g,p} \big\uparrow & & \big\uparrow \tau'_{g,p} \\
p_* R_{gp} & \xrightarrow[p_* \hat{\psi}_{gp}]{} & p_* R'_{gp};
\end{array}
$$

and in view of the of the commutative diagram in (4.10.2)(b), and of the definition of ψ_f for proper f, application of the functor g_* to the preceding diagram shows that $\hat{\psi}_{gp} = \psi_{gp}$, whence ③ commutes.

We have now defined ψ_f for all f. The commutativity in (4.10.2.2) shows that no other family (ψ_f) can satisfy (4.10.3). It remains to be proved that with the present (ψ_f), commutativity does hold for the two diagrams in (4.10.2.2).

For the first of those diagrams, the problem is to show, given a sequence $U \xrightarrow{u} X \xrightarrow{v} Z \xrightarrow{g} Y$ with u and v open immersions and g proper, that the following natural diagram commutes:

$$
\begin{array}{ccccccc}
u^* R_{gv} & \xrightarrow{\sim} & u^* v^* R_g & \longrightarrow & u^* v^* R'_g & \xrightarrow{\sim} & u^* R'_{gv} \\
\big\downarrow & & \big\downarrow & & \big\downarrow & & \big\downarrow \\
R_{gvu} & \longrightarrow & (vu)^* R_g & \longrightarrow & (vu)^* R'_g & \longrightarrow & R'_{gvu};
\end{array}
$$

but this is an immediate consequence of (4.10.2)(a).

For the second diagram in (4.10.2.2), suppose there is given a sequence $X' \xrightarrow{g} X \xrightarrow{v} Z \xrightarrow{h} Y$ with u an open immersion and g, h both proper. As above, there are maps $X' \xrightarrow{w} W \xrightarrow{\bar{g}} Z$ such that w maps X' isomorphically onto a dense open subscheme of W, $\bar{g}$ is proper, and $\bar{g}w = vg$:

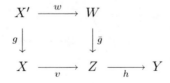

The proper map g factors naturally as $X' \to \bar{g}^{-1}(vX) \to X$, whence $w(X')$ is open, closed and dense in—hence equal to—$\bar{g}^{-1}(vX)$, and so there is a natural isomorphism $\theta\colon v^* \bar{g}_* \xrightarrow{\sim} g_* w^*$.

The problem is to show commutativity of the natural diagram

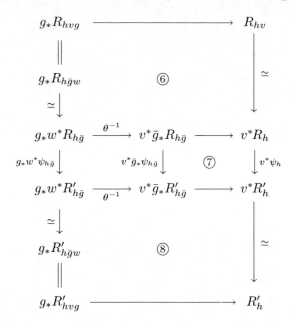

Commutativity of subdiagrams ⑥ and ⑧ is given by (4.10.2)(c). The argument that subdiagram ⑦ commutes is similar to that used above for ③. Commutativity of the remaining subdiagram is obvious.   Q.E.D.

Here is the main result of this section.[18]

Theorem 4.10.4 *Let* **S** *be a category of noetherian schemes such that if* $Y \in \mathbf{S}$ *and* $f\colon X \to Y$ *is a separated finite-type map then* $X \in \mathbf{S}$. *Suppose every scheme in* **S** *has a Dualizing Complex.*

Then there exists on **S** *a* $\overline{\mathbf{D}}_{\mathsf{c}}^{+}$*-valued pseudofunctor* $^!$ *which is uniquely determined up to isomorphism by the properties that it restricts to the inverse-image pseudofunctor* $*$ *on the subcategory of open immersions, that for a proper* $f \in \mathbf{S}$, *the functor* $f^!$ *is right-adjoint to* $f_*\colon \overline{\mathbf{D}}_{\mathsf{c}}^{+}(X) \to \overline{\mathbf{D}}_{\mathsf{c}}^{+}(Y)$ *(see* (3.9.2.6)(c)), *and that for any fiber square* σ *in* **S**

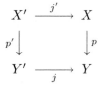

with j *an open immersion and* p *proper, the base-change map* β_σ *of* (4.4.3) *is the natural composite isomorphism*

$$j'^* p^! = j'^! p^! \xrightarrow{\;\sim\;} (pj')^! = (jp')^! \xrightarrow{\;\sim\;} p'^! j^! = p'^! j^*.$$

[18] Cf. [**H**, p. 383, Cor. 3.4], and its proof.

With this[1], *each Dualizing Complex* $(\bar{R}, \bar{\gamma}, \bar{\tau})$ *on* Y *is isomorphic to the one in* (4.10.2.1)(A) *with* $R := \bar{R}_{(\text{identity of } Y)}$.

Remarks. This says less than Theorem (4.8.1): the restriction to **S** of the pseudofunctor in that Theorem satisfies this one. The point is, however, that Theorem (4.10.4) captures Grothendieck's strategy for constructing a duality pseudofunctor by means of Dualizing Complexes. Indeed, showing the existence of Dualizing Complexes is a major theme of the second half of [**H**]. (See also the discussion and clarification of this material in [**C**, §§3.1–3.4].)[19]

Let us add a few words, in passing, about noetherian *formal schemes*. Applying his results about pasting pseudofunctors to the duality theory in [**AJL′**], Nayak gets the existence of a duality pseudofunctor for composites of any number of pseudo-proper maps and open immersions [**Nk**, §7.1]. (As of 2009, one doesn't know whether or not any pseudo-finite separated map of formal schemes is such a composite.) On the other hand, using an analog of Theorem (4.10.4), Sastry constructs a duality pseudofunctor on the category of all formal schemes admitting a dualizing complex (suitably defined for formal schemes), with "essentially pseudo-finite type" maps; and he shows that this pseudofunctor agrees with Nayak's whenever both are defined [**S′**, §9].

Sastry's approach has some resemblance to the one in [**H**], but there are a number of new techniques involved in the construction of Dualizing Complexes. In short, Chapter 6 of [**H**] is localized, generalized, and extended to the context of formal schemes in [**LNS**]; and then, among other things, the main results of Chapter 7 of [**H**], are extended to this context in [**S′**].

Thus, at the present time (2008), the theory of Dualizing Complexes for formal schemes gives rise in certain situations to the *only* way to construct dualizing pseudofunctors.

Proof of (4.10.4) (Outline only). For each $Y \in \mathbf{S}$ choose a Dualizing Complex $\mathcal{R}^Y = (R^Y, \gamma^Y, \tau^Y)$. For any **S**-map $f \colon X \to Y$ let $\mathcal{D}_f^Y$ be the functor from $\mathbf{D_c}(X)$ to $\mathbf{D_c}(X)$ given by

$$\mathcal{D}_f^Y(E) := \mathcal{H}_X(E, R_f^Y) \qquad (\mathcal{H}_X := \mathbf{R}\mathcal{H}\mathrm{om}_X^\bullet).$$

We set $R_Y := R_{1_Y}^Y$ and $\mathcal{D}^Y := \mathcal{D}_{1_Y}^Y$ where 1_Y is the identity map of Y.

For any **S**-map $f \colon X \to Y$, the functor $f^! \colon \mathbf{D_c}(Y) \to \mathbf{D_c}(X)$ is defined to be

$$f^! := \mathcal{D}_f^Y f^* \mathcal{D}^Y.$$

This functor has the following properties.

---

[19] Recently, Yekutieli and Zhang have exploited the notion of "rigid dualizing complex," introduced by Van den Bergh in the context of noncommutative algebra, to give an elegant new approach to the existence question, at least for finite tor-dimension maps of schemes of finite type over a regular scheme. See [**YZ**] for a preliminary account.

(1) If f is an open immersion, there is a natural functorial isomorphism $f^! \xrightarrow{\sim} f^*$, namely, the natural composition, with $E \in \mathbf{D_c}(Y)$,

$$f^! E = \mathcal{H}_X\big(f^* \mathcal{D}^Y E, R_f^Y\big) \xrightarrow[\text{via (4.6.7)}]{\sim} \mathcal{H}_X\big(\mathcal{H}_X(f^*E, f^*R_Y), R_f^Y\big)$$

$$\xrightarrow[\text{via } \gamma_{1_Y,u}^{-1}]{\sim} \mathcal{H}_X\big(\mathcal{H}_X(f^*E, R_f^Y), R_f^Y\big) \xrightarrow{\sim} f^*E$$

(the last isomorphism resulting from R_f^Y being a dualizing $\mathcal{O}_X$-complex).

(2) If f is proper then $f^!$ is right-adjoint to $f_* \colon \mathbf{D_c}(X) \to \mathbf{D_c}(Y)$. Indeed, for $E \in \mathbf{D_c}(X)$, $F \in \mathbf{D_c}(Y)$ we have, in view of (4.10.2)(b), natural functorial isomorphisms

$$\mathrm{Hom}_{\mathbf{D}(X)}\big(E, f^!F\big) \xrightarrow[(2.6.1)']{\sim} \mathrm{Hom}_{\mathbf{D}(X)}\big(E \otimes f^*\mathcal{H}_Y(F, R_Y), R_f^Y\big)$$

$$\xrightarrow{\sim} \mathrm{Hom}_{\mathbf{D}(Y)}\big(f_*(E \otimes f^*\mathcal{H}_Y(F, R_Y)), R_Y\big)$$

$$\xrightarrow[(3.9.4)]{\sim} \mathrm{Hom}_{\mathbf{D}(Y)}\big(f_*E \otimes \mathcal{H}_Y(F, R_Y), R_Y\big)$$

$$\xrightarrow[(2.6.1)']{\sim} \mathrm{Hom}_{\mathbf{D}(Y)}\big(f_*E, \mathcal{H}_Y(\mathcal{H}_Y(F, R_Y), R_Y)\big)$$

$$\xrightarrow{\sim} \mathrm{Hom}_{\mathbf{D}(Y)}\big(f_*E, R_Y\big).$$

(3) There is a natural isomorphism $f^! R_Y \xrightarrow{\sim} R_f^Y$. This follows easily from the natural isomorphism $\mathcal{D}^Y R_Y \xrightarrow{\sim} \mathcal{O}_Y$.

(4) The functor $^!$ extends to a pseudofunctor.
For the proof, we need:

Lemma 4.10.4.1 *For any sequence* $V \xrightarrow{h} W \xrightarrow{g} X \xrightarrow{f} Y$ *in* $\mathbf{S}$ *there is a natural isomorphism*

$$\phi_{f,g,h} \colon \mathcal{D}_{gh}^X h^* \mathcal{D}_g^X \xrightarrow{\sim} \mathcal{D}_{fgh}^Y h^* \mathcal{D}_{fg}^Y,$$

such that

$$\phi_{f,g,h} \circ \phi_{g,1_W,h} = \phi_{fg,1_W,h} \colon \mathcal{D}_h^W h^* \mathcal{D}^W \xrightarrow{\sim} \mathcal{D}_{fgh}^Y h^* \mathcal{D}_{fg}^Y.$$

Proof. By (4.10.3.1) there is an invertible $\mathcal{O}_X$-module $\mathcal{L}_f$, a locally constant, integer-valued function n_f, and an isomorphism of Dualizing Complexes

$$\alpha \colon \mathcal{R}^X \xrightarrow{\sim} (\mathcal{R}^Y \times_Y X) \otimes \mathcal{L}_f[n_f],$$

see (4.10.2.1), (B) and (C). Set $\mathcal{I} := \mathcal{L}_f[n_f]$ and $\mathcal{I}^{-1} := \mathcal{H}_X(\mathcal{I}, \mathcal{O}_X)$, so that there is a canonical isomorphism $\mathcal{I} \otimes \mathcal{I}^{-1} \xrightarrow{\sim} \mathcal{O}_X$. For any map $e \colon Z \to X$ and $F, G \in \mathbf{D}(Z)$, the map coming from (3.5.3)(g) is an *isomorphism* $\mathcal{H}_Z(F, G) \otimes e^*\mathcal{I}^{-1} \xrightarrow{\sim} \mathcal{H}_Z(F \otimes e^*\mathcal{I}, G)$. (The question being local, the proof reduces easily to the simple case $\mathcal{I} = \mathcal{O}_X$.)

There results, for any $E \in \mathbf{D_c}(W)$, a composite isomorphism

$$\varphi_{\alpha,\mathcal{L}} \colon \mathcal{D}_{gh}^X h^* \mathcal{D}_g^X E \xrightarrow{\sim} \mathcal{D}_{fgh}^Y(h^* \mathcal{D}_g^X E) \otimes (gh)^* \mathfrak{I}$$

$$\xrightarrow{\sim} \mathcal{D}_{fgh}^Y(h^* \mathcal{D}_{fg}^Y E \otimes (gh)^* \mathfrak{I}) \otimes (gh)^* \mathfrak{I}$$

$$\xrightarrow{\sim} \mathcal{D}_{fgh}^Y h^* \mathcal{D}_{fg}^Y E \otimes (gh)^* \mathfrak{I}^{-1} \otimes (gh)^* \mathfrak{I}$$

$$\xrightarrow{\sim} \mathcal{D}_{fgh}^Y h^* \mathcal{D}_{fg}^Y E \otimes (gh)^* (\mathfrak{I}^{-1} \otimes \mathfrak{I})$$

$$\xrightarrow{\sim} \mathcal{D}_{fgh}^Y h^* \mathcal{D}_{fg}^Y E.$$

It is easily checked that $\varphi_{\alpha,\mathcal{L}}$ is independent of the choice of α and of $\mathcal{L}$, i.e., if μ is a unit in $\mathrm{H}^0(X, \mathcal{O}_X)$, and if $\mathcal{L}' \cong \mathcal{L}$, then $\varphi_{\alpha,\mathcal{L}} = \varphi_{\mu\alpha,\mathcal{L}'}$. So we can set $\phi_{f,g,h} = \varphi_{\alpha,\mathcal{L}}$.

The final assertion is left to the very patient reader. (A direct approach seems to involve a formidable diagram—although the analogous statement (3.3.13) in [**C**, p. 135] is said there to be "easy to check.") Q.E.D.

Next, with f, g, h as in (4.10.4.1), we define the functorial isomorphism $d_{g,f} \colon g^! f^! \xrightarrow{\sim} (fg)^!$ to be the natural composition

$$g^! f^! = \mathcal{D}_g^X g^* \mathcal{D}^X \mathcal{D}_f^Y f^* \mathcal{D}^Y \xrightarrow[\phi_{f,1_V,g}]{\sim} \mathcal{D}_{fg}^Y g^* \mathcal{D}_f^Y \mathcal{D}_f^Y f^* \mathcal{D}^Y$$

$$\xrightarrow{\sim} \mathcal{D}_{fg}^Y g^* f^* \mathcal{D}^Y$$

$$\xrightarrow{\sim} \mathcal{D}_{fg}^Y (fg)^* \mathcal{D}^Y = (fg)^!$$

Pseudofunctoriality requires the following diagram to commute:[20]

$$
\begin{array}{ccc}
(fgh)^! & \xleftarrow{\ d_{h,fg}\ } & h^!(fg)^! \\
{\scriptstyle d_{gh,f}}\big\uparrow & & \big\uparrow{\scriptstyle h^! d_{f,g}} \\
(gh)^! f^! & \xleftarrow{\ \ d_{g,h}\ \ } & h^! g^! f^!
\end{array}
$$

Expanding this diagram according to the definition of $d_{g,f}$, one finds quickly that the problem is to show commutativity of the following diagram of natural isomorphisms:

$$
\begin{array}{ccccc}
\mathcal{D}_{fgh}^Y (gh)^* \mathcal{D}_f^Y & \longrightarrow & \mathcal{D}_{fgh}^Y h^* g^* \mathcal{D}_f^Y & \longleftarrow & \mathcal{D}_{fgh}^Y h^* \mathcal{D}_{fg}^Y \mathcal{D}_{fg}^Y g^* \mathcal{D}_f^Y \\
{\scriptstyle \phi_{f,1_X,gh}}\big\uparrow & & & & \big\uparrow{\scriptstyle \phi_{fg,1_W,h}} \\
\mathcal{D}_{gh}^X (gh)^* \mathcal{D}^X & \longleftarrow & \mathcal{D}_{gh}^X h^* g^* \mathcal{D}^X & & \mathcal{D}_h^W h^* \mathcal{D}^W \mathcal{D}_{fg}^Y g^* \mathcal{D}_f^Y \\
& & \big\uparrow & & \mathcal{D}_h^W h^* \mathcal{D}^W\big\uparrow{\scriptstyle (\phi_{f,1_X,g})} \\
& & \mathcal{D}_{gh}^X h^* \mathcal{D}_g^X \mathcal{D}_g^X g^* \mathcal{D}^X & \xleftarrow{\ \ \phi_{g,1_W,h}\ \ } & \mathcal{D}_h^W h^* \mathcal{D}^W \mathcal{D}_g^X g^* \mathcal{D}^X
\end{array}
$$

---

[20] Strictly speaking, we need also to "normalize" $^!$, i.e., to replace $(1_Y)^!$ by the identity functor of $\mathbf{D_c}(Y)$ for every $Y \in \mathbf{S}$.

Using the equality in (4.10.4.1), one transforms the question to commutativity of

$$
\begin{array}{ccc}
\mathcal{D}^Y_{fgh}(gh)^*\mathcal{D}^Y_f & \longrightarrow & \mathcal{D}^Y_{fgh}h^*g^*\mathcal{D}^Y_f & \longleftarrow & \mathcal{D}^Y_{fgh}h^*\mathcal{D}^Y_{fg}\mathcal{D}^Y_{fg}g^*\mathcal{D}^Y_f \\[2ex]
\phi_{f,1_X,gh}\uparrow & & & & \uparrow \\[2ex]
\mathcal{D}^X_{gh}(gh)^*\mathcal{D}^X & \longleftarrow & \mathcal{D}^X_{gh}h^*g^*\mathcal{D}^X & & \mathcal{D}^Y_{fgh}h^*\mathcal{D}^Y_{fg}\Big|(\phi_{f,1_X,g}) \\[2ex]
& & \uparrow & & \\[2ex]
& & \mathcal{D}^X_{gh}h^*\mathcal{D}^X_g\mathcal{D}^X_g g^*\mathcal{D}^X & \xleftarrow[\phi_{f,g,h}]{} & \mathcal{D}^Y_{fgh}h^*\mathcal{D}^Y_{fg}\mathcal{D}^X_g g^*\mathcal{D}^X
\end{array}
$$

Checking this commutativity is left to the few (if any) extremely patient readers who might be willing to do it. Again, the complete expansion according to definitions is intimidating—but the analogous associativity statement is said in [**C**, p. 136] to be "straightforward to check."

Pseudofunctoriality being thus established, one must now verify that the isomorphism in (1) above is *pseudofunctorial;* that on proper maps, $*$ and $^!$ are adjoint as *pseudofunctors* (see (2) and (3.6.7(d))); that the isomorphism in (3) extends to an isomorphism of Dualizing Complexes; and that β_σ is as described in Theorem (4.10.4). And finally, the *uniqueness* (up to isomorphism) of the pseudofunctor $^!$ can be verified as at the beginning of the proof of (4.8.4).

Each of these verifications amounts, upon expansion according to definitions, to checking commutativity of a rather unpleasant diagram.

For the purposes of these Notes, Thm. (4.10.4) is not one of the "main results" referred to in Section (0.3) of the Introduction; so I leave it at that.

References

[AIL]   L. Avramov, S. Iyengar, J. Lipman, Reflexivity and rigidity for complexes, preprint.

[AJL]   L. Alonso Tarrío, A. Jeremías López, J. Lipman, Local homology and cohomology on schemes, *Ann. Scient. Éc. Norm. Sup.*, 30 (1997), pp. 1–39.

[AJL']   ———, A. Jeremías López, J. Lipman, Duality and flat base change on formal schemes, *Contemporary Math.*, 244 (1999), pp. 3–90.

[AJS]   L. Alonso Tarrío, A. Jeremías López, M. J. Souto Salorio, Localization in categories of complexes and unbounded resolutions, *Canadian Math. J.*, 52 (2000), pp. 225–247.

[AHK]   A. Altman, R. Hoobler, S. L. Kleiman, A note on the base change map for cohomology, *Compositio Math.*, 27 (1973), pp. 25–37.

[Ay]   J. Ayoub, *Les six opérations de Grothendieck et le formalisme des cycles évanescents dans le monde motivique*, Astérisque, Soc. Math. de France, 2007. <http://www.math.uiuc.edu/K-theory/0761/THESE.pdf>.

[B]   N. Bourbaki, *Algèbre, Chap. 10, Algèbre Homologique*, Masson, Paris, 1980.

[B']   ———, *Algèbre Commutative, Chap. 1, Modules Plats*, Hermann, Paris, 1961.

[BB]   A. Bondal, M. van den Bergh, Generators and representability of functors in commutative and noncommutative geometry, *Moscow Math. J.*, 3 (2003), pp. 1–36.

[BN]   M. Bökstedt, A. Neeman, Homotopy limits in triangulated categories, *Compositio Math.*, 86 (1993), pp. 209–234.

[Bt]   P. Berthelot, Immersions régulières et calcul du $K^\bullet$ d'un schéma éclaté, in *Théorie des Intersections et Théorème de Riemann-Roch (SGA 6)*, Lecture Notes in Math., no. 225, Springer-Verlag, New York, 1971, pp. 416–465.

[C]   B. Conrad, *Grothendieck Duality and Base Change*, Lecture Notes in Math., no. 1750, Springer-Verlag, New York, 2000.

[C']   B. Conrad, Deligne's notes on Nagata compactifications, *J. Ramanujan Math. Soc.* 22 (2007), 250–257.

[De]   P. Deligne, Cohomologie à supports propres, in *Théorie des Topos et Cohomologie Étale des Schémas (SGA 4) Tome 3*, Lecture Notes in Math., no. 305, Springer-Verlag, New York, 1973, pp. 250–461.

[De']   ———, Cohomologie à support propre et construction du foncteur $f^!$, in *Residues and Duality*, Lecture Notes in Math., no. 20, Springer-Verlag, New York, 1966, pp. 404–421.

[Do]   A. Dold, Zur Homotopie theorie der Kettenkomplexe, *Math. Annalen*, 140 (1960), pp. 278–298.

[EK]   S. Eilenberg, G. M. Kelly, Closed categories, in *Proceedings of the Conference on Categorical Algebra, La Jolla 1965*, edited by S. Eilenberg, D. K. Harrison, S. Mac Lane and H. Röhrl, Springer-Verlag, New York, 1966, pp. 421–562.

[E]   F. Elzein, Complexe dualisant et applications à la classe fondamentale d'un cycle, *Bull. Soc. Math. France*, Mémoire 58 (1978).

[G]   R. Godement, *Théorie des faisceaux*, Act. Sci. et Industrielles no. 1252, Hermann, Paris, 1964.

[Gl]   P.-P. Grivel, Catégories dérivées et foncteurs dérivées, in *Algebraic D-modules*, by A. Borel et al., Academic Press, Orlando, Florida, 1987, pp. 1–108.

[Gl']   ———, Une démonstration du théorème de dualité de Verdier, *L'Enseignement Math.*, 31, 1985, pp. 227–247.

[Gr]   A. Grothendieck, Sur quelques points d'algèbre homologique, *Tôhoku Math. Journal*, 9 (1957), pp. 119–221.

[Gr']   ———, The cohomology theory of abstract algebraic varieties, in *Proceedings of the International Congress of Mathematicians (Edinburgh, 1958)*, Cambridge Univ. Press, New York, 1960, pp. 103–118.

[EGA]   A. Grothendieck, J. Dieudonné, *Éléments de Géométrie Algébrique*, **II**, Publ. Math. IHES, 8 (1961); **III**, *ibid.*, 11 (1961), 17 (1963); **IV**, *ibid.*, 20 (1964), 24 (1965), 28 (1966), 32 (1967).

[GD]   ———, *Éléments de Géométrie Algébrique I*, Springer-Verlag, New York, 1971.

[GV]   A. Grothendieck, J.-L. Verdier, Préfaisceaux, in *Théorie des Topos et Cohomologie Étale des Schémas (SGA 4)*, Lecture Notes in Math., no. 269, Springer-Verlag, New York, 1972, pp. 1–184.

[H]   R. Hartshorne, *Residues and Duality*, Lecture Notes in Math., no. 20, Springer-Verlag, New York, 1966.

[HS]   R. Hübl, P. Sastry, Regular differential forms and relative duality, *Amer. J. Math.*, 115 (1993), pp. 749–787.

[Hsh]   M. Hashimoto, Equivariant twisted inverses, in *Foundations of Grothendieck Duality for Diagrams of Schemes*, Lecture Notes in Math., no 1960, Springer-Verlag, New York, 2009, 263–474.

[I]   L. Illusie, Généralités sur les conditions de finitude dans les catégories dérivées, etc., in *Théorie des Intersections et Théorème de Riemann-Roch (SGA 6)*, Lecture Notes in Math., no. 225, Springer-Verlag, New York, 1971, pp. 78–273.

[I']   ———, Catégories dérivées et dualité, travaux de J.-L. Verdier, *L'Enseignement Math.*, 36 (1990), pp. 369–391.

[Iv]   B. Iversen, *Cohomology of Sheaves*, Springer-Verlag, New York, 1986.

[K]   G. M. Kelly, *Basic Concepts of Enriched Category Theory*, Cambridge Univ. Press, 1982.

[K']   ———, Coherence theorems for lax algebras and for distributive laws, in *Category Seminar, Sydney 1972/73*, Lecture Notes in Math., no. 420, Springer-Verlag, New York, 1974.

[Kf]   G. R. Kempf, Some elementary proofs of basic theorems in the cohomology of quasi-coherent sheaves, *Rocky Mountain J. Math*, 10 (1980), pp. 637–645.

[Kl]   R. Kiehl, Ein "Descente"-Lemma und Grothendiecks Projektionssatz für nichtnoethersche Schemata, *Math. Annalen*, 198 (1972), pp. 287–316.

[Kn]   S. L. Kleiman, Relative duality for quasi-coherent sheaves, *Compositio Math.*, 41 (1980), pp. 39–60.

[KS]   M. Kashiwara, P. Schapira, *Sheaves on Manifolds*, Springer-Verlag, New York, 1990.

[L]   M. Laplaza, Coherence for distributivity, in *Coherence in Categories*, Lecture Notes in Math., no. 281, Springer-Verlag, New York, 1972, pp. 29–65.

[L′]   ——, A new result of coherence for distributivity, *ibid.*, pp. 214–235.

[LN]   J. Lipman, A. Neeman, Quasi-perfect scheme-maps and boundedness of the twisted inverse image functor, *Illinois J. Math.*, 51 (2007), pp. 209–236.

[LNS]   J. Lipman, S. Nayak, P. Sastry, Pseudofunctorial behavior of Cousin complexes on formal schemes, *Contemporary Math.*, 375 (2005), pp. 3–133.

[Lp]   J. Lipman, *Dualizing sheaves, Differentials, and Residues on Algebraic Varieties*, Astérisque, vol. 117, Soc. Math. de France, 1984.

[Lp′]   ——, Residues and Traces of Differential Forms via Hochschild Homology, Contemporary Math., vol. 61, Amer. Math. Soc., 1987.

[LO]   Y. Laszlo, M. Olsson, The six operations for sheaves on Artin stacks I: finite coefficients, preprint, arXiv:math.AG/0512097 v1, 5 Dec 2005.

[Lt]   W. Lütkebohmert, On compactification of schemes, *Manuscr. Math.*, 80 (1993), pp. 95–111.

[Lw]   G. Lewis, Coherence for a closed functor, in *Coherence in Categories*, Lecture Notes in Math., no. 281, Springer-Verlag, New York, 1972, pp. 148–195.

[M]   S. Mac Lane, *Categories for the Working Mathematician*, Second Edition, Springer-Verlag, New York, 1998.

[Mb]   Z. Mebkhout, *Le formalisme des six opérations de Grothendieck pour les $\mathcal{D}_X$-modules. cohérents*, Travaux en Cours 35, Hermann, Paris, 1989.

[N]   A. Neeman, The Grothendieck duality theorem via Bousfield's techniques and Brown representability, *J. Amer. Math. Soc.*, 9 (1996), pp. 205–236.

[N′]   ——, Triangulated Categories, Princeton Univ. Press, Princeton NJ, 2001.

[N″]   ——, Derived Categories and Grothendieck Duality, preprint.

[Nk]   S. Nayak, Pasting pseudofunctors, *Contemporary Math.* 375 (2005), pp. 195–271.

[Nk′]   ——, Compactification for essentially finite-type maps, arXiv:0809.1201.

[R]   M. Raussen, C. Skau, Interview with Jean-Pierre Serre, *Notices Amer. Math. Soc.*, 51 (2004), pp. 210–214.

[S]   P. Sastry, Base change and Grothendieck duality for Cohen-Macaulay maps, *Compositio Math.*, 140 (2004), pp. 729–777.

[S′]   P. Sastry, Duality for Cousin complexes, *Contemporary Math.*, 375 (2005), pp. 137–192.

[Sc]   H. Schubert, *Categories*, Springer-Verlag, New York, 1972.

[Sm]   C. Simpson, Explaining Gabriel-Zisman localization to the computer. *J. Automat. Reason*, 36 (2006), pp. 259–285.

[Sp]   N. Spaltenstein, Resolutions of unbounded complexes, *Compositio Math.*, 65 (1988), pp. 121–154.

[Sv]   S. V. Soloviev, On the conditions of full coherence in closed categories, *J. Pure and Applied Algebra*, 69 (1990), pp. 301–329.

[TT]   R. W. Thomason, T. Trobaugh, Higher algebraic K-theory of schemes and of derived categories, in *The Grothendieck Festschrift*, Vol. III, Progr. Math., 88, Birkhäuser Boston, Boston, MA, 1990, pp. 247–435.

[V]   J.-L. Verdier, Catégories dérivées, état 0, in *Cohomologie Étale (SGA $4\frac{1}{2}$)*, Lecture Notes in Math., no. 569, Springer-Verlag, New York, 1977, pp. 262–311.

[V′]   ——, Base change for twisted inverse image of coherent sheaves, in *Algebraic Geometry (Bombay, 1968)*, Oxford Univ. Press, London, 1969, pp. 393–408.

[Vj]   P. Vojta, Nagata's embedding theorem, arXiv:0706.1907v1.

[W]   C. Weibel, *An introduction to homological algebra*, Cambridge Univ. Press, Cambridge, 1994.

[YZ]   A. Yekutieli, J. Zhang, Rigid dualizing complexes on schemes, (preliminary version of a series of papers), arXiv:math.AG/0405570 v3, 7 Apr 2005.

Index

Part II
Mitsuyasu Hashimoto:
Equivariant Twisted Inverses

To Tomoko

Abstract

An equivariant version of the twisted inverse pseudofunctor is defined, and equivariant versions of some important properties, including the Grothendieck duality of proper morphisms and flat base change are proved. As an application, a generalized version of Watanabe's theorem on the Gorenstein property of the ring of invariants is proved.

Introduction

Let S be a scheme, G a flat S-group scheme of finite type, X and Y noetherian S-schemes with G-actions, and $f\colon X \to Y$ a finite-type separated G-morphism.

The purpose of these notes is to construct an equivariant version of the twisted inverse functor $f^!$ and study its basic properties.

One of the main motivations of the work is applications to invariant theory. As an example, we give a short proof of a generalized version of Watanabe's theorem on the Gorenstein property of invariant subrings [45]. Also, there might be some meaning in formulating the equivariant duality theorem, of which Serre duality for representations of reductive groups (see [21, (II.4.2)]) is a special case, in a reasonably general form. As a byproduct, we give some foundations for G-equivariant sheaf theory. More generally, we discuss sheaves over diagrams of schemes.

In the case where G is trivial, $f^!$ is defined as follows. For a scheme Z, we denote the category of $\mathcal{O}_Z$-modules by $\mathrm{Mod}(Z)$. By definition, a *plump subcategory* of an abelian category is a non-empty full subcategory which is closed under kernels, cokernels, and extensions [26, (1.9.1)]. We denote the plump subcategory of $\mathrm{Mod}(Z)$ consisting of quasi-coherent $\mathcal{O}_Z$-modules by $\mathrm{Qch}(Z)$.

By Nagata's compactification theorem [34], [27], there exists some factorization

$$X \xrightarrow{i} \bar{X} \xrightarrow{p} Y$$

such that p is proper and i an open immersion. We call such a factorization a *compactification*. We define $f^! \colon D^+_{\mathrm{Qch}(Y)}(\mathrm{Mod}(Y)) \to D^+_{\mathrm{Qch}(X)}(\mathrm{Mod}(X))$ to be the composite $i^* p^\times$, where $p^\times \colon D^+_{\mathrm{Qch}(Y)}(\mathrm{Mod}(Y)) \to D^+_{\mathrm{Qch}(\bar{X})}(\mathrm{Mod}(\bar{X}))$ is the right adjoint of Rp_*, and i^* is the restriction. This definition of $f^!$ is independent of the choice of compactification.

In order to consider a non-trivial G, we need to replace $\mathrm{Qch}(X)$ and $\mathrm{Mod}(X)$ by some appropriate categories which respect G-actions. The category $\mathrm{Qch}(G, X)$ which corresponds to $\mathrm{Qch}(X)$ is fairly well-known. It is

J. Lipman, M. Hashimoto, *Foundations of Grothendieck Duality for Diagrams* 267
of Schemes, Lecture Notes in Mathematics 1960,
© Springer-Verlag Berlin Heidelberg 2009

the category of G-linearized quasi-coherent $\mathcal{O}_X$-modules defined by Mumford
[32]. The category $\mathrm{Qch}(G, X)$ is equivalent to the category of quasi-coherent
sheaves over the diagram of schemes

$$B_G^M(X) := \left(G \times_S G \times_S X \xrightarrow[\substack{\xrightarrow{1_G \times a} \\ \xrightarrow{\mu \times 1_X} \\ \xrightarrow{p_{23}}}]{} G \times_S X \xrightarrow[\substack{\xrightarrow{a} \\ \xrightarrow{p_2}}]{} X \right),$$

where $a : G \times X \to X$ is the action, $\mu : G \times G \to G$ the product, and p_{23}
and p_2 are appropriate projections. Thus it is natural to embed the cate-
gory $\mathrm{Qch}(G, X)$ into the category of all $\mathcal{O}_{B_G^M(X)}$-modules $\mathrm{Mod}(B_G^M(X))$, and
$\mathrm{Mod}(B_G^M(X))$ is a good substitute of $\mathrm{Mod}(X)$. As G is flat, $\mathrm{Qch}(G, X)$ is a
plump subcategory of $\mathrm{Mod}(B_G^M(X))$, and we may consider the triangulated
subcategory $D_{\mathrm{Qch}(G,X)}(\mathrm{Mod}(B_G^M(X)))$. However, our construction utilizes
an intermediate category $\mathrm{Lqc}(G, X)$ (the category of locally quasi-coherent
sheaves), and is not an obvious interpretation of the non-equivariant case.

Note that there is a natural restriction functor $\mathrm{Mod}(B_G^M(X)) \to \mathrm{Mod}(X)$,
which sends $\mathrm{Qch}(G, X)$ to $\mathrm{Qch}(X)$. This functor is regarded as the forgetful
functor, forgetting the G-action. The equivariant duality theorem which we
are going to establish must be compatible with this restriction functor, otherwise the theory would be something different from the usual scheme theory
and probably useless.

Most of the discussion in these notes treats more general diagrams of
schemes. This makes the discussion easier, as some of the important properties are proved by induction on the number of objects in the diagram. Our
main construction and theorems are only for the class of finite diagrams of
schemes of certain type, which contains the diagrams of the form $B_G^M(X)$.

In Chapters 1–3, we review some general facts on homological algebra. In
Chapter 1, we give some basic facts on commutativity of various diagrams
of functors derived from an adjoint pair of almost-pseudofunctors over closed
symmetric monoidal categories. In Chapter 2, we give basics about sheaves
on ringed sites. In Chapter 3, we give some basics about unbounded derived
categories.

The construction of $f^!$ is divided into five steps. The first is to study the
functoriality of sheaves over diagrams of schemes. Chapters 4–7 are devoted
to this step. The second is the derived version of the first step. This will be
done in Chapters 8, 13, and 14. Note that not only the categories of all mod-
ule sheaves $\mathrm{Mod}(X_\bullet)$ and the category of quasi-coherent sheaves $\mathrm{Qch}(X_\bullet)$,
but also the category of locally quasi-coherent sheaves $\mathrm{Lqc}(X_\bullet)$ plays an important role in our construction.

The third is to prove the existence of the right adjoint $p_\bullet^\times$ of $R(p_\bullet)_*$ for
(componentwise) proper morphism $p_\bullet$ of diagrams of schemes. This is not
so difficult, and is done in Chapter 17. We use Neeman's existence theorem
on the right adjoint of triangulated functors. Not only to utilize Neeman's
theorem, but to calculate composites of various left and right derived functors,

it is convenient to utilize unbounded derived functors. A short survey on unbounded derived functors is given in Chapter 3.

The fourth step is to prove various commutativities related to the well-definedness of the twisted inverse pseudofunctors, Chapters 16, 18, and 19. Among them, the compatibility with restrictions (Proposition 18.14) is the key to our construction. Given a separated G-morphism of finite type $f : X \to Y$ between noetherian G-schemes, the associated morphism $B_G^M(f) : B_G^M(X) \to B_G^M(Y)$ is cartesian, see (4.2) for the definition. If we could find a compactification

$$B_G^M(X) \xrightarrow{i_\bullet} Z_\bullet \xrightarrow{p_\bullet} B_G^M(Y)$$

such that $p_\bullet$ is proper and cartesian, and $i_\bullet$ an image-dense open immersion, then the construction of $f^!$ and the proof of commutativity of various diagrams would be very easy. However, it seems that this is almost the same as the problem of equivariant compactifications. Equivariant compactifications are known to exist only in very restricted cases, see [40]. We avoid this difficult open problem, and prove the commutativity of various diagrams without assuming that $p_\bullet$ is cartesian.

The fifth part is the existence of a factorization $f_\bullet = p_\bullet i_\bullet$, where $p_\bullet$ is proper and $i_\bullet$ an image-dense open immersion. This is easily done utilizing Nagata's compactification theorem, and is done in Chapter 20. This completes the basic construction of the equivariant twisted inverse pseudofunctor. Theorem 20.4 is our main theorem.

In Chapters 21–28, we prove equivariant versions of most of the known results on twisted inverses including equivariant Grothendieck duality and the flat base change, except that equivariant dualizing complexes are treated later. We also prove that the twisted inverse functor preserves quasi-coherence of cohomology groups. As we already know the corresponding results on single schemes and the commutativity with restrictions, this consists in straightforward (but not easy) checking of commutativity of various diagrams of functors.

Almost all results above are valid for any diagram of noetherian schemes with flat arrows over a finite ordered category. Although our construction can be done using the diagram $B_G^M(X)$, some readers might ask why this were not done over the simplicial scheme $B_G(X)$ associated with the action of G on X. We explain the simplicial method and the related descent theory in Chapters 9 and 10. In the literature, it seems that equivariant sheaves with respect to the action of G on X has been regarded as equivariant sheaves on the diagram $B_G(X)$, see for example, [6, Appendix B]. The relation between $X_\bullet := B_G^M(X)$ and $Y_\bullet := B_G(X)$ is subtle. The category $\mathrm{Qch}(X_\bullet)$ and $\mathrm{Qch}(Y_\bullet)$ are equivalent, and the category of equivariant modules $\mathrm{EM}(X_\bullet)$ and $\mathrm{EM}(Y_\bullet)$ are equivalent. However, I do not know anything about the relationship between $\mathrm{Mod}(X_\bullet)$ and $\mathrm{Mod}(Y_\bullet)$. We use $B_G^M(X)$ only because it is a diagram over a finite ordered category.

In Chapter 11, we prove that if $X_\bullet$ is a simplicial groupoid of schemes, $d_0(1)$ and $d_1(1)$ are concentrated, and X_0 is concentrated, then $\mathrm{Qch}(X_\bullet)$ is Grothendieck. If, moreover, X_0 is noetherian, then $\mathrm{Qch}(X_\bullet)$ is locally noetherian, and $\mathcal{M} \in \mathrm{Qch}(X_\bullet)$ is a noetherian object if and only if $\mathcal{M}_0$ is coherent. In Chapter 12, we study groupoids of schemes and their relations with simplicial groupoids of schemes. In Chapter 15, we compare the two derived functors of $\underline{\mathrm{Hom}}^\bullet_{\mathcal{O}_{X_\bullet}}(\mathcal{M}, \mathcal{N})$ for $\mathcal{M} \in D^-_{\mathrm{Coh}}(\mathrm{Qch}(X_\bullet))$ and $\mathcal{N} \in D^+(\mathrm{Qch}(X_\bullet))$. One is the derived functor taken in $D(\mathrm{Mod}(X_\bullet))$ and the other is the derived functor taken in $D(\mathrm{Qch}(X_\bullet))$. They coincide under mild noetherian hypothesis.

Finally, we consider the group actions on schemes. In Chapter 29, we give a groupoid of schemes associated with a group action. The equivariant duality theorem for group actions is established. In Chapter 30, we prove that the equivariant twisted inverse is compatible with the derived G-invariance. In Chapter 31, we give a definition of the equivariant dualizing complexes. As an application, we give a short proof of a generalized version of Watanabe's theorem on the Gorenstein property of invariant subrings in Chapter 32. In Chapter 33, we give some other examples of diagrams of schemes.

Acknowledgement The author is grateful to Professor Luchezar Avramov, Professor Ryoshi Hotta, Professor Joseph Lipman, and Professor Jun-ichi Miyachi for valuable advice. Special thanks are due to Professor Joseph Lipman for correcting English of this introduction.

Chapter 1
Commutativity of Diagrams Constructed from a Monoidal Pair of Pseudofunctors

(1.1) Let $\mathcal{S}$ be a category. A (covariant) *almost-pseudofunctor* $\#$ on $\mathcal{S}$ assigns to each object $X \in \mathcal{S}$ a category $X_\#$, to each morphism $f \colon X \to Y$ in $\mathcal{S}$ a functor $f_\# \colon X_\# \to Y_\#$, and for each $X \in \mathcal{S}$, a natural isomorphism $\mathfrak{e}_X \colon \mathrm{Id}_{X_\#} \to (\mathrm{id}_X)_\#$ is assigned, and for each composable pair of morphisms $X \xrightarrow{f} Y \xrightarrow{g} Z$, a natural isomorphism

$$c = c_{f,g} \colon (gf)_\# \xrightarrow{\cong} g_\# f_\#$$

is given, and the following conditions are satisfied.

1. For any $f \colon X \to Y$, the map $f_\# \, \mathrm{Id}_{X_\#} = f_\# = (f\mathrm{id}_X)_\# \xrightarrow{c_{f,\mathrm{id}}} f_\#(\mathrm{id}_X)_\#$ agrees with $f_\# \mathfrak{e}_X$.
2. For any $f \colon X \to Y$, the map $\mathrm{Id}_{Y_\#} \, f_\# = f_\# = (\mathrm{id}_Y f)_\# \xrightarrow{c_{\mathrm{id},f}} (\mathrm{id}_Y)_\# f_\#$ agrees with $\mathfrak{e}_Y f_\#$.
3. For any composable triple of morphisms $X \xrightarrow{f} Y \xrightarrow{g} Z \xrightarrow{h} W$, the diagram

$$
\begin{array}{ccc}
(hgf)_\# & \xrightarrow{\ c_{f,hg}\ } & (hg)_\# f_\# \\
\downarrow{\scriptstyle c_{gf,h}} & & \downarrow{\scriptstyle c_{g,h}} \\
h_\#(gf)_\# & \xrightarrow{\ c_{f,g}\ } & h_\# g_\# f_\#
\end{array}
$$

commutes.

If $(?)_\#$ is an almost-pseudofunctor on $\mathcal{S}$, then $(?)_\#^{\mathrm{op}}$ given by $(X)_\#^{\mathrm{op}} = X_\#^{\mathrm{op}}$ for $X \in \mathcal{S}$ and $(f)_\#^{\mathrm{op}} = f_\#^{\mathrm{op}} \colon X_\#^{\mathrm{op}} \to Y_\#^{\mathrm{op}}$ for a morphism $f \colon X \to Y$ of $\mathcal{S}$ together with $\mathfrak{e}^{-1}$ and c^{-1} is again an almost-pseudofunctor on $\mathcal{S}$. Letting $\mathfrak{e}_X = \mathrm{id}$ for each X, a pseudofunctor [26, (3.6.5)] is an almost-pseudofunctor.

(1.2) Let $*$ be an almost-pseudofunctor on $\mathcal{S}$. Let $gf = f'g'$ be a commutative diagram in $\mathcal{S}$. The composite isomorphism

$$g_* f_* \xrightarrow{\ c^{-1}\ } (gf)_* = (f'g')_* \xrightarrow{\ c\ } f'_* g'_*$$

is also denoted by $c = c(gf = f'g')$, by abuse of notation.

J. Lipman, M. Hashimoto, *Foundations of Grothendieck Duality for Diagrams*   271
of Schemes, Lecture Notes in Mathematics 1960,
© Springer-Verlag Berlin Heidelberg 2009

Lemma 1.3. *Let* $fg' = gf'$ *and* $hg'' = g'h'$ *be commutative squares in* S. *Then the diagram*

$$(fh)_*g''_* \xrightarrow{c} f_*h_*g''_* \xrightarrow{c} f_*g'_*h'_*$$
$$\downarrow c \qquad\qquad\qquad\qquad \downarrow c$$
$$g_*(f'h')_* \xrightarrow{\qquad\qquad c \qquad\qquad} g_*f'_*h'_*$$

is commutative.

Proof. Easy.   □

Lemma 1.4. *Let* $fg' = gf'$ *and* $f'h' = hf''$ *be commutative squares in* S. *Then the diagram*

$$f_*(g'h')_* \xrightarrow{c} f_*g'_*h'_* \xrightarrow{c} g_*f'_*h'_*$$
$$\downarrow c \qquad\qquad\qquad\qquad \downarrow c$$
$$(gh)_*f''_* \xrightarrow{\qquad\qquad c \qquad\qquad} g_*h_*f''_*$$

is commutative.

Proof. Follows from Lemma 1.3.   □

(1.5) A *contravariant almost-pseudofunctor* $(?)^\#$ is defined similarly. For $X \in S$, a category $X^\#$ is assigned, and for a morphism $f\colon X \to Y$ in S, a functor $f^\#\colon Y^\# \to X^\#$ is assigned, and for $X \in S$, a natural isomorphism $\mathfrak{f} = \mathfrak{f}_X\colon \mathrm{id}_X^\# \to \mathrm{Id}_{X^\#}$ is assigned, and for a composable pair of morphisms $X \xrightarrow{f} Y \xrightarrow{g} Z$, a natural isomorphism

$$d_{f,g}\colon f^\# g^\# \to (gf)^\#$$

is given, and $((?)^\#)^{\mathrm{op}}$ together with $(\mathfrak{f}_X)_{X \in S}$ and $(d_{g,f})$ is a covariant almost-pseudofunctor on S^{op}. If $(?)^\#$ is a contravariant almost-pseudofunctor on S, then $((?)^\#)^{\mathrm{op}}$ together with $\mathfrak{f}^{-1}$ and $(d_{f,g}^{-1})$ is again a contravariant almost-pseudofunctor on S.

(1.6) Let $(?)^*$ be a contravariant almost-pseudofunctor on S. For a commutative diagram $gf = f'g'$ in S, the composite map

$$(g')^*(f')^* \xrightarrow{d} (f'g')^* = (gf)^* \xrightarrow{d^{-1}} f^*g^*$$

is also denoted by $d = d(gf = f'g')$, by abuse of notation.

(1.7) Let $*$ and $\#$ be almost-pseudofunctors on S such that $X_* = X_\#$. A *morphism of almost-pseudofunctors* $\upsilon\colon * \to \#$ is a family of natural maps $\upsilon_f\colon f_* \to f_\#$ (one for each $f \in \mathrm{Mor}(S)$) such that for any $X \in S$ the diagram

$$\mathrm{Id}_{X_*} \xrightarrow{\mathrm{id}} \mathrm{Id}_{X_\#}$$
$$\downarrow \epsilon \qquad\qquad \downarrow \epsilon$$
$$(\mathrm{id}_X)_* \xrightarrow{\upsilon} (\mathrm{id}_X)_\#$$

commutes, and for any composable pair of morphisms $X \xrightarrow{f} Y \xrightarrow{g} Z$, the diagram

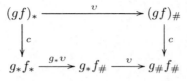

commutes.

If v is a morphism of almost-pseudofunctors, and v_f is a natural isomorphism for each f, then we say that v is an isomorphism of almost-pseudofunctors.

(1.8) Let $\#$ be an almost-pseudofunctor on $\mathcal{S}$. We define $*$ by $X_* = X_\#$ for $X \in \mathcal{S}$, $(\mathrm{id}_X)_* = \mathrm{Id}_{X_\#}$ for $X \in \mathcal{S}$, and $f_* = f_\#$ if $f \neq \mathrm{id}_X$ for any X. For a composable pair of morphisms $X \xrightarrow{f} Y \xrightarrow{g} Z$, we define $c_{f,g} \colon (gf)_* \to g_* f_*$ to be the identity map if $f = \mathrm{id}_X$ or $g = \mathrm{id}_Y$. If $Z = X$, $g = f^{-1}$ and $f \neq \mathrm{id}_X$, then $c_{f,f^{-1}} \colon (\mathrm{id}_X)_* \to f_*^{-1} f_*$ is defined to be the composite

$$\mathrm{Id}_{X_\#} \xrightarrow{\varepsilon_X} (\mathrm{id}_X)_\# \xrightarrow{c_{f,f^{-1}}} f_\#^{-1} f_\#.$$

Otherwise, we define $c_{f,g}$ to be the original $c_{f,g}$ of $\#$. It is easy to see that $*$ is a pseudofunctor on $\mathcal{S}$.

We define $v \colon f_* \to f_\#$ to be ε_X if $f = \mathrm{id}_X$, and the identity of $f_\#$ otherwise. Then v is an isomorphism of almost-pseudofunctors. Thus any almost-pseudofunctor is isomorphic to a pseudofunctor. We call $*$ the *associated pseudofunctor* of the almost-pseudofunctor $\#$.

Similarly, any contravariant almost-pseudofunctor is isomorphic to a contravariant pseudofunctor, and the associated contravariant pseudofunctor of a contravariant almost-pseudofunctor is defined.

(1.9) In this paper, various (different) adjoint pairs appears almost everywhere. By abuse of notation, the unit (resp. the counit) of adjunction is usually simply denoted by the same symbol u (resp. ε). When we mention an adjunction of functors, we implicitly (or occasionally explicitly) fix the unit u and the counit ε.

(1.10) We need the notion of the conjugation from [29, (IV.7)] and [26, (3.3.5)]. Let X and Y be categories, and f_* and g_* functors $X \to Y$ with respective left adjoints f^* and g^*. By Hom, we denote the set of natural transformations. Then $\Phi \colon \mathrm{Hom}(f_*, g_*) \to \mathrm{Hom}(g^*, f^*)$ given by

$$\Phi(\alpha) \colon g^* \xrightarrow{u} g^* f_* f^* \xrightarrow{\alpha} g^* g_* f^* \xrightarrow{\varepsilon} f^*$$

and $\Psi \colon \mathrm{Hom}(g^*, f^*) \to \mathrm{Hom}(f_*, g_*)$ given by

$$\Psi(\beta) \colon f_* \xrightarrow{u} g_* g^* f_* \xrightarrow{\beta} g_* f^* f_* \xrightarrow{\varepsilon} g_*$$

are inverse each other. We say that $\Phi(\alpha)$ is *left conjugate* to α, $\Psi(\beta)$ is *right conjugate* to β, and α and $\Phi(\alpha)$ are *conjugate*. The identity map in $\mathrm{Hom}(f_*, f_*)$ is conjugate to the identity in $\mathrm{Hom}(f^*, f^*)$. Let $h_* : X \to Y$ be a functor with the left adjoint h^*. If $\alpha \in \mathrm{Hom}(f_*, g_*)$ and $\alpha' \in \mathrm{Hom}(g_*, h_*)$, and $\beta \in \mathrm{Hom}(g^*, f^*)$ and $\beta' \in \mathrm{Hom}(h^*, g^*)$ are their respective conjugates, then $\alpha' \circ \alpha$ and $\beta \circ \beta'$ are conjugate. In particular, $\alpha \in \mathrm{Hom}(f_*, g_*)$ is an isomorphism if and only if its conjugate $\beta \in \mathrm{Hom}(g^*, f^*)$ is an isomorphism, and if this is the case, α^{-1} and β^{-1} are conjugate.

Lemma 1.11. *Let X, Y and f_*, g_*, f^*, g^* be as above. Let $\alpha \in \mathrm{Hom}(f_*, g_*)$, and $\beta \in \mathrm{Hom}(g^*, f^*)$. Then the following are equivalent.*

1 α *and* β *are conjugate.*
2 *One of the following diagrams commutes.*

$$
\begin{array}{ccc}
g^* f_* & \xrightarrow{\ \alpha\ } & g^* g_* \\
\downarrow{\scriptstyle \beta} & & \downarrow{\scriptstyle \varepsilon} \\
f^* f_* & \xrightarrow{\ \varepsilon\ } & 1
\end{array}
\qquad
\begin{array}{ccc}
1 & \xrightarrow{\ u\ } & f_* f^* \\
\downarrow{\scriptstyle u} & & \downarrow{\scriptstyle \alpha} \\
g_* g^* & \xrightarrow{\ \beta\ } & g_* f^*
\end{array}
$$

Proof. See [29, (IV.7), Theorem 2]. $\qquad\qquad\square$

(1.12) Let $\mathcal{S}$ be a category, $(?)_*$ be an almost-pseudofunctor on $\mathcal{S}$. Let $(?)^*$ be a left adjoint of $(?)_*$. Namely, for each morphism f of $\mathcal{S}$, we have a left adjoint f^* of f_* (and the explicitly given unit $u : 1 \to f_* f^*$ and counit $\varepsilon : f^* f_* \to 1$).

For $X \in \mathcal{S}$, X^* is defined to be X_*.

For composable two morphisms f and g in $\mathcal{S}$, we denote the map $f^* g^* \to (gf)^*$ conjugate to $c : (gf)_* \to g_* f_*$ by $d = d_{f,g}$. Thus $d_{f,g}$ is the composite

$$
f^* g^* \xrightarrow{u} f^* g^* (gf)_* (gf)^* \xrightarrow{c} f^* g^* g_* f_* (gf)^* \xrightarrow{\varepsilon} f^* f_* (gf)^* \xrightarrow{\varepsilon} (gf)^*.
$$

Being the conjugate of an isomorphism, d is an isomorphism. As d^{-1} is conjugate to c^{-1}, it is the composite

$$
(gf)^* \xrightarrow{u} (gf)^* g_* g^* \xrightarrow{u} (gf)^* g_* f_* f^* g^* \xrightarrow{c^{-1}} (gf)^* (gf)_* f^* g^* \xrightarrow{\varepsilon} f^* g^*.
$$

Lemma 1.13. *Let f and g be morphisms in $\mathcal{S}$, and assume that gf is defined. Then the composite*

$$
1 \xrightarrow{u} g_* g^* \xrightarrow{u} g_* f_* f^* g^* \xrightarrow{c^{-1}} (gf)_* f^* g^* \xrightarrow{d} (gf)_* (gf)^*
$$

is u.

Proof. Follows immediately from Lemma 1.11. $\qquad\qquad\square$

Lemma 1.14. *Let f and g be morphisms in $\mathcal{S}$, and assume that gf is defined. Then the composite*

$$(gf)^*(gf)_* \xrightarrow{c} (gf)^*g_*f_* \xrightarrow{d^{-1}} f^*g^*g_*f_* \xrightarrow{\varepsilon} f^*f_* \xrightarrow{\varepsilon} 1$$

is ε.

Proof. Follows immediately from Lemma 1.11. $\qquad\qquad\square$

(1.15) For $X \in \mathcal{S}$, Id_{X^*} is left adjoint to Id_{X_*} (with $u = \mathrm{id}$ and $\varepsilon = \mathrm{id}$). The morphism left conjugate to $\mathfrak{e}_X \colon \mathrm{Id}_{X_*} \to (\mathrm{id}_X)_*$ is denoted by $\mathfrak{f}_X \colon (\mathrm{id}_X)^* \to \mathrm{Id}_{X^*}$. Namely, $\mathfrak{f}_X$ is the composite

$$(\mathrm{id}_X)^* \xrightarrow{\mathrm{id}} (\mathrm{id}_X)^* \, \mathrm{Id}_{X_*} \, \mathrm{Id}_{X^*} \xrightarrow{\mathfrak{e}} (\mathrm{id}_X)^*(\mathrm{id}_X)_* \, \mathrm{Id}_{X^*} \xrightarrow{\varepsilon} \mathrm{Id}_{X^*}.$$

(1.16) Let $\mathcal{S}$, $(?)_*$, and $(?)^*$ be as above. Then it is easy to see that $(?)^*$ together with d and $\mathfrak{f}$ defined above forms a contravariant almost-pseudofunctor. We say that $((?)^*, (?)_*)$ is an *adjoint pair* of almost-pseudofunctors on $\mathcal{S}$, with this situation. For a commutative diagram $gf = f'g'$ in $\mathcal{S}$, the composite maps $c(gf = f'g')$ and $d(gf = f'g')$ are conjugate.

The opposite $((?)_*^{\mathrm{op}}, ((?)^*)^{\mathrm{op}})$ of $((?)^*, (?)_*)$ is an adjoint pair of almost-pseudofunctors on $\mathcal{S}^{\mathrm{op}}$. $c_{f,g}$, $d_{f,g}$, $u : 1 \to f_*f^*$, and $\varepsilon : f^*f_* \to 1$ of $((?)^*, (?)_*)$ correspond to $d_{g,f}$, $c_{g,f}$, ε, and u of $((?)_*^{\mathrm{op}}, ((?)^*)^{\mathrm{op}})$, respectively.

(1.17) Let $\mathcal{S}$ be as above, and $(?)^*$ a given contravariant almost-pseudofunctor, and $(?)_*$ its right adjoint. Then $((?)^*)^{\mathrm{op}}$ is a covariant almost-pseudofunctor on $\mathcal{S}^{\mathrm{op}}$ as in (1.5). Then $((?)_*^{\mathrm{op}}, ((?)^*)^{\mathrm{op}})$ is an adjoint pair of almost-pseudofunctors. So $((?)^*, (?)_*) = (((?)^*)^{\mathrm{opop}}, (?)_*^{\mathrm{opop}})$ is also an adjoint pair of almost-pseudofunctors.

(1.18) Let $\mathcal{S}$ be as above, and $(?)_*$ a given covariant almost-pseudofunctor, and $(?)^!$ its *right* adjoint. For composable morphisms f and g, define $d_{f,g} : f^!g^! \to (gf)^!$ to be the map right conjugate to $c_{f,g} : (gf)_* \to g_*f_*$. For $X \in \mathcal{S}$, define $\mathfrak{f}_X : (\mathrm{id}_X)^! \to \mathrm{Id}_{X^!}$ to be the map right conjugate to $\mathfrak{e}_X : \mathrm{Id}_{X_*} \to (\mathrm{id}_X)_*$. Then it is straightforward to check that $(?)^!$ is a contravariant almost-pseudofunctor on $\mathcal{S}$. We say that $((?)_*, (?)^!)$ is an *opposite adjoint pair* of almost-pseudofunctors on $\mathcal{S}$. Opposite adjoint pair is also obtained from a given contravariant almost-pseudofunctor $(?)^!$ and its left adjoint $(?)_*$. Note that $((?)^\#, (?)_\#)$ is an adjoint pair of almost-pseudofunctors on $\mathcal{S}$ if and only if $((?)_\#^{\mathrm{op}}, ((?)^\#)^{\mathrm{op}})$ is an opposite adjoint pair of almost-pseudofunctors on $\mathcal{S}$.

(1.19) Let $((?)^*, (?)_*)$ be an adjoint pair of pseudofunctors on $\mathcal{S}$. Let $\sigma = (fg' = gf')$ be a commutative diagram in $\mathcal{S}$.

Lemma 1.20. *The following composite maps agree:*

1 $g^*f_* \xrightarrow{u} g^*f_*g'_*(g')^* \xrightarrow{c} g^*g_*f'_*(g')^* \xrightarrow{\varepsilon} f'_*(g')^*$;

2 $g^*f_* \xrightarrow{u} f'_*(f')^*g^*f_* \xrightarrow{d} f'_*(g')^*f^*f_* \xrightarrow{\varepsilon} f'_*(g')^*$.

For the proof and more information, see [26, (3.7.2)].

(1.21)   We denote the composite map in the lemma by $\theta(\sigma)$ or θ, and call it *Lipman's theta*. Note that $\theta(fg' = gf')$ of $((?)^*, (?)_*)$ is $\theta(g'f = f'g)$ in the opposite $((?)^{\mathrm{op}}_*, ((?)^*)^{\mathrm{op}})$.

Lemma 1.22. *Let $fg' = gf'$ and $hg'' = g'h'$ be commutative squares in $\mathcal{S}$. Then the diagram*

$$
\begin{array}{ccc}
g^*(fh)_* & \xrightarrow{c} g^* f_* h_* \xrightarrow{\theta} f'_*(g')^* h_* \\
\downarrow \theta & \downarrow \theta \\
(f'h')_*(g'')^* & \xrightarrow{\hspace{3cm} c} f'_* h'_*(g'')^*
\end{array}
$$

is commutative.

Proof. See [26, (3.7.2)].   □

Lemma 1.23. *Let $\sigma = (fg' = gf')$ and $\tau = (f'h' = hf'')$ be commutative diagrams in $\mathcal{S}$. Then the composite*

$$
(gh)^* f_* \xrightarrow{d^{-1}} h^* g^* f_* \xrightarrow{\theta} h^* f'_*(g')^* \xrightarrow{\theta} f''_*(h')^*(g')^* \xrightarrow{d} f''_*(g'h')^*
$$

agrees with θ for $f(g'h') = (gh)f''$.

Proof. This is the 'opposite assertion' of Lemma 1.22. Namely, Lemma 1.22 applied to the opposite pair $((?)^{\mathrm{op}}_*, ((?)^*)^{\mathrm{op}})$ is this lemma.   □

Lemma 1.24. *Let $fg' = gf'$ be a commutative diagram in $\mathcal{S}$. Then the composite*

$$
f_* \xrightarrow{u} g_* g^* f_* \xrightarrow{\theta} g_* f'_*(g')^* \xrightarrow{c} f_* g'_*(g')^*
$$

is u.

Proof. Obvious by the commutativity of the diagram

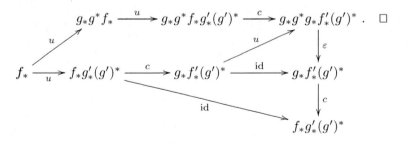

Lemma 1.25. *Let $fg' = gf'$ be a commutative diagram in $\mathcal{S}$. Then the composite*

$$
g^* g_* f'_* \xrightarrow{c} g^* f_* g'_* \xrightarrow{\theta} f'_*(g')^* g'_* \xrightarrow{\varepsilon} f'_*
$$

is ε.

Proof. Obvious by the commutativity of the diagram

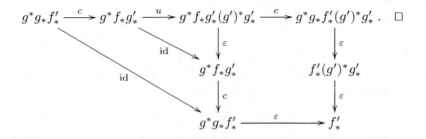

Lemma 1.26. *Let $fg' = gf'$ be a commutative diagram in $\mathcal{S}$. Then the composite*

$$g^* \xrightarrow{u} g^* f_* f^* \xrightarrow{\theta} f'_*(g')^* f^* \xrightarrow{d} f'_*(f')^* g^*$$

is u.

Proof. This is the opposite version of Lemma 1.25. $\square$

Lemma 1.27. *Let $fg' = gf'$ be a commutative diagram in $\mathcal{S}$. Then the composite*

$$(g')^* f^* f_* \xrightarrow{d} (f')^* g^* f_* \xrightarrow{\theta} (f')^* f'_*(g')^* \xrightarrow{\varepsilon} (g')^*$$

is ε.

Proof. This is the opposite assertion of Lemma 1.24. $\square$

(1.28) We say that $(?)_*$ is a covariant symmetric monoidal almost-pseudofunctor on a category $\mathcal{S}$, if $(?)_*$ is an almost-pseudofunctor on $\mathcal{S}$, and the following conditions are satisfied. For each $X \in \mathcal{S}$,

$$X_* = (X_*, \otimes, \mathcal{O}_X, \alpha, \lambda, \gamma, [?, *], \pi)$$

is a (symmetric monoidal) closed category (see e.g., [26, (3.5.1)]), where X_* on the right-hand side is the underlying category, $\otimes : X_* \times X_* \to X_*$ the product structure, $\mathcal{O}_X \in X_*$ the unit object, $\alpha : (a \otimes b) \otimes c \cong a \otimes (b \otimes c)$ the associativity isomorphism, $\lambda : \mathcal{O}_X \otimes a \cong a$ the left unit isomorphism, $\gamma : a \otimes b \cong b \otimes a$ the twisting (symmetry) isomorphism, $[?, -] : X_*^{\mathrm{op}} \times X_* \to X_*$ the internal hom, and

$$\pi : X_*(a \otimes b, c) \cong X_*(a, [b, c]) \tag{1.29}$$

the associative adjunction isomorphism of X, respectively. For a morphism $f : X \to Y$ in $\mathcal{S}$, $f_* : X_* \to Y_*$ is a symmetric monoidal functor [26, (3.4.2)], and $\mathfrak{e}_X : \mathrm{Id}_{X_*} \to (\mathrm{id}_X)_*$ and $c_{f,g}$ are morphisms of symmetric monoidal functors, see [26, (3.6.7)].

(1.30)   The unit map and the counit map arising from the adjunction (1.29) are denoted (by less worse abuse of notation, not using u or ε) by (the same symbol)

$$\mathrm{tr} : a \to [b, a \otimes b] \qquad \text{(the trace map)}$$

and

$$\mathrm{ev} : [b, c] \otimes b \to c \qquad \text{(the evaluation map)},$$

respectively.

(1.31)   By the definition of closed categories, the associative adjunction isomorphism (1.29) is natural on a, b and c. So not only that tr and ev are natural transformations, we have the following.

Lemma 1.32. *For $a, b, b', c \in X_*$ and a morphism $\varphi \colon b \to b'$, the diagrams*

$$
\begin{array}{ccc}
a & \xrightarrow{\;\;\mathrm{tr}\;\;} & [b, a \otimes b] \\
{\scriptstyle \mathrm{tr}}\downarrow & & \downarrow{\scriptstyle [1_b, 1_a \otimes \varphi]} \\
[b', a \otimes b'] & \xrightarrow{[\varphi, 1_a \otimes 1_b]} & [b, a \otimes b']
\end{array}
\qquad
\begin{array}{ccc}
[b', c] \otimes b & \xrightarrow{[\varphi, 1_c] \otimes 1_b} & [b, c] \otimes b \\
{\scriptstyle [1_{b'}, 1_c] \otimes \varphi}\downarrow & & \downarrow{\scriptstyle \mathrm{ev}} \\
[b', c] \otimes b' & \xrightarrow{\;\;\mathrm{ev}\;\;} & c
\end{array}
$$

are commutative.

Proof. We only prove the commutativity of the first diagram. By the naturality of π, the diagram

$$
\begin{array}{ccc}
X_*(a \otimes b, a \otimes b) & \xrightarrow{(1_a \otimes \varphi)_*} X_*(a \otimes b, a \otimes b') \xleftarrow{(1_a \otimes \varphi)^*} & X_*(a \otimes b', a \otimes b') \\
\downarrow{\scriptstyle \pi} & \downarrow{\scriptstyle \pi} & \downarrow{\scriptstyle \pi} \\
X_*(a, [b, a \otimes b]) & \xrightarrow{[1_b, 1_a \otimes \varphi]_*} X_*(a, [b, a \otimes b']) \xleftarrow{[\varphi, 1_a \otimes 1_{b'}]_*} & X_*(a, [b', a \otimes b'])
\end{array}
$$

is commutative. Considering the image of $1_{a \otimes b} \in X_*(a \otimes b, a \otimes b)$ and $1_{a \otimes b'} \in X_*(a \otimes b', a \otimes b')$ in $X_*(a, [b, a \otimes b'])$, we have

$$[1_b, 1_a \otimes \varphi] \circ \mathrm{tr} = \pi(1_a \otimes \varphi) = [\varphi, 1_a \otimes 1_b] \circ \mathrm{tr}.$$

This is what we wanted to prove.   □

(1.33)   Let $f : X \to Y$ be a morphism. Then f_* is a symmetric monoidal functor. The natural map

$$f_* a \otimes f_* b \to f_*(a \otimes b)$$

is denoted by $m = m(f)$, and the map

$$\mathcal{O}_Y \to f_* \mathcal{O}_X$$

is denoted by $\eta = \eta(f)$.

A covariant symmetric monoidal almost-pseudofunctor which is a pseudo-functor is called a covariant symmetric monoidal pseudofunctor. Let $\star$ be the associated pseudofunctor of the symmetric monoidal almost-pseudofunctor $*$. Then letting the closed structure of $X_\star$ be the same as that of X_*, and letting

$$m\colon (\mathrm{id}_X)_\star a \otimes (\mathrm{id}_X)_\star b \to (\mathrm{id}_X)_\star (a \otimes b)$$

and

$$\eta\colon \mathcal{O}_X \to (\mathrm{id}_X)_\star \mathcal{O}_X$$

to be the identity morphisms, $\star$ is a symmetric monoidal pseudofunctor which is isomorphic to $*$ as a symmetric monoidal almost-pseudofunctor.

(1.34) Let $f : X \to Y$ be a morphism in $\mathcal{S}$. The composite natural map

$$f_*[a,b] \xrightarrow{\mathrm{tr}} [f_*a, f_*[a,b] \otimes f_*a] \xrightarrow{\text{via } m} [f_*a, f_*([a,b] \otimes a)] \xrightarrow{\text{via } \mathrm{ev}} [f_*a, f_*b]$$

is denoted by H.

(1.35) G. Lewis proved a theorem which guarantee that some diagrams involving two symmetric monoidal closed categories and one symmetric monoidal functor commute [25].

By Lewis's result, we have that the following diagrams are commutative for any morphism $f : X \to Y$ (also checked by a direct computation).

$$
\begin{array}{ccc}
f_*[a,b] \otimes f_*a & \xrightarrow{H \otimes 1} & [f_*a, f_*b] \otimes f_*a \\
\downarrow{m} & & \downarrow{\mathrm{ev}} \\
f_*([a,b] \otimes a) & \xrightarrow{f_* \mathrm{ev}} & f_*b
\end{array}
\tag{1.36}
$$

$$
\begin{array}{ccc}
f_*a & \xrightarrow{f_* \mathrm{tr}} & f_*[b, a \otimes b] \\
\downarrow{\mathrm{tr}} & & \downarrow{H} \\
[f_*b, f_*a \otimes f_*b] & \xrightarrow{m} & [f_*b, f_*(a \otimes b)]
\end{array}
\tag{1.37}
$$

$$
\begin{array}{ccccccc}
f_*a & \xrightarrow{\lambda^{-1}} & \mathcal{O}_Y \otimes f_*a & \xrightarrow{\eta} & f_*\mathcal{O}_X \otimes f_*a & \xrightarrow{\text{via } \mathrm{tr}} & f_*[a, \mathcal{O}_X \otimes a] \otimes f_*a \\
\downarrow{1} & & & & & & \downarrow{\text{via } \lambda} \\
f_*a & \xleftarrow{\mathrm{ev}} & & [f_*a, f_*a] \otimes f_*a & \xleftarrow{H \otimes 1} & & f_*[a,a] \otimes f_*a
\end{array}
\tag{1.38}
$$

Lemma 1.39. *Let $f : X \to Y$ and $g : Y \to Z$ be morphisms in $\mathcal{S}$. Then the diagram*

$$
\begin{array}{ccccc}
g_*f_*[a,b] & \xrightarrow{H} & g_*[f_*a, f_*b] & \xrightarrow{H} & [g_*f_*a, g_*f_*b] \\
\uparrow{c} & & & & \uparrow{[c^{-1}, c]} \\
(gf)_*[a,b] & & \xrightarrow{\hspace{2cm} H \hspace{2cm}} & & [(gf)_*a, (gf)_*b]
\end{array}
$$

is commutative.

Proof. Consider the diagram

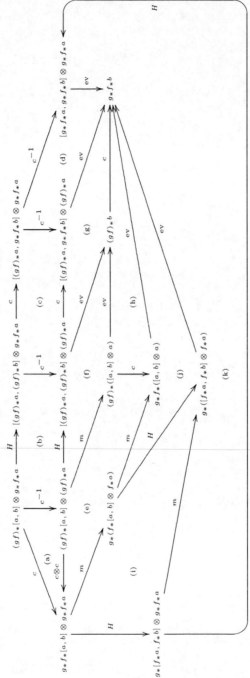

The commutativity of (a), (b), (c), (g), (h) and (i) is trivial. The commutativity of (d) is Lemma 1.32. (e) is commutative, since c is assumed to be a morphism of symmetric monoidal functors, see [26, (3.6.7.2)]. The commutativity of (f), (j), and (k) is the commutativity of (1.36). Thus the whole diagram is commutative. Taking the adjoint, we get the commutativity of the diagram in the lemma. $\square$

(1.40) Let $(?)_*$ be a covariant symmetric monoidal almost-pseudofunctor on $\mathcal{S}$. Let $(?)^*$ be its left adjoint. Namely, for each morphism f of $\mathcal{S}$, a left adjoint f^* of f_* (and the unit map $1 \to f_* f^*$ and the counit map $f^* f_* \to 1$) is given. For a morphism $f : X \to Y$, the map $f^* \mathcal{O}_Y \to \mathcal{O}_X$ adjoint to $\eta : \mathcal{O}_Y \to f_* \mathcal{O}_X$ is denoted by C. The composite map

$$f^*(a \otimes b) \xrightarrow{u \otimes u} f^*(f_* f^* a \otimes f_* f^* b) \xrightarrow{m} f^* f_*(f^* a \otimes f^* b) \xrightarrow{\varepsilon} f^* a \otimes f^* b \quad (1.41)$$

is denoted by Δ.

Almost by definition, the diagrams

$$
\begin{array}{ccc}
a \otimes b & \xrightarrow{\ u \otimes u\ } & f_* f^* a \otimes f_* f^* b \\
\downarrow u & & \downarrow m \\
f_* f^*(a \otimes b) & \xrightarrow{\ \Delta\ } & f_*(f^* a \otimes f^* b)
\end{array}
\qquad (1.42)
$$

and

$$
\begin{array}{ccc}
f^*(f_* a \otimes f_* b) & \xrightarrow{\ m\ } & f^* f_*(a \otimes b) \\
\downarrow \Delta & & \downarrow \varepsilon \\
f^* f_* a \otimes f^* f_* b & \xrightarrow{\ \varepsilon \otimes \varepsilon\ } & a \otimes b
\end{array}
\qquad (1.43)
$$

are commutative.

If $(?)_*$ is the associated pseudofunctor of $(?)_*$ and $(?)^*$ is the associated contravariant pseudofunctor of $(?)^*$, then $((?)_*, (?)^\star)$ is a monoidal adjoint pair. Note that $u \colon \mathrm{Id} \to f_* f^*$ is the identity if $f = \mathrm{id}_X$ for some X, and u agrees with the unit map for the original adjoint pair $((?)_*, (?)^*)$ otherwise. Similarly for the counit map ε.

Lemma 1.44. *Let $((?)^*, (?)_*)$ be a monoidal adjoint pair on $\mathcal{S}$. Let $\sigma = (fg' = gf')$ be a commutative diagram in $\mathcal{S}$. Then the diagram*

$$
\begin{array}{ccccc}
f^*(g_* a \otimes g_* b) & \xrightarrow{\ m\ } & f^* g_*(a \otimes b) & \xrightarrow{\ \theta\ } & g'_*(f')^*(a \otimes b) \\
\downarrow \Delta & & & & \downarrow \Delta \\
f^* g_* a \otimes f^* g_* b & \xrightarrow{\ \theta \otimes \theta\ } & g'_*(f')^* a \otimes g'_*(f')^* b & \xrightarrow{\ m\ } & g'_*((f')^* a \otimes (f')^* b)
\end{array}
$$

is commutative.

Proof. Utilize the commutativity of (1.42) and (1.43). $\square$

(1.45) For a symmetric monoidal category X_*, the composite

$$a \otimes \mathcal{O}_X \xrightarrow{\gamma} \mathcal{O}_X \otimes a \xrightarrow{\lambda} a$$

is called the *right unit isomorphism*, and is denoted by ρ. Let $f_* : X_* \to Y_*$ be a symmetric monoidal functor. Then it is easy to see that the diagram

$$
\begin{array}{ccc}
f_*a \otimes \mathcal{O}_Y & \xrightarrow{\ \ \rho\ \ } & f_*a \\
\downarrow{\scriptstyle 1 \otimes \eta} & & \uparrow{\scriptstyle \rho} \\
f_*a \otimes f_*\mathcal{O}_X & \xrightarrow{\ m\ } & f_*(a \otimes \mathcal{O}_X)
\end{array}
\tag{1.46}
$$

is commutative for $a \in X_*$.

Lemma 1.47. *Let $f_* : X_* \to Y_*$ be a symmetric monoidal functor between closed categories. Then the diagram*

$$
\begin{array}{ccccc}
f_*[\mathcal{O}_X, c \otimes \mathcal{O}_X] & \xrightarrow{\ \rho\ } & f_*[\mathcal{O}_X, c] & \xrightarrow{\ H\ } & [f_*\mathcal{O}_X, f_*c] \\
\uparrow{\scriptstyle \mathrm{tr}} & & & & \downarrow{\scriptstyle \eta} \\
f_*c & \xrightarrow{\ \mathrm{tr}\ } & [\mathcal{O}_Y, f_*c \otimes \mathcal{O}_Y] & \xrightarrow{\ \rho\ } & [\mathcal{O}_Y, f_*c]
\end{array}
$$

is commutative.

Proof. Consider the diagram

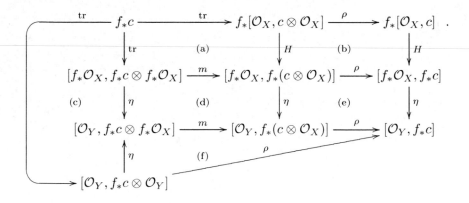

By the commutativity of (1.37), (a) commutes. Clearly, (b), (d), and (e) commute. By Lemma 1.32, (c) commutes. (f) is nothing but (1.46), and commutes. So the whole diagram commutes, and this is what we want to prove. $\qquad\square$

Definition 1.48. A monoidal adjoint pair $((?)^*, (?)_*)$ is said to be *Lipman* if $\Delta : f^*(a \otimes b) \to f^*a \otimes f^*b$ and $C : f^*\mathcal{O}_Y \to \mathcal{O}_X$ are isomorphisms for any morphism $f : X \to Y$ in $\mathcal{S}$ and any $a, b \in Y_*$.

(1.49) Note that $\Delta : f^*(a \otimes b) \to f^*a \otimes f^*b$ is a natural isomorphism if and only if its right conjugate (see (1.10)) is an isomorphism. The right conjugate is the composite

$$f_*[f^*b, c] \xrightarrow{H} [f_*f^*b, f_*c] \xrightarrow{u} [b, f_*c].$$

Let $((?)^*, (?)_*)$ be a Lipman adjoint pair of monoidal almost-pseudofunctors over $\mathcal{S}$. Then $(?)^*$ together with Δ^{-1} and C^{-1} form a *covariant* symmetric monoidal almost-pseudofunctor on $\mathcal{S}^{\mathrm{op}}$.

(1.50) For a morphism $f : X \to Y$ in $\mathcal{S}$, the composite map

$$f^*[a, b] \xrightarrow{\text{via tr}} [f^*a, f^*[a, b] \otimes f^*a] \xrightarrow{\text{via } \Delta^{-1}} [f^*a, f^*([a, b] \otimes a)] \xrightarrow{\text{via ev}} [f^*a, f^*b]$$

is denoted by P. We can apply Lewis's theorem to f^*. In particular, the following diagrams are commutative by (1.35) for a morphism $f : X \to Y$.

$$
\begin{array}{ccc}
f^*[a, b] \otimes f^*a & \xrightarrow{P \otimes 1} & [f^*a, f^*b] \otimes f^*a \\
\downarrow \Delta^{-1} & & \downarrow \text{ev} \\
f^*([a, b] \otimes a) & \xrightarrow{f^* \text{ev}} & f_*b
\end{array}
\qquad (1.51)
$$

$$
\begin{array}{ccc}
f^*a & \xrightarrow{f^* \text{tr}} & f^*[b, a \otimes b] \\
\downarrow \text{tr} & & \downarrow P \\
[f^*b, f^*a \otimes f^*b] & \xrightarrow{\Delta^{-1}} & [f^*b, f^*(a \otimes b)]
\end{array}
\qquad (1.52)
$$

$$
\begin{array}{ccccc}
f^*a & \xrightarrow{\lambda^{-1}} \mathcal{O}_X \otimes f^*a \xrightarrow{C^{-1}} & f^*\mathcal{O}_Y \otimes f^*a & \xrightarrow{\text{via tr}} & f^*[a, \mathcal{O}_Y \otimes a] \otimes f^*a \\
\downarrow 1 & & & & \downarrow \text{via } \lambda \\
f^*a \longleftarrow & \xrightarrow{\text{ev}} [f^*a, f^*a] \otimes f^*a & \xleftarrow{P \otimes 1} & f^*[a, a] \otimes f^*a
\end{array}
\qquad (1.53)
$$

Lemma 1.54. *Let $f : X \to Y$ and $g : Y \to Z$ be morphisms in $\mathcal{S}$. Then the diagram*

$$
\begin{array}{ccccc}
f^*g^*[a, b] & \xrightarrow{P} & f^*[g^*a, g^*b] & \xrightarrow{P} & [f^*g^*a, f^*g^*b] \\
\uparrow d^{-1} & & & & \uparrow [d, d^{-1}] \\
(gf)^*[a, b] & & \xrightarrow{P} & & [(gf)^*a, (gf)^*b]
\end{array}
$$

is commutative.

Proof. Follows instantly by Lemma 1.39.   □

Lemma 1.55. *The diagram*

$$[a,b] \otimes a \xrightarrow{\quad\quad\text{ev}\quad\quad} b$$
$$\downarrow u \otimes u \qquad\qquad\qquad\qquad\qquad\qquad \downarrow u$$
$$f_* f^*[a,b] \otimes f_* f^* a \xrightarrow{HP \otimes 1} [f_* f^* a, f_* f^* b] \otimes f_* f^* a \xrightarrow{\text{ev}} f_* f^* b$$

is commutative.

Proof. Consider the diagram

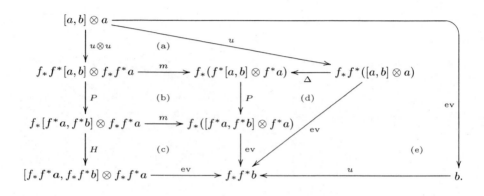

Then (a), (c), and (d) are commutative by the commutativity of (1.42), (1.36), and (1.51), respectively. The commutativity of (b) and (e) is trivial. Thus the whole diagram is commutative, and the lemma follows.   □

Lemma 1.56. *The following diagrams are commutative.*

$$[a,b] \xrightarrow{\;u\;} [a, f_* f^* b]$$
$$\downarrow u \qquad\qquad \uparrow u$$
$$f_* f^*[a,b] \xrightarrow{HP} [f_* f^* a, f_* f^* b]$$

$$f^* f_*[a,b] \xrightarrow{PH} [f^* f_* a, f^* f_* b]$$
$$\downarrow \varepsilon \qquad\qquad\qquad \downarrow \varepsilon$$
$$[a,b] \xrightarrow{\;\varepsilon\;} [f^* f_* a, b]$$

Proof. We prove the commutativity of the first diagram. Consider the diagram

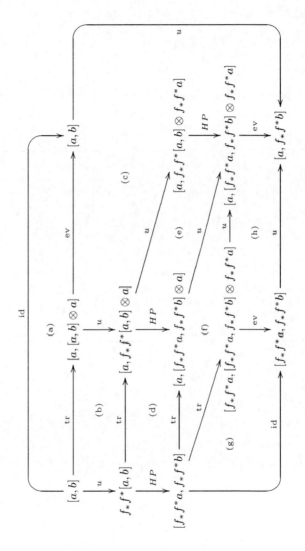

The commutativity of (a), (b), (d), (e), (g), and (h) is trivial. The commutativity of (c) is Lemma 1.55. The commutativity of (f) is Lemma 1.32. Thus the whole diagram is commutative, and the first part of the lemma follows.

The commutativity of the second diagram is proved by a similar diagram drawing. The details are left to the reader.   □

(1.57) Let X be an object of $\mathcal{S}$. We denote the composite isomorphism

$$X_*(a,b) \xrightarrow{\text{via } \lambda} X_*(\mathcal{O}_X \otimes a, b) \xrightarrow{\pi} X_*(\mathcal{O}_X, [a,b])$$

by h_X.

Lemma 1.58. *Let $f : X \to Y$ be a morphism in $\mathcal{S}$. Then the composite map*

$$X_*(a,b) \xrightarrow{h_X} X_*(\mathcal{O}_X, [a,b]) \xrightarrow{C} X_*(f^*\mathcal{O}_Y, [a,b]) \cong Y_*(\mathcal{O}_Y, f_*[a,b])$$

$$\xrightarrow{H} Y_*(\mathcal{O}_Y, [f_*a, f_*b]) \xrightarrow{h_Y^{-1}} Y_*(f_*a, f_*b)$$

agrees with the map given by $\varphi \mapsto f_*\varphi$. The composite map

$$Y_*(a',b') \xrightarrow{h_Y} Y_*(\mathcal{O}_Y, [a',b']) \xrightarrow{f^*} X_*(f^*\mathcal{O}_Y, f^*[a',b'])$$

$$\xrightarrow{X_*(C^{-1},P)} X_*(\mathcal{O}_X, [f^*a', f^*b']) \xrightarrow{h_X^{-1}} X_*(f^*a', f^*b')$$

agrees with f^*.

Proof. We prove the first assertion. The all maps are natural on a. By Yoneda's lemma, we may assume that $a = b$ and it suffices to show that the identity map 1_b is mapped to 1_{f_*b} by the map. It is straightforward to check that 1_b goes to the composite map

$$f_*b \xrightarrow{\lambda^{-1}} \mathcal{O}_Y \otimes f_*b \xrightarrow{u} f_*f^*\mathcal{O}_Y \otimes f_*b \xrightarrow{C} f_*\mathcal{O}_X \otimes f_*b$$

$$\xrightarrow{\text{via tr}} f_*[b, \mathcal{O}_X \otimes b] \otimes f_*b \xrightarrow{\lambda} f_*[b,b] \otimes f_*b \xrightarrow{\text{ev}} f_*b.$$

By the commutativity of (1.38), we are done.

The second assertion is proved similarly, utilizing the commutativity of (1.53).   □

Lemma 1.59. *Let $\sigma = (fg' = gf')$ be a commutative diagram in $\mathcal{S}$. Then the diagram*

$$
\begin{array}{ccc}
g^*f_*[a,b] & \xrightarrow{\;\;\;\;\;\;\theta\;\;\;\;\;\;} & f'_*(g')^*[a,b] \\
\downarrow {\scriptstyle PH} & & \downarrow {\scriptstyle HP} \\
[g^*f_*a, g^*f_*b] \xrightarrow{\theta} & [g^*f_*a, f'_*(g')^*b] & \xleftarrow{\theta} [f'_*(g')^*a, f'_*(g')^*b]
\end{array}
$$

is commutative.

Proof. Follows from Lemma 1.56.   □

Chapter 2
Sheaves on Ringed Sites

(2.1)   We fix a universe [16] $\mathcal{U}$ and a universe $\mathcal{V}$ such that $\mathcal{U} \in \mathcal{V}$ or $\mathcal{U} = \mathcal{V}$. A set is said to be *small* or $\mathcal{U}$-*small* if it is an element of $\mathcal{V}$, and is bijective to an element of $\mathcal{U}$. A category is said to be small if both the object set and the set of morphisms are small. Ringed spaces (including schemes) are required to be small, unless otherwise specified.

The categories of small sets and small abelian groups are respectively denoted by <u>Set</u> and <u>Ab</u>. For example, M is an object of <u>Ab</u> if and only if M is a group whose underlying set is in $\mathcal{V}$, and M is bijective to some set in $\mathcal{U}$.

A category $\mathcal{C}$ is said to be a $\mathcal{U}$-*category* if for any objects x, y of $\mathcal{C}$, the set $\mathcal{C}(x, y)$ is small. Note that <u>Set</u> and <u>Ab</u> are $\mathcal{U}$-categories.

(2.2)   For categories I and $\mathcal{C}$, we denote the functor category $\mathrm{Func}(I^{\mathrm{op}}, \mathcal{C})$ by $\mathcal{P}(I, \mathcal{C})$. An object of $\mathcal{P}(I, \mathcal{C})$ is sometimes referred as a *presheaf* over I with values in $\mathcal{C}$.

If $\mathcal{C}$ is a $\mathcal{U}$-category and I is small, then $\mathcal{P}(I, \mathcal{C})$ is a $\mathcal{U}$-category. For a small category $\mathbb{X}$, we denote $\mathcal{P}(\mathbb{X}, \underline{\mathrm{Ab}})$ by $\mathrm{PA}(\mathbb{X})$.

In these notes, a site (i.e., a category with a Grothendieck topology, in the sense of [42]) is required to be a small category whose topology is defined by a pretopology (see [42]).

Let $\mathcal{C}$ be a category with small products. If $\mathbb{X}$ is a site, then the category of sheaves on $\mathbb{X}$ with values in $\mathcal{C}$ is denoted by $\mathcal{S}(\mathbb{X}, \mathcal{C})$. The inclusion $q : \mathcal{S}(\mathbb{X}, \mathcal{C}) \to \mathcal{P}(\mathbb{X}, \mathcal{C})$ is fully faithful. If $\mathcal{C}$ is a $\mathcal{U}$-category, then $\mathcal{S}(\mathbb{X}, \mathcal{C})$ is a $\mathcal{U}$-category. The category $\mathcal{S}(\mathbb{X}, \underline{\mathrm{Ab}})$ is denoted by $\mathrm{AB}(\mathbb{X})$.

Let $\mathbb{X}$ be a site. The inclusion $\mathrm{AB}(\mathbb{X}) \to \mathrm{PA}(\mathbb{X})$ is denoted by $q(\mathbb{X}, \mathrm{AB})$. We denote the *sheafification* functor $\mathrm{PA}(\mathbb{X}) \to \mathrm{AB}(\mathbb{X})$ by $a(\mathbb{X}, \mathrm{AB})$. Namely, $a = a(\mathbb{X}, \mathrm{AB})$ is the left adjoint of $q(\mathbb{X}, \mathrm{AB})$. Note that a is exact.

We review the construction of the sheafification described in [2, (II.1)].

For $\mathcal{M} \in \mathrm{PA}(\mathbb{X})$, $x \in \mathbb{X}$, and a covering $\mathcal{U} = (x_i \to x)_{i \in I}$ of x, we denote the kernel of the map

$$\prod_{i \in I} \Gamma(x_i, \mathcal{M}) \xrightarrow{\varphi} \prod_{(i,j) \in I \times I} \Gamma(x_i \times_x x_j, \mathcal{M})$$

J. Lipman, M. Hashimoto, *Foundations of Grothendieck Duality for Diagrams*   287
of Schemes, Lecture Notes in Mathematics 1960,
© Springer-Verlag Berlin Heidelberg 2009

by $\check{H}^0(\mathcal{U}, \mathcal{M})$, where $\varphi((m_i)_{i \in I}) = (\mathrm{res}_{x_i \times_x x_j, x_j}\, m_j - \mathrm{res}_{x_i \times_x x_j, x_i}\, m_i)_{(i,j) \in I \times I}$.
Note that $\underline{\check{H}}^0(\mathcal{U}, ?)$ is a functor from PA($\mathbb{X}$) to $\underline{\mathrm{Ab}}$, which is compatible with arbitrary limits.

The set of all coverings of x is a directed set. Let $\mathcal{U} = (\phi_i \colon x_i \to x)_{i \in I}$ and $\mathcal{V} = (\psi_j \colon y_j \to x)_{j \in J}$ be coverings of x. We say that $\mathcal{V}$ is a *refinement* of $\mathcal{U}$ if there are a map $\tau \colon J \to I$ and a collection of morphisms $(\eta_j \colon y_j \to x_{\tau j})_{j \in J}$ such that $\phi_{\tau j} \circ \eta_j = \psi_j$ for $j \in J$. If $(\tau, (\eta_j))$ makes $\mathcal{V}$ a refinement of $\mathcal{U}$, then we define

$$\mathbb{L} = \mathbb{L}(\mathcal{V}, \mathcal{U}; (\tau, (\eta_j)))(\mathcal{M}) \colon \check{H}^0(\mathcal{U}, \mathcal{M}) \to \check{H}^0(\mathcal{V}, \mathcal{M})$$

by $\mathbb{L}((m_i)_{i \in I}) = (\mathrm{res}_{\eta_j}(m_{\tau j}))_{j \in J}$. It is easy to see that $\mathbb{L}$ is independent of the choice of τ or η_j, and depends only on $\mathcal{U}$ and $\mathcal{V}$, see [31, Lemma III.2.1]. If $\mathcal{W}$ is a refinement of $\mathcal{V}$, then $\mathbb{L}(\mathcal{W}, \mathcal{U}) = \mathbb{L}(\mathcal{W}, \mathcal{V}) \circ \mathbb{L}(\mathcal{V}, \mathcal{U})$. Thus we get an inductive system $(\check{H}^0(\mathcal{U}, \mathcal{M}))_{\mathcal{U}}$, where $\mathcal{U}$ runs through the all coverings of x. We denote $\varinjlim \check{H}^0(\mathcal{U}, \mathcal{M})$ by $\check{H}^0(x, \mathcal{M})$. This is a small abelian group, and $\check{H}^0(x, ?)$ is a left exact functor from PA($\mathbb{X}$) to $\underline{\mathrm{Ab}}$.

Let $x' \to x$ be a morphism. Then a covering $\mathcal{U} = (x_i \to x)_{i \in I}$ gives a covering $x' \times_x \mathcal{U} = (x' \times_x x_i \to x')_{i \in I}$ in a natural way. This correspondence induces a map $\check{H}^0(\mathcal{U}, \mathcal{M}) \to \check{H}^0(x' \times_x \mathcal{U}, \mathcal{M})$. So we have a canonical map $\check{H}^0(x, \mathcal{M}) \to \check{H}^0(x', \mathcal{M})$. So we have a presheaf of abelian groups $\underline{\check{H}}^0(\mathcal{M})$ such that $\Gamma(x, \underline{\check{H}}^0(\mathcal{M})) = \check{H}^0(x, \mathcal{M})$. Note that $\underline{\check{H}}^0$ is an endofunctor of PA($\mathbb{X}$).

Note that there is a natural map

$$Y = Y(\mathcal{M}) \colon \mathcal{M} \to \underline{\check{H}}^0(\mathcal{M}).$$

The map Y at the object x

$$Y(x) \colon \Gamma(x, \mathcal{M}) \to \Gamma(x, \underline{\check{H}}^0(\mathcal{M})) = \check{H}^0(x, \mathcal{M})$$

is given by $Y(x)(m) = m \in \check{H}^0(\mathrm{id}_x, \mathcal{M}) \to \check{H}^0(x, \mathcal{M})$ for $m \in \Gamma(x, \mathcal{M})$, where id_x is the covering $(x \to x)$ consisting of the one morphism id_x. $Y = Y(\mathcal{M})$ is an isomorphism if and only if $\mathcal{M}$ is a sheaf.

It is known that $\underline{\check{H}}^0(\underline{\check{H}}^0(\mathcal{M}))$ is a sheaf, and it is the sheafification $a\mathcal{M}$. The composite map

$$u \colon \mathcal{M} \xrightarrow{Y(\mathcal{M})} \underline{\check{H}}^0(\mathcal{M}) \xrightarrow{Y(\underline{\check{H}}^0(\mathcal{M}))} \underline{\check{H}}^0(\underline{\check{H}}^0(\mathcal{M})) = a\mathcal{M} = qa\mathcal{M}$$

is the unit of adjunction. By the naturality of Y, u also agrees with the composite map

$$\mathcal{M} \xrightarrow{Y(\mathcal{M})} \underline{\check{H}}^0(\mathcal{M}) \xrightarrow{\underline{\check{H}}^0(Y(\mathcal{M}))} \underline{\check{H}}^0(\underline{\check{H}}^0(\mathcal{M})).$$

Note that the counit of adjunction $\varepsilon : aq \to \mathrm{Id}$ is given as the unique natural map such that $q\varepsilon : qaq \to q$ is the inverse of uq.

(2.3)   Let $\mathbb{X} = (\mathbb{X}, \mathcal{O}_{\mathbb{X}})$ be a ringed site. Namely, let $\mathbb{X}$ be a site and $\mathcal{O}_{\mathbb{X}}$ a sheaf of commutative rings on $\mathbb{X}$. We denote the category of presheaves (resp. sheaves) of $\mathcal{O}_{\mathbb{X}}$-modules by $\mathrm{PM}(\mathbb{X})$ (resp. $\mathrm{Mod}(\mathbb{X})$). The inclusion $\mathrm{Mod}(\mathbb{X}) \to \mathrm{PM}(\mathbb{X})$ is denoted by $q(\mathbb{X}, \mathrm{Mod})$. The sheafification $\mathrm{PM}(\mathbb{X}) \to \mathrm{Mod}(\mathbb{X})$ is denoted by $a(\mathbb{X}, \mathrm{Mod})$. Note that $a(\mathbb{X}, \mathrm{Mod})$ is constructed in the same way as in (2.2), since $\check{\underline{H}}^0(\mathcal{M})$ is in $\mathrm{PM}(\mathbb{X})$ in a natural way for $\mathcal{M} \in \mathrm{PM}(\mathbb{X})$. Since q is fully faithful, $\varepsilon : aq \to \mathrm{Id}$ is an isomorphism.

The forgetful functor $\mathrm{Mod}(\mathbb{X}) \to \mathrm{AB}(\mathbb{X})$ is denoted by $F(\mathbb{X})$. The forgetful functor $\mathrm{PM}(\mathbb{X}) \to \mathrm{PA}(\mathbb{X})$ is denoted by $F'(\mathbb{X})$. Thus $F'(\mathbb{X}) \circ q(\mathbb{X}, \mathrm{Mod}) = q(\mathbb{X}, \mathrm{AB}) \circ F(\mathbb{X})$ and $a(\mathbb{X}, \mathrm{AB}) \circ F'(\mathbb{X}) = F(\mathbb{X}) \circ a(\mathbb{X}, \mathrm{Mod})$.

We say that a category $\mathcal{A}$ is *Grothendieck* if it is an abelian $\mathcal{U}$-category with a generator which satisfies the (AB5) condition in [13] (the existence of arbitrary small coproducts, and the exactness of small filtered inductive limits), see [37]. The categories $\mathrm{AB}(\mathbb{X})$ and $\mathrm{Mod}(\mathbb{X})$ are Grothendieck. In general, a Grothendieck category satisfies (AB3*), see [37, Corollary 7.10]. A ringed category $(\mathbb{X}, \mathcal{O}_{\mathbb{X}})$ is a pair such that $\mathbb{X}$ is a small category, and $\mathcal{O}_{\mathbb{X}}$ is a presheaf of commutative rings on $\mathbb{X}$. If $\mathbb{X}$ is a ringed category, then $\mathrm{PA}(\mathbb{X})$ and $\mathrm{PM}(\mathbb{X})$ are Grothendieck with (AB4*).

(2.4)   Let $f : \mathbb{Y} \to \mathbb{X}$ be a functor between small categories. Then the pull-back $\mathrm{PA}(\mathbb{X}) \to \mathrm{PA}(\mathbb{Y})$ is denoted by $f_{\mathrm{PA}}^{\#}$. Note that $f_{\mathrm{PA}}^{\#}(\mathcal{F}) := \mathcal{F} \circ f^{\mathrm{op}}$. In general, the pull-back $\mathcal{P}(\mathbb{X}, \mathcal{C}) \to \mathcal{P}(\mathbb{Y}, \mathcal{C})$ is defined in a similar way, and is denoted by $f^{\#}$. If f is a continuous functor (i.e., $f_{\mathrm{Set}}^{\#}$ carries sheaves to sheaves) between sites, then $f_{\mathrm{AB}}^{\#} : \mathrm{AB}(\mathbb{X}) \to \mathrm{AB}(\mathbb{Y})$ is defined to be the restriction of $f_{\mathrm{PA}}^{\#}$. Throughout these notes, we require that a continuous functor $f : \mathbb{Y} \to \mathbb{X}$ between sites satisfies the following condition. For $y \in \mathbb{Y}$, a covering $(y_i \to y)_{i \in I}$, and any $i, j \in I$, the morphisms $f(y_i \times_y y_j) \to f(y_i)$ and $f(y_i \times_y y_j) \to f(y_j)$ make $f(y_i \times_y y_j)$ the fiber product $f(y_i) \times_{f(y)} f(y_j)$. The identity functor is continuous. A composite of continuous functors is again continuous.

Thanks to the re-definition of sites and continuous functors, we have the following.

Lemma 2.5. *Let* $f : \mathbb{Y} \to \mathbb{X}$ *be a functor between sites. Then* f *is continuous if and only if the following holds.*

If $(\varphi_i : y_i \to y)_{i \in I}$ *is a covering, then* $(f\varphi_i : fy_i \to fy)_{i \in I}$ *is a covering, and for any* $i, j \in I$, *the morphisms* $f(y_i \times_y y_j) \to f(y_i)$ *and* $f(y_i \times_y y_j) \to f(y_j)$ *make* $f(y_i \times_y y_j)$ *the fiber product* $f(y_i) \times_{f(y)} f(y_j)$.

For the proof, see [43, (1.6)].

(2.6)   Let $f : \mathbb{Y} \to \mathbb{X}$ be a functor between small categories. The left adjoint of $f_{\mathrm{PA}}^{\#}$, which exists by Kan's lemma (see e.g., [2, Theorem I.2.1]), is denoted by $f_{\#}^{\mathrm{PA}}$.

For $x \in \mathbb{X}$, we define the small category I_x^f as follows. An object of I_x^f is a pair (y, ϕ) with $y \in \mathbb{Y}$ and $\phi \in \mathbb{X}(x, f(y))$. A morphism $h : (y, \phi) \to (y', \phi')$ is a morphism $h \in \mathbb{Y}(y, y')$ such that $f(h) \circ \phi = \phi'$. Note that $\Gamma(x, f_\#^{\mathrm{PA}}(\mathcal{F})) = \varinjlim \Gamma(y, \mathcal{F})$, where the colimit is taken over $(I_x^f)^{\mathrm{op}}$.

The left adjoint of $f_{\mathcal{C}}^{\#} : \mathcal{P}(\mathbb{X}, \mathcal{C}) \to \mathcal{P}(\mathbb{Y}, \mathcal{C})$ is constructed similarly, provided $\mathcal{C}$ has arbitrary small colimits. The left adjoint is denoted by $f_\#^{\mathcal{C}}$ or simply by $f_\#$.

For a continuous functor $f : \mathbb{Y} \to \mathbb{X}$ between sites, the left adjoint $f_\#^{\mathrm{AB}}$ of $f_{\mathrm{AB}}^{\#}$ is given by $f_\#^{\mathrm{AB}} = a(\mathbb{X}, \mathrm{AB}) \circ f_\#^{\mathrm{PA}} \circ q(\mathbb{Y}, \mathrm{AB})$.

Lemma 2.7. *If $(I_x^f)^{\mathrm{op}}$ is pseudofiltered (see e.g., [16, 31]) for each $x \in \mathbb{X}$, then $f_\#^{\mathrm{PA}}$ is exact.*

Proof. This is a consequence of [16, Corollaire 2.10].   □

(2.8) We say that $f : \mathbb{Y} \to \mathbb{X}$ is *admissible* if f is continuous and the functor $f_\#^{\mathrm{PA}}$ is exact.

(2.9) Let $f : \mathbb{Y} \to \mathbb{X}$ be an admissible functor. Then $f_\#^{\mathrm{AB}}$ is exact. Indeed, $f_\#^{\mathrm{AB}}$ is right exact, since it is a left adjoint of $f_{\mathrm{AB}}^{\#}$. On the other hand, being a composite of left exact functors, $f_\#^{\mathrm{AB}} = a f_\#^{\mathrm{PA}} q$ is left exact.

(2.10) If $\mathbb{Y}$ has finite limits and f preserves finite limits, then $f_\#^{\mathrm{PA}}$ is exact by Lemma 2.7. It follows that a continuous map between topological spaces induces an admissible continuous functor between the corresponding sites.

(2.11) The right adjoint functor of $f_{\mathrm{PA}}^{\#}$, which we denote by $f_\flat^{\mathrm{PA}}$ also exists, as $\underline{\mathrm{Ab}}^{\mathrm{op}}$ has arbitrary small colimits (i.e., $\underline{\mathrm{Ab}}$ has small limits). The functor $f_\flat^{\mathrm{PA}}$ is the composite

$$\mathrm{Func}(\mathbb{Y}^{\mathrm{op}}, \underline{\mathrm{Ab}}) \xrightarrow{\mathrm{op}} \mathrm{Func}(\mathbb{Y}, \underline{\mathrm{Ab}}^{\mathrm{op}}) \xrightarrow{(f^{\mathrm{op}})_\#} \mathrm{Func}(\mathbb{X}, \underline{\mathrm{Ab}}^{\mathrm{op}}) \xrightarrow{\mathrm{op}} \mathrm{Func}(\mathbb{X}^{\mathrm{op}}, \underline{\mathrm{Ab}}),$$

where $(f^{\mathrm{op}})_\#$ is the left adjoint of

$$(f^{\mathrm{op}})^{\#} : \mathrm{Func}(\mathbb{X}, \underline{\mathrm{Ab}}^{\mathrm{op}}) \to \mathrm{Func}(\mathbb{Y}, \underline{\mathrm{Ab}}^{\mathrm{op}}),$$

where $f^{\mathrm{op}} = f$ is the opposite of f, namely, f viewed as a functor $\mathbb{Y}^{\mathrm{op}} \to \mathbb{X}^{\mathrm{op}}$.

(2.12) For $\mathcal{M}, \mathcal{N} \in \mathrm{PM}(\mathbb{X})$, the *presheaf tensor product* is denoted by $\otimes_{\mathcal{O}_\mathbb{X}}^p$. It is defined by

$$\Gamma(x, \mathcal{M} \otimes_{\mathcal{O}_\mathbb{X}}^p \mathcal{N}) := \Gamma(x, \mathcal{M}) \otimes_{\Gamma(x, \mathcal{O}_\mathbb{X})} \Gamma(x, \mathcal{N})$$

for $x \in \mathbb{X}$.

The *sheaf tensor product* $a(q\mathcal{M} \otimes_{\mathcal{O}_\mathbb{X}}^p q\mathcal{N})$ of $\mathcal{M}, \mathcal{N} \in \mathrm{Mod}(\mathbb{X})$ is denoted by $\mathcal{M} \otimes_{\mathcal{O}_\mathbb{X}} \mathcal{N}$.

Let $\mathcal{M}, \mathcal{N} \in \mathrm{PM}(\mathbb{X})$, $x \in \mathbb{X}$, and $\mathcal{U} = (x_i \to x)_{i \in I}$ and $\mathcal{V} = (x'_j \to x)_{j \in J}$ be coverings of x. We define a map

$$Z = Z(\mathcal{U}, \mathcal{V}; \mathcal{M}, \mathcal{N}) \colon \check{H}^0(\mathcal{U}, \mathcal{M}) \otimes_{\Gamma(x, \mathcal{O}_{\mathbb{X}})} \check{H}^0(\mathcal{V}, \mathcal{N}) \to \check{H}^0(\mathcal{U} \times \mathcal{V}, \mathcal{M} \otimes^p \mathcal{N})$$

by $Z((m_i)_{i \in I} \otimes (n_j)_{j \in J}) = (m_i \otimes n_j)_{(i,j) \in I \times J}$, where $\mathcal{U} \times \mathcal{V}$ denotes the covering $(x_i \times_x x'_j \to x)_{(i,j) \in I \times J}$ of x. Note that Z induces

$$Z = Z(\mathcal{M}, \mathcal{N}) \colon \underline{\check{H}}^0(\mathcal{M}) \otimes^p \underline{\check{H}}^0(\mathcal{N}) \to \underline{\check{H}}^0(\mathcal{M} \otimes^p \mathcal{N}).$$

Lemma 2.13. *The composite*

$$\mathcal{M} \otimes^p \mathcal{N} \xrightarrow{Y(\mathcal{M}) \otimes Y(\mathcal{N})} \underline{\check{H}}^0 \mathcal{M} \otimes^p \underline{\check{H}}^0 \mathcal{N} \xrightarrow{Z} \underline{\check{H}}^0(\mathcal{M} \otimes^p \mathcal{N})$$

agrees with $Y(\mathcal{M} \otimes^p \mathcal{N})$.

Proof. This is straightforward, and we omit it. □

Lemma 2.14. *The composite*

$$\underline{\check{H}}^0 \mathcal{M} \otimes^p \underline{\check{H}}^0 \mathcal{N} \xrightarrow{Z} \underline{\check{H}}^0(\mathcal{M} \otimes^p \mathcal{N}) \xrightarrow{\underline{\check{H}}^0(Y \otimes Y)} \underline{\check{H}}^0(\underline{\check{H}}^0 \mathcal{M} \otimes^p \underline{\check{H}}^0 \mathcal{N})$$

agrees with $Y = Y(\underline{\check{H}}^0 \mathcal{M} \otimes^p \underline{\check{H}}^0 \mathcal{N})$.

Proof. Note that a section of $\underline{\check{H}}^0(\underline{\check{H}}^0 \mathcal{M} \otimes^p \underline{\check{H}}^0 \mathcal{N})$ at x is represented by data as follows. A covering $\mathcal{V} = (x_i \to x)_{i \in I}$, a collection of coverings $\mathcal{V}_i = (y^i_j \to x_i)_{j \in J_i}$ $(i \in I)$, and a collection of elements $(\sum_l (m^{i,l}_j)_{j \in J_i} \otimes (n^{i,l}_j)_{j \in J_i})_{i \in I}$ subject to the patching conditions, where $m^{i,l}_j \in \Gamma(y^i_j, \mathcal{M})$ and $n^{i,l}_j \in \Gamma(y^i_j, \mathcal{N})$.

Let $\mathcal{U} = (z_l \to x)_{l \in L}$ and $\mathcal{U}' = (z'_{l'} \to x)_{l' \in L'}$ be coverings of x, and $(m_l)_{l \in L}$ and $(n_{l'})_{l' \in L'}$ elements of $\check{H}^0(\mathcal{U}, \mathcal{M})$ and $\check{H}^0(\mathcal{U}', \mathcal{N})$, respectively. Then $Y((m_l) \otimes (n_{l'}))$ is represented by the collection $I = \{\mathrm{id}_x\}$, $\mathcal{V} = (\mathrm{id}_x)$, $J_{\mathrm{id}_x} = L \times L'$, $\mathcal{V}_{\mathrm{id}_x} = (z_l \times_x z'_{l'} \to x)_{(l,l') \in L \times L'}$, and $((\mathrm{res}_{z_l \times_x z'_{l'}, z_l} m_l) \otimes (\mathrm{res}_{z_l \times_x z'_{l'}, z'_{l'}} n_{l'})) \in \check{H}^0(\mathcal{V}, \underline{\check{H}}^0 \mathcal{M} \otimes^p \underline{\check{H}}^0 \mathcal{N})$. As an element of $\check{H}^0(x, \underline{\check{H}}^0 \mathcal{M} \otimes^p \underline{\check{H}}^0 \mathcal{N})$, this element is the same as the element represented by the collection $I = L \times L'$, $\mathcal{V} = \mathcal{U} \times \mathcal{U}'$, $J_{l_1, l'_1} = L \times L'$ for any $(l_1, l'_1) \in L \times L'$, $\mathcal{V}_{l_1, l'_1} = z_{l_1} \times_x \mathcal{U} \times_x z'_{l'_1} \times_x \mathcal{U}'$ for $(l_1, l'_1) \in L \times L'$, and $((\mathrm{res}_{z_{l_1,l,l'_1,l'}, z_l} m_l) \otimes (\mathrm{res}_{z_{l_1,l,l'_1,l'}, z'_{l'}} n_{l'}))_{(l_1, l'_1) \in L \times L'}$, where $z_{l_1, l, l'_1, l'} := z_{l_1} \times_x z_l \times_x z'_{l'_1} \times_x z'_{l'}$. Since $\mathrm{res}_{z_{l_1} \times_x z_l, z_l} m_l = \mathrm{res}_{z_{l_1} \times_x z_l, z_{l_1}} m_{l_1}$ and $\mathrm{res}_{z'_{l'} \times_x z'_{l'_1}, z'_{l'}} n_{l'} = \mathrm{res}_{z'_{l'} \times_x z'_{l'_1}, z'_{l'_1}} n_{l'_1}$, this element agrees with the element represented by the collection $I = L \times L'$, $\mathcal{V} = \mathcal{U} \times \mathcal{U}'$, $J_{l_1, l'_1} = L \times L'$ for $(l_1, l'_1) \in L \times L'$, $\mathcal{V}_{l_1, l'_1} = z_{l_1} \times_x \mathcal{U} \times_x z'_{l'_1} \times_x \mathcal{U}'$ for $(l_1, l'_1) \in L \times L'$, and $((\mathrm{res}_{z_{l_1,l,l'_1,l'}, z_{l_1}} m_{l_1}) \otimes (\mathrm{res}_{z_{l_1,l,l'_1,l'}, z'_{l'_1}} n_{l'_1}))_{(l_1, l'_1) \in L \times L'}$.

It also agrees with the element represented by the collection $I = L \times L'$, $\mathcal{V} = \mathcal{U} \times \mathcal{U}'$, $J_{l_1,l_1'}$ is the singleton $\{\mathrm{id}_{z_{l_1} \times_x z'_{l_1'}}\}$ for $(l_1, l_1') \in L \times L'$, $\mathcal{V}_{l_1,l_1'} = (\mathrm{id}_{z_{l_1} \times_x z'_{l_1'}})$ for $(l_1, l_1') \in L \times L'$, and $((\mathrm{res}_{z_{l_1} \times_x z'_{l_1'}, z_{l_1}} m_{l_1}) \otimes (\mathrm{res}_{z_{l_1} \times_x z'_{l_1'}, z'_{l_1'}} n_{l_1'}))_{(l_1,l_1') \in L \times L'}$, which agrees with the image of $(m_l) \otimes (n_{l'})$ by $\check{\underline{H}}^0(Y \otimes Y) \circ Z$. This shows that $Y = \check{\underline{H}}^0(Y \otimes Y) \circ Z$.   □

(2.15)   We define a natural map

$$m' : qa\mathcal{M} \otimes^p qa\mathcal{N} \to qa(\mathcal{M} \otimes^p \mathcal{N})$$

as the composite

$$qa\mathcal{M} \otimes^p qa\mathcal{N} = \check{\underline{H}}^0 \check{\underline{H}}^0 \mathcal{M} \otimes^p \check{\underline{H}}^0 \check{\underline{H}}^0 \mathcal{N} \xrightarrow{Z} \check{\underline{H}}^0(\check{\underline{H}}^0 \mathcal{M} \otimes^p \check{\underline{H}}^0 \mathcal{N})$$

$$\xrightarrow{\check{\underline{H}}^0 Z} \check{\underline{H}}^0 \check{\underline{H}}^0 (\mathcal{M} \otimes^p \mathcal{N}) = qa(\mathcal{M} \otimes^p \mathcal{N}).$$

Lemma 2.16. *The composite*

$$\mathcal{M} \otimes^p \mathcal{N} \xrightarrow{u \otimes u} qa\mathcal{M} \otimes^p qa\mathcal{N} \xrightarrow{m'} qa(\mathcal{M} \otimes^p \mathcal{N})$$

agrees with the unit map u.

Proof. Consider the diagram

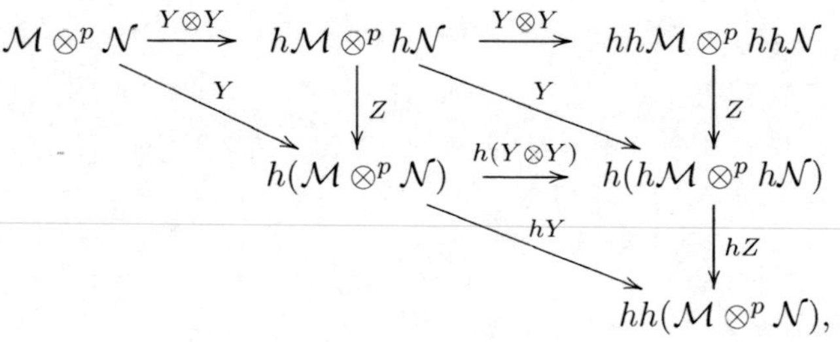

where $h = \check{\underline{H}}^0$. Then the four triangles in the diagram commutes by Lemma 2.13 and Lemma 2.14. So the whole diagram commutes, and the lemma follows.   □

Lemma 2.17. *The composite*

$$qa\mathcal{M} \otimes^p qa\mathcal{N} \xrightarrow{m'} qa(\mathcal{M} \otimes^p \mathcal{N}) \xrightarrow{qa(u \otimes^p u)} qa(qa\mathcal{M} \otimes^p qa\mathcal{N})$$

agrees with u.

Proof. Consider the diagram

$$hh\mathcal{M} \otimes^p hh\mathcal{N} \xrightarrow{\;Z\;} h(h\mathcal{M} \otimes^p h\mathcal{N}) \xrightarrow{\;hZ\;} hh(\mathcal{M} \otimes^p \mathcal{N})$$

with maps Y, $h(Y \otimes^p Y)$, hY, $hh(Y \otimes^p Y)$,

$$h(hh\mathcal{M} \otimes^p hh\mathcal{N}) \xrightarrow{\;hZ\;} hh(h\mathcal{M} \otimes^p h\mathcal{N})$$

hY, $hh(Y \otimes^p Y)$,

$$hh(hh\mathcal{M} \otimes^p hh\mathcal{N}),$$

where $h = \check{\underline{H}}^0$. The four triangles in the diagram are commutative by Lemma 2.13 and Lemma 2.14. So the whole diagram is commutative, and the lemma follows. □

Lemma 2.18. *For* $\mathcal{M}, \mathcal{N} \in \mathrm{PM}(\mathbb{X})$, *the natural map*

$$\bar{\Delta} := a(u \otimes^p u) \colon a(\mathcal{M} \otimes^p \mathcal{N}) \to a(qa\mathcal{M} \otimes^p qa\mathcal{N})$$

is an isomorphism.

Proof. Consider the diagram

$$\mathcal{M} \otimes^p \mathcal{N} \xrightarrow{\;\varphi\;} qa\mathcal{M} \otimes^p qa\mathcal{N}$$

with maps τ, m', β, $q(\bar{\Delta})$,

$$qa(\mathcal{M} \otimes^p \mathcal{N}) \underset{q(\psi)}{\overset{q(\bar{\Delta})}{\rightleftarrows}} qa(qa\mathcal{M} \otimes^p qa\mathcal{N}),$$

where $\varphi = u \otimes^p u$, $\tau = u$, $\beta = u$, $\bar{\Delta} = a(u \otimes^p u)$, and $\psi \colon a(qa\mathcal{M} \otimes^p qa\mathcal{N}) \to a(\mathcal{M} \otimes^p \mathcal{N})$ is the unique map of sheaves such that $m' = q(\psi)\beta$ (this map exists by the universality of the sheafification). By Lemma 2.16, $\tau = m'\varphi$. By Lemma 2.17, $\beta = q(\bar{\Delta})m'$.

So

$$q(\psi\bar{\Delta})\tau = q(\psi)q(\bar{\Delta})m'\varphi = q(\psi)\beta\varphi = m'\varphi = \tau = q(\mathrm{id})\tau.$$

By the universality of the sheafification τ, we have that $\psi\bar{\Delta} = \mathrm{id}$. Moreover,

$$q(\bar{\Delta}\psi)\beta = q(\bar{\Delta})q(\psi)\beta = q(\bar{\Delta})m' = \beta = q(\mathrm{id})\beta.$$

By the universality of the sheafification β, we have that $\bar{\Delta}\psi = \mathrm{id}$. This shows that $\bar{\Delta}$ is an isomorphism. □

(2.19) Let $(\mathbb{Y}, \mathcal{O}_\mathbb{Y})$ and $(\mathbb{X}, \mathcal{O}_\mathbb{X})$ be ringed categories. We say that $f : (\mathbb{Y}, \mathcal{O}_\mathbb{Y}) \to (\mathbb{X}, \mathcal{O}_\mathbb{X})$ is a *ringed functor* if $f : \mathbb{Y} \to \mathbb{X}$ is a functor, and a morphism of presheaves of rings $\eta : \mathcal{O}_\mathbb{Y} \to f^\# \mathcal{O}_\mathbb{X}$ is given.

If, moreover, both $(\mathbb{Y}, \mathcal{O}_\mathbb{Y})$ and $(\mathbb{X}, \mathcal{O}_\mathbb{X})$ are ringed sites and f is continuous, then we call f a *ringed continuous functor*.

The pull-back $\mathrm{PM}(\mathbb{X}) \to \mathrm{PM}(\mathbb{Y})$ is denoted by $f_{\mathrm{PM}}^\#$, and its left adjoint is denoted by $f_\#^{\mathrm{PM}}$. The left adjoint $f_\#^{\mathrm{PM}}$ is defined by

$$\Gamma(x, f_\#^{\mathrm{PM}} \mathcal{M}) := \varinjlim \Gamma(x, \mathcal{O}_\mathbb{X}) \otimes_{\Gamma(y, \mathcal{O}_\mathbb{Y})} \Gamma(y, \mathcal{M})$$

for $x \in \mathbb{X}$ and $\mathcal{M} \in \mathrm{PM}(\mathbb{Y})$, where the colimit is taken over the category $(I_x^f)^{\mathrm{op}}$. Similarly, $f_{\mathrm{Mod}}^\# : \mathrm{Mod}(\mathbb{X}) \to \mathrm{Mod}(\mathbb{Y})$ and its left adjoint $f_\#^{\mathrm{Mod}} = a f_{\mathrm{PM}}^\# q$ is defined. Note that $q f_{\mathrm{Mod}}^\# = f_{\mathrm{PM}}^\# q$ and $q f_{\mathrm{AB}}^\# = f_{\mathrm{PA}}^\# q$. We sometimes denote the identity map $q f_{\mathrm{Mod}}^\# = f_{\mathrm{PM}}^\# q$ and $q f_{\mathrm{AB}}^\# = f_{\mathrm{PA}}^\# q$ and their inverses by $c = c(f)$. If $(I_x^f)^{\mathrm{op}}$ is filtered for any $x \in \mathbb{X}$, then $f_\#^{\mathrm{PA}} \mathcal{O}_\mathbb{Y}$ has a structure of a presheaf of rings in a natural way, and there is a canonical isomorphism $f_\#^{\mathrm{PM}} \mathcal{M} \cong \mathcal{O}_\mathbb{X} \otimes_{f_\#^{\mathrm{PA}} \mathcal{O}_\mathbb{Y}}^p f_\#^{\mathrm{PA}} \mathcal{M}$. The right adjoint of $f_{\mathrm{PM}}^\#$, which exists as in (2.11), is denoted by $f_\flat^{\mathrm{PM}}$.

(2.20) Let $f : (\mathbb{Y}, \mathcal{O}_\mathbb{Y}) \to (\mathbb{X}, \mathcal{O}_\mathbb{X})$ be a ringed continuous functor. For later use, we need the explicit description of the unit $u : \mathrm{Id} \to f_\heartsuit^\# f_\#^\heartsuit$ and the counit $\varepsilon : f_\#^\heartsuit f_\heartsuit^\# \to \mathrm{Id}$, where $\heartsuit$ denotes either PM or Mod. The unit u for the case $\heartsuit = \mathrm{PM}$ is induced by the map

$$\Gamma(y, \mathcal{M}) \to \Gamma(fy, \mathcal{O}_\mathbb{X}) \otimes_{\Gamma(y, \mathcal{O}_\mathbb{Y})} \Gamma(y, \mathcal{M}) \to$$
$$\varinjlim \Gamma(fy, \mathcal{O}_\mathbb{X}) \otimes_{\Gamma(y', \mathcal{O}_\mathbb{Y})} \Gamma(y', \mathcal{M}) = \Gamma(fy, f_\# \mathcal{M}) = \Gamma(y, f^\# f_\# \mathcal{M}),$$

where the first map sends m to $1 \otimes m$, and the second map is the obvious map. The counit ε for the case $\heartsuit = \mathrm{PM}$ is induced by the map

$$\Gamma(x, f_\# f^\# \mathcal{M}) = \varinjlim \Gamma(x, \mathcal{O}_\mathbb{X}) \otimes_{\Gamma(y, \mathcal{O}_\mathbb{Y})} \Gamma(fy, \mathcal{M}) \to \Gamma(x, \mathcal{M}),$$

where the colimit is taken over $(I_x^f)^{\mathrm{op}}$, and the last map is given by $a \otimes m \mapsto a \, \mathrm{res}_{x, fy}(m)$. It is easy to verify that the composite

$$f_\# \xrightarrow{u} f_\# f^\# f_\# \xrightarrow{\varepsilon} f_\#$$

is the identity, and the composite

$$f^\# \xrightarrow{u} f^\# f_\# f^\# \xrightarrow{\varepsilon} f^\#$$

is the identity, and thus certainly $(f_\#, f^\#)$ is an adjoint pair.

The unit $u : \mathrm{Id} \to f_{\mathrm{Mod}}^\# f_\#^{\mathrm{Mod}}$ is the composite

$$\mathrm{Id} \xrightarrow{\varepsilon^{-1}} aq \xrightarrow{u} a f_{\mathrm{PM}}^\# f_\#^{\mathrm{PM}} q \xrightarrow{\theta} f_{\mathrm{Mod}}^\# a f_\#^{\mathrm{PM}} q = f_{\mathrm{Mod}}^\# f_\#^{\mathrm{Mod}},$$

where θ is the composite

$$af^{\#} \xrightarrow{u} af^{\#}qa \xrightarrow{c} aqf^{\#}a \xrightarrow{\varepsilon} f^{\#}a,$$

see Lemma 2.32 below.

The counit $\varepsilon : f_{\#}^{\mathrm{Mod}} f_{\mathrm{Mod}}^{\#} \to \mathrm{Id}$ is the composite

$$f_{\#}^{\mathrm{Mod}} f_{\mathrm{Mod}}^{\#} = af_{\#}^{\mathrm{PM}} q f_{\mathrm{Mod}}^{\#} \xrightarrow{c} af_{\#}^{\mathrm{PM}} f_{\mathrm{PM}}^{\#} q \xrightarrow{\varepsilon} aq \xrightarrow{\varepsilon} \mathrm{Id}.$$

(2.21)   If there is no confusion, the $\heartsuit$ attached to the functors of sheaves defined above are omitted. For example, $f^{\#}$ stands for $f_{\heartsuit}^{\#}$. Note that $f_{\heartsuit}^{\#}(\mathcal{F})$ viewed as a presheaf of abelian groups is independent of $\heartsuit$.

(2.22)   Let $\mathbb{X}$ be a ringed site, and $x \in \mathbb{X}$. The category $\mathbb{X}/x$ is a site with the same topology as that of $\mathbb{X}$. The canonical functor $\mathfrak{R}_x : \mathbb{X}/x \to \mathbb{X}$ is a ringed continuous functor, and yields the pull-backs $(\mathfrak{R}_x)_{\mathrm{AB}}^{\#}$ and $(\mathfrak{R}_x)_{\mathrm{PA}}^{\#}$, which we denote by $(?)|_x^{\mathrm{AB}}$ and $(?)|_x^{\mathrm{PA}}$, respectively. Their left adjoints are denoted by L_x^{AB} and L_x^{PA}, respectively. Note that $\mathfrak{R}_x$ is admissible, see [31, p.78].

(2.23)   Note that $\mathbb{X}/x$ is a ringed site with the structure sheaf $\mathcal{O}_{\mathbb{X}}|_x$. Thus, $(?)|_x^{\mathrm{Mod}}$ and $(?)|_x^{\mathrm{PM}}$ are defined in an obvious way, and their left adjoints L_x^{Mod} and L_x^{PM} are also defined. Note that L_x^{Mod} and L_x^{PM} are faithful and exact.

(2.24)   For a morphism $\phi : x \to y$, we have an obvious admissible ringed continuous functor $\mathfrak{R}_{\phi} : \mathbb{X}/x \to \mathbb{X}/y$. The corresponding pull-back is denoted by $\phi_{\heartsuit}^{\star}$, and its left adjoint is denoted by $\phi_{\star}^{\heartsuit}$, where $\heartsuit$ is AB, PA, Mod or PM.

For $\mathcal{M}, \mathcal{N} \in \heartsuit(\mathbb{X})$, we define $\underline{\mathrm{Hom}}_{\heartsuit(\mathbb{X})}(\mathcal{M}, \mathcal{N})$ to be the object of $\heartsuit(\mathbb{X})$ given by

$$\Gamma(x, \underline{\mathrm{Hom}}_{\heartsuit(\mathbb{X})}(\mathcal{M}, \mathcal{N})) := \mathrm{Hom}_{\heartsuit(\mathbb{X}/x)}(\mathcal{M}|_x^{\heartsuit}, \mathcal{N}|_x^{\heartsuit}),$$

where $\heartsuit = \mathrm{PA}, \mathrm{AB}, \mathrm{PM}$, or Mod. For $\phi : x \to y$, the restriction map

$$\mathrm{Hom}_{\heartsuit(\mathbb{X}/y)}(\mathcal{M}|_y^{\heartsuit}, \mathcal{N}|_y^{\heartsuit}) \to \mathrm{Hom}_{\heartsuit(\mathbb{X}/x)}(\mathcal{M}|_x^{\heartsuit}, \mathcal{N}|_x^{\heartsuit})$$

is given by $\phi_{\heartsuit}^{\star}$. It is easy to see that if $\mathcal{N}$ is a sheaf, then $\underline{\mathrm{Hom}}_{\heartsuit(\mathbb{X})}(\mathcal{M}, \mathcal{N})$ is a sheaf. Note that $\underline{\mathrm{Hom}}_{\heartsuit(\mathbb{X})}(\mathcal{M}, \mathcal{N})$ is a functor from $\heartsuit(\mathbb{X})^{\mathrm{op}} \times \heartsuit(\mathbb{X})$ to $\heartsuit(\mathbb{X})$.

(2.25)   Let $f : \mathbb{Y} \to \mathbb{X}$ be a ringed continuous functor, $y \in \mathbb{Y}$, and $\mathcal{U} = (y_i \to y)_{i \in I}$ a covering of y. Then $f\mathcal{U} = (fy_i \to fy)_{i \in I}$ is a covering of fy, since f is continuous, see Lemma 2.5. Let $\mathcal{M} \in \mathrm{PM}(\mathbb{X})$. Then we have a canonical isomorphism

$$\nu : \check{H}^0(\mathcal{U}, f^{\#}\mathcal{M}) \cong \check{H}^0(f\mathcal{U}, \mathcal{M}),$$

since the canonical map $f(y_i \times_y y_j) \to fy_i \times_{fy} fy_j$ is an isomorphism. This induces a natural map

$$\nu \colon \underline{\check{H}}^0 f^\# \mathcal{M} \to f^\# \underline{\check{H}}^0 \mathcal{M}.$$

Lemma 2.26. *Let f and $\mathcal{M}$ be as above. Then the composite*

$$f^\# \mathcal{M} \xrightarrow{Y} \underline{\check{H}}^0 f^\# \mathcal{M} \xrightarrow{\nu} f^\# \underline{\check{H}}^0 \mathcal{M}$$

is $f^\# Y$.

Proof. This is straightforward, and left to the reader.   □

Lemma 2.27. *Let f and $\mathcal{M}$ be as above. Then the composite*

$$\underline{\check{H}}^0 f^\# \mathcal{M} \xrightarrow{\nu} f^\# \underline{\check{H}}^0 \mathcal{M} \xrightarrow{Y} \underline{\check{H}}^0 f^\# \underline{\check{H}}^0 \mathcal{M}$$

agrees with $\underline{\check{H}}^0 f^\# Y$.

Proof. This is proved quite similarly to Lemma 2.14, and we omit the proof.
                                                                                      □

(2.28)   Let $f \colon \mathbb{Y} \to \mathbb{X}$ be a ringed continuous functor, and $\mathcal{M} \in \mathrm{PM}(\mathbb{X})$. Then we define the natural map $\bar{\theta} \colon af^\# \mathcal{M} \to f^\# a\mathcal{M}$ to be the unique map such that $q\bar{\theta}$ is the composite

$$q\bar{\theta} \colon qaf^\# \mathcal{M} = \underline{\check{H}}^0 \underline{\check{H}}^0 f^\# \mathcal{M} \xrightarrow{\underline{\check{H}}^0 \nu} \underline{\check{H}}^0 f^\# \underline{\check{H}}^0 \mathcal{M}$$
$$\xrightarrow{\nu} f^\# \underline{\check{H}}^0 \underline{\check{H}}^0 \mathcal{M} = f^\# qa\mathcal{M} \xrightarrow{c} qf^\# a\mathcal{M}.$$

Lemma 2.29. *Let f and $\mathcal{M}$ be as above. Then the composite*

$$f^\# \mathcal{M} \xrightarrow{u} qaf^\# \mathcal{M} \xrightarrow{q\bar{\theta}} qf^\# a\mathcal{M} \xrightarrow{c} f^\# qa\mathcal{M}$$

is $f^\# u$.

Proof. Consider the diagram

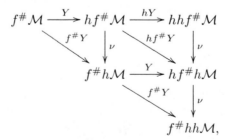

$$f^\# hh\mathcal{M},$$

where $h = \check{\underline{H}}^0$. The four triangles in the diagram commutes by Lemma 2.26 and Lemma 2.27. So the whole diagram commutes, and the lemma follows. $\square$

(2.30)   Let $f : \mathbb{Y} \to \mathbb{X}$ be a ringed continuous functor between ringed sites. The following is a restricted version of the results on cocontinuous functors in [43]. We give a proof for convenience of readers.

Lemma 2.31. *Assume that for any $y \in \mathbb{Y}$ and any covering $(x_\lambda \to fy)_{\lambda \in \Lambda}$ of fy, there is a covering $(y_\mu \to y)_{\mu \in M}$ of y such that there is a map $\phi : M \to \Lambda$ such that $fy_\mu \to fy$ factors through $x_{\phi(\mu)} \to fy$ for each μ. Then the pullback $f^\#$ is compatible with the sheafification in the sense that the canonical natural transformation*

$$\bar{\theta} : a(\mathbb{Y}, \mathrm{Mod}) f_{\mathrm{PM}}^\# \to f_{\mathrm{Mod}}^\# a(\mathbb{X}, \mathrm{Mod})$$

is a natural isomorphism. If this is the case, $f_{\mathrm{Mod}}^\#$ has the right adjoint $f_\flat^{\mathrm{Mod}}$, and in particular, it preserves arbitrary limits and arbitrary colimits.

Proof. Let $\mathcal{M} \in \mathrm{PM}(\mathbb{X})$ and $y \in \mathbb{Y}$. Recall that

$$\nu : \check{H}^0((y_i \to y)_{i \in I}, f^\# \mathcal{M}) \to \check{H}^0((fy_i \to fy)_{i \in I}, \mathcal{M})$$

is an isomorphism, and induces

$$\nu : \check{H}^0(y, f^\# \mathcal{M}) = \varinjlim \check{H}^0((y_i \to y)_{i \in I}, f^\# \mathcal{M}) \xrightarrow{\nu}$$
$$\varinjlim \check{H}^0((fy_i \to fy)_{i \in I}, \mathcal{M}) \to \varinjlim \check{H}^0((x_j \to fy)_{j \in J}, \mathcal{M}) = \check{H}^0(fy, \mathcal{M}).$$

This is also an isomorphism, since coverings of the form $(fy_i \to fy)$ is final in the category of all coverings of fy. By the definition of $\bar{\theta}$, $\bar{\theta}$ is an isomorphism.

Next we show that $f_{\mathrm{Mod}}^\#$ has a right adjoint. To prove this, it suffices to show that $f_\flat^{\mathrm{PM}}(\mathcal{M})$ is a sheaf if so is $\mathcal{M}$ for $\mathcal{M} \in \mathrm{PM}(\mathbb{X})$. Let $u : \mathrm{Id}_{\mathrm{PM}(\mathbb{X})} \to f_\flat^{\mathrm{PM}} f_{\mathrm{PM}}^\#$ be the unit of adjunction, $\varepsilon : f_{\mathrm{PM}}^\# f_\flat^{\mathrm{PM}} \to \mathrm{Id}_{\mathrm{PM}(\mathbb{Y})}$ the counit of adjunction, $v(\mathbb{X}) : \mathrm{Id}_{\mathrm{PM}(\mathbb{X})} \to q(\mathbb{X}, \mathrm{Mod}) a(\mathbb{X}, \mathrm{Mod})$ the unit of adjunction, and $v(\mathbb{Y}) : \mathrm{Id}_{\mathrm{PM}(\mathbb{Y})} \to q(\mathbb{Y}, \mathrm{Mod}) a(\mathbb{Y}, \mathrm{Mod})$ the unit of adjunction. Then the diagram of functors

$$
\begin{array}{ccccccc}
f_\flat^{\mathrm{PM}} & \xrightarrow{u f_\flat^{\mathrm{PM}}} & f_\flat^{\mathrm{PM}} f_{\mathrm{PM}}^\# f_\flat^{\mathrm{PM}} & \xrightarrow{\mathrm{id}} & f_\flat^{\mathrm{PM}} f_{\mathrm{PM}}^\# f_\flat^{\mathrm{PM}} & \xrightarrow{f_\flat^{\mathrm{PM}} \varepsilon} & f_\flat^{\mathrm{PM}} \\
\downarrow{\scriptstyle v(\mathbb{X}) f_\flat^{\mathrm{PM}}} & & \downarrow{\scriptstyle f_\flat^{\mathrm{PM}} f_{\mathrm{PM}}^\# v(\mathbb{X}) f_\flat^{\mathrm{PM}}} & & \downarrow{\scriptstyle f_\flat^{\mathrm{PM}} v(\mathbb{Y}) f_{\mathrm{PM}}^\# f_\flat^{\mathrm{PM}}} & & \downarrow{\scriptstyle f_\flat^{\mathrm{PM}} v(\mathbb{Y})} \\
q a f_\flat^{\mathrm{PM}} & \xrightarrow{u q a f_\flat^{\mathrm{PM}}} & f_\flat^{\mathrm{PM}} f_{\mathrm{PM}}^\# q a f_\flat^{\mathrm{PM}} & \xrightarrow{\cong} & f_\flat^{\mathrm{PM}} q a f_{\mathrm{PM}}^\# f_\flat^{\mathrm{PM}} & \xrightarrow{f_\flat^{\mathrm{PM}} q a \varepsilon} & f_\flat^{\mathrm{PM}} q a
\end{array}
$$

is commutative, where $\cong$ is the *inverse* of the canonical map caused by $q a f_{\mathrm{PM}}^\# \xrightarrow{q\bar{\theta}} q f_{\mathrm{Mod}}^\# a \xrightarrow{c} f_{\mathrm{PM}}^\# q a$, which exists by the first part. As $(f_{\mathrm{PM}}^\#, f_\flat^{\mathrm{PM}})$ is an adjoint pair, the composite of the first row of the diagram is the identity. As $v(\mathbb{Y})(\mathcal{M})$ is an isomorphism, the right-most vertical arrow evaluated at $\mathcal{M}$ is an isomorphism. Hence, $v(\mathbb{X}) f_\flat^{\mathrm{PM}}(\mathcal{M})$, which is the left-most vertical arrow evaluated at $\mathcal{M}$, is a split monomorphism. As it is a direct summand of a sheaf, $f_\flat^{\mathrm{PM}}(\mathcal{M})$ is a sheaf, as desired.

As it is a right adjoint of $f_\#^{\mathrm{Mod}}$, the functor $f_{\mathrm{Mod}}^{\#}$ preserves arbitrary limits. As it is a left adjoint of $f_\flat^{\mathrm{Mod}}$, the functor $f_{\mathrm{Mod}}^{\#}$ preserves arbitrary colimits.
$\square$

Lemma 2.32. *Let* $f \colon \mathbb{Y} \to \mathbb{X}$ *be a ringed continuous functor. Then* $\bar{\theta} \colon a f_{\mathrm{PM}}^{\#} \to f_{\mathrm{Mod}}^{\#} a$ *agrees with the composite*

$$\theta \colon a f^{\#} \xrightarrow{u} a f^{\#} q a \xrightarrow{c} a q f^{\#} a \xrightarrow{\varepsilon} f^{\#} a.$$

Proof. Consider the diagram

$$
\begin{array}{ccccc}
f^{\#} & \xrightarrow{\ u\ } & f^{\#} q a & \xrightarrow{\ c\ } & q f^{\#} a \\
\downarrow u & \text{(a)} & \downarrow u & \text{(b)} & \downarrow u \ \text{(c)} \quad \searrow^{\mathrm{id}} \\
q a f^{\#} & \xrightarrow{\ u\ } & q a f^{\#} q a & \xrightarrow{\ c\ } & q a q f^{\#} a & \xrightarrow{\ \varepsilon\ } & q f^{\#} a \ .
\end{array}
$$

(a) and (b) are commutative by the naturality of u. The commutativity of (c) is basics on adjunction. So the adjoint $q\theta \circ u$ of θ agrees with the composite

$$f^{\#} \xrightarrow{u} f^{\#} q a \xrightarrow{c} q f^{\#} a.$$

By Lemma 2.29, this agrees with the adjoint $q\bar{\theta} \circ u$ of $\bar{\theta}$. Since the adjoint maps agree, we have $\theta = \bar{\theta}$.
$\square$

Lemma 2.33. *Let* $f \colon \mathbb{Y} \to \mathbb{X}$ *be a ringed continuous functor. Then the composite*

$$f_{\mathrm{PM}}^{\#} \xrightarrow{u} q a f_{\mathrm{PM}}^{\#} \xrightarrow{\theta} q f_{\mathrm{Mod}}^{\#} a \xrightarrow{c} f_{\mathrm{PM}}^{\#} q a$$

agrees with u.

Proof. Follows from Lemma 2.29 and Lemma 2.32.
$\square$

Lemma 2.34. *Let* $f \colon \mathbb{Y} \to \mathbb{X}$ *be a ringed continuous functor. Then the conjugate of* $c \colon q f_{\mathrm{Mod}}^{\#} \to f_{\mathrm{PM}}^{\#} q$ *agrees with*

$$a f_{\#}^{\mathrm{PM}} \xrightarrow{u} a f_{\#}^{\mathrm{PM}} q a = f_{\#}^{\mathrm{Mod}} a. \tag{2.35}$$

In particular, being a conjugate of an isomorphism, (2.35) *is an isomorphism.*

Proof. Straightforward.
$\square$

(2.36) Let $\mathbb{X}$ be a ringed site, and $x \in \mathbb{X}$. It is easy to see that $\mathfrak{R}_x \colon \mathbb{X}/x \to \mathbb{X}$ satisfies the condition in Lemma 2.31. So $(?)|_x^{\mathrm{Mod}}$ preserves arbitrary limits and colimits. In particular, $(?)|_x^{\mathrm{Mod}}$ is exact. Similarly, for a morphism $\phi \colon x \to y$ in $\mathbb{X}$, $\phi_{\mathrm{Mod}}^{\star}$ preserves arbitrary limits and colimits.

(2.37) Let $\mathbb{X}$ be a ringed site, $\mathcal{M} \in \mathrm{PM}(\mathbb{X})$, and $\mathcal{N} \in \mathrm{Mod}(\mathbb{X})$. We define an isomorphism

$$\mathbb{V}\colon q\,\underline{\mathrm{Hom}}_{\mathrm{Mod}(\mathbb{X})}(a\mathcal{M},\mathcal{N}) \to \underline{\mathrm{Hom}}_{\mathrm{PM}(\mathbb{X})}(\mathcal{M},q\mathcal{N})$$

as follows. For $x \in \mathbb{X}$, the map $\mathbb{V}$ at x is the composite

$$\mathbb{V}(x)\colon \mathrm{Hom}_{\mathrm{Mod}(\mathbb{X}/x)}((a\mathcal{M})|_x,\mathcal{N}|_x) \xrightarrow{\bar{\theta}} \mathrm{Hom}_{\mathrm{Mod}(\mathbb{X}/x)}(a(\mathcal{M}|_x),\mathcal{N}|_x)$$

$$\cong \mathrm{Hom}_{\mathrm{PM}(\mathbb{X}/x)}(\mathcal{M}|_x,q(\mathcal{N}|_x)) \xrightarrow{c} \mathrm{Hom}_{\mathrm{PM}(\mathbb{X}/x)}(\mathcal{M}|_x,(q\mathcal{N})|_x).$$

The $\cong$ is an isomorphism coming from the adjunction. Note that $\bar{\theta}\colon a(\mathcal{M}|_x) \to (a\mathcal{M})|_x$ is an isomorphism by Lemma 2.31.

A morphism $\varphi\colon (a\mathcal{M})|_x \to \mathcal{N}|_x$ is mapped to the composite

$$\mathcal{M}|_x \xrightarrow{u|_x} (qa\mathcal{M})|_x \xrightarrow{c} q((a\mathcal{M})|_x) \xrightarrow{q\varphi} q(\mathcal{N}|_x) \xrightarrow{c} (q\mathcal{N})|_x.$$

Note that $\mathbb{V}$ is a map of presheaves, that is, $\mathbb{V}$ is compatible with the restriction. Indeed, for $\varphi \in \mathrm{Hom}_{\mathrm{Mod}(\mathbb{X}/x)}((a\mathcal{M})|_x,\mathcal{N}|_x)$ and $\phi\colon y \to x$, $\phi^\star\mathbb{V}\varphi$ is the composite

$$\phi^\star(\mathcal{M}|_x) \xrightarrow{\phi^\star(u|_x)} \phi^\star((qa\mathcal{M})|_x) \xrightarrow{c}$$

$$\phi^\star(q((a\mathcal{M})|_x)) \xrightarrow{\phi^\star(q\varphi)} \phi^\star(q(\mathcal{N}|_x)) \xrightarrow{c} \phi^\star((q\mathcal{N})|_x),$$

which can be identified with the composite

$$\mathcal{M}|_y \xrightarrow{u|_y} (qa\mathcal{M})|_y \xrightarrow{c} q((a\mathcal{M})|_y) \xrightarrow{q(\phi^\star\varphi)} q(\mathcal{N}|_y) \xrightarrow{c} (q\mathcal{N})|_y.$$

This map is $\mathbb{V}\phi^\star\varphi$, and $\mathbb{V}$ is compatible with the restriction maps.

Lemma 2.38. *Let $\mathbb{X}$ be a ringed site, and $\mathcal{M},\mathcal{N} \in \mathrm{Mod}(\mathbb{X})$. Then the composite*

$$\bar{H}\colon q\,\underline{\mathrm{Hom}}_{\mathrm{Mod}(\mathbb{X})}(\mathcal{M},\mathcal{N}) \xrightarrow{\varepsilon} q\,\underline{\mathrm{Hom}}_{\mathrm{Mod}(\mathbb{X})}(aq\mathcal{M},\mathcal{N}) \xrightarrow{\mathbb{V}} \underline{\mathrm{Hom}}_{\mathrm{PM}(\mathbb{X})}(q\mathcal{M},q\mathcal{N})$$

is given as follows. For $x \in \mathbb{X}$ and $\varphi \in \mathrm{Hom}_{\mathrm{Mod}(\mathbb{X}/x)}(\mathcal{M}|_x,\mathcal{N}|_x)$, $\bar{H}(x)(\varphi) \in \mathrm{Hom}_{\mathrm{PM}(\mathbb{X}/x)}((q\mathcal{M})|_x,(q\mathcal{N})|_x)$ is the composite

$$(q\mathcal{M})|_x \xrightarrow{c} q(\mathcal{M}|_x) \xrightarrow{q\varphi} q(\mathcal{N}|_x) \xrightarrow{c} (q\mathcal{N})|_x.$$

Proof. By the definition of $\mathbb{V}$, the map in question is the composite

$$(q\mathcal{M})|_x \xrightarrow{u} (qaq\mathcal{M})|_x \xrightarrow{c} q((aq\mathcal{M})|_x) \xrightarrow{\varepsilon} q(\mathcal{M}|_x) \xrightarrow{q\varphi} q(\mathcal{N}|_x) \xrightarrow{c} (q\mathcal{N})|_x.$$

It agrees with the composite

$$(q\mathcal{M})|_x \xrightarrow{u} (qaq\mathcal{M})|_x \xrightarrow{\varepsilon} (q\mathcal{M})|_x \xrightarrow{c} q(\mathcal{M}|_x) \xrightarrow{q\varphi} q(\mathcal{N}|_x) \xrightarrow{c} (q\mathcal{N})|_x.$$

Since $\varepsilon u = \mathrm{id}$, the assertion follows. $\qquad\qquad\qquad\qquad\qquad\square$

(2.39)   Let $\mathcal{M}, \mathcal{N} \in \mathrm{PM}(\mathbb{X})$. The composite

$$a\,\underline{\mathrm{Hom}}_{\mathrm{PM}(\mathbb{X})}(\mathcal{M}, \mathcal{N}) \xrightarrow{u} a\,\underline{\mathrm{Hom}}_{\mathrm{PM}(\mathbb{X})}(\mathcal{M}, qa\mathcal{N}) \xrightarrow{\mathrm{V}^{-1}}$$

$$aq\,\underline{\mathrm{Hom}}_{\mathrm{Mod}(\mathbb{X})}(a\mathcal{M}, a\mathcal{N}) \xrightarrow{\varepsilon} \underline{\mathrm{Hom}}_{\mathrm{Mod}(\mathbb{X})}(a\mathcal{M}, a\mathcal{N})$$

is denoted by $\bar{P}$.

Lemma 2.40. *Let* $(\mathbb{X}, \mathcal{O}_{\mathbb{X}})$ *be a ringed site. The category* $\mathrm{PM}(\mathbb{X})$ *is a closed symmetric monoidal category (see* [29, (VII.7)]*) with* $\otimes^p_{\mathcal{O}_{\mathbb{X}}}$ *the multiplication,* $q\mathcal{O}_{\mathbb{X}}$ *the unit object,* $\underline{\mathrm{Hom}}_{\mathrm{PM}(\mathbb{X})}(?, ?)$ *the internal hom, etc., etc.*

The proof of the lemma (including the precise statement) is straightforward, but we give some remarks on non-trivial natural maps.

(2.41)   The evaluation map

$$\mathrm{ev}\colon \underline{\mathrm{Hom}}_{\mathrm{PM}(\mathbb{X})}(\mathcal{M}, \mathcal{N}) \otimes^p \mathcal{M} \to \mathcal{N}$$

at the section $\Gamma(x, ?)$,

$$\Gamma(x, \underline{\mathrm{Hom}}_{\mathrm{PM}(\mathbb{X})}(\mathcal{M}, \mathcal{N}) \otimes^p \mathcal{M}) =$$

$$\mathrm{Hom}_{\mathrm{Mod}(\mathbb{X}/x)}(\mathcal{M}|_x, \mathcal{N}|_x) \otimes_{\Gamma(x, \mathcal{O}_{\mathbb{X}})} \Gamma(x, \mathcal{M}) \to \Gamma(x, \mathcal{N}),$$

is given by $\varphi \otimes a \mapsto \varphi(\mathrm{id}_x)(a)$ for $\varphi \in \mathrm{Hom}_{\mathrm{Mod}(\mathbb{X}/x)}(\mathcal{M}|_x, \mathcal{N}|_x)$ and $a \in \Gamma(x, \mathcal{M})$.

(2.42)   The trace map

$$\mathrm{tr}\colon \mathcal{M} \to \underline{\mathrm{Hom}}_{\mathrm{PM}(\mathbb{X})}(\mathcal{N}, \mathcal{M} \otimes^p_{\mathcal{O}_{\mathbb{X}}} \mathcal{N})$$

at the section $\Gamma(x, ?)$,

$$\Gamma(x, \mathcal{M}) \to \mathrm{Hom}_{\mathrm{PM}(\mathbb{X}/x)}(\mathcal{N}|_x, (\mathcal{M} \otimes^p_{\mathcal{O}_{\mathbb{X}}} \mathcal{N})|_x),$$

maps $\alpha \in \Gamma(x, \mathcal{M})$ to the map

$$\mathrm{tr}(x)(\alpha) \in \mathrm{Hom}_{\mathrm{PM}(\mathbb{X}/x)}(\mathcal{N}|_x, (\mathcal{M} \otimes^p_{\mathcal{O}_{\mathbb{X}}} \mathcal{N})|_x)$$

as follows. For $\phi\colon x' \to x$, $\beta \in \Gamma(\phi, \mathcal{N}|_x) = \Gamma(x', \mathcal{N})$ is mapped to

$$\mathrm{res}_{x', x}(\alpha) \otimes \beta \in \Gamma(x', \mathcal{M}) \otimes_{\Gamma(x', \mathcal{O}_{\mathbb{X}})} \Gamma(x', \mathcal{N}) = \Gamma(\phi, (\mathcal{M} \otimes^p_{\mathcal{O}_{\mathbb{X}}} \mathcal{N})|_x)$$

by $\mathrm{tr}(x)(\alpha)$.

Lemma 2.43. *The category* $\mathrm{Mod}(\mathbb{X})$ *is a closed symmetric monoidal category with* $\otimes_{\mathcal{O}_{\mathbb{X}}}$ *the multiplication,* $\mathcal{O}_{\mathbb{X}}$ *the unit object,* $\underline{\mathrm{Hom}}_{\mathrm{Mod}(\mathbb{X})}(?,?)$ *the internal hom, etc., etc.*

The precise statement and the proof is left to the reader. We only remark the following.

(2.44) Let $\mathcal{M}, \mathcal{N}, \mathcal{P} \in \mathrm{Mod}(\mathbb{X})$. Then the associativity morphism

$$\alpha\colon (\mathcal{M} \otimes \mathcal{N}) \otimes \mathcal{P} \to \mathcal{M} \otimes (\mathcal{N} \otimes \mathcal{P})$$

is the composite

$$(\mathcal{M} \otimes \mathcal{N}) \otimes \mathcal{P} = a(qa(q\mathcal{M} \otimes^p q\mathcal{N}) \otimes^p q\mathcal{P}) \xrightarrow{u} a(qa(q\mathcal{M} \otimes^p q\mathcal{N}) \otimes^p qaq\mathcal{P})$$

$$\xrightarrow{\bar{\Delta}^{-1}} a((q\mathcal{M} \otimes^p q\mathcal{N}) \otimes^p q\mathcal{P}) \xrightarrow{a\alpha'} a(q\mathcal{M} \otimes^p (q\mathcal{N} \otimes^p q\mathcal{P}))$$

$$\xrightarrow{u} a(q\mathcal{M} \otimes^p qa(q\mathcal{N} \otimes^p q\mathcal{P})) = \mathcal{M} \otimes (\mathcal{N} \otimes \mathcal{P}),$$

where $\bar{\Delta}^{-1}$ is the inverse of the map $\bar{\Delta}$, see Lemma 2.18, and α' is the associativity morphism for presheaves.

(2.45) The left unit isomorphism $\lambda\colon \mathcal{O}_{\mathbb{X}} \otimes \mathcal{M} \to \mathcal{M}$ is defined to be the composite

$$\mathcal{O}_{\mathbb{X}} \otimes \mathcal{M} = a(q\mathcal{O}_{\mathbb{X}} \otimes^p q\mathcal{M}) \xrightarrow{a\lambda'} aq\mathcal{M} \xrightarrow{\varepsilon} \mathcal{M},$$

where λ' is the left unit isomorphism for presheaves.

(2.46) The twisting isomorphism

$$\gamma\colon \mathcal{M} \otimes \mathcal{N} \to \mathcal{N} \otimes \mathcal{M}$$

is nothing but

$$\mathcal{M} \otimes \mathcal{N} = a(q\mathcal{M} \otimes^p q\mathcal{N}) \xrightarrow{a\gamma'} a(q\mathcal{N} \otimes^p q\mathcal{M}) = \mathcal{N} \otimes \mathcal{M},$$

where γ' is the twisting map for presheaves.

(2.47) The natural map ev is the composite

$$\underline{\mathrm{Hom}}_{\mathrm{Mod}(\mathbb{X})}(\mathcal{M}, \mathcal{N}) \otimes_{\mathcal{O}_{\mathbb{X}}} \mathcal{M} = a(q\,\underline{\mathrm{Hom}}_{\mathrm{Mod}(\mathbb{X})}(\mathcal{M}, \mathcal{N}) \otimes^p q\mathcal{M}) \xrightarrow{\bar{H}}$$

$$a(\underline{\mathrm{Hom}}_{\mathrm{PM}(\mathbb{X})}(q\mathcal{M}, q\mathcal{N}) \otimes^p q\mathcal{M}) \xrightarrow{a\,\mathrm{ev}'} aq\mathcal{M} \xrightarrow{\varepsilon} \mathcal{M},$$

where ev' is the evaluation map for presheaves.

(2.48) The natural map tr is the composite

$$\mathcal{M} \xrightarrow{\varepsilon^{-1}} aq\mathcal{M} \xrightarrow{\text{tr}'} a\,\underline{\text{Hom}}_{\text{PM}(\mathbb{X})}(q\mathcal{N}, q\mathcal{M} \otimes^p q\mathcal{N})$$
$$\xrightarrow{\bar{P}} \underline{\text{Hom}}_{\text{Mod}(\mathbb{X})}(aq\mathcal{N}, a(q\mathcal{M} \otimes^p q\mathcal{N}))$$
$$= \underline{\text{Hom}}_{\text{Mod}(\mathbb{X})}(aq\mathcal{N}, \mathcal{M} \otimes \mathcal{N}) \xrightarrow{\varepsilon^{-1}} \underline{\text{Hom}}_{\text{Mod}(\mathbb{X})}(\mathcal{N}, \mathcal{M} \otimes \mathcal{N}),$$

where tr′ is the trace map for presheaves.

Lemma 2.49. *The inclusion* $q: \text{Mod}(\mathbb{X}) \to \text{PM}(\mathbb{X})$ *and the natural transformations*

$$m\colon q\mathcal{M} \otimes^p_{\mathcal{O}_{\mathbb{X}}} q\mathcal{N} \xrightarrow{u} qa(q\mathcal{M} \otimes^p_{\mathcal{O}_{\mathbb{X}}} q\mathcal{N}) = q(\mathcal{M} \otimes_{\mathcal{O}_{\mathbb{X}}} \mathcal{N})$$

and

$$\eta\colon q\mathcal{O}_{\mathbb{X}} \xrightarrow{\text{id}} q\mathcal{O}_{\mathbb{X}}$$

form a symmetric monoidal functor, see [26, (3.4.2)].

    Letting $\mathcal{S}$ *be the connected category with two objects and one non-trivial morphism,* (a, q) *forms a Lipman monoidal adjoint pair. The map* H *(see for the definition, (1.34)) agrees with* $\bar{H}$ *(see Lemma 2.38). The map* P *(see for the definition, (1.50)) agrees with* $\bar{P}$ *(see (2.39)). The map* Δ *(see (1.40)) agrees with* $\bar{\Delta}$ *(see Lemma 2.18).*

Proof. We prove the first assertion. First we prove that the diagram

$$\begin{array}{ccc}
q\mathcal{O}_{\mathbb{X}} \otimes^p q\mathcal{M} & \xrightarrow{m} & q(\mathcal{O}_{\mathbb{X}} \otimes \mathcal{M}) \\
\uparrow{\scriptstyle \eta \otimes^p 1} & & \downarrow{\scriptstyle q\lambda} \\
q\mathcal{O}_{\mathbb{X}} \otimes^p q\mathcal{M} & \xrightarrow{\lambda'} & q\mathcal{M}
\end{array}$$

is commutative for $\mathcal{M} \in \text{Mod}(\mathbb{X})$. This is trivial from the definition of the sheaf tensor product (2.12), the definition of λ (2.45), and the commutativity of the diagram

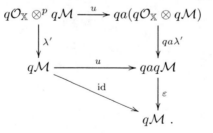

Next we prove that the diagram

$$\begin{array}{ccc}
q\mathcal{M} \otimes^p q\mathcal{N} & \xrightarrow{m} & q(\mathcal{M} \otimes \mathcal{N}) \\
\downarrow{\scriptstyle \gamma'} & & \downarrow{\scriptstyle q\gamma} \\
q\mathcal{N} \otimes^p q\mathcal{M} & \xrightarrow{m} & q(\mathcal{N} \otimes \mathcal{M})
\end{array}$$

is commutative for $\mathcal{M}, \mathcal{N} \in \text{Mod}(\mathbb{X})$. By the definition of m and γ (2.46), the diagram is nothing but

$$
\begin{array}{ccc}
q\mathcal{M} \otimes^p q\mathcal{N} & \xrightarrow{u} & qa(q\mathcal{M} \otimes^p q\mathcal{N}) \\
\downarrow \gamma' & & \downarrow qa\gamma' \\
q\mathcal{N} \otimes^p q\mathcal{M} & \xrightarrow{u} & qa(q\mathcal{N} \otimes^p q\mathcal{M})
\end{array} ,
$$

which is commutative by the naturality of u.

To prove that q is a symmetric monoidal functor, it remains to prove that

$$
\begin{array}{ccc}
(q\mathcal{M} \otimes^p q\mathcal{N}) \otimes^p q\mathcal{P} & \xrightarrow{\alpha'} & q\mathcal{M} \otimes^p (q\mathcal{N} \otimes^p q\mathcal{P}) \\
\downarrow m & & \downarrow m \\
q(\mathcal{M} \otimes \mathcal{N}) \otimes^p q\mathcal{P} & & q\mathcal{M} \otimes^p q(\mathcal{N} \otimes \mathcal{P}) \\
\downarrow m & & \downarrow m \\
q((\mathcal{M} \otimes \mathcal{N}) \otimes \mathcal{P}) & \xrightarrow{q\alpha} & q(\mathcal{M} \otimes (\mathcal{N} \otimes \mathcal{P}))
\end{array}
$$

is commutative, where α' is the associativity map for presheaves. By the definition of m, the diagram equals

$$
\begin{array}{ccc}
(q\mathcal{M} \otimes^p q\mathcal{N}) \otimes^p q\mathcal{P} & \xrightarrow{\alpha'} & q\mathcal{M} \otimes^p (q\mathcal{N} \otimes^p q\mathcal{P}) \\
\downarrow u & & \downarrow u \\
qa(q\mathcal{M} \otimes^p q\mathcal{N}) \otimes^p q\mathcal{P} & & q\mathcal{M} \otimes^p qa(q\mathcal{N} \otimes^p q\mathcal{P}) \\
\downarrow u & & \downarrow u \\
qa(qa(q\mathcal{M} \otimes^p q\mathcal{N}) \otimes^p q\mathcal{P}) & \xrightarrow{q\alpha} & qa(q\mathcal{M} \otimes^p qa(q\mathcal{N} \otimes^p q\mathcal{P}))
\end{array} .
$$

By the naturality of u, the commutativity of this diagram is reduced to the commutativity of

$$
\begin{array}{ccc}
(q\mathcal{M} \otimes^p q\mathcal{N}) \otimes^p q\mathcal{P} & \xrightarrow{\alpha'} & q\mathcal{M} \otimes^p (q\mathcal{N} \otimes^p q\mathcal{P}) \\
\downarrow u & \text{(a)} & \downarrow u \\
qa((q\mathcal{M} \otimes^p q\mathcal{N}) \otimes^p q\mathcal{P}) & \xrightarrow{qa\alpha'} & qa(q\mathcal{M} \otimes^p (q\mathcal{N} \otimes^p q\mathcal{P})) \\
\downarrow u & \text{(b)} & \downarrow u \\
qa(qa(q\mathcal{M} \otimes^p q\mathcal{N}) \otimes^p q\mathcal{P}) & \xrightarrow{q\alpha} & qa(q\mathcal{M} \otimes^p qa(q\mathcal{N} \otimes^p q\mathcal{P}))
\end{array} .
$$

The commutativity of (a) is obvious by the naturality of u. The commutativity of (b) follows from the definition of α (2.44). We have proved that q is a symmetric monoidal functor.

Next we prove that Δ agrees with $\bar{\Delta}$. By definition (1.40), Δ is the composite

$$a(\mathcal{M} \otimes^p \mathcal{N}) \xrightarrow{\bar{\Delta}} a(qa\mathcal{M} \otimes^p qa\mathcal{N}) \xrightarrow{u} aqa(qa\mathcal{M} \otimes^p qa\mathcal{N}) \xrightarrow{\varepsilon} a(qa\mathcal{M} \otimes^p qa\mathcal{N}),$$

which agrees with $\bar{\Delta}$.

Next we prove that $H = \bar{H}$. For $b, c \in \mathrm{Mod}(\mathbb{X})$, the diagram

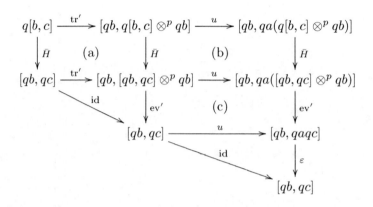

is commutative, where $[d, e]$ stands for $\underline{\mathrm{Hom}}(d, e)$. Indeed, (a) is commutative by the naturality of tr', and (b) and (c) are commutative by the naturality of u. The commutativity of the two triangles are obvious.

By the definition of H (1.34) and ev (2.47), the composite $\varepsilon\,\mathrm{ev}'\,\bar{H}u\,\mathrm{tr}'$ agrees with H. By the commutativity of the diagram, we have that $H = \bar{H}$.

Now we prove that (a, q) forms a Lipman adjoint pair. That is, Δ and C are isomorphisms. Since $\bar{\Delta}$ is an isomorphism by Lemma 2.18 and $\Delta = \bar{\Delta}$, Δ is an isomorphism. Note that $C \colon aq\mathcal{O}_{\mathbb{X}} \to \mathcal{O}_{\mathbb{X}}$ is nothing but ε by definition. Since q is fully faithful, ε is an isomorphism (apply [19, Lemma I.1.2.6, 4] for the adjoint pair $(q^{\mathrm{op}}, a^{\mathrm{op}})$), and we are done. So the definition of P makes sense.

We prove that $P = \bar{P}$. For $b, c \in \mathrm{PM}(\mathbb{X})$, the diagram

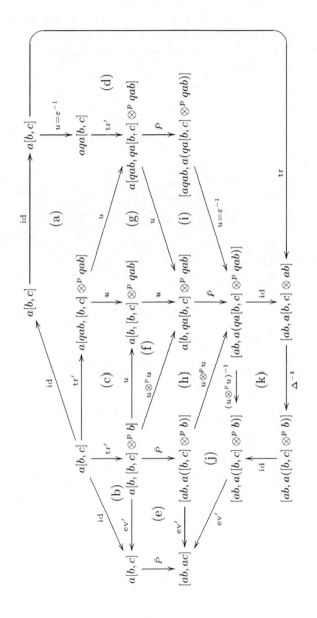

is commutative. Note that $\varepsilon\colon aqa \to a$ has an inverse, which must agree with u. The commutativity of (a) is the naturality of tr'. The commutativity of (b) is the basics on adjunction. The commutativity of (c) is Lemma 1.32. The commutativity of (d) is the definition of tr, see (2.48). The commutativity of (e), (h) and (i) is the naturality of $\bar{P}$. The commutativity of (f), (g) and (j) is trivial. The commutativity of (k) is the definition of $\bar{\Delta}$. Thus the diagram is commutative, and we have $\bar{P} = P$, by the definition of P. $\qquad\square$

(2.50)   Let $f\colon \mathbb{Y} \to \mathbb{X}$ be a ringed continuous functor. For $\mathcal{M}, \mathcal{N} \in \mathrm{PM}(\mathbb{X})$, we define

$$m = m_{\mathrm{PM}}(f)\colon f_{\mathrm{PM}}^{\#}\mathcal{M} \otimes_{\mathcal{O}_\mathbb{Y}}^{p} f_{\mathrm{PM}}^{\#}\mathcal{N} \to f_{\mathrm{PM}}^{\#}(\mathcal{M} \otimes_{\mathcal{O}_\mathbb{X}}^{p} \mathcal{N})$$

by

$$\Gamma(y, f_{\mathrm{PM}}^{\#}\mathcal{M} \otimes_{\mathcal{O}_\mathbb{Y}}^{p} f_{\mathrm{PM}}^{\#}\mathcal{N}) = \Gamma(fy, \mathcal{M}) \otimes_{\Gamma(y, \mathcal{O}_\mathbb{Y})} \Gamma(fy, \mathcal{N})$$

$$\to \Gamma(fy, \mathcal{M}) \otimes_{\Gamma(fy, \mathcal{O}_\mathbb{X})} \Gamma(fy, \mathcal{N}) = \Gamma(y, f_{\mathrm{PM}}^{\#}(\mathcal{M} \otimes_{\mathcal{O}_\mathbb{X}}^{p} \mathcal{N}))$$

$$(m \otimes n \mapsto m \otimes n)$$

for each $y \in \mathbb{Y}$.

We also define $\eta = \eta_{\mathrm{PM}}(f)$ to be the canonical map

$$q\mathcal{O}_\mathbb{Y} \to qf_{\mathrm{Mod}}^{\#}\mathcal{O}_\mathbb{X} \xrightarrow{c} f_{\mathrm{PM}}^{\#}q\mathcal{O}_\mathbb{X}.$$

Lemma 2.51. *The functor $f_{\mathrm{PM}}^{\#}$ together with m and η above forms a symmetric monoidal functor.*

Proof. Consider the diagram

$$
\begin{array}{ccc}
f^{\#}q\mathcal{O}_\mathbb{X} \otimes^{p} f^{\#}\mathcal{M} & \xrightarrow{m} & f^{\#}(q\mathcal{O}_\mathbb{X} \otimes^{p} \mathcal{M}) \\
\uparrow{\scriptstyle \eta \otimes^{p} 1} & & \downarrow{\scriptstyle f^{\#}\lambda} \\
q\mathcal{O}_\mathbb{Y} \otimes^{p} f^{\#}\mathcal{M} & \xrightarrow{\lambda} & f^{\#}\mathcal{M}
\end{array}
,
$$

whose commutativity we need to prove. For $y \in \mathbb{Y}$, applying $\Gamma(y, ?)$ to this diagram, we get

$$
\begin{array}{ccc}
\Gamma(fy, \mathcal{O}_\mathbb{X}) \otimes_{\Gamma(y, \mathcal{O}_\mathbb{Y})} \Gamma(fy, \mathcal{M}) & \to & \Gamma(fy, \mathcal{O}_\mathbb{X}) \otimes_{\Gamma(fy, \mathcal{O}_\mathbb{X})} \Gamma(fy, \mathcal{M}) \\
\uparrow & & \downarrow \\
\Gamma(y, \mathcal{O}_\mathbb{Y}) \otimes_{\Gamma(y, \mathcal{O}_\mathbb{Y})} \Gamma(fy, \mathcal{M}) & \to & \Gamma(fy, \mathcal{M})
\end{array}
.
$$

By the bottom horizontal arrow, $a \otimes m \in \Gamma(y, \mathcal{O}_\mathbb{Y}) \otimes_{\Gamma(y, \mathcal{O}_\mathbb{Y})} \Gamma(fy, \mathcal{M})$ goes to $\eta(a)m$. We get the same result when we keep track the other path in the diagram. So the diagram in question is commutative.

The rest of the proof is similar, and we leave it to the reader. $\qquad\square$

(2.52) Let $f \colon \mathbb{Y} \to \mathbb{X}$ be a ringed continuous functor as above. We define $m = m_{\mathrm{Mod}}(f)$ to be the composite map

$$f^{\#}_{\mathrm{Mod}}\mathcal{M} \otimes_{\mathcal{O}_{\mathbb{Y}}} f^{\#}_{\mathrm{Mod}}\mathcal{N} = a(qf^{\#}_{\mathrm{Mod}}\mathcal{M} \otimes^{p}_{\mathcal{O}_{\mathbb{Y}}} qf^{\#}_{\mathrm{Mod}}\mathcal{N}) \xrightarrow{c} a(f^{\#}_{\mathrm{PM}}q\mathcal{M} \otimes^{p}_{\mathcal{O}_{\mathbb{Y}}} f^{\#}_{\mathrm{PM}}q\mathcal{N})$$

$$\xrightarrow{\text{via } m_{\mathrm{PM}}} af^{\#}_{\mathrm{PM}}(q\mathcal{M} \otimes^{p}_{\mathcal{O}_{\mathbb{X}}} q\mathcal{N}) \xrightarrow{\text{via } \theta} f^{\#}_{\mathrm{Mod}}a(q\mathcal{M} \otimes^{p}_{\mathcal{O}_{\mathbb{X}}} q\mathcal{N}) = f^{\#}_{\mathrm{Mod}}(\mathcal{M} \otimes_{\mathcal{O}_{\mathbb{X}}} \mathcal{N}),$$

where $\theta \colon af^{\#}_{\mathrm{PM}} \to f^{\#}_{\mathrm{Mod}}a$ is the composite map

$$af^{\#}_{\mathrm{PM}} \xrightarrow{\text{via } u} af^{\#}_{\mathrm{PM}}qa \xrightarrow{c} aqf^{\#}_{\mathrm{Mod}}a \xrightarrow{\text{via } \varepsilon} f^{\#}_{\mathrm{Mod}}a.$$

We define $\eta = \eta_{\mathrm{Mod}}(f)$ to be the given map of sheaves of rings $\mathcal{O}_{\mathbb{Y}} \to f^{\#}_{\mathrm{Mod}}\mathcal{O}_{\mathbb{X}}$.

Lemma 2.53. *Let the notation be as above. Then the functor* $f^{\#}_{\mathrm{Mod}}\colon \mathrm{Mod}(\mathbb{X}) \to \mathrm{Mod}(\mathbb{Y})$, *together with m and η, is a symmetric monoidal functor.*

Proof. The diagram

$$
\begin{array}{ccccccc}
af^{\#}q & \xrightarrow{u} & af^{\#}qaq & \xrightarrow{c} & aqf^{\#}aq & \xrightarrow{\varepsilon} & f^{\#}aq \\
& \searrow \text{ id} & \downarrow \varepsilon & & \downarrow \varepsilon & & \downarrow \varepsilon \\
& & af^{\#}q & \xrightarrow{c} & aqf^{\#} & \xrightarrow{\varepsilon} & f^{\#}
\end{array}
\tag{2.54}
$$

is commutative. For $\mathcal{M} \in \mathrm{Mod}(\mathbb{X})$, consider the diagram

$$
\begin{array}{ccccc}
a(f^{\#}q\mathcal{O}_{\mathbb{X}} \otimes^{p} f^{\#}q\mathcal{M}) & \xrightarrow{m_{\mathrm{PM}}} & af^{\#}(q\mathcal{O}_{\mathbb{X}} \otimes^{p} q\mathcal{M}) & \xrightarrow{\theta} & f^{\#}a(q\mathcal{O}_{\mathbb{X}} \otimes^{p} q\mathcal{M}) \\
\uparrow \eta_{\mathrm{PM}} & \text{(a)} & \downarrow \lambda & \text{(b)} & \downarrow \lambda \\
a(q\mathcal{O}_{\mathbb{Y}} \otimes^{p} f^{\#}q\mathcal{M}) & \xrightarrow{\lambda} & af^{\#}q\mathcal{M} & \xrightarrow{\theta} & f^{\#}aq\mathcal{M} \\
\downarrow c & \text{(c)} & \downarrow c & \text{(d)} & \downarrow \varepsilon \\
a(q\mathcal{O}_{\mathbb{Y}} \otimes^{p} qf^{\#}\mathcal{M}) & \xrightarrow{\lambda} & aqf^{\#}\mathcal{M} & \xrightarrow{\varepsilon} & f^{\#}\mathcal{M}
\end{array}
.
$$

(a) is commutative by Lemma 2.51. (b) is commutative by the naturality of θ. (c) is commutative by the naturality of λ. (d) is commutative by the commutativity of (2.54). So the whole diagram is commutative.

This shows that the diagram

$$
\begin{array}{ccc}
f^{\#}_{\mathrm{Mod}}\mathcal{O}_{\mathbb{X}} \otimes f^{\#}_{\mathrm{Mod}}\mathcal{M} & \xrightarrow{m_{\mathrm{Mod}}} & f^{\#}_{\mathrm{Mod}}(\mathcal{O}_{\mathbb{X}} \otimes \mathcal{M}) \\
\uparrow \eta & & \downarrow \lambda \\
\mathcal{O}_{\mathbb{Y}} \otimes f^{\#}_{\mathrm{Mod}}\mathcal{M} & \xrightarrow{\lambda} & f^{\#}_{\mathrm{Mod}}\mathcal{M}
\end{array}
$$

is commutative.

The other axioms are checked similarly. The details are left to the reader. $\square$

Now the following is easy to prove.

Lemma 2.55. *Let $\mathcal{S}'$ denote the category of ringed categories and ringed functors. Then $((?)^{\mathrm{PM}}_{\#}, (?)^{\#}_{\mathrm{PM}})$ is an adjoint pair of monoidal almost-pseudofunctors on $(\mathcal{S}')^{\mathrm{op}}$. Let $\mathcal{S}$ denote the category of ringed sites and ringed continuous functors. Then $((?)^{\mathrm{Mod}}_{\#}, (?)^{\#}_{\mathrm{Mod}})$ is an adjoint pair of monoidal almost-pseudofunctors on $\mathcal{S}^{\mathrm{op}}$, see (1.40).*

(2.56) We give some comments on Lemma 2.55. Let $g : \mathbb{Z} \to \mathbb{Y}$ and $f : \mathbb{Y} \to \mathbb{X}$ be morphisms in $\mathcal{S}$. Then $c : (fg)^{\#}_{\heartsuit} \to g^{\#}_{\heartsuit} f^{\#}_{\heartsuit}$ is the identity map for $\heartsuit = \mathrm{PM}, \mathrm{Mod}$. For $z \in \mathbb{Z}$,

$$\Gamma(z, (fg)^{\#}\mathcal{M}) = \Gamma(fgz, \mathcal{M}) = \Gamma(gz, f^{\#}\mathcal{M}) = \Gamma(z, g^{\#}f^{\#}\mathcal{M}).$$

In particular, the diagram

$$
\begin{array}{ccc}
q(fg)^{\#} & \xrightarrow{\ \ c\ \ } & (fg)^{\#}q \\
\downarrow{c} & & \downarrow{c} \\
qf^{\#}g^{\#} \xrightarrow{c} f^{\#}qg^{\#} & \xrightarrow{c} & f^{\#}g^{\#}q
\end{array}
$$

is commutative.

Note also that $(\mathrm{id}_X)^{\#} : \heartsuit(X) \to \heartsuit(X)$ is the identity functor, and $\mathfrak{e}_X : \mathrm{Id} \to (\mathrm{id}_X)^{\#}$ is the identity.

A straightforward computation shows that $d : f^{\mathrm{PM}}_{\#} g^{\mathrm{PM}}_{\#} \to (fg)^{\mathrm{PM}}_{\#}$ is given by

$$\Gamma(x, f_{\#}g_{\#}\mathcal{M}) = \varinjlim \Gamma(x, \mathcal{O}_{\mathbb{X}}) \otimes_{\Gamma(y,\mathcal{O}_{\mathbb{Y}})} \varinjlim \Gamma(y, \mathcal{O}_{\mathbb{Y}}) \otimes_{\Gamma(z,\mathcal{O}_{\mathbb{Z}})} \Gamma(z, \mathcal{M}) \to$$

$$\varinjlim \Gamma(x, \mathcal{O}_{\mathbb{X}}) \otimes_{\Gamma(z,\mathcal{O}_{\mathbb{Z}})} \Gamma(z, \mathcal{M}) = \Gamma(x, (fg)_{\#}\mathcal{M}),$$

where $\to$ is given by $a \otimes b \otimes m \mapsto ab \otimes m$.

A straightforward diagram chasing (using Lemma 2.33) shows that $d : f^{\mathrm{Mod}}_{\#} g^{\mathrm{Mod}}_{\#} \to (fg)^{\mathrm{Mod}}_{\#}$ agrees with the composite

$$f^{\mathrm{Mod}}_{\#} g^{\mathrm{Mod}}_{\#} = a f^{\mathrm{PM}}_{\#} qa g^{\mathrm{PM}}_{\#} q \xrightarrow{u^{-1}} a f^{\mathrm{PM}}_{\#} g^{\mathrm{PM}}_{\#} q \xrightarrow{d} a(fg)^{\mathrm{PM}}_{\#} q = (fg)_{\#},$$

where u^{-1} is the inverse of the isomorphism $u : af_{\#} \to af_{\#}qa$, see Lemma 2.34.

(2.57) Let

$$
\begin{array}{ccc}
\mathbb{X} & \xrightarrow{g'} & \mathbb{X}' \\
\uparrow{f} & & \uparrow{f'} \\
\mathbb{Y} & \xrightarrow{g} & \mathbb{Y}'
\end{array}
\tag{2.58}
$$

be a commutative diagram of ringed sites and ringed (continuous) functors. Then the natural map $\theta : g_{\#}f^{\#} \to (f')^{\#}g'_{\#}$ is defined, see (1.21). Using the explicit description of the unit map $u : 1 \to (g')^{\#}g'_{\#}$ and the counit map $\varepsilon : g_\# g^\# \to 1$ in (2.20) and $c : f^\#(g')^\# \cong g^\#(f')^\#$ in (2.56), it is straightforward to check the following. For $\mathcal{M} \in \mathrm{PM}(\mathbb{X})$ and $y' \in \mathbb{Y}$,

$$\Gamma(y', \theta) : \Gamma(y', g_\# f^\# \mathcal{M}) = \varinjlim_{y' \to gy} \Gamma(y', \mathcal{O}_{\mathbb{Y}'}) \otimes_{\Gamma(y, \mathcal{O}_{\mathbb{Y}})} \Gamma(fy, \mathcal{M})$$

$$\to \Gamma(y', (f')^\# g'_\# \mathcal{M}) = \varinjlim_{f'y' \to g'x} \Gamma(f'y', \mathcal{O}_{\mathbb{X}'}) \otimes_{\Gamma(x, \mathcal{O}_{\mathbb{X}})} \Gamma(x, \mathcal{M})$$

is induced by the map

$$\Gamma(y', \mathcal{O}_{\mathbb{Y}'}) \otimes_{\Gamma(y, \mathcal{O}_{\mathbb{Y}})} \Gamma(fy, \mathcal{M}) \to \Gamma(f'y', \mathcal{O}_{\mathbb{X}'}) \otimes_{\Gamma(fy, \mathcal{O}_{\mathbb{X}})} \Gamma(fy, \mathcal{M})$$
$$(a \otimes m \mapsto a \otimes m)$$

for $(y' \to gy) \in (I_{y'}^g)^{\mathrm{op}}$.

θ_{Mod} is described by θ_{PM} as follows.

Lemma 2.59. *Let (2.58) be a commutative diagram of ringed sites and ringed continuous functors. Then $\theta_{\mathrm{Mod}} : g_\# f^\# \to (f')^\# g'_\#$ is the composite*

$$g_\# f^\# = ag_\# qf^\# \xrightarrow{c} qg_\# f^\# q \xrightarrow{\theta_{\mathrm{PM}}} a(f')^\# g'_\# q \xrightarrow{\theta} (f')^\# ag'_\# q = (f')^\# g'_\#.$$

Proof. Left to the reader as an exercise (utilize Lemma 2.60). $\qquad\square$

Lemma 2.60. *Let $f : \mathbb{Y} \to \mathbb{X}$ be a ringed continuous functor. Then the diagram of functors $\mathrm{PM}(\mathbb{X}) \to \mathrm{PM}(\mathbb{Y})$*

$$\begin{array}{ccc} f^\# & \xrightarrow{u} & f^\# qa \\ \downarrow u & & \downarrow c(f) \\ qaf^\# & \xrightarrow{\theta} & qf^\# a \end{array}$$

is commutative.

Proof. Left to the reader as an exercise. $\qquad\square$

(2.61) Let $\mathcal{S}'$ and $\mathcal{S}$ be as in Lemma 2.55. Then the monoidal adjoint pairs $((?)_\#^{\mathrm{PM}}, (?)_{\mathrm{PM}}^\#)$ and $((?)_\#^{\mathrm{Mod}}, (?)_{\mathrm{Mod}}^\#)$ are not Lipman, see (6.9).

Lemma 2.62. *Let $\mathcal{A}$ be an abelian category which satisfies the (AB3) condition, I a small category, and $((a_\lambda)_{\lambda \in I}, (\varphi_f)_{f \in \mathrm{Mor}(I)})$ a direct system in $\mathcal{A}$. Assume that I has an initial object λ_0, and φ_f is an isomorphism for any $f \in \mathrm{Mor}(I)$. Then $a_\lambda \to \varinjlim a_\lambda$ is an isomorphism for any λ.*

Proof. It suffices to show that $a_{\lambda_0} \to \varinjlim a_\lambda$ has an inverse. For each λ, consider $\varphi_{f(\lambda)}^{-1} : a_\lambda \to a_{\lambda_0}$, where $f(\lambda)$ is the unique map $\lambda_0 \to \lambda$. Then the collection $(\varphi_{f(\lambda)}^{-1})$ induces a morphism $\varinjlim a_\lambda \to a_{\lambda_0}$. This gives the desired inverse. $\cdot \qquad\square$

Lemma 2.63. *Let* $\mathbb{Y}$ *and* $\mathbb{X}$ *be ringed categories, and* $f : \mathbb{Y} \to \mathbb{X}$ *be a ringed functor. Assume that for each* $x \in \mathbb{X}$, I_x^f *has a terminal object. Then* $C : f_{\#}\mathcal{O}_{\mathbb{Y}} \to \mathcal{O}_{\mathbb{X}}$ *is an isomorphism.*

Proof. Let $x \in \mathbb{X}$. Then $\Gamma(x, f_{\#}\mathcal{O}_{\mathbb{Y}}) = \varinjlim \Gamma(x, \mathcal{O}_{\mathbb{X}}) \otimes_{\Gamma(y,\mathcal{O}_{\mathbb{Y}})} \Gamma(y, \mathcal{O}_{\mathbb{Y}})$, where the colimit is taken over $(I_x^f)^{\mathrm{op}}$, which has an initial object $(y_0, (h : x \to fy_0))$ by assumption. By Lemma 2.62, the canonical maps

$$\Gamma(x, \mathcal{O}_{\mathbb{X}}) \cong \Gamma(x, \mathcal{O}_{\mathbb{X}}) \otimes_{\Gamma(y_0,\mathcal{O}_{\mathbb{Y}})} \Gamma(y_0, \mathcal{O}_{\mathbb{Y}}) \to \Gamma(x, f_{\#}\mathcal{O}_{\mathbb{Y}})$$

are isomorphisms. So C is an isomorphism, as can be seen easily. □

Corollary 2.64. *Let* $f : \mathbb{Y} \to \mathbb{X}$ *be a morphism of ringed sites. If* I_x^f *has a terminal object for* $x \in \mathbb{X}$, *then* $C : f_{\#}^{\mathrm{Mod}}\mathcal{O}_{\mathbb{Y}} \to \mathcal{O}_{\mathbb{X}}$ *is an isomorphism.*

Proof. Note that C is the composite

$$f_{\#}^{\mathrm{Mod}}\mathcal{O}_{\mathbb{Y}} = af_{\#}q\mathcal{O}_{\mathbb{Y}} \xrightarrow{C'} aq\mathcal{O}_{\mathbb{X}} \xrightarrow{\varepsilon} \mathcal{O}_{\mathbb{X}},$$

where $C' : f_{\#}q\mathcal{O}_{\mathbb{Y}} \to q\mathcal{O}_{\mathbb{X}}$ is the C associated with the ringed functor $(\mathbb{Y}, q\mathcal{O}_{\mathbb{Y}}) \to (\mathbb{X}, q\mathcal{O}_{\mathbb{X}})$. Note that C' is an isomorphism by the lemma. As $\varepsilon : aq \to 1$ is also an isomorphism, C is also an isomorphism, as desired. □

Corollary 2.65. *Let* $f : X \to Y$ *be a morphism of ringed spaces. Then* $C : f^*\mathcal{O}_Y \to \mathcal{O}_X$ *is an isomorphism.*

Proof. For each open subset U of X, $I_U^{f^{-1}}$ has the terminal object $(Y, (U \hookrightarrow X))$. □

Corollary 2.66. *Let* $f : \mathbb{Y} \to \mathbb{X}$ *be a morphism of ringed sites, and* $y \in \mathbb{Y}$. *Then we may consider the induced morphism of ringed sites* $f/y : \mathbb{Y}/y \to \mathbb{X}/fy$. *The canonical map* $C : (f/y)_{\#}(\mathcal{O}_{\mathbb{Y}}|_y) \to \mathcal{O}_{\mathbb{X}}|_x$ *is an isomorphism.*

Proof. For $\varphi : x \to fy$ in $\mathbb{X}/fy$, $I_\varphi^{f/y}$ has a terminal object (id_y, φ). □

Chapter 3
Derived Categories and Derived Functors of Sheaves on Ringed Sites

We utilize the notation and terminology on triangulated categories in [44]. However, we usually write the suspension (translation) functor of a triangulated category by Σ or $(?)[1]$.

Let $\mathcal{T}$ be a triangulated category.

Lemma 3.1. *Let*

$$(a_\lambda \xrightarrow{f_\lambda} b_\lambda \xrightarrow{g_\lambda} c_\lambda \xrightarrow{h_\lambda} \Sigma a_\lambda)$$

be a small family of distinguished triangles in $\mathcal{T}$. Assume that the coproducts $\bigoplus a_\lambda$, $\bigoplus b_\lambda$, and $\bigoplus c_\lambda$ exist. Then the triangle

$$\bigoplus a_\lambda \xrightarrow{\oplus f_\lambda} \bigoplus b_\lambda \xrightarrow{\oplus g_\lambda} \bigoplus c_\lambda \xrightarrow{H \circ \oplus h_\lambda} \Sigma \left(\bigoplus a_\lambda \right)$$

is distinguished, where $H : \bigoplus \Sigma a_\lambda \to \Sigma(\bigoplus a_\lambda)$ is the canonical isomorphism. Similarly, a product of distinguished triangles is a distinguished triangle.

We refer the reader to [36, Proposition 1.2.1] for the proof of the second assertion. The proof for the first assertion is similar [36, Remark 1.2.2].

(3.2) Let $\mathcal{A}$ be an abelian category. The category of unbounded (resp. bounded below, bounded above, bounded) complexes in $\mathcal{A}$ is denoted by $C(\mathcal{A})$ (resp. $C^+(\mathcal{A})$, $C^-(\mathcal{A})$, $C^b(\mathcal{A})$). The corresponding homotopy category and the derived category are denoted by $K^?(\mathcal{A})$ and $D^?(\mathcal{A})$, where ? is either $\emptyset$ (i.e., nothing), $+$, $-$ or b. The localization $K^?(\mathcal{A}) \to D^?(\mathcal{A})$ is denoted by Q. We denote the homotopy category of complexes in $\mathcal{A}$ with unbounded (resp. bounded below, bounded above, bounded) cohomology groups by $\bar{K}^?(\mathcal{A})$. The corresponding derived category is denoted by $\bar{D}^?(\mathcal{A})$.

For a plump subcategory $\mathcal{A}'$ of $\mathcal{A}$, we denote by $K^?_{\mathcal{A}'}(\mathcal{A})$ (resp. $\bar{K}^?_{\mathcal{A}'}(\mathcal{A})$) the full subcategory of $K^?(\mathcal{A})$ (resp. $\bar{K}^?(\mathcal{A})$) consisting of complexes whose cohomology groups are objects of $\mathcal{A}'$. The localization of $K^?_{\mathcal{A}'}(\mathcal{A})$ by the épaisse subcategory (see for the definition, [44, Chapitre 1, §2, (1.1)]) of exact complexes is denoted by $D^?_{\mathcal{A}'}(\mathcal{A})$. The category $\bar{D}^?_{\mathcal{A}'}(\mathcal{A})$ is defined similarly. Note that the canonical functor $D^?_{\mathcal{A}'}(\mathcal{A}) \to D(\mathcal{A})$ is fully faithful,

J. Lipman, M. Hashimoto, *Foundations of Grothendieck Duality for Diagrams of Schemes*, Lecture Notes in Mathematics 1960,
© Springer-Verlag Berlin Heidelberg 2009

and hence $D^?_{\mathcal{A}'}(\mathcal{A})$ is identified with the full subcategory of $D(\mathcal{A})$ consisting of unbounded (resp. bounded below, bounded above, bounded) complexes whose cohomology groups are in $\mathcal{A}'$. Note also that the canonical functor $D^?_{\mathcal{A}'}(\mathcal{A}) \to \bar{D}^?_{\mathcal{A}'}(\mathcal{A})$ is an equivalence.

(3.3)   Let $\mathcal{A}$ and $\mathcal{B}$ be abelian categories, and $F : K^?(\mathcal{A}) \to K^*(\mathcal{B})$ a triangulated functor. Let $\mathcal{C}$ be a triangulated subcategory of $K^?(\mathcal{A})$ such that

1 If $c \in \mathcal{C}$ is exact, then Fc is exact.
2 For any $a \in K^?(\mathcal{A})$, there exists some quasi-isomorphism $a \to c$.

The condition **2** implies that the canonical functor

$$i(\mathcal{C}) : \mathcal{C}/(\mathcal{E} \cap \mathcal{C}) \to K^?(\mathcal{A})/\mathcal{E} = D^?(\mathcal{A})$$

is an equivalence, where $\mathcal{E}$ denotes the épaisse subcategory of exact complexes in $K^?(\mathcal{A})$, see [44, Chapitre 2, §1, (2.3)]. We fix a quasi-inverse $p(\mathcal{C}) : D^?(\mathcal{A}) \to \mathcal{C}/(\mathcal{E} \cap \mathcal{C})$. On the other hand, the composite

$$\mathcal{C} \hookrightarrow K^?(\mathcal{A}) \xrightarrow{F} K^*(\mathcal{B}) \xrightarrow{Q} D^*(\mathcal{B}) \tag{3.4}$$

is factorized as

$$\mathcal{C} \xrightarrow{Q_{\mathcal{C}}} \mathcal{C}/(\mathcal{E} \cap \mathcal{C}) \xrightarrow{\mathfrak{F}} D^*(\mathcal{B}) \tag{3.5}$$

up to a unique natural isomorphism, by the universality of localization and the condition **1** above. Under the setting above, we have the following [17, (I.5.1)].

Lemma 3.6. *The composite functor*

$$RF : D^?(\mathcal{A}) \xrightarrow{p(\mathcal{C})} \mathcal{C}/(\mathcal{E} \cap \mathcal{C}) \xrightarrow{\mathfrak{F}} D^*(\mathcal{B})$$

is a right derived functor of F.

For more about the existence of a derived functor, see [26, (2.2)]. We denote the map $QF \to (RF)Q$ in the definition of RF (see [17, p.51]) by Ξ or $\Xi(F)$.

(3.7)   Here we are going to review Spaltenstein's work on unbounded derived categories [39].

A chain complex I of $\mathcal{A}$ is called *K-injective* if for any exact sequence E of $\mathcal{A}$, the complex of abelian groups $\mathrm{Hom}^\bullet_{\mathcal{A}}(E, I)$ is also exact.

A morphism $f : C \to I$ in $K(\mathcal{A})$ is called a *K-injective resolution* of C, if I is K-injective and f is a quasi-isomorphism.

The following is pointed out in [9].

Lemma 3.8. *Let $\mathcal{A}$ be an abelian category, and $\mathbb{I} \in C(\mathcal{A})$. Then the following are equivalent.*

1 $\mathbb{I}$ *is K-injective, and $\mathbb{I}^n$ is an injective object of $\mathcal{A}$ for each $n \in \mathbb{Z}$.*

2 *For an exact sequence $0 \to \mathbb{A} \to \mathbb{B} \to \mathbb{C} \to 0$ in $C(\mathcal{A})$ with $\mathbb{C}$ exact, any chain map $\mathbb{A} \to \mathbb{I}$ lifts to $\mathbb{B}$.*

Proof. **1$\Rightarrow$2.** The sequence of complexes of abelian groups

$$0 \to \operatorname{Hom}^{\bullet}_{\mathcal{A}}(\mathbb{C}, \mathbb{I}) \to \operatorname{Hom}^{\bullet}_{\mathcal{A}}(\mathbb{B}, \mathbb{I}) \to \operatorname{Hom}^{\bullet}_{\mathcal{A}}(\mathbb{A}, \mathbb{I}) \to 0$$

is exact, since each term of $\mathbb{I}$ is injective. So

$$H^0(\operatorname{Hom}^{\bullet}_{\mathcal{A}}(\mathbb{B}, \mathbb{I})) \to H^0(\operatorname{Hom}^{\bullet}_{\mathcal{A}}(\mathbb{A}, \mathbb{I})) \to H^1(\operatorname{Hom}^{\bullet}_{\mathcal{A}}(\mathbb{C}, \mathbb{I}))$$

is exact. But $H^1(\operatorname{Hom}^{\bullet}_{\mathcal{A}}(\mathbb{C}, \mathbb{I})) = 0$, since $\mathbb{C}$ is exact and $\mathbb{I}$ is K-injective. So $H^0(\operatorname{Hom}^{\bullet}_{\mathcal{A}}(\mathbb{B}, \mathbb{I})) \to H^0(\operatorname{Hom}^{\bullet}_{\mathcal{A}}(\mathbb{A}, \mathbb{I}))$ is surjective. Since $\operatorname{Hom}^{-1}_{\mathcal{A}}(\mathbb{B}, \mathbb{I}) \to \operatorname{Hom}^{-1}_{\mathcal{A}}(\mathbb{A}, \mathbb{I})$ is surjective, the commutative diagram with exact rows

$$
\begin{array}{ccccccc}
\operatorname{Hom}^{-1}_{\mathcal{A}}(\mathbb{B}, \mathbb{I}) & \xrightarrow{\partial} & Z^0(\operatorname{Hom}^{\bullet}_{\mathcal{A}}(\mathbb{B}, \mathbb{I})) & \longrightarrow & H^0(\operatorname{Hom}^{\bullet}_{\mathcal{A}}(\mathbb{B}, \mathbb{I})) & \longrightarrow & 0 \\
\downarrow & & \downarrow & & \downarrow & & \\
\operatorname{Hom}^{-1}_{\mathcal{A}}(\mathbb{A}, \mathbb{I}) & \xrightarrow{\partial} & Z^0(\operatorname{Hom}^{\bullet}_{\mathcal{A}}(\mathbb{A}, \mathbb{I})) & \longrightarrow & H^0(\operatorname{Hom}^{\bullet}_{\mathcal{A}}(\mathbb{A}, \mathbb{I})) & \longrightarrow & 0
\end{array}
$$

shows that $Z^0(\operatorname{Hom}^{\bullet}_{\mathcal{A}}(\mathbb{B}, \mathbb{I})) \to Z^0(\operatorname{Hom}^{\bullet}_{\mathcal{A}}(\mathbb{A}, \mathbb{I}))$ is surjective. This is what we wanted to prove.

2$\Rightarrow$1 First we prove that $\mathbb{I}$ is K-injective. It suffices to show that for any exact complex $\mathbb{F}$, any chain map $\varphi : \mathbb{F} \to \mathbb{I}$ is null-homotopic. Let $\mathbb{C} = \operatorname{Cone}(\varphi)$, where Cone denotes the mapping cone. Consider the exact sequence

$$0 \to \mathbb{I} \to \mathbb{C} \to \mathbb{F}[1] \to 0.$$

So the identity map $\mathbb{I} \to \mathbb{I}$ lifts to $\psi : \mathbb{C} \to \mathbb{I}$. Let s be the restriction of ψ to $\mathbb{F}[1] \subset \mathbb{C}$. It is easy to see that $\varphi = sd + ds$. So φ is null-homotopic, as desired.

Next we show that $\mathbb{I}^n$ is injective for any n. To prove this, let $f : A \to B$ be a monomorphism in $\mathcal{A}$, and $\varphi : A \to \mathbb{I}^n$ a morphism. Let C be the cokernel of f. Define a complex $\mathbb{A}$ by $\mathbb{A}^n = \mathbb{A}^{n+1} = A$, $d^n_{\mathbb{A}} = \operatorname{id}$, and $\mathbb{A}^i = 0$ ($i \neq n, n+1$). Replacing A by B and C, we define the complexes $\mathbb{B}$ and $\mathbb{C}$, respectively. Define $f^{\bullet} : \mathbb{A} \to \mathbb{B}$ by $f^n = f^{n+1} = f$ and $f^i = 0$ for $i \neq n, n+1$. Obviously, $\operatorname{Coker} f^{\bullet} \cong \mathbb{C}$ is exact.

Define a chain map $\Phi : \mathbb{A} \to \mathbb{I}$ by $\Phi^n = \varphi$ and $\Phi^{n+1} = d^n_{\mathbb{I}} \circ \varphi$. By assumption, there is a chain map $\Psi : \mathbb{B} \to \mathbb{I}$ such that $\Phi = \Psi f^{\bullet}$. So $\varphi = \Phi^n = \Psi^n \circ f^n = \Psi^n \circ f$, and Ψ^n lifts φ. $\square$

For $\mathbb{I} \in C(\mathcal{A})$, we say that $\mathbb{I}$ is *strictly injective* if $\mathbb{I}$ satisfies the equivalent conditions in the lemma. A *strictly injective resolution* is a quasi-isomorphism $\mathbb{F} \to \mathbb{I}$ with $\mathbb{I}$ strictly injective. The following is proved in [9]. See also [39] and [1].

Lemma 3.9. *If $\mathcal{A}$ is Grothendieck, then for any chain complex $\mathbb{F} \in C(\mathcal{A})$ admits a strictly injective resolution $\mathbb{F} \to \mathbb{I}$ which is a monomorphism.*

A chain complex I is K-injective if and only if $K(\mathcal{A})(E, I) = 0$ for any exact sequence E. It is easy to see that the K-injective complexes form an épaisse subcategory $I(\mathcal{A})$ of $K(\mathcal{A})$.

(3.10)   Let $F : K(\mathcal{A}) \to K(\mathcal{B})$ be a triangulated functor, and assume that $\mathcal{A}$ is Grothendieck. Let $\mathcal{I}$ be the full subcategory of K-injective complexes of $K(\mathcal{A})$. It is easy to see that $\mathcal{I}$ is triangulated, and $\mathcal{I} \cap \mathcal{E} = 0$. By Lemma 3.6, the composite

$$D(\mathcal{A}) \xrightarrow{p(\mathcal{I})} \mathcal{I} \xrightarrow{\tilde{F}} D(\mathcal{B})$$

is a right derived functor RF of F. Note that to fix $p(\mathcal{I})$ and the isomorphism $\mathrm{Id}_{D(\mathcal{A})} \to i(\mathcal{I})p(\mathcal{I})$ is nothing but to fix a functorial K-injective resolution $\mathbb{F} \to pQ\mathbb{F} = \mathbb{I}_{\mathbb{F}}$ in $K(\mathcal{A})$.

(3.11)   Let $F : K(\mathcal{A}) \to K(\mathcal{B})$ be a triangulated functor. Assume that $\mathcal{A}$ is Grothendieck. For $\mathbb{F} \in K(\mathcal{A})$, $\mathbb{F}$ is (right) F-*acyclic* (more precisely, $Q \circ F$-acyclic, where $Q : K(\mathcal{B}) \to D(\mathcal{B})$ is the localization. See for the definition, [26, (2.2.5)]) if and only if for some K-injective resolution $\mathbb{F} \to \mathbb{I}$, $F(\mathbb{F}) \to F(\mathbb{I})$ is a quasi-isomorphism, if and only if for any K-injective resolution $\mathbb{F} \to \mathbb{I}$, $F(\mathbb{F}) \to F(\mathbb{I})$ is a quasi-isomorphism. Note that the set of F-acyclic objects in $K(\mathcal{A})$ forms a localizing subcategory of $K(\mathcal{A})$, see [26, (2.2.5.1)].

Lemma 3.12. *Let $\mathcal{A}$ and $\mathcal{B}$ be abelian categories, and $F : \mathcal{A} \to \mathcal{B}$ an exact functor with the right adjoint G. Assume that $\mathcal{B}$ is Grothendieck. Then $KG : K(\mathcal{B}) \to K(\mathcal{A})$ preserves K-injective complexes. Moreover, $RG : D(\mathcal{B}) \to D(\mathcal{A})$ is the right adjoint of $\dot{L}F = F$.*

Proof. Let $\mathbb{M} \in K(\mathcal{A})$, and $\mathbb{I}$ a K-injective complex of $K(\mathcal{B})$. Then

$$\mathrm{Hom}_{K(\mathcal{A})}(\mathbb{M}, (KG)\mathbb{I}) \cong H^0(\mathrm{Hom}^{\bullet}_{\mathcal{A}}(\mathbb{M}, G\mathbb{I}))$$
$$\cong H^0(\mathrm{Hom}^{\bullet}_{\mathcal{B}}(F\mathbb{M}, \mathbb{I})) = \mathrm{Hom}_{K(\mathcal{B})}(F\mathbb{M}, \mathbb{I}).$$

If $\mathbb{M}$ is exact, then the last group is zero. This shows $(KG)\mathbb{I}$ is K-injective.

Now let $\mathbb{M} \in D(\mathcal{A})$ and $\mathbb{N} \in D(\mathcal{B})$ be arbitrary. Then by the first part, we have a functorial isomorphism

$$\mathrm{Hom}_{D(\mathcal{A})}(\mathbb{M}, (RG)\mathbb{N}) \cong \mathrm{Hom}_{K(\mathcal{A})}(\mathbb{M}, (KG)\mathbb{I}_{\mathbb{N}})$$
$$\cong \mathrm{Hom}_{K(\mathcal{B})}(F\mathbb{M}, \mathbb{I}_{\mathbb{N}}) \cong \mathrm{Hom}_{D(\mathcal{B})}(F\mathbb{M}, \mathbb{N}).$$

This proves the last assertion. $\square$

Remark 3.13. Note that for an abelian category $\mathcal{A}$, we have $\mathrm{ob}(C(\mathcal{A})) = \mathrm{ob}(K(\mathcal{A})) = \mathrm{ob}(D(\mathcal{A}))$. Thus, an object of one of the three categories is sometimes viewed as an object of another.

(3.14) Let $\mathcal{A}$ be a closed symmetric monoidal abelian category which satisfies the (AB3) and (AB3*) conditions. Let $\otimes$ be the multiplication and $[?, ?]$ be the internal hom. For a fixed $b \in \mathcal{A}$, $(? \otimes b, [b, ?])$ is an adjoint pair. In particular, $? \otimes b$ preserves colimits, and $[b, ?]$ preserves limits. By symmetry, $a \otimes ?$ also preserves colimits. As we have an isomorphism

$$\mathcal{A}(a, [b, c]) \cong \mathcal{A}(a \otimes b, c) \cong \mathcal{A}(b \otimes a, c) \cong \mathcal{A}(b, [a, c]) \cong \mathcal{A}^{\mathrm{op}}([a, c], b),$$

we have $[?, c] : \mathcal{A}^{\mathrm{op}} \to \mathcal{A}$ is right adjoint to $[?, c] : \mathcal{A} \to \mathcal{A}^{\mathrm{op}}$. This shows that $[?, c]$ changes colimits to limits.

As in [17], we define the tensor product $\mathbb{F} \otimes^{\bullet} \mathbb{G}$ of $\mathbb{F}, \mathbb{G} \in C(\mathcal{A})$ by

$$(\mathbb{F} \otimes^{\bullet} \mathbb{G})^n := \bigoplus_{p+q=n} \mathbb{F}^p \otimes \mathbb{G}^q.$$

The differential d^n on $\mathbb{F}^p \otimes \mathbb{G}^q$ is defined to be

$$d^n = d_{\mathbb{F}} \otimes 1 + (-1)^p 1 \otimes d_{\mathbb{G}}$$

(the sign convention is slightly different from [17], but this is not essential). We have $\mathbb{F} \otimes^{\bullet} \mathbb{G} \in C(\mathcal{A})$. Similarly, $[\mathbb{F}, \mathbb{G}]^{\bullet} \in C(\mathcal{A})$ is defined by

$$[\mathbb{F}, \mathbb{G}]^n := \prod_{p \in \mathbb{Z}} [\mathbb{F}^p, \mathbb{G}^{n+p}]$$

and

$$d^n := [d_{\mathbb{F}}, 1] + (-1)^{n+1} [1, d_{\mathbb{G}}].$$

It is straightforward to prove the following.

Lemma 3.15. *Let $\mathcal{A}$ be as above. Then the category of chain complexes $C(\mathcal{A})$ is closed symmetric monoidal with $\otimes^{\bullet}$ the multiplication and $[?, ?]^{\bullet}$ the internal hom. The bi-triangulated functors*

$$\otimes^{\bullet} : K(\mathcal{A}) \times K(\mathcal{A}) \to K(\mathcal{A})$$

and

$$[?, ?]^{\bullet} : K(\mathcal{A})^{\mathrm{op}} \times K(\mathcal{A}) \to K(\mathcal{A}),$$

are induced, and $K(\mathcal{A})$ is a closed symmetric monoidal triangulated category (see [26, (3.5), (3.6)]).

(3.16) Let $\mathcal{A}$ be an abelian category, and $\mathfrak{P}$ a full subcategory of $C(\mathcal{A})$. An inverse system $(\mathbb{F}_i)_{i \in I}$ in $C(\mathcal{A})$ is said to be $\mathfrak{P}$-*special* if the following conditions are satisfied.

i I is well-ordered.

ii If $i \in I$ has no predecessor, then the canonical map $I_i \to \varprojlim_{j<i} I_j$ is an isomorphism (in particular, $I_{i_0} = 0$ if i_0 is the minimum element of I).

iii If $i \in I$ has a predecessor $i - 1$, then the natural chain map $I_i \to I_{i-1}$ is an epimorphism, the kernel C_i is isomorphic to some object of $\mathfrak{P}$, and the exact sequence

$$0 \to C_i \to I_i \to I_{i-1} \to 0$$

is semi-split.

Similarly, $\mathfrak{P}$-special direct systems are also defined, see [39].

The full subcategory of $C(\mathcal{A})$ consisting of inverse (resp. direct) limits of $\mathfrak{P}$-special inverse (resp. direct) systems is denoted by $\underleftarrow{\mathfrak{P}}$ (resp. $\underrightarrow{\mathfrak{P}}$).

(3.17) Let $(\mathbb{X}, \mathcal{O}_{\mathbb{X}})$ be a ringed site. Various definitions and results on unbounded complexes of sheaves over a ringed space by Spaltenstein [39] is generalized to those for ringed sites. However, note that we can not utilize the notion related to closed subsets, points, or stalks of sheaves.

(3.18) We say that a complex $\mathbb{F} \in C(\mathrm{Mod}(\mathbb{X}))$ is *K-flat* if $\mathbb{G} \otimes^{\bullet} \mathbb{F}$ is exact whenever $\mathbb{G}$ is an exact complex in $\mathrm{Mod}(\mathbb{X})$. We say that $\mathbb{A} \in C(\mathrm{Mod}(\mathbb{X}))$ is *weakly K-injective* if $\mathbb{A}$ is $\mathrm{Hom}^{\bullet}_{\mathrm{Mod}(\mathbb{X})}(\mathbb{F}, ?)$-acyclic for any K-flat complex $\mathbb{F}$.

(3.19) Let $(\mathbb{X}, \mathcal{O}_{\mathbb{X}})$ be a ringed site. For $x \in \mathbb{X}$, we define $\mathcal{O}^p_x$ to be $L^{\mathrm{PM}}_x((q\mathcal{O}_{\mathbb{X}})|_x)$, and $\mathcal{O}_x := L^{\mathrm{Mod}}_x(\mathcal{O}_{\mathbb{X}}|_x) \cong a\mathcal{O}^p_x$. We denote by $\mathfrak{P}_0 = \mathfrak{P}_0(\mathbb{X}, \mathcal{O}_{\mathbb{X}})$ the full subcategory of $C(\mathrm{Mod}(\mathbb{X}))$ consisting of complexes of the form $\mathcal{O}_x[n]$ with $x \in \mathbb{X}$. We define $\mathfrak{P} = \mathfrak{P}(\mathbb{X}, \mathcal{O}_{\mathbb{X}})$ to be $\underrightarrow{\mathfrak{P}_0}$. We call an object of $\mathfrak{P}$ a *strongly K-flat* complex. We also define $\mathfrak{Q}$ to be the full subcategory of $C(\mathrm{Mod}(\mathbb{X}))$ consisting of bounded above complexes whose terms are direct sums of copies of $\mathcal{O}_x$.

We say that $\mathbb{A} \in C(\mathrm{Mod}(\mathbb{X}))$ is *K-limp* if $\mathbb{A}$ is $\mathrm{Hom}^{\bullet}_{\mathrm{Mod}(\mathbb{X})}(\mathbb{F}, ?)$-acyclic for any strongly K-flat complex $\mathbb{F}$.

Lemma 3.20. *Let* $f : (\mathbb{Y}, \mathcal{O}_{\mathbb{Y}}) \to (\mathbb{X}, \mathcal{O}_{\mathbb{X}})$ *be a ringed continuous functor. Then we have an isomorphism*

$$f^{\mathrm{Mod}}_{\#}(\mathcal{O}_y) \cong \mathcal{O}_{fy}$$

for $y \in \mathbb{Y}$. *In particular, if* $\mathbb{F} \in \mathfrak{P}(\mathbb{Y})$, *then* $f_{\#}\mathbb{F} \in \mathfrak{P}(\mathbb{X})$.

Proof. For $y \in \mathbb{Y}$, we denote the canonical continuous ringed functor

$$(\mathbb{Y}/y, \mathcal{O}_{\mathbb{Y}}|_y) \to (\mathbb{X}/fy, \mathcal{O}_{\mathbb{X}}|_{fy})$$

by f/y. We have $\mathfrak{R}_{fy} \circ (f/y) = f \circ \mathfrak{R}_y$. Hence by Corollary 2.66,

$$f_{\#}\mathcal{O}_y = f_{\#}L_y(\mathcal{O}_{\mathbb{Y}}|_y) \cong L_{fy}(f/y)_{\#}(\mathcal{O}_{\mathbb{Y}}|_y) \cong L_{fy}(\mathcal{O}_{\mathbb{X}}|_{fy}) = \mathcal{O}_{fy}.$$

$\square$

Lemma 3.21. *Let* $(\mathbb{X}, \mathcal{O}_{\mathbb{X}})$ *be a ringed site, and* $\mathbb{F}, \mathbb{G} \in C(\mathrm{Mod}(\mathbb{X}))$. *Then the following hold:*

1 $\mathbb{F}$ *is K-flat if and only if* $\underline{\mathrm{Hom}}^{\bullet}_{\mathrm{Mod}(\mathbb{X})}(\mathbb{F}, \mathbb{I})$ *is K-injective for any K-injective complex $\mathbb{I}$.*

2 *If $\mathbb{F}$ is K-flat exact, then $\mathbb{G} \otimes^{\bullet}_{\mathcal{O}_{\mathbb{X}}} \mathbb{F}$ is exact.*

3 *The inductive limit of a pseudo-filtered inductive system of K-flat complexes is again K-flat.*

4 *The tensor product of two K-flat complexes is again K-flat.*

See [39] for the proof. For **2**, utilize **3** of Lemma 3.25 and Corollary 3.23 below.

Proposition 3.22. *Let $(\mathbb{X}, \mathcal{O}_{\mathbb{X}})$ be a ringed site, and $x \in \mathbb{X}$. Then $\mathcal{O}_x$ is K-flat.*

Proof. It suffices to show that for any exact complex $\mathcal{E}$, $\mathcal{E} \otimes \mathcal{O}_x$ is exact. To verify this, it suffices to show that for any K-injective complex $\mathcal{I}$, the complex $\mathrm{Hom}^{\bullet}_{\mathcal{O}_{\mathbb{X}}}(\mathcal{E} \otimes \mathcal{O}_x, \mathcal{I})$ is exact. Indeed, then if we consider the K-injective resolution $\mathcal{E} \otimes \mathcal{O}_x \to \mathcal{I}$, it must be null-homotopic and thus $\mathcal{E} \otimes \mathcal{O}_x$ must be exact.

Note that we have

$$\mathrm{Hom}^{\bullet}_{\mathcal{O}_{\mathbb{X}}}(\mathcal{E} \otimes \mathcal{O}_x, \mathcal{I}) \cong \mathrm{Hom}^{\bullet}_{\mathcal{O}_{\mathbb{X}}}(\mathcal{O}_x, \underline{\mathrm{Hom}}^{\bullet}_{\mathcal{O}_{\mathbb{X}}}(\mathcal{E}, \mathcal{I})) \cong$$
$$\mathrm{Hom}^{\bullet}_{\mathrm{Mod}(\mathbb{X}/x)}(\mathcal{E}|_x, \mathcal{I}|_x) \cong \mathrm{Hom}^{\bullet}_{\mathcal{O}_{\mathbb{X}}}(L_x(\mathcal{E}|_x), \mathcal{I}).$$

As $(?)|_x$ and L_x are exact (2.36), (2.23), the last complex is exact, and we are done. □

Corollary 3.23. *A strongly K-flat complex is K-flat.*

Proof. Follows immediately from the proposition. □

(3.24) Let $\mathcal{A}$ be an abelian category. For an object

$$\mathbb{F} : \cdots \to F^n \xrightarrow{d^n} F^{n+1} \xrightarrow{d^{n+1}} \to \cdots$$

in $C(\mathcal{A})$, we denote the truncated complex

$$0 \to F^n / \mathrm{Im}\, d^{n-1} \xrightarrow{d^n} F^{n+1} \xrightarrow{d^{n+1}} \to \cdots$$

by $\tau_{\geq n}\mathbb{F}$. Similarly, the truncated complex

$$\cdots \to F^{n-2} \xrightarrow{d^{n-2}} F^{n-1} \xrightarrow{d^{n-1}} \mathrm{Ker}\, d^n \to 0$$

is denoted by $\tau_{\leq n}\mathbb{F}$.

Lemma 3.25. *Let $(\mathbb{X}, \mathcal{O}_{\mathbb{X}})$ be a ringed site and $\mathbb{F} \in C(\mathrm{Mod}(\mathbb{X}))$.*

1 *We have $\mathfrak{Q} \subset \mathfrak{P}$.*

2 *A K-injective complex is weakly K-injective, and a weakly K-injective complex is K-limp.*

3 *For any $\mathbb{H} \in C(\mathrm{Mod}(\mathbb{X}))$, there is a $\mathfrak{Q}$-special direct system $(\mathbb{F}_n)$ and a direct system of chain maps $(f_n : \mathbb{F}_n \to \tau_{\leq n}\mathbb{H})$ such that f_n is a quasi-isomorphism for each $n \in \mathbb{N}$, and $(\mathbb{F}_n)_l = 0$ for $l \geq n+1$. We have $\varinjlim \mathbb{F}_n \to \mathbb{H}$ is a quasi-isomorphism, and $\varinjlim \mathbb{F}_n \in \mathfrak{P}$. Moreover, $Q\mathbb{H}$ is the homotopy colimit of the inductive system $(\tau_{\leq n}Q\mathbb{H})$ in the category $D(\mathrm{Mod}(\mathbb{X}))$.*

4 *The following are equivalent.*

  i $\mathbb{F}$ *is K-limp.*
  ii $\mathbb{F}$ *is K-limp as a complex of sheaves of abelian groups.*
  iii $\mathbb{F}$ *is $\mathrm{Hom}^{\bullet}_{\mathrm{Mod}(\mathbb{X})}(\mathcal{O}_x, ?)$-acyclic for $x \in \mathbb{X}$.*
  iv $\mathbb{F}$ *is $\Gamma(x, ?)$-acyclic for $x \in \mathbb{X}$.*
  v *If $\mathbb{G} \in \mathfrak{P}$ and $\mathbb{G}$ is exact, then $\mathrm{Hom}^{\bullet}_{\mathrm{Mod}(\mathbb{X})}(\mathbb{G}, \mathbb{F})$ is exact.*

5 *The following are equivalent.*

  i $\mathbb{F}$ *is weakly K-injective.*
  ii *If $\mathbb{G}$ is K-flat exact, then $\mathrm{Hom}^{\bullet}_{\mathrm{Mod}(\mathbb{X})}(\mathbb{G}, \mathbb{F})$ is exact.*
  iii *For any K-flat complex $\mathbb{G}$, $\underline{\mathrm{Hom}}^{\bullet}_{\mathrm{Mod}(\mathbb{X})}(\mathbb{G}, \mathbb{F})$ is weakly K-injective.*

Proof. **1** is trivial. **2** follows from the definition and Corollary 3.23. The proof of **3** and **4** are left to the reader, see [39, (3.2), (3.3), (5.16), (5.17), (5.21)]. **5** is similar. $\square$

Lemma 3.26. *Let $(\mathbb{X}, \mathcal{O}_{\mathbb{X}})$ be a ringed site and $\mathbb{F}, \mathbb{G} \in C(\mathrm{Mod}(\mathbb{X}))$. If $\mathbb{F}$ is weakly K-injective and $\mathbb{G}$ is K-flat, then $\mathbb{F}$ is $\underline{\mathrm{Hom}}^{\bullet}_{\mathrm{Mod}(\mathbb{X})}(\mathbb{G}, ?)$-acyclic.*

Proof. Let $\mathbb{F} \to \mathbb{I}$ be the K-injective resolution, and $\mathbb{J}$ the mapping cone. Let $\varphi : \mathbb{H} \to \underline{\mathrm{Hom}}^{\bullet}_{\mathrm{Mod}(\mathbb{X})}(\mathbb{G}, \mathbb{J})$ be a $\mathfrak{P}$-resolution. As $\mathbb{H} \otimes^{\bullet} \mathbb{G}$ is K-flat and $\mathbb{J}$ is weakly K-injective exact,

$$\mathrm{Hom}^{\bullet}_{\mathrm{Mod}(\mathbb{X})}(\mathbb{H}, \underline{\mathrm{Hom}}^{\bullet}_{\mathrm{Mod}(\mathbb{X})}(\mathbb{G}, \mathbb{J})) \cong \mathrm{Hom}^{\bullet}_{\mathrm{Mod}(\mathbb{X})}(\mathbb{H} \otimes^{\bullet} \mathbb{G}, \mathbb{J})$$

is exact. So φ must be null-homotopic, and hence $\underline{\mathrm{Hom}}^{\bullet}_{\mathrm{Mod}(\mathbb{X})}(\mathbb{G}, \mathbb{J})$ is exact. This is what we wanted to prove. $\square$

(3.27) Let $(\mathbb{X}, \mathcal{O}_{\mathbb{X}})$ be a ringed site. For $\mathbb{G} \in C(\mathrm{Mod}(\mathbb{X}))$, it is easy to see that $\mathbb{G} \otimes^{\bullet}_{\mathcal{O}_{\mathbb{X}}} ?$ induces a functor from $K(\mathrm{Mod}(\mathbb{X}))$ to itself. By Lemma 3.25, **3** and the dual assertion of [17, Theorem I.5.1], the derived functor $L(\mathbb{G} \otimes^{\bullet}_{\mathcal{O}_{\mathbb{X}}} ?)$ is induced, and it is calculated using any K-flat resolution of $?$. If we fix $?$, then $L(\mathbb{G} \otimes^{\bullet}_{\mathcal{O}_{\mathbb{X}}} ?)$ is a functor on $\mathbb{G}$, and it induces a bifunctor

$$* \otimes^{\bullet, L}_{\mathcal{O}_{\mathbb{X}}} ? : D(\mathrm{Mod}(\mathbb{X})) \times D(\mathrm{Mod}(\mathbb{X})) \to D(\mathrm{Mod}(\mathbb{X})).$$

$\mathbb{G} \otimes^{\bullet, L}_{\mathcal{O}_{\mathbb{X}}} \mathbb{F}$ is calculated using any K-flat resolution of $\mathbb{F}$ or any K-flat resolution of $\mathbb{G}$. Note that $\otimes^{\bullet, L}_{\mathcal{O}_{\mathbb{X}}}$ is a $\triangle$-functor as in [26, (2.5.7)].

We define the *hyperTor* functor as follows:

$$\underline{\mathrm{Tor}}_i^{\mathcal{O}_{\mathbb{X}}}(\mathbb{F}, \mathbb{G}) := H^{-i}(\mathbb{F} \otimes_{\mathcal{O}_{\mathbb{X}}}^{\bullet, L} \mathbb{G}).$$

(3.28) Let $(\mathbb{X}, \mathcal{O}_{\mathbb{X}})$ be a ringed site. For $\mathbb{F} \in C(\mathrm{Mod}(\mathbb{X}))$, the functor $\underline{\mathrm{Hom}}_{\mathcal{O}_{\mathbb{X}}}^{\bullet}(\mathbb{F}, ?)$ induces a functor from $K(\mathrm{Mod}(\mathbb{X}))$ to itself. As $\mathrm{Mod}(\mathbb{X})$ is Grothendieck, we can take K-injective resolutions, and hence the right derived functor $R\underline{\mathrm{Hom}}_{\mathcal{O}_{\mathbb{X}}}^{\bullet}(\mathbb{F}, ?)$ is induced. Thus a bifunctor

$$R\underline{\mathrm{Hom}}_{\mathcal{O}_{\mathbb{X}}}(*, ?) : D(\mathrm{Mod}(\mathbb{X}))^{\mathrm{op}} \times D(\mathrm{Mod}(\mathbb{X})) \to D(\mathrm{Mod}(\mathbb{X}))$$

is induced. For $\mathbb{F}, \mathbb{G} \in D(\mathrm{Mod}(\mathbb{X}))$, we define the *hyperExt* sheaf of $\mathbb{F}$ and $\mathbb{G}$ by

$$\underline{\mathrm{Ext}}_{\mathcal{O}_{\mathbb{X}}}^i(\mathbb{F}, \mathbb{G}) := H^i(R\underline{\mathrm{Hom}}_{\mathcal{O}_{\mathbb{X}}}^{\bullet}(\mathbb{F}, \mathbb{G})).$$

Similarly, the functor $\mathrm{Hom}_{\mathcal{O}_{\mathbb{X}}}^{\bullet}(*, ?)$ induces

$$R\mathrm{Hom}_{\mathcal{O}_{\mathbb{X}}}^{\bullet}(*, ?) : D(\mathrm{Mod}(\mathbb{X}))^{\mathrm{op}} \times D(\mathrm{Mod}(\mathbb{X})) \to D(\underline{\mathrm{Ab}}).$$

Almost by definition, we have

$$H^i(R\mathrm{Hom}_{\mathcal{O}_{\mathbb{X}}}^{\bullet}(\mathbb{F}, \mathbb{G})) \cong \mathrm{Hom}_{D(\mathrm{Mod}(\mathbb{X}))}(\mathbb{F}, \mathbb{G}[i]).$$

Sometimes we denote these groups by $\mathrm{Ext}_{\mathcal{O}_{\mathbb{X}}}^i(\mathbb{F}, \mathbb{G})$.

Lemma 3.29. *Let $(\mathbb{X}, \mathcal{O}_{\mathbb{X}})$ be a ringed site. Then $D(\mathrm{Mod}(\mathbb{X}))$ is a closed symmetric monoidal triangulated category with $\otimes_{\mathcal{O}_{\mathbb{X}}}^{\bullet, L}$ its product and $R\underline{\mathrm{Hom}}_{\mathcal{O}_{\mathbb{X}}}^{\bullet}(*, ?)$ its internal hom.*

Proof. This is straightforward. □

Lemma 3.30. *Let $f : \mathbb{Y} \to \mathbb{X}$ be a continuous functor between sites. If $\mathbb{I} \in C$ $(\mathrm{AB}(\mathbb{X}))$ is K-limp and exact, then $f_{\mathrm{AB}}^{\#}\mathbb{I}$ is exact.*

Proof. Let $\xi : \mathbb{F} \to f_{\mathrm{AB}}^{\#}\mathbb{I}$ be a $\mathfrak{P}(\mathrm{AB})$-resolution of $f_{\mathrm{AB}}^{\#}\mathbb{I}$. It suffices to show $\mathrm{Hom}_{\mathrm{AB}(\mathbb{Y})}^{\bullet}(\mathbb{F}, f_{\mathrm{AB}}^{\#}\mathbb{I})$ is exact (if so, then ξ must be null-homotopic). Since $f_{\#}^{\mathrm{AB}}\mathbb{F} \in \mathfrak{P}(\mathrm{AB})$ and $\mathbb{I}$ is K-limp exact, this is obvious. □

By the lemma, a K-limp complex is $f^{\#}$-acyclic.

Lemma 3.31. *Let $f : (\mathbb{Y}, \mathcal{O}_{\mathbb{Y}}) \to (\mathbb{X}, \mathcal{O}_{\mathbb{X}})$ be an admissible ringed continuous functor. Then the following hold:*

1. *If $\mathbb{I} \in C(\mathrm{AB}(\mathbb{X}))$ is a K-injective (resp. K-limp) complex of sheaves of abelian groups, then so is $f_{\mathrm{AB}}^{\#}\mathbb{I}$.*
2. *If $\mathbb{F} \in C(\mathrm{Mod}(\mathbb{Y}))$ is strongly K-flat and exact, then $f_{\#}^{\mathrm{Mod}}\mathbb{F}$ is strongly K-flat and exact.*

Proof. As $f_{\mathrm{AB}}^{\#}$ has an exact left adjoint $f_{\#}^{\mathrm{AB}}$, the assertion for K-injectivity in **1** is obvious.

We prove the assertion for the K-limp property in **1**. Let $\mathbb{P} \in \mathfrak{P}(\mathbb{Y})$ be an exact complex. As $f_{\#}^{\mathrm{AB}}$ is exact, $f_{\#}^{\mathrm{AB}}\mathbb{P}$ is exact and a complex in $\mathfrak{P}(\mathbb{X})$ by Lemma 3.20. Hence,

$$\mathrm{Hom}^{\bullet}_{\mathrm{AB}(\mathbb{Y})}(\mathbb{P}, f_{\mathrm{AB}}^{\#}\mathbb{I}) \cong \mathrm{Hom}^{\bullet}_{\mathrm{AB}(\mathbb{X})}(f_{\#}^{\mathrm{AB}}\mathbb{P}, \mathbb{I})$$

is exact. This shows $f_{\mathrm{AB}}^{\#}\mathbb{I}$ is K-limp.

We prove **2**. We already know that $f_{\#}\mathbb{F}$ is strongly K-flat by Lemma 3.20. We prove that $\mathrm{Hom}^{\bullet}_{\mathrm{Mod}(\mathbb{X})}(f_{\#}^{\mathrm{Mod}}\mathbb{F}, \mathbb{I})$ is exact, where $\eta : f_{\#}^{\mathrm{Mod}}\mathbb{F} \to \mathbb{I}$ is a K-injective resolution. Then η must be null-homotopic, and we have $f_{\#}^{\mathrm{Mod}}\mathbb{F}$ is exact and the proof is complete. Clearly, $\mathbb{I}$ is K-limp, and hence so is $f_{\mathrm{Mod}}^{\#}\mathbb{I}$ by **1** and Lemma 3.25, **4**. The assertion follows immediately by adjunction. □

(3.32) By the lemma, there is a derived functor $Lf_{\#}^{\mathrm{Mod}} : D(\mathrm{Mod}(\mathbb{Y})) \to D(\mathrm{Mod}(\mathbb{X}))$ of $f_{\#}^{\mathrm{Mod}}$ for an admissible ringed continuous functor f. It is calculated via strongly K-flat resolutions.

Now as in [39, section 6] and [26], the following is proved.

Lemma 3.33. *Let $\mathcal{S}$ be the category of ringed sites and admissible ringed continuous functors. Then $(L(?)_{\#}^{\mathrm{Mod}}, R(?)_{\mathrm{Mod}}^{\#})$ is a monoidal adjoint pair of Δ-almost-pseudofunctors (defined appropriately as in [26, (3.6.7)]) on $\mathcal{S}^{\mathrm{op}}$.*

The proof is basically the same as that in [26, Chapter 1–3], and left to the reader.

Remark 3.34. Later we will treat ringed continuous functor f which may not be admissible. In this case, we may use $Rf^{\#}$ and related functorialities, but not $Lf_{\#}$.

Chapter 4
Sheaves over a Diagram of S-Schemes

(4.1) Let S be a (small) scheme, and I a small category. We call an object of $\mathcal{P}(I^{\mathrm{op}}, \underline{\mathrm{Sch}}/S)$ an I-diagram of S-schemes, where $\underline{\mathrm{Sch}}/S$ denotes the category of (small) S-schemes. We denote $\underline{\mathrm{Sch}}/\operatorname{Spec}\mathbb{Z}$ simply by $\underline{\mathrm{Sch}}$. So an object of $\mathcal{P}(I, \underline{\mathrm{Sch}}/S)$ is referred as an I^{op}-diagram of S-schemes. Let $X_\bullet \in \mathcal{P}(I, \underline{\mathrm{Sch}}/S)$. We denote $X_\bullet(i)$ by X_i for $i \in I$, and $X_\bullet(\phi)$ by X_ϕ for $\phi \in \mathrm{Mor}(I)$. Let $\mathbb{P}$ be a property of schemes (e.g., quasi-compact, locally noetherian, regular). We say that $X_\bullet$ satisfies $\mathbb{P}$ if X_i satisfies $\mathbb{P}$ for any $i \in I$. Let $\mathbb{Q}$ be a property of morphisms of schemes (e.g., quasi-compact, locally of finite type, smooth). We say that $X_\bullet$ is $\mathbb{Q}$ over S if the structure map $X_i \to S$ satisfies $\mathbb{Q}$ for any $i \in I$. We say that $X_\bullet$ has $\mathbb{Q}$ arrows if X_ϕ satisfies $\mathbb{Q}$ for any $\phi \in \mathrm{Mor}(I)$.

(4.2) Let $f_\bullet : X_\bullet \to Y_\bullet$ be a morphism in $\mathcal{P}(I, \underline{\mathrm{Sch}}/S)$. For $i \in I$, we denote $f_\bullet(i) : X_i \to Y_i$ by f_i. For a property $\mathbb{Q}$ of morphisms of schemes, we say that $f_\bullet$ satisfies $\mathbb{Q}$ if so does f_i for any $i \in I$. We say that $f_\bullet$ is *cartesian* if the canonical map $(f_j, X_\phi) : X_j \to Y_j \times_{Y_i} X_i$ is an isomorphism for any morphism $\phi : i \to j$ of I.

(4.3) Let S, I and $X_\bullet$ be as above. We define the *Zariski site* of $X_\bullet$, denoted by $\mathrm{Zar}(X_\bullet)$, as follows. An object of $\mathrm{Zar}(X_\bullet)$ is a pair (i, U) such that $i \in I$ and U is an open subset of X_i. A morphism $(\phi, h) : (j, V) \to (i, U)$ is a pair (ϕ, h) such that $\phi \in I(i, j)$ and $h : V \to U$ is the restriction of X_ϕ. For a given morphism $\phi : i \to j$, U, and V, such an h exists if and only if $V \subset X_\phi^{-1}(U)$, and it is unique. We denote this h by $h(\phi; U, V)$. The composition of morphisms is defined in an obvious way. Thus $\mathrm{Zar}(X_\bullet)$ is a small category. For $(i, U) \in \mathrm{Zar}(X_\bullet)$, a covering of (i, U) is a family of morphisms of the form

$$((\mathrm{id}_i, h(\mathrm{id}_i; U, U_\lambda)) : (i, U_\lambda) \to (i, U))_{\lambda \in \Lambda}$$

such that $\bigcup_{\lambda \in \Lambda} U_\lambda = U$. This defines a pretopology of $\mathrm{Zar}(X_\bullet)$, and $\mathrm{Zar}(X_\bullet)$ is a site. As we will consider only the Zariski topology, a presheaf or sheaf on $\mathrm{Zar}(X_\bullet)$ will be sometimes referred as a presheaf or sheaf on $X_\bullet$, if there

J. Lipman, M. Hashimoto, *Foundations of Grothendieck Duality for Diagrams of Schemes*, Lecture Notes in Mathematics 1960,

is no danger of confusion. Thus $\mathcal{P}(X_\bullet, \mathcal{C})$ and $\mathcal{S}(X_\bullet, \mathcal{C})$ mean $\mathcal{P}(\mathrm{Zar}(X_\bullet), \mathcal{C})$ and $\mathcal{S}(\mathrm{Zar}(X_\bullet), \mathcal{C})$, respectively.

(4.4) Let S and I be as above, and let $\sigma : J \hookrightarrow I$ be a subcategory of I. Then we have an obvious restriction functor $\sigma^\# : \mathcal{P}(I, \underline{\mathrm{Sch}}/S) \to \mathcal{P}(J, \underline{\mathrm{Sch}}/S)$, which we denote by $(?)|_J$. If $\mathrm{ob}(J)$ is finite and $I(j, i)$ is finite for each $j \in J$ and $i \in I$, then $(?)|_J$ has a right adjoint functor $\mathrm{cosk}_J^I = (?)^{\mathrm{op}} \sigma_\#^{\mathrm{op}} (?)^{\mathrm{op}}$, because $\underline{\mathrm{Sch}}/S$ has finite limits. Note that $(\mathrm{cosk}_J^I X_\bullet)_i = \varprojlim X_j$, where the limit is taken over $I_i^{\sigma^{\mathrm{op}}}$, where $\sigma^{\mathrm{op}} : J^{\mathrm{op}} \to I^{\mathrm{op}}$ is the opposite of σ. See [10, pp. 9–12].

(4.5) Let $X_\bullet \in \mathcal{P}(I, \underline{\mathrm{Sch}}/S)$. Then we have an obvious continuous functor $Q(X_\bullet, J) : \mathrm{Zar}((X_\bullet)|_J) \hookrightarrow \mathrm{Zar}(X_\bullet)$. Note that $Q(X_\bullet, J)$ may not be admissible. The restriction functors $Q(X_\bullet, J)_{\mathrm{AB}}^\#$ and $Q(X_\bullet, J)_{\mathrm{PA}}^\#$ are denoted by $(?)_J^{\mathrm{AB}}$ and $(?)_J^{\mathrm{PA}}$, respectively. For $i \in I$, we consider that i is the subcategory of I whose object set is $\{i\}$ with $\mathrm{Hom}_i(i, i) = \{\mathrm{id}\}$. The restrictions $(?)_i^\heartsuit$ for $\heartsuit = \mathrm{AB}, \mathrm{PA}$ are defined.

(4.6) Let $\mathcal{F} \in \mathrm{PA}(X_\bullet)$ and $i \in I$. Then $\mathcal{F}_i \in \mathrm{PA}(X_i)$, and thus we have a family of sheaves $(\mathcal{F}_i)_{i \in I}$. Moreover, for $(i, U) \in \mathrm{Zar}(X_\bullet)$ and $\phi : i \to j$, we have the restriction map

$$\Gamma(U, \mathcal{F}_i) = \Gamma((i, U), \mathcal{F}) \xrightarrow{\mathrm{res}} \Gamma((j, X_\phi^{-1}(U)), \mathcal{F}) = \Gamma(X_\phi^{-1}(U), \mathcal{F}_j)$$
$$= \Gamma(U, (X_\phi)_* \mathcal{F}_j),$$

which induces

$$\beta_\phi(\mathcal{F}) \in \mathrm{Hom}_{\mathrm{PA}(X_i)}(\mathcal{F}_i, (X_\phi)_* \mathcal{F}_j). \tag{4.7}$$

The corresponding map in $\mathrm{Hom}_{\mathrm{PA}(X_j)}((X_\phi)_{\mathrm{PA}}^*(\mathcal{F}_i), \mathcal{F}_j)$ is denoted by $\alpha_\phi^{\mathrm{PA}}(\mathcal{F})$. If $\mathcal{F}$ is a sheaf, then (4.7) yields

$$\alpha_\phi^{\mathrm{AB}}(\mathcal{F}) \in \mathrm{Hom}_{\mathrm{AB}(X_j)}((X_\phi)_{\mathrm{AB}}^*(\mathcal{F}_i), \mathcal{F}_j).$$

It is straightforward to check the following.

Lemma 4.8 ([10]). *Let $\heartsuit$ be either* AB *or* PA. *The following hold:*

1 *For any $i \in I$, we have $\alpha_{\mathrm{id}_i}^\heartsuit : (X_{\mathrm{id}_i})_\heartsuit^*(\mathcal{F}_i) \to \mathcal{F}_i$ is the canonical identification $\int_{X_i}$.*

2 *If $\phi \in I(i, j)$ and $\psi \in I(j, k)$, then the composite map*

$$(X_{\psi\phi})_\heartsuit^*(\mathcal{F}_i) \xrightarrow{d^{-1}} (X_\psi)_\heartsuit^*(X_\phi)_\heartsuit^*(\mathcal{F}_i) \xrightarrow{(X_\psi)_\heartsuit^* \alpha_\phi^\heartsuit} (X_\psi)_\heartsuit^*(\mathcal{F}_j) \xrightarrow{\alpha_\psi^\heartsuit} \mathcal{F}_k \tag{4.9}$$

agrees with $\alpha_{\psi\phi}^\heartsuit$.

3 *Conversely, a family $((\mathcal{G}_i)_{i \in I}, (\alpha_\phi)_{\phi \in \mathrm{Mor}(I)})$ such that $\mathcal{G}_i \in \heartsuit(X_i)$, $\alpha_\phi \in \mathrm{Hom}_{\heartsuit(X_j)}((X_\phi)_\heartsuit^*(\mathcal{G}_i), \mathcal{G}_j)$ for $\phi \in I(i, j)$, and that the conditions corresponding to **1,2** are satisfied yields $\mathcal{G} \in \heartsuit(X_\bullet)$, and this correspondence gives an equivalence.*

(4.10)   Similarly, a family $((\mathcal{G}_i)_{i \in \mathrm{ob}(I)}, (\beta_\phi)_{\phi \in \mathrm{Mor}(I)})$ with

$$\mathcal{G}_i \in \heartsuit(X_i) \text{ and } \beta_\phi \in \mathrm{Hom}_{\heartsuit(X_i)}(\mathcal{G}_i, (X_\phi)_*^\heartsuit \mathcal{G}_j)$$

satisfying the conditions

1' For $i \in \mathrm{ob}(I)$, $\beta_{\mathrm{id}_i} : \mathcal{G}_i \to (X_{\mathrm{id}_i})_* \mathcal{G}_i$ is the canonical identification $\mathfrak{e}_{X_i}$;
2' For $\phi \in I(i,j)$ and $\psi \in I(j,k)$, the composite

$$\mathcal{F}_i \xrightarrow{\beta_\phi} (X_\phi)_*(\mathcal{F}_j) \xrightarrow{(X_\phi)_*\beta_\psi} (X_\phi)_*(X_\psi)_*(\mathcal{F}_k) \xrightarrow{c^{-1}} (X_{\psi\phi})_*(\mathcal{F}_k)$$

agrees with $\beta_{\psi\phi}$

is in one to one correspondence with $\mathcal{G} \in \heartsuit(X_\bullet)$.

(4.11)   Let $\mathcal{F} \in \mathrm{AB}(X_\bullet)$. We say that $\mathcal{F}$ is an *equivariant* abelian sheaf if $\alpha_\phi^{\mathrm{AB}}$ are isomorphisms for all $\phi \in \mathrm{Mor}(I)$. For $\mathcal{F} \in \mathrm{PA}(X_\bullet)$, we say that $\mathcal{F}$ is an equivariant abelian presheaf if $\alpha_\phi^{\mathrm{PA}}$ are isomorphisms for all $\phi \in \mathrm{Mor}(I)$. An equivariant sheaf may not be an equivariant presheaf. However, an equivariant presheaf which is a sheaf is an equivariant sheaf. We denote the category of equivariant sheaves and presheaves by $\mathrm{EqAB}(X_\bullet)$ and $\mathrm{EqPA}(X_\bullet)$, respectively. As $(X_\phi)_\heartsuit^*$ is exact for $\heartsuit = \mathrm{AB}, \mathrm{PA}$ and any ϕ, we have that $\mathrm{EqAB}(X_\bullet)$ is plump in $\mathrm{AB}(X_\bullet)$, and $\mathrm{EqPA}(X_\bullet)$ is plump in $\mathrm{PA}(X_\bullet)$.

(4.12)   Let $X_\bullet \in \mathcal{P}(I, \underline{\mathrm{Sch}}/S)$. The data

$$((\mathcal{O}_{X_i})_{i \in I}, (\beta_\phi = \eta : \mathcal{O}_{X_i} \to (X_\phi)_*\mathcal{O}_{X_j})_{\phi \in \mathrm{Mor}(I)})$$

gives a sheaf of commutative rings on $X_\bullet$, which we denote by $\mathcal{O}_{X_\bullet}$, and thus $\mathrm{Zar}(X_\bullet)$ is a ringed site. The categories $\mathrm{PM}(\mathrm{Zar}(X_\bullet))$ and $\mathrm{Mod}(\mathrm{Zar}(X_\bullet))$ are denoted by $\mathrm{PM}(X_\bullet)$ and $\mathrm{Mod}(X_\bullet)$, respectively. Let $\heartsuit = \mathrm{PM}, \mathrm{Mod}$. Note that for $\mathcal{M} \in \heartsuit(X_\bullet)$ and $\phi : i \to j$, $\beta_\phi : \mathcal{M}_i \to (X_\phi)_*\mathcal{M}_j$ is a morphism in $\heartsuit(X_i)$, which we denote by $\beta_\phi^\heartsuit$. The adjoint morphism $X_\phi^*\mathcal{M}_i \to \mathcal{M}_j$ is denoted by $\alpha_\phi^\heartsuit$. α is not compatible with the forgetful functors in general.

(4.13)   For $J \subset I$, we have $\mathcal{O}_{X_\bullet|_J} = (\mathcal{O}_{X_\bullet})_J$ by definition. The continuous functor

$$Q(X_\bullet, J) : (\mathrm{Zar}(X_\bullet|_J), \mathcal{O}_{X_\bullet|_J}) \to (\mathrm{Zar}(X_\bullet), \mathcal{O}_{X_\bullet})$$

is actually a ringed continuous functor.

The corresponding restriction $Q(X_\bullet, J)_\heartsuit^\#$ is denoted by $(?)_J^\heartsuit$ for $\heartsuit = \mathrm{PM}, \mathrm{Mod}$. For subcategories $J_1 \subset J \subset I$ of I, we denote the restriction $(?)_{J_1}^\heartsuit : \heartsuit(X_\bullet|_J) \to \heartsuit(X_\bullet|_{J_1})$ by $(?)_{J_1, J}^\heartsuit$, to emphasize J.

(4.14)   Let $\heartsuit$ be PM or Mod. Note that $\mathcal{M} \in \heartsuit(X_\bullet)$ is nothing but a family

$$\mathrm{Dat}(\mathcal{M}) := ((\mathcal{M}_i)_{i \in I}, (\alpha_\phi^\heartsuit)_{\phi \in \mathrm{Mor}(I)})$$

such that $\mathcal{M}_i \in \heartsuit(X_i)$, $\alpha_\phi^\heartsuit : (X_\phi)_\heartsuit^*(\mathcal{M}_i) \to \mathcal{M}_j$ is a morphism of $\heartsuit(X_\bullet)$ for any $\phi : i \to j$, and the conditions corresponding to **1,2** in Lemma 4.8 are satisfied.

We say that $\mathcal{M} \in \heartsuit(X_\bullet)$ is *equivariant* if $\alpha_\phi^\heartsuit$ is an isomorphism for any $\phi \in \mathrm{Mor}(I)$. Note that equivariance depends on $\heartsuit$, and is not preserved by the forgetful functors in general. We denote the full subcategory of $\mathrm{Mod}(X_\bullet)$ consisting of equivariant objects by $\mathrm{EM}(X_\bullet)$.

(4.15) Let $\heartsuit$ be Mod or AB, and $\mathcal{M} \in \heartsuit(X_\bullet)$. For a morphism $\phi : i \to j$, the diagram

$$
\begin{array}{ccc}
q\mathcal{M}_i & \xrightarrow{\ \ \beta\ \ } & q(X_\phi)_*\mathcal{M}_j \\
\downarrow c & & \downarrow c \\
(q\mathcal{M})_i \xrightarrow{\beta} (X_\phi)_*(q\mathcal{M})_j & \xrightarrow{c} & (X_\phi)_*q\mathcal{M}_j
\end{array}
\tag{4.16}
$$

is commutative. This is checked at the section level directly. Utilizing this fact, we have the following.

Lemma 4.17. *Let $\mathcal{M} \in \mathrm{PM}(X_\bullet)$. For $\phi : i \to j$, the diagram*

$$
\begin{array}{ccc}
a\mathcal{M}_i \xrightarrow{\beta} a(X_\phi)_*\mathcal{M}_j \xrightarrow{\theta} (X_\phi)_*a\mathcal{M}_j \\
\downarrow \theta \qquad\qquad\qquad\qquad \downarrow \theta \\
(a\mathcal{M})_i \xrightarrow{\qquad\qquad \beta \qquad\qquad} (X_\phi)_*(a\mathcal{M})_j
\end{array}
$$

is commutative.

Proof. Straightforward diagram drawing. $\qquad\qquad\qquad\qquad\qquad\qquad$ $\square$

(4.18) Let $X_\bullet \in \mathcal{P}(I, \underline{\mathrm{Sch}}/S)$, and $\phi : i \to j$ be a morphism of I. Let $\mathcal{M} \in \mathrm{PM}(X_\bullet)$. Then $\alpha_\phi : X_\phi^*\mathcal{M}_i \to \mathcal{M}_j$ is the composite

$$
X_\phi^*\mathcal{M}_i \xrightarrow{\beta} X_\phi^*(X_\phi)_*\mathcal{M}_j \xrightarrow{\varepsilon} \mathcal{M}_j.
\tag{4.19}
$$

Thus for $U \in \mathrm{Zar}(X_j)$, α_ϕ is given by

$$
\Gamma(U, X_\phi^*\mathcal{M}_i) = \varinjlim \Gamma(U, \mathcal{O}_{X_j}) \otimes_{\Gamma(V, \mathcal{O}_{X_i})} \Gamma((i, V), \mathcal{M})
$$
$$
\to \Gamma((j, U), \mathcal{M}) = \Gamma(U, \mathcal{M}_j),
$$

where the colimit is taken over the open subsets V of X_i such that $U \subset X_\phi^{-1}(V)$, and the arrow is given by $a \otimes m \mapsto a\,\mathrm{res}_{(j,U),(i,V)}\,m$, see (2.20).

(4.20) Let $X_\bullet$ and $\phi : i \to j$ be as in (4.18). Let $\mathcal{M} \in \mathrm{Mod}(X_\bullet)$. Then $\alpha_\phi : X_\phi^*\mathcal{M}_i \to \mathcal{M}_j$ is also given by the composite (4.19). By (2.20), it is given by the composite

$$
X_\phi^*(?)_i = aX_\phi^*q(?)_i \xrightarrow{\beta} aX_\phi^*q(X_\phi)_*(?)_j \xrightarrow{c} aX_\phi^*(X_\phi)_*q(?)_j \xrightarrow{\varepsilon} aq(?)_j \xrightarrow{\varepsilon} (?)_j.
$$

By the commutativity of (4.16), it is easy to see that it agrees with

$$X_\phi^*(?)_i = aX_\phi^*q(?)_i \xrightarrow{c} aX_\phi^*(?)_i q \xrightarrow{\alpha_\phi} a(?)_j q \xrightarrow{c} aq(?)_j \xrightarrow{\varepsilon} (?)_j.$$

Thus if $\mathcal{M} \in \mathrm{Mod}(X_\bullet)$ is equivariant as an object of $\mathrm{PM}(X_\bullet)$, then it is equivariant as an object of $\mathrm{Mod}(X_\bullet)$.

Chapter 5
The Left and Right Inductions and the Direct and Inverse Images

Let I be a small category, S a scheme, and $X_\bullet \in \mathcal{P}(I, \underline{\mathrm{Sch}}/S)$.

(5.1) Let J be a subcategory of I. The left adjoint $Q(X_\bullet, J)_\#^\heartsuit$ of $(?)_J^\heartsuit$ (see (4.13)) is denoted by $L_J^\heartsuit$ for $\heartsuit = \mathrm{PA}, \mathrm{AB}, \mathrm{PM}, \mathrm{Mod}$. The right adjoint $Q(X_\bullet, J)_\flat^\heartsuit$ of $(?)_J^\heartsuit$, which exists by Lemma 2.31, is denoted by $R_J^\heartsuit$ for $\heartsuit = \mathrm{PA}, \mathrm{AB}, \mathrm{PM}, \mathrm{Mod}$. We call $L_J^\heartsuit$ and $R_J^\heartsuit$ the left and right induction functor, respectively.

Let $J_1 \subset J \subset I$ be subcategories of I. The left and right adjoints of $(?)_{J_1, J}^\heartsuit$ are denoted by $L_{J, J_1}^\heartsuit$ and $R_{J, J_1}^\heartsuit$, respectively. As $(?)_J^\heartsuit$ has both a left adjoint and a right adjoint, we have

Lemma 5.2. *The functor $(?)_J^\heartsuit$ preserves arbitrary limits and colimits (hence is exact) for $\heartsuit = \mathrm{PA}, \mathrm{AB}, \mathrm{PM}, \mathrm{Mod}$.*

The functor $\heartsuit(X_\bullet) \to \prod_{i \in I} \heartsuit(X_i)$ given by $\mathcal{F} \mapsto (\mathcal{F}_i)_{i \in I}$ is faithful for $\heartsuit = \mathrm{PA}, \mathrm{AB}, \mathrm{PM}, \mathrm{Mod}$.

(5.3) Let $f_\bullet : X_\bullet \to Y_\bullet$ be a morphism in $\mathcal{P}(I, \underline{\mathrm{Sch}}/S)$. This induces an obvious ringed continuous functor

$$f_\bullet^{-1} : (\mathrm{Zar}(Y_\bullet), \mathcal{O}_{Y_\bullet}) \to (\mathrm{Zar}(X_\bullet), \mathcal{O}_{X_\bullet}).$$

We have $\mathrm{id}^{-1} = \mathrm{id}$, and $(g_\bullet \circ f_\bullet)^{-1} = f_\bullet^{-1} \circ g_\bullet^{-1}$ for $g_\bullet : Y_\bullet \to Z_\bullet$.

We define the *direct image* $(f_\bullet)_*^\heartsuit$ to be $(f_\bullet^{-1})_*^\#$, and the *inverse image* $(f_\bullet)_\heartsuit^*$ to be $(f_\bullet^{-1})_\#^\heartsuit$ for $\heartsuit = \mathrm{Mod}, \mathrm{PM}, \mathrm{AB}, \mathrm{PA}$.

Lemma 5.4. *Let $f_\bullet : X_\bullet \to Y_\bullet$ be a morphism in $\mathcal{P}(I, \underline{\mathrm{Sch}}/S)$, and $K \subset J \subset I$. Then we have*

1 $Q(X_\bullet, J) \circ Q(X_\bullet|_J, K) = Q(X_\bullet, K)$
2 $f_\bullet^{-1} \circ Q(Y_\bullet, J) = Q(X_\bullet, J) \circ (f_\bullet|_J)^{-1}$.

J. Lipman, M. Hashimoto, *Foundations of Grothendieck Duality for Diagrams*   327
of Schemes, Lecture Notes in Mathematics 1960,
© Springer-Verlag Berlin Heidelberg 2009

(5.5)   Let us fix I and S. By Lemma 2.43 and Lemma 2.55, we have various natural maps between functors on sheaves arising from the closed structures and the monoidal pairs, involving various J-diagrams of schemes, where J varies subcategories of I. In the sequel, many of the natural maps are referred as 'the canonical maps' or 'the canonical isomorphisms' without any explicit definitions. Many of them are defined in [26] and Chapter 1, and various commutativity theorems are proved there.

Example 5.6. Let I be a small category, S a scheme, and $f_\bullet : X_\bullet \to Y_\bullet$ and $g_\bullet : Y_\bullet \to Z_\bullet$ are morphisms in $\mathcal{P}(I, \underline{\mathrm{Sch}}/S)$. Let $K \subset J \subset I$ be subcategories, and $\heartsuit$ denote PM, Mod, PA, or AB.

1 There is a natural isomorphism

$$c^\heartsuit_{I,J,K} : (?)^\heartsuit_{K,I} \cong (?)^\heartsuit_{K,J} \circ (?)^\heartsuit_{J,I}.$$

Taking the conjugate,

$$d^\heartsuit_{I,J,K} : L^\heartsuit_{I,J} \circ L^\heartsuit_{J,K} \cong L^\heartsuit_{I,K}$$

is induced.

2 There is a natural isomorphism

$$c^\heartsuit_{J,f_\bullet} : (?)^\heartsuit_J \circ (f_\bullet)^\heartsuit_* \cong (f_\bullet|_J)^\heartsuit_* \circ (?)^\heartsuit_J$$

and its conjugate

$$d^\heartsuit_{J,f_\bullet} : L^\heartsuit_J \circ (f_\bullet|_J)^*_\heartsuit \cong (f_\bullet)^*_\heartsuit \circ L^\heartsuit_J.$$

3 We have

$$(c^\heartsuit_{K,f_\bullet|_J}(?)^\heartsuit_J) \circ ((?)^\heartsuit_{K,J} c^\heartsuit_{J,f_\bullet}) = ((f_\bullet|_K)^\heartsuit_* c^\heartsuit_{I,J,K}) \circ c^\heartsuit_{K,f_\bullet} \circ ((c^\heartsuit_{I,J,K})^{-1}(f_\bullet)^\heartsuit_*).$$

4 We have

$$((g_\bullet|_J)^\heartsuit_* c^\heartsuit_{J,f_\bullet}) \circ (c^\heartsuit_{J,g_\bullet}(f_\bullet)^\heartsuit_*) = (c^\heartsuit_{f_\bullet|_J,g_\bullet|_J}(?)^\heartsuit_J) \circ c^\heartsuit_{J,g_\bullet \circ f_\bullet} \circ ((?)^\heartsuit_J (c^\heartsuit_{f_\bullet,g_\bullet})^{-1}),$$

where $c^\heartsuit_{f_\bullet,g_\bullet} : (g_\bullet \circ f_\bullet)^\heartsuit_* \cong (g_\bullet)^\heartsuit_* \circ (f_\bullet)^\heartsuit_*$ is the canonical isomorphism, and similarly for $c^\heartsuit_{f_\bullet|_J,g_\bullet|_J}$.

5 The canonical map

$$m_J : \mathcal{M}_J \otimes_{\mathcal{O}_{X_\bullet|_J}} \mathcal{N}_J \to (\mathcal{M} \otimes_{\mathcal{O}_{X_\bullet}} \mathcal{N})_J$$

is an isomorphism, as can be seen easily (the corresponding assertion for PM is obvious. Utilize Lemma 5.7 below to show the case of Mod). The canonical map

$$\Delta : L_J(\mathcal{M} \otimes_{\mathcal{O}_{X_\bullet|_J}} \mathcal{N}) \cong (L_J\mathcal{M}) \otimes_{\mathcal{O}_{X_\bullet}} (L_J\mathcal{N}).$$

is defined, which may not be an isomorphism.

Lemma 5.7. *Let I be a small category, S a scheme, and $X_\bullet \in \mathcal{P}(I, \underline{\mathrm{Sch}}/S)$. Let J be a subcategory of I. Then the natural map*

$$\theta = \bar{\theta} : a(?)_J^{\mathrm{PM}} \to (?)_J^{\mathrm{Mod}} a$$

is an isomorphism.

Proof. Obvious by Lemma 2.31. $\qquad\square$

Chapter 6
Operations on Sheaves Via the Structure Data

Let I be a small category, S a scheme, and $\mathcal{P} := \mathcal{P}(I, \underline{\mathrm{Sch}}/S)$. To study sheaves on objects of $\mathcal{P}$, it is convenient to utilize the structure data of them, and then utilize the usual sheaf theory on schemes.

(6.1) Let $X_\bullet \in \mathcal{P}$. Let $\heartsuit$ be any of $\mathrm{PA}, \mathrm{AB}, \mathrm{PM}, \mathrm{Mod}$, and $\mathcal{M}, \mathcal{N} \in \heartsuit(X_\bullet)$. An element (φ_i) in $\prod \mathrm{Hom}_{\heartsuit(X_i)}(\mathcal{M}_i, \mathcal{N}_i)$ is given by some $\varphi \in \mathrm{Hom}_{\heartsuit(X_\bullet)}(\mathcal{M}, \mathcal{N})$ (by the canonical faithful functor $\heartsuit(X_\bullet) \to \prod \heartsuit(X_i)$), if and only if

$$\varphi_j \circ \alpha_\phi(\mathcal{M}) = \alpha_\phi(\mathcal{N}) \circ (X_\phi)_\heartsuit^*(\varphi_i) \tag{6.2}$$

holds (or equivalently, $\beta_\phi(\mathcal{N}) \circ \varphi_i = (X_\phi)_* \varphi_j \circ \beta_\phi(\mathcal{M})$ holds) for any $(\phi : i \to j) \in \mathrm{Mor}(I)$.

We say that a family of morphisms $(\varphi_i)_{i \in I}$ between structure data

$$\varphi_i : \mathcal{M}_i \to \mathcal{N}_i$$

is a morphism of structure data if φ_i is a morphism in $\heartsuit(X_i)$ for each i, and (6.2) is satisfied for any ϕ. Thus the categories of structure data of sheaves, presheaves, modules, and premodules on $X_\bullet$, denoted by $\mathfrak{D}_\heartsuit(X_\bullet)$ are defined, and the equivalence $\mathrm{Dat}_\heartsuit : \heartsuit(X_\bullet) \cong \mathfrak{D}_\heartsuit(X_\bullet)$ are given. This is the precise meaning of Lemma 4.8.

(6.3) Let $X_\bullet \in \mathcal{P}$ and $\mathcal{M}, \mathcal{N} \in \mathrm{Mod}(X_\bullet)$. As in Example 5.6, **5**, we have an isomorphism

$$m_i : \mathcal{M}_i \otimes_{\mathcal{O}_{X_i}} \mathcal{N}_i \cong (\mathcal{M} \otimes_{\mathcal{O}_{X_\bullet}} \mathcal{N})_i.$$

This is trivial for presheaves, and utilize the fact the sheafification is compatible with $(?)_i$ for sheaves. At the section level, for $\mathcal{M}, \mathcal{N} \in \mathrm{PM}(X_\bullet)$, $i \in I$, and $U \in \mathrm{Zar}(X_i)$,

$$m_i^p : \Gamma(U, \mathcal{M}_i \otimes_{\mathcal{O}_{X_i}}^p \mathcal{N}_i) \to \Gamma(U, (\mathcal{M} \otimes_{\mathcal{O}_{X_\bullet}}^p \mathcal{N})_i)$$

J. Lipman, M. Hashimoto, *Foundations of Grothendieck Duality for Diagrams of Schemes*, Lecture Notes in Mathematics 1960,
© Springer-Verlag Berlin Heidelberg 2009

is nothing but the identification

$$\Gamma(U, \mathcal{M}_i) \otimes_{\Gamma(U, \mathcal{O}_{X_i})} \Gamma(U, \mathcal{N}_i) = \Gamma((i, U), \mathcal{M}) \otimes_{\Gamma((i,U), \mathcal{O}_{X_\bullet})} \Gamma((i, U), \mathcal{N})$$
$$= \Gamma((i, U), \mathcal{M} \otimes^p_{\mathcal{O}_{X_\bullet}} \mathcal{N}).$$

For $\mathcal{M}, \mathcal{N} \in \mathrm{Mod}(X_\bullet)$ and $i \in I$, m_i is given as the composite

$$\mathcal{M}_i \otimes_{\mathcal{O}_{X_i}} \mathcal{N}_i = a(q\mathcal{M}_i \otimes^p_{\mathcal{O}_{X_i}} q\mathcal{N}_i) \xrightarrow{c} a((q\mathcal{M})_i \otimes^p_{\mathcal{O}_{X_i}} (q\mathcal{N})_i) \xrightarrow{m_i^p}$$
$$a(q\mathcal{M} \otimes^p_{\mathcal{O}_{X_\bullet}} q\mathcal{N})_i \xrightarrow{\theta} (a(q\mathcal{M} \otimes^p_{\mathcal{O}_{X_\bullet}} q\mathcal{N}))_i = (\mathcal{M} \otimes_{\mathcal{O}_{X_\bullet}} \mathcal{N})_i,$$

see (2.52). Utilizing this identification, the structure map α_ϕ of $\mathcal{M} \otimes \mathcal{N}$ can be completely described via those of $\mathcal{M}$ and $\mathcal{N}$. Namely,

Lemma 6.4. *Let* $X_\bullet \in \mathcal{P}(I, \underline{\mathrm{Sch}}/S)$, *and* $\mathcal{M}, \mathcal{N} \in \heartsuit(X_\bullet)$, *where* $\heartsuit$ *is* PM *or* Mod. *For* $\phi \in I(i, j)$, $\alpha_\phi(\mathcal{M} \otimes \mathcal{N})$ *agrees with the composite map*

$$X_\phi^*(\mathcal{M} \otimes \mathcal{N})_i \xrightarrow{m_i^{-1}} X_\phi^*(\mathcal{M}_i \otimes \mathcal{N}_i) \xrightarrow{\Delta} X_\phi^* \mathcal{M}_i \otimes X_\phi^* \mathcal{N}_i \xrightarrow{\alpha_\phi \otimes \alpha_\phi} \mathcal{M}_j \otimes \mathcal{N}_j \xrightarrow{m_j} (\mathcal{M} \otimes \mathcal{N})_j,$$

where $\otimes$ *should be replaced by* $\otimes^p$ *when* $\heartsuit = $ PM.

Proof (sketch). It is not so difficult to show that it suffices to show that $\beta_\phi(\mathcal{M} \otimes \mathcal{N})$ agrees with the composite

$$(\mathcal{M} \otimes \mathcal{N})_i \xrightarrow{m_i^{-1}} \mathcal{M}_i \otimes \mathcal{N}_i \xrightarrow{\beta \otimes \beta} (X_\phi)_* \mathcal{M}_j \otimes (X_\phi)_* \mathcal{N}_j \xrightarrow{m}$$
$$(X_\phi)_*(\mathcal{M}_j \otimes \mathcal{N}_j) \xrightarrow{m_j} (X_\phi)_*(\mathcal{M} \otimes \mathcal{N})_j. \quad (6.5)$$

First we prove this for the case that $\heartsuit = $ PM. For an open subset U of X_i, this composite map evaluated at U is

$$\Gamma((i, U), (\mathcal{M} \otimes \mathcal{N})) = \Gamma((i, U), \mathcal{M}) \otimes_{\Gamma((i,U), \mathcal{O}_{X_\bullet})} \Gamma((i, U), \mathcal{N}) \xrightarrow{\mathrm{res} \otimes \mathrm{res}}$$
$$\Gamma((j, X_\phi^{-1}(U)), \mathcal{M}) \otimes_{\Gamma((i,U), \mathcal{O}_{X_\bullet})} \Gamma((j, X_\phi^{-1}(U)), \mathcal{N}) \xrightarrow{p}$$
$$\Gamma((j, X_\phi^{-1}(U)), \mathcal{M}) \otimes_{\Gamma((j, X_\phi^{-1}(U)), \mathcal{O}_{X_\bullet})} \Gamma((j, X_\phi^{-1}(U)), \mathcal{N})$$
$$= \Gamma((j, X_\phi^{-1}(U)), \mathcal{M} \otimes \mathcal{N}),$$

where $p(m \otimes n) = m \otimes n$. This composite map is nothing but the restriction map of $\mathcal{M} \otimes \mathcal{N}$. So by definition, it agrees with

$$\beta_\phi : \Gamma(U, (\mathcal{M} \otimes \mathcal{N})_i) \to \Gamma(U, (X_\phi)_*(\mathcal{M} \otimes \mathcal{N})_j).$$

Next we consider the case $\heartsuit = $ Mod. First note that the diagram

$$(a(q\mathcal{M} \otimes^p q\mathcal{N}))_i \xrightarrow{\quad\beta\quad} (X_\phi)_*(a(q\mathcal{M} \otimes^p q\mathcal{N}))_j$$
$$\uparrow \theta \qquad\qquad\qquad\qquad\qquad\qquad\qquad\qquad \uparrow \theta$$
$$a(q\mathcal{M} \otimes^p q\mathcal{N})_i \xrightarrow{\beta} a(X_\phi)_*(q\mathcal{M} \otimes^p q\mathcal{N})_j \xrightarrow{\theta} (X_\phi)_*a(q\mathcal{M} \otimes^p q\mathcal{N})_j$$

is commutative by Lemma 4.17. By the presheaf version of the lemma, which has been proved in the last paragraph, the diagram

$$a(q\mathcal{M} \otimes^p q\mathcal{N})_i \xleftarrow{\quad m_i \quad} a((q\mathcal{M})_i \otimes^p (q\mathcal{N})_i)$$

with vertical maps β on the left, $\beta\otimes\beta$ then m on the right:

$$a((X_\phi)_*(q\mathcal{M})_j \otimes^p (X_\phi)_*(q\mathcal{N})_j)$$

$$a(X_\phi)_*(q\mathcal{M} \otimes^p q\mathcal{N})_j \xleftarrow{\quad m_j \quad} a(X_\phi)_*((q\mathcal{M})_j \otimes^p (q\mathcal{N})_j)$$

is commutative. By the commutativity of the diagram (4.16), the diagram

$$a((q\mathcal{M})_i \otimes^p (q\mathcal{N})_i) \xrightarrow{\quad c\otimes c \quad} a(q\mathcal{M}_i \otimes^p q\mathcal{N}_i)$$

with $\beta\otimes\beta$ on the left and $\beta\otimes\beta$ then $c\otimes c$ on the right:

$$a(q(X_\phi)_*\mathcal{M}_j \otimes^p q(X_\phi)_*\mathcal{N}_j)$$

$$a((X_\phi)_*(q\mathcal{M})_j \otimes^p (X_\phi)_*(q\mathcal{N})_j) \xrightarrow{c\otimes c} a((X_\phi)_*q\mathcal{M}_j \otimes^p (X_\phi)_*q\mathcal{N}_j)$$

is commutative. Combining the commutativity of these three diagrams (and some other easy commutativity), it is not so difficult to show that the map

$$\beta : (\mathcal{M}\otimes\mathcal{N})_i = (a(q\mathcal{M}\otimes^p q\mathcal{N}))_i \to (X_\phi)_*(a(q\mathcal{M}\otimes^p q\mathcal{N}))_j = (X_\phi)_*(\mathcal{M}\otimes\mathcal{N})_j$$

agrees with the composite

$$(\mathcal{M}\otimes\mathcal{N})_i = (a(q\mathcal{M}\otimes^p q\mathcal{N}))_i \xrightarrow{\theta^{-1}} a(q\mathcal{M}\otimes^p q\mathcal{N})_i \xrightarrow{m_i^{-1}} a((q\mathcal{M})_i \otimes^p (q\mathcal{N})_i)$$

$$\xrightarrow{c\otimes c} a(q\mathcal{M}_i \otimes^p q\mathcal{N}_i) \xrightarrow{\beta\otimes\beta} a(q(X_\phi)_*\mathcal{M}_j \otimes^p q(X_\phi)_*\mathcal{N}_j) \xrightarrow{c\otimes c}$$

$$a((X_\phi)_*q\mathcal{M}_j \otimes^p (X_\phi)_*q\mathcal{N}_j) \xrightarrow{m} a(X_\phi)_*(q\mathcal{M}_j \otimes^p q\mathcal{N}_j) \xrightarrow{\theta} (X_\phi)_*a(q\mathcal{M}_j \otimes^p q\mathcal{N}_j)$$

$$\xrightarrow{c\otimes c} (X_\phi)_*a((q\mathcal{M})_j \otimes^p (q\mathcal{N})_j) \xrightarrow{m_j} (X_\phi)_*a((q\mathcal{M} \otimes^p q\mathcal{N})_j) \xrightarrow{\theta}$$

$$(X_\phi)_*(a(q\mathcal{M} \otimes^p q\mathcal{N}))_j = (X_\phi)_*(\mathcal{M} \otimes \mathcal{N})_j.$$

This composite map agrees with the composite map (6.5). This proves the lemma. $\square$

(6.6) Let $X_\bullet \in \mathcal{P}$, and J a subcategory of I. The left adjoint functor $L_J^\heartsuit = Q(X_\bullet, J)_\#^\heartsuit$ of $(?)_J^\heartsuit$ is given by the structure data as follows explicitly. For $\mathcal{M} \in \heartsuit(X_\bullet|_J)$ and $i \in I$, we have

Lemma 6.7. *There is an isomorphism*

$$\lambda_{J,i} : (L_J^\heartsuit(\mathcal{M}))_i^\heartsuit \cong \varinjlim(X_\phi)_\heartsuit^*(\mathcal{M}_j),$$

where the colimit is taken over the subcategory $(I_i^{(J^{\mathrm{op}} \to I^{\mathrm{op}})})^{\mathrm{op}}$ of I/i whose objects are $(\phi : j \to i) \in I/i$ with $j \in \mathrm{ob}(J)$ and morphisms are morphisms φ of I/i such that $\varphi \in \mathrm{Mor}(J)$. The translation map of the direct system is given as follows. For morphisms $\phi : j \to i$ and $\psi : j' \to j$, the translation map $X_{\phi\psi}^* \mathcal{M}_{j'} \to X_\phi^* \mathcal{M}_j$ is the composite

$$X_{\phi\psi}^* \mathcal{M}_{j'} \xrightarrow{d} X_\phi^* X_\psi^* \mathcal{M}_{j'} \xrightarrow{\alpha_\psi} X_\phi^* \mathcal{M}_j.$$

Proof. We prove the lemma for the case that $\heartsuit = \mathrm{PM}, \mathrm{Mod}$. The case that $\heartsuit = \mathrm{PA}, \mathrm{AB}$ is similar and easier.

Consider the case $\heartsuit = \mathrm{PM}$ first. For any object $(\phi, h) : (i, U) \to (j, V)$ of $I_{(i,U)}^{\mathrm{Zar}(X_\bullet|_J) \hookrightarrow \mathrm{Zar}(X_\bullet)}$, consider the obvious map

$$\Gamma((i, U), \mathcal{O}_{X_\bullet}) \otimes_{\Gamma((j,V), \mathcal{O}_{X_\bullet|_J})} \Gamma((j, V), \mathcal{M}) = \Gamma(U, \mathcal{O}_{X_i}) \otimes_{\Gamma(V, \mathcal{O}_{X_j})} \Gamma(V, \mathcal{M}_j)$$

$$\to \varinjlim_{X_\phi^{-1}(V') \supset U} \Gamma(U, \mathcal{O}_{X_i}) \otimes_{\Gamma(V', \mathcal{O}_{X_j})} \Gamma(V', \mathcal{M}_j)$$

$$= \Gamma(U, X_\phi^* \mathcal{M}_j) \to \varinjlim \Gamma(U, X_{\phi'}^* \mathcal{M}_{j'}),$$

where the last $\varinjlim$ is taken over $(\phi' : j' \to i) \in (I_i^{(J^{\mathrm{op}} \to I^{\mathrm{op}})})^{\mathrm{op}}$. This map induces a unique map

$$\Gamma(U, (L_J \mathcal{M})_i) = \Gamma((i, U), L_J \mathcal{M}) =$$
$$\varinjlim \Gamma((i, U), \mathcal{O}_{X_\bullet}) \otimes_{\Gamma((j,V), \mathcal{O}_{X_\bullet|_J})} \Gamma((j, V), \mathcal{M}) \to \varinjlim \Gamma(U, X_{\phi'}^* \mathcal{M}_{j'}).$$

It is easy to see that this defines $\lambda_{J,i}$.

We define the inverse of $\lambda_{J,i}$ explicitly. Let $(\phi : j \to i) \in (I_i^{(J^{\mathrm{op}} \to I^{\mathrm{op}})})^{\mathrm{op}}$. Let $U \in \mathrm{Zar}(X_i)$ and $V \in \mathrm{Zar}(X_j)$ such that $U \subset X_\phi^{-1}(V)$. We have an obvious map

$$\Gamma(U, \mathcal{O}_{X_i}) \otimes_{\Gamma(V, \mathcal{O}_{X_j})} \Gamma(V, \mathcal{M}_j) = \Gamma((i, U), \mathcal{O}_{X_\bullet}) \otimes_{\Gamma((j,V), \mathcal{O}_{X_\bullet|_J})} \Gamma((j, V), \mathcal{M})$$

$$\to \varinjlim \Gamma((i, U), \mathcal{O}_{X_\bullet}) \otimes_{\Gamma((j,V), \mathcal{O}_{X_\bullet|_J})} \Gamma((j, V), \mathcal{M})$$

$$= \Gamma((i, U), L_J \mathcal{M}) = \Gamma(U, (L_J \mathcal{M})_i),$$

which induces

$$\Gamma(U, X_\phi^* \mathcal{M}) = \varinjlim \Gamma(U, \mathcal{O}_{X_i}) \otimes_{\Gamma(V, \mathcal{O}_{X_j})} \Gamma(V, \mathcal{M}_j) \to \Gamma(U, (L_J \mathcal{M})_i).$$

This gives a morphism $X_\phi^* \mathcal{M} \to (L_J \mathcal{M})_i$. It is easy to see that this defines $\varinjlim X_\phi^* \mathcal{M} \to (L_J \mathcal{M})_i$, which is the inverse of $\lambda_{J,i}$. This completes the proof for the case that $\heartsuit = \mathrm{PM}$.

Now consider the case $\heartsuit = \mathrm{Mod}$. Define $\lambda_{J,i}^{\mathrm{Mod}}$ to be the composite

$$(L_J^{\mathrm{Mod}} \mathcal{M})_i = (?)_i a L_J^{\mathrm{PM}} q \mathcal{M} \xrightarrow{\theta^{-1}} a(?)_i L_J^{\mathrm{PM}} q \mathcal{M} \xrightarrow{\lambda_{J,i}^{\mathrm{PM}}} a \varinjlim X_\phi^* (q\mathcal{M})_j$$

$$\cong \varinjlim a X_\phi^* (q\mathcal{M})_j \xrightarrow{c} \varinjlim a X_\phi^* q \mathcal{M}_j = \varinjlim X_\phi^* \mathcal{M}_j.$$

As the morphisms appearing in the composition are all isomorphisms, $\lambda_{J,i}^{\mathrm{Mod}}$ is an isomorphism. □

In particular, we have an isomorphism

$$\lambda_{j,i} : (L_j^\heartsuit(\mathcal{M}))_i^\heartsuit \cong \bigoplus_{\phi \in I(j,i)} (X_\phi)_\heartsuit^*(\mathcal{M}). \tag{6.8}$$

(6.9) As announced in (2.61), we show that the monoidal adjoint pair $((?)_\#^{\mathrm{Mod}}, (?)_{\mathrm{Mod}}^\#)$ in Lemma 2.55 is not Lipman.

We define a finite category $\mathcal{K}$ by $\mathrm{ob}(\mathcal{K}) = \{s, t\}$, and $\mathcal{K}(s, t) = \{u, v\}$, $\mathcal{K}(s, s) = \{\mathrm{id}_s\}$, and $\mathcal{K}(t, t) = \{\mathrm{id}_t\}$. Pictorially, $\mathcal{K}$ looks like $t \overset{u}{\underset{v}{\rightleftarrows}} s$. Let k be a field, and define $X_\bullet \in \mathcal{P}(\mathcal{K}, \underline{\mathrm{Sch}})$ by $X_s = X_t = \mathrm{Spec}\, k$, and $X_u = X_v = \mathrm{id}$. Then $\Gamma(X_t, (L_s \mathcal{O}_{X_s})_t)$ is two-dimensional by (6.8). So $L_s \mathcal{O}_{X_s}$ and $\mathcal{O}_{X_\bullet}$ are not isomorphic by the dimension reason. Similarly, $L_s(\mathcal{O}_{X_s} \otimes_{\mathcal{O}_{X_s}} \mathcal{O}_{X_s})$ cannot be isomorphic to $L_s \mathcal{O}_{X_s} \otimes_{\mathcal{O}_{X_\bullet}} L_s \mathcal{O}_{X_s}$.

Similarly, $((?)_\#^{\mathrm{PM}}, (?)_{\mathrm{PM}}^\#)$ in Lemma 2.55 is not Lipman.

(6.10) Let $\psi : i \to i'$ be a morphism. The structure map

$$\alpha_\psi : (X_\psi)_\heartsuit^*((L_J^\heartsuit(\mathcal{M}))_i^\heartsuit) \to (L_J^\heartsuit(\mathcal{M}))_{i'}^\heartsuit$$

is induced by

$$(X_\psi)_\heartsuit^*((X_\phi)_\heartsuit^*(\mathcal{M}_j)) \cong (X_{\psi\phi})_\heartsuit^*(\mathcal{M}_j).$$

More precisely, for $\psi : i \to i'$, the diagram

$$
\begin{array}{ccc}
X_\psi^*((L_J\mathcal{M})_i) & \xrightarrow{\lambda_{J,i}} X_\psi^* \varinjlim X_\phi^* \mathcal{M}_j \cong \varinjlim X_\psi^* X_\phi^* \mathcal{M}_j \\
\downarrow \alpha_\psi & \downarrow h \\
(L_J\mathcal{M})_{i'} & \xrightarrow{\lambda_{J,i'}} \varinjlim X_{\phi'}^* \mathcal{M}_{j'}
\end{array}
$$

is commutative, where $\phi : i \to j$ runs through $(I_i^f)^{\mathrm{op}}$, and $\phi' : i' \to j'$ runs through $(I_{i'}^f)^{\mathrm{op}}$, where $f : J^{\mathrm{op}} \to I^{\mathrm{op}}$ is the inclusion. The map h is induced by $d : X_\psi^* X_\phi^* \to (X_\phi X_\psi)^* = X_{\psi\phi}^*$. This is checked at the section level directly when $\heartsuit = \mathrm{PM}$.

We consider the case that $\heartsuit = \mathrm{Mod}$. Then the composite

$$X_\psi^*(?)_i L_J \xrightarrow{\lambda_{J,i}} X_\psi^* \varinjlim X_\phi^*(?)_j \cong \varinjlim X_\psi^* X_\phi^*(?)_j \xrightarrow{h} \varinjlim X_{\phi'}^*(?)_{j'}$$

agrees with the composite

$$X_\psi^*(?)_i L_J = a X_\psi^* q(?)_i a L_J q \xrightarrow{\theta^{-1}} a X_\psi^* q a(?)_i L_J q \xrightarrow{\lambda_{J,i}^{\mathrm{PM}}} a X_\psi^* q a \varinjlim X_\phi^*(?)_j q$$

$$\xrightarrow{\cong} a X_\psi^* q \varinjlim a X_\phi^*(?)_j q \xrightarrow{c} a X_\psi^* q \varinjlim a X_\phi^* q(?)_j \xrightarrow{\cong} \varinjlim a X_\psi^* q a X_\phi^* q(?)_j \xrightarrow{u^{-1}}$$

$$\varinjlim a X_\psi^* X_\phi^* q(?)_j \xrightarrow{d} \varinjlim a X_{\psi\phi}^* q(?)_j \to \varinjlim a X_{\phi'}^* q(?)_{j'} = \varinjlim X_{\phi'}^*(?)_{j'}.$$

Using Lemma 2.60, it is straightforward to show that this map agrees with

$$X_\psi^*(?)_i L_J = a X_\psi^* q(?)_i a L_J q \xrightarrow{c} a X_\psi^*(?)_i q a L_J q \xrightarrow{\alpha_\psi} a(?)_{i'} q a L_J q \xrightarrow{c}$$

$$a q(?)_{i'} a L_J q \xrightarrow{\varepsilon} (?)_{i'} a L_J q \xrightarrow{\theta^{-1}} a(?)_{i'} L_J q \xrightarrow{\lambda_{J,i'}} a \varinjlim X_{\phi'}^*(?)_{j'} q \xrightarrow{\cong}$$

$$\varinjlim a X_{\phi'}^*(?)_{j'} q \xrightarrow{c} \varinjlim a X_{\phi'}^* q(?)_{j'} = \varinjlim X_{\phi'}^*(?)_{j'}.$$

This composite map agrees with

$$X_\psi^*(?)_i L_J \xrightarrow{\alpha_\psi} (?)_{i'} L_J \xrightarrow{\lambda_{J,i'}} \varinjlim X_{\phi'}^*(?)_{j'}$$

by (4.20) and the definition of $\lambda_{J,i'}$ for sheaves (see the proof of Lemma 6.7). This is what we wanted to prove.

The case that $\heartsuit = \mathrm{PA}, \mathrm{AB}$ is proved similarly.

(6.11) In the remainder of this chapter, we do not give detailed proofs, since the strategy is similar to the above (just check the commutativity at the section level for presheaves, and sheafify it).

(6.12) The counit map $\varepsilon : L_J(?)_J \to \mathrm{Id}$ is given as a morphism of structure data as follows.

$$\varepsilon_i : (?)_i L_J(?)_J \to (?)_i$$

agrees with

$$(?)_i L_J(?)_J \xrightarrow{\lambda_{J,i}} \varinjlim X_\phi^*(?)_j(?)_J \xrightarrow{c} \varinjlim X_\phi^*(?)_j \xrightarrow{\alpha} (?)_i,$$

where α is induced by $\alpha_\phi : X_\phi^*(?)_j \to (?)_i$.

(6.13) The unit map $u : \mathrm{Id} \to (?)_J L_J$ is also described, as follows.

$$u_j : (?)_j \to (?)_j(?)_J L_J$$

agrees with

$$(?)_j \xrightarrow{\mathrm{f}^{-1}} X^*_{\mathrm{id}_j}(?)_j \to \varinjlim X^*_\phi(?)_k \xrightarrow{\lambda^{-1}_{J,j}} (?)_j L_J \cong (?)_j(?)_J L_J,$$

where the colimit is taken over $(\phi : k \to j) \in (I_j^{(J^{\mathrm{op}} \subset I^{\mathrm{op}})})^{\mathrm{op}}$.

(6.14) Let $X_\bullet \in \mathcal{P}$, and J a subcategory of I. The right adjoint functor $R_J^\heartsuit$ of $(?)_J^\heartsuit$ is given as follows explicitly. For $\mathcal{M} \in \heartsuit(X_\bullet|_J)$ and $i \in I$, we have

$$\rho^{J,i} : (R_J^\heartsuit(\mathcal{M}))_i^\heartsuit \cong \varprojlim(X_\phi)_*^\heartsuit(\mathcal{M}_j),$$

where the limit is taken over $I_i^{(J \to I)}$, see (2.6) for the notation. The descriptions of α, u, and ε for the right induction are left to the reader.

Lemma 6.15. *Let $X_\bullet \in \mathcal{P}$, and J a full subcategory of I. Then we have the following.*

1 *The counit of adjunction $\varepsilon : (?)_J^\heartsuit \circ R_J^\heartsuit \to \mathrm{Id}$ is an isomorphism. In particular, $R_J^\heartsuit$ is full and faithful.*
2 *The unit of adjunction $u : \mathrm{Id} \to (?)_J^\heartsuit \circ L_J^\heartsuit$ is an isomorphism. In particular, $L_J^\heartsuit$ is full and faithful.*

Proof. **1** For $i \in J$, the restriction

$$\varepsilon_i : (?)_i^\heartsuit (?)_J^\heartsuit R_J^\heartsuit \mathcal{M} = \varprojlim(X_\phi)_*^\heartsuit(\mathcal{M}_j) \to (X_{\mathrm{id}_i})_* \mathcal{M}_i = \mathcal{M}_i = (?)_i \mathcal{M}$$

is nothing but the canonical map from the projective limit, where the limit is taken over $(\phi : i \to j) \in I_i^{(J \to I)}$. As J is a full subcategory, we have $I_i^{(J \to I)}$ equals i/J, and hence id_i is its initial object. So the limit is equal to $\mathcal{M}_i$, and ε_i is the Identity map. Since ε_i is an Isomorphism for each $i \in J$, we have that ε is an isomorphism.

The proof of **2** is similar, and we omit it.   □

Let $\mathcal{C}$ be a small category. A connected component of $\mathcal{C}$ is a full subcategory of $\mathcal{C}$ whose object set is one of the equivalence classes of $\mathrm{ob}(\mathcal{C})$ with respect to the transitive symmetric closure of the relation $\sim$ given by

$$c \sim c' \iff \mathcal{C}(c, c') \neq \emptyset.$$

Definition 6.16. We say that a subcategory J of I is *admissible* if

1 For $i \in I$, the category $(I_i^{(J^{\mathrm{op}} \subset I^{\mathrm{op}})})^{\mathrm{op}}$ is pseudofiltered.
2 For $j \in J$, we have id_j is the initial object of one of the connected components of $I_j^{(J^{\mathrm{op}} \subset I^{\mathrm{op}})}$ (i.e., id_j is the terminal object of one of the connected components of $(I_j^{(J^{\mathrm{op}} \subset I^{\mathrm{op}})})^{\mathrm{op}}$).

Note that for $j \in I$, the subcategory $j = (\{j\}, \{\mathrm{id}_j\})$ of I is admissible.

In Lemma 6.7, the colimit in the right hand side is pseudo-filtered and hence it preserves exactness, if **1** is satisfied. In particular, if **1** is satisfied, then $Q(X_\bullet, J) : \mathrm{Zar}(X_\bullet|_J) \to \mathrm{Zar}(X_\bullet)$ is an admissible functor. As in the proof of Lemma 6.15, $(?)_j$ is a direct summand of $(?)_j \circ L_J$ for $j \in J$ so that L_J is faithful, if **2** is satisfied. We have the following.

Lemma 6.17. *Let $X_\bullet \in \mathcal{P}(I, \underline{\mathrm{Sch}}/S)$, and $K \subset J \subset I$ be admissible subcategories of I. Then $L_{J,K}^{\mathrm{PA}}$ is faithful and exact. The morphism of sites $Q(X_\bullet|_J, K)$ is admissible. If, moreover, X_ϕ is flat for any $\phi \in I(k, j)$ with $j \in J$ and $k \in K$, then $L_{J,K}^\heartsuit$ is faithful and exact for $\heartsuit = \mathrm{Mod}$.*

Proof. Assume that $\mathcal{M} \in \heartsuit(X_\bullet|_K)$, $\mathcal{M} \neq 0$, and $L_{J,K}\mathcal{M} = 0$. There exists some $k \in K$ such that $M_k \neq 0$. Since $L_{J,K}\mathcal{M} = 0$, we have that $0 \cong (?)_k L_{I,J} L_{J,K} \mathcal{M} \cong (?)_k L_{I,K} \mathcal{M}$. This contradicts the fact that $\mathcal{M}_k$ is a direct summand of $(L_{I,K}\mathcal{M})_k$. Hence $L_{J,K}$ is faithful.

We prove that $L_{J,K}^\heartsuit$ is exact. It suffices to show that for any $j \in J$, $(?)_j L_{J,K}$ is exact. As J is admissible, $(?)_j$ is a direct summand of $(?)_j L_{I,J}$. Hence it suffices to show that $(?)_j L_{I,K} \cong (?)_j L_{I,J} L_{J,K}$ is exact. By Lemma 6.7, $(?)_j L_{I,K} \cong \varinjlim (X_\phi)_\heartsuit^* (?)_k$, where the colimit is taken over $(\phi \colon k \to j) \in (I_j^{K^{\mathrm{op}} \subset I^{\mathrm{op}}})^{\mathrm{op}}$. By assumption, $(X_\phi)_\heartsuit^*$ is exact for any ϕ in the colimit. As $(I_j^{K^{\mathrm{op}} \subset I^{\mathrm{op}}})^{\mathrm{op}}$ is pseudo-filtered by assumption, $(?)_j L_{I,K}$ is exact, as desired. $\square$

(6.18) As in Example 5.6, **2**, we have an isomorphism

$$c_{i,f_\bullet} : (?)_i \circ (f_\bullet)_* \cong (f_i)_* \circ (?)_i. \tag{6.19}$$

The translation α_ϕ is described as follows.

Lemma 6.20. *Let $f_\bullet \colon X_\bullet \to Y_\bullet$ be a morphism in $\mathcal{P}(I, \underline{\mathrm{Sch}}/S)$. For $\phi \in I(i, j)$,*

$$\alpha_\phi(f_\bullet)_* : Y_\phi^*(?)_i (f_\bullet)_* \to (?)_j (f_\bullet)_*$$

agrees with

$$Y_\phi^*(?)_i(f_\bullet)_* \xrightarrow{c_{i,f_\bullet}} Y_\phi^*(f_i)_*(?)_i \xrightarrow{\text{via } \theta} (f_j)_* X_\phi^*(?)_i$$

$$\xrightarrow{(f_j)_* \alpha_\phi} (f_j)_*(?)_j \xrightarrow{c_{j,f_\bullet}^{-1}} (?)_j(f_\bullet)_*, \tag{6.21}$$

where θ is Lipman's theta [26, (3.7.2)].

One of the definitions of θ is the composite

$$\theta : Y_\phi^*(f_i)_* \xrightarrow{\text{via } u} Y_\phi^*(f_i)_*(X_\phi)_* X_\phi^* \xrightarrow{c} Y_\phi^*(Y_\phi)_*(f_j)_* X_\phi^* \xrightarrow{\text{via } \varepsilon} (f_j)_* X_\phi^*.$$

Proof. Note that the diagram

$$
\begin{array}{ccc}
(?)_i(f_\bullet)_* & \xrightarrow{\beta} (Y_\phi)_*(?)_j(f_\bullet)_* \xrightarrow{c} (Y_\phi)_*(f_j)_*(?)_j \\
\downarrow c & \downarrow c \\
(f_i)_*(?)_i & \xrightarrow{\quad (f_i)_*\beta \quad} (f_i)_*(X_\phi)_*(?)_j
\end{array}
\tag{6.22}
$$

is commutative. Indeed, when we apply the functor $\Gamma(U, ?)$ for an open subset U of Y_i, then we get an obvious commutative diagram

$$
\begin{array}{ccc}
\Gamma((i, f_i^{-1}(U)), ?) & \xrightarrow{\text{res}} \Gamma((j, f_j^{-1}(Y_\phi^{-1}(U))), ?) & \xrightarrow{\text{id}} \Gamma((j, f_j^{-1}(Y_\phi^{-1}(U))), ?) \\
\downarrow \text{id} & & \downarrow \text{id} \\
\Gamma((i, f_i^{-1}(U)), ?) & \xrightarrow{\qquad\qquad \text{res} \qquad\qquad} & \Gamma((j, X_\phi^{-1}(f_i^{-1}(U))), ?).
\end{array}
$$

Now the assertion of the lemma follows from the commutativity of the diagram

$$
\begin{array}{ccccccc}
Y_\phi^*(?)_i(f_\bullet)_* & \xrightarrow{\alpha} & (?)_j(f_\bullet)_* & \xrightarrow{c} & (f_j)_*(?)_j & \xrightarrow{\text{id}} & (f_j)_*(?)_j \\
\downarrow \text{id} \quad (a) & & \uparrow \varepsilon \quad (b) & & \uparrow \varepsilon & & \\
Y_\phi^*(?)_i(f_\bullet)_* & \xrightarrow{\beta} & Y_\phi^*(Y_\phi)_*(?)_j(f_\bullet)_* & \xrightarrow{c} & Y_\phi^*(Y_\phi)_*(f_j)_*(?)_j & & \\
\downarrow c & & (c) & & \downarrow c & & \\
Y_\phi^*(f_i)_*(?)_i & & \xrightarrow{\quad\quad \beta \quad\quad} & & Y_\phi^*(f_i)_*(X_\phi)_*(?)_j \quad (f) & & \downarrow \text{id} \\
\downarrow \theta & & (d) & & \downarrow \theta & & \\
(f_j)_*X_\phi^*(?)_i & & \xrightarrow{\quad\quad \beta \quad\quad} & & (f_j)_*X_\phi^*(X_\phi)_*(?)_j & & \\
\downarrow \text{id} & & (e) & & \downarrow \varepsilon & & \\
(f_j)_*X_\phi^*(?)_i & & \xrightarrow{\quad\quad \alpha \quad\quad} & & (f_j)_*(?)_j & \xleftarrow{\text{id}} & (f_j)_*(?)_j.
\end{array}
$$

Indeed, the commutativity of (a) and (e) is the definition of α. The commutativity of (b) follows from the naturality of ε. The commutativity of (c) follows from the commutativity of (6.22). The commutativity of (d) is the naturality of θ. The commutativity of (f) follows from the definition of θ and the fact that the composite

$$
(X_\phi)_* \xrightarrow{u} (X_\phi)_*X_\phi^*(X_\phi)_* \xrightarrow{\varepsilon} (X_\phi)_*
$$

is the identity.   □

Proposition 6.23. *Let* $f_\bullet : X_\bullet \to Y_\bullet$ *be a morphism in* $\mathcal{P}$, J *a subcategory of* I, *and* $i \in I$. *Then the composite map*

$$
(?)_i L_J(f_\bullet|_J)_* \xrightarrow{\text{via } \theta} (?)_i(f_\bullet)_* L_J \xrightarrow{\text{via } c_{i,f_\bullet}} (f_i)_*(?)_i L_J
$$

agrees with the composite map

$$(?)_i L_J(f_\bullet|_J)_* \xrightarrow{\text{via }\lambda_{J,i}} \varinjlim Y_\phi^*(?)_j(f_\bullet|_J)_* \xrightarrow{\text{via }c_{j,f_\bullet|_J}} \varinjlim Y_\phi^*(f_j)_*(?)_j$$

$$\xrightarrow{\text{via }\theta} \varinjlim (f_i)_* X_\phi^*(?)_j \rightarrow (f_i)_* \varinjlim X_\phi^*(?)_j \xrightarrow{\text{via }\lambda_{J,i}^{-1}} (f_i)_*(?)_i L_J.$$

Proof. Note that θ in the first composite map is the composite

$$\theta = \theta(J, f_\bullet) : L_J(f_\bullet|_J)_* \xrightarrow{\text{via }u} L_J(f_\bullet|_J)_*(?)_J L_J \xrightarrow{c} L_J(?)_J(f_\bullet)_* L_J \xrightarrow{\varepsilon} (f_\bullet)_* L_J.$$

The description of u and ε are already given, and the proof is reduced to the iterative use of (6.10), (6.12), (6.13), and Lemma 6.20. The detailed argument is left to a patient reader. The reason why the second map involves θ is Lemma 6.20. $\qquad\square$

Similarly, we have the following.

Proposition 6.24. *Let $f_\bullet : X_\bullet \to Y_\bullet$ be a morphism in $\mathcal{P}$, J a subcategory of I, and $i \in I$. Then the composite map*

$$(f_i)^*(?)_i L_J \xrightarrow{\text{via }\theta(f_\bullet,i)} (?)_i (f_\bullet)^* L_J \xrightarrow{\text{via }d_{f_\bullet,J}} (?)_i L_J(f_\bullet|_J)^*$$

agrees with the composite map

$$(f_i)^*(?)_i L_J \xrightarrow{\text{via }\lambda_{J,i}} (f_i)^* \varinjlim Y_\phi^*(?)_j \cong \varinjlim (f_i)^* Y_\phi^*(?)_j$$

$$\xrightarrow{d} \varinjlim X_\phi^*(f_j)^*(?)_j \xrightarrow{\text{via }\theta(f_\bullet|_J,j)} \varinjlim X_\phi^*(?)_j(f_\bullet|_J)^* \xrightarrow{\text{via }\lambda_{J,i}^{-1}} (?)_i L_J(f_\bullet|_J)^*.$$

The proof is left to the reader. The proof of Proposition 6.23 and Proposition 6.24 are formal, and the propositions are valid for $\heartsuit = $ PM, Mod, PA, and AB.

Let $f_\bullet : X_\bullet \to Y_\bullet$ be a morphism in $\mathcal{P}$, and $J \subset I$ a subcategory. The inverse image $(f_\bullet)_\heartsuit^*$ is compatible with the restriction $(?)_J$.

Lemma 6.25. *The natural map*

$$\theta_\heartsuit = \theta_\heartsuit(f_\bullet, J) : ((f_\bullet)|_J)_\heartsuit^* \circ (?)_J \to (?)_J \circ (f_\bullet)_\heartsuit^*$$

is an isomorphism for $\heartsuit = $ PA, AB, PM, Mod. In particular, $f_\bullet^{-1} : \text{Zar}(Y_\bullet) \to \text{Zar}(X_\bullet)$ is an admissible continuous functor.

Proof. We consider the case where $\heartsuit = $ PM.

Let $\mathcal{M} \in \text{PM}(Y_\bullet)$, and $(j, U) \in \text{Zar}(X_\bullet|_J)$. We have

$$\Gamma((j, U), (f_\bullet|_J)^* \mathcal{M}_J) = \varinjlim \Gamma((j, U), \mathcal{O}_{X_\bullet}) \otimes_{\Gamma((j',V),\mathcal{O}_{Y_\bullet})} \Gamma((j', V), \mathcal{M}),$$

where the colimit is taken over $(j', V) \in (I_{(j,U)}^{(f_\bullet|_J)^{-1}})^{\text{op}}$. On the other hand, we have

$$\Gamma((j, U), (?)_J f_\bullet^* \mathcal{M}) = \varinjlim \Gamma((j, U), \mathcal{O}_{X_\bullet}) \otimes_{\Gamma((i,V),\mathcal{O}_{Y_\bullet})} \Gamma((i, V), \mathcal{M}),$$

where the colimit is taken over $(i, V) \in (I^{f^{-1}}_{(j,U)})^{\mathrm{op}}$. There is an obvious map from the first to the second. This obvious map is θ, see (2.57).

To verify that this is an isomorphism, it suffices to show that the category $(I^{(f_\bullet|_J)^{-1}}_{(j,U)})^{\mathrm{op}}$ is final in the category $(I^{f^{-1}}_{(j,U)})^{\mathrm{op}}$. In fact, any $(\phi, h) : (j, U) \to (i, f_i^{-1}(V))$ with $(i, V) \in \mathrm{Zar}(Y_\bullet)$ factors through

$$(\mathrm{id}_j, h) : (j, U) \to (j, f_j^{-1} Y_\phi^{-1}(V)).$$

Hence, $\theta_\heartsuit$ is an isomorphism for $\heartsuit = \mathrm{PM}$. The construction for the case where $\heartsuit = \mathrm{PA}$ is similar.

As $(?)_J$ is compatible with the sheafification by Lemma 2.31, we have that θ is an isomorphism for $\heartsuit = \mathrm{Mod}, \mathrm{AB}$ by Lemma 2.59. $\square$

Corollary 6.26. *The conjugate*

$$\xi_\heartsuit = \xi_\heartsuit(f_\bullet, J) : (f_\bullet)_*^\heartsuit R_J \to R_J(f_\bullet|_J)_*^\heartsuit$$

of $\theta_\heartsuit(f_\bullet, J)$ is an isomorphism for $\heartsuit = \mathrm{PA}, \mathrm{AB}, \mathrm{PM}, \mathrm{Mod}$.

Proof. Obvious by Lemma 6.25. $\square$

(6.27) By Corollary 6.26, we may define the composite

$$\mu_\heartsuit = \mu_\heartsuit(f_\bullet, J) : f_\bullet^* R_J \xrightarrow{u} f_\bullet^* R_J(f_\bullet|_J)_*(f_\bullet|_J)^*$$

$$\xrightarrow{\xi^{-1}} f_\bullet^*(f_\bullet)_* R_J(f_\bullet|_J)^* \xrightarrow{\varepsilon} R_J(f_\bullet|_J)^*.$$

Observe that the diagram

$$(?)_i f_\bullet^* R_J \xrightarrow{\theta^{-1}} f_i^*(?)_i R_J \xrightarrow{\rho} f_i^* \varprojlim(Y_\phi)_*(?)_j \longrightarrow \varprojlim f_i^*(Y_\phi)_*(?)_j$$

$$\downarrow \mu \qquad\qquad\qquad\qquad\qquad\qquad\qquad\qquad \theta \qquad\qquad\qquad$$

$$(?)_i R_J f_\bullet|_J^* \xrightarrow{\rho} \varprojlim(X_\phi)_*(?)_j f_\bullet|_J^* \xrightarrow{\theta^{-1}} \varprojlim(X_\phi)_* f_j^*(?)_j$$

is commutative.

Lemma 6.28. *Let the notation be as above, and $\mathcal{M}, \mathcal{N} \in \heartsuit(Y_\bullet)$. Then the diagram*

$$(f_\bullet|_J)_\heartsuit^*(\mathcal{M}_J \otimes \mathcal{N}_J) \xrightarrow{m} (f_\bullet|_J)_\heartsuit^*((\mathcal{M} \otimes \mathcal{N})_J) \xrightarrow{\theta} ((f_\bullet)_\heartsuit^*(\mathcal{M} \otimes \mathcal{N}))_J$$

$$\downarrow \Delta \qquad\qquad\qquad\qquad\qquad\qquad\qquad\qquad \downarrow (?)_J \Delta$$

$$(f_\bullet|_J)_\heartsuit^* \mathcal{M}_J \otimes (f_\bullet|_J)_\heartsuit^* \mathcal{N}_J \xrightarrow{\theta \otimes \theta} ((f_\bullet)_\heartsuit^* \mathcal{M})_J \otimes ((f_\bullet)_\heartsuit^* \mathcal{N})_J \xrightarrow{m} ((f_\bullet)_\heartsuit^* \mathcal{M} \otimes (f_\bullet)_\heartsuit^* \mathcal{N})_J$$

$$(6.29)$$

is commutative.

Proof. This is an immediate consequence of Lemma 1.44. $\square$

Corollary 6.30. *The adjoint pair* $((?)^*_{\mathrm{Mod}}, (?)^{\mathrm{Mod}}_*)$ *over the category* $\mathcal{P}(I, \underline{\mathrm{Sch}}/S)$ *is Lipman.*

Proof. Let $f_\bullet : X_\bullet \to Y_\bullet$ be a morphism of $\mathcal{P}(I, \underline{\mathrm{Sch}}/S)$. It is easy to see that the diagram

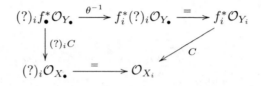

is commutative. So utilizing Lemma 1.25, it is easy to see that

$$(?)_i f_\bullet^* \mathcal{O}_{Y_\bullet} \xrightarrow{\theta^{-1}} f_i^*(?)_i \mathcal{O}_{Y_\bullet} \xrightarrow{=} f_i^* \mathcal{O}_{Y_i}$$

is also commutative. Since $C \colon f_i^* \mathcal{O}_{Y_i} \to \mathcal{O}_{X_i}$ is an isomorphism by Corollary 2.65, $(?)_i C$ is an isomorphism for any $i \in I$. Hence $C \colon f_\bullet^* \mathcal{O}_{Y_\bullet} \to \mathcal{O}_{X_\bullet}$ is also an isomorphism.

Let us consider $\mathcal{M}, \mathcal{N} \in \heartsuit(Y_\bullet)$. To verify that Δ is an isomorphism, it suffices to show that

$$(?)_i \Delta : (f_\bullet^*(\mathcal{M} \otimes \mathcal{N}))_i \to (f_\bullet^* \mathcal{M} \otimes f_\bullet^* \mathcal{N})_i$$

is an isomorphism for any $i \in \mathrm{ob}(I)$. Now consider the diagram (6.29) for $J = i$. Horizontal maps in the diagram are isomorphisms by (6.3) and Lemma 6.25. The left Δ is an isomorphism, since f_i is a morphism of single schemes. By Lemma 6.28, $(?)_i \Delta$ is also an isomorphism. $\quad\square$

(6.31) The description of the translation map α_ϕ for $f_\bullet^*$ is as follows. For $\phi \in I(i, j)$,

$$\alpha_\phi : X_\phi^*(?)_i f_\bullet^* \to (?)_j f_\bullet^*$$

is the composite

$$X_\phi^*(?)_i f_\bullet^* \xrightarrow{X_\phi^* \theta^{-1}} X_\phi^* f_i^*(?)_i \xrightarrow{d} f_j^* Y_\phi^*(?)_i \xrightarrow{f_j^* \alpha_\phi} f_j^*(?)_j \xrightarrow{\theta} (?)_j f_\bullet^*.$$

(6.32) Let $X_\bullet \in \mathcal{P}$, and $\mathcal{M}, \mathcal{N} \in \heartsuit(X_\bullet)$. Although there is a canonical map

$$H_i : \underline{\mathrm{Hom}}_{\heartsuit(X_\bullet)}(\mathcal{M}, \mathcal{N})_i \to \underline{\mathrm{Hom}}_{\heartsuit(X_i)}(\mathcal{M}_i, \mathcal{N}_i)$$

arising from the closed structure for $i \in I$, this may not be an isomorphism. However, we have the following.

Lemma 6.33. *Let $i \in I$. If $\mathcal{M}$ is equivariant, then the canonical map*

$$H_i : \underline{\mathrm{Hom}}_{\heartsuit(X_\bullet)}(\mathcal{M}, \mathcal{N})_i \to \underline{\mathrm{Hom}}_{\heartsuit(X_i)}(\mathcal{M}_i, \mathcal{N}_i)$$

is an isomorphism of presheaves. In particular, it is an isomorphism in
$\heartsuit(X_i)$.

Proof. It suffices to prove that

$$H_i : \mathrm{Hom}_{\heartsuit(\mathrm{Zar}(X_\bullet)/(i,U))}(\mathcal{M}|_{(i,U)}, \mathcal{N}|_{(i,U)}) \to \mathrm{Hom}_{\heartsuit(U)}(\mathcal{M}_i|_U, \mathcal{N}_i|_U)$$

is an isomorphism for any Zariski open set U in X_i.

To give an element of $\varphi \in \mathrm{Hom}_{\heartsuit(\mathrm{Zar}(X_\bullet)/(i,U))}(\mathcal{M}|_{(i,U)}, \mathcal{N}|_{(i,U)})$ is the same as to give a family $(\varphi_\phi)_{\phi:i\to j}$ with

$$\varphi_\phi \in \mathrm{Hom}_{\heartsuit(X_\phi^{-1}(U))}(\mathcal{M}_j|_{X_\phi^{-1}(U)}, \mathcal{N}_j|_{X_\phi^{-1}(U)})$$

such that for any $\phi: i \to j$ and $\psi: j \to j'$,

$$\varphi_{\psi\phi} \circ (\alpha_\psi(\mathcal{M}))|_{X_{\psi\phi}^{-1}(U)} = (\alpha_\psi(\mathcal{N}))|_{X_{\psi\phi}^{-1}(U)} \circ ((X_\psi)|_{X_{\psi\phi}^{-1}(U)})^*_\heartsuit(\varphi_\phi). \quad (6.34)$$

As $\alpha_\phi(\mathcal{M})$ is an isomorphism for any $\phi : i \to j$, we have that such a (φ_ϕ) is uniquely determined by φ_{id_i} by the formula

$$\varphi_\phi = (\alpha_\phi(\mathcal{N}))|_{X_\phi^{-1}(U)} \circ ((X_\phi)|_{X_\phi^{-1}(U)})^*_\heartsuit(\varphi_{\mathrm{id}_i}) \circ (\alpha_\phi(\mathcal{M}))|^{-1}_{X_\phi^{-1}(U)}. \quad (6.35)$$

Conversely, fix φ_{id_i}, and define φ_ϕ by (6.35). Consider the diagram

$$
\begin{array}{ccccccc}
X_{\psi\phi}^* \mathcal{M}_i & \xrightarrow{d^{-1}} & X_\psi^* X_\phi^* \mathcal{M}_i & \xrightarrow{\alpha_\phi} & X_\psi^* \mathcal{M}_j & \xrightarrow{\alpha_\psi} & \mathcal{M}_{j'} \\
\varphi_{\mathrm{id}_i} \downarrow & (a) & \downarrow \varphi_{\mathrm{id}_i} & (b) & \downarrow \varphi_\phi & (c) & \downarrow \varphi_{\psi\varphi} \\
X_{\psi\varphi}^* \mathcal{N}_i & \xrightarrow{d^{-1}} & X_\psi^* X_\phi^* \mathcal{N}_i & \xrightarrow{\alpha_\phi} & X_\psi^* \mathcal{N}_j & \xrightarrow{\alpha_\psi} & \mathcal{N}_{j'} \, .
\end{array}
$$

The diagram (a) is commutative by the naturality of d^{-1}. The diagram (b) and (a)+(b)+(c) are commutative, by the definition of φ_ϕ and $\varphi_{\psi\phi}$ (6.35), respectively. Since d^{-1} and $\alpha_\phi(\mathcal{M})$ are isomorphisms, the diagram (c) is commutative, and hence (6.34) holds. Hence H_i is bijective, as desired. $\qquad\square$

Lemma 6.36. *Let J be a subcategory of I. If $\mathcal{M}$ is equivariant, then the canonical map*

$$H_J : \underline{\mathrm{Hom}}_{\heartsuit(X_\bullet)}(\mathcal{M}, \mathcal{N})_J \to \underline{\mathrm{Hom}}_{\heartsuit(X_\bullet|_J)}(\mathcal{M}_J, \mathcal{N}_J)$$

is an isomorphism of presheaves. In particular, it is an isomorphism in
$\heartsuit(X_\bullet|_J)$.

Proof. It suffices to show that

$$(H_J)_i : (\underline{\mathrm{Hom}}_{\heartsuit(X_\bullet)}(\mathcal{M},\mathcal{N})_J)_i \to \underline{\mathrm{Hom}}_{\heartsuit(X_J)}(\mathcal{M}_J,\mathcal{N}_J)_i$$

is an isomorphism for each $i \in J$. By Lemma 1.39, the composite map

$$\underline{\mathrm{Hom}}_{\heartsuit(X_\bullet)}(\mathcal{M},\mathcal{N})_i \cong (\underline{\mathrm{Hom}}_{\heartsuit(X_\bullet)}(\mathcal{M},\mathcal{N})_J)_i$$

$$\xrightarrow{(H_J)_i} \underline{\mathrm{Hom}}_{\heartsuit(X_J)}(\mathcal{M}_J,\mathcal{N}_J)_i \xrightarrow{H_i} \underline{\mathrm{Hom}}_{\heartsuit(X_i)}(\mathcal{M}_i,\mathcal{N}_i)$$

agrees with H_i. As $\mathcal{M}_J$ is also equivariant, we have that the two H_i are isomorphisms by Lemma 6.33, and hence $(H_J)_i$ is an isomorphism for any $i \in J$. $\square$

(6.37)   By the lemma, the sheaf $\underline{\mathrm{Hom}}_{\heartsuit(X_\bullet)}(\mathcal{M},\mathcal{N})$ is given by the collection

$$(\underline{\mathrm{Hom}}_{\heartsuit(X_i)}(\mathcal{M}_i,\mathcal{N}_i))_{i \in I}$$

provided $\mathcal{M}$ is equivariant. The structure map is the canonical composite map

$$\alpha_\phi : (X_\phi)^*_\heartsuit \underline{\mathrm{Hom}}_{\heartsuit(X_i)}(\mathcal{M}_i,\mathcal{N}_i) \xrightarrow{P} \underline{\mathrm{Hom}}_{\heartsuit(X_j)}((X_\phi)^*_\heartsuit \mathcal{M}_i, (X_\phi)^*_\heartsuit \mathcal{N}_i)$$

$$\xrightarrow{\underline{\mathrm{Hom}}_{\heartsuit(X_j)}(\alpha_\phi^{-1},\alpha_\phi)} \underline{\mathrm{Hom}}_{\heartsuit(X_j)}(\mathcal{M}_j,\mathcal{N}_j).$$

Similarly, the following is also easy to prove.

Lemma 6.38. *Let $i \in I$ be an initial object of I. Then the following hold:*

1 *If $\mathcal{M} \in \heartsuit(X_\bullet)$ is equivariant, then*

$$(?)_i : \mathrm{Hom}_{\heartsuit(X_\bullet)}(\mathcal{M},\mathcal{N}) \to \mathrm{Hom}_{\heartsuit(X_i)}(\mathcal{M}_i,\mathcal{N}_i)$$

is an isomorphism.

2 $(?)_i : \mathrm{EM}(X_\bullet) \to \mathrm{Mod}(X_i)$ *is an equivalence, whose quasi-inverse is L_i.*

The fact that $L_i(\mathcal{M})$ is equivariant for $\mathcal{M} \in \mathrm{Mod}(X_i)$ is checked directly from the definition.

Chapter 7
Quasi-Coherent Sheaves Over a Diagram of Schemes

Let I be a small category, S a scheme, and $X_\bullet \in \mathcal{P}(I, \underline{\text{Sch}}/S)$.

(7.1) Let $\mathcal{M} \in \text{Mod}(X_\bullet)$. We say that $\mathcal{M}$ is *locally quasi-coherent* (resp. *locally coherent*) if $\mathcal{M}_i$ is quasi-coherent (resp. coherent) for any $i \in I$. We say that $\mathcal{M}$ is *quasi-coherent* if for any $(i, U) \in \text{Zar}(X_\bullet)$ with $U = \text{Spec} A$ being affine, there exists an exact sequence in $\text{Mod}(\text{Zar}(X_\bullet)/(i, U))$ of the form

$$(\mathcal{O}_{X_\bullet}|_{(i,U)})^{(T)} \to (\mathcal{O}_{X_\bullet}|_{(i,U)})^{(\Sigma)} \to \mathcal{M}|_{(i,U)} \to 0, \qquad (7.2)$$

where T and Σ are arbitrary small sets.

Lemma 7.3. *Let $\mathcal{M} \in \text{Mod}(X_\bullet)$. Then the following are equivalent.*

1 *$\mathcal{M}$ is quasi-coherent.*
2 *$\mathcal{M}$ is locally quasi-coherent and equivariant.*
3 *For any morphism $(\phi, h) : (j, V) \to (i, U)$ in $\text{Zar}(X_\bullet)$ such that $V = \text{Spec} B$ and $U = \text{Spec} A$ are affine, the canonical map $B \otimes_A \Gamma((i, U), \mathcal{M}) \to \Gamma((j, V), \mathcal{M})$ is an isomorphism.*

Proof. **1$\Rightarrow$2** Let $i \in I$ and U an affine open subset of X_i. Then there is an exact sequence of the form (7.2). Applying the restriction functor $\text{Mod}(\text{Zar}(X_\bullet)/(i, U)) \to \text{Mod}(U)$, we get an exact sequence

$$\mathcal{O}_U^{(T)} \to \mathcal{O}_U^{(\Sigma)} \to (\mathcal{M}_i)|_U \to 0,$$

which shows that $\mathcal{M}_i$ is quasi-coherent for any $i \in I$. We prove that $\alpha_\phi(\mathcal{M})$ is an isomorphism for any $\phi : i \to j$, to show that $\mathcal{M}$ is equivariant. Take an affine open covering (U_λ) of X_i, and we prove that $\alpha_\phi(\mathcal{M})$ is an isomorphism over $X_\phi^{-1}(U_\lambda)$ for each λ. But this is obvious by the existence of an exact sequence of the form (7.2) and the five lemma.

2$\Rightarrow$3 Set $W := X_\phi^{-1}(U)$, and let $\iota : V \hookrightarrow W$ be the inclusion map. Obviously, we have $h = (X_\phi)|_W \circ \iota$. As $\mathcal{M}$ is equivariant, we have that the canonical map

J. Lipman, M. Hashimoto, *Foundations of Grothendieck Duality for Diagrams* 345
of Schemes, Lecture Notes in Mathematics 1960,
© Springer-Verlag Berlin Heidelberg 2009

$$\alpha_\phi|_W(\mathcal{M}) : (X_\phi)|_W{}^*_{\mathrm{Mod}}(\mathcal{M}_i)|_U \to (\mathcal{M}_j)|_W$$

is an isomorphism. Applying ι^*_{Mod} to the isomorphism, we have that $h^*_{\mathrm{Mod}}((\mathcal{M}_i)|_U) \cong (\mathcal{M}_j)|_V$. The assertion follows from the assumption that $\mathcal{M}_i$ is quasi-coherent.

3⇒1 Let $(i, U) \in \mathrm{Zar}(X_\bullet)$ with $U = \mathrm{Spec}\, A$ affine. There is a presentation of the form

$$A^{(T)} \to A^{(\Sigma)} \to \Gamma((i, U), \mathcal{M}) \to 0.$$

It suffices to prove that the induced sequence (7.2) is exact. To verify this, it suffices to prove that the sequence is exact after taking the section at $((\phi, h) : (j, V) \to (i, U)) \in \mathrm{Zar}(X_\bullet)/(i, U)$ with $V = \mathrm{Spec}\, B$ being affine. We have a commutative diagram

$$
\begin{array}{ccccccc}
B \otimes_A A^{(T)} & \to & B \otimes_A A^{(\Sigma)} & \to & B \otimes_A \Gamma((i, U), \mathcal{M}) & \to & 0 \\
\downarrow \cong & & \downarrow \cong & & \downarrow \cong & & \\
B^{(T)} & \to & B^{(\Sigma)} & \to & \Gamma((j, V), \mathcal{M}) & & \to 0
\end{array}
$$

whose first row is exact and vertical arrows are isomorphisms. Hence, the second row is also exact, and (7.2) is exact. ☐

Definition 7.4. We say that $\mathcal{M} \in \mathrm{Mod}(X_\bullet)$ is *coherent* if it is equivariant and locally coherent. We denote the full subcategory of $\mathrm{Mod}(X_\bullet)$ consisting of coherent objects by $\mathrm{Coh}(X_\bullet)$.

(7.5) Let $J \subset I$ be a subcategory. We say that J is *big* in I if for any $(\psi : j \to k) \in \mathrm{Mor}(I)$, there exists some $(\phi : i \to j) \in \mathrm{Mor}(J)$ such that $\psi \circ \phi \in \mathrm{Mor}(J)$. Note that $\mathrm{ob}(J) = \mathrm{ob}(I)$ if J is big in I. Let $\mathbb{Q}$ be a property of morphisms of schemes. We say that $X_\bullet$ has $\mathbb{Q}$ J-arrows if $(X_\bullet)|_J$ has $\mathbb{Q}$-arrows.

Lemma 7.6. *Let $J \subset I$ be a subcategory, and $\mathcal{M} \in \mathrm{Mod}(X_\bullet)$.*

1 *The full subcategory $\mathrm{Lqc}(X_\bullet)$ of $\mathrm{Mod}(X_\bullet)$ consisting of locally quasi-coherent objects is a plump subcategory.*
2 *If $\mathcal{M}$ is equivariant (resp. locally quasi-coherent, quasi-coherent), then so is $\mathcal{M}_J^{\mathrm{Mod}}$.*
3 *If J is big in I and $\mathcal{M}_J$ is equivariant (resp. locally quasi-coherent, quasi-coherent), then so is $\mathcal{M}$.*
4 *If J is big in I and $X_\bullet$ has flat J-arrows, then the full subcategory $\mathrm{EM}(X_\bullet)$ (resp. $\mathrm{Qch}(X_\bullet)$) of $\mathrm{Mod}(X_\bullet)$ consisting of equivariant (resp. quasi-coherent) objects is a plump subcategory.*
5 *If J is big in I, then $(?)_J$ is faithful and exact.*

Proof. **1** and **2** are trivial.

We prove **3**. The assertion for the local quasi-coherence is obvious, because we have $\mathrm{ob}(J) = \mathrm{ob}(I)$. By Lemma 7.3, it remains to show the assertion for the equivariance. Let us assume that $\mathcal{M}_J$ is equivariant and $\psi : j \to k$ is a morphism in I, and take $\phi : i \to j$ such that $\phi, \psi\phi \in \mathrm{Mor}(J)$. Then the composite map

$$(X_{\psi\phi})^*_{\mathrm{Mod}}(\mathcal{M}_i) \cong (X_\psi)^*_{\mathrm{Mod}}(X_\phi)^*_{\mathrm{Mod}}(\mathcal{M}_i) \xrightarrow{(X_\psi)^*_{\mathrm{Mod}}\alpha^{\mathrm{Mod}}_\phi} (X_\psi)^*_{\mathrm{Mod}}(\mathcal{M}_j) \xrightarrow{\alpha^{\mathrm{Mod}}_\psi} \mathcal{M}_k,$$

which agrees with $\alpha^{\mathrm{Mod}}_{\psi\phi}$, is an isomorphism by assumption. As we have $\alpha^{\mathrm{Mod}}_\phi$ is also an isomorphism, we have that $\alpha^{\mathrm{Mod}}_\psi$ is an isomorphism. Thus $\mathcal{M}$ is equivariant.

We prove **4**. By **1** and Lemma 7.3, it suffices to prove the assertion only for the equivariance. Let

$$\mathcal{M}_1 \to \mathcal{M}_2 \to \mathcal{M}_3 \to \mathcal{M}_4 \to \mathcal{M}_5$$

be an exact sequence in $\mathrm{Mod}(X_\bullet)$, and assume that $\mathcal{M}_i$ is equivariant for $i = 1, 2, 4, 5$. We prove that $\mathcal{M}_3$ is equivariant. The sequence remains exact after applying the functor $(?)^{\mathrm{Mod}}_J$. By **3**, replacing I by J and $X_\bullet$ by $(X_\bullet)|_J$, we may assume that $X_\bullet$ has flat arrows. Now the assertion follows easily from the five lemma.

The assertion **5** is obvious, because $\mathrm{ob}(J) = \mathrm{ob}(I)$. □

Lemma 7.7. *Let $(\mathcal{M}_\lambda)$ be a diagram in $\mathrm{Mod}(X_\bullet)$. If each $\mathcal{M}_\lambda$ is locally quasi-coherent (resp. equivariant, quasi-coherent), then so is $\varinjlim \mathcal{M}_\lambda$.*

Proof. As $(?)_i$ preserves colimits, the assertion for local quasi-coherence is trivial. Assume that each $\mathcal{M}_\lambda$ is equivariant. For $(\phi : i \to j) \in \mathrm{Mor}(I)$, $\alpha_\phi(\mathcal{M}_\lambda)$ is an isomorphism. As $\alpha_\phi(\varinjlim \mathcal{M}_\lambda)$ is nothing but the composite

$$(X_\phi)^*_{\mathrm{Mod}}((\varinjlim \mathcal{M}_\lambda)_i) \cong \varinjlim (X_\phi)^*_{\mathrm{Mod}}(\mathcal{M}_\lambda)_i \xrightarrow{\varinjlim \alpha_\phi(\mathcal{M}_\lambda)} \varinjlim(\mathcal{M}_\lambda)_j \cong (\varinjlim \mathcal{M}_\lambda)_j,$$

it is an isomorphism. The rest of the assertions follow. □

By Lemma 6.7, we have the following.

Lemma 7.8. *Let $J \subset I$ be a subcategory, and $\mathcal{M} \in \mathrm{Lqc}(X_\bullet|_J)$. Then we have $L^{\mathrm{Mod}}_J(\mathcal{M}) \in \mathrm{Lqc}(X_\bullet)$.*

Similarly, we have the next lemma. We say that a morphism $f : X \to Y$ of schemes is *quasi-separated* if the diagonal map $X \to X \times_Y X$ is quasi-compact. A quasi-compact quasi-separated morphism is said to be *concentrated*. If $f : X \to Y$ is concentrated, and $\mathcal{M} \in \mathrm{Qch}(X)$, then $f_*\mathcal{M} \in \mathrm{Qch}(Y)$ [14, (9.2.1)], where $\mathrm{Qch}(X)$ and $\mathrm{Qch}(Y)$ denote the category of quasi-coherent sheaves on X and Y, respectively.

Lemma 7.9. *Let $j \in I$. Assume that $X_\bullet$ has concentrated arrows, and that $I(i, j)$ is finite for any $i \in I$. If $\mathcal{M} \in \mathrm{Qch}(X_j)$, then we have $R_j\mathcal{M} \in \mathrm{Lqc}(X_\bullet)$.*

The following is also proved easily, using (6.3) and Lemma 6.4.

Lemma 7.10. *Let $\mathcal{M}$ and $\mathcal{N}$ be locally quasi-coherent (resp. equivariant, quasi-coherent) $\mathcal{O}_{X_\bullet}$-modules. Then $\mathcal{M} \otimes_{\mathcal{O}_{X_\bullet}} \mathcal{N}$ is also locally quasi-coherent (resp. equivariant, quasi-coherent).*

The following is a consequence of the observation in (6.37).

Lemma 7.11. *Let $\mathcal{M}$ be a coherent $\mathcal{O}_{X_\bullet}$-module, and $\mathcal{N}$ a locally quasi-coherent $\mathcal{O}_{X_\bullet}$-module. Then $\underline{\mathrm{Hom}}_{\mathrm{Mod}(X_\bullet)}(\mathcal{M}, \mathcal{N})$ is locally quasi-coherent. If, moreover, there is a big subcategory J of I such that $X_\bullet$ has flat J-arrows and $\mathcal{N}$ is quasi-coherent, then $\underline{\mathrm{Hom}}_{\mathrm{Mod}(X_\bullet)}(\mathcal{M}, \mathcal{N})$ is quasi-coherent.*

Lemma 7.12. *Let $f : X \to Y$ be a concentrated morphism of schemes, and $g_Y : Y' \to Y$ a flat morphism of schemes. Set $X' := X \times_Y Y'$, $g_X : X' \to X$ the first projection, and $f' : X' \to Y'$ the second projection. Then for $\mathcal{M} \in \mathrm{Qch}(X)$, the canonical morphism*

$$\theta : g_Y^* f_* \mathcal{M} \to f'_* g_X^* \mathcal{M}$$

is an isomorphism.

Proof. First note that the assertion is true if g_Y is an open immersion. Indeed, it is easy to check that θ_{PM} and θ in the composition in Lemma 2.59 are isomorphisms in this case.

Using Lemma 1.23, we may assume that both Y and Y' are affine. Thus X is quasi-compact. Let (U_i) be a finite affine open covering of X, which exists. Set $\tilde{X} = \coprod_i U_i$, and let $p : \tilde{X} \to X$ be the obvious map. Since f is quasi-separated and Y is affine, $U_i \cap U_j$ is quasi-compact for any i, j. Thus p is quasi-compact. Note also that p is separated, since $\tilde{X}$ is affine. Let $p_i : \tilde{X} \times_X \tilde{X} \to \tilde{X}$ be the ith projection for $i = 1, 2$, and set $q = pp_1 = pp_2$. Note that p_1, p_2 and q are quasi-compact separated. Almost by the definition of a sheaf, there is an exact sequence of the form

$$0 \to \mathcal{M} \xrightarrow{u} p_* p^* \mathcal{M} \to q_* q^* \mathcal{M}.$$

Since $q_* q^* \mathcal{M} \cong p_*((p_1)_* q^* \mathcal{M})$, and $p^* \mathcal{M}$ and $(p_1)_* q^* \mathcal{M}$ are quasi-coherent, we may assume that $\mathcal{M} = p_* \mathcal{N}$ for some $\mathcal{N} \in \mathrm{Qch}(\tilde{X})$ by the five lemma. By Lemma 1.22, replacing f by p and fp, we may assume that f is quasi-compact separated. Then repeating the same argument as above, we may assume that p is affine now. Replacing f by p and fp again, we may assume that f is affine. That is, X is affine. But this case is trivial. $\square$

(7.13) Let $f_\bullet : X_\bullet \to Y_\bullet$ be a morphism in $\mathcal{P} = \mathcal{P}(I, \underline{\mathrm{Sch}}/S)$.
As θ in (6.21) is not an isomorphism in general, $(f_\bullet)^\heartsuit_*(\mathcal{M})$ need not be equivariant even if $\mathcal{M}$ is equivariant. However, we have

Lemma 7.14. *Let $f_\bullet : X_\bullet \to Y_\bullet$ be a morphism in $\mathcal{P}$, and J a big subcategory of I. Then we have the following:*

1 $f_\bullet$ is cartesian if and only if $(f_\bullet)|_J$ is cartesian.

2 If $f_\bullet$ is concentrated and $\mathcal{M} \in \mathrm{Lqc}(X_\bullet)$, then $(f_\bullet)_*(\mathcal{M}) \in \mathrm{Lqc}(Y_\bullet)$.

3 If $f_\bullet$ is cartesian concentrated, $Y_\bullet$ has flat J-arrows, and $\mathcal{M} \in \mathrm{Qch}(X_\bullet)$, then we have $(f_\bullet)_*(\mathcal{M}) \in \mathrm{Qch}(Y_\bullet)$.

Proof. **1** Assume that $f_\bullet|_J$ is cartesian, and let $\psi : j \to k$ be a morphism in I. Take $\phi : i \to j$ such that $\phi, \psi\phi \in \mathrm{Mor}(J)$. Consider the commutative diagram

$$
\begin{array}{ccccc}
X_k & \xrightarrow{X_\psi} & X_j & \xrightarrow{X_\phi} & X_i \\
\downarrow f_k(a) & & \downarrow f_j(b) & & \downarrow f_i \\
Y_k & \xrightarrow{Y_\psi} & Y_j & \xrightarrow{Y_\phi} & Y_i.
\end{array}
$$

By assumption, the square (b) and the whole rectangle ((a)+(b)) are fiber squares. Hence (a) is also a fiber square. This shows that $f_\bullet$ is cartesian. The converse is obvious.

The assertion **2** is obvious by the isomorphism $((f_\bullet)_*\mathcal{M})_i \cong (f_i)_*(\mathcal{M}_i)$ for $i \in I$.

We prove **3**. By Lemma 7.6, we may assume that $J = I$. Then $(f_\bullet)_*(\mathcal{M})$ is locally quasi-coherent by **2**. As $\mathcal{M}$ is equivariant and θ in (6.21) is an isomorphism by Lemma 7.12, we have that $(f_\bullet)_*(\mathcal{M})$ is equivariant. Hence by Lemma 7.3, $(f_\bullet)_*(\mathcal{M})$ is quasi-coherent. $\qquad\square$

(7.15) Let the notation be as in Lemma 7.14. If $f_\bullet$ is concentrated, then $(f_\bullet)_*^{\mathrm{Lqc}} : \mathrm{Lqc}(X_\bullet) \to \mathrm{Lqc}(Y_\bullet)$ is defined as the restriction of $(f_\bullet)_*^{\mathrm{Mod}}$. If $f_\bullet$ is concentrated cartesian and $Y_\bullet$ has flat J-arrows, then $(f_\bullet)_*^{\mathrm{Qch}} : \mathrm{Qch}(X_\bullet) \to \mathrm{Qch}(Y_\bullet)$ is induced.

Lemma 7.16. *Let* $f_\bullet : X_\bullet \to Y_\bullet$ *and* $g_\bullet : Y_\bullet \to Z_\bullet$ *be morphisms in* $\mathcal{P}$. *Then the following hold.*

0 *An isomorphism is a cartesian morphism.*

1 *If* $f_\bullet$ *and* $g_\bullet$ *are cartesian, then so is* $g_\bullet \circ f_\bullet$.

2 *If* $g_\bullet$ *and* $g_\bullet \circ f_\bullet$ *are cartesian, then so is* $f_\bullet$.

3 *If* $f_\bullet$ *is faithfully flat cartesian and* $g_\bullet \circ f_\bullet$ *is cartesian, then* $g_\bullet$ *is cartesian.*

Proof. Trivial. $\qquad\square$

Lemma 7.17. *Let* $f_\bullet : X_\bullet \to Y_\bullet$ *and* $g_\bullet : Y'_\bullet \to Y_\bullet$ *be morphisms in* $\mathcal{P}$. *Let* $f'_\bullet : X'_\bullet \to Y'_\bullet$ *be the base change of* $f_\bullet$ *by* $g_\bullet$.

1 *If* $f_\bullet$ *is cartesian, then so is* $f'_\bullet$.

2 *If* $f'_\bullet$ *is cartesian and* $g_\bullet$ *is faithfully flat, then* $f_\bullet$ *is cartesian.*

Proof. Obvious. $\qquad\square$

(7.18) Let $f : X \to Y$ be a morphism of schemes. If f is concentrated, then f_* is compatible with pseudo-filtered inductive limits.

Lemma 7.19 ([23, p. 641, Proposition 6]). *Let* $f : X \to Y$ *be a concentrated morphism of schemes, and* $(\mathcal{M}_i)$ *a pseudo-filtered inductive system of* $\mathcal{O}_X$*-modules. Then the canonical map*

$$\varinjlim f_* \mathcal{M}_i \to f_* \varinjlim \mathcal{M}_i$$

is an isomorphism.

By the lemma, the following follows immediately.

Lemma 7.20. *Let* $f_\bullet : X_\bullet \to Y_\bullet$ *be a morphism in* $\mathcal{P}(I, \underline{\mathrm{Sch}}/S)$. *If* $f_\bullet$ *is concentrated, then* $(f_\bullet)_*^{\mathrm{Mod}}$ *and* $(f_\bullet)_*^{\mathrm{Lqc}}$ *preserve pseudo-filtered inductive limits. If, moreover,* $f_\bullet$ *is cartesian and* $Y_\bullet$ *has flat arrows, then* $(f_\bullet)_*^{\mathrm{Qch}}$ *preserves pseudo-filtered inductive limits.*

Lemma 7.21. *Let* $f_\bullet : X_\bullet \to Y_\bullet$ *be a morphism in* $\mathcal{P}$. *Let* J *be an admissible subcategory of* I. *If* $Y_\bullet$ *has flat arrows and* $f_\bullet$ *is cartesian and concentrated, then the canonical map*

$$\theta(J, f_\bullet) : L_J \circ (f_\bullet|_J)_* \to (f_\bullet)_* \circ L_J$$

is an isomorphism of functors from $\mathrm{Lqc}(X_\bullet|_J)$ *to* $\mathrm{Lqc}(Y_\bullet)$.

Proof. This is obvious by Proposition 6.23, Lemma 7.19, and Lemma 7.12. □

The following is obvious by Lemma 6.25 and (6.31).

Lemma 7.22. *Let* $f_\bullet : X_\bullet \to Y_\bullet$ *be a morphism in* $\mathcal{P}$. *If* $\mathcal{M} \in \mathrm{Mod}(Y_\bullet)$ *is equivariant (resp. locally quasi-coherent, quasi-coherent), then so is* $(f_\bullet)_{\mathrm{Mod}}^*(\mathcal{M})$. *If* $\mathcal{M} \in \mathrm{Mod}(Y_\bullet)$, $f_\bullet$ *is faithfully flat, and* $(f_\bullet)_{\mathrm{Mod}}^*(\mathcal{M})$ *is equivariant, then we have* $\mathcal{M}$ *is equivariant.*

The restriction $(f_\bullet)^* : \mathrm{Qch}(Y_\bullet) \to \mathrm{Qch}(X_\bullet)$ is sometimes denoted by $(f_\bullet)_{\mathrm{Qch}}^*$.

Chapter 8
Derived Functors of Functors on Sheaves of Modules Over Diagrams of Schemes

(8.1) Let I be a small category, and S a scheme. Set $\mathcal{P} := \mathcal{P}(I, \underline{\mathrm{Sch}}/S)$, and let $X_\bullet \in \mathcal{P}$. In these notes, we use some abbreviated notation for derived categories of modules over diagrams of schemes. In the sequel, $D(\mathrm{Mod}(X_\bullet))$ may be denoted by $D(X_\bullet)$. $D^+_{\mathrm{EM}(X_\bullet)}(\mathrm{Mod}(X_\bullet))$ may be denoted by $D^+_{\mathrm{EM}}(X_\bullet)$. $D^b_{\mathrm{Coh}(X_\bullet)}(\mathrm{Qch}(X_\bullet))$ may be denoted by $D^b_{\mathrm{Coh}}(\mathrm{Qch}(X_\bullet))$, and so on. This notation will be also used for a single scheme. For a scheme X, $D^+_{\mathrm{Qch}(X)}(\mathrm{Mod}(X))$ will be denoted by $D^+_{\mathrm{Qch}}(X)$, where $\mathrm{Mod}(X)$ is the category of $\mathcal{O}_X$-modules.

Proposition 8.2. *Let $X_\bullet \in \mathcal{P}$, and $\mathbb{I} \in K(\mathrm{Mod}(X_\bullet))$. We have $\mathbb{I}$ is K-limp if and only if so is $\mathbb{I}_i$ for $i \in I$.*

Proof. The only if part follows from Lemma 3.31 and Lemma 3.25, **4**.

We prove the if part. Let $\mathbb{I} \to \mathbb{J}$ be a K-injective resolution, and let $\mathbb{C}$ be the mapping cone. Note that $\mathbb{C}_i$ is exact for each i.

Let $(U, i) \in \mathrm{Zar}(X_\bullet)$. We have an isomorphism

$$\Gamma((U, i), \mathbb{C}) \cong \Gamma(U, \mathbb{C}_i).$$

As $\mathbb{C}_i$ is K-limp by the only if part, these are exact for each (U, i). It follows that $\mathbb{I}$ is K-limp. $\qquad\square$

Corollary 8.3. *Let J be a subcategory of I, and $f_\bullet : X_\bullet \to Y_\bullet$ a morphism in $\mathcal{P}(I, \underline{\mathrm{Sch}}/S)$. Then there is a canonical isomorphism*

$$c(J, f_\bullet) : (?)_J R(f_\bullet)_* \cong R(f_\bullet|_J)_* (?)_J.$$

Lemma 8.4. *Let J be an admissible subcategory of I. Assume that $X_\bullet$ has flat arrows. If $\mathbb{I}$ is a K-injective complex in $\mathrm{Mod}(X_\bullet)$, then $\mathbb{I}_J$ is K-injective.*

Proof. This is simply because $(?)_J$ has an exact left adjoint L_J. $\qquad\square$

J. Lipman, M. Hashimoto, *Foundations of Grothendieck Duality for Diagrams of Schemes*, Lecture Notes in Mathematics 1960,
© Springer-Verlag Berlin Heidelberg 2009

Lemma 8.5. *Let* $f_\bullet\colon X_\bullet \to Y_\bullet$ *be a concentrated morphism in* $\mathcal{P}(I, \underline{\mathrm{Sch}}/S)$. *Then* $R(f_\bullet)_*$ *takes* $D_{\mathrm{Lqc}}(X_\bullet)$ *to* $D_{\mathrm{Lqc}}(Y_\bullet)$. $R(f_\bullet)_* \colon D_{\mathrm{Lqc}}(X_\bullet) \to D_{\mathrm{Lqc}}(Y_\bullet)$ *is way-out in both directions if* $Y_\bullet$ *is quasi-compact and* I *is finite.*

Proof. Follows from [26, (3.9.2)] and Corollary 8.3 easily. □

Lemma 8.6. *Let* $X_\bullet \in \mathcal{P}$. *Assume that* $X_\bullet$ *has flat arrows. For a complex* $\mathbb{F}$ *in* $\mathrm{Mod}(X_\bullet)$, $\mathbb{F}$ *has equivariant cohomology groups if and only if* $\alpha_\phi \colon X_\phi^* \mathbb{F}_i \to \mathbb{F}_j$ *is a quasi-isomorphism for any morphism* $\phi\colon i \to j$ *in* I.

Proof. This is easy, since X_ϕ^* is an exact functor. □

Lemma 8.7. *Let* $f_\bullet\colon X_\bullet \to Y_\bullet$ *be a morphism in* $\mathcal{P}$. *Assume that* $f_\bullet$ *is concentrated and cartesian, and* $Y_\bullet$ *has flat arrows. If* $\mathbb{F} \in D_{\mathrm{Qch}}(X_\bullet)$, *then* $R(f_\bullet)_*\mathbb{F} \in D_{\mathrm{Qch}}(Y_\bullet)$.

Proof. By the derived version of Lemma 6.20,

$$\alpha_\phi \colon Y_\phi^*(?)_i R(f_\bullet)_* \mathbb{F} \to (?)_j R(f_\bullet)_* \mathbb{F} \tag{8.8}$$

agrees with the composite

$$Y_\phi^*(?)_i R(f_\bullet)_* \mathbb{F} \xrightarrow{c} Y_\phi^* R(f_i)_* \mathbb{F}_i \xrightarrow{\theta} R(f_j)_* X_\phi^* \mathbb{F}_i \xrightarrow{\alpha_\phi} R(f_j)_* \mathbb{F}_j \xrightarrow{c} (?)_j R(f_\bullet)_* \mathbb{F}.$$

The first and the fourth map c's are isomorphisms. The second map θ is an isomorphism by [26, (3.9.5)]. The third map α_ϕ is an isomorphism by assumption and Lemma 8.6. Thus (8.8) is an isomorphism. Again by Lemma 8.6, we have the desired assertion. □

(8.9) Let X be a scheme, $x \in X$, and M an $\mathcal{O}_{X,x}$-module. We define $\xi_x(M) \in \mathrm{Mod}(X)$ by $\Gamma(U, \xi_x(M)) = M$ if $x \in U$, and zero otherwise. The restriction maps are defined in an obvious way. For an exact complex $\mathbb{H}$ of $\mathcal{O}_{X,x}$-modules, $\xi_x(\mathbb{H})$ is exact not only as a complex of sheaves, but also as a complex of presheaves. For a morphism of schemes $f\colon X \to Y$, we have that $f_* \xi_x(M) \cong \xi_{f(x)}(M)$.

Lemma 8.10. *Let* $\mathbb{F} \in C(\mathrm{Mod}(X_\bullet))$. *The following are equivalent.*

1 $\mathbb{F}$ *is* K-*flat.*
2 $\mathbb{F}_i$ *is* K-*flat for* $i \in \mathrm{ob}(I)$.
3 $\mathbb{F}_{i,x}$ *is a* K-*flat complex of* $\mathcal{O}_{X_i,x}$-*modules for any* $i \in \mathrm{ob}(I)$ *and* $x \in X_i$.

Proof. **3⇒1** Let $\mathbb{G} \in C(\mathrm{Mod}(X_\bullet))$ be exact. We are to prove that $\mathbb{F} \otimes_{\mathcal{O}_{X_\bullet}} \mathbb{G}$ is exact. For $i \in \mathrm{ob}(I)$ and $x \in X_i$, we have

$$(\mathbb{F} \otimes_{\mathcal{O}_{X_\bullet}} \mathbb{G})_{i,x} \cong (\mathbb{F}_i \otimes_{\mathcal{O}_{X_i}} \mathbb{G}_i)_x \cong \mathbb{F}_{i,x} \otimes_{\mathcal{O}_{X_i,x}} \mathbb{G}_{i,x}.$$

Since $\mathbb{G}_{i,x}$ is exact, $(\mathbb{F} \otimes_{\mathcal{O}_{X_i,x}} \mathbb{G})_{i,x}$ is exact. So $\mathbb{F} \otimes_{\mathcal{O}_{X_\bullet}} \mathbb{G}$ is exact.

1⇒3 Let $\mathbb{H} \in C(\mathrm{Mod}(\mathcal{O}_{X_{i},x}))$ be an exact complex, and we are to prove that $\mathbb{F}_{i,x} \otimes_{\mathcal{O}_{X_{i},x}} \mathbb{H}$ is exact. For each $j \in \mathrm{ob}(I)$,

$$(?)_{j} R_{i} \xi_{x}(\mathbb{H}) \cong \prod_{\phi \in I(j,i)} (X_{\phi})_{*} \xi_{x}(\mathbb{H}) \cong \prod_{\phi \in I(j,i)} \xi_{X_{\phi}(x)}(\mathbb{H})$$

is exact, since a direct product of exact complexes of *presheaves* is exact. So $R_{i}\xi_{x}(\mathbb{H})$ is exact. It follows that $\mathbb{F} \otimes_{\mathcal{O}_{X_{\bullet}}} R_{i}\xi_{x}(\mathbb{H})$ is exact. Hence

$$(\mathbb{F} \otimes_{\mathcal{O}_{X_{\bullet}}} R_{i}\xi_{x}(\mathbb{H}))_{i} \cong \mathbb{F}_{i} \otimes_{\mathcal{O}_{X_{i}}} \prod_{\phi \in I(i,i)} \xi_{X_{\phi}(x)} \mathbb{H}$$

is also exact. So $\mathbb{F}_{i} \otimes_{\mathcal{O}_{X_{i}}} \xi_{\mathrm{id}_{X_{i}}(x)}(\mathbb{H}) = \mathbb{F}_{i} \otimes_{\mathcal{O}_{X_{i}}} \xi_{x}\mathbb{H}$ is exact. So

$$(\mathbb{F}_{i} \otimes_{\mathcal{O}_{X_{i}}} \xi_{x}\mathbb{H})_{x} \cong \mathbb{F}_{i,x} \otimes_{\mathcal{O}_{X_{i},x}} (\xi_{x}\mathbb{H})_{x} \cong \mathbb{F}_{i,x} \otimes_{\mathcal{O}_{X_{i},x}} \mathbb{H}$$

is also exact.

Applying **1⇔3**, which has already been proved, to the complex $\mathbb{F}_{i}$ over the single scheme X_{i}, we get **2⇔3**. □

Hence by [39], we have the following.

Lemma 8.11. *Let $f_{\bullet} : X_{\bullet} \to Y_{\bullet}$ be a morphism in $\mathcal{P}$. Then we have the following.*

1 *If $\mathbb{F} \in C(\mathrm{Mod}(Y_{\bullet}))$ is K-flat, then so is $f_{\bullet}^{*}\mathbb{F}$.*
2 *If $\mathbb{F} \in C(\mathrm{Mod}(Y_{\bullet}))$ is K-flat exact, then so is $f_{\bullet}^{*}\mathbb{F}$.*
3 *If $\mathbb{I} \in C(\mathrm{Mod}(X_{\bullet}))$ is weakly K-injective, then so is $(f_{\bullet})_{*}\mathbb{I}$.*

(8.12) By the lemma, the left derived functor $Lf_{\bullet}^{*}$, which we already know its existence by Lemma 6.25, can also be calculated by K-flat resolutions.

Lemma 8.13. *Let J be a subcategory of I, and $f_{\bullet} : X_{\bullet} \to Y_{\bullet}$ a morphism in $\mathcal{P}$. Then we have the following.*

1 *The canonical map*

$$\theta(f_{\bullet}, J) : L(f_{\bullet}|_{J})^{*}(?)_{J} \to (?)_{J} Lf_{\bullet}^{*}$$

is an isomorphism.
2 *The diagram*

$$
\begin{array}{ccc}
(?)_{J} & \xrightarrow{\quad\mathrm{id}\quad} & (?)_{J} \\
\downarrow{\scriptstyle u} & & \downarrow{\scriptstyle u} \\
R(f_{\bullet}|_{J})_{*}L(f_{\bullet}|_{J})^{*}(?)_{J} \xrightarrow{\theta} R(f_{\bullet}|_{J})_{*}(?)_{J}Lf_{\bullet}^{*} \xrightarrow{c^{-1}} & (?)_{J}R(f_{\bullet})_{*}Lf_{\bullet}^{*}
\end{array}
$$

is commutative.

3 *The diagram*

$$
\begin{array}{ccc}
(?)_J & \xrightarrow{\quad\quad\quad\quad \text{id} \quad\quad\quad\quad} & (?)_J \\
\uparrow{\scriptstyle\varepsilon} & & \uparrow{\scriptstyle\varepsilon} \\
L(f_\bullet|_J)^* R(f_\bullet|_J)_*(?)_J \xrightarrow{\ c^{-1}\ } L(f_\bullet|_J)^*(?)_J R(f_\bullet)_* \xrightarrow{\ \theta\ } (?)_J L(f_\bullet)^* R(f_\bullet)_*
\end{array}
$$

is commutative.

Proof. Since $(?)_J$ preserves K-flat complexes by Lemma 8.10, we have

$$
L(f_\bullet|_J)^*(?)_J \cong L((f_\bullet|_J) \circ (?)_J).
$$

On the other hand, it is obvious that we have $(?)_J L f_\bullet^* \cong L((?)_J f_\bullet^*)$. By Lemma 6.25, we have a composite isomorphism

$$
\theta : L(f_\bullet|_J)^*(?)_J \cong L((f_\bullet|_J)^* \circ (?)_J) \xrightarrow{L\theta} L((?)_J \circ f_\bullet^*) \cong (?)_J L(f_\bullet)^*,
$$

and **1** is proved.

2 and **3** follow from the proofs of Lemma 1.24 and Lemma 1.25, respectively. $\qquad\square$

Lemma 8.14. *Let $X_\bullet \in \mathcal{P}$, and $\mathbb{F}, \mathbb{G} \in D(X_\bullet)$. Then we have the following.*

1 $\mathbb{F}_J \otimes^{\bullet,L}_{\mathcal{O}_{X_\bullet|_J}} \mathbb{G}_J \cong (\mathbb{F} \otimes^{\bullet,L}_{\mathcal{O}_{X_\bullet}} \mathbb{G})_J$ *for any subcategory $J \subset I$.*

2 *If $\mathbb{F}$ and $\mathbb{G}$ have locally quasi-coherent cohomology groups, then $\underline{\mathrm{Tor}}_i^{\mathcal{O}_{X_\bullet}}(\mathbb{F}, \mathbb{G})$ is also locally quasi-coherent for any $i \in \mathbb{Z}$.*

3 *Assume that there exists some big subcategory J of I such that $X_\bullet$ has flat J-arrows. If both $\mathbb{F}$ and $\mathbb{G}$ have equivariant (resp. quasi-coherent) cohomology groups, then $\underline{\mathrm{Tor}}_i^{\mathcal{O}_{X_\bullet}}(\mathbb{F}, \mathbb{G})$ is also equivariant (resp. quasi-coherent).*

Proof. The assertion **1** is an immediate consequence of Lemma 8.10 and Example 5.6, **5**.

2 In view of **1**, we may assume that $X = X_\bullet$ is a single scheme. As the question is local, we may assume that X is even affine.

We may assume that $\mathbb{F} = \varinjlim \mathbb{F}_n$, where $(\mathbb{F}_n)$ is the $\mathfrak{P}(X_\bullet)$-special direct system such that each $\mathbb{F}_n$ is bounded above and has locally quasi-coherent cohomology groups as in Lemma 3.25, **3**. Similarly, we may assume that $\mathbb{G} = \varinjlim \mathbb{G}_n$. As filtered inductive limits are exact and compatible with tensor products, and the colimit of locally quasi-coherent sheaves is locally quasi-coherent, we may assume that both $\mathbb{F}$ and $\mathbb{G}$ are bounded above, flat, and has locally quasi-coherent cohomology groups. By [17, Proposition I.7.3], we may assume that both $\mathbb{F}$ and $\mathbb{G}$ are single quasi-coherent sheaves. This case is trivial.

3 In view of **1**, we may assume that $J = I$ and $X_\bullet$ has flat arrows. By **2**, it suffices to show the assertion for equivariance. Assuming that $\mathbb{F}$ and $\mathbb{G}$

are K-flat with equivariant cohomology groups, we prove that $\mathbb{F} \otimes \mathbb{G}$ has equivariant cohomology groups. This is enough.

Let $\phi : i \to j$ be a morphism of I. As X_ϕ is flat and $\mathbb{F}$ and $\mathbb{G}$ have equivariant cohomology groups, $\alpha_\phi : X_\phi^* \mathbb{F}_i \to \mathbb{F}_j$ and $\alpha_\phi : X_\phi^* \mathbb{G}_i \to \mathbb{G}_j$ are quasi-isomorphisms. The composite

$$X_\phi^*(\mathbb{F} \otimes_{\mathcal{O}_{X_\bullet}}^\bullet \mathbb{G})_i \cong X_\phi^* \mathbb{F}_i \otimes_{\mathcal{O}_{X_j}}^\bullet X_\phi^* \mathbb{G}_i \xrightarrow{\alpha_\phi \otimes \alpha_\phi} \mathbb{F}_j \otimes_{\mathcal{O}_{X_j}}^\bullet \mathbb{G}_j \cong (\mathbb{F} \otimes_{\mathcal{O}_{X_\bullet}}^\bullet \mathbb{G})_j$$

is a quasi-isomorphism, since $X_\phi^* \mathbb{G}_i$ and $\mathbb{F}_j$ are K-flat. By (6.3), $\alpha_\phi(\mathbb{F} \otimes_{\mathcal{O}_{X_\bullet}}^\bullet \mathbb{G})$ is a quasi-isomorphism.

As $X_\bullet$ has flat arrows, this shows that $\mathbb{F} \otimes_{\mathcal{O}_{X_\bullet}}^\bullet \mathbb{G}$ has equivariant cohomology groups. $\qquad\square$

(8.15) Let $X_\bullet \in \mathcal{P}$, and J an admissible subcategory of I. By Lemma 6.17, the left derived functor

$$LL_J^{\mathrm{Mod}} : D(X_\bullet|_J) \to D(X_\bullet)$$

of L_J^{Mod} is defined, since $Q(X_\bullet, J)$ is admissible. This is also calculated using K-flat resolutions. Namely,

Lemma 8.16. *Let $X_\bullet$ and J be as above. If $\mathbb{F} \in K(\mathrm{Mod}(X_\bullet|_J))$ is K-flat, then so is $L_J \mathbb{F}$. If $\mathbb{F}$ is K-flat exact, then so is $L_J \mathbb{F}$.*

Proof. This is trivial by Lemma 6.7. $\qquad\square$

Corollary 8.17. *Let $X_\bullet \in \mathcal{P}(I, \underline{\mathrm{Sch}}/S)$, J an admissible subcategory of I, and $\mathbb{I} \in K(\mathrm{Mod}(X_\bullet))$. If $\mathbb{I}$ is weakly K-injective, then $\mathbb{I}_J$ is weakly K-injective.*

Proof. Let $\mathbb{F}$ be a K-flat exact complex in $K(\mathrm{Mod}(X_\bullet|_J))$. Then,

$$\mathrm{Hom}_{\mathrm{Mod}(X_\bullet|_J)}^\bullet(\mathbb{F}, \mathbb{I}_J) \cong \mathrm{Hom}_{\mathrm{Mod}(X_\bullet)}^\bullet(L_J \mathbb{F}, \mathbb{I})$$

is exact by the lemma. By Lemma 3.25, **5**, we are done. $\qquad\square$

Lemma 8.18. *Let $f : X \to Y$ be a morphism of schemes, and $\mathcal{M} \in \mathrm{Qch}(Y)$. Then for any $i \geq 0$, $L_i f^* \mathcal{M} \in \mathrm{Qch}(X)$.*

Proof. Note that Lf^* is computed using a flat resolution, and a flat object is preserved by f^*. If g is a flat morphism of schemes, then g^* is exact. Thus using a spectral sequence argument, it is easy to see that the question is local both on Y and X. So we may assume that $X = \mathrm{Spec}\, B$ and $Y = \mathrm{Spec}\, A$ are affine. If $\Gamma(Y, \mathcal{M}) = M$ and $\mathbb{F} \to M$ is an A-projective resolution, then

$$L_i f^* \mathcal{M} = H_i(f^* \tilde{\mathbb{F}}) = H_i((B \otimes_A \mathbb{F})^\sim) = \mathrm{Tor}_i^A(B, M)^\sim.$$

Thus $L_i f^* \mathcal{M}$ is quasi-coherent for any $i \geq 0$, as desired. $\qquad\square$

Lemma 8.19. *Let $X_\bullet \in \mathcal{P}$ and J an admissible subcategory of I. Let $\mathbb{F} \in D_{\mathrm{Lqc}}(X_\bullet|_J)$. Then, $LL_J\mathbb{F} \in D_{\mathrm{Lqc}}(X_\bullet)$.*

Proof. First we consider the case that $\mathbb{F} = \mathcal{M}$ is a single locally quasi-coherent sheaf. Then by the uniqueness of the derived functor,

$$(?)_i H^{-n}(LL_J\mathcal{M}) = L_n((?)_i L_J)\mathcal{M} = \varinjlim L_n X_\phi^* \mathcal{M}_j$$

for $i \in I$. Thus $LL_J\mathcal{M} \in D_{\mathrm{Lqc}}(X_\bullet)$ by Lemma 8.18.

Now using the standard spectral sequence argument (or the way-out lemma [17, (I.7.3)]), the case that $\mathbb{F}$ is bounded above follows. The general case follows immediately by Lemma 3.25, **3**. □

Lemma 8.20. *Let $f_\bullet : X_\bullet \to Y_\bullet$ be a morphism in $\mathcal{P}$. If $\mathbb{F} \in D_{\mathrm{Lqc}}(Y_\bullet)$, then $Lf_\bullet^*\mathbb{F} \in D_{\mathrm{Lqc}}(X_\bullet)$. If $Y_\bullet$ and $X_\bullet$ have flat arrows and $\mathbb{F} \in D_{\mathrm{EM}}(Y_\bullet)$, then $Lf_\bullet^*\mathbb{F} \in D_{\mathrm{EM}}(X_\bullet)$.*

Proof. For the first assertion, we may assume that $f : X \to Y$ is a morphism of single schemes by Lemma 8.13. If $\mathbb{F}$ is a single quasi-coherent sheaf, this is obvious by Lemma 8.18. So the case that $\mathbb{F}$ is bounded above follows from the way-out lemma. The general case follows from Lemma 3.25, **3**.

We prove the second assertion. If $\mathbb{F}$ is a K-flat complex in $\mathrm{Mod}(Y_\bullet)$ with equivariant cohomology groups, then $\alpha_\phi : Y_\phi^*\mathbb{F}_i \to \mathbb{F}_j$ is a quasi-isomorphism for any morphism $\phi : i \to j$ of I by Lemma 8.6. As the mapping cone $\mathrm{Cone}(\alpha_\phi)$ is K-flat exact by Lemma 8.10, $f_j^* \mathrm{Cone}(\alpha_\phi)$ is also exact. Thus $f_j^*\alpha_\phi : f_j^*Y_\phi^*\mathbb{F}_i \to f_j^*\mathbb{F}_j$ is a quasi-isomorphism. This shows that $\alpha_\phi : X_\phi^*(f_\bullet^*\mathbb{F})_i \to (f_\bullet^*\mathbb{F})_j$ is a quasi-isomorphism for any ϕ. So $f_\bullet^*\mathbb{F}$ has equivariant cohomology groups by Lemma 8.6. This is what we wanted to prove. □

Lemma 8.21. *Let $f_\bullet : X_\bullet \to Y_\bullet$ be a flat morphism in $\mathcal{P}$. If $\mathbb{F} \in D_{\mathrm{EM}}(Y_\bullet)$, then $Lf_\bullet^*\mathbb{F} \in D_{\mathrm{EM}}(X_\bullet)$.*

Proof. Let $\mathbb{F}$ be a K-flat complex in $\mathrm{Mod}(Y_\bullet)$ with equivariant cohomology groups. Then $H^n(f_\bullet^*\mathbb{F}) \cong f_\bullet^*(H^n\mathbb{F})$ is equivariant by Lemma 7.22. This is what we wanted to prove. □

(8.22) Let I be a small category, and S a scheme. Set $\mathcal{P} := \mathcal{P}(I, \underline{\mathrm{Sch}}/S)$. As we have seen, for a morphism $f_\bullet : X_\bullet \to Y_\bullet$, $f_\bullet^{-1} : \mathrm{Zar}(Y_\bullet) \to \mathrm{Zar}(X_\bullet)$ is an admissible ringed continuous functor by Lemma 6.25. Moreover, if J and K are admissible subcategories of I such that $J \subset K$, then $Q(X_\bullet|_J, K) : \mathrm{Zar}(X_\bullet|_K) \to \mathrm{Zar}(X_\bullet|_J)$ is also admissible. Utilizing Lemma 3.33 and Lemma 5.4, we have the following.

Example 8.23. Let I be a small category, S a scheme, and $f_\bullet : X_\bullet \to Y_\bullet$ and $g_\bullet : Y_\bullet \to Z_\bullet$ are morphisms in $\mathcal{P}(I, \underline{\mathrm{Sch}}/S)$. Let $K \subset J \subset I$ be admissible subcategories. Then we have the following.

1 There is a natural isomorphism

$$c_{I,J,K} : (?)_{K,I} \cong (?)_{K,J} \circ (?)_{J,I}.$$

Taking the conjugate,

$$d_{I,J,K} : LL_{I,J} \circ LL_{J,K} \cong LL_{I,K}$$

is induced.

2 There are natural isomorphism

$$c_{J,f_\bullet} : (?)_J \circ R(f_\bullet)_* \cong R(f_\bullet|_J)_* \circ (?)_J$$

and its conjugate

$$d_{J,f_\bullet} : LL_J \circ L(f_\bullet|_J)^* \cong L(f_\bullet)^* \circ LL_J.$$

3 We have

$$(c_{K,f_\bullet|_J}(?)_J) \circ ((?)_{K,J} c_{J,f_\bullet}) = (R(f_\bullet|_K)_* c_{I,J,K}) \circ c_{K,f_\bullet} \circ (c_{I,J,K}^{-1} R(f_\bullet)_*).$$

4 We have

$$(R(g_\bullet|_J)_* c_{J,f_\bullet}) \circ (c_{J,g_\bullet} R(f_\bullet)_*) = (c_{f_\bullet|_J,g_\bullet|_J}(?)_J) \circ c_{J,g_\bullet \circ f_\bullet} \circ ((?)_J c_{f_\bullet,g_\bullet}^{-1}),$$

where $c_{f_\bullet,g_\bullet} : R(g_\bullet \circ f_\bullet)_* \cong R(g_\bullet)_* \circ R(f_\bullet)_*$ is the canonical isomorphism, and similarly for $c_{f_\bullet|_J,g_\bullet|_J}$.

5 The adjoint pair $(L(?)_{\mathrm{Mod}}^*, R(?)_*^{\mathrm{Mod}})$ over the category $\mathcal{P}(I, \underline{\mathrm{Sch}}/S)$ is Lipman.

Chapter 9
Simplicial Objects

(9.1) For $n \in \mathbb{Z}$ with $n \geq -1$, we define $[n]$ to be the totally ordered finite set $\{0 < 1 < \ldots < n\}$. Thus, $[-1] = \emptyset$, $[0] = \{0\}$, $[1] = \{0 < 1\}$, and so on. We define (Δ^+) to be the small category given by $\mathrm{ob}(\Delta^+) := \{[n] \mid n \in \mathbb{Z},\ n \geq -1\}$ and

$$\mathrm{Mor}(\Delta^+) := \{\text{monotone maps}\}.$$

For a subset S of $\{-1, 0, 1, \ldots\}$, we define $(\Delta^+)_S$ to be the full subcategory of (Δ^+) such that $\mathrm{ob}((\Delta^+)_S) = \{[n] \mid n \in S\}$. We define $(\Delta) := (\Delta^+)_{[0,\infty)}$. If $-1 \notin S$, then $(\Delta^+)_S$ is also denoted by $(\Delta)_S$.

We define $(\Delta^+)^{\mathrm{mon}}$ to be the subcategory of (Δ^+) by $\mathrm{ob}((\Delta^+)^{\mathrm{mon}}) := \mathrm{ob}(\Delta^+)$ and

$$\mathrm{Mor}((\Delta^+)^{\mathrm{mon}}) := \{\text{injective monotone maps}\}.$$

For $S \subset \{-1, 0, 1, \ldots\}$, the full subcategories $(\Delta^+)^{\mathrm{mon}}_S$ and $(\Delta)^{\mathrm{mon}}_S$ of $(\Delta^+)^{\mathrm{mon}}$ are defined similarly.

We denote $(\Delta)^{\mathrm{mon}}_{\{0,1,2\}}$ and $(\Delta^+)^{\mathrm{mon}}_{\{-1,0,1,2\}}$ by Δ_M and Δ^+_M, respectively.

Let $\mathcal{C}$ be a category. We call an object of $\mathcal{P}((\Delta^+), \mathcal{C})$ (resp. $\mathcal{P}((\Delta), \mathcal{C})$, an *augmented simplicial object* (resp. *simplicial object*) of $\mathcal{C}$.

For a subcategory $\mathcal{D}$ of (Δ^+) and an object $X_\bullet \in \mathcal{P}(\mathcal{D}, \mathcal{C})$, we denote $X_{[n]}$ by X_n.

As $[-1]$ is the initial object of (Δ^+), an augmented simplicial object $X_\bullet$ of $\mathcal{C}$ with $X_{-1} = c$ is identified with a simplicial object of $\mathcal{C}/c$.

We define some particular morphisms in (Δ^+). The unique map $[-1] \to [n]$ is denoted by $\varepsilon(n)$. The unique injective monotone map $[n-1] \to [n]$ such that i is not in the image is denoted by $\delta_i(n)$ for $i \in [n]$. The unique surjective monotone map $[n+1] \to [n]$ such that i has two inverse images is denoted by $\sigma_i(n)$ for $i \in [n]$. The unique map $[0] \to [n]$ such that i is in the image is denoted by $\rho_i(n)$. The unique map $[n] \to [0]$ is denoted by λ_n.

Let $\mathcal{D}$ be a subcategory of (Δ^+). For $X_\bullet \in \mathcal{P}(\mathcal{D}, \mathcal{C})$, we denote $X_\bullet(\varepsilon(n))$ (resp. $X_\bullet(\delta_i(n))$, $X_\bullet(\sigma_i(n))$, $X_\bullet(\rho_i(n))$, and $X_\bullet(\lambda_n)$) by $e(n, X_\bullet)$ (resp.

J. Lipman, M. Hashimoto, *Foundations of Grothendieck Duality for Diagrams* 359
of Schemes, Lecture Notes in Mathematics 1960,
© Springer-Verlag Berlin Heidelberg 2009

$d_i(n, X_\bullet)$, $s_i(n, X_\bullet)$, $r_i(n, X_\bullet)$, and $l_n(X_\bullet)$), or simply by $e(n)$ (resp. $d_i(n)$, $s_i(n)$, $r_i(n)$, l_n), if there is no danger of confusion.

Note that (Δ) is generated by $\delta_i(n)$, $\sigma_i(n)$ for various i and n.

(9.2) Note that $(\Delta^+)([m], [n])$ is a finite set for any m, n. Assume that $\mathcal{C}$ has finite limits and let $f : X \to Y$ be a morphism in $\mathcal{C}$. Then the *Čech nerve* is defined to be $\mathrm{Nerve}(f) := \mathrm{cosk}^{(\Delta^+)}_{(\Delta^+)_{\{-1,0\}}}(f)$, where $\mathrm{cosk}^{(\Delta^+)}_{(\Delta^+)_{\{-1,0\}}}$ is the right adjoint of the restriction. It is described as follows. $\mathrm{Nerve}(f)_n = X \times_Y \times \cdots \times_Y X$ ($(n+1)$-fold fiber product) for $n \geq 0$, and $\mathrm{Nerve}(f)_{-1} = Y$. Note that $d_i(n)$ is given by

$$d_i(n)(x_n, \ldots, x_1, x_0) = (x_n, \cdots \overset{i}{\check{\cdots}} \cdots, x_1, x_0),$$

and $s_i(n)$ is given by

$$s_i(n)(x_n, \ldots, x_1, x_0) = (x_n, \ldots, x_{i+1}, x_i, x_i, x_{i-1}, \ldots, x_1, x_0)$$

if $\mathcal{C} = \underline{\mathrm{Set}}$.

(9.3) Let S be a scheme. A simplicial object (resp. augmented simplicial object) in $\underline{\mathrm{Sch}}/S$, in other words, an object of $\mathcal{P}((\Delta), \underline{\mathrm{Sch}}/S)$ (resp. $\mathcal{P}((\Delta^+), \underline{\mathrm{Sch}}/S)$), is called a simplicial (resp. augmented simplicial) S-scheme.

If I is a subcategory of (Δ^+), $X_\bullet \in \mathcal{P}(I, \underline{\mathrm{Sch}}/\mathbb{Z})$, $\heartsuit = \mathrm{Mod}, \mathrm{PM}, \mathrm{AB}, \mathrm{PA}$, $\mathcal{M} \in \heartsuit(X_\bullet)$ and $[n] \in I$, then we sometimes denote $\mathcal{M}_{[n]}$ by $\mathcal{M}_n$.

The following is well-known.

Lemma 9.4. *Let* $X_\bullet \in \mathcal{P}((\Delta), \underline{\mathrm{Sch}}/S)$. *Then the restriction* $(?)_{\Delta_M} : \mathrm{EM}(X_\bullet)$ $\to \mathrm{EM}(X_\bullet|_{\Delta_M})$ *is an equivalence. With the equivalence, quasi-coherent sheaves correspond to quasi-coherent sheaves.*

Proof. We define a third category $\mathcal{A}$ as follows. An object of $\mathcal{A}$ is a pair $(\mathcal{M}_0, \varphi)$ such that, $\mathcal{M}_0 \in \mathrm{Mod}(X_0)$, $\varphi \in \mathrm{Hom}_{\mathrm{Mod}(X_1)}(d_0^*(\mathcal{M}_0), d_1^*(\mathcal{M}_0))$, φ an isomorphism, and that $d_1^*(\varphi) = d_2^*(\varphi) \circ d_0^*(\varphi)$ (more precisely, the composite map

$$r_2^* \mathcal{M}_0 \xrightarrow{d^{-1}} d_1^* d_0^* \mathcal{M}_0 \xrightarrow{d_1^* \varphi} d_1^* d_1^* \mathcal{M}_0 \xrightarrow{d} r_0^* \mathcal{M}_0$$

agrees with the composite map

$$r_2^* \mathcal{M}_0 \xrightarrow{d^{-1}} d_0^* d_0^* \mathcal{M}_0 \xrightarrow{d_0^* \varphi} d_0^* d_1^* \mathcal{M}_0 \xrightarrow{d} d_2^* d_0^* \mathcal{M}_0 \xrightarrow{d_2^* \varphi} d_2^* d_1^* \mathcal{M}_0 \xrightarrow{d} r_0^* \mathcal{M}_0.$$

We use such a simplified notation throughout the proof of this lemma). Note that applying l_2^* to the last equality, we get $l_1^*(\varphi) = l_1^*(\varphi) \circ l_1^*(\varphi)$. As φ is an isomorphism, we get $l_1^*(\varphi) = \mathrm{id}$.

A morphism $\gamma_0 : (\mathcal{M}_0, \varphi) \to (\mathcal{N}_0, \psi)$ is an element

$$\gamma_0 \in \mathrm{Hom}_{\mathrm{Mod}(X_0)}(\mathcal{M}_0, \mathcal{N}_0)$$

such that

$$\psi \circ d_0^*(\gamma_0) = d_1^*(\gamma_0) \circ \varphi.$$

We define a functor $\Phi : \mathrm{EM}(X_\bullet|_{\Delta_M}) \to \mathcal{A}$ by

$$\Phi(\mathcal{M}) := (\mathcal{M}_0, \alpha_{d_1(1)}^{-1} \circ \alpha_{d_0(1)}).$$

It is easy to verify that this gives a well-defined functor.

Now we define a functor $\Psi : \mathcal{A} \to \mathrm{EM}(X_\bullet)$. Note that an object $\mathcal{M}$ of $\mathrm{EM}(X_\bullet)$ is identified with a family $(\mathcal{M}_n, \alpha_w)_{[n] \in (\Delta),\ w \in \mathrm{Mor}((\Delta))}$ such that $\mathcal{M}_n \in \mathrm{Mod}(X_n)$,

$$\alpha_w \in \mathrm{Hom}_{\mathrm{Mod}(X_n)}((X_w)_{\mathrm{Mod}}^*(\mathcal{M}_m), \mathcal{M}_n)$$

for $w \in \Delta(m, n)$, α_w is an isomorphism, and

$$\alpha_{ww'} = \alpha_w \circ X_w^* \alpha_{w'} \circ d^{-1} \tag{9.5}$$

whenever ww' is defined, see (4.6).

For $(\mathcal{M}_0', \varphi) \in \mathcal{A}$, we define $\mathcal{M}_{n,i} := (r_i(n))^*(\mathcal{M}_0')$, and $\mathcal{M}_n := \mathcal{M}_{n,0}$ for $n \geq 0$ and $0 \leq i \leq n$. We define $\psi_i(n) : \mathcal{M}_{n,i+1} \to \mathcal{M}_{n,i}$ to be $(X_{q(i,n)})^*(\varphi)$ for $n \geq 1$ and $0 \leq i < n$, where $q(i,n) : [1] \to [n]$ is the unique injective monotone map with $\{i, i+1\} = \mathrm{Im}\, q(i,n)$. We define $\varphi_i(n) : \mathcal{M}_{n,i} \cong \mathcal{M}_n$ to be the composite map

$$\varphi_i(n) := \psi_0(n) \circ \psi_1(n) \circ \cdots \circ \psi_{i-1}(n)$$

for $n \geq 0$ and $0 \leq i \leq n$.

Now we define

$$\alpha_w \in \mathrm{Hom}_{\mathrm{Mod}(X_n)}(X_w^*(\mathcal{M}_m), \mathcal{M}_n)$$

to be the map

$$X_w^* \mathcal{M}_m = X_w^* r_0(m)^* \mathcal{M}_0' \xrightarrow{d} r_{w(0)}(n)^* \mathcal{M}_0' = \mathcal{M}_{n,w(0)} \xrightarrow{\varphi_{w(0)}(n)} \mathcal{M}_n$$

for $w \in \Delta([m], [n])$.

Thus $(\mathcal{M}_0', \varphi)$ yields a family $(\mathcal{M}_n, \alpha_w)$, and this gives the definition of $\Psi : \mathcal{A} \to \mathrm{EM}(X_\bullet)$. The details of the proof of the well-definedness is left to the reader.

It is also straightforward to check that $(?)_{\Delta_M}$, Φ, and Ψ give the equivalence of these three categories. The proof is also left to the reader.

The last assertion is obvious from the construction. □

Chapter 10
Descent Theory

Let S be a scheme.

(10.1) Consider the functor shift : $(\Delta^+) \to (\Delta)$ given by $\text{shift}[n] := [n+1]$, $\text{shift}(\delta_i(n)) := \delta_{i+1}(n+1)$, $\text{shift}(\sigma_i(n)) := \sigma_{i+1}(n+1)$, and $\text{shift}(\varepsilon(0)) := \delta_1(1)$. We have a natural transformation $(\delta_0^+) : \text{Id}_{(\Delta^+)} \to \iota \circ \text{shift}$ given by $(\delta_0^+)_n := \delta_0(n+1)$ for $n \geq 0$ and $(\delta_0^+)_{-1} := \varepsilon(0)$, where $\iota : (\Delta) \hookrightarrow (\Delta^+)$ is the inclusion. We denote $(\delta_0^+)\iota$ by (δ_0). Note that (δ_0) can be viewed as a natural map $(\delta_0) : \text{Id}_{(\Delta)} \to \text{shift } \iota$.

Let $X_\bullet \in \mathcal{P}((\Delta), \underline{\text{Sch}}/S)$. We define $X_\bullet'$ to be the augmented simplicial scheme $\text{shift}^\#(X_\bullet) = X_\bullet \text{ shift}$. The natural map

$$X_\bullet(\delta_0) : X_\bullet'|_{(\Delta)} = X_\bullet \text{ shift } \iota \to X_\bullet$$

is denoted by $(d_0)(X_\bullet)$ or (d_0). Similarly, if $Y_\bullet \in \mathcal{P}((\Delta^+), \underline{\text{Sch}}/S)$, then

$$(d_0^+)(Y_\bullet) : (Y_\bullet|_{(\Delta)})' = Y_\bullet \iota \text{ shift } \xrightarrow{Y_\bullet(\delta_0^+)} Y_\bullet$$

is defined as well.

(10.2) We say that $X_\bullet \in \mathcal{P}((\Delta), \underline{\text{Sch}}/S)$ is a *simplicial groupoid* of S-schemes if there is a faithfully flat morphism of S-schemes $g : Z \to Y$ such that there is a faithfully flat cartesian morphism $f_\bullet : Z_\bullet \to X_\bullet$ of $\mathcal{P}((\Delta), \underline{\text{Sch}}/S)$, where $Z_\bullet = \text{Nerve}(g)|_{(\Delta)}$.

Lemma 10.3. *Let* $X_\bullet \in \mathcal{P}((\Delta), \underline{\text{Sch}}/S)$.

1 *If* $X_\bullet \cong \text{Nerve}(g)|_{(\Delta)}$ *for some faithfully flat morphism g of S-schemes, then $X_\bullet$ is a simplicial groupoid.*

2 *If* $f_\bullet : Z_\bullet \to X_\bullet$ *is a faithfully flat cartesian morphism of simplicial S-schemes and $Z_\bullet$ is a simplicial groupoid, then we have $X_\bullet$ is also a simplicial groupoid.*

J. Lipman, M. Hashimoto, *Foundations of Grothendieck Duality for Diagrams of Schemes*, Lecture Notes in Mathematics 1960,
© Springer-Verlag Berlin Heidelberg 2009

3 $X_\bullet$ *is a simplicial groupoid if and only if* $(d_0) : X'_\bullet|_{(\Delta)} \to X_\bullet$ *is cartesian,*
the canonical unit map

$$X'_\bullet \to \mathrm{Nerve}(d_1(1)) = \mathrm{cosk}^{(\Delta^+)}_{(\Delta^+)^{\mathrm{mon}}_{\{-1,0\}}}(X'_\bullet|_{(\Delta^+)^{\mathrm{mon}}_{\{-1,0\}}})$$

is an isomorphism, and $d_0(1)$ *and* $d_1(1)$ *are flat.*
4 *If* $f_\bullet : Z_\bullet \to X_\bullet$ *is a cartesian morphism of simplicial S-schemes and* $X_\bullet$
is a simplicial groupoid, then $Z_\bullet$ *is a simplicial groupoid.*
5 *A simplicial groupoid has faithfully flat* $(\Delta)^{\mathrm{mon}}$*-arrows.*
6 *If* $X_\bullet$ *is a simplicial groupoid of S-schemes such that* $d_0(1)$ *and* $d_1(1)$
are separated (*resp. quasi-compact, quasi-separated, of finite type, smooth,*
étale), *then* $X_\bullet$ *has separated* (*resp. quasi-compact, quasi-separated, of fi-*
nite type, smooth, étale) $(\Delta)^{\mathrm{mon}}$*-arrows, and* $(d_0) : X'_\bullet|_{(\Delta)} \to X_\bullet$ *is sepa-*
rated (*resp. quasi-compact, quasi-separated, of finite type, smooth, étale*).

Proof. **1** and **2** are obvious by definition.

We prove **3**. We prove the 'if' part. As $d_0(1)s_0(0) = \mathrm{id} = d_1(1)s_0(0)$, we
have that $d_0(1)$ and $d_1(1)$ are faithfully flat by assumption. As $d_0(1) = (d_0)_0$
is faithfully flat and (d_0) is cartesian, it is easy to see that (d_0) is also faithfully
flat. So this direction is obvious.

We prove the 'only if' part. As $X_\bullet$ is a simplicial groupoid, there is a
faithfully flat S-morphism $g : Z \to Y$ and a faithfully flat cartesian morphism
$f_\bullet : Z_\bullet \to X_\bullet$ of simplicial S-schemes, where $Z_\bullet = \mathrm{Nerve}(g)|_{(\Delta)}$. It is easy
to see that $(d_0) : Z'_\bullet|_{(\Delta)} \to Z_\bullet$ is nothing but the base change by g, and it
is faithfully flat cartesian. It is also obvious that $Z'_\bullet \cong \mathrm{Nerve}(d_1(1)(Z_\bullet))$ and
$d_0(1)(Z_\bullet)$ and $d_1(1)(Z_\bullet)$ are flat. It is obvious that $f'_\bullet : Z'_\bullet \to X'_\bullet$ is faithfully
flat cartesian. Now by Lemma 7.16, $(d_0)(X_\bullet)$ is cartesian. As $f_\bullet$ is faithfully
flat cartesian and $d_0(1)(Z_\bullet)$ and $d_1(1)(Z_\bullet)$ are flat, we have that $d_0(1)(X_\bullet)$
and $d_1(1)(X_\bullet)$ are flat. When we base change $X'_\bullet \to \mathrm{Nerve}(d_1(1)(X_\bullet))$ by
$f_0 : Z_0 \to X_0$, then we have the isomorphism $Z'_\bullet \cong \mathrm{Nerve}(d_1(1)(Z_\bullet))$. As f_0
is faithfully flat, we have that $X'_\bullet \to \mathrm{Nerve}(d_1(1))$ is also an isomorphism.

The assertions **4**, **5** and **6** are proved easily. $\square$

(10.4) Let $X_\bullet \in \mathcal{P}((\Delta), \underline{\mathrm{Sch}}/S)$. Then we define $F : \mathrm{Zar}(X'_\bullet) \to \mathrm{Zar}(X_\bullet)$
by $F(([n], U)) = (\mathrm{shift}[n], U)$ and $F((w, h)) = (\mathrm{shift}\, w, h)$. The corresponding
pull-back $F^\#_{\mathrm{Mod}}$ is denoted by $(?)'$. It is easy to see that $(?)'$ has a left and a
right adjoint. It also preserves equivariant and locally quasi-coherent sheaves.

Let $\mathcal{M} \in \mathrm{Mod}(X_\bullet)$. Then we define $(\alpha) : (d_0)^* \mathcal{M} \to \mathcal{M}'_{(\Delta)}$ by

$$(\alpha)_n : ((d_0)^* \mathcal{M})_n \xrightarrow{\theta^{-1}} d_0(n)^* \mathcal{M}_n \xrightarrow{\alpha_{\delta_0(n)}} \mathcal{M}_{n+1} = \mathcal{M}'_n.$$

It is easy to see that $(\alpha) : (d_0)^* \to (?)_{(\Delta)} \circ (?)'$ is a natural map. Similarly,
for $Y_\bullet \in \mathcal{P}((\Delta^+), \underline{\mathrm{Sch}}/S)$,

$$(\alpha^+) : (d_0^+)^* \to (?)' \circ (?)_{(\Delta)}$$

is defined.

(10.5) Let $X_\bullet \in \mathcal{P}((\Delta^+), \underline{\mathrm{Sch}}/S)$, and $\mathcal{M} \in \mathrm{Mod}(X_\bullet|_{(\Delta)})$. Then, we have a cosimplicial object $\mathrm{Cos}(\mathcal{M})$ of $\mathrm{Mod}(X_{-1})$ (i.e., a simplicial object of $\mathrm{Mod}(X_{-1})^{\mathrm{op}}$). We have $\mathrm{Cos}(\mathcal{M})_n := e(n)_*(\mathcal{M}_n)$, and

$$\mathrm{Cos}(\mathcal{M})_w : e(m)_*(\mathcal{M}_m) \xrightarrow{\beta_w} e(m)_*(w_*(\mathcal{M}_n)) \xrightarrow{c^{-1}} e(n)_*(\mathcal{M}_n)$$

for a morphism $w : [m] \to [n]$ in Δ. Similarly, the augmented cosimplicial object $\mathrm{Cos}^+(\mathcal{N})$ of $\mathrm{Mod}(X_{-1})$ is defined for $\mathcal{N} \in \mathrm{Mod}(X_\bullet)$.

By (6.14), it is easy to see that $\mathcal{M}^+ := R^{\mathrm{Mod}}_{(\Delta)}(\mathcal{M})$ is $\mathcal{M}$ on $X_\bullet|_{(\Delta)}$, $\mathcal{M}^+_{-1}$ is $\varprojlim \mathrm{Cos}(\mathcal{M})$, and $\beta_{\bar{\varepsilon}(n)}(\mathcal{M}^+)$ is nothing but the canonical map

$$\varprojlim \mathrm{Cos}(\mathcal{M}) \to \mathrm{Cos}(\mathcal{M})_n = e(n)_*(\mathcal{M}_n) = e(n)_*(\mathcal{M}^+_n).$$

Note that $\mathrm{Cos}(\mathcal{M})$ can be viewed as a (co)chain complex such that $\mathrm{Cos}(\mathcal{M})^n = e(n)_*(\mathcal{M}_n)$ for $n \geq 0$, and the boundary map $\partial^n : \mathrm{Cos}(\mathcal{M})^n \to \mathrm{Cos}(\mathcal{M})^{n+1}$ is given by $\partial^n = d_0 - d_1 + \cdots + (-1)^{n+1} d_{n+1}$, where $d_i = d_i(n+1) = \mathrm{Cos}(\mathcal{M})_{\delta_i(n+1)}$. Similarly, for $\mathcal{N} \in \mathrm{Mod}(X_\bullet)$, $\mathrm{Cos}^+(\mathcal{N})$ can be viewed as an augmented cochain complex.

Note also that for $\mathcal{M} \in \mathrm{Mod}(X_\bullet|_{(\Delta)})$, we have

$$\varprojlim \mathrm{Cos}(\mathcal{M}) = \mathrm{Ker}(d_0(1) - d_1(1)) = H^0(\mathrm{Cos}(\mathcal{M})), \qquad (10.6)$$

which is determined only by $\mathcal{M}_{(\Delta)_{\{0,1\}}}$.

Lemma 10.7. *Let $f_\bullet : X_\bullet \to Y_\bullet$ be a morphism of $\mathcal{P}((\Delta^+), \underline{\mathrm{Sch}}/S)$. If $f_\bullet|_{(\Delta^+)_{\{-1,0,1\}}}$ is flat cartesian, and $Y_\bullet$ has concentrated $(\Delta^+)^{\mathrm{mon}}_{\{-1,0,1\}}$-arrows, then the canonical map*

$$\mu : f^*_\bullet \circ R_{(\Delta)} \to R_{(\Delta)} \circ (f_\bullet|_{(\Delta)})^*$$

(see (6.27)) is an isomorphism of functors from $\mathrm{Lqc}(Y_\bullet|_{(\Delta)})$ to $\mathrm{Lqc}(X_\bullet)$.

Proof. To prove that the map in question is an isomorphism, it suffices to show that the map is an isomorphism after applying the functor $(?)_n$ for $n \geq -1$. This is trivial if $n \geq 0$. On the other hand, if $n = -1$, the map restricted at -1 and evaluated at $\mathcal{M} \in \mathrm{Lqc}(Y_\bullet|_{(\Delta)})$ is nothing but

$$f^*_{-1}(H^0(\mathrm{Cos}(\mathcal{M}))) \cong H^0(f^*_{-1}(\mathrm{Cos}(\mathcal{M}))) \to H^0(\mathrm{Cos}((f_\bullet|_{(\Delta)})^*(\mathcal{M}))).$$

The first map is an isomorphism as f_{-1} is flat. Although the map

$$f^*_{-1}(\mathrm{Cos}(\mathcal{M})) \to \mathrm{Cos}((f_\bullet|_{(\Delta)})^*(\mathcal{M}))$$

may not be a chain isomorphism, it is an isomorphism at the degrees $-1, 0, 1$, and it induces the isomorphism of H^0. $\qquad\square$

Lemma 10.8. *Let $X_\bullet \in \mathcal{P}((\Delta), \underline{\mathrm{Sch}}/S)$, and $\mathcal{M} \in \mathrm{Mod}(X_\bullet)$. Then the (associated chain complex of the) augmented cosimplicial object $\mathrm{Cos}^+(\mathcal{M}')$ of $\mathcal{M}' \in \mathrm{Mod}(X_\bullet')$ is split exact. In particular, the unit map $u : \mathcal{M}' \to R_{(\Delta)}\mathcal{M}'_{(\Delta)}$ is an isomorphism.*

Proof. Define $s_n : \mathrm{Cos}^+(\mathcal{M}')_n \to \mathrm{Cos}^+(\mathcal{M}')_{n-1}$ to be

$$\mathrm{Cos}^+(\mathcal{M}')_n = (r_0)(n+1)_*(\mathcal{M}_{n+1}) \xrightarrow{(r_0)(n+1)_*\beta_{\sigma_0(n)}}$$

$$(r_0)(n+1)_*s_0(n)_*(\mathcal{M}_n) \xrightarrow{c^{-1}} (r_0)(n)_*(\mathcal{M}_n) = \mathrm{Cos}^+(\mathcal{M}')_{n-1}$$

for $n \geq 0$, and $s_{-1} : \mathrm{Cos}^+(\mathcal{M}')_{-1} \to 0$ to be 0. It is easy to verify that s is a chain deformation of $\mathrm{Cos}^+(\mathcal{M}')$.   □

Corollary 10.9. *Let the notation be as in the lemma. Then there is a functorial isomorphism*

$$R_{(\Delta)}(d_0)^*(\mathcal{M}) \to \mathcal{M}' \tag{10.10}$$

for $\mathcal{M} \in \mathrm{EM}(X_\bullet)$. In particular, there is a functorial isomorphism

$$(R_{(\Delta)}(d_0)^*(\mathcal{M}))_{-1} \to \mathcal{M}_0. \tag{10.11}$$

Proof. The first map (10.10) is defined to be the composite

$$R_{(\Delta)}(d_0)^* \xrightarrow{R_{(\Delta)}(\alpha)} R_{(\Delta)}(?)_{(\Delta)}(?)' \xrightarrow{u^{-1}} (?)'.$$

As $(\alpha)(\mathcal{M})$ is an isomorphism if $\mathcal{M}$ is equivariant, this is an isomorphism. The second map (10.11) is obtained from (10.10), applying $(?)_{-1}$.   □

The following well-known theorem in descent theory contained in [33] is now easy to prove.

Proposition 10.12. *Let $f : X \to Y$ be a morphism of S-schemes, and set $X_\bullet^+ := \mathrm{Nerve}(f)$, and $X_\bullet := X_\bullet^+|_{(\Delta)}$. Let $\mathcal{M} \in \mathrm{Mod}(X_\bullet)$. Then we have the following.*

0 *The counit of adjunction*

$$\varepsilon : (R_{(\Delta)}\mathcal{M})_{(\Delta)} \to \mathcal{M}$$

is an isomorphism.

1 *If f is concentrated and $\mathcal{M} \in \mathrm{Lqc}(X_\bullet)$, then $R_{(\Delta)}\mathcal{M} \in \mathrm{Lqc}(X_\bullet^+)$.*

2 *If f is faithfully flat concentrated and $\mathcal{M} \in \mathrm{Qch}(X_\bullet)$, then we have $R_{(\Delta)}\mathcal{M} \in \mathrm{Qch}(X_\bullet^+)$.*

3 *If f is faithfully flat concentrated, $\mathcal{N} \in \mathrm{EM}(X_\bullet^+)$, and $\mathcal{N}_{(\Delta)} \in \mathrm{Qch}(X_\bullet)$, then the unit of adjunction*

$$u : \mathcal{N} \to R_{(\Delta)}(\mathcal{N}_{(\Delta)})$$

is an isomorphism. In particular, $\mathcal{N}$ is quasi-coherent.

4 *If f is faithfully flat concentrated, then the restriction functor*

$$(?)_{(\Delta)} : \mathrm{Qch}(X_\bullet^+) \to \mathrm{Qch}(X_\bullet)$$

is an equivalence, with $R_{(\Delta)}$ its quasi-inverse.

Proof. The assertion **0** follows from Lemma 6.15.

We prove **1**. By **0**, it suffices to prove that

$$(R_{(\Delta)}\mathcal{M})_{-1} = \mathrm{Ker}(e(0)_*\beta_{\delta_0(1)} - e(0)_*\beta_{\delta_1(1)})$$

is quasi-coherent. This is obvious by [14, (9.2.2)].

Now we assume that f is faithfully flat concentrated, to prove the assertions **2**, **3**, and **4**.

We prove **2**. As we already know that $R_{(\Delta)}\mathcal{M}$ is locally quasi-coherent, it suffices to show that it is equivariant. As $(d_0^+) : (X_\bullet|_{(\Delta)})' \to X_\bullet$ is faithfully flat, it suffices to show that $(d_0^+)^*R_{(\Delta)}\mathcal{M}$ is equivariant, by Lemma 7.22. Now the assertion is obvious by Lemma 10.7 and Corollary 10.9, as $\mathcal{M}'$ is quasi-coherent.

We prove **3**. Note that the composite map

$$(d_0^+)^*\mathcal{N} \xrightarrow{(d_0^+)^*u} (d_0^+)^*R_{(\Delta)}(\mathcal{N}_{(\Delta)}) \xrightarrow{\mu} R_{(\Delta)}(d_0)^*\mathcal{N}_{(\Delta)} \cong R_{(\Delta)}((d_0^+)^*\mathcal{N})_{(\Delta)} \tag{10.13}$$

is nothing but the unit of adjunction $u((d_0^+)^*\mathcal{N})$. As $(\alpha^+) : (d_0^+)^*\mathcal{N} \to (\mathcal{N}_{(\Delta)})'$ is an isomorphism since $\mathcal{N}$ is equivariant, we have that $u((d_0^+)^*\mathcal{N})$ is an isomorphism by Lemma 10.8. As μ in (10.13) is an isomorphism by Lemma 10.7, we have that $(d_0^+)^*u$ is an isomorphism. As (d_0^+) is faithfully flat, we have that $u : \mathcal{N} \to R_{(\Delta)}(\mathcal{N}_{(\Delta)})$ is an isomorphism, as desired. The last assertion is obvious by **2**, and **3** is proved.

The assertion **4** is a consequence of **0**, **2** and **3**. □

Corollary 10.14. *Let $f : X \to Y$ be a faithfully flat quasi-compact morphism of schemes, and $\mathcal{M} \in \mathrm{Mod}(Y)$. Then $\mathcal{M}$ is quasi-coherent if and only if $f^*\mathcal{M}$ is.*

Proof. The 'only if' part is trivial.

We prove the 'if' part. We may assume that Y is affine. So X is quasi-compact, and has a finite affine open covering (U_i). Replacing X by $\coprod_i U_i$, we may assume that X is also affine. Thus f is faithfully flat concentrated. If $f^*\mathcal{M}$ is quasi-coherent, then $\mathcal{N} := L_{-1}\mathcal{M}$ satisfies the assumption of **3** of the proposition, as can be seen easily. So $\mathcal{M} \cong (\mathcal{N})_{-1}$ is quasi-coherent. □

Corollary 10.15. *Let the notation be as in the proposition, and assume that f is faithfully flat concentrated. The composite functor*

$$\mathbb{A} := (?)_{(\Delta)} \circ L_{-1} : \mathrm{Qch}(Y) \to \mathrm{Qch}(X_\bullet)$$

is an equivalence with

$$\mathbb{D} := (?)_{-1} \circ R_{(\Delta)}$$

its quasi-inverse.

Proof. Follows immediately by the proposition and Lemma 6.38, **2**, since $[-1]$ is the initial object of (Δ^+). □

We call $\mathbb{A}$ in the corollary the *ascent functor*, and $\mathbb{D}$ the *descent functor*.

Corollary 10.16. *Let the notation be as in the proposition. Then the composite functor*

$$\mathbb{A} \circ \mathbb{D} : \mathrm{Lqc}(X_\bullet) \to \mathrm{Qch}(X_\bullet)$$

is the right adjoint functor of the inclusion $\mathrm{Qch}(X_\bullet) \hookrightarrow \mathrm{Lqc}(X_\bullet)$.

Proof. Note that $\mathbb{D} : \mathrm{Lqc}(X_\bullet) \to \mathrm{Qch}(Y)$ is a well-defined functor, and hence $\mathbb{A} \circ \mathbb{D}$ is a functor from $\mathrm{Lqc}(X_\bullet)$ to $\mathrm{Qch}(X_\bullet)$.

For $\mathcal{M} \in \mathrm{Qch}(X_\bullet)$ and $\mathcal{N} \in \mathrm{Lqc}(X_\bullet)$, we have

$$\mathrm{Hom}_{\mathrm{Qch}(X_\bullet)}(\mathcal{M}, \mathbb{A}\mathbb{D}\mathcal{N}) \cong \mathrm{Hom}_{\mathrm{Qch}(Y)}(\mathbb{D}\mathcal{M}, \mathbb{D}\mathcal{N})$$
$$\cong \mathrm{Hom}_{\mathrm{Lqc}(X_\bullet^+)}(R_{(\Delta)}\mathcal{M}, R_{(\Delta)}\mathcal{N})$$
$$\cong \mathrm{Hom}_{\mathrm{Lqc}(X_\bullet)}((R_{(\Delta)}\mathcal{M})_{(\Delta)}, \mathcal{N}) \cong \mathrm{Hom}_{\mathrm{Lqc}(X_\bullet)}(\mathcal{M}, \mathcal{N})$$

by the proposition, Corollary 10.15, and Lemma 6.38, **1**. □

Corollary 10.17. *Let $X_\bullet$ be a simplicial groupoid of S-schemes, and assume that $d_0(1)$ and $d_1(1)$ are concentrated. Then*

$$(d_0)_*^{\mathrm{Qch}} \circ \mathbb{A} : \mathrm{Qch}(X_0) \to \mathrm{Qch}(X_\bullet)$$

is a right adjoint of $(?)_0 : \mathrm{Qch}(X_\bullet) \to \mathrm{Qch}(X_0)$, *where* $\mathbb{A} : \mathrm{Qch}(X_0) \to \mathrm{Qch}(X_\bullet'|_{(\Delta)})$ *is the ascent functor defined in Corollary 10.15.*

Proof. Note that $(d_0)_*^{\mathrm{Qch}}$ is well-defined, because (d_0) is concentrated cartesian, and the simplicial groupoid $X_\bullet$ has flat $(\Delta)^{\mathrm{mon}}$-arrows, see (7.15) and Lemma 10.3. It is obvious that $\mathbb{D} \circ (d_0)_{\mathrm{Qch}}^*$ is the left adjoint of $(d_0)_*^{\mathrm{Qch}} \circ \mathbb{A}$ by Corollary 10.15. On the other hand, we have $(?)_0 \cong \mathbb{D} \circ (d_0)_{\mathrm{Qch}}^*$ by Corollary 10.9. Hence, $(d_0)_*^{\mathrm{Qch}} \circ \mathbb{A}$ is a right adjoint of $(?)_0$, as desired. □

(10.18) Let $f : X \to Y$ be a morphism of S-schemes, and set $X_\bullet^+ := \mathrm{Nerve}(f)$, and $X_\bullet = X_\bullet^+|_{(\Delta)}$. It seems that even if f is concentrated and faithfully flat, the canonical descent functor $\mathrm{EM}(X_\bullet) \to \mathrm{Mod}(Y)$ may not be an isomorphism. However, we have this kind of isomorphism for special morphisms.

Let $f : X \to Y$ be a morphism of schemes. We say that f is a *locally an open immersion* if there exists some open covering (U_i) of X such that $f|_{U_i}$ is an open immersion for any i. Assume that f is locally an open immersion.

Lemma 10.19. *Let $f : X \to Y$ be locally an open immersion. Let $g : Y' \to Y$ be any morphism, $X' := Y' \times_Y X$, $g' : X' \to X$ the second projection, and $f' : X' \to Y'$ the first projection. Then the canonical map $\theta : f^* g_* \to g'_* (f')^*$ between the functors from $\mathrm{Mod}(Y')$ to $\mathrm{Mod}(X)$ is an isomorphism.*

Proof. Use Lemma 2.59. □

Lemma 10.20. *Let $f : X \to Y$, $X_\bullet^+$, and $X_\bullet$ be as in (10.18). Assume that f is faithfully flat and locally an open immersion. Then the descent functor $\mathbb{D} = (?)_{-1} R_{(\Delta)} : \mathrm{EM}(X_\bullet) \to \mathrm{Mod}(Y)$ is an equivalence with $\mathbb{A} = (?)_{(\Delta)} L_{-1}$ its quasi-inverse.*

Proof. Similar to the proof of Proposition 10.12. □

Lemma 10.21. *Let $f : X \to Y$, $X_\bullet^+$, and $X_\bullet$ be as in (10.18). Assume that f is faithfully flat and locally an open immersion. Then for $\mathcal{M} \in \mathrm{EM}(X'_\bullet)$, the direct image $(d_0)_* \mathcal{M}$ is equivariant. The restriction $\mathrm{EM}(X_\bullet) \to \mathrm{Mod}(X_0) = \mathrm{Mod}(X)$ has the right adjoint $(d_0)_* \mathbb{A}$.*

Proof. Easy. □

Chapter 11
Local Noetherian Property

An abelian category $\mathcal{A}$ is called *locally noetherian* if it is a $\mathcal{U}$-category, satisfies the (AB5) condition, and has a small set of noetherian generators [11]. For a locally noetherian category $\mathcal{A}$, we denote the full subcategory of $\mathcal{A}$ consisting of its noetherian objects by $\mathcal{A}_f$.

Lemma 11.1. *Let $\mathcal{A}$ be an abelian $\mathcal{U}$-category which satisfies the (AB3) condition, and $\mathcal{B}$ a locally noetherian category. Let $F : \mathcal{A} \to \mathcal{B}$ be a faithful exact functor, and G its right adjoint. If G preserves filtered inductive limits, then the following hold.*

1 *$\mathcal{A}$ is locally noetherian.*
2 *$a \in \mathcal{A}$ is a noetherian object if and only if Fa is.*

Proof. The 'if' part of **2** is obvious, as F is faithful and exact. Note that $\mathcal{A}$ satisfies the (AB5) condition, as F is faithful exact and colimit preserving, and $\mathcal{B}$ satisfies the (AB5) condition.

Note also that, for $a \in \mathcal{A}$, the set of subobjects of a is small, because the set of subobjects of Fa is small [13] and F is faithful exact.

Let S be a small set of noetherian generators of $\mathcal{B}$. As any noetherian object is a quotient of a finite sum of objects in S, we may assume that any noetherian object in $\mathcal{B}$ is isomorphic to an element of S, replacing S by some larger small set, if necessary. For each $s \in S$, the set of subobjects of Gs is small by the last paragraph. Hence, there is a small subset T of $\mathrm{ob}(\mathcal{A})$ such that, any element $t \in T$ admits a monomorphism $t \to Gs$ for some $s \in S$, Ft is noetherian, and that if $a \in \mathcal{A}$ admits a monomorphism $a \to Gs$ for some $s \in S$ and Fa noetherian then $a \cong t$ for some $t \in T$.

We claim that any $a \in \mathcal{A}$ is a filtered inductive limit $\varinjlim a_\lambda$ of subobjects a_λ of a, with each a_λ is isomorphic to some element in T.

If the claim is true, then **1** is obvious, as T is a small set of noetherian generators of $\mathcal{A}$, and $\mathcal{A}$ satisfies the (AB5) condition, as we have already seen.

The 'only if' part of **2** is also true if the claim is true, since if $a \in \mathcal{A}$ is noetherian, then it is a quotient of a finite sum of elements of T, and hence Fa is noetherian.

J. Lipman, M. Hashimoto, *Foundations of Grothendieck Duality for Diagrams* 371
of Schemes, Lecture Notes in Mathematics 1960,
© Springer-Verlag Berlin Heidelberg 2009

It suffices to prove the claim. As $\mathcal{B}$ is locally noetherian, we have $Fa = \varinjlim b_\lambda$, where (b_λ) is the filtered inductive system of noetherian subobjects of Fa.

Let $u : \mathrm{Id} \to GF$ be the unit of adjunction, and $\varepsilon : FG \to \mathrm{Id}$ be the counit of adjunction. It is well-known that we have $(\varepsilon F) \circ (Fu) = \mathrm{id}_F$. As Fu is a split monomorphism, u is also a monomorphism. We define $a_\lambda := u(a)^{-1}(Gb_\lambda)$. As G preserves filtered inductive limits and $\mathcal{A}$ satisfies the (AB5) condition, we have

$$\varinjlim a_\lambda = u(a)^{-1}(G\varinjlim b_\lambda) = u(a)^{-1}(GFa) = a.$$

Note that $a_\lambda \to Gb_\lambda$ is a monomorphism, with b_λ being noetherian.

It remains to show that Fa_λ is noetherian. Let $i_\lambda : a_\lambda \hookrightarrow a$ be the inclusion map, and $j_\lambda : b_\lambda \to Fa$ the inclusion. Then the diagram

$$
\begin{array}{ccccc}
Fa & \xrightarrow{Fu(a)} & FGFa & \xrightarrow{\varepsilon F(a)} & Fa \\
{\scriptstyle Fi_\lambda} \uparrow & & \uparrow {\scriptstyle FGj_\lambda} & & \uparrow {\scriptstyle j_\lambda} \\
Fa_\lambda & \longrightarrow & FGb_\lambda & \xrightarrow{\varepsilon(b_\lambda)} & b_\lambda
\end{array}
$$

is commutative. As the composite of the first row is the identity map and Fi_λ is a monomorphism, we have that the composite of the second row $Fa_\lambda \to b_\lambda$ is a monomorphism. As b_λ is noetherian, we have that Fa_λ is also noetherian, as desired. $\square$

Lemma 11.2. *Let $\mathcal{A}$ be an abelian $\mathcal{U}$-category which satisfies the (AB3) condition, and $\mathcal{B}$ a Grothendieck category. Let $\mathcal{A} \to \mathcal{B}$ be a faithful exact functor, and G its right adjoint. If G preserves filtered inductive limits, then $\mathcal{A}$ is Grothendieck.*

Proof. Similar. $\square$

(11.3) Let S be a scheme, and $X_\bullet \in \mathcal{P}((\Delta), \underline{\mathrm{Sch}}/S)$.

Lemma 11.4. *The restriction functor $(?)_0 : \mathrm{EM}(X_\bullet) \to \mathrm{Mod}(X_0)$ is faithful exact.*

Proof. This is obvious, because for any $[n] \in (\Delta)$, there is a morphism $[0] \to [n]$.

Lemma 11.5. *Let $X_\bullet$ be a simplicial groupoid of S-schemes, and assume that $d_0(1)$ and $d_1(1)$ are concentrated. If $\mathrm{Qch}(X_0)$ is Grothendieck, then $\mathrm{Qch}(X_\bullet)$ is Grothendieck. Assume moreover that $\mathrm{Qch}(X_0)$ is locally noetherian. Then we have*

1 $\mathrm{Qch}(X_\bullet)$ *is locally noetherian.*
2 $\mathcal{M} \in \mathrm{Qch}(X_\bullet)$ *is a noetherian object if and only if $\mathcal{M}_0$ is a noetherian object.*

Proof. Let $F := (?)_0 : \mathrm{Qch}(X_\bullet) \to \mathrm{Qch}(X_0)$ be the restriction. By Lemma 11.4, F is faithful exact. Let $G := (d_0)_*^{\mathrm{Qch}} \circ \mathbb{A}$ be the right adjoint of F, see Corollary 10.17. As $\mathbb{A}$ is an equivalence and $(d_0)_*^{\mathrm{Qch}}$ preserves filtered inductive limits by Lemma 7.20, G preserves filtered inductive limits. As $\mathrm{Qch}(X_\bullet)$ satisfies (AB3) by Lemma 7.7, the assertion is obvious by Lemma 11.1 and Lemma 11.2. $\qquad\qquad\qquad\qquad\qquad\qquad\qquad\qquad\qquad\qquad\square$

The following is well-known, see [18, pp. 126–127].

Corollary 11.6. *Let Y be a noetherian scheme. Then $\mathrm{Qch}(Y)$ is locally noetherian, and $\mathcal{M} \in \mathrm{Qch}(Y)$ is a noetherian object if and only if it is coherent.*

Proof. This is obvious if $Y = \mathrm{Spec}\, A$ is affine. Now consider the general case. Let $(U_i)_{1 \le i \le r}$ be an affine open covering of Y, and set $X := \coprod_i U_i$. Let $p : X \to Y$ be the canonical map, and set $X_\bullet := \mathrm{Nerve}(f)|_{(\Delta)}$. Note that p is faithfully flat quasi-compact separated. By assumption and the lemma, we have that $\mathrm{Qch}(X_\bullet)$ is locally noetherian, and $\mathcal{M} \in \mathrm{Qch}(X_\bullet)$ is a noetherian object if and only if $\mathcal{M}_0$ is noetherian, i.e., coherent. As $\mathbb{A} : \mathrm{Qch}(Y) \to \mathrm{Qch}(X_\bullet)$ is an equivalence, we have that $\mathrm{Qch}(Y)$ is locally noetherian, and $\mathcal{M} \in \mathrm{Qch}(Y)$ is a noetherian object if and only if $(\mathbb{A}\mathcal{M})_0 = p^*\mathcal{M}$ is coherent if and only if $\mathcal{M}$ is coherent. $\qquad\qquad\qquad\qquad\qquad\qquad\qquad\square$

A scheme X is said to be *concentrated* if the structure map $X \to \mathrm{Spec}\,\mathbb{Z}$ is concentrated.

Corollary 11.7. *Let Y be a concentrated scheme. Then $\mathrm{Qch}(Y)$ is Grothendieck.*

Proof. Similar. $\qquad\qquad\qquad\qquad\qquad\qquad\qquad\qquad\qquad\qquad\qquad\qquad\qquad\square$

Now the following is obvious.

Corollary 11.8. *Let $X_\bullet$ be a simplicial groupoid of S-schemes, with $d_0(1)$ and $d_1(1)$ concentrated. If X_0 is concentrated, then $\mathrm{Qch}(X_\bullet)$ is Grothendieck. If, moreover, X_0 is noetherian, then $\mathrm{Qch}(X_\bullet)$ is locally noetherian, and $\mathcal{M} \in \mathrm{Qch}(X_\bullet)$ is a noetherian object if and only if $\mathcal{M}_0$ is coherent.*

Lemma 11.9. *Let I be a finite category, S a scheme, and $X_\bullet \in \mathcal{P}(I, \underline{\mathrm{Sch}}/S)$. If $X_\bullet$ is noetherian, then $\mathrm{Mod}(X_\bullet)$ and $\mathrm{Lqc}(X_\bullet)$ are locally noetherian. $\mathcal{M} \in \mathrm{Lqc}(X_\bullet)$ is a noetherian object if and only if $\mathcal{M}$ is locally coherent.*

Proof. Let J be the discrete subcategory of I such that $\mathrm{ob}(J) = \mathrm{ob}(I)$. Obviously, the restriction $(?)_J$ is faithful and exact. For $i \in \mathrm{ob}(I)$, there is an isomorphism of functors $(?)_i R_J \cong \prod_{j \in \mathrm{ob}(J)} \prod_{\phi \in I(i,j)} (X_\phi)_*(?)_j$. The product is a finite product, as I is finite. As each X_ϕ is concentrated, $(X_\phi)_*(?)_j$ preserves filtered inductive limits by Lemma 7.19. Hence R_J preserves filtered inductive limits. Note also that R_J preserves local quasi-coherence.

Hence we may assume that I is a discrete finite category, which case is trivial by [17, Theorem II.7.8] and Corollary 11.6. $\qquad\qquad\qquad\square$

Lemma 11.10. *Let I be a finite category, S a scheme, and $X_\bullet \in \mathcal{P}(I, \underline{\mathrm{Sch}}/S)$. If $X_\bullet$ is concentrated, then $\mathrm{Lqc}(X_\bullet)$ is Grothendieck.*

Proof. Similar.

Chapter 12
Groupoid of Schemes

(12.1) Let $\mathcal{C}$ be a category with finite limits. A $\mathcal{C}$-*groupoid* X_* is a functor from $\mathcal{C}^{\mathrm{op}}$ to the category of groupoids $\in \mathcal{U}$ (i.e., category $\in \mathcal{U}$ all of whose morphisms are isomorphisms) such that the set valued functors $X_0 := \mathrm{ob} \circ X_*$ and $X_1 := \mathrm{Mor} \circ X_*$ are representable.

Let X_* be a $\mathcal{C}$-groupoid. Let us denote the source (resp. target) $X_1 \to X_0$ by d_1 (resp. d_0). Then $X_2 := X_1 {}_{d_1}\!\times_{d_0} X_1$ represents the functor of pairs (f, g) of morphisms of X_* such that $f \circ g$ is defined. Let $d'_0 : X_2 \to X_1$ (resp. $d'_2 : X_2 \to X_1$) be the first (resp. second) projection , and $d'_1 : X_2 \to X_1$ the composition.

By Yoneda's lemma, d_0, d_1, d'_0, d'_1, and d'_2 are morphisms of $\mathcal{C}$. $d_1 : X_1(T) \to X_0(T)$ is surjective for any $T \in \mathrm{ob}(\mathcal{C})$. Note that the squares

$$
\begin{array}{ccc}
X_2 \xrightarrow{d'_0} X_1 & \quad X_2 \xrightarrow{d'_1} X_1 & \quad X_2 \xrightarrow{d'_1} X_1 \\
\downarrow{\scriptstyle d'_2} \quad \downarrow{\scriptstyle d_1} & \quad \downarrow{\scriptstyle d'_2} \quad \downarrow{\scriptstyle d_1} & \quad \downarrow{\scriptstyle d'_0} \quad \downarrow{\scriptstyle d_0} \\
X_1 \xrightarrow{d_0} X_0 & \quad X_1 \xrightarrow{d_1} X_0 & \quad X_1 \xrightarrow{d_0} X_0
\end{array}
\qquad (12.2)
$$

are fiber squares. In particular,

$$
X_* := X_2 \; \begin{smallmatrix} \xrightarrow{\ d'_0\ } \\ \xrightarrow{\ d'_1\ } \\ \xrightarrow{\ d'_2\ } \end{smallmatrix} \; X_1 \; \begin{smallmatrix} \xrightarrow{\ d_0\ } \\ \xrightarrow{\ d_1\ } \end{smallmatrix} \; X_0
\qquad (12.3)
$$

forms an object of $\mathcal{P}(\Delta_M, \mathcal{C})$. Finally, by the associativity,

$$
\circ (\circ \times 1) = \circ (1 \times \circ),
\qquad (12.4)
$$

where $\circ : X_1 {}_{d_1}\!\times_{d_0} X_1 \to X_1$ denotes the composition, or $\circ$ is the composite

$$
X_1 {}_{d_1}\!\times_{d_0} X_1 \cong X_2 \xrightarrow{d'_1} X_1.
$$

J. Lipman, M. Hashimoto, *Foundations of Grothendieck Duality for Diagrams of Schemes*, Lecture Notes in Mathematics 1960,
© Springer-Verlag Berlin Heidelberg 2009

Conversely, a diagram $X_* \in \mathcal{P}(\Delta_M, \mathcal{C})$ as in (12.3) such that the squares in (12.2) are fiber squares, $d_1(T) : X_1(T) \to X_0(T)$ are surjective for all $T \in \mathrm{ob}(\mathcal{C})$, and the associativity (12.4) holds gives a $\mathcal{C}$-groupoid [12]. In the sequel, we mainly consider that a $\mathcal{C}$-groupoid is an object of $\mathcal{P}(\Delta_M, \mathcal{C})$.

Let S be a scheme. We say that $X_\bullet \in \mathcal{P}(\Delta_M, \underline{\mathrm{Sch}}/S)$ is an S-*groupoid*, if $X_\bullet$ is a $(\underline{\mathrm{Sch}}/S)$-groupoid with flat arrows.

(12.5) Let $X_\bullet$ be a $(\underline{\mathrm{Sch}}/S)$-groupoid, and set $X_n := X_1 \,_{d_1}\times_{d_0} X_1 \,_{d_1}\times_{d_0} \cdots \,_{d_1}\times_{d_0} X_1$ (X_1 appears n times) for $n \geq 2$. For $n \geq 2$, $d_i : X_n \to X_{n-1}$ is defined by $d_0(x_{n-1}, \ldots, x_1, x_0) = (x_{n-1}, \ldots, x_1)$, $d_n(x_{n-1}, \ldots, x_1, x_0) = (x_{n-2}, \ldots, x_0)$, and $d_i(x_{n-1}, \ldots, x_1, x_0) = (x_{n-1}, \ldots, x_i \circ x_{i-1}, \ldots, x_0)$ for $0 < i < n$. $s_i : X_n \to X_{n+1}$ is defined by

$$s_i(x_{n-1}, \ldots, x_1, x_0) = (x_{n-1}, \ldots, x_i, \mathrm{id}, x_{i-1}, \ldots, x_0).$$

It is easy to see that this gives a simplicial S-scheme $\Sigma(X_\bullet)$ such that $\Sigma(X_\bullet)|_{\Delta_M} = X_\bullet$.

For any simplicial S-scheme $Z_\bullet$ and $\psi_\bullet : Z_\bullet|_{\Delta_M} \to X_\bullet$, there exists some unique $\varphi_\bullet : Z_\bullet \to \Sigma(X_\bullet)$ such that $\varphi|_{\Delta_M} : Z_\bullet|_{\Delta_M} \to \Sigma(X_\bullet)|_{\Delta_M} = X_\bullet$ equals ψ. Indeed, φ is given by

$$\varphi_n(z) = (\psi_1(Q_{n-1}(z)), \ldots, \psi_1(Q_0(z))),$$

where $q_i : [1] \to [n]$ is the injective monotone map such that $\mathrm{Im}\, q_i = \{i, i+1\}$ for $0 \leq i < n$, and $Q_i : Z_n \to Z_1$ is the associated morphism. This shows that $\Sigma(X_\bullet) \cong \mathrm{cosk}_{\Delta_M}^{(\Delta)} X_\bullet$, and the counit map $(\mathrm{cosk}_{\Delta_M}^{(\Delta)} X_\bullet)|_{\Delta_M} \to X_\bullet$ is an isomorphism.

Note that under the identification $\Sigma(X_\bullet)_{n+1} \cong X_n \,_{r_0}\times_{d_0} X_1$, the morphism $d_0 : \Sigma(X_\bullet)_{n+1} \to \Sigma(X_\bullet)_n$ is nothing but the first projection. So $(d_0) : \Sigma(X_\bullet)' \to \Sigma(X_\bullet)$ is cartesian. If, moreover, $d_0(1)$ is flat, then (d_0) is faithfully flat.

We construct an isomorphism $h_\bullet : \Sigma(X_\bullet)' \to \mathrm{Nerve}(d_1(1))$. Define $h_{-1} = \mathrm{id}$ and $h_0 = \mathrm{id}$. Define h_1 to be the composite

$$X_1 \,_{d_1}\times_{d_0} X_1 \xrightarrow{(d_0 \boxtimes d_2)^{-1}} X_2 \xrightarrow{d_1 \boxtimes d_2} X_1 \,_{d_1}\times_{d_1} X_1.$$

Now define h_n to be the composite

$$X_1 \,_{d_1}\times_{d_0} X_1 \,_{d_1}\times_{d_0} \cdots \,_{d_1}\times_{d_0} X_1 \,_{d_1}\times_{d_0} X_1 \xrightarrow{1 \times h_1} X_1 \,_{d_1}\times_{d_0} X_1 \,_{d_1}\times_{d_0} \cdots \,_{d_1}\times_{d_0} X_1 \,_{d_1}\times_{d_1} X_1$$
$$\xrightarrow{\text{via } h_1} \cdots \xrightarrow{h_1 \times 1} X_1 \,_{d_1}\times_{d_1} X_1 \,_{d_1}\times_{d_1} \cdots \,_{d_1}\times_{d_1} X_1 \,_{d_1}\times_{d_1} X_1.$$

It is straightforward to check that this gives a well-defined isomorphism $h_\bullet : \Sigma(X_\bullet)' \to \mathrm{Nerve}(d_1(1))$. In conclusion, we have

Lemma 12.6. *If $X_\bullet$ is an S-groupoid, then $\mathrm{cosk}_{\Delta_M}^{(\Delta)} X_\bullet$ is a simplicial S-groupoid, and the counit $\varepsilon : (\mathrm{cosk}_{\Delta_M}^{(\Delta)} X_\bullet)|_{\Delta_M} \to X_\bullet$ is an isomorphism.*

Conversely, the following holds.

Lemma 12.7. *If $Y_\bullet$ is a simplicial S-groupoid, then $Y_\bullet|_{\Delta_M}$ is an S-groupoid, and the unit map $u : Y_\bullet \to \mathrm{cosk}_{\Delta_M}^{(\Delta)}(Y_\bullet|_{\Delta_M})$ is an isomorphism.*

Proof. It is obvious that $Y_\bullet|_{\Delta_M}$ has flat arrows. So it suffices to show that $Y_\bullet|_{\Delta_M}$ is a $(\underline{\mathrm{Sch}}/S)$-groupoid. Since $Y_\bullet'$ is isomorphic to $\mathrm{Nerve}\, d_1$, the square

$$
\begin{array}{ccc}
Y_2 & \xrightarrow{d_2} & Y_1 \\
\downarrow d_1 & & \downarrow d_1 \\
Y_1 & \xrightarrow{d_1} & Y_0
\end{array}
$$

is a fiber square. Since $(d_0) : Y_\bullet'|_{(\Delta)} \to Y_\bullet$ is a cartesian morphism, the squares

$$
\begin{array}{ccc}
Y_2 & \xrightarrow{d_1} & Y_1 \\
\downarrow d_0 & & \downarrow d_0 \\
Y_1 & \xrightarrow{d_0} & Y_0
\end{array}
\qquad\qquad
\begin{array}{ccc}
Y_2 & \xrightarrow{d_2} & Y_1 \\
\downarrow d_0 & & \downarrow d_0 \\
Y_1 & \xrightarrow{d_1} & Y_0
\end{array}
$$

are fiber squares.

As $d_1 s_0 = \mathrm{id}$, $d_1(T) : Y_1(T) \to Y_0(T)$ is surjective for any S-scheme T. Let us denote the composite

$$
Y_1\ {}_{d_1}\!\times_{d_0} Y_1 \cong Y_2 \xrightarrow{d_1} Y_1
$$

by $\circ$. It remains to show the associativity.

As the three squares in the diagram

$$
\begin{array}{ccccc}
Y_3 & \xrightarrow{d_3} & Y_2 & \xrightarrow{d_2} & Y_1 \\
\downarrow d_0 & & \downarrow d_0 & & \downarrow d_0 \\
Y_2 & \xrightarrow{d_2} & Y_1 & \xrightarrow{d_1} & Y_0 \\
\downarrow d_0 & & \downarrow d_0 & & \\
Y_1 & \xrightarrow{d_1} & Y_0 & &
\end{array}
$$

are all fiber squares, the canonical map $Q := Q_2 \boxtimes Q_1 \boxtimes Q_0 : Y_3 \to Y_1\ {}_{d_1}\!\times_{d_0} Y_1\ {}_{d_1}\!\times_{d_0} Y_1$ is an isomorphism. So it suffices to show that the maps

$$
Y_3 \xrightarrow{Q} Y_1\ {}_{d_1}\!\times_{d_0} Y_1\ {}_{d_1}\!\times_{d_0} Y_1 \xrightarrow{\circ\times 1} Y_1\ {}_{d_1}\!\times_{d_0} Y_1 \xrightarrow{\circ} Y_1
$$

and

$$
Y_3 \xrightarrow{Q} Y_1\ {}_{d_1}\!\times_{d_0} Y_1\ {}_{d_1}\!\times_{d_0} Y_1 \xrightarrow{1\times\circ} Y_1\ {}_{d_1}\!\times_{d_0} Y_1 \xrightarrow{\circ} Y_1
$$

agree. But it is not so difficult to show that the first map is $d_1 d_2$, while the second one is $d_1 d_1$. So $Y_\bullet|_{\Delta_M}$ is an S-groupoid.

Set $Z_\bullet := \mathrm{cosk}_{\Delta_M}^{(\Delta)}(Y_\bullet|_{\Delta_M})$, and we are to show that the unit $u_\bullet : Y_\bullet \to Z_\bullet$ is an isomorphism. Since $Y_\bullet|_{\Delta_M}$ is an S-groupoid, $\varepsilon_\bullet : Z_\bullet|_{\Delta_M} \to Y_\bullet|_{\Delta_M}$ is an isomorphism. It follows that $u_\bullet|_{\Delta_M} : Y_\bullet|_{\Delta_M} \to Z_\bullet|_{\Delta_M}$ is also an isomorphism. Hence $\mathrm{Nerve}(d_1(1)(u_\bullet)) : \mathrm{Nerve}(d_1(1)(Y_\bullet)) \to \mathrm{Nerve}(d_1(1)(Z_\bullet))$ is also an isomorphism. As both $Y_\bullet$ and $Z_\bullet$ are simplicial S-groupoids by Lemma 12.6, $u'_\bullet : Y'_\bullet \to Z'_\bullet$ is an isomorphism. So $u_n : Y_n \to Z_n$ are all isomorphisms, and we are done. $\square$

Lemma 12.8. *Let S be a scheme, and $X_\bullet$ an S-groupoid, with $d_0(1)$ and $d_1(1)$ concentrated. If X_0 is concentrated, then $\mathrm{Qch}(X_\bullet)$ is Grothendieck. If, moreover, X_0 is noetherian, then $\mathrm{Qch}(X_\bullet)$ is locally noetherian, and $\mathcal{M} \in \mathrm{Qch}(X_\bullet)$ is a noetherian object if and only if $\mathcal{M}_0$ is coherent.*

Proof. This is immediate by Corollary 11.8 and Lemma 9.4. $\square$

(12.9) Let $f : X \to Y$ be a faithfully flat concentrated S-morphism. Set $X_\bullet^+ := (\mathrm{Nerve}(f))|_{\Delta_M^+}$ and $X_\bullet := (X_\bullet^+)|_{\Delta_M}$. We define the descent functor

$$\mathbb{D} : \mathrm{Lqc}(X_\bullet) \to \mathrm{Qch}(Y)$$

to be the composite $(?)_{[-1]} R_{\Delta_M}$. The left adjoint $(?)_{\Delta_M} L_{[-1]}$ is denoted by $\mathbb{A}$, and called the ascent functor.

Lemma 12.10. *Let the notation be as above. Then $\mathbb{D} : \mathrm{Qch}(X_\bullet) \to \mathrm{Qch}(Y)$ is an equivalence, with $\mathbb{A}$ its quasi-inverse. The composite*

$$\mathbb{A} \circ \mathbb{D} : \mathrm{Lqc}(X_\bullet) \to \mathrm{Qch}(X_\bullet)$$

is the right adjoint of the inclusion $\mathrm{Qch}(X_\bullet) \hookrightarrow \mathrm{Lqc}(X_\bullet)$.

Proof. Follows easily from Lemma 9.4, Corollary 10.15, and Corollary 10.16. $\square$

Lemma 12.11. *Let $X_\bullet$ be an S-groupoid, and assume that $d_0(1)$ and $d_1(1)$ are concentrated. Set $Y_\bullet^+ := ((\mathrm{cosk}_{\Delta_M}^{(\Delta)} X_\bullet)')|_{\Delta_M^+}$, and $Y_\bullet := (Y_\bullet^+)|_{\Delta_M}$. Let $(d_0) : Y_\bullet \to X_\bullet$ be the canonical map*

$$Y_\bullet = ((\mathrm{cosk}_{\Delta_M}^{(\Delta)} X_\bullet)')|_{\Delta_M} \xrightarrow{\;(d_0)|_{\Delta_M}\;} (\mathrm{cosk}_{\Delta_M}^{(\Delta)} X_\bullet)|_{\Delta_M} \cong X_\bullet.$$

Then (d_0) is concentrated faithfully flat cartesian, and

$$(d_0)_*^{\mathrm{Qch}} \circ \mathbb{A} : \mathrm{Qch}(X_0) \to \mathrm{Qch}(X_\bullet)$$

is a right adjoint of $(?)_0 : \mathrm{Qch}(X_\bullet) \to \mathrm{Qch}(X_0)$, where $\mathbb{A} : \mathrm{Qch}(X_0) \to \mathrm{Qch}(Y_\bullet)$ is the ascent functor.

Proof. Follows easily from Corollary 10.17. $\square$

Utilizing Lemma 9.4, Lemma 10.20 and Lemma 10.21, we have the following easily.

Lemma 12.12. *Let $f : X \to Y$ be a faithfully flat locally an open immersion of schemes. Set $X_{\bullet} = \mathrm{Nerve}(f)|_{\Delta_M}$. Then the descent functor $\mathbb{D} = (?)_{-1} R_{(\Delta_M)} : \mathrm{EM}(X_{\bullet}) \to \mathrm{Mod}(Y)$ is an equivalence with the ascent functor $\mathbb{A} = (?)_{\Delta_M} L_{-1}$ its quasi-inverse. The restriction $(?)_0 : \mathrm{EM}(X_{\bullet}) \to \mathrm{Mod}(X_0) = \mathrm{Mod}(X)$ is faithfully exact with $(d_0)_* \mathbb{A}$ its right adjoint.*

Proof. Easy. □

(12.13) We say that $X_{\bullet} \in \mathcal{P}(\Delta_M, \underline{\mathrm{Sch}}/S)$ is an *almost-S-groupoid* if the three squares in (12.2) are cartesian, (12.4) holds, and $d_0(1)$ and $d_1(1)$ are faithfully flat. By definition, an S-groupoid is an almost-S-groupoid. If $X_{\bullet}$ is an almost-S-groupoid, then there is a faithfully flat cartesian morphism $p_{\bullet} : \mathrm{Nerve}(d_1(1))|_{\Delta_M} \to X_{\bullet}$ such that $p_0 : X_1 \to X_0$ is $d_0(1)$ (prove it). So Lemma 12.8 is true when we replace S-groupoid by almost-S-groupoid.

Chapter 13
Bökstedt–Neeman Resolutions and HyperExt Sheaves

(13.1) Let $\mathcal{T}$ be a triangulated category with small direct products. Note that a direct product of distinguished triangles is again a distinguished triangle (Lemma 3.1).

Let

$$\cdots \to t_3 \xrightarrow{s_3} t_2 \xrightarrow{s_2} t_1 \tag{13.2}$$

be a sequence of morphisms in $\mathcal{T}$. We define $d : \prod_{i \geq 1} t_i \to \prod_{i \geq 1} t_i$ by $p_i \circ d = p_i - s_{i+1} \circ p_{i+1}$, where $p_i : \prod_i t_i \to t_i$ is the projection. Consider a distinguished triangle of the form

$$M \xrightarrow{m} \prod_{i \geq 1} t_i \xrightarrow{d} \prod_{i \geq 1} t_i \xrightarrow{q} \Sigma M,$$

where Σ denotes the suspension.

We call M, which is determined uniquely up to isomorphisms, the *homotopy limit* of (13.2) and denote it by $\operatorname{holim} t_i$.

(13.3) Dually, *homotopy colimit* is defined and denoted by hocolim, if $\mathcal{T}$ has small coproducts.

(13.4) Let $\mathcal{A}$ be an abelian category which satisfies (AB3*). Let $(\mathbb{F}_\lambda)_{\lambda \in \Lambda}$ be a small family of objects in $K(\mathcal{A})$. Then for any $\mathbb{G} \in K(\mathcal{A})$, we have that

$$\operatorname{Hom}_{K(\mathcal{A})}(\mathbb{G}, \prod_\lambda \mathbb{F}_\lambda) = H^0(\operatorname{Hom}_{\mathcal{A}}^\bullet(\mathbb{G}, \prod_\lambda \mathbb{F}_\lambda)) \cong H^0(\prod_\lambda \operatorname{Hom}_{\mathcal{A}}^\bullet(\mathbb{G}, \mathbb{F}_\lambda))$$

$$\cong \prod_\lambda H^0(\operatorname{Hom}_{\mathcal{A}}^\bullet(\mathbb{G}, \mathbb{F}_\lambda)) = \prod_\lambda \operatorname{Hom}_{K(\mathcal{A})}(\mathbb{G}, \mathbb{F}_\lambda).$$

That is, the direct product $\prod_\lambda \mathbb{F}_\lambda$ in $C(\mathcal{A})$ is also a direct product in $K(\mathcal{A})$.

(13.5) Let $\mathcal{A}$ be a Grothendieck abelian category, and (t_λ) a small family of objects of $D(\mathcal{A})$. Let $(\mathbb{F}_\lambda)$ be a family of K-injective objects of $K(\mathcal{A})$ such that $\mathbb{F}_\lambda$ represents t_λ for each λ. Then $Q(\prod_\lambda \mathbb{F}_\lambda)$ is a direct product of t_λ in

J. Lipman, M. Hashimoto, *Foundations of Grothendieck Duality for Diagrams*   381
of Schemes, Lecture Notes in Mathematics 1960,
© Springer-Verlag Berlin Heidelberg 2009

$D(\mathcal{A})$ (note that the direct product $\prod_\lambda \mathbb{F}_\lambda$ exists, see [37, Corollary 7.10]). Hence $D(\mathcal{A})$ has small products.

Lemma 13.6. *Let I be a small category, S be a scheme, and let $X_\bullet \in \mathcal{P}(I, \underline{\mathrm{Sch}}/S)$. Let $\mathbb{F}$ be an object of $C(\mathrm{Mod}(X_\bullet))$. Assume that $\mathbb{F}$ has locally quasi-coherent cohomology groups. Then the following hold.*

1 *Let $\mathfrak{I}$ denote the full subcategory of $C(\mathrm{Mod}(X_\bullet))$ consisting of bounded below complexes of injective objects of $\mathrm{Mod}(X_\bullet)$ with locally quasi-coherent cohomology groups. There is an $\mathfrak{I}$-special inverse system $(I_n)_{n \in \mathbb{N}}$ with the index set $\mathbb{N}$ and an inverse system of chain maps $(f_n : \tau_{\geq -n}\mathbb{F} \to I_n)$ such that*

    i *f_n is a quasi-isomorphism for any $n \in \mathbb{N}$.*
    ii *$I_n^i = 0$ for $i < -n$.*

2 *If (I_n) and (f_n) are as in **1**, then the following hold.*

    i *For each $i \in \mathbb{Z}$, the canonical map $H^i(\varprojlim I_n) \to H^i(I_n)$ is an isomorphism for $n \geq \max(1, -i)$, where the projective limit is taken in the category $C(\mathrm{Mod}(X_\bullet))$, and $H^i(?)$ denotes the ith cohomology sheaf of a complex of sheaves.*
    ii *$\varprojlim f_n : \mathbb{F} \to \varprojlim I_n$ is a quasi-isomorphism.*
    iii *The projective limit $\varprojlim I_n$, viewed as an object of $K(\mathrm{Mod}(X))$, is the homotopy limit of (I_n).*
    iv *$\varprojlim I_n$ is K-injective.*

Proof. The assertion **1** is [39, (3.7)].

We prove **2**, **i**. Let $j \in \mathrm{ob}(I)$ and U an affine open subset of X_j. Then for any $n \geq 1$, I_n^i and $H^i(I_n)$ are $\Gamma((j, U), ?)$-acyclic for each $i \in \mathbb{Z}$. As I_n is bounded below, each $Z^i(I_n)$ and $B^i(I_n)$ are also $\Gamma((j, U), ?)$-acyclic, and the sequence

$$0 \to \Gamma((j, U), Z^i(I_n)) \to \Gamma((j, U), I_n^i) \to \Gamma((j, U), B^{i+1}(I_n)) \to 0 \quad (13.7)$$

and

$$0 \to \Gamma((j, U), B^i(I_n)) \to \Gamma((j, U), Z^i(I_n)) \to \Gamma((j, U), H^i(I_n)) \to 0 \quad (13.8)$$

are exact for each i, as can be seen easily, where B^i and Z^i respectively denote the ith coboundary and the cocycle sheaves.

In particular, the inverse system $(\Gamma((j, U), B^i(I_n)))$ is a Mittag-Leffler inverse system of abelian groups by (13.7), since $(\Gamma((j, U), I_n^i))$ is. On the other hand, as we have $H^i(I_n) \cong H^i(\mathbb{F})$ for $n \geq \max(1, -i)$, the inverse system $(\Gamma((j, U), H^i(I_n)))$ stabilizes, and hence we have $(\Gamma((j, U), Z^i(I_n)))$ is also Mittag-Leffler.

Passing through the projective limit,

$$0 \to \Gamma((j, U), Z^i(\varprojlim I_n)) \to \Gamma((j, U), \varprojlim I_n) \to \Gamma((j, U), \varprojlim B^{i+1}(I_n)) \to 0$$

is exact. Hence, the canonical map $B^i(\varprojlim I_n) \to \varprojlim B^i(I_n)$ is an isomorphism, since (j, U) with U an affine open subset of X_j generates the topology of $\mathrm{Zar}(X_\bullet)$.

Taking the projective limit of (13.8), we have

$$0 \to \Gamma((j,U), B^i(\varprojlim I_n)) \to \Gamma((j,U), Z^i(\varprojlim I_n)) \to \Gamma((j,U), \varprojlim H^i(I_n)) \to 0$$

is an exact sequence for any j and any affine open subset U of X_j.

Hence, the canonical maps

$$\Gamma((j,U), H^i(I_n)) \cong \Gamma((j,U), \varprojlim H^i(I_n)) \leftarrow \Gamma((j,U), H^i(\varprojlim I_n))$$

are all isomorphisms for $n \geq \max(1, -i)$, and we have $H^i(I_n) \cong H^i(\varprojlim I_n)$ for $n \geq \max(1, -i)$.

The assertion **ii** is now trivial.

The assertion **iii** is now a consequence of [7, Remark 2.3] (one can work at the presheaf level where we have the (AB4*) property). The assertion **iv** is now obvious. $\square$

Let I be a small category, S a scheme, and $X_\bullet \in \mathcal{P}(I, \underline{\mathrm{Sch}}/S)$.

Lemma 13.9. *Assume that $X_\bullet$ has flat arrows. Let J be a subcategory of I, and let $\mathbb{F} \in D_{\mathrm{EM}}(X_\bullet)$ and $\mathbb{G} \in D(X_\bullet)$. Assume one of the following.*

a $\mathbb{G} \in D^+(X_\bullet)$.
b $\mathbb{F} \in D_{\mathrm{EM}}^+(X_\bullet)$.
c $\mathbb{G} \in D_{\mathrm{Lqc}}(X_\bullet)$.

Then the canonical map

$$H_J : (?)_J R\underline{\mathrm{Hom}}^\bullet_{\mathrm{Mod}(X_\bullet)}(\mathbb{F}, \mathbb{G}) \to R\underline{\mathrm{Hom}}^\bullet_{\mathrm{Mod}(X_\bullet|_J)}(\mathbb{F}_J, \mathbb{G}_J)$$

is an isomorphism of functors to $D(\mathrm{PM}(X_\bullet|_J))$ (here $\underline{\mathrm{Hom}}^\bullet_{\mathrm{Mod}(X_\bullet)}(?, *)$ is viewed as a functor to $\mathrm{PM}(X_\bullet)$, and similarly for $\underline{\mathrm{Hom}}_{\mathrm{Mod}(X_\bullet|_J)}(?, *)$). In particular, it is an isomorphism of functors to $D(X_\bullet|_J)$.

Proof. By Lemma 1.39, we may assume that $J = i$ for an object i of I.

So what we want to prove is for any complex in $\mathrm{Mod}(X_\bullet)$ with equivariant cohomology groups $\mathbb{F}$ and any K-injective complex $\mathbb{G}$ in $\mathrm{Mod}(X_\bullet)$,

$$H_i : \underline{\mathrm{Hom}}_{\mathrm{Mod}(X_\bullet)}(\mathbb{F}, \mathbb{G})_i \to \underline{\mathrm{Hom}}_{\mathrm{Mod}(X_i)}(\mathbb{F}_i, \mathbb{G}_i)$$

is a quasi-isomorphism of complexes in $\mathrm{PM}(X_i)$ (in particular, it is a quasi-isomorphism of complexes in $\mathrm{Mod}(X_i)$), under the additional assumptions corresponding to **a**, **b**, or **c**. Indeed, if so, $\mathbb{G}_i$ is K-injective by Lemma 8.4.

First consider the case that $\mathbb{F}$ is a single equivariant object. Then the assertion is true by Lemma 6.36. By the way-out lemma [17, Proposition I.7.1], the case that $\mathbb{F}$ is bounded holds. Under the assumption of **a**, the case that $\mathbb{F}$ is bounded above holds.

Now consider the general case for **a**. As the functors in question on $\mathbb{F}$ changes coproducts to products, the map in question is a quasi-isomorphism if $\mathbb{F}$ is a direct sum of complexes bounded above with equivariant cohomology groups. Indeed, a direct product of quasi-isomorphisms of complexes of $\mathrm{PM}(X_i)$ is again quasi-isomorphic. In particular, the lemma holds if $\mathbb{F}$ is a homotopy colimit of objects of $D^-_{\mathrm{EM}}(X_\bullet)$. As any object $\mathbb{F}$ of $D_{\mathrm{EM}}(X_\bullet)$ is the homotopy colimit of $(\tau_{\leq n}\mathbb{F})$, we are done.

The proof for the case **b** is similar. As $\mathbb{F}$ has bounded below cohomology groups, $\tau_{\leq n}\mathbb{F}$ has bounded cohomology groups for each n.

We prove the case **c**. By Lemma 13.6, we may assume that $\mathbb{G}$ is a homotopy limit of K-injective complexes with locally quasi-coherent bounded below cohomology groups. As the functors on $\mathbb{G}$ in consideration commute with homotopy limits, the problem is reduced to the case **a**.   □

Lemma 13.10. *Let I be a small category, S a scheme, and $X_\bullet \in \mathcal{P}(I, \underline{\mathrm{Sch}}/S)$. Assume that $X_\bullet$ has flat arrows and is locally noetherian. Let $\mathbb{F} \in D^-_{\mathrm{Coh}}(X_\bullet)$ and $\mathbb{G} \in D^+_{\mathrm{Lqc}}(X_\bullet)$ (resp. $D^+_{\mathrm{Lch}}(X_\bullet)$), where Lch denotes the plump subcategory of* Mod *consisting of locally coherent sheaves. Then $\underline{\mathrm{Ext}}^i_{\mathcal{O}_{X_\bullet}}(\mathbb{F}, \mathbb{G})$ is locally quasi-coherent (resp. locally coherent) for $i \in \mathbb{Z}$. If, moreover, $\mathbb{G}$ has quasi-coherent (resp. coherent) cohomology groups, then $\underline{\mathrm{Ext}}^i_{\mathcal{O}_{X_\bullet}}(\mathbb{F}, \mathbb{G})$ is quasi-coherent (resp. coherent) for $i \in \mathbb{Z}$.*

Proof. We prove the assertion for the local quasi-coherence and the local coherence. By Lemma 13.9, we may assume that $X_\bullet$ is a single scheme. This case is [17, Proposition II.3.3].

We prove the assertion for the quasi-coherence (resp. coherence), assuming that $\mathbb{G}$ has quasi-coherent (resp. coherent) cohomology groups. By [17, Proposition I.7.3], we may assume that $\mathbb{F}$ is a single coherent sheaf, and $\mathbb{G}$ is an injective resolution of a single quasi-coherent (resp. coherent) sheaf.

As $X_\bullet$ has flat arrows and the restrictions are exact, it suffices to show that

$$\alpha_\phi : X^*_\phi(?)_i \underline{\mathrm{Hom}}^\bullet_{\mathrm{Mod}(X_\bullet)}(\mathbb{F}, \mathbb{G}) \to (?)_j \underline{\mathrm{Hom}}^\bullet_{\mathrm{Mod}(X_\bullet)}(\mathbb{F}, \mathbb{G})$$

is a quasi-isomorphism for any morphism $\phi : i \to j$ in I.

As X_ϕ is flat, $\alpha_\phi : X^*_\phi \mathbb{F}_i \to \mathbb{F}_j$ and $\alpha_\phi : X^*_\phi \mathbb{G}_i \to \mathbb{G}_j$ are quasi-isomorphisms. In particular, the latter is a K-injective resolution.

By the derived version of (6.37), it suffices to show that

$$P : X^*_\phi R\underline{\mathrm{Hom}}^\bullet_{\mathcal{O}_{X_i}}(\mathbb{F}_i, \mathbb{G}_i) \to R\underline{\mathrm{Hom}}^\bullet_{\mathcal{O}_{X_j}}(X^*_\phi \mathbb{F}_i, X^*_\phi \mathbb{G}_i)$$

is an isomorphism. This is [17, Proposition II.5.8].   □

Chapter 14
The Right Adjoint of the Derived Direct Image Functor

(14.1) Let X be a scheme. A right adjoint of the inclusion $F_X : \mathrm{Qch}(X) \hookrightarrow \mathrm{Mod}(X)$ is called the *quasi-coherator* of X, and is denoted by $\mathrm{qch} = \mathrm{qch}(X)$.

If $f : Y \to X$ is a concentrated morphism of schemes and $\mathrm{qch}(X)$ and $\mathrm{qch}(Y)$ exist, then there is a canonical isomorphism $f_* \,\mathrm{qch}(Y) \cong \mathrm{qch}(X) f_*$, which is the conjugate to $f^* F_X \cong F_Y f^*$. Note also that if $\mathrm{qch}(X)$ exists, then the unit $u : \mathrm{Id} \to \mathrm{qch}(X) F_X$ is an isomorphism, see [19, (I.1.2.6)].

(14.2) Let S be a scheme, I a small category, and $X_\bullet \in \mathcal{P}(I, \underline{\mathrm{Sch}}/S)$. Assume that for each $i \in I$, there exists some $\mathrm{qch}(X_i)$ and that $X_\bullet$ has concentrated arrows. Then we define $\mathrm{lqc}(X_\bullet) : \mathrm{Mod}(X_\bullet) \to \mathrm{Lqc}(X_\bullet)$ as follows. Let $\mathcal{M} \in \mathrm{Mod}(X_\bullet)$. Then $\mathcal{M}$ is expressed in terms of the data $((\mathcal{M}_i)_{i \in I}, (\beta_\phi)_{\phi \in \mathrm{Mor}(I)})$. $\mathrm{lqc}(\mathcal{M})$ is then defined in terms of the data as follows. $(\mathrm{lqc}(\mathcal{M}))_i = \mathrm{qch}\,\mathcal{M}_i$ for $i \in I$, and β_ϕ is the composite

$$\mathrm{qch}\,\mathcal{M}_i \xrightarrow{\;\mathrm{qch}\,\beta_\phi\;} \mathrm{qch}(X_\phi)_* \mathcal{M}_j \cong (X_\phi)_* \,\mathrm{qch}\,\mathcal{M}_j$$

for $\phi : i \to j$. It is easy to see that $\mathrm{lqc}(X_\bullet)$ is the right adjoint of the inclusion $F_X : \mathrm{Lqc}(X_\bullet) \to \mathrm{Mod}(X_\bullet)$. We call lqc the *local quasi-coherator*.

Lemma 14.3. *Let X be a concentrated scheme.*

1 *There is a right adjoint $\mathrm{qch}(X) : \mathrm{Mod}(X) \to \mathrm{Qch}(X)$ of the canonical inclusion $F_X : \mathrm{Qch}(X) \hookrightarrow \mathrm{Mod}(X)$. $\mathrm{qch}(X)$ preserves filtered inductive limits.*

1' *$\mathrm{qch}(X)$ preserves K-injective complexes. $R\,\mathrm{qch}(X) : D(\mathrm{Mod}(X)) \to D(\mathrm{Qch}(X))$ is right adjoint to $F_X : D(\mathrm{Qch}(X)) \to D(\mathrm{Mod}(X))$.*

2 *Assume that X is separated or noetherian. Then the functor $F_X : D(\mathrm{Qch}(X)) \to D(\mathrm{Mod}(X))$ is full and faithful, and induces an equivalence $D(\mathrm{Qch}(X)) \to D_{\mathrm{Qch}(X)}(\mathrm{Mod}(X))$.*

3 *Assume that X is separated or noetherian. Then the unit of adjunction $u : \mathrm{Id} \to R\,\mathrm{qch}(X) F_X$ is an isomorphism, and $\varepsilon : F_X R\,\mathrm{qch}(X) \to \mathrm{Id}$ is an isomorphism on $D_{\mathrm{Qch}(X)}(\mathrm{Mod}(X))$.*

J. Lipman, M. Hashimoto, *Foundations of Grothendieck Duality for Diagrams*   385
of Schemes, Lecture Notes in Mathematics 1960,
© Springer-Verlag Berlin Heidelberg 2009

Proof. The existence assertion of **1** is proved in [20, Lemme 3.2]. If Spec $A = X$ is affine, then it is easy to see that the functor $\mathcal{M} \mapsto \Gamma(X, \mathcal{M})^{\sim}$ is a desired qch(X). In fact, for a quasi-coherent $\mathcal{N}$, a morphism $\mathcal{N} \to \mathcal{M}$ is uniquely determined by the A-linear map $\Gamma(X, \mathcal{N}) \to \Gamma(X, \mathcal{M})$. So qch$(X)$ preserves filtered inductive limits by [23, Proposition 6].

Next consider the general case. As X is quasi-compact, there is a finite affine open covering (U_i) of X. Set $Y = \coprod_i U_i$, and let $p \colon Y \to X$ be the canonical map. Note that p is locally an open immersion and faithfully flat. Let $X_{\bullet} = \mathrm{Nerve}(p)|_{\Delta_M}$.

Assume that each X_i admits qch(X_i). This is the case if X is separated (and hence X_i is affine for each i). Note that the inclusion $\mathrm{Qch}(X) \hookrightarrow \mathrm{Mod}(X)$ is equivalent to the composite

$$\mathrm{Qch}(X) \xrightarrow{\mathbb{A}} \mathrm{Qch}(X_{\bullet}) \hookrightarrow \mathrm{Lqc}(X_{\bullet}) \hookrightarrow \mathrm{Mod}(X_{\bullet}) \xrightarrow{\mathbb{D}} \mathrm{Mod}(X).$$

By Lemma 12.10, $\mathbb{A}$ is an equivalence. So it suffices to show that $\mathbb{D} : \mathrm{Qch}(X_{\bullet}) \to \mathrm{Mod}(X)$ has a right adjoint. By Lemma 12.12 and Lemma 12.10, for $\mathcal{M} \in \mathrm{Qch}(X_{\bullet})$ and $\mathcal{N} \in \mathrm{Mod}(X)$, we have

$$\mathrm{Hom}_{\mathrm{Mod}(X)}(\mathbb{D}\mathcal{M}, \mathcal{N}) \cong \mathrm{Hom}_{\mathrm{EM}(X_{\bullet})}(\mathcal{M}, \mathbb{A}\mathcal{N}) \cong \mathrm{Hom}_{\mathrm{Mod}(X_{\bullet})}(\mathcal{M}, \mathbb{A}\mathcal{N})$$

$$\cong \mathrm{Hom}_{\mathrm{Lqc}(X_{\bullet})}(\mathcal{M}, \mathrm{lqc}\,\mathbb{A}\mathcal{N}) \cong \mathrm{Hom}_{\mathrm{Qch}(X_{\bullet})}(\mathcal{M}, \mathbb{A}\mathbb{D}\,\mathrm{lqc}\,\mathbb{A}\mathcal{N}).$$

Thus $\mathbb{D} : \mathrm{Qch}(X_{\bullet}) \to \mathrm{Mod}(X)$ has a right adjoint $\mathbb{A}\mathbb{D}\,\mathrm{lqc}\,\mathbb{A}$, as desired (so qch$(X) = \mathbb{D}\,\mathrm{lqc}\,\mathbb{A}$, as can be seen easily).

So the case that X is quasi-compact separated is done. Now repeating the same argument for the general X (then X_i is quasi-compact separated for each i), the construction of qch is done.

We prove that qch is compatible with filtered inductive limits. Assume first that X is separated. Then lqc : $\mathrm{Mod}(X_{\bullet}) \to \mathrm{Lqc}(X_{\bullet})$ preserves filtered inductive limits by the affine case. As $\mathbb{A}$, lqc, and $\mathbb{D}$ preserves filtered inductive limits, so is qch$(X) = \mathbb{D}\,\mathrm{lqc}\,\mathbb{A}$. Now repeating the same argument, the general case follows.

The assertion **1'** follows from **1** and Lemma 3.12.

Clearly, **2** and **3** are equivalent. **2** for the case that X is separated is proved in [7]. We remark that Verdier's example [20, Appendice I] shows that the assertions are *not* true for a general concentrated scheme X which is not separated or noetherian.

We give a proof for **3** for the case that X is noetherian, using the result for the case that X is separated. It suffices to show that if $\mathbb{I}$ is a K-injective complex in $K(\mathrm{Mod}(X))$ with quasi-coherent cohomology groups, then qch$(\mathbb{I}) \to \mathbb{I}$ is a quasi-isomorphism. Since $\mathrm{Mod}(X)$ is Grothendieck, there is a strictly injective resolution $\mathbb{I} \to \mathbb{J}$ by Lemma 3.9. As $\mathrm{Cone}(\mathbb{I} \to \mathbb{J})$ is null-homotopic, replacing $\mathbb{I}$ by $\mathbb{J}$, we may assume that $\mathbb{I}$ is strictly injective.

Let $\mathfrak{U} = (U_i)_{1 \le i \le m}$ be a finite affine open covering of X. For a finite subset I of $\{1, \ldots, m\}$, we denote $\bigcap_{i \in I} U_i$ by U_I. Note that each U_I is noetherian

and separated. Let $g_I : U_I \hookrightarrow X$ be the inclusion. For $M \in \mathrm{Mod}(X)$, the
Čech complex $\mathrm{Čech}(M) = \mathrm{Čech}_{\mathfrak{U}}(M)$ of M is defined to be

$$0 \to \bigoplus_{\#I=1} (g_I)_* g_I^* M \to \bigoplus_{\#I=2} (g_I)_* g_I^* M \to \cdots \to \bigoplus_{\#I=m} (g_I)_* g_I^* M \to 0,$$

where $(g_I)_* g_I^* M \to (g_J)_* g_J^* M$ is the $\pm$ of the unit of adjunction if $J \supset I$, and
zero if $J \not\supset I$. The augmented Čech complex $0 \to M \to \mathrm{Čech}(M)$ is denoted
by $\mathrm{Čech}^+(M) = \mathrm{Čech}_{\mathfrak{U}}^+(M)$. The lth term $\bigoplus_{\#I=l+1} (g_I)_* g_I^* M$ is denoted by
$\mathrm{Čech}^l(M)$.

Note that if $U_i = X$ for some i, then $\mathrm{Čech}_{\mathfrak{U}}^+(M)$ is split exact. In par-
ticular, $g_i^*(\mathrm{Čech}_{\mathfrak{U}}^+(M)) \cong \mathrm{Čech}_{U_i \cap \mathfrak{U}}^+(g_i^* M)$ is split exact, where $U_i \cap \mathfrak{U} = (U_i \cap U_j)_{1 \le j \le m}$ is the open covering of U_i. Let $g : Y = \coprod_i U_i \to X$ be the
canonical map. Since g is faithfully flat and $g^*(\mathrm{Čech}_{\mathfrak{U}}^+(M))$ is split exact,
$\mathrm{Čech}_{\mathfrak{U}}^+(M)$ is exact. Note also that if $M = (g_i)_* N$ for some $N \in \mathrm{Mod}(U_i)$,
then $\mathrm{Čech}_{\mathfrak{U}}^+(M) \cong (g_i)_*(\mathrm{Čech}_{U_i \cap \mathfrak{U}}^+(N))$ is split exact. In particular, if M is
a direct summand of $g_* N$ for some $N \in \mathrm{Mod}(Y)$, then $\mathrm{Čech}^+(M)$ is split
exact. This is the case if M is injective, since $M \hookrightarrow g_* g^* M$ splits.

Now we want to prove that $\mathrm{qch}(\mathbb{I}) \to \mathbb{I}$ is a quasi-isomorphism. Since
$\mathbb{I}$ is strictly injective, $\mathrm{Čech}^+(\mathbb{I})$ is split exact. So $\mathbb{I} \to \mathrm{Čech}(\mathbb{I})$ is a quasi-
isomorphism. As $\mathrm{qch}(\mathrm{Čech}^+(\mathbb{I}))$ is split exact, $\mathrm{qch}(\mathbb{I}) \to \mathrm{qch}(\mathrm{Čech}(\mathbb{I}))$
is a quasi-isomorphism. So it suffices to show that $\mathrm{qch}(\mathrm{Čech}(\mathbb{I})) \to$
$\mathrm{Čech}(\mathbb{I})$ is a quasi-isomorphism. To verify this, it suffices to show that
$\mathrm{qch}(\mathrm{Čech}^l(\mathbb{I})) \to \mathrm{Čech}^l(\mathbb{I})$ is a quasi-isomorphism for $l = 0, \ldots, m-1$. To
verify this, it suffices to show that for each non-empty subset I of $\{1, \ldots, m\}$,
$\varepsilon : F_X \mathrm{qch}((g_I)_* g_I^* \mathbb{I}) \to (g_I)_* g_I^* \mathbb{I}$ is a quasi-isomorphism. This map can be
identified with $(g_I)_* \varepsilon : (g_I)_* F_{U_I} \mathrm{qch}\, g_I^* \mathbb{I} \to (g_I)_* g_I^* \mathbb{I}$. By the case that X
is separated, $\varepsilon : F_{U_I} \mathrm{qch}\, g_I^* \mathbb{I} \to g_I^* \mathbb{I}$ is a quasi-isomorphism, since $g_I^* \mathbb{I}$ is a
K-injective complex and U_I is noetherian separated. Note that $g_I^* \mathbb{I}$ is $(g_I)_*$-
acyclic simply because it is K-injective. On the other hand, since each term
of $\mathrm{qch}\, g_I^* \mathbb{I}$ is an injective object of $\mathrm{Qch}(U_I)$, it is also an injective object
of $\mathrm{Mod}(U_I)$, see [17, (II.7)]. In particular, each term of $\mathrm{qch}\, g_I^* \mathbb{I}$ is quasi-
coherent and $(g_I)_*$-acyclic. By [26, (3.9.3.5)], $\mathrm{qch}\, g_I^* \mathbb{I}$ is $(g_I)_*$-acyclic. Hence
$\varepsilon : F_X \mathrm{qch}((g_I)_* g_I^* \mathbb{I}) \to (g_I)_* g_I^* \mathbb{I}$ is a quasi-isomorphism, as desired.   □

By the lemma, the following follows immediately.

Corollary 14.4. *Let X be a concentrated scheme. Then $\mathrm{Qch}(X)$ has arbitrary small direct products.*

This also follows from Corollary 11.7 and [37, Corollary 7.10].

(14.5) Let $f : X \to Y$ be a concentrated morphism of schemes. Then $f_*^{\mathrm{Qch}} :$
$\mathrm{Qch}(X) \to \mathrm{Qch}(Y)$ is defined, and we have $F_Y \circ f_*^{\mathrm{Qch}} \cong f_*^{\mathrm{Mod}} \circ F_X$, where
F_Y and F_X are the forgetful functors. Note that $Rf_*(D_{\mathrm{Qch}}(X)) \subset D_{\mathrm{Qch}}(Y)$,

see [26, (3.9.2)]. If, moreover, X is concentrated, then there is a right derived functor Rf_*^{Qch} of f_*^{Qch} by Corollary 11.7.

Lemma 14.6. *Let $f : X \to Y$ be a morphism of schemes. Then, we have the following.*

1 *If X is noetherian or both Y and f are quasi-compact separated, then the canonical maps*

$$F_Y \circ Rf_*^{\text{Qch}} \cong R(F_Y \circ f_*^{\text{Qch}}) \cong R(f_*^{\text{Mod}} \circ F_X) \to Rf_*^{\text{Mod}} \circ F_X$$

are all isomorphisms.

2 *Assume that both X and Y are either noetherian or quasi-compact separated. Then there are a left adjoint F of Rf_*^{Qch} and an isomorphism $F_X F \cong Lf_{\text{Mod}}^* F_Y$.*

3 *Let X and Y be as in **2**. Then there is an isomorphism*

$$Rf_*^{\text{Qch}} R\,\text{qch} \cong R\,\text{qch}\, Rf_*^{\text{Mod}}.$$

Proof. We prove **1**. It suffices to show that, if $\mathbb{I} \in K(\text{Qch}(X))$ is K-injective, then $F_X \mathbb{I}$ is f_*^{Mod}-acyclic. By Corollary 11.7 and Lemma 3.9, we may assume that each term of $\mathbb{I}$ is injective. By [26, (3.9.3.5)] (applied to the plump subcategory $\mathcal{A}_\# = \text{Qch}(X)$ of $\text{Mod}(X)$), it suffices to show that an injective object $\mathcal{I}$ of $\text{Qch}(X)$ is f_*^{Mod}-acyclic. This is trivial if X is noetherian, since then $\mathcal{I}$ is injective in $\text{Mod}(X)$, see [17, (II.7)].

Now assume that both Y and f are quasi-compact separated. Let $g : Z \to X$ be a faithfully flat morphism such that Z is affine. Such a morphism exists, as X is quasi-compact. Then it is easy to see that $\mathcal{I}$ is a direct summand of $g_* \mathcal{J}$ for some injective object $\mathcal{J}$ in $\text{Qch}(Z)$. So it suffices to show that $F_X g_* \mathcal{J}$ is f_*^{Mod}-acyclic. Let $F_Z \mathcal{J} \to \mathbb{J}$ be an injective resolution. As g is affine (since X is separated and Z is affine) and $\mathcal{J}$ is quasi-coherent, we have $R^i g_*^{\text{Mod}} F_Z \mathcal{J} = 0$ for $i > 0$. Hence $g_* F_Z \mathcal{J} \to g_* \mathbb{J}$ is a quasi-isomorphism, and hence is a K-limp resolution of $F_X g_* \mathcal{J} \cong g_* F_Z \mathcal{J}$. As $f \circ g$ is also affine, $f_* g_* F_Z \mathcal{J} \to f_* g_* \mathbb{J}$ is still a quasi-isomorphism, and this shows that $F_X g_* \mathcal{J}$ is f_*-acyclic.

2 Define $F : D(\text{Qch}(Y)) \to D(\text{Qch}(X))$ by $F := R\,\text{qch}\, Lf_{\text{Mod}}^* F_Y$. Note that $Lf_{\text{Mod}}^* F_Y (D(\text{Qch}(Y))) \subset D_{\text{Qch}}(X)$, see [26, (3.9.1)]. So we have

$$F_X F = F_X R\,\text{qch}\, Lf_{\text{Mod}}^* F_Y \xrightarrow{\text{via }\varepsilon} Lf_{\text{Mod}}^* F_Y$$

is an isomorphism by Lemma 14.3. Hence

$$\text{Hom}_{D(\text{Qch}(X))}(F\mathbb{F}, \mathbb{G}) \cong \text{Hom}_{D(X)}(F_X F\mathbb{F}, F_X \mathbb{G}) \cong$$
$$\text{Hom}_{D(X)}(Lf_{\text{Mod}}^* F_Y \mathbb{F}, F_X \mathbb{G}) \cong \text{Hom}_{D(Y)}(F_Y \mathbb{F}, Rf_*^{\text{Mod}} F_X \mathbb{G}) \cong$$
$$\text{Hom}_{D(Y)}(F_Y \mathbb{F}, F_Y Rf_*^{\text{Qch}} \mathbb{G}) \cong \text{Hom}_{D(\text{Qch}(Y))}(\mathbb{F}, Rf_*^{\text{Qch}} \mathbb{G}).$$

This shows that F is left adjoint to Rf_*^{Qch}.

3 Take the conjugate of **2**. $\square$

Lemma 14.7. *Let I be a small category, and $f_\bullet : X_\bullet \to Y_\bullet$ an affine morphism in $\mathcal{P}(I, \underline{\mathrm{Sch}})$. Let $\mathbb{F} \in D_{\mathrm{Lqc}}(X_\bullet)$. Then $R^0(f_\bullet)_* \mathbb{F} \cong (f_\bullet)_*(H^0(\mathbb{F}))$.*

Proof. Clearly, $(f_\bullet)_*(H^0(\mathbb{F})) \cong R^0(f_\bullet)_*(H^0(\mathbb{F}))$. Since $R^i(f_\bullet)_*(\tau_{>0}\mathbb{F}) = 0$ for $i \leq 0$ is obvious, $R^i(f_\bullet)_*(\tau_{\leq 0}\mathbb{F}) \cong R^i(f_\bullet)_* \mathbb{F}$ for $i \leq 0$. So it suffices to show that $R^i(f_\bullet)_*(\tau_{<0}\mathbb{F}) = 0$ for $i \geq 0$.

To verify this, we may assume that $f_\bullet = f : X \to Y$ is a map of single schemes, and Y is affine. By Lemma 14.3, we may assume that $\mathbb{F} = F_X \mathbb{G}$ for some $D(\mathrm{Qch}(X))$. By Lemma 14.6, it suffices to show that $R^i f_*^{\mathrm{Qch}}(\tau_{<0}\mathbb{G}) = 0$ for $i \geq 0$. But this is trivial, since f_*^{Qch} is an exact functor. □

Corollary 14.8. *Let I and $f_\bullet$ be as in the lemma. If $\mathbb{F} \in D_{\mathrm{Lqc}}(X_\bullet)$, then $R^n(f_\bullet)_* \mathbb{F} \cong (f_\bullet)_*(H^n(\mathbb{F}))$.*

(14.9) Let $\mathcal{C}$ be an additive category, and $c \in \mathcal{C}$. We say that c is a *compact object*, if for any small family of objects $(t_\lambda)_{\lambda \in \Lambda}$ of $\mathcal{C}$ such that the coproduct (direct sum) $\bigoplus_{\lambda \in \Lambda} t_\lambda$ exists, the canonical map

$$\bigoplus_\lambda \mathrm{Hom}_\mathcal{C}(c, t_\lambda) \to \mathrm{Hom}_\mathcal{C}(c, \bigoplus_\lambda t_\lambda)$$

is an isomorphism.

A triangulated category $\mathcal{T}$ is said to be *compactly generated*, if $\mathcal{T}$ has small coproducts, and there is a small set C of compact objects of $\mathcal{T}$ such that $\mathrm{Hom}_\mathcal{T}(c, t) = 0$ for all $c \in C$ implies $t = 0$. The following was proved by A. Neeman [35].

Theorem 14.10. *Let $\mathcal{S}$ be a compactly generated triangulated category, $\mathcal{T}$ any triangulated category, and $F : \mathcal{S} \to \mathcal{T}$ a triangulated functor. Suppose that F preserves coproducts, that is to say, for any small family of objects (s_λ) of $\mathcal{S}$, the canonical maps $F(s_\lambda) \to F(\bigoplus_\lambda s_\lambda)$ make $F(\bigoplus_\lambda s_\lambda)$ the coproduct of $F(s_\lambda)$. Then F has a right adjoint $G : \mathcal{T} \to \mathcal{S}$.*

For the definition of triangulated category and triangulated functor, see [36]. Related to the theorem, we remark the following.

Lemma 14.11 (Keller and Vossieck [22]). *Let $\mathcal{S}$ and $\mathcal{T}$ be triangulated categories, and $F : \mathcal{S} \to \mathcal{T}$ a triangulated functor. If G is a right adjoint of F, then G is also a triangulated functor.*

Proof. By the opposite assertion of [29, (IV.1), Theorem 3], G is additive.

Let $\phi_F : F\Sigma \to \Sigma F$ be the canonical isomorphism. Then its inverse induces an isomorphism

$$\psi : F\Sigma^{-1} = \Sigma^{-1}\Sigma F\Sigma^{-1} \xrightarrow{\phi_F^{-1}} \Sigma^{-1}F\Sigma\Sigma^{-1} = \Sigma^{-1}F.$$

Taking the conjugate of ψ, we get an isomorphism $\phi_G : G\Sigma \cong \Sigma G$.

Let $A \xrightarrow{a} B \xrightarrow{b} C \xrightarrow{c} \Sigma A$ be a distinguished triangle in $\mathcal{T}$. Then there exists some distinguished triangle of the form

$$GA \xrightarrow{Ga} GB \xrightarrow{\alpha} X \xrightarrow{\beta} \Sigma(GA).$$

Then there exists some $d \colon FX \to C$ such that

$$
\begin{array}{ccccccc}
FGA & \xrightarrow{FGa} & FGB & \xrightarrow{F\alpha} & FX & \xrightarrow{\phi \circ F\beta} & \Sigma(FGA) \\
\downarrow{\scriptstyle \varepsilon} & & \downarrow{\scriptstyle \varepsilon} & & \downarrow{\scriptstyle d} & & \downarrow{\scriptstyle \Sigma\varepsilon} \\
A & \xrightarrow{a} & B & \xrightarrow{b} & C & \xrightarrow{c} & \Sigma A
\end{array}
$$

is a map of triangles. Then taking the adjoint, we get a commutative diagram

$$
\begin{array}{ccccccc}
GA & \xrightarrow{Ga} & GB & \xrightarrow{\alpha} & X & \xrightarrow{\beta} & \Sigma GA \\
\downarrow{\scriptstyle \mathrm{id}} & & \downarrow{\scriptstyle \mathrm{id}} & & \downarrow{\scriptstyle \delta} & & \downarrow \\
GA & \xrightarrow{Ga} & GB & \xrightarrow{Gb} & GC & \xrightarrow{\phi_G \circ Gc} & \Sigma GA,
\end{array}
\tag{14.12}
$$

where δ is the adjoint of d, and the right most vertical arrow is the composite

$$\Sigma GA \xrightarrow{u} GF\Sigma GA \xrightarrow{\phi_F} G\Sigma FGA \xrightarrow{\varepsilon} G\Sigma A \xrightarrow{\phi_G} \Sigma GA,$$

which agrees with id, as can be seen easily. This induces a commutative diagram

$$
\begin{array}{ccccccccc}
\mathcal{S}(?,GA) & \xrightarrow{(Ga)_*} & \mathcal{S}(?,GB) & \xrightarrow{\alpha_*} & \mathcal{S}(?,X) & \xrightarrow{\beta_*} & \mathcal{S}(?,\Sigma GA) & \xrightarrow{(\Sigma Ga)_*} & \mathcal{S}(?,\Sigma GB) \\
\downarrow{\scriptstyle \mathrm{id}} & & \downarrow{\scriptstyle \mathrm{id}} & & \downarrow{\scriptstyle \delta_*} & & \downarrow{\scriptstyle \mathrm{id}} & & \downarrow{\scriptstyle \mathrm{id}} \\
\mathcal{S}(?,GA) & \xrightarrow{(Ga)_*} & \mathcal{S}(?,GB) & \xrightarrow{(Gb)_*} & \mathcal{S}(?,GC) & \xrightarrow{(\phi_G \circ Gc)_*} & \mathcal{S}(?,\Sigma GA) & \xrightarrow{(\Sigma Ga)_*} & \mathcal{S}(?,\Sigma GB).
\end{array}
$$

The first row is exact, since it comes from a distinguished triangle, see [17, Proposition I.1.1]. As the second row is isomorphic to the sequence

$$\mathcal{T}(F?,A) \xrightarrow{a_*} \mathcal{T}(F?,B) \xrightarrow{b_*} \mathcal{T}(F?,C) \xrightarrow{c_*} \mathcal{T}(F?,\Sigma A) \xrightarrow{(\Sigma a)_*} \mathcal{T}(F?,\Sigma B)$$

and $A \xrightarrow{a} B \xrightarrow{b} C \xrightarrow{c} \Sigma A$ is a distinguished triangle, it is also exact. By the five lemma, δ_* is an isomorphism. By Yoneda's lemma, δ is an isomorphism. This shows that (14.12) is an isomorphism, and hence the second row of (14.12) is distinguished. This is what we wanted to show. $\qquad\square$

Lemma 14.13. *Let $\mathcal{S}$ and $\mathcal{T}$ be triangulated categories, and $F \colon \mathcal{S} \to \mathcal{T}$ a triangulated functor with a right adjoint G. If both $\mathcal{S}$ and $\mathcal{T}$ have t-structures* and F is way-out left (i.e., $F(\tau_{\leq 0}\mathcal{S}) \subset \tau_{\leq d}\mathcal{T}$ for some d), then G is way-out right.

Proof. For the definition of t-structures on triangulated categories, see [5]. Assuming that $F(\tau_{\leq 0}\mathcal{S}) \subset \tau_{\leq d}\mathcal{T}$, we show $G(\tau_{\geq 0}\mathcal{T}) \subset \tau_{\geq -d}\mathcal{S}$.

Let $t \in \tau_{\geq 0}\mathcal{T}$ and $s \in \tau_{\leq -d-1}\mathcal{S}$. Then since $Fs \in \tau_{\leq -1}\mathcal{T}$, we have $\mathcal{S}(s, Gt) \cong \mathcal{T}(Fs, t) = 0$. By [5, (1.3.4)], $Gt \in \tau_{\geq -d}\mathcal{S}$. □

The following was proved by A. Neeman [35] for the quasi-compact separated case, and was proved generally by A. Bondal and M. van den Bergh [8].

Theorem 14.14. *Let X be a concentrated scheme. Then $c \in D_{\mathrm{Qch}}(X)$ is a compact object if and only if c is isomorphic to a perfect complex, where we say that $\mathbb{C} \in C(\mathrm{Qch}(X))$ is perfect if $\mathbb{C}$ is bounded, and each term of $\mathbb{C}$ is locally free of finite rank. Moreover, $D_{\mathrm{Qch}}(X)$ is compactly generated.*

Lemma 14.15. *Let $f : X \to Y$ be a concentrated morphism of schemes. Then $Rf_*^{\mathrm{Mod}} : D_{\mathrm{Qch}}(X) \to D(Y)$ preserves coproducts.*

Proof. See [35] or [26, (3.9.3.2), Remark (b)]. □

(14.16) Let $f : X \to Y$ be a concentrated morphism of schemes such that X is concentrated. Then by Theorem 14.10, Theorem 14.14 and Lemma 14.15, there is a right adjoint

$$f^\times : D(Y) \to D_{\mathrm{Qch}}(X)$$

of Rf_*^{Mod}.

By restriction, $f^\times : D_{\mathrm{Qch}}(Y) \to D_{\mathrm{Qch}}(X)$ is a right adjoint of $Rf_*^{\mathrm{Mod}} : D_{\mathrm{Qch}}(X) \to D_{\mathrm{Qch}}(Y)$. Note that $(R(?)_*, (?)^\times)$ is an adjoint pair of Δ-pseudofunctors on the opposite of the category of concentrated schemes. In other words, $(R(?)_*, (?)^\times)$ is an opposite adjoint pair (of Δ-pseudofunctors) on the category of concentrated schemes, see (1.18).

Chapter 15
Comparison of Local Ext Sheaves

(15.1) Let S be a scheme, and $X_\bullet$ an almost-S-groupoid. Assume that $d_0(1)$ and $d_1(1)$ are affine, and X_0 is locally noetherian.

Lemma 15.2. *Let* $\mathbb{F} \in K_{\mathrm{Coh}}^-(\mathrm{Qch}(X_\bullet))$ *and* $\mathbb{G} \in K^+(\mathrm{Qch}(X_\bullet))$. *If* $\mathbb{G}$ *is a bounded below complex consisting of injective objects of* $\mathrm{Qch}(X_\bullet)$, *then* $\mathbb{G}$ *is* $\mathrm{Hom}^\bullet_{\mathcal{O}_{X_\bullet}}(\mathbb{F}, ?)$-*acyclic as a complex of* $\mathrm{Mod}(X_\bullet)$.

Proof. It is easy to see that we may assume that $\mathbb{F}$ is a single coherent sheaf, and $\mathbb{G}$ is a single injective object of $\mathrm{Qch}(X_\bullet)$. To prove this case, it suffices to show that

$$\underline{\mathrm{Ext}}^i_{\mathcal{O}_{X_\bullet}}(\mathbb{F}, \mathbb{G}) = 0$$

for $i > 0$.

Set $X'_\bullet := \mathrm{Nerve}(d_1(1))|_{\Delta_M}$, and let $p_\bullet : X'_\bullet \to X_\bullet$ be a cartesian morphism such that $p_0 : X_1 \to X_0$ is $d_0(1)$, see (12.13). In particular, $p_\bullet$ is affine and faithfully flat.

Let $\mathbb{A} : \mathrm{Mod}(X_0) \to \mathrm{Mod}(X'_\bullet)$ be the ascent functor, and $\mathbb{D} : \mathrm{Mod}(X'_\bullet) \to \mathrm{Mod}(X_0)$ be the descent functor.

As $(p_\bullet)_* : \mathrm{Qch}(X'_\bullet) \to \mathrm{Qch}(X_\bullet)$ has a faithful exact left adjoint $p_\bullet^*$, there exists some injective object $\mathbb{I}$ of $\mathrm{Qch}(X'_\bullet)$ such that $\mathbb{G}$ is a direct summand of $(p_\bullet)_*\mathbb{I}$. We may assume that $\mathbb{G} = (p_\bullet)_*\mathbb{I}$. As

$$R\underline{\mathrm{Hom}}^\bullet_{\mathcal{O}_{X_\bullet}}(\mathbb{F}, R(p_\bullet)_*\mathbb{I}) \cong R(p_\bullet)_* R\underline{\mathrm{Hom}}^\bullet_{\mathcal{O}_{X'_\bullet}}(p_\bullet^*\mathbb{F}, \mathbb{I}),$$

$p_\bullet$ is affine by assumption, and $R\underline{\mathrm{Hom}}^\bullet_{\mathcal{O}_{X'_\bullet}}(p_\bullet^*\mathbb{F}, \mathbb{I})$ has quasi-coherent cohomology groups, we may assume that $X_\bullet = \mathrm{Nerve}(f)|_{\Delta_M}$ for some faithfully flat affine morphism $f : X \to Y$ between S-schemes with Y locally noetherian (but we may lose the assumption that X_0 is locally noetherian). For each l, we have that

$$(?)_l R\underline{\mathrm{Hom}}^\bullet_{\mathcal{O}_{X_\bullet}}(\mathbb{F}, \mathbb{G}) \cong R\underline{\mathrm{Hom}}^\bullet_{\mathcal{O}_{X_l}}((?)_l\mathbb{A}\mathbb{D}\mathbb{F}, (?)_l\mathbb{A}\mathbb{D}\mathbb{G}) \cong$$
$$R\underline{\mathrm{Hom}}^\bullet_{\mathcal{O}_{X_l}}(e(l)^*\mathbb{D}\mathbb{F}, e(l)^*\mathbb{D}\mathbb{G}) \cong e(l)^* R\underline{\mathrm{Hom}}^\bullet_{\mathcal{O}_Y}(\mathbb{D}\mathbb{F}, \mathbb{D}\mathbb{G})$$

J. Lipman, M. Hashimoto, *Foundations of Grothendieck Duality for Diagrams of Schemes*, Lecture Notes in Mathematics 1960,
© Springer-Verlag Berlin Heidelberg 2009

by [17, (II.5.8)] (note that $\mathbb{D}\mathbb{F}$ is coherent). As $\mathbb{D} : \mathrm{Qch}(X_\bullet) \to \mathrm{Qch}(Y)$ is an equivalence, $\mathbb{D}\mathbb{G}$ is an injective object of $\mathrm{Qch}(Y)$. Hence it is also an injective object of $\mathrm{Mod}(Y)$ by [17, (II.7)]. Hence $\underline{\mathrm{Ext}}^i_{\mathcal{O}_Y}(\mathbb{D}\mathbb{F}, \mathbb{D}\mathbb{G}) = 0$ for $i > 0$, as desired.   □

The following is a generalization of [19, Theorem II.1.1.12].

Corollary 15.3. *Let the notation be as in the lemma. Then $\mathbb{G}_0$ is* $\underline{\mathrm{Hom}}^\bullet_{\mathcal{O}_{X_0}}(\mathbb{F}_0, ?)$-*acyclic as a complex of* $\mathcal{O}_{X_0}$-*modules.*

Proof. Let $\mathbb{G} \to \mathbb{I}$ be a K-injective resolution in $\mathrm{Mod}(X_\bullet)$ such that $\mathbb{I}$ is bounded below. Let $\mathbb{C}$ be the mapping cone of this. Since $(?)_0$ has an exact left adjoint, $\mathbb{G}_0 \to \mathbb{I}_0$ is a K-injective resolution in $K(\mathrm{Mod}(X_0))$. So it suffices to show that $\underline{\mathrm{Hom}}^\bullet_{\mathcal{O}_{X_0}}(\mathbb{F}_0, \mathbb{C}_0)$ is exact. As each term of $\mathbb{F}$ is equivariant, this complex is isomorphic to $\underline{\mathrm{Hom}}^\bullet_{\mathcal{O}_{X_\bullet}}(\mathbb{F}, \mathbb{C})_0$, which is exact by the lemma.   □

Chapter 16
The Composition of Two Almost-Pseudofunctors

Definition 16.1. We say that $\mathcal{C} = (\mathcal{A}, \mathcal{F}, \mathcal{P}, \mathcal{I}, \mathcal{D}, \mathcal{D}^+, (?)^\#, (?)^\flat, \zeta)$ is a *composition data* of contravariant almost-pseudofunctors if the following eighteen conditions are satisfied:

1 $\mathcal{A}$ is a category with fiber products.
2 $\mathcal{P}$ and $\mathcal{I}$ are sets of morphisms of $\mathcal{A}$.
3 Any isomorphism in $\mathcal{A}$ is in $\mathcal{P} \cap \mathcal{I}$.
4 The composite of two morphisms in $\mathcal{P}$ is again a morphism in $\mathcal{P}$.
5 The composite of two morphisms in $\mathcal{I}$ is again a morphism in $\mathcal{I}$.
6 A base change of a morphism in $\mathcal{P}$ is again a morphism in $\mathcal{P}$.
7 Any $f \in \mathrm{Mor}(\mathcal{A})$ admits a factorization $f = pi$ such that $p \in \mathcal{P}$ and $i \in \mathcal{I}$.

Before stating the remaining conditions, we give some definitions for convenience.

i Let $\mathcal{C}$ be a set of morphisms in $\mathcal{A}$ containing all identity maps and being closed under compositions. We define $\mathcal{A}_\mathcal{C}$ by $\mathrm{ob}(\mathcal{A}_\mathcal{C}) := \mathrm{ob}(\mathcal{A})$ and $\mathrm{Mor}(\mathcal{A}_\mathcal{C}) := \mathcal{C}$. In particular, the subcategories $\mathcal{A}_\mathcal{P}$ and $\mathcal{A}_\mathcal{I}$ of $\mathcal{A}$ are defined.
ii We call a commutative diagram of the form $p \circ i = i' \circ p'$ with $p, p' \in \mathcal{P}$ and $i, i' \in \mathcal{I}$ a *pi-square*. We denote the set of all pi-squares by Π.

8 $\mathcal{D} = (\mathcal{D}(X))_{X \in \mathrm{ob}(\mathcal{A})}$ is a family of categories.
9 $(?)^\#$ is a contravariant almost-pseudofunctor on $\mathcal{A}_\mathcal{P}$, $(?)^\flat$ is a contravariant almost-pseudofunctor on $\mathcal{A}_\mathcal{I}$, and we have $X^\# = X^\flat = \mathcal{D}(X)$ for each $X \in \mathrm{ob}(\mathcal{A})$.
10 $\zeta = (\zeta(\sigma))_{\sigma=(pi=jq)\in\Pi}$ is a family of natural transformations

$$\zeta(\sigma) : i^\flat p^\# \to q^\# j^\flat.$$

J. Lipman, M. Hashimoto, *Foundations of Grothendieck Duality for Diagrams of Schemes*, Lecture Notes in Mathematics 1960,

11 If

$$U_1 \xrightarrow{j_1} V_1 \xrightarrow{i_1} X_1$$
$$\downarrow p_U\sigma' \quad \downarrow p_V\sigma \quad \downarrow p_X$$
$$U \xrightarrow{j} V \xrightarrow{i} X$$

is a commutative diagram in $\mathcal{A}$ such that $\sigma, \sigma' \in \Pi$, then the composite map

$$(i_1 j_1)^b p_X^\# \xrightarrow{d^{-1}} j_1^b i_1^b p_X^\# \xrightarrow{\zeta(\sigma)} j_1^b p_V^\# i^b \xrightarrow{\zeta(\sigma')} p_U^\# j^b i^b \xrightarrow{d} p_U^\# (ij)^b$$

agrees with $\zeta(\sigma'\sigma)$, where $\sigma'\sigma$ is the pi-square $p_X(i_1 j_1) = (ij)p_U$.

12 For any morphism $p : X \to Y$ in $\mathcal{P}$, the composite

$$p^\# \xrightarrow{f^{-1}} 1_X^b p^\# \xrightarrow{\zeta(p1_X = 1_Y p)} p^\# 1_Y^b \xrightarrow{f} p^\#$$

is the identity.

13 If

$$U_2 \xrightarrow{i_2} X_2$$
$$\downarrow q_U\sigma' \quad \downarrow q_X$$
$$U_1 \xrightarrow{i_1} X_1$$
$$\downarrow p_U\sigma \quad \downarrow p_X$$
$$U \xrightarrow{i} X$$

is a commutative diagram in $\mathcal{A}$ such that $\sigma, \sigma' \in \Pi$, then the composite

$$i_2^b(p_X q_X)^\# \xrightarrow{d^{-1}} i_2^b q_X^\# p_X^\# \xrightarrow{\zeta(\sigma')} q_U^\# i_1^b p_X^\# \xrightarrow{\zeta(\sigma)} q_U^\# p_U^\# i^b \xrightarrow{d} (p_U q_U)^\# i^b$$

agrees with $\zeta((p_X q_X)i_2 = i(p_U q_U))$.

14 For any morphism $i : U \to X$ in $\mathcal{I}$, the composite

$$i^b \xrightarrow{f^{-1}} i^b 1_X^\# \xrightarrow{\zeta} 1_U^\# i^b \xrightarrow{f} i^b$$

is the identity.

15 $\mathcal{F}$ is a subcategory of $\mathcal{A}$, and any isomorphism in $\mathcal{A}$ between objects of $\mathcal{F}$ is in Mor($\mathcal{F}$).

16 $\mathcal{D}^+ = (\mathcal{D}^+(X))_{X \in \mathrm{ob}(\mathcal{F})}$ is a family of categories such that $\mathcal{D}^+(X)$ is a full subcategory of $\mathcal{D}(X)$ for each $X \in \mathrm{ob}(\mathcal{F})$.

17 If $f : X \to Y$ is a morphism in $\mathcal{F}$, $f = p \circ i$, $p \in \mathcal{P}$ and $i \in \mathcal{I}$, then we have $i^b p^!(\mathcal{D}^+(Y)) \subset \mathcal{D}^+(X)$.

18 If

$$V \xrightarrow{j} U_1 \xrightarrow{i_1} X_1$$
$$\downarrow p_U\sigma \quad \downarrow p_X$$
$$U \xrightarrow{i} X \xrightarrow{q} Y$$

is a diagram in $\mathcal{A}$ such that $\sigma \in \Pi$, $V, U, Y \in \mathrm{ob}(\mathcal{F})$, $p_U j \in \mathrm{Mor}(\mathcal{F})$ and $qi \in \mathrm{Mor}(\mathcal{F})$, then

$$j^{\flat}\zeta(\sigma)q^{\#} : j^{\flat}i_1^{\flat}p_X^{\#}q^{\#} \to j^{\flat}p_U^{\#}i^{\flat}q^{\#}$$

is an isomorphism between functors from $\mathcal{D}^+(Y)$ to $\mathcal{D}^+(V)$.

(16.2) Let $\mathcal{C} = (\mathcal{A}, \mathcal{F}, \mathcal{P}, \mathcal{I}, \mathcal{D}, \mathcal{D}^+, (?)^{\#}, (?)^{\flat}, \zeta)$ be a composition data of contravariant almost-pseudofunctors. We call a commutative diagram of the form $f = pi$ with $p \in \mathcal{P}$, $i \in \mathcal{I}$ and $f \in \mathrm{Mor}(\mathcal{A})$ a *compactification*. We call a commutative diagram of the form $pi = qj$ with $p, q \in \mathcal{P}$, $i, j \in \mathcal{I}$ and $pi = qj \in \mathrm{Mor}(\mathcal{F})$ an *independence diagram*.

Lemma 16.3. *Let*

$$
\begin{array}{ccc}
U & \xrightarrow{i_1} & X_1 \\
\downarrow{i} & \tau & \downarrow{p_1} \\
X & \xrightarrow{p} & Y
\end{array}
$$

be an independence diagram. Then the following hold:

1 *There is a diagram of the form*

$$
\begin{array}{ccccc}
U & \xrightarrow{j} & Z & \xrightarrow{q_1} & X_1 \\
 & & \downarrow{q} & & \downarrow{p_1} \\
 & & X & \xrightarrow{p} & Y
\end{array}
$$

such that $qj = i$, $q_1 j = i_1$, $pq = p_1 q_1$, $q, q_1 \in \mathcal{P}$, and $j \in \mathcal{I}$.

2 *$\zeta(qj = i1_U)p^{\#} : j^{\flat}q^{\#}p^{\#} \to i^{\flat}p^{\#}$ is an isomorphism between functors from $\mathcal{D}^+(Y)$ to $\mathcal{D}^+(U)$.*

3 *$\zeta(q_1 j_1 = i_1 1_U)p_1^{\#}$ is also an isomorphism between functors from $\mathcal{D}^+(Y)$ to $\mathcal{D}^+(U)$.*

4 *The composite isomorphism*

$$\Upsilon(\tau) : i^{\flat}p^{\#} \xrightarrow{\mathfrak{f}^{-1}} 1_U^{\#}i^{\flat}p^{\#} \xrightarrow{(\zeta(qj=i1_U)p^{\#})^{-1}} j^{\flat}q^{\#}p^{\#}$$

$$\xrightarrow{d} j^{\flat}q_1^{\#}p_1^{\#} \xrightarrow{\zeta(q_1 j_1 = i_1 1_U)p_1^{\#}} 1_U^{\#}i_1^{\flat}p_1^{\#} \xrightarrow{\mathfrak{f}} i_1^{\flat}p_1^{\#}$$

(between functors defined over $\mathcal{D}^+(Y)$, not over $\mathcal{D}(Y)$) depends only on τ.

5 *If $\tau' = (p_1 i_1 = p_2 i_2)$ is an independence diagram, then we have*

$$\Upsilon(\tau') \circ \Upsilon(\tau) = \Upsilon(pi = p_2 i_2).$$

The proof is left to the reader. We call $\Upsilon(\tau)$ the *independence isomorphism* of τ.

(16.4) Any $f \in \mathrm{Mor}(\mathcal{A})$ has a compactification by assumption. We fix a family of compactifications $\mathcal{T} := (\tau(f) : (f = p(f) \circ i(f)))_{f \in \mathrm{Mor}(\mathcal{A})}$.

For $X \in \mathrm{ob}(\mathcal{F})$, we define $X^! := \mathcal{D}^+(X)$. For a morphism $f : X \to Y$ in $\mathcal{F}$, we define $f^! := i(f)^{\flat} p(f)^{\#}$, which is a functor from $Y^!$ to $X^!$ by assumption.

Let $f : X \to Y$ and $g : Y \to Z$ be morphisms in $\mathcal{F}$. Let $i(g)p(f) = qj$ be a compactification of $i(g)p(f)$. Then by **18** in Definition 16.1 and Lemma 16.3, the composite map

$$(gf)^! = i(gf)^{\flat} p(gf)^{\#} \xrightarrow{\Upsilon(p(gf)i(gf)=(p(g)q)(ji(f)))} (j \circ i(f))^{\flat} (p(g) \circ q)^{\#}$$

$$\cong i(f)^{\flat} j^{\flat} q^{\#} p(g)^{\#} \xrightarrow{i(f)^{\flat} \zeta(qj=i(g)p(f))p(g)^{\#}} i(f)^{\flat} p(f)^{\#} i(g)^{\flat} p(g)^{\#} = f^! g^!$$

is an isomorphism. We define $d_{f,g} : f^! g^! \to (gf)^!$ to be the inverse of this composite.

Lemma 16.5. *The definition of $d_{f,g}$ is independent of choice of q and j above.*

The proof of the lemma is left to the reader.

For $X \in \mathrm{ob}(\mathcal{F})$, we define $\mathfrak{f}_X : \mathrm{id}_X^! \to \mathrm{Id}_{X^!}$ to be the composite

$$\mathrm{id}_X^! = i(\mathrm{id}_X)^{\flat} p(\mathrm{id}_X)^{\#} \xrightarrow{\zeta} \mathrm{id}_X^{\#} \mathrm{id}_X^{\flat} \xrightarrow{\mathfrak{f}} \mathrm{id}_X^{\flat} \xrightarrow{\mathfrak{f}} \mathrm{Id}_{X^!} .$$

Proposition 16.6. *Let the notation be as above.*

1 *$(?)^!$ together with $(d_{f,g})$ and $(\mathfrak{f}_X)$ form a contravariant almost-pseudofunctor on $\mathcal{F}$.*

2 *For $j \in \mathcal{I} \cap \mathrm{Mor}(\mathcal{F})$, define $\psi : j^! \to j^{\flat}$ to be the composite*

$$j^! = i(j)^{\flat} p(j)^{\#} \xrightarrow{\Upsilon} j^{\flat} \mathrm{id}^{\#} \xrightarrow{\mathfrak{f}} j^{\flat}.$$

Then $\psi : (?)^! \to (?)^{\flat}$ is an isomorphism of contravariant almost-pseudofunctors on $\mathcal{A}_{\mathcal{I}} \cap \mathcal{F}$.

3 *For $q \in \mathcal{P} \cap \mathrm{Mor}(\mathcal{F})$, define $\psi : q^! \to q^{\#}$ to be the composite*

$$q^! = i(q)^{\flat} p(q)^{\#} \xrightarrow{\Upsilon} \mathrm{id}^{\flat} q^{\#} \xrightarrow{\mathfrak{f}} q^{\#}.$$

Then $\psi : (?)^! \to (?)^{\#}$ is an isomorphism of contravariant almost-pseudofunctors on $\mathcal{A}_{\mathcal{P}} \cap \mathcal{F}$.

4 *Let us take another family of compactifications $(f = p_1(f)i_1(f))_{f \in \mathrm{Mor}(\mathcal{F})}$, and let $(?)^{\star}$ be the resulting contravariant almost-pseudofunctor defined by $f^{\star} = i_1(f)^{\flat} p_1(f)^{\#}$. Then $\Upsilon : f^! \to f^{\star}$ induces an isomorphism of contravariant almost-pseudofunctors $(?)^! \cong (?)^{\star}$.*

The proof is left to the reader. We call $(?)^!$ the *composite* of $(?)^{\#}$ and $(?)^{\flat}$. The composite is uniquely defined up to isomorphisms of almost-pseudofunctors on $\mathcal{F}$. The discussion above has an obvious triangulated version. Composition data of contravariant triangulated almost-pseudofunctors are defined appropriately, and the composition of two contravariant triangulated almost-pseudofunctors is obtained as a contravariant triangulated almost-pseudofunctor.

(16.7) Let S be a scheme. Let $\mathcal{A}$ be the category whose objects are noetherian S-schemes and morphisms are morphisms separated of finite type. Set $\mathcal{F} = \mathcal{A}$. Let $\mathcal{I}$ be the class of open immersions. Let $\mathcal{P}$ be the class of proper morphisms. Set $\mathcal{D}(X) = \mathcal{D}^+(X) = D^+_{\mathrm{Qch}}(X)$ for $X \in \mathcal{A}$. Let $(?)^\flat := (?)^*$, the (derived) inverse image almost-pseudofunctor for morphisms in $\mathcal{I}$, where $X^\flat := \mathcal{D}(X)$. Let $(?)^\# := (?)^\times$, the *twisted inverse* almost-pseudofunctor (see [26, Chapter 4]) for morphisms in $\mathcal{P}$, where $X^\# := \mathcal{D}(X)$ again. Note that the left adjoint $R(?)_*$ is way-out left for morphisms in $\mathcal{P}$ so that $(?)^\times$ is way-out right by Lemma 14.13, and $(?)^\#$ is well-defined. The conditions **1–9** in Definition 16.1 hold. Note that **7** is nothing but Nagata's compactification theorem [34]. The conditions **15–17** are trivial.

Let $\sigma_0 : pi = jq$ be a pi-diagram, which is also a fiber square. Then the canonical map

$$\theta : j^*(Rp_*) \to (Rq_*)i^*$$

is an isomorphism of triangulated functors, see [26, (3.9.5)]. Hence, taking the inverse of the conjugate, we have an isomorphism

$$\xi = \xi(\sigma_0) : (Ri_*)q^\times \cong p^\times(Rj_*). \tag{16.8}$$

So we have a morphism of triangulated functors

$$\zeta_0(\sigma_0) : i^*p^\times \xrightarrow{\text{via } u} i^*p^\times(Rj_*)j^* \xrightarrow{\text{via } \xi^{-1}} i^*(Ri_*)q^\times j^* \xrightarrow{\text{via } \varepsilon} q^\times j^*,$$

which is an *isomorphism*, see [41]. The statements **10, 12** and **14**, and corresponding statements to **11, 13** only for *fiber square pi-diagrams*, are readily proved.

In particular, for a closed open immersion $\eta : U \to X$, we have an isomorphism

$$v(\eta) : \eta^* \xrightarrow{\text{f}^{-1}} 1^\times_U \eta^* \xrightarrow{\zeta_0(\eta 1_U = \eta 1_U)^{-1}} 1^*_U \eta^\times \xrightarrow{\text{f}} \eta^\times.$$

Let $\sigma = (pi = qj)$ be an arbitrary pi-diagram. Let j_1 be the base change of j by p, and let p_1 be the base change of p by j. Let η be the unique morphism such that $q = p_1\eta$ and $i = j_1\eta$. Note that η is a closed open immersion. Define $\zeta(\sigma)$ to be the composite isomorphism

$$i^*p^\times \cong \eta^*j_1^*p^\times \xrightarrow{\text{via } \zeta_0} \eta^*p_1^\times j^* \xrightarrow{\text{via } v(\eta)} \eta^\times p_1^\times j^* \cong q^\times j^*.$$

Now the proof of conditions **11, 13** consists in diagram chasing arguments, while **18** is trivial, since $\zeta(\sigma)$ is always an isomorphism. Thus the twisted inverse triangulated almost-pseudofunctor $(?)^!$ on $\mathcal{A}$ is defined to be the composite of $(?)^\times$ and $(?)^*$.

Chapter 17
The Right Adjoint of the Derived Direct Image Functor of a Morphism of Diagrams

Let I be a small category, S a scheme, and $X_\bullet \in \mathcal{P}(I, \underline{\mathrm{Sch}}/S)$.

Lemma 17.1. *Assume that $X_\bullet$ is concentrated. That is, X_i is concentrated for each $i \in I$. Let C_i be a small set of compact generators of $D_{\mathrm{Qch}}(X_i)$, which exists by* **Theorem 14.14.** *Then*

$$C := \{LL_i c \mid i \in I, \, c \in C_i\}$$

is a small set of compact generators of $D_{\mathrm{Lqc}}(X_\bullet)$. In particular, the category $D_{\mathrm{Lqc}}(X_\bullet)$ is compactly generated.

Proof. Let $t \in D_{\mathrm{Lqc}}(X_\bullet)$ and assume that

$$\mathrm{Hom}_{D(X_\bullet)}(LL_i c, t) \cong \mathrm{Hom}_{D(X_i)}(c, t_i) = 0$$

for any $i \in I$ and any $c \in C_i$. Then, $t_i = 0$ for all i. This shows $t = 0$. It is easy to see that $LL_i c$ is compact, and C is small. So C is a small set of compact generators. As $D_{\mathrm{Lqc}}(X_\bullet)$ has coproducts, it is compactly generated. $\square$

Lemma 17.2. *Let $f_\bullet : X_\bullet \to Y_\bullet$ be a concentrated morphism in $\mathcal{P}(I, \underline{\mathrm{Sch}}/S)$. Then*

$$R(f_\bullet)_* : D_{\mathrm{Lqc}}(X_\bullet) \to D_{\mathrm{Lqc}}(Y_\bullet)$$

preserves coproducts.

Proof. Let (t_λ) be a small family of objects in $D_{\mathrm{Lqc}}(X_\bullet)$, and consider the canonical map

$$\bigoplus_\lambda R(f_\bullet)_* t_\lambda \to R(f_\bullet)_* \left(\bigoplus_\lambda t_\lambda \right).$$

For each $i \in I$, apply $(?)_i$ to the map. As $(?)_i$ obviously preserves coproducts and we have a canonical isomorphism

$$(?)_i R(f_\bullet)_* \cong R(f_i)_* (?)_i,$$

J. Lipman, M. Hashimoto, *Foundations of Grothendieck Duality for Diagrams of Schemes*, Lecture Notes in Mathematics 1960,

the result is the canonical map

$$\bigoplus_\lambda R(f_i)_*(t_\lambda)_i \to R(f_i)_*(\bigoplus_\lambda (t_\lambda)_i),$$

which is an isomorphism by Lemma 14.15. Hence, $R(f_\bullet)_*$ preserves coproducts.   □

By Theorem 14.10, we have the first (original) theorem of these notes:

Theorem 17.3. *Let I be a small category, S a scheme, and $f_\bullet : X_\bullet \to Y_\bullet$ a morphism in $\mathcal{P}(I, \underline{\mathrm{Sch}}/S)$. If $X_\bullet$ and $f_\bullet$ are concentrated, then*

$$R(f_\bullet)_* : D_{\mathrm{Lqc}}(X_\bullet) \to D_{\mathrm{Lqc}}(Y_\bullet)$$

has a right adjoint $f_\bullet^\times$.

(17.4) Let I be a small category. Let $\mathcal{S}$ be the category of concentrated I^{op} diagrams of schemes. Note that any morphism of $\mathcal{S}$ is concentrated (follows easily from [15, (1.2.3), (1.2.4)]). For $X_\bullet \in \mathcal{S}$, set $R(X_\bullet)_* := D_{\mathrm{Lqc}}(X_\bullet)$. For a morphism $f_\bullet$ of $\mathcal{S}$, set $R(f_\bullet)_*$ be the derived direct image. Then $R(?)_*$ is a covariant almost-pseudofunctor on $\mathcal{S}$. Thus its right adjoint $(?)^\times$ is a contravariant almost-pseudofunctor on $\mathcal{S}$, and $(R(?)_*, (?)^\times)$ is an opposite adjoint pair of Δ-pseudofunctors on $\mathcal{S}$, see (1.18).

For composable morphisms $f_\bullet$ and $g_\bullet$, $d_{f_\bullet, g_\bullet} : f_\bullet^\times g_\bullet^\times \to (g_\bullet f_\bullet)^\times$ is the composite

$$f_\bullet^\times g_\bullet^\times \xrightarrow{u} (g_\bullet f_\bullet)^\times R(g_\bullet f_\bullet)_* f_\bullet^\times g_\bullet^\times \xrightarrow{c} (g_\bullet f_\bullet)^\times R(g_\bullet)_* R(f_\bullet)_* f_\bullet^\times g_\bullet^\times \xrightarrow{\varepsilon}$$
$$(g_\bullet f_\bullet)^\times R(g_\bullet)_* g_\bullet^\times \xrightarrow{\varepsilon} (g_\bullet f_\bullet)^\times.$$

For $X_\bullet \in \mathcal{S}$, $\mathfrak{f} : \mathrm{id}_{X_\bullet}^\times \to \mathrm{Id}_{X_\bullet^\times}$ is the composite

$$\mathrm{id}_{X_\bullet}^\times \xrightarrow{\mathfrak{e}} R(\mathrm{id}_{X_\bullet})_* \mathrm{id}_{X_\bullet}^\times \xrightarrow{\varepsilon} \mathrm{Id}.$$

Lemma 17.5. *Let S be a scheme, I a small category, and $f_\bullet : X_\bullet \to Y_\bullet$ and $g_\bullet : Y_\bullet' \to Y_\bullet$ be morphisms in $\mathcal{P}(I, \underline{\mathrm{Sch}}/S)$. Set $X_\bullet' := Y_\bullet' \times_{Y_\bullet} X_\bullet$. Let $f_\bullet' : X_\bullet' \to Y_\bullet'$ be the first projection, and $g_\bullet' : X_\bullet' \to X_\bullet$ the second projection. Assume that $f_\bullet$ is concentrated, and $g_\bullet$ is flat. Then the canonical map*

$$\theta(g_\bullet, f_\bullet) : (g_\bullet)^* R(f_\bullet)_* \to R(f_\bullet')_* (g_\bullet')^*$$

is an isomorphism of functors from $D_{\mathrm{Lqc}}(X_\bullet)$ to $D_{\mathrm{Lqc}}(Y_\bullet')$.

Proof. It suffices to show that for each $i \in I$,

$$(?)_i \theta : (?)_i (g_\bullet)^* R(f_\bullet)_* \to (?)_i R(f_\bullet')_* (g_\bullet')^*$$

is an isomorphism. By Lemma 1.22, it is easy to verify that the diagram

$$g_i^* R(f_i)_*(?)_i \xrightarrow{\ c^{-1}\ } g_i^*(?)_i R(f_\bullet)_* \xrightarrow{\ \theta\ } (?)_i g_\bullet^* R(f_\bullet)_*$$
$$\downarrow \theta(g_i, f_i) \qquad\qquad\qquad\qquad\qquad \downarrow (?)_i \theta(g_\bullet, f_\bullet) \qquad (17.6)$$
$$R(f_i')_*(g_i')^*(?)_i \xrightarrow{\ \theta\ } R(f_i')_*(?)_i(g_\bullet')^* \xrightarrow{\ c^{-1}\ } (?)_i R(f_\bullet')_*(g_\bullet')^*$$

is commutative. As the horizontal maps and $\theta(g_i, f_i)$ are isomorphisms by [26, (3.9.5)], $(?)_i \theta(g_\bullet, f_\bullet)$ is also an isomorphism. $\square$

Chapter 18
Commutativity of Twisted Inverse with Restrictions

(18.1) Let S be a scheme, I a small category, and $f_\bullet : X_\bullet \to Y_\bullet$ a morphism in $\mathcal{P}(I, \underline{\mathrm{Sch}}/S)$. Let J be an admissible subcategory of I. Assume that $f_\bullet$ is concentrated. Then there is a natural map

$$\theta(J, f_\bullet) : LL_J \circ R(f_\bullet|_J)_* \to R(f_\bullet)_* \circ LL_J \qquad (18.2)$$

between functors from $D_{\mathrm{Lqc}}(X_\bullet|_J)$ to $D_{\mathrm{Lqc}}(Y_\bullet)$, see [26, (3.7.2)].

(18.3) Let S, I and $f_\bullet$ be as in (18.1). We assume that $X_\bullet$ and $f_\bullet$ are concentrated, so that the right adjoint functor

$$f_\bullet^\times : D_{\mathrm{Lqc}}(Y_\bullet) \to D_{\mathrm{Lqc}}(X_\bullet)$$

of

$$R(f_\bullet)_* : D_{\mathrm{Lqc}}(X_\bullet) \to D_{\mathrm{Lqc}}(Y_\bullet)$$

exists. Let J be a subcategory of I which may not be admissible.
  We define the natural transformation

$$\xi(J, f_\bullet) : (?)_J \circ f_\bullet^\times \to (f_\bullet|_J)^\times \circ (?)_J$$

to be the composite

$$(?)_J f_\bullet^\times \xrightarrow{u} (f_\bullet|_J)^\times R(f_\bullet|_J)_* (?)_J f_\bullet^\times \xrightarrow{c^{-1}} (f_\bullet|_J)^\times (?)_J R(f_\bullet)_* f_\bullet^\times \xrightarrow{\varepsilon} (f_\bullet|_J)^\times (?)_J.$$

By definition, ξ is the conjugate map of $\theta(J, f_\bullet)$ in (18.2) if J is admissible. Do not confuse $\xi(J, f_\bullet)$ with $\xi(f_\bullet, J)$ (see Corollary 6.26).

Lemma 18.4. *Let $J_2 \subset J_1 \subset I$ be subcategories of I. Let S be a scheme, $f_\bullet : X_\bullet \to Y_\bullet$ be a morphism in $\mathcal{P}(I, \underline{\mathrm{Sch}}/S)$. Assume both $X_\bullet$ and $f_\bullet$ are concentrated. Then the composite map*

J. Lipman, M. Hashimoto, *Foundations of Grothendieck Duality for Diagrams of Schemes*, Lecture Notes in Mathematics 1960,
© Springer-Verlag Berlin Heidelberg 2009

$$(?)_{J_2} f_\bullet^\times \xrightarrow{c} (?)_{J_2} (?)_{J_1} f_\bullet^\times \xrightarrow{\xi(J_1, f_\bullet)} (?)_{J_2} (f_\bullet|_{J_1})^\times (?)_{J_1}$$

$$\xrightarrow{\xi(J_2, f_\bullet|_{J_1})} (f_\bullet|_{J_2})^\times (?)_{J_2} (?)_{J_1} \xrightarrow{c^{-1}} (f_\bullet|_{J_2})^\times (?)_{J_2}$$

is equal to $\xi(J_2, f_\bullet)$.

Proof. Straightforward (and tedious) diagram drawing.   □

Lemma 18.5. *Let S and $f_\bullet : X_\bullet \to Y_\bullet$ be as in* Lemma 18.4. *Let J be a subcategory of I. Assume that $Y_\bullet$ has flat arrows and $f_\bullet$ is cartesian. Then $\xi(J, f_\bullet) : (?)_J f_\bullet^\times \to (f_\bullet|_J)^\times (?)_J$ is an isomorphism between functors from $D_{\mathrm{Lqc}}(Y_\bullet)$ to $D_{\mathrm{Lqc}}(X_\bullet|_J)$.*

Proof. In view of Lemma 18.4, we may assume that $J = i$ for an object i of I. Then, as i is an admissible subcategory of I and $\xi(i, f_\bullet)$ is a conjugate map of $\theta(i, f_\bullet)$, it suffices to show that $(?)_j \theta(i, f_\bullet)$ is an isomorphism for any $j \in \mathrm{ob}(I)$. As $Y_\bullet$ has flat arrows, $L_i : \mathrm{Mod}(Y_i) \to \mathrm{Mod}(Y_\bullet)$ is exact. As $f_\bullet$ is cartesian, $L_i : \mathrm{Mod}(X_i) \to \mathrm{Mod}(X_\bullet)$ is also exact.

By Proposition 6.23, the composite

$$(?)_j L_i R(f_i)_* \xrightarrow{(?)_j \theta} (?)_j R(f_\bullet)_* L_i \xrightarrow{c} R(f_j)_* (?)_j L_i$$

agrees with the composite

$$(?)_j L_i R(f_i)_* \xrightarrow{\lambda_{i,j}} \bigoplus_{\phi \in I(i,j)} Y_\phi^* R(f_i)_* \xrightarrow{\oplus \theta} \bigoplus_\phi R(f_j)_* X_\phi^*$$

$$\xrightarrow{C} R(f_j)_* \left(\bigoplus_\phi X_\phi^* \right) \xrightarrow{\lambda_{i,j}^{-1}} R(f_j)_* (?)_j L_i,$$

where C is the canonical map. By Lemma 14.15, C is an isomorphism. As $f_\bullet$ is cartesian and $Y_\bullet$ has flat arrows, $\theta : Y_\phi^* R(f_i)_* \to R(f_j)_* X_\phi^*$ is an isomorphism for each $\phi \in I(i, j)$ by [26, (3.9.5)]. Hence the second composite is an isomorphism. As the first composite is an isomorphism and c is also an isomorphism, we have that $(?)_j \theta(i, f_\bullet)$ is an isomorphism.   □

(18.6) Let I be a small category, S a scheme, and $f_\bullet : X_\bullet \to Y_\bullet$ a morphism in $\mathcal{P}(I, \underline{\mathrm{Sch}}/S)$. Assume that $X_\bullet$ and $f_\bullet$ are concentrated.

Lemma 18.7. *Let J be a subcategory of I. Then the following hold:*

1 *The composite map*

$$(?)_J \xrightarrow{u} (?)_J f_\bullet^\times R(f_\bullet)_* \xrightarrow{\xi(J, f_\bullet)} (f_\bullet|_J)^\times (?)_J R(f_\bullet)_* \xrightarrow{c_{J, f_\bullet}} (f_\bullet|_J)^\times R(f_\bullet|_J)_* (?)_J$$

agrees with u.

2 *The composite map*

$$(?)_J R(f_\bullet)_* f_\bullet^\times \xrightarrow{c_{J,f_\bullet}} R(f_\bullet|_J)_* (?)_J f_\bullet^\times \xrightarrow{\xi(J,f_\bullet)} R(f_\bullet|_J)_* (f_\bullet|_J)^\times (?)_J \xrightarrow{\varepsilon} (?)_J$$

agrees with ε.

Proof. The proof consists in straightforward diagram drawings. □

(18.8) Let I be a small category. For $i, j \in \mathrm{ob}(I)$, we say that $i \leq j$ if $I(i, j) \neq \emptyset$. This definition makes $\mathrm{ob}(I)$ a pseudo-ordered set. We say that I is *ordered* if $\mathrm{ob}(I)$ is an ordered set with the pseudo-order structure above, and $I(i, i) = \{\mathrm{id}\}$ for $i \in I$.

Lemma 18.9. *Let I be an ordered small category. Let J_0 and J_1 be full subcategories of I, such that $\mathrm{ob}(J_0) \cup \mathrm{ob}(J_1) = \mathrm{ob}(I)$, $\mathrm{ob}(J_0) \cap \mathrm{ob}(J_1) = \emptyset$, and $I(j_1, j_0) = \emptyset$ for $j_1 \in J_1$ and $j_0 \in J_0$. Let $X_\bullet \in \mathcal{P}(I, \underline{\mathrm{Sch}}/S)$. Then, we have the following.*

1 *The unit of adjunction $u : \mathrm{Id}_{\mathrm{Mod}(X_\bullet|_{J_1})} \to (?)_{J_1} \circ L_{J_1}$ is an isomorphism.*
2 *$(?)_{J_0} \circ L_{J_1}$ is zero.*
3 *L_{J_1} is exact, and J_1 is an admissible subcategory of I.*
4 *For $\mathcal{M} \in \mathrm{Mod}(X_\bullet)$, $\mathcal{M}_{J_0} = 0$ if and only if $\varepsilon : L_{J_1} \mathcal{M}_{J_1} \to \mathcal{M}$ is an isomorphism.*
5 *The counit of adjunction $(?)_{J_0} \circ R_{J_0} \to \mathrm{Id}_{\mathrm{Mod}(X_\bullet|_{J_0})}$ is an isomorphism.*
6 *$(?)_{J_1} \circ R_{J_0}$ is zero.*
7 *R_{J_0} is exact and preserves local-quasi-coherence.*
8 *For $\mathcal{M} \in \mathrm{Mod}(X_\bullet)$, $\mathcal{M}_{J_1} = 0$ if and only if $u : \mathcal{M} \to R_{J_0} \mathcal{M}_{J_0}$ is an isomorphism.*
9 *The sequence*

$$0 \to L_{J_1}(?)_{J_1} \xrightarrow{\varepsilon} \mathrm{Id} \xrightarrow{u} R_{J_0}(?)_{J_0} \to 0$$

is exact, and induces a distinguished triangle in $D(X_\bullet)$.

Proof. **1** This is obvious by Lemma 6.15.

2 The category $I_{j_0}^{(J_1^{\mathrm{op}} \hookrightarrow I^{\mathrm{op}})}$ is empty, if $j_0 \in J_0$, since $I^{\mathrm{op}}(j_0, j_1) = \emptyset$ if $j_1 \in J_1$ and $j_0 \in J_0$. It follows that $(?)_{j_0} \circ L_{J_1} = 0$ if $j_0 \in J_0$. Hence, $(?)_{J_0} \circ L_{J_1} = 0$.

3 This is trivial by **1,2** and their proof.

4 The 'if' part is trivial by **2**. We prove the 'only if' part. By assumption and **2**, both $\mathcal{M}_{J_0}$ and $(?)_{J_0} L_{J_1} \mathcal{M}_{J_1}$ are zero, and $(?)_{J_0} \varepsilon$ is an isomorphism. On the other hand, $((?)_{J_1} \varepsilon)(u(?)_{J_1}) = \mathrm{id}$, and u is an isomorphism by **1**. Hence,

$$(?)_{J_1} \varepsilon : (?)_{J_1} L_{J_1} \mathcal{M}_{J_1} \to (?)_{J_1} \mathcal{M}$$

is an isomorphism. Hence, ε is an isomorphism.

The assertions **5,6,7,8,9** are similar, and we omit the proof. □

Lemma 18.10. *Let I, S, J_1 and J_0 be as in Lemma 18.9. Let $f_\bullet : X_\bullet \to Y_\bullet$ be a morphism in $\mathcal{P}(I, \underline{\mathrm{Sch}}/S)$. Assume that $X_\bullet$ and $f_\bullet$ are concentrated. Then we have that $\theta(J_1, f_\bullet)$ and $\xi(J_1, f_\bullet)$ are isomorphisms.*

Proof. Note that J_1 is admissible by Lemma 18.9, **3**, and hence $\theta(J_1, f_\bullet)$ and $\xi(J_1, f_\bullet)$ are defined. Since we have $\xi(J_1, f_\bullet)$ is the conjugate of $\theta(J_1, f_\bullet)$ by definition, it suffices to show that $\theta(J_1, f_\bullet)$ is an isomorphism. It suffices to show that

$$(?)_i LL_{J_1} R(f_\bullet|_{J_1})_* \xrightarrow{(?)_i \theta} (?)_i R(f_\bullet)_* LL_{J_1} \cong R(f_i)_* (?)_i LL_{J_1}$$

is an isomorphism for any $i \in I$.

If $i \in J_0$, then both hand sides are zero functors, and it is an isomorphism. On the other hand, if $i \in J_1$, then the map in question is equal to the composite isomorphism

$$(?)_i LL_{J_1} R(f_\bullet|_{J_1})_* \cong (?)_i R(f_\bullet|_{J_1})_* \xrightarrow{c_{i, f_\bullet|_{J_1}}} R(f_i)_* (?)_i \cong R(f_i)_* (?)_i LL_{J_1}$$

by Proposition 6.23. Hence $\theta(J_1, f_\bullet)$ is an isomorphism, as desired. $\square$

(18.11)   Let S be a scheme, I an ordered small category, $i \in I$, and $X_\bullet \in \mathcal{P}(I, \underline{\mathrm{Sch}}/S)$. Let J_1 be a filter of $\mathrm{ob}(I)$ such that i is a minimal element of J_1 (e.g., $[i, \infty)$), and set $\Gamma_i := L_{I, J_1} \circ R_{J_1, i}$. Then we have $(?)_j \Gamma_i = 0$ if $j \neq i$ and $(?)_i \Gamma_i = \mathrm{Id}$. Hence Γ_i does not depend on the choice of J_1, and depends only on i. Note that Γ_i preserves arbitrary limits and colimits (hence is exact). Assume that X_i is concentrated. Then $D_{\mathrm{Qch}}(X_i)$ is compactly generated, and the derived functor

$$\Gamma_i : D_{\mathrm{Qch}}(X_i) \to D_{\mathrm{Lqc}}(X_\bullet)$$

preserves coproducts. It follows that there is a right adjoint

$$\Sigma_i : D_{\mathrm{Lqc}}(X_\bullet) \to D_{\mathrm{Qch}}(X_i).$$

As Γ_i is obviously way-out left, we have Σ_i is way-out right by Lemma 14.13.

(18.12)   Let S be a scheme, I a small category, and $X_\bullet \in \mathcal{P}(I, \underline{\mathrm{Sch}}/S)$. We define $\mathcal{D}^+(X_\bullet)$ (resp. $\mathcal{D}^-(X_\bullet)$) to be the full subcategory of $D(X_\bullet)$ consisting of $\mathbb{F} \in D(X_\bullet)$ such that $\mathbb{F}_i$ is bounded below (resp. above) and has quasi-coherent cohomology groups for each $i \in I$. For a plump full subcategory $\mathcal{A}$ of $\mathrm{Lqc}(X_\bullet)$, we denote the triangulated subcategory of $\mathcal{D}^+(X_\bullet)$ (resp. $\mathcal{D}^-(X_\bullet)$) consisting of objects all of whose cohomology groups belong to $\mathcal{A}$ by $\mathcal{D}_\mathcal{A}^+(X_\bullet)$ (resp. $\mathcal{D}_\mathcal{A}^-(X_\bullet)$).

(18.13)   Let P be an ordered set. We say that P is *upper Jordan-Dedekind* (*UJD* for short) if for any $p \in P$, the subset

$$[p, \infty) := \{q \in P \mid q \geq p\}$$

is finite. We say that an ordered small category I is UJD if the ordered set $\text{ob}(I)$ is UJD, and $I(i,j)$ is finite for $i, j \in I$.

Proposition 18.14. *Let I be an ordered UJD small category. Let S be a scheme, and $g_\bullet : U_\bullet \to X_\bullet$ and $f_\bullet : X_\bullet \to Y_\bullet$ be morphisms in $\mathcal{P}(I, \underline{\text{Sch}}/S)$. Assume that $Y_\bullet$ is noetherian with flat arrows, $f_\bullet$ is proper, $g_\bullet$ is an open immersion such that $g_i(U_i)$ is dense in X_i for each $i \in I$, and $f_\bullet \circ g_\bullet$ is cartesian. Then $g_\bullet$ is cartesian, and for any $i \in I$ the composite natural map*

$$(?)_i g_\bullet^* f_\bullet^\times \xrightarrow{\text{via } \theta^{-1}} g_i^*(?)_i f_\bullet^\times \xrightarrow{\text{via } \xi(i)} g_i^* f_i^\times (?)_i$$

is an isomorphism between functors $\mathcal{D}^+(Y_\bullet) \to D_{\text{Qch}}(U_i)$, where $\theta : g_i^*(?)_i \to (?)_i g_\bullet^*$ is the canonical isomorphism.

Proof. Note that $U_\bullet$ has flat arrows, since $Y_\bullet$ has flat arrows and $f_\bullet \circ g_\bullet$ is cartesian.

We prove that $g_\bullet$ is cartesian. Let $\phi : i \to j$ be a morphism in I. Then, the canonical map $(U_\phi, f_j g_j) : U_j \to U_i \times_{Y_i} Y_j$ is an isomorphism by assumption. This map factors through $(U_\phi, g_j) : U_j \to U_i \times_{X_i} X_j$, and it is easy to see that (U_ϕ, g_j) is a closed immersion. On the other hand, it is an image dense open immersion, as can be seen easily, and hence it is an isomorphism. So $g_\bullet$ is cartesian.

Set $J_1 := [i, \infty)$ and $J_0 := \text{ob}(I) \setminus J_1$. By Lemma 18.10, $\xi(J_1, f_\bullet)$ is an isomorphism. By Lemma 18.4, we may replace I by J_1, and we may assume that I is an ordered finite category, and i is a minimal element of $\text{ob}(I)$. Now we have $\mathcal{D}^+(Y_\bullet)$ agrees with $D_{\text{Lqc}}^+(Y_\bullet)$. Since we have $\text{ob}(I)$ is finite, it is easy to see that $R(f_\bullet)_*$ is way-out in both directions. It follows that $f_\bullet^\times$ is way-out right by Lemma 14.13.

It suffices to show that

$$g_i^* \xi(i) : g_i^*(?)_i f_\bullet^\times \to g_i^* f_i^\times (?)_i$$

is an isomorphism of functors from $D_{\text{Lqc}}^+(Y_\bullet)$ to $D_{\text{Qch}}^+(U_i)$. As $g_i^* R(g_i)_* \cong \text{Id}$, it suffices to show that $R(g_i)_* g_i^* \xi(i)$ is an isomorphism. This is equivalent to say that for any perfect complex $\mathbb{P} \in C(\text{Qch}(X_i))$, we have

$$R(g_i)_* g_i^* \xi(i) : \text{Hom}_{D(X_i)}(\mathbb{P}, R(g_i)_* g_i^*(?)_i f_\bullet^\times)$$
$$\to \text{Hom}_{D(X_i)}(\mathbb{P}, R(g_i)_* g_i^* f_i^\times (?)_i)$$

is an isomorphism. By [41, Lemma 2], this is equivalent to say that the canonical map

$$\varinjlim \text{Hom}_{D(Y_\bullet)}(R(f_\bullet)_* LL_i(\mathbb{P} \otimes_{\mathcal{O}_{X_i}}^{\bullet, L} \mathcal{J}^n), ?)$$
$$\to \varinjlim \text{Hom}_{D(Y_\bullet)}(LL_i R(f_i)_*(\mathbb{P} \otimes_{\mathcal{O}_{X_i}}^{\bullet, L} \mathcal{J}^n), ?)$$

induced by the conjugate $\theta(i, f_\bullet)$ of $\xi(i)$ is an isomorphism, where $\mathcal{J}$ is a defining ideal sheaf of the closed subset $X_i \setminus U_i$ in X_i.

As I is ordered and $\mathrm{ob}(I)$ is finite, we may label

$$\mathrm{ob}(I) = \{i = i(0), i(1), i(2), \ldots\}$$

so that $I(i(s), i(t)) \neq \emptyset$ implies that $s \leq t$. Let $J(r)$ denote the full subcategory of I whose object set is $\{i(r), i(r+1), \ldots\}$.

By descending induction on t, we prove that the map

$$\text{via } \theta(i, f_\bullet) : \varinjlim \mathrm{Hom}_{D(Y_\bullet)}(L_{J(t)}(?)_{J(t)} R(f_\bullet)_* LL_i(\mathbb{P} \otimes_{\mathcal{O}_{X_i}}^{\bullet, L} \mathcal{J}^n), ?)$$

$$\to \varinjlim \mathrm{Hom}_{D(Y_\bullet)}(L_{J(t)}(?)_{J(t)} LL_i R(f_i)_* (\mathbb{P} \otimes_{\mathcal{O}_{X_i}}^{\bullet, L} \mathcal{J}^n), ?)$$

is an isomorphism. This is enough to prove the proposition, since $L_{J(1)}(?)_{J(1)} = \mathrm{Id}$.

Since the sequence

$$0 \to L_{J(t+1)}(?)_{J(t+1)} \xrightarrow{\text{via } \varepsilon} L_{J(t)}(?)_{J(t)} \to \Gamma_{i(t)}(?)_{i(t)} \to 0$$

is an exact sequence of exact functors, it suffices to prove that the map

$$\text{via } \theta(i, f_\bullet) : \varinjlim \mathrm{Hom}_{D(Y_\bullet)}(\Gamma_{i(t)}(?)_{i(t)} R(f_\bullet)_* LL_i(\mathbb{P} \otimes_{\mathcal{O}_{X_i}}^{\bullet, L} \mathcal{J}^n), ?)$$

$$\to \varinjlim \mathrm{Hom}_{D(Y_\bullet)}(\Gamma_{i(t)}(?)_{i(t)} LL_i R(f_i)_* (\mathbb{P} \otimes_{\mathcal{O}_{X_i}}^{\bullet, L} \mathcal{J}^n), ?)$$

is an isomorphism by induction assumption and the five lemma. By Proposition 6.23, this is equivalent to say that the map

$$\text{via } \theta : \varinjlim \mathrm{Hom}_{D(Y_{i(t)})}(R(f_{i(t)})_* \bigoplus_\phi LX_\phi^*(\mathbb{P} \otimes_{\mathcal{O}_{X_i}}^{\bullet, L} \mathcal{J}^n), \Sigma_i(?))$$

$$\to \varinjlim \mathrm{Hom}_{D(Y_{i(t)})}(\bigoplus_\phi Y_\phi^* R(f_i)_* (\mathbb{P} \otimes_{\mathcal{O}_{X_i}}^{\bullet, L} \mathcal{J}^n), \Sigma_i(?))$$

is an isomorphism, where the sum is taken over the *finite* set $I(i, i(t))$. It suffices to prove that the map

$$\text{via } \xi : \varinjlim \mathrm{Hom}_{D(X_i)}(\mathbb{P} \otimes_{\mathcal{O}_{X_i}}^{\bullet, L} \mathcal{J}^n, R(X_\phi)_* f_{i(t)}^\times \Sigma_i(?))$$

$$\to \varinjlim \mathrm{Hom}_{D(X_i)}(\mathbb{P} \otimes_{\mathcal{O}_{X_i}}^{\bullet, L} \mathcal{J}^n, f_i^\times R(Y_\phi)_* \Sigma_i(?))$$

induced by the map $\xi : R(X_\phi)_* f_{i(t)}^\times \to f_i^\times R(Y_\phi)_*$, which is conjugate to

$$\theta : Y_\phi^* R(f_i)_* \to R(f_{i(t)})_* LX_\phi^*,$$

is an isomorphism for $\phi \in I(i, i(t))$. Since Σ_i, $R(X_\phi)_* f_{i(t)}^\times \Sigma_i$, and $f_i^\times R(Y_\phi)_* \Sigma_i$ are way-out right, it suffices to show the canonical map

$$g_i^* \xi(Y_\phi f_{i(t)} = f_i X_\phi) : g_i^* R(X_\phi)_* f_{i(t)}^\times \to g_i^* f_i^\times R(Y_\phi)_*$$

is an isomorphism between functors from $D_{\mathrm{Qch}}^+(Y_{i(t)})$ to $D_{\mathrm{Qch}}^+(U_i)$. Let $X' := X_i \times_{Y_i} Y_{i(t)}$, $p_1 : X' \to X_i$ be the first projection, $p_2 : X' \to Y_{i(t)}$ the second projection, and $\pi : X_{i(t)} \to X'$ be the map $(X_\phi, f_{i(t)})$. It is easy to see that $\xi(Y_\phi f_{i(t)} = f_i X_\phi)$ equals the composite map

$$R(X_\phi)_* f_{i(t)}^\times \cong R(p_1)_* R\pi_* \pi^\times p_2^\times \xrightarrow{\varepsilon} R(p_1)_* p_2^\times \cong f_i^\times R(Y_\phi)_*.$$

Note that the last map is an isomorphism since Y_ϕ is flat. As we have

$$U_{i(t)} \to U_i \times_{Y_i} Y_{i(t)} \cong U_i \times_{X_i} X'$$

is an isomorphism and g_i is an open immersion by assumption, the canonical map

$$g_i^* R(p_1)_* \to R(U_\phi)_* (\pi \circ g_{i(t)})^*$$

is an isomorphism. So it suffices to prove that

$$(\pi \circ g_{i(t)})^* R\pi_* \pi^\times \xrightarrow{\text{via } \varepsilon} (\pi \circ g_{i(t)})^*$$

is an isomorphism.

Consider the fiber square

$$
\begin{array}{ccc}
U_{i(t)} & \xrightarrow{g_{i(t)}} & X_{i(t)} \\
\downarrow \mathrm{id} & \sigma & \downarrow \pi \\
U_{i(t)} & \xrightarrow{\pi \circ g_{i(t)}} & X'.
\end{array}
$$

By [41, Theorem 2], $\zeta_0(\sigma) : g_{i(t)}^* \pi^\times \to (\pi \circ g_{i(t)})^*$ is an isomorphism (in [41], schemes are assumed to have finite Krull dimension, but this assumption is not used in the proof there and unnecessary). By definition (16.7), ζ_0 is the composite map

$$g_{i(t)}^* \pi^\times \xrightarrow{u} \mathrm{id}^\times R\mathrm{id}_* g_{i(t)}^* \pi^\times \xrightarrow{\cong} \mathrm{id}^\times (\pi \circ g_{i(t)})^* R\pi_* \pi^\times \xrightarrow{\varepsilon} (\pi \circ g_{i(t)})^*.$$

Since the first and the second maps are isomorphisms, the third map is an isomorphism. This was what we wanted to prove. □

Corollary 18.15. *Under the same assumption as in the proposition, we have* $g_\bullet^* f_\bullet^\times (\mathcal{D}^+(Y_\bullet)) \subset \mathcal{D}^+(U_\bullet)$.

Proof. This is because $g_i^* f_i^\times (D_{\mathrm{Qch}}^+(Y_i)) \subset D_{\mathrm{Qch}}^+(U_i)$ for each $i \in I$.

Chapter 19
Open Immersion Base Change

(19.1) Let S be a scheme, I a small category, and

$$
\begin{array}{ccc}
X'_\bullet & \xrightarrow{g'_\bullet} & X_\bullet \\
\downarrow f'_\bullet \;\; \sigma & & \downarrow f_\bullet \\
Y'_\bullet & \xrightarrow{g_\bullet} & Y_\bullet
\end{array}
$$

a fiber square in $\mathcal{P}(I, \underline{\mathrm{Sch}}/S)$. Assume that $X_\bullet$ and $f_\bullet$ are concentrated, and $g_\bullet$ is flat. By Lemma 17.5, the canonical map

$$
\theta(g_\bullet, f_\bullet) : g_\bullet^* R(f_\bullet)_* \to R(f'_\bullet)_* (g'_\bullet)^*
$$

is an isomorphism of functors from $D_{\mathrm{Lqc}}(X_\bullet)$ to $D_{\mathrm{Lqc}}(Y'_\bullet)$. We define $\zeta(\sigma) = \zeta(g_\bullet, f_\bullet)$ to be the composite map

$$
\zeta(\sigma) : (g'_\bullet)^* f_\bullet^\times \xrightarrow{u} (f'_\bullet)^\times R(f'_\bullet)_* (g'_\bullet)^* f_\bullet^\times \xrightarrow{\theta^{-1}} (f'_\bullet)^\times g_\bullet^* R(f_\bullet)_* f_\bullet^\times \xrightarrow{\varepsilon} (f'_\bullet)^\times g_\bullet^*.
$$

Lemma 19.2. *Let σ be as above, and J a subcategory of I. Then the diagram*

$$
\begin{array}{ccc}
(?)_J (g'_\bullet)^* f_\bullet^\times & \xrightarrow{\theta^{-1}} (g'_\bullet|_J)^* (?)_J f_\bullet^\times \xrightarrow{\xi} (g'_\bullet|_J)^* (f_\bullet|_J)^\times (?)_J \\
\downarrow \zeta & & \downarrow \zeta \\
(?)_J (f'_\bullet)^\times g_\bullet^* & \xrightarrow{\xi} (f'_\bullet|_J)^\times (?)_J g_\bullet^* \xrightarrow{\theta^{-1}} (f'_\bullet|_J)^\times (g_\bullet|_J)^* (?)_J
\end{array}
$$

is commutative.

Proof. Follows immediately from Lemma 18.7 and the commutativity of (17.6). □

Lemma 19.3. *Let σ be as above. Then the composite*

$$
(g'_\bullet)^* \xrightarrow{u} (g'_\bullet)^* f_\bullet^\times R(f_\bullet)_* \xrightarrow{\zeta} (f'_\bullet)^\times g_\bullet^* R(f_\bullet)_* \xrightarrow{\theta} (f'_\bullet)^\times R(f'_\bullet)_* (g'_\bullet)^*
$$

is u.

J. Lipman, M. Hashimoto, *Foundations of Grothendieck Duality for Diagrams of Schemes*, Lecture Notes in Mathematics 1960,

Proof. Follows from the commutativity of the diagram

$$
\begin{array}{ccc}
(g'_\bullet)^* & \xrightarrow{\;\;u\;\;} & (g'_\bullet)^* f_\bullet^\times R(f_\bullet)_* \\
\end{array}
$$

$$
(f'_\bullet)^\times R(f'_\bullet)_* (g'_\bullet)^* \xrightarrow{\;u\;} (f'_\bullet)^\times R(f'_\bullet)_* (g'_\bullet)^* (f_\bullet)^\times R(f_\bullet)_*
$$

$$
\theta^{-1}
$$

$$
\mathrm{id} \qquad (f'_\bullet)^\times g_\bullet^* R(f_\bullet)_* \xrightarrow{\;u\;} (f'_\bullet)^\times g_\bullet^* R(f_\bullet)_* f_\bullet^\times R(f_\bullet)_*
$$

$$
\mathrm{id} \qquad \qquad \varepsilon
$$

$$
(f'_\bullet)^\times R(f'_\bullet)_* (g'_\bullet)^* \xleftarrow{\;\;\theta\;\;} (f'_\bullet)^\times g_\bullet^* R(f_\bullet)_* \;.
$$

□

Theorem 19.4. *Let S be a scheme, I an ordered UJD small category, and*

$$
\begin{array}{ccc}
V_\bullet \xrightarrow{j_\bullet} U'_\bullet \xrightarrow{i'_\bullet} X'_\bullet \\
p_\bullet^U \downarrow \quad \sigma \quad \downarrow p_\bullet^X \\
U_\bullet \xrightarrow{i_\bullet} X_\bullet \xrightarrow{q_\bullet} Y_\bullet
\end{array}
$$

be a diagram in $\mathcal{P}(I, \underline{\mathrm{Sch}}/S)$. Assume the following.

1 *$Y_\bullet$ is noetherian with flat arrows.*
2 *$j_\bullet$, $i'_\bullet$ and $i_\bullet$ are image dense open immersions.*
3 *$q_\bullet$, $p_\bullet^X$ and $p_\bullet^U$ are proper.*
4 *$p_\bullet^X i'_\bullet = i_\bullet p_\bullet^U$.*
5 *$q_\bullet i_\bullet$ and $p_\bullet^U j_\bullet$ are cartesian.*

Then σ is a fiber square, and

$$
j_\bullet^* \zeta(\sigma) q_\bullet^\times : j_\bullet^* (i'_\bullet)^* (p_\bullet^X)^\times q_\bullet^\times \to j_\bullet^* (p_\bullet^U)^\times i_\bullet^* q_\bullet^\times
$$

is an isomorphism of functors from $\mathcal{D}^+(Y_\bullet)$ to $\mathcal{D}^+(V_\bullet)$.

Proof. The square σ is a fiber square, since the canonical map $U'_\bullet \to U_\bullet \times_{X_\bullet} X'_\bullet$ is an image dense closed open immersion, and is an isomorphism.

To prove the theorem, it suffices to show that the map in question is an isomorphism after applying $(?)_i$ for any $i \in I$. By Proposition 18.14, Lemma 19.2, and [26, (3.7.2), (iii)], the problem is reduced to the flat base change theorem (in fact open immersion base change theorem is enough) for schemes [41, Theorem 2], and we are done.   □

Chapter 20
The Existence of Compactification and Composition Data for Diagrams of Schemes Over an Ordered Finite Category

(20.1) Let I be an ordered finite category which is non-empty. Let $\mathcal{A}$ denote the category of noetherian I^{op}-diagrams of schemes as its objects and morphisms separated of finite type as its morphisms. Let $\mathcal{P}$ denote the class of proper morphisms in $\mathrm{Mor}(\mathcal{A})$. Let $\mathcal{I}$ denote the class of image dense open immersions in $\mathrm{Mor}(\mathcal{A})$.

Define $\mathcal{D}(X_\bullet) := D_{\mathrm{Lqc}}(X_\bullet)$ for $X_\bullet \in \mathrm{ob}(\mathcal{A})$. Define a pseudofunctor $(?)^{\#}$ on $\mathcal{A}_{\mathcal{P}}$ to be $(?)^\times$, where $X_\bullet^{\#} = \mathcal{D}(X_\bullet)$ for $X_\bullet \in \mathrm{ob}(\mathcal{A}_{\mathcal{P}})$. Define a pseudofunctor $(?)^{\flat}$ on $\mathcal{A}_{\mathcal{I}}$ to be $(?)^*$, where $X_\bullet^{\flat} = \mathcal{D}(X_\bullet)$ for $X_\bullet \in \mathrm{ob}(\mathcal{A}_{\mathcal{I}})$.

For a pi-square σ, define $\zeta(\sigma)$ to be the natural map defined in (19.1).

Lemma 20.2. *Let the notation be as above. Conditions* 1–6 *and* 8–14 *in Definition 16.1 are satisfied. Moreover, any pi-square is a fiber square.*

Proof. This is easy. $\square$

Proposition 20.3. *Let the notation be as in* (20.1). *Then the condition* **7** *in Definition 16.1 is satisfied. That is, for any morphism $f_\bullet$ in $\mathcal{A}$, there is a factorization $f_\bullet = p_\bullet j_\bullet$ with $p_\bullet \in \mathcal{P}$ and $j_\bullet \in \mathcal{I}$.*

Proof. Label the object set $\mathrm{ob}(I)$ of I as $\{i(1), \cdots, i(n)\}$ so that $I(i(s), i(t)) = \emptyset$ if $s > t$. Set $J(r)$ to be the full subcategory of I with $\mathrm{ob}(J(r)) = \{i(1), \ldots, i(r)\}$. By induction on r, we construct morphisms $j_\bullet(r) : X_\bullet|_{J(r)} \to Z_\bullet(r)$ and $p_\bullet(r) : Z_\bullet(r) \to Y_\bullet|_{J(r)}$ such that

1 $j_\bullet(r)$ is an open immersion whose scheme theoretic image is $Z_\bullet(r)$ (i.e., for any j, $j_\bullet(r)_j$ is an open immersion whose scheme theoretic image is $Z_\bullet(r)_j$). In particular, $j_\bullet(r)$ is an image dense open immersion.
2 $p_\bullet(r)$ is proper.
3 $p_\bullet(r) j_\bullet(r) = f_\bullet|_{J(r)}$.
4 $Z_\bullet(r)|_{J(j)} = Z_\bullet(j)$, $j_\bullet(r)|_{J(j)} = j_\bullet(j)$, $p_\bullet(r)|_{J(j)} = p_\bullet(j)$ for $j < r$.

The proposition follows from this construction for $r = n$. We may assume that the construction is done for $j < r$.

J. Lipman, M. Hashimoto, *Foundations of Grothendieck Duality for Diagrams of Schemes*, Lecture Notes in Mathematics 1960,
© Springer-Verlag Berlin Heidelberg 2009

First, consider the case where $i(r)$ is a minimal element in $\mathrm{ob}(I)$. By Nagata's compactification theorem [34] (see also [27]), there is a factorization

$$X_{i(r)} \xrightarrow{k} Z \xrightarrow{p} Y_{i(r)}$$

such that p is proper, k is an open immersion whose scheme theoretic image is Z, and $pk = f_{i(r)}$. Now define $Z_\bullet(r)_{i(r)} := Z$ and $Z_\bullet(r)_{\mathrm{id}_{i(r)}} = \mathrm{id}_Z$. Defining the other structures after **4**, we get $Z_\bullet(r)$, since $I(i(r), i(s)) = \emptyset = I(i(s), i(r))$ for $s < r$ and $I(i(r), i(r)) = \{\mathrm{id}\}$ by assumption. Define $j_\bullet(r)_{i(r)} := k$ and by **4**, we get a morphism $j_\bullet(r) : X_\bullet|_{J(r)} \to Z_\bullet(r)$ by the same reason. Similarly, $p_\bullet(r)_{i(r)} := p$ and **4** define $p_\bullet(r) : Z_\bullet(r) \to Y_\bullet|_{J(r)}$. and **1**–**4** are satisfied by the induction assumption. So this case is OK.

Now assume that $i(r)$ is not minimal so that $\bigcup_{j<r} I(i(j), i(r)) \neq \emptyset$. For simplicity, set $Y_j = Y_{i(j)}$, $X_j = X_{i(j)}$ for $1 \leq j \leq n$ and $Z_j = Z_\bullet(j)_j$ for $j < r$. Similarly, $f_j := f_{i(j)}$ $(1 \leq j \leq n)$, $j_j := j_\bullet(j)_{i(j)}$, and $p_j := p_\bullet(j)_{i(j)}$ $(1 \leq j < r)$.

Consider the direct product of Y_r-schemes

$$W := \prod_{j<r} \prod_{\phi \in I(i(j), i(r))} Y_r \, {}_{Y_\phi}{\times}_{Y_j} Z_j.$$

Note that each $Y_r \, {}_{Y_\phi}{\times}_{Y_j} Z_j$ is proper over Y_r, and hence W is proper over Y_r. There is a unique Y_r-morphism $h : X_r \to W$ induced by

$$(f_r, j_j \circ X_\phi) : X_r \to Y_r \, {}_{Y_\phi}{\times}_{Y_j} Z_j.$$

Since h is separated of finite type, there is a factorization

$$X_r \xrightarrow{k} Z \xrightarrow{p} W$$

such that k is an open immersion whose scheme theoretic image is Z, p is proper, and $pk = h$.

Now define $Z_{i(r)} = Z$, $Z_{\mathrm{id}_{i(r)}} = \mathrm{id}_Z$, and Z_ϕ to be the composite

$$Z \xrightarrow{p} W \xrightarrow{\text{projection}} Z_j$$

for $j < r$ and $\phi \in I(i(j), i(r))$. We define $Z_\bullet(r)$ by these data and by **4**. Set $j_r = j_\bullet(r)_r = k$, and $p_r = p_\bullet(r)_r$ to be the composite

$$Z \xrightarrow{p} W \to Y_r,$$

where the second map is the structure map of W as a Y_r-scheme.

Note that $Z_\psi j_j = j_{j'} X_\psi$ and $Y_\psi p_j = p_{j'} Z_\psi$ hold for $1 \leq j' \leq j \leq r$ and $\psi \in I(i(j'), i(j))$. Indeed, this is trivial by induction assumption if $j < r$, also trivial if $\psi = \mathrm{id}_{i(r)}$, and follows easily from the construction if $j' < j = r$. In particular, $j_\bullet(r)$ and $p_\bullet(r)$ are defined so that **4** is satisfied and are morphisms of diagrams of schemes provided that $Z_\bullet(r)$ is a diagram of schemes.

We need to check that $Z_\bullet(r)$ is certainly a diagram of schemes. To verify this, it suffices to show that, for any $j' < j < r$ and any $\phi \in I(i(j), i(r))$ and $\psi \in I(i(j'), i(j))$, $Z_{\phi\psi} = Z_\psi Z_\phi$ holds. Let A be the locus in Z such that $Z_{\phi\psi}$ and $Z_\psi Z_\phi$ agree. Note that the diagrams

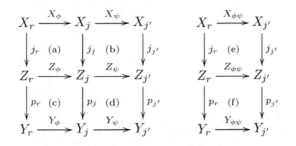

are commutative. Since (c), (d) and (f) are commutative and $Y_{\phi\psi} = Y_\psi Y_\phi$, we have that $p_{j'} Z_{\phi\psi} = p_{j'} Z_\psi Z_\phi$. Since there is a cartesian square

$$
\begin{array}{ccc}
A & \longrightarrow & Z_{j'} \\
\downarrow & & \downarrow \Delta \\
Z_r & \xrightarrow{(Z_{\phi\psi}, Z_\psi Z_\phi)} & Z_{j'} \times_{Y_{j'}} Z_{j'}
\end{array}
$$

and $p_{j'} : Z_{j'} \to Y_{j'}$ is separated, we have that A is a closed subscheme of Z_r. Since the scheme theoretic image of the open immersion $j_r : X_r \hookrightarrow Z_r$ is Z_r, it suffices to show that j_r factors through A. That is, it suffices to show that $j_r Z_{\phi\psi} = j_r Z_\psi Z_\phi$. But this is trivial by the commutativity of (a), (b), and (e), and the fact $X_{\phi\psi} = X_\psi X_\phi$. So $Z_\bullet(r)$ is a diagram of schemes, and $j_\bullet(r)$ and $p_\bullet(r)$ are morphisms of diagrams of schemes.

The conditions 1–4 are now easy to verify, and the proof is complete. □

Theorem 20.4. *Let the notation be as in* (20.1). *Set $\mathcal{F}$ to be the subcategory of $\mathcal{A}$ whose objects are objects of $\mathcal{A}$ with flat arrows, and whose morphisms are cartesian morphisms in $\mathcal{A}$. Define $\mathcal{D}^+$ by $\mathcal{D}^+(X_\bullet) := D^+_{\mathrm{Lqc}}(X_\bullet)$. Then $(\mathcal{A}, \mathcal{F}, \mathcal{P}, \mathcal{I}, \mathcal{D}, \mathcal{D}^+, (?)^\#, (?)^\flat, \zeta)$ is a composition data of contravariant almost-pseudofunctors.*

Proof. Conditions **1–14** in Definition 16.1 have already been checked. **15** follows from Lemma 7.16. **16** is trivial. Since I is finite, the definition of $\mathcal{D}^+(X_\bullet)$ is consistent with that in (18.12). Hence **17** is Corollary 18.15. **18** is Theorem 19.4. □

(20.5) We call the composite of $(?)^\#$ and $(?)^\flat$ defined by the composition data in the theorem the *equivariant twisted inverse* almost-pseudofunctor, and denote it by $(?)^!$.

Chapter 21
Flat Base Change

Let the notation be as in Theorem 20.4. Let $f_\bullet : X_\bullet \to Y_\bullet$ be a morphism in $\mathcal{F}$, and J a subcategory of I. Let $f_\bullet = p_\bullet i_\bullet$ be a compactification.

Lemma 21.1. *The composite map*

$$(?)_J f_\bullet^! \xrightarrow{\text{via } \Upsilon} (?)_J i_\bullet^* p_\bullet^\times \xrightarrow{\theta^{-1}} i_J^*(?)_J p_\bullet^\times \xrightarrow{\xi} i_J^* p_J^\times (?)_J \xrightarrow{\text{via } \Upsilon} f_\bullet|_J^!(?)_J$$

is independent of choice of compactification $f_\bullet = p_\bullet i_\bullet$, *where* Υ's *are the independence isomorphisms.*

The proof utilizes Lemma 16.3, and left to the reader. We denote by $\bar{\xi} = \bar{\xi}(J, f_\bullet)$ the composite map in the lemma.

Lemma 21.2. *Let* $f_\bullet : X_\bullet \to Y_\bullet$ *be a morphism in* $\mathcal{F}$, *and* $K \subset J \subset I$ *be subcategories. Then the composite map*

$$(?)_K f_\bullet^! \cong (?)_K (?)_J f_\bullet^! \xrightarrow{\bar{\xi}} (?)_K f_\bullet|_J^!(?)_J \xrightarrow{\bar{\xi}} f_\bullet|_K^!(?)_K (?)_J \cong f_K^!(?)_K$$

agrees with $\bar{\xi}(K, f_\bullet)$.

Proof. Follows easily from Lemma 18.4. □

Lemma 21.3. *Let* $f_\bullet : X_\bullet \to Y_\bullet$ *be a morphism in* $\mathcal{F}$, *and* J *a subcategory of* I. *Then* $\bar{\xi}(J, f_\bullet)$ *is an isomorphism.*

Proof. It suffices to show that $(?)_i \bar{\xi}(J, f_\bullet)$ is an isomorphism for any $i \in \text{ob}(J)$. By Lemma 21.2, we have

$$\bar{\xi}(i, f_\bullet|_J) \circ ((?)_i \bar{\xi}(J, f_\bullet)) = \bar{\xi}(i, f_\bullet).$$

By Proposition 18.14, we have $\bar{\xi}(i, f_\bullet|_J)$ and $\bar{\xi}(i, f_\bullet)$ are isomorphisms. Hence the natural map $(?)_i \bar{\xi}(J, f_\bullet)$ is also an isomorphism. □

J. Lipman, M. Hashimoto, *Foundations of Grothendieck Duality for Diagrams of Schemes*, Lecture Notes in Mathematics 1960,
© Springer-Verlag Berlin Heidelberg 2009

Lemma 21.4. *Let* $f : X \to Y$ *be a flat morphism of locally noetherian schemes, and* U *a dense open subset of* Y. *Then* $f^{-1}(U)$ *is a dense open subset of* X.

Proof. The question is local both on Y and X, and hence we may assume that both $Y = \operatorname{Spec} A$ and $X = \operatorname{Spec} B$ are affine. Let I be the radical ideal of A defining the closed subset $Y \setminus U$. By assumption, I is not contained in any minimal prime of A. Assume that $f^{-1}(U)$ is not dense in X. Then, there is a minimal prime P of B which contains IB. As we have $I \subset IB \cap A \subset P \cap A$ and $P \cap A$ is minimal by the going-down theorem (see [30, Theorem 9.5]), this is a contradiction. $\square$

(21.5) Let the notation be as in Theorem 20.4. Let

$$
\begin{array}{ccc}
X'_{\bullet} & \xrightarrow{f'_{\bullet}} & Y'_{\bullet} \\
g^X_{\bullet} \downarrow & \sigma & \downarrow g_{\bullet} \\
X_{\bullet} & \xrightarrow{f_{\bullet}} & Y_{\bullet}
\end{array}
$$

be a diagram in $\mathcal{P}(I, \underline{\mathrm{Sch}})$ such that

1 All objects lie in $\mathcal{F}$;
2 $f_{\bullet}$ and $f'_{\bullet}$ are morphisms in $\mathcal{F}$;
3 σ is a fiber square;
4 $g_{\bullet}$ is flat (not necessarily a morphism of $\mathcal{A}$).

By assumption, there is a diagram

$$
\begin{array}{ccccc}
X'_{\bullet} & \xrightarrow{i'_{\bullet}} & Z'_{\bullet} & \xrightarrow{p'_{\bullet}} & Y'_{\bullet} \\
\downarrow g^X_{\bullet}\, \sigma_1 & & \downarrow g^Z_{\bullet}\, \sigma_2 & & \downarrow g_{\bullet} \\
X_{\bullet} & \xrightarrow{i_{\bullet}} & Z_{\bullet} & \xrightarrow{p_{\bullet}} & Y_{\bullet}
\end{array} \tag{21.6}
$$

such that $f_{\bullet} = p_{\bullet} i_{\bullet}$ is a compactification, σ_1 and σ_2 are fiber squares, and the whole rectangle $\sigma_1 \sigma_2$ equals σ. By Lemma 21.4, we have that $f'_{\bullet} = p'_{\bullet} i'_{\bullet}$ is a compactification.

Lemma 21.7. *The composite map*

$$
(g^X_{\bullet})^* f^!_{\bullet} \xrightarrow{\Upsilon} (g^X_{\bullet})^* i^*_{\bullet} p^{\times}_{\bullet} \xrightarrow{d} (i'_{\bullet})^* (g^Z_{\bullet})^* p^{\times}_{\bullet} \xrightarrow{\zeta} (i'_{\bullet})^* (p'_{\bullet})^{\times} g^*_{\bullet} \xrightarrow{\Upsilon} (f'_{\bullet})^! g^*_{\bullet}
$$

is independent of choice of the diagram (21.6), *and depends only on* σ, *where* Υ's *are independence isomorphisms.*

Proof. Obvious by Lemma 16.3. $\square$

We denote the composite map in the lemma by $\bar{\zeta} = \bar{\zeta}(\sigma)$.

Theorem 21.8. *Let the notation be as above. Then we have:*

1   *Let J be a subcategory of I. Then the diagram*

$$
\begin{array}{ccc}
(?)_J(g_\bullet^X)^* f_\bullet^! & \xrightarrow{\theta^{-1}} (g_J^X)^*(?)_J f_\bullet^! & \xrightarrow{\bar{\xi}} (g_J^X)^* f_J^!(?)_J \\
\downarrow (?)_J \bar{\zeta}(\sigma) & & \downarrow \bar{\zeta}(\sigma_J)(?)_J \\
(?)_J(f_\bullet')^!(g_\bullet)^* & \xrightarrow{\bar{\xi}} (f_J')^!(?)_J(g_\bullet)^* & \xrightarrow{\theta^{-1}} (f_J')^!(g_J)^*(?)_J
\end{array}
\qquad (21.9)
$$

is commutative.

2   $\bar{\zeta}(\sigma)$ *is an isomorphism.*

Proof. **1** is an immediate consequence of Lemma 19.2 and [26, (3.7.2)].

2 Let i be an object of I. By Lemma 21.3, the horizontal arrows in the diagram (21.9) for $J = i$ are isomorphisms. By Verdier's flat base change theorem [41, Theorem 2], we have that $\bar{\zeta}(\sigma_i)$ is an isomorphism. Hence, we have that $(?)_i \bar{\zeta}(\sigma)$ is an isomorphism for any $i \in I$ by **1** applied to $J = i$, and the assertion follows.   □

Chapter 22
Preservation of Quasi-Coherent Cohomology

(22.1) Let the notation be as in (20.1). Let $\mathcal{F}$ be as in Theorem 20.4.

Lemma 22.2. *Let $f : X \to Y$ be a separated morphism of finite type between noetherian schemes. If $\mathbb{F} \in \mathcal{D}^+_{\mathrm{Coh}(Y)}(\mathrm{Mod}(Y))$, then $f^! \mathbb{F} \in \mathcal{D}^+_{\mathrm{Coh}(X)}(\mathrm{Mod}(X))$.*

Proof. We may assume that both Y and X are affine. So we may assume that f is either smooth or a closed immersion. The case where f is smooth is obvious by [41, Theorem 3]. The case where f is a closed immersion is also obvious by Proposition III.6.1 and Theorem III.6.7 in [17]. $\qquad\square$

Proposition 22.3. *Let $f_\bullet : X_\bullet \to Y_\bullet$ be a morphism in $\mathcal{F}$, and $\phi : i \to j$ a morphism in I. Then the composite map*

$$X^*_\phi(?)_i f^!_\bullet \xrightarrow{\alpha_\phi} (?)_j f^!_\bullet \xrightarrow{\bar{\xi}} f^!_j(?)_j$$

agrees with the composite map

$$X^*_\phi(?)_i f^!_\bullet \xrightarrow{\bar{\xi}} X^*_\phi f^!_i(?)_i \xrightarrow{\bar{\zeta}} f^!_j Y^*_\phi(?)_i \xrightarrow{\alpha_\phi} f^!_j(?)_j.$$

Proof. By Lemma 21.2, we may assume that I is the ordered category given by $\mathrm{ob}(I) = \{i, j\}$ and $I(i, j) = \{\phi\}$.

Then it is easy to see that there is a compactification

$$X_\bullet \xrightarrow{i_\bullet} Z_\bullet \xrightarrow{p_\bullet} Y_\bullet$$

of $f_\bullet$ such that $p_\bullet$ is cartesian. Note that $i_\bullet$ is cartesian, and $Z_\bullet$ has flat arrows.

By the definition of $\bar{\xi}$ and $\bar{\zeta}$, it suffices to prove that the composite map

$$X^*_\phi(?)_i i^*_\bullet p^\times_\bullet \xrightarrow{\theta^{-1}} X^*_\phi i^*_i(?)_i p^\times_\bullet \xrightarrow{d} i^*_j Z^*_\phi(?)_i p^\times_\bullet \xrightarrow{\xi} i^*_j Z^*_\phi p^\times_i(?)_i \xrightarrow{\zeta} i^*_j p^\times_j Y^*_\phi(?)_i \xrightarrow{\alpha_\phi} i^*_j p^\times_j(?)_j$$

agrees with

$$X_\phi^*(?)_i i_\bullet^* p_\bullet^\times \xrightarrow{\alpha_\phi} (?)_j i_\bullet^* p_\bullet^\times \xrightarrow{\theta^{-1}} i_j^*(?)_j p_\bullet^\times \xrightarrow{\xi} i_j^* p_j^\times(?)_j.$$

By the "derived version" of (6.31), the composite map

$$X_\phi^*(?)_i i_\bullet^* p_\bullet^\times \xrightarrow{\theta^{-1}} X_\phi^* i_i^*(?)_i p_\bullet^\times \xrightarrow{d} i_j^* Y_\phi^*(?)_i p_\bullet^\times \xrightarrow{\alpha_\phi} i_j^*(?)_j p_\bullet^\times$$

agrees with

$$X_\phi^*(?)_i i_\bullet^* p_\bullet^\times \xrightarrow{\alpha_\phi} (?)_j i_\bullet^* p_\bullet^\times \xrightarrow{\theta^{-1}} i_j^*(?)_j p_\bullet^\times.$$

Hence it suffices to prove the map

$$Z_\phi^*(?)_i p_\bullet^\times \xrightarrow{\xi} Z_\phi^* p_i^\times(?)_i \xrightarrow{\zeta} p_j^\times Y_\phi^*(?)_i \xrightarrow{\alpha_\phi} p_j^\times(?)_j$$

agrees with

$$Z_\phi^*(?)_i p_\bullet^\times \xrightarrow{\alpha_\phi} (?)_j p_\bullet^\times \xrightarrow{\xi} p_j^\times(?)_j.$$

Now the proof consists in a straightforward diagram drawing utilizing Lemma 19.3 and the derived version of Lemma 6.20.   □

Corollary 22.4. *Let $f_\bullet : X_\bullet \to Y_\bullet$ be a morphism in $\mathcal{F}$. Then we have $f_\bullet^!(\mathcal{D}_{\mathrm{Qch}}^+(Y_\bullet)) \subset \mathcal{D}_{\mathrm{Qch}}^+(X_\bullet)$ and $f_\bullet^!(\mathcal{D}_{\mathrm{Coh}}^+(Y_\bullet)) \subset \mathcal{D}_{\mathrm{Coh}}^+(X_\bullet)$.*

Proof. Let $\phi : i \to j$ be a morphism in I. By the flat base change theorem, Lemma 21.3, and the proposition, we have $\alpha_\phi : X_\phi^*(?)_i f_\bullet^! \to (?)_j f_\bullet^!$ is an isomorphism if $\alpha_\phi : Y_\phi^*(?)_i \to (?)_j$ is an isomorphism. So $f_\bullet^!$ preserves equivariance of cohomology groups, and the first assertion follows.

On the other hand, by Lemma 22.2 and Proposition 18.14, $f^!$ preserves local coherence of cohomology groups. Hence it also preserves the coherence of cohomology groups, by the first paragraph.   □

Chapter 23
Compatibility with Derived Direct Images

(23.1) Let the notation be as in (21.5). Consider the diagram (21.6). Lipman's theta $\theta(\sigma_2) : g_\bullet^* R(p_\bullet)_* \to R(p_\bullet')_*(g_\bullet^Z)^*$ induces the conjugate map

$$\xi(\sigma_2) : R(g_\bullet^Z)_*(p_\bullet')^\times \to p_\bullet^\times R(g_\bullet)_*.$$

As σ_2 is a fiber square, $\theta(\sigma_2)$ is an isomorphism. Hence $\xi(\sigma_2)$ is also an isomorphism. Note that

$$\theta : i_\bullet^* R(g_\bullet^Z)_* \to R(g_\bullet^X)_*(i_\bullet')^*$$

is an isomorphism, since σ_1 is a fiber square. We define $\bar{\xi} : R(g_\bullet^X)_*(f_\bullet')^! \to f_\bullet^! R(g_\bullet)_*$ to be the composite

$$R(g_\bullet^X)_*(f_\bullet')^! \xrightarrow{\Upsilon} R(g_\bullet^X)_*(i_\bullet')^*(p_\bullet')^\times \xrightarrow{\theta^{-1}} i_\bullet^* R(g_\bullet^Z)_*(p_\bullet')^\times \xrightarrow{\xi} i_\bullet^* p_\bullet^\times R(g_\bullet)_* \xrightarrow{\Upsilon} f_\bullet^! R(g_\bullet)_*.$$

As the all maps in the composition are isomorphisms, we have

Lemma 23.2. $\bar{\xi}$ *is an isomorphism.*

Lemma 23.3. *For any subcategory J, the composite*

$$(?)_J R(g_\bullet^Z)_*(p_\bullet')^\times \xrightarrow{\xi} (?)_J p_\bullet^\times R(g_\bullet)_* \xrightarrow{\xi} (p_\bullet|_J)^\times (?)_J R(g_\bullet)_* \xrightarrow{c} (p_\bullet|_J)^\times R(g_\bullet|_J)_*(?)_J$$

agrees with the composite

$$(?)_J R(g_\bullet^Z)_*(p_\bullet')^\times \xrightarrow{c} R(g_\bullet^Z|_J)_*(?)_J (p_\bullet')^\times \xrightarrow{\xi}$$

$$R(g_\bullet^Z|_J)_*(p_\bullet'|_J)^\times (?)_J \xrightarrow{\xi} (p_\bullet|_J)^\times R(g_\bullet|_J)_*(?)_J.$$

Proof. Follows from Lemma 18.7. $\qquad\qquad\square$

J. Lipman, M. Hashimoto, *Foundations of Grothendieck Duality for Diagrams*   425
of Schemes, Lecture Notes in Mathematics 1960,
© Springer-Verlag Berlin Heidelberg 2009

Chapter 24
Compatibility with Derived Right Inductions

(24.1) Let I be a finite category, and $X_\bullet \in \mathcal{P}(I, \underline{\text{Sch}})$. Assume that $X_\bullet$ has concentrated arrows. Let J be a subcategory of I. For $i \in I$, $I_i^{(J \to I)}$ is finite, since I is finite. So for any $\mathcal{M} \in \text{Lqc}(X_\bullet|_J)$, we have $R_J \mathcal{M} \in \text{Lqc}(X_\bullet)$ by (6.14). So $R_J : \text{Lqc}(X_\bullet|_J) \to \text{Lqc}(X_\bullet)$ is a right adjoint of $(?)_J : \text{Lqc}(X_\bullet) \to \text{Lqc}(X_\bullet|_J)$.

Lemma 24.2. *Let I be a finite category, and $X_\bullet \in \mathcal{P}(I, \underline{\text{Sch}})$. Assume that $X_\bullet$ is noetherian. If $\mathcal{I} \in \text{Lqc}(X_\bullet)$ is an injective object, then it is injective as an object of $\text{Mod}(X_\bullet)$.*

Proof. Let J be the discrete subcategory of I such that $\text{ob}(J) = \text{ob}(I)$. Let $\mathcal{I}_J \hookrightarrow \mathcal{J}$ be the injective hull in $\text{Lqc}(X_\bullet|_J)$. Since $(?)_J$ is faithful, the composite $\mathcal{I} \to R_J(?)_J \mathcal{I} \hookrightarrow R_J \mathcal{J}$ is a monomorphism, and hence it splits. For each $i \in I$, $\mathcal{J}_i$ is injective as an object of $\text{Qch}(X_i)$, since $\text{Lqc}(X_\bullet|_J) \cong \prod_i \text{Qch}(X_i)$ in a natural way. So it is also injective as an object of $\text{Mod}(X_i)$ by [17, (II.7)]. So $\mathcal{J}$ is injective as an object of $\text{Mod}(X_\bullet|_J)$. Since R_J preserves injectives, $R_J \mathcal{J}$ is an injective object of $\text{Mod}(X_\bullet)$. Hence its direct summand $\mathcal{I}$ is also injective in $\text{Mod}(X_\bullet)$. $\square$

Corollary 24.3. *Let I and $X_\bullet$ be as in the lemma. Then for any subcategory $J \subset I$,*

$$RR_J : D(X_\bullet|_J) \to D(X_\bullet)$$

takes $D_{\text{Lqc}}^+(X_\bullet|_J)$ to $D_{\text{Lqc}}^+(X_\bullet)$, and RR_J is right adjoint to $(?)_J$: $D_{\text{Lqc}}^+(X_\bullet) \to D_{\text{Lqc}}^+(X_\bullet|_J)$.

Proof. By the way-out lemma, it suffices to prove that for a single object $\mathcal{M} \in \text{Lqc}(X_\bullet|_J)$, $R^n R_J \mathcal{M} \in \text{Lqc}(X_\bullet)$. Let $\mathcal{M} \to \mathbb{I}$ be an injective resolution in the category $\text{Lqc}(X_\bullet|_J)$, which exists by Lemma 11.9. Then $\mathbb{I}$ is also an injective resolution in $\text{Mod}(X_\bullet|_J)$ by the lemma. So $R^n R_J \mathcal{M} \cong H^n(R_J \mathbb{I})$ lies in $\text{Lqc}(X_\bullet)$. $\square$

J. Lipman, M. Hashimoto, *Foundations of Grothendieck Duality for Diagrams*   427
of Schemes, Lecture Notes in Mathematics 1960,
ⓒ Springer-Verlag Berlin Heidelberg 2009

(24.4) Let the notation be as in Theorem 20.4. Let $f_\bullet : X_\bullet \to Y_\bullet$ be a morphism in $\mathcal{A}$. Let J be a subcategory of I. As

$$c : (?)_J R(f_\bullet)_* \to R(f_\bullet|_J)_*(?)_J$$

is an isomorphism of functors from $D^+_{\mathrm{Lqc}}(X_\bullet)$ to $D^+_{\mathrm{Lqc}}(X_\bullet|_J)$, its conjugate map

$$c' : RR_J(f_\bullet|_J)^\times \to f_\bullet^\times RR_J$$

is also an isomorphism.

Let $g_\bullet : U_\bullet \to X_\bullet$ be a cartesian image dense open immersion in $\mathcal{A}$. Let $\mu = \mu(g_\bullet, J)$ be the canonical map

$$g_\bullet^* RR_J \xrightarrow{u} g_\bullet^* RR_J R(g_\bullet|_J)_*(g_\bullet|_J)^* \xrightarrow{\xi^{-1}} g_\bullet^* R(g_\bullet)_* RR_J(g_\bullet|_J)^* \xrightarrow{\varepsilon} RR_J(g_\bullet|_J)^*,$$

where $\xi : R(g_\bullet)_* RR_J \to RR_J R(g_\bullet|_J)_*$ is the conjugate of the isomorphism $\theta : (g_\bullet)|_J^*(?)_J \to (?)_J g_\bullet^*$.

Lemma 24.5. *Let the notation be as above. Then* $\mu : g_\bullet^* RR_J \to RR_J(g_\bullet|_J)^*$ *is an isomorphism of functors from* $D(X_\bullet|_J)$ *to* $D(U_\bullet)$.

Proof. As $g_\bullet^*$, $(g_\bullet|_J)^*$, R_J, $(g_\bullet)_*$, and $(g_\bullet|_J)_*$ have exact left adjoints, it suffices to show that

$$(?)_i \mu : (?)_i g_\bullet^* R_J \mathbb{I} \to (?)_i R_J(g_\bullet|_J)^* \mathbb{I}$$

is an isomorphism for any K-injective complex in $C(\mathrm{Mod}(X_\bullet|_J))$ and $i \in \mathrm{ob}(I)$. This map agrees with

$$(?)_i g_\bullet^* R_J \mathbb{I} \xrightarrow{\theta^{-1}} g_i^*(?)_i R_J \mathbb{I} \cong g_i^* \varprojlim(X_\phi)_* \mathbb{I}_j \cong \varprojlim g_i^*(X_\phi)_* \mathbb{I}_j$$
$$\xrightarrow{\theta} \varprojlim(U_\phi)_* g_j^* \mathbb{I}_j \xrightarrow{\theta} \varprojlim(U_\phi)_*(?)_j(g_\bullet|_J)^* \mathbb{I} \cong (?)_i R_J(g_\bullet|_J)^* \mathbb{I},$$

which is obviously an isomorphism, where the limit is taken over $\phi \in I_i^{(J \to I)}$.
$\square$

(24.6) Let $f_\bullet : X_\bullet \to Y_\bullet$ be a morphism in $\mathcal{F}$, and $f_\bullet = p_\bullet i_\bullet$ a compactification. We define $\bar{c} : f_\bullet^! RR_J \to RR_J(f_\bullet|_J)^!$ to be the composite

$$f_\bullet^! RR_J \xrightarrow{\Upsilon} i_\bullet^* p_\bullet^\times RR_J \xrightarrow{c'} i_\bullet^* RR_J(p_\bullet|_J)^\times \xrightarrow{\mu} RR_J(i_\bullet|_J)^*(p_\bullet|_J)^\times \xrightarrow{\Upsilon} RR_J(f_\bullet|_J)^!.$$

By Lemma 24.5, we have

Lemma 24.7. $\bar{c} : f_\bullet^! RR_J \to RR_J(f_\bullet|_J)^!$ *is an isomorphism of functors from* $D^+_{\mathrm{Lqc}}(Y_\bullet|_J)$ *to* $D^+_{\mathrm{Lqc}}(X_\bullet)$.

Chapter 25
Equivariant Grothendieck's Duality

Theorem 25.1 (Grothendieck's duality). *Let $f : X \to Y$ be a proper morphism of noetherian schemes. For $\mathbb{F} \in D_{\mathrm{Qch}}(X)$ and $\mathbb{G} \in D_{\mathrm{Qch}}^{+}(Y)$, The canonical map*

$$\Theta(f) : Rf_* R\underline{\mathrm{Hom}}_{\mathcal{O}_X}^{\bullet}(\mathbb{F}, f^{\times}\mathbb{G}) \xrightarrow{H} R\underline{\mathrm{Hom}}_{\mathcal{O}_Y}^{\bullet}(Rf_*\mathbb{F}, Rf_* f^{\times}\mathbb{G})$$
$$\xrightarrow{\varepsilon} R\underline{\mathrm{Hom}}_{\mathcal{O}_Y}^{\bullet}(Rf_*\mathbb{F}, \mathbb{G})$$

is an isomorphism.

Proof. As pointed out in [35, section 6], this is an immediate consequence of the open immersion base change [41, Theorem 2]. □

Theorem 25.2 (Equivariant Grothendieck's duality). *Let I be a small category, and $f_{\bullet} : X_{\bullet} \to Y_{\bullet}$ a morphism in $\mathcal{P}(I, \underline{\mathrm{Sch}}/\mathbb{Z})$. If $Y_{\bullet}$ is noetherian with flat arrows and $f_{\bullet}$ is proper cartesian, then the composite*

$$\Theta(f_{\bullet}) : R(f_{\bullet})_* R\underline{\mathrm{Hom}}_{\mathrm{Mod}(X_{\bullet})}^{\bullet}(\mathbb{F}, f_{\bullet}^{\times}\mathbb{G}) \xrightarrow{H} R\underline{\mathrm{Hom}}_{\mathrm{Mod}(Y_{\bullet})}^{\bullet}(R(f_{\bullet})_*\mathbb{F}, R(f_{\bullet})_* f_{\bullet}^{\times}\mathbb{G})$$
$$\xrightarrow{\varepsilon} R\underline{\mathrm{Hom}}_{\mathrm{Mod}(Y_{\bullet})}^{\bullet}(R(f_{\bullet})_*\mathbb{F}, \mathbb{G})$$

is an isomorphism for $\mathbb{F} \in D_{\mathrm{Qch}}(X_{\bullet})$ and $\mathbb{G} \in \mathcal{D}^{+}(Y_{\bullet})$.

Proof. It suffices to show that $(?)_i \Theta(f_{\bullet})$ is an isomorphism for $i \in \mathrm{ob}(I)$. By Lemma 1.39 and Lemma 18.7, **2**, it is easy to see that the composite

$$(?)_i R(f_{\bullet})_* R\underline{\mathrm{Hom}}_{\mathrm{Mod}(X_{\bullet})}^{\bullet}(\mathbb{F}, f_{\bullet}^{\times}\mathbb{G}) \xrightarrow{c} R(f_i)_*(?)_i R\underline{\mathrm{Hom}}_{\mathrm{Mod}(X_{\bullet})}^{\bullet}(\mathbb{F}, f_{\bullet}^{\times}\mathbb{G})$$
$$\xrightarrow{H_i} R(f_i)_* R\underline{\mathrm{Hom}}_{\mathrm{Mod}(X_i)}^{\bullet}(\mathbb{F}_i, (?)_i f_{\bullet}^{\times}\mathbb{G}) \xrightarrow{\xi} R(f_i)_* R\underline{\mathrm{Hom}}_{\mathrm{Mod}(X_i)}^{\bullet}(\mathbb{F}_i, f_i^{\times}\mathbb{G}_i)$$
$$\xrightarrow{\Theta(f_i)} R\underline{\mathrm{Hom}}_{\mathrm{Mod}(Y_i)}^{\bullet}(R(f_i)_*\mathbb{F}_i, \mathbb{G}_i)$$

J. Lipman, M. Hashimoto, *Foundations of Grothendieck Duality for Diagrams*   429
of Schemes, Lecture Notes in Mathematics 1960,
© Springer-Verlag Berlin Heidelberg 2009

agrees with the composite

$$(?)_i R(f_\bullet)_* R \underline{\mathrm{Hom}}^\bullet_{\mathrm{Mod}(X_\bullet)}(\mathbb{F}, f_\bullet^\times \mathbb{G}) \xrightarrow{(?)_i \Theta(f_\bullet)} (?)_i R \underline{\mathrm{Hom}}^\bullet_{\mathrm{Mod}(Y_\bullet)}(R(f_\bullet)_* \mathbb{F}, \mathbb{G})$$
$$\xrightarrow{H_i} R \underline{\mathrm{Hom}}^\bullet_{\mathrm{Mod}(Y_i)}((?)_i R(f_\bullet)_* \mathbb{F}, \mathbb{G}_i) \xrightarrow{c} R \underline{\mathrm{Hom}}^\bullet_{\mathrm{Mod}(Y_i)}(R(f_i)_* \mathbb{F}_i, \mathbb{G}_i).$$

Consider the first composite map. By Lemma 13.9, H_i is an isomorphism. By Lemma 18.5, ξ is an isomorphism. By Theorem 25.1, $\Theta(f_i)$ is an isomorphism. Hence the first composite is an isomorphism, and so is the second.

Consider the second composite map. By Lemma 8.7, $R(f_\bullet)_* \mathbb{F} \in D_{\mathrm{Qch}}(Y_\bullet)$. So the second map H_i is an isomorphism by Lemma 13.9. So the first map $(?)_i \Theta(f_\bullet)$ must be an isomorphism. This is what we wanted to prove. $\square$

Chapter 26
Morphisms of Finite Flat Dimension

(26.1) Let $((?)^*, (?)_*)$ be a monoidal adjoint pair of almost-pseudofunctors over a category $\mathcal{S}$. For a morphism $f : X \to Y$ in $\mathcal{S}$, we define the *projection morphism* $\Pi = \Pi(f)$ to be the composite

$$f_*a \otimes b \xrightarrow{u} f_*f^*(f_*a \otimes b) \xrightarrow{\Delta} f_*(f^*f_*a \otimes f^*b) \xrightarrow{\varepsilon} f_*(a \otimes f^*b),$$

where $a \in X_*$ and $b \in Y_*$ (see Chapter 1 for the notation).

Lemma 26.2. *Let the notation be as above, and $f : X \to Y$ and $g : Y \to Z$ be morphisms in $\mathcal{S}$. For $x \in X_*$ and $z \in Z_*$, the composite*

$$(gf)_*x \otimes z \xrightarrow{c} g_*(f_*x) \otimes z \xrightarrow{\Pi(g)} g_*(f_*x \otimes g^*z) \xrightarrow{\Pi(f)} g_*f_*(x \otimes f^*g^*z)$$

$$\xrightarrow{c^{-1}} (gf)_*(x \otimes f^*g^*z) \xrightarrow{d} (gf)_*(x \otimes (gf)^*z)$$

agrees with $\Pi(gf)$.

Proof. Left to the reader. □

(26.3) Let I be a small category, S a scheme, and $f_\bullet : X_\bullet \to Y_\bullet$ a morphism in $\mathcal{P}(I, \underline{\mathrm{Sch}}/S)$.

Lemma 26.4 (Projection Formula). *Assume that $f_\bullet$ is concentrated. Then the natural map*

$$\Pi = \Pi(f_\bullet) : (Rf_\bullet)_*\mathbb{F} \otimes^{\bullet,L}_{\mathcal{O}_{Y_\bullet}} \mathbb{G} \to (Rf_\bullet)_*(\mathbb{F} \otimes^{\bullet,L}_{\mathcal{O}_{X_\bullet}} Lf_\bullet^*\mathbb{G})$$

is an isomorphism for $\mathbb{F} \in D_{\mathrm{Lqc}}(X_\bullet)$ and $\mathbb{G} \in D_{\mathrm{Lqc}}(Y_\bullet)$.

Proof. For each $i \in \mathrm{ob}(I)$, the composite

$$R(f_i)_*\mathbb{F}_i \otimes^{\bullet,L}_{\mathcal{O}_{Y_i}} \mathbb{G}_i \xrightarrow{\Pi(f_i)} R(f_i)_*(\mathbb{F}_i \otimes^{\bullet,L}_{\mathcal{O}_{X_i}} Lf_i^*\mathbb{G}_i) \xrightarrow{\theta(f_\bullet, i)} R(f_i)_*(\mathbb{F}_i \otimes^{\bullet,L}_{\mathcal{O}_{X_i}} (?)_i Lf_\bullet^*\mathbb{G})$$

$$\xrightarrow{m} R(f_i)_*(?)_i(\mathbb{F} \otimes^{\bullet,L}_{\mathcal{O}_{X_\bullet}} Lf_\bullet^*\mathbb{G}) \xrightarrow{c} (?)_i R(f_\bullet)_*(\mathbb{F} \otimes^{\bullet,L}_{\mathcal{O}_{X_\bullet}} Lf_\bullet^*\mathbb{G})$$

J. Lipman, M. Hashimoto, *Foundations of Grothendieck Duality for Diagrams of Schemes*, Lecture Notes in Mathematics 1960,
© Springer-Verlag Berlin Heidelberg 2009

is an isomorphism by [26, (3.9.4)] and Lemma 8.13, **1**. On the other hand, it is straightforward to check that this composite isomorphism agrees with the composite

$$R(f_i)_*\mathbb{F}_i \otimes^L_{\mathcal{O}_{Y_i}} \mathbb{G}_i \xrightarrow{c^{-1}} (?)_i R(f_\bullet)_* \mathbb{F} \otimes^L_{\mathcal{O}_{Y_i}} \mathbb{G}_i \xrightarrow{m} (?)_i (R(f_\bullet)_* \mathbb{F} \otimes^{\bullet,L}_{\mathcal{O}_{Y_\bullet}} \mathbb{G})$$

$$\xrightarrow{(?)_i \Pi(f_\bullet)} (?)_i R(f_\bullet)_* (\mathbb{F} \otimes^{\bullet,L}_{\mathcal{O}_{X_\bullet}} Lf_\bullet^* \mathbb{G}).$$

It follows that $(?)_i \Pi(f_\bullet)$ is an isomorphism for any $i \in \mathrm{ob}(I)$. Hence, $\Pi(f_\bullet)$ is an isomorphism. $\square$

(26.5) Let $f_\bullet : X_\bullet \to Y_\bullet$ be a morphism in $\mathcal{P}(I, \underline{\mathrm{Sch}}/S)$, and assume that both $X_\bullet$ and $f_\bullet$ are concentrated. Define $\chi = \chi(f_\bullet)$ to be the composite

$$f_\bullet^\times \mathbb{F} \otimes^{\bullet,L}_{\mathcal{O}_{X_\bullet}} Lf_\bullet^* \mathbb{G} \xrightarrow{u} f_\bullet^\times R(f_\bullet)_* (f_\bullet^\times \mathbb{F} \otimes^{\bullet,L}_{\mathcal{O}_{X_\bullet}} Lf_\bullet^* \mathbb{G})$$

$$\xrightarrow{\Pi(f_\bullet)^{-1}} f_\bullet^\times (R(f_\bullet)_* f_\bullet^\times \mathbb{F} \otimes^{\bullet,L}_{\mathcal{O}_{Y_\bullet}} \mathbb{G}) \xrightarrow{\varepsilon} f_\bullet^\times (\mathbb{F} \otimes^{\bullet,L}_{\mathcal{O}_{Y_\bullet}} \mathbb{G}),$$

where $\mathbb{F}, \mathbb{G} \in D_{\mathrm{Lqc}}(Y_\bullet)$.

Utilizing the commutativity as in the proof of Lemma 26.4 and Lemma 18.7, it is not so difficult to show the following.

Lemma 26.6. *Let $f_\bullet : X_\bullet \to Y_\bullet$ be as in (26.5). For a subcategory J of I, the composite*

$$(?)_J f_\bullet^\times \mathbb{F} \otimes^{\bullet,L}_{\mathcal{O}_{X_\bullet|_J}} (?)_J Lf_\bullet^* \mathbb{G} \xrightarrow{\xi \otimes \theta^{-1}} (f_\bullet|_J)^\times \mathbb{F}_J \otimes^{\bullet,L}_{\mathcal{O}_{X_\bullet|_J}} L(f_\bullet|_J)^* \mathbb{G}_J$$

$$\xrightarrow{\chi(f_\bullet|_J)} (f_\bullet|_J)^\times (\mathbb{F}_J \otimes^{\bullet,L}_{\mathcal{O}_{Y_\bullet|_J}} \mathbb{G}_J) \xrightarrow{m} (f_\bullet|_J)^\times (?)_J (\mathbb{F} \otimes^{\bullet,L}_{\mathcal{O}_{Y_\bullet}} \mathbb{G})$$

agrees with

$$(?)_J f_\bullet^\times \mathbb{F} \otimes^{\bullet,L}_{\mathcal{O}_{X_\bullet|_J}} (?)_J Lf_\bullet^* \mathbb{G} \xrightarrow{m} (?)_J (f_\bullet^\times \mathbb{F} \otimes^{\bullet,L}_{\mathcal{O}_{X_\bullet}} Lf_\bullet^* \mathbb{G})$$

$$\xrightarrow{\chi(f_\bullet)} (?)_J f_\bullet^\times (\mathbb{F} \otimes^{\bullet,L}_{\mathcal{O}_{Y_\bullet}} \mathbb{G}) \xrightarrow{\xi} (f_\bullet|_J)^\times (?)_J (\mathbb{F} \otimes^{\bullet,L}_{\mathcal{O}_{Y_\bullet}} \mathbb{G}).$$

Lemma 26.7. *Let $f_\bullet : X_\bullet \to Y_\bullet$ be as in (26.5). The composite*

$$f_\bullet^\times \mathbb{F} \otimes^{\bullet,L}_{\mathcal{O}_{X_\bullet}} Lf_\bullet^* (\mathbb{G} \otimes^{\bullet,L}_{\mathcal{O}_{Y_\bullet}} \mathbb{H}) \xrightarrow{\Delta} f_\bullet^\times \mathbb{F} \otimes^{\bullet,L}_{\mathcal{O}_{X_\bullet}} (Lf_\bullet^* \mathbb{G} \otimes^{\bullet,L}_{\mathcal{O}_{X_\bullet}} Lf_\bullet^* \mathbb{H}) \xrightarrow{\alpha^{-1}}$$

$$(f_\bullet^\times \mathbb{F} \otimes^{\bullet,L}_{\mathcal{O}_{X_\bullet}} Lf_\bullet^* \mathbb{G}) \otimes^{\bullet,L}_{\mathcal{O}_{X_\bullet}} Lf_\bullet^* \mathbb{H} \xrightarrow{\chi} f_\bullet^\times (\mathbb{F} \otimes^{\bullet,L}_{\mathcal{O}_{Y_\bullet}} \mathbb{G}) \otimes^{\bullet,L}_{\mathcal{O}_{X_\bullet}} Lf_\bullet^* \mathbb{H} \xrightarrow{\chi}$$

$$f_\bullet^\times ((\mathbb{F} \otimes^{\bullet,L}_{\mathcal{O}_{Y_\bullet}} \mathbb{G}) \otimes^{\bullet,L}_{\mathcal{O}_{Y_\bullet}} \mathbb{H}) \xrightarrow{\alpha} f_\bullet^\times (\mathbb{F} \otimes^{\bullet,L}_{\mathcal{O}_{Y_\bullet}} (\mathbb{G} \otimes^{\bullet,L}_{\mathcal{O}_{Y_\bullet}} \mathbb{H}))$$

agrees with χ.

Lemma 26.8. *Let S, I and σ be as in* (19.1). *For $\mathbb{F} \in D_{\mathrm{Lqc}}(Y_\bullet)$, the composite*

$$(g'_\bullet)^* f_\bullet^\times \mathbb{F} \otimes^{\bullet,L}_{\mathcal{O}_{X'_\bullet}} (g'_\bullet)^* L f_\bullet^* \mathbb{G} \xrightarrow{\Delta^{-1}} (g'_\bullet)^* (f_\bullet^\times \mathbb{F} \otimes^{\bullet,L}_{\mathcal{O}_{X_\bullet}} L f_\bullet^* \mathbb{G})$$

$$\xrightarrow{\chi} (g'_\bullet)^* f_\bullet^\times (\mathbb{F} \otimes^{\bullet,L}_{\mathcal{O}_{Y_\bullet}} \mathbb{G}) \xrightarrow{\zeta(\sigma)} (f'_\bullet)^\times g_\bullet^* (\mathbb{F} \otimes^{\bullet,L}_{\mathcal{O}_{Y_\bullet}} \mathbb{G})$$

agrees with

$$(g'_\bullet)^* f_\bullet^\times \mathbb{F} \otimes^{\bullet,L}_{\mathcal{O}_{X'_\bullet}} (g'_\bullet)^* L f_\bullet^* \mathbb{G} \xrightarrow{\zeta(\sigma) \otimes d} (f'_\bullet)^\times g_\bullet^* \mathbb{F} \otimes^{\bullet,L}_{\mathcal{O}_{X'_\bullet}} L(f'_\bullet)^* g_\bullet^* \mathbb{G}$$

$$\xrightarrow{\chi} (f'_\bullet)^\times (g_\bullet^* \mathbb{F} \otimes^{\bullet,L}_{\mathcal{O}_{Y'_\bullet}} g_\bullet^* \mathbb{G}) \xrightarrow{\Delta^{-1}} (f'_\bullet)^\times g_\bullet^* (\mathbb{F} \otimes^{\bullet,L}_{\mathcal{O}_{Y_\bullet}} \mathbb{G}).$$

Lemma 26.9. *Let $f_\bullet : X_\bullet \to Y_\bullet$ and $g_\bullet : Y_\bullet \to Z_\bullet$ be morphisms in $\mathcal{P}(I, \underline{\mathrm{Sch}}/S)$. Assume that $X_\bullet$, $Y_\bullet$ and $g_\bullet$ are concentrated. Then the composite*

$$(g_\bullet f_\bullet)^\times \mathbb{F} \otimes^{\bullet,L}_{\mathcal{O}_{X_\bullet}} (g_\bullet f_\bullet)^* \mathbb{G} \cong f_\bullet^\times g_\bullet^\times \mathbb{F} \otimes^{\bullet,L}_{\mathcal{O}_{X_\bullet}} f_\bullet^* g_\bullet^* \mathbb{G} \xrightarrow{\chi(f_\bullet)} f_\bullet^\times (g_\bullet^\times \mathbb{F} \otimes^{\bullet,L}_{\mathcal{O}_{Y_\bullet}} g_\bullet^* \mathbb{G})$$

$$\xrightarrow{\chi(g_\bullet)} f_\bullet^\times g_\bullet^\times (\mathbb{F} \otimes^{\bullet,L}_{\mathcal{O}_{Z_\bullet}} \mathbb{G}) \cong (g_\bullet f_\bullet)^\times (\mathbb{F} \otimes^{\bullet,L}_{\mathcal{O}_{Z_\bullet}} \mathbb{G})$$

agrees with $\chi(g_\bullet f_\bullet)$.

The proof of the lemmas above are left to the reader.

(26.10)   Let the notation be as in Theorem 20.4. Let $i_\bullet : U_\bullet \to X_\bullet$ be a morphism in $\mathcal{I}$, and $p_\bullet : X_\bullet \to Y_\bullet$ a morphism in $\mathcal{P}$.

We define $\bar{\chi} = \bar{\chi}(p_\bullet, i_\bullet)$ to be the composite

$$\bar{\chi} : i_\bullet^* p_\bullet^\times \mathbb{F} \otimes^{\bullet,L}_{\mathcal{O}_{U_\bullet}} L(p_\bullet i_\bullet)^* \mathbb{G} \xrightarrow{d^{-1}} i_\bullet^* p_\bullet^\times \mathbb{F} \otimes^{\bullet,L}_{\mathcal{O}_{U_\bullet}} i_\bullet^* L p_\bullet^* \mathbb{G}$$

$$\xrightarrow{\Delta^{-1}} i_\bullet^* (p_\bullet^\times \mathbb{F} \otimes^{\bullet,L}_{\mathcal{O}_{X_\bullet}} L p_\bullet^* \mathbb{G}) \xrightarrow{i_\bullet^* \chi} i_\bullet^* p_\bullet^\times (\mathbb{F} \otimes^{\bullet,L}_{\mathcal{O}_{Y_\bullet}} \mathbb{G}).$$

Lemma 26.11. *Let $f_\bullet$ be a morphism in $\mathcal{F}$, and $f_\bullet = p_\bullet i_\bullet = q_\bullet j_\bullet$ an independence square. Then the composite*

$$i_\bullet^* p_\bullet^\times \mathbb{F} \otimes^{\bullet,L}_{\mathcal{O}_{U_\bullet}} L(p_\bullet i_\bullet)^* \mathbb{G} \xrightarrow{\bar{\chi}} i_\bullet^* p_\bullet^\times (\mathbb{F} \otimes^{\bullet,L}_{\mathcal{O}_{Y_\bullet}} \mathbb{G}) \xrightarrow{\Upsilon(p_\bullet i_\bullet = q_\bullet j_\bullet)} j_\bullet^* q_\bullet^\times (\mathbb{F} \otimes^{\bullet,L}_{\mathcal{O}_{Y_\bullet}} \mathbb{G})$$

agrees with

$$i_\bullet^* p_\bullet^\times \mathbb{F} \otimes^{\bullet,L}_{\mathcal{O}_{U_\bullet}} L(p_\bullet i_\bullet)^* \mathbb{G} \xrightarrow{\Upsilon \otimes 1} j_\bullet^* q_\bullet^\times \mathbb{F} \otimes^{\bullet,L}_{\mathcal{O}_{U_\bullet}} L(q_\bullet j_\bullet)^* \mathbb{G} \xrightarrow{\bar{\chi}} j_\bullet^* q_\bullet^\times (\mathbb{F} \otimes^{\bullet,L}_{\mathcal{O}_{Y_\bullet}} \mathbb{G}).$$

Proof. As Υ is constructed from ζ and d by definition, the assertion follows easily from Lemma 26.8 and Lemma 26.9. $\square$

(26.12)   Let $f_{\bullet} : X_{\bullet} \to Y_{\bullet}$ be a morphism in $\mathcal{F}$. We define $\bar{\chi}(f_{\bullet})$ to be $\bar{\chi}(p_{\bullet}, i_{\bullet})$, where $f_{\bullet} = p_{\bullet} i_{\bullet}$ is the (fixed) compactification of $f_{\bullet}$.

By Lemma 26.11, $\bar{\chi}(f_{\bullet})$ is an isomorphism if and only if there exists some compactification $f_{\bullet} = q_{\bullet} j_{\bullet}$ such that $\bar{\chi}(q_{\bullet}, j_{\bullet})$ is an isomorphism.

Lemma 26.13. *Let the notation be as in* Theorem 20.4, *and* $f_{\bullet} : X_{\bullet} \to Y_{\bullet}$ *and* $g_{\bullet} : Y_{\bullet} \to Z_{\bullet}$ *morphisms in* $\mathcal{F}$. *Then the composite*

$$(g_{\bullet} f_{\bullet})^! \mathbb{F} \otimes_{\mathcal{O}_{X_{\bullet}}}^{\bullet, L} L(g_{\bullet} f_{\bullet})^* \mathbb{G} \cong f_{\bullet}^! g_{\bullet}^! \mathbb{F} \otimes_{\mathcal{O}_{X_{\bullet}}}^{\bullet, L} L f_{\bullet}^* L g_{\bullet}^* \mathbb{G} \xrightarrow{\bar{\chi}(f_{\bullet})} f_{\bullet}^! (g_{\bullet}^! \mathbb{F} \otimes_{\mathcal{O}_{Y_{\bullet}}}^{\bullet, L} L g_{\bullet}^* \mathbb{G})$$

$$\xrightarrow{\bar{\chi}(g_{\bullet})} f_{\bullet}^! g_{\bullet}^! (\mathbb{F} \otimes_{\mathcal{O}_{Z_{\bullet}}}^{\bullet, L} \mathbb{G}) \cong (g_{\bullet} f_{\bullet})^! (\mathbb{F} \otimes_{\mathcal{O}_{Z_{\bullet}}}^{\bullet, L} \mathbb{G})$$

agrees with $\bar{\chi}(g_{\bullet} f_{\bullet})$.

Theorem 26.14. *Let the notation be as in* Theorem 20.4, *and* $f_{\bullet} : X_{\bullet} \to Y_{\bullet}$ *a morphism in* $\mathcal{F}$. *If* $f_{\bullet}$ *is of finite flat dimension, then*

$$\bar{\chi}(f_{\bullet}) : f_{\bullet}^! \mathbb{F} \otimes_{\mathcal{O}_{Y_{\bullet}}}^{\bullet, L} L f_{\bullet}^* \mathbb{G} \to f_{\bullet}^! (\mathbb{F} \otimes_{\mathcal{O}_{X_{\bullet}}}^{\bullet, L} \mathbb{G})$$

is an isomorphism for $\mathbb{F}, \mathbb{G} \in D_{\mathrm{Lqc}}^+(Y_{\bullet})$.

Proof. Let $f_{\bullet} = p_{\bullet} i_{\bullet}$ be a compactification of $f_{\bullet}$. It suffices to show that $\bar{\chi}(p_{\bullet}, i_{\bullet})$ is an isomorphism. In view of Lemma 26.7, we may assume that $\mathbb{F} = \mathcal{O}_{Y_{\bullet}}$. Then in view of Proposition 18.14 and Lemma 26.6, it suffices to show that

$$\bar{\chi}(f_j) : i_j^* p_j^\times \mathcal{O}_{Y_j} \otimes_{\mathcal{O}_{X_j}}^{\bullet, L} L f_j^* \mathbb{G}_i \to i_j^* p_j^\times (\mathcal{O}_{Y_j} \otimes_{\mathcal{O}_{Y_j}}^{\bullet, L} \mathbb{G}_j)$$

is an isomorphism for any $j \in \mathrm{ob}(I)$. So we may assume that $I = j$.

By the flat base change theorem and Lemma 26.8, the question is local on Y_j. Clearly, the question is local on X_j. Hence we may assume that Y_j and X_j are affine. Set $f = f_j$, $Y = Y_j$, and $X = X_j$. Note that f is a closed immersion defined by an ideal of finite projective dimension, followed by an affine n-space.

By Lemma 26.13, it suffices to prove that $\bar{\chi}(f)$ is an isomorphism if f is a closed immersion defined by an ideal of finite projective dimension or an affine n-space. Both cases are proved easily, using [35, Theorem 5.4] (note that an affine n-space is an open subscheme of a projective n-space).   □

Chapter 27
Cartesian Finite Morphisms

(27.1) Let I be a small category, S a scheme, and $f_\bullet : X_\bullet \to Y_\bullet$ a morphism in $\mathcal{P}(I, \underline{\mathrm{Sch}}/S)$. Let Z denote the ringed site $(\mathrm{Zar}(Y_\bullet), (f_\bullet)_*(\mathcal{O}_{X_\bullet}))$. Assume that $Y_\bullet$ is locally noetherian. There are obvious admissible ringed continuous functors $i : \mathrm{Zar}(Y_\bullet) \to Z$ and $g : Z \to \mathrm{Zar}(X_\bullet)$ such that $gi = f_\bullet^{-1}$. If $f_\bullet$ is affine, then $g_\# : \mathrm{Mod}(Z) \to \mathrm{Mod}(X_\bullet)$ is an exact functor, as can be seen easily.

Lemma 27.2. *If $f_\bullet$ is affine, then the counit*

$$\varepsilon : g_\# R g^\# \mathbb{F} \to \mathbb{F}$$

is an isomorphism for $\mathbb{F} \in D^+_{\mathrm{Lqc}}(X_\bullet)$.

Proof. The construction of ε is compatible with restrictions. So we may assume that $f_\bullet = f : X \to Y$ is an affine morphism of single schemes. Further, the question is local on Y, and hence we may assume that $Y = \mathrm{Spec}\, A$ is affine. As f is affine, $X = \mathrm{Spec}\, B$ is affine. By Lemma 14.3, we may assume that $\mathbb{F} = F_X \mathbb{G}$ for some $\mathbb{G} \in D(\mathrm{Qch}(X))$. In view of Lemma 14.6, it suffices to show that $\varepsilon : g_\# g^\# \mathbb{G} \to \mathbb{G}$ is an isomorphism if $\mathbb{G}$ is a K-injective complex in $\mathrm{Qch}(X)$.

To verify this, it suffices to show that $\varepsilon : g_\# g^\# \mathcal{M} \to \mathcal{M}$ is an isomorphism for $\mathcal{M} \in \mathrm{Qch}(X)$. By Lemma 7.19, $f_* = i^\# g^\#$ on $\mathrm{Qch}(X)$ respects coproducts and is exact. Since $i^\#$ respects coproducts and is faithful exact, $g^\#$ respects coproducts and is exact. So $g_\# g^\# : \mathrm{Qch}(X) \to \mathrm{Qch}(X)$ respects coproducts and is exact.

Since X is affine, there is an exact sequence of the form

$$\mathcal{O}_X^{(J)} \to \mathcal{O}_X^{(I)} \to \mathcal{M} \to 0.$$

So we may assume that $\mathcal{M} = \mathcal{O}_X$. But this case is trivial. $\qquad\square$

J. Lipman, M. Hashimoto, *Foundations of Grothendieck Duality for Diagrams of Schemes*, Lecture Notes in Mathematics 1960,
© Springer-Verlag Berlin Heidelberg 2009

(27.3) Let I, S, $f_\bullet : X_\bullet \to Y_\bullet$, Z, g, and i be as in (27.1). Assume that $f_\bullet$ is finite cartesian.

We say that an $\mathcal{O}_Z$-module $\mathcal{M}$ is locally quasi-coherent (resp. quasi-coherent, coherent) if $i^\#\mathcal{M}$ is. The corresponding full subcategory of $\mathrm{Mod}(Z)$ is denoted by $\mathrm{Lqc}(Z)$ (resp. $\mathrm{Qch}(Z)$, $\mathrm{Coh}(Z)$).

Lemma 27.4. *Let the notation be as above. Then an $\mathcal{O}_Z$-module $\mathcal{M}$ is locally quasi-coherent if and only if for any $j \in \mathrm{ob}(I)$ and any affine open subscheme U of Y_j, there exists an exact sequence of $((\mathcal{O}_Z)_j)|_U$-modules*

$$(((\mathcal{O}_Z)_j)|_U)^{(T)} \to (((\mathcal{O}_Z)_j)|_U)^{(\Sigma)} \to \mathcal{M}_j|_U \to 0.$$

Proof. As we assume that $f_\bullet$ is finite cartesian, $\mathcal{O}_Z$ is coherent. Hence the existence of such exact sequences implies that $\mathcal{M}$ is locally quasi-coherent.

We prove the converse. Let $j \in \mathrm{ob}(I)$ and U an affine open subset of Y_j. Set $C := \Gamma(U, (\mathcal{O}_Z)_j) = \Gamma(f_j^{-1}(U), \mathcal{O}_{X_j})$ and $M := \Gamma(U, \mathcal{M}_j)$. There is a canonical map $(g_j|_{f_j^{-1}(U)})^\#(\tilde{M}) \to \mathcal{M}_j|_U$, where $\tilde{M}$ is the quasi-coherent sheaf over $\mathrm{Spec}\, C \subset X_j$ associated with the C-module M. When we apply $(i_j|_U)^\#$ to this map, we get $\tilde{M}_0 \to ((i^\#\mathcal{M})_j)|_U$, where M_0 is M viewed as a $\Gamma(U, \mathcal{O}_{Y_j})$-module. This is an isomorphism, since $(i^\#\mathcal{M})_j|_U$ is quasi-coherent and U is affine. As $(i_j|_U)^\#$ is faithful and exact, we have $(g_j|_{f_j^{-1}(U)})^\#(\tilde{M}) \cong \mathcal{M}_j|_U$.

Take an exact sequence of the form

$$C^{(T)} \to C^{(\Sigma)} \to M \to 0.$$

Applying the exact functor $(g_j|_{f_j^{-1}(U)})^\# \circ \tilde{?}$, we get an exact sequence of the desired type. $\square$

Corollary 27.5. *Under the same assumption as in the lemma, the functor $g_\#$ preserves local quasi-coherence.*

Proof. As $g_\#$ is compatible with restrictions, we may assume that I consists of one object and one morphism. Further, as the question is local, we may assume that $Y_\bullet = Y$ is an affine scheme. By the lemma, it suffices to show that $g_\# \mathcal{O}_Z$ is quasi-coherent, since $g_\#$ is exact and preserves direct sums. As $g_\# \mathcal{O}_Z = g_\# g^\# \mathcal{O}_X \cong \mathcal{O}_X$, we are done. $\square$

Lemma 27.6. *Let the notation be as above. The unit of adjunction $u : \mathbb{F} \to Rg_\# g_\# \mathbb{F}$ is an isomorphism for $\mathbb{F} \in D^+_{\mathrm{Lqc}}(Z)$.*

Proof. We may assume that I consists of one object and one morphism, and $Y_\bullet = Y$ is affine. By the way-out lemma, we may assume that $\mathbb{F}$ is a single quasi-coherent sheaf. Then by Corollary 27.5, $g_\# \mathbb{F}$ is quasi-coherent, and is $g^\#$-acyclic. So it suffices to show that $u : \mathcal{M} \to g^\# g_\# \mathcal{M}$ is an isomorphism for a quasi-coherent sheaf $\mathcal{M}$ on Z. Note that $g^\# g_\# : \mathrm{Qch}(Z) \to \mathrm{Qch}(Z)$

respects coproducts. By Lemma 27.4 and the five lemma, we may assume that $\mathbb{F} = \mathcal{O}_Z = g^\# \mathcal{O}_X$ and it suffices to prove that $ug^\# : g^\# \mathcal{O}_X \to g^\# g_\# g^\# \mathcal{O}_X$ is an isomorphism. As $\mathrm{id} = (g^\# \varepsilon)(ug^\#)$ and ε is an isomorphism, we are done. $\square$

For $\mathcal{N} \in \mathrm{Mod}(Y_\bullet)$ and $\mathcal{M} \in \mathrm{Mod}(Z)$, the sheaf $\underline{\mathrm{Hom}}_{\mathcal{O}_{Y_\bullet}}(\mathcal{M}, \mathcal{N})$ on $Y_\bullet$ has a structure of $\mathcal{O}_Z$-module, and it belongs to $\mathrm{Mod}(Z)$. There is an obvious isomorphism of functors

$$\kappa : i^\# \underline{\mathrm{Hom}}_{\mathcal{O}_{Y_\bullet}}(\mathcal{M}, \mathcal{N}) \cong \underline{\mathrm{Hom}}_{\mathcal{O}_{Y_\bullet}}(i^\# \mathcal{M}, \mathcal{N}).$$

For $\mathcal{M}, \mathcal{M}' \in \mathrm{Mod}(Z)$, there is a natural map

$$\upsilon : \underline{\mathrm{Hom}}_{\mathrm{Mod}(Z)}(\mathcal{M}, \mathcal{M}') \to \underline{\mathrm{Hom}}_{\mathcal{O}_{Y_\bullet}}(\mathcal{M}, i^\# \mathcal{M}').$$

Note that the composite

$$i^\# \underline{\mathrm{Hom}}_{\mathrm{Mod}(Z)}(\mathcal{M}, \mathcal{M}') \xrightarrow{i^\# \upsilon} i^\# \underline{\mathrm{Hom}}_{\mathcal{O}_{Y_\bullet}}(\mathcal{M}, i^\# \mathcal{M}') \xrightarrow{\kappa} \underline{\mathrm{Hom}}_{\mathcal{O}_{Y_\bullet}}(i^\# \mathcal{M}, i^\# \mathcal{M}')$$

agrees with H.

(27.7) Let I, S, $f_\bullet : X_\bullet \to Y_\bullet$, Z, g, and i be as in (27.1). Assume that $f_\bullet$ is finite cartesian, and $Y_\bullet$ has flat arrows. Define $f_\bullet^\natural : \mathcal{D}^+(Y_\bullet) \to D(X_\bullet)$ by

$$f_\bullet^\natural(\mathbb{F}) := g_\# R\,\underline{\mathrm{Hom}}^\bullet_{\mathcal{O}_{Y_\bullet}}(\mathcal{O}_Z, \mathbb{F}).$$

As $f_\bullet$ is finite cartesian, $\mathcal{O}_Z$ is coherent. By Lemma 13.10, $i^\# R\,\underline{\mathrm{Hom}}^\bullet_{\mathcal{O}_{Y_\bullet}}(\mathcal{O}_Z, \mathbb{F}) \in \mathcal{D}^+(Y_\bullet)$. It follows that $f_\bullet^\natural(\mathbb{F}) \in \mathcal{D}^+(X_\bullet)$, and $f_\bullet^\natural$ is a functor from $\mathcal{D}^+(Y_\bullet)$ to $\mathcal{D}^+(X_\bullet)$.

Define $\varepsilon : R(f_\bullet)_* f_\bullet^\natural \to \mathrm{Id}_{\mathcal{D}^+(Y_\bullet)}$ by

$$R(f_\bullet)_* f_\bullet^\natural \mathbb{F} = i^\# Rg^\# g_\# R\,\underline{\mathrm{Hom}}^\bullet_{\mathcal{O}_{Y_\bullet}}(\mathcal{O}_Z, \mathbb{F}) \xrightarrow{u^{-1}} i^\# R\,\underline{\mathrm{Hom}}^\bullet_{\mathcal{O}_{Y_\bullet}}(\mathcal{O}_Z, \mathbb{F})$$

$$\xrightarrow{\kappa} R\,\underline{\mathrm{Hom}}^\bullet_{\mathcal{O}_{Y_\bullet}}(i^\# \mathcal{O}_Z, \mathbb{F}) = R\,\underline{\mathrm{Hom}}^\bullet_{\mathcal{O}_{Y_\bullet}}((f_\bullet)_* \mathcal{O}_{X_\bullet}, \mathbb{F}) \xrightarrow{\eta} R\,\underline{\mathrm{Hom}}^\bullet_{\mathcal{O}_{Y_\bullet}}(\mathcal{O}_{Y_\bullet}, \mathbb{F}) \cong \mathbb{F}.$$

Define $u : \mathrm{Id}_{\mathcal{D}^+(X_\bullet)} \to f_\bullet^\natural R(f_\bullet)_*$ by

$$\mathbb{F} \cong R\,\underline{\mathrm{Hom}}^\bullet_{\mathcal{O}_{X_\bullet}}(\mathcal{O}_{X_\bullet}, \mathbb{F}) \xrightarrow{\varepsilon^{-1}} g_\# Rg^\# R\,\underline{\mathrm{Hom}}^\bullet_{\mathcal{O}_{X_\bullet}}(\mathcal{O}_{X_\bullet}, \mathbb{F})$$

$$\xrightarrow{H} g_\# R\,\underline{\mathrm{Hom}}^\bullet_{\mathrm{Mod}(Z)}(g^\# \mathcal{O}_{X_\bullet}, Rg^\# \mathbb{F}) \xrightarrow{\upsilon} g_\# R\,\underline{\mathrm{Hom}}^\bullet_{\mathcal{O}_{Y_\bullet}}(\mathcal{O}_Z, R(f_\bullet)_* \mathbb{F}) = f_\bullet^\natural R(f_\bullet)_* \mathbb{F}.$$

Theorem 27.8. *Let the notation be as above. Then $f_\bullet^\natural$ is right adjoint to $R(f_\bullet)_*$, and ε and u defined above are the counit and unit of adjunction, respectively. In particular, if, moreover, $X_\bullet$ is quasi-compact, then $f_\bullet^\natural$ is isomorphic to $f_\bullet^\times$.*

Proof. It is easy to see that the composite

$$R\underline{\mathrm{Hom}}^\bullet_{\mathcal{O}_{Y_\bullet}}(\mathcal{O}_Z,\mathbb{F}) \cong R\underline{\mathrm{Hom}}^\bullet_{\mathrm{Mod}(Z)}(\mathcal{O}_Z, R\underline{\mathrm{Hom}}^\bullet_{\mathcal{O}_{Y_\bullet}}(\mathcal{O}_Z,\mathbb{F}))$$

$$\xrightarrow{v} R\underline{\mathrm{Hom}}^\bullet_{\mathcal{O}_{Y_\bullet}}(\mathcal{O}_Z, i^\# R\underline{\mathrm{Hom}}^\bullet_{\mathcal{O}_{Y_\bullet}}(\mathcal{O}_Z,\mathbb{F})) \xrightarrow{\kappa} R\underline{\mathrm{Hom}}^\bullet_{\mathcal{O}_{Y_\bullet}}(\mathcal{O}_Z, R\underline{\mathrm{Hom}}^\bullet_{\mathcal{O}_{Y_\bullet}}(i^\#\mathcal{O}_Z,\mathbb{F}))$$

$$\xrightarrow{\eta} R\underline{\mathrm{Hom}}^\bullet_{\mathcal{O}_{Y_\bullet}}(\mathcal{O}_Z, R\underline{\mathrm{Hom}}^\bullet_{\mathcal{O}_{Y_\bullet}}(\mathcal{O}_{Y_\bullet},\mathbb{F})) \cong R\underline{\mathrm{Hom}}^\bullet_{\mathcal{O}_{Y_\bullet}}(\mathcal{O}_Z,\mathbb{F})$$

is the identity. Utilizing this and Lemma 1.47, $(f_\bullet^\flat \varepsilon) \circ (u f_\bullet^\flat) = \mathrm{id}$ and $(\varepsilon R(f_\bullet)_*) \circ (R(f_\bullet)_* u) = \mathrm{id}$ are checked directly.

The last assertion is obvious, as the right adjoint functor is unique.   □

Chapter 28
Cartesian Regular Embeddings and Cartesian Smooth Morphisms

(28.1) Let I be a small category, S a scheme, and $X_\bullet \in \mathcal{P}(I, \underline{\mathrm{Sch}}/S)$. An $\mathcal{O}_{X_\bullet}$-module sheaf $\mathcal{M} \in \mathrm{Mod}(X_\bullet)$ is said to be *locally free* (resp. *invertible*) if $\mathcal{M}$ is coherent and $\mathcal{M}_i$ is locally free (resp. invertible) for any $i \in \mathrm{ob}(I)$. A *perfect complex* of $X_\bullet$ is a bounded complex in $C^b(\mathrm{Mod}(X_\bullet))$ each of whose terms is locally free.

A point of $X_\bullet$ is a pair (i, x) such that $i \in \mathrm{ob}(I)$ and $x \in X_i$. A *stalk* of a sheaf $\mathcal{M} \in \mathrm{AB}(X_\bullet)$ at the point (i, x) is defined to be $(\mathcal{M}_i)_x$, and we denote it by $\mathcal{M}_{i,x}$.

A *connected component* of $X_\bullet$ is an equivalence class with respect to the equivalence relation of the set of points of $X_\bullet$ generated by the following relations.

1 (i, x) and (i', x') are equivalent if $i = i'$ and x and x' belong to the same connected component of X_i.

2 (i, x) and (i', x') are equivalent if there exists some $\phi : i \to i'$ such that $X_\phi(x') = x$.

We say that $X_\bullet$ is *d-connected* if $X_\bullet$ consists of one connected component (note that the word 'connected' is reserved for componentwise connectedness). If $X_\bullet$ is locally noetherian, then a connected component of $X_\bullet$ is a closed open subdiagram of schemes in a natural way. If this is the case, the rank function $(i, x) \mapsto \mathrm{rank}_{\mathcal{O}_{X_i, x}} \mathcal{F}_{i,x}$ of a locally free sheaf $\mathcal{F}$ is constant on a connected component of $X_\bullet$.

Lemma 28.2. *Let I be a small category, S a scheme, and $X_\bullet \in \mathcal{P}(I, \underline{\mathrm{Sch}}/S)$. Let $\mathbb{F}$ be a perfect complex of $X_\bullet$. Then we have*

1 *The canonical map*

$$H_J : (?)_J R\underline{\mathrm{Hom}}^\bullet_{\mathcal{O}_{X_\bullet}}(\mathbb{F}, \mathbb{G}) \to R\underline{\mathrm{Hom}}^\bullet_{\mathcal{O}_{X_\bullet|_J}}(\mathbb{F}_J, \mathbb{G}_J)$$

is an isomorphism for $\mathbb{G} \in D(X_\bullet)$.

J. Lipman, M. Hashimoto, *Foundations of Grothendieck Duality for Diagrams of Schemes*, Lecture Notes in Mathematics 1960,
© Springer-Verlag Berlin Heidelberg 2009

2 *The canonical map*

$$R\underline{\operatorname{Hom}}_{\mathcal{O}_{X_\bullet}}^\bullet(\mathbb{F}, \mathbb{G}) \otimes_{\mathcal{O}_{X_\bullet}}^{\bullet, L} \mathbb{H} \to R\underline{\operatorname{Hom}}_{\mathcal{O}_{X_\bullet}}^\bullet(\mathbb{F}, \mathbb{G} \otimes_{\mathcal{O}_{X_\bullet}}^{\bullet, L} \mathbb{H})$$

is an isomorphism for $\mathbb{G}, \mathbb{H} \in D(X_\bullet)$.

Proof. **1** It suffices to show that

$$H_J : (?)_J \underline{\operatorname{Hom}}_{\mathcal{O}_{X_\bullet}}^\bullet(\mathbb{F}, \mathbb{G}) \to \underline{\operatorname{Hom}}_{\mathcal{O}_{X_\bullet|_J}}^\bullet(\mathbb{F}_J, \mathbb{G}_J)$$

is an isomorphism of complexes if $\mathbb{G}$ is a K-injective complex in $C(\operatorname{Mod}(X_\bullet))$, since $\mathbb{F}_J$ is K-flat and $\mathbb{G}_J$ is weakly K-injective. The assertion follows immediately by Lemma 6.36.

2 We may assume that $\mathbb{F}$ is a single locally free sheaf. By **1**, we may assume that $X = X_\bullet$ is a single scheme. We may assume that X is affine and $\mathbb{F} = \mathcal{O}_X^n$ for some n. This case is trivial. $\qquad\square$

(28.3) Let I be a small category, S a scheme, and $X_\bullet \in \mathcal{P}(I, \underline{\operatorname{Sch}}/S)$. An $\mathcal{O}_{X_\bullet}$-module $\mathcal{M}$ is said to be *locally of finite projective dimension* if $\mathcal{M}_{i,x}$ is of finite projective dimension as an $\mathcal{O}_{X_i,x}$-module for any point (i,x) of $X_\bullet$. We say that $\mathcal{M}$ has *finite projective dimension* if there exists some non-negative integer d such that $\operatorname{proj.dim}_{\mathcal{O}_{X_i,x}} \mathcal{M}_{i,x} \leq d$ for any point (i,x) of $X_\bullet$.

Lemma 28.4. *Let I be a small category, S a scheme, and $X_\bullet \in \mathcal{P}(I, \underline{\operatorname{Sch}}/S)$. Assume that $X_\bullet$ has flat arrows and is locally noetherian. If $\mathbb{F}$ is a complex in $C(\operatorname{Mod}(X_\bullet))$ with bounded coherent cohomology groups which have finite projective dimension, then the canonical map*

$$R\underline{\operatorname{Hom}}_{\mathcal{O}_{X_\bullet}}^\bullet(\mathbb{F}, \mathbb{G}) \otimes_{\mathcal{O}_{X_\bullet}}^{\bullet, L} \mathbb{H} \to R\underline{\operatorname{Hom}}_{\mathcal{O}_{X_\bullet}}^\bullet(\mathbb{F}, \mathbb{G} \otimes_{\mathcal{O}_{X_\bullet}}^{\bullet, L} \mathbb{H})$$

is an isomorphism for $\mathbb{G}, \mathbb{H} \in D(X_\bullet)$.

Proof. We may assume that $\mathbb{G} = \mathcal{O}_{X_\bullet}$. By the way-out lemma, we may assume that $\mathbb{F}$ is a single coherent sheaf which has finite projective dimension, say d. By Lemma 13.9, it is easy to see that $\underline{\operatorname{Ext}}_{\mathcal{O}_{X_\bullet}}^i(\mathbb{F}, G) = 0$ $(i > d)$ for $G \in \operatorname{Mod}(X_\bullet)$. In particular, $R\underline{\operatorname{Hom}}_{\mathcal{O}_{X_\bullet}}^\bullet(\mathbb{F}, ?)$ is way-out in both directions. On the other hand, as $R\underline{\operatorname{Hom}}_{\mathcal{O}_{X_\bullet}}^\bullet(\mathbb{F}, \mathcal{O}_{X_\bullet})$ has finite flat dimension, and hence $R\underline{\operatorname{Hom}}_{\mathcal{O}_{X_\bullet}}(\mathbb{F}, \mathcal{O}_{X_\bullet}) \otimes_{\mathcal{O}_{X_\bullet}}^{\bullet, L} ?$ is also way-out in both directions. By the way-out lemma, we may assume that $\mathbb{H}$ is a single $\mathcal{O}_{X_\bullet}$-module. By Lemma 13.9, we may assume that $X = X_\bullet$ is a single scheme. The question is local, and we may assume that $X = \operatorname{Spec} A$ is affine. Moreover, we may assume that $\mathbb{F}$ is a complex of sheaves associated with a finite projective resolution of a single finitely generated module. As $\mathbb{F}$ is perfect, the result follows from Lemma 28.2. $\qquad\square$

(28.5) Let S, I, and $X_\bullet$ be as above. For a locally free sheaf $\mathcal{F}$ over $X_\bullet$, we denote $\underline{\mathrm{Hom}}_{\mathcal{O}_{X_\bullet}}(\mathcal{F}, \mathcal{O}_{X_\bullet})$ by $\mathcal{F}^\vee$. It is easy to see that $\mathcal{F}^\vee$ is again locally free. If $\mathcal{L}$ is an invertible sheaf, then

$$\mathcal{O}_{X_\bullet} \xrightarrow{\mathrm{tr}} \underline{\mathrm{Hom}}_{\mathcal{O}_{X_\bullet}}(\mathcal{L}, \mathcal{L}) \cong \mathcal{L}^\vee \otimes_{\mathcal{O}_{X_\bullet}} \mathcal{L}$$

are isomorphisms.

(28.6) Let I be a small category, S a scheme, and $i_\bullet : Y_\bullet \to X_\bullet$ a closed immersion in $\mathcal{P}(I, \underline{\mathrm{Sch}}/S)$. Then the canonical map $\eta : \mathcal{O}_{X_\bullet} \to (i_\bullet)_* \mathcal{O}_{Y_\bullet}$ is an epimorphism in $\mathrm{Lqc}(X_\bullet)$. Set $\mathcal{I} := \mathrm{Ker}\, \eta$. Then $\mathcal{I}$ is a locally quasi-coherent ideal of $\mathcal{O}_{X_\bullet}$. Conversely, if $\mathcal{I}$ is a given locally quasi-coherent ideal of $\mathcal{O}_{X_\bullet}$, then

$$Y_\bullet := \underline{\mathrm{Spec}}_\bullet \, \mathcal{O}_{X_\bullet}/\mathcal{I} \xrightarrow{i_\bullet} X_\bullet$$

is defined appropriately, and $i_\bullet$ is a closed immersion. Thus the isomorphism classes of closed immersions to $X_\bullet$ in the category $\mathcal{P}(I, \underline{\mathrm{Sch}}/S)/X_\bullet$ and locally quasi-coherent ideals of $\mathcal{O}_{X_\bullet}$ are in one-to-one correspondence. We call $\mathcal{I}$ the *defining ideal sheaf* of $Y_\bullet$.

Note that $i_\bullet$ is cartesian if and only if $(i_\bullet)_* \mathcal{O}_{Y_\bullet}$ is equivariant. If $X_\bullet$ has flat arrows, this is equivalent to say that $\mathcal{I}$ is equivariant.

(28.7) Let $X_\bullet$ be locally noetherian. A morphism $i_\bullet : Y_\bullet \to X_\bullet$ is said to be a *regular embedding*, if $i_\bullet$ is a closed immersion such that $i_j : Y_j \to X_j$ is a regular embedding for each $j \in \mathrm{ob}(I)$, or equivalently, $\mathcal{I}$ is locally coherent and $\mathcal{I}_{j,x}$ is a complete intersection ideal of $\mathcal{O}_{X_j,x}$ for any $j \in \mathrm{ob}(I)$ and $x \in X_j$. If this is the case, we say that $\mathcal{I}$ is a *local complete intersection* ideal sheaf.

A cartesian closed immersion $i_\bullet : Y_\bullet \to X_\bullet$ with $X_\bullet$ locally noetherian with flat arrows is a cartesian regular embedding if and only if $i_\bullet^* \mathcal{I}$ is locally free and $\mathcal{I}_{i,x}$ is of finite projective dimension as an $\mathcal{O}_{X_i,x}$-module for any $i \in I$ and $x \in X_i$.

Note that $i_\bullet^* \mathcal{I} \cong \mathcal{I}/\mathcal{I}^2$, and we have

$$\mathrm{ht}_{\mathcal{O}_{X_i,x}} \mathcal{I}_x = \mathrm{rank}_{\mathcal{O}_{Y_i,y}} (i_\bullet^* \mathcal{I})_{i,y}$$

for any point (i, y) of $Y_\bullet$, where $x = i_i(y)$. We call these numbers the codimension of $\mathcal{I}$ at (i, y).

Proposition 28.8. *Let I be a small category, S a scheme, and $i_\bullet : Y_\bullet \to X_\bullet$ a morphism in $\mathcal{P}(I, \underline{\mathrm{Sch}}/S)$. Assume that $X_\bullet$ is locally noetherian with flat arrows and $i_\bullet$ is a cartesian regular embedding. Let $\mathcal{I}$ be the defining ideal of $Y_\bullet$, and assume that $Y_\bullet$ has a constant codimension d. Then we have the following.*

1 $\underline{\mathrm{Ext}}^i_{\mathcal{O}_{X_\bullet}}((i_\bullet)_*\mathcal{O}_{Y_\bullet}, \mathcal{O}_{X_\bullet}) = 0$ *for* $i \neq d$.

2 *The canonical map*

$$\underline{\mathrm{Ext}}^d_{\mathcal{O}_{X_\bullet}}((i_\bullet)_*\mathcal{O}_{Y_\bullet}, \mathcal{O}_{X_\bullet}) \to \underline{\mathrm{Ext}}^d_{\mathcal{O}_{X_\bullet}}((i_\bullet)_*\mathcal{O}_{Y_\bullet}, (i_\bullet)_*\mathcal{O}_{Y_\bullet})$$

is an isomorphism.

3 *The Yoneda algebra*

$$\underline{\mathrm{Ext}}^\bullet_{\mathcal{O}_{Y_\bullet}}((i_\bullet)_*\mathcal{O}_{Y_\bullet}, (i_\bullet)_*\mathcal{O}_{Y_\bullet}) := \bigoplus_{j \geq 0} \underline{\mathrm{Ext}}^j_{\mathcal{O}_{Y_\bullet}}((i_\bullet)_*\mathcal{O}_{Y_\bullet}, (i_\bullet)_*\mathcal{O}_{Y_\bullet})$$

is isomorphic to the exterior algebra $(i_\bullet)_* \bigwedge^\bullet (i_\bullet^* \mathcal{I})^\vee$ *as graded* $\mathcal{O}_{X_\bullet}$-*algebras.*

4 *There is an isomorphism*

$$i_\bullet^\flat \mathcal{O}_{X_\bullet} \cong \bigwedge^d (i_\bullet^* \mathcal{I})^\vee [-d].$$

5 *For* $\mathbb{F} \in \mathcal{D}^+(X_\bullet)$, *there is a functorial isomorphism*

$$i_\bullet^\flat \mathbb{F} \cong \bigwedge^d (i_\bullet^* \mathcal{I})^\vee \otimes^{\bullet, L}_{\mathcal{O}_{Y_\bullet}} Li_\bullet^* \mathbb{F}[-d].$$

Proof. **1** is trivial, since $\mathcal{I}_{i,x}$ is a complete intersection ideal of the local ring $\mathcal{O}_{X_i, x}$ of codimension d for any point (i, x) of $X_\bullet$.

2 Note that $(i_\bullet)_* \mathcal{O}_{Y_\bullet} \cong \mathcal{O}_{X_\bullet}/\mathcal{I}$. From the short exact sequence

$$0 \to \mathcal{I} \to \mathcal{O}_{X_\bullet} \to \mathcal{O}_{X_\bullet}/\mathcal{I} \to 0,$$

we get an isomorphism

$$\underline{\mathrm{Ext}}^1_{\mathcal{O}_{X_\bullet}}((i_\bullet)_*\mathcal{O}_{Y_\bullet}, (i_\bullet)_*\mathcal{O}_{Y_\bullet}) \cong \underline{\mathrm{Hom}}_{\mathcal{O}_{X_\bullet}}(\mathcal{I}, \mathcal{O}_{X_\bullet}/\mathcal{I}) \cong (i_\bullet)_*(i_\bullet^* \mathcal{I})^\vee.$$

The canonical map

$$(i_\bullet)_*(i_\bullet^* \mathcal{I})^\vee \cong \underline{\mathrm{Ext}}^1_{\mathcal{O}_{X_\bullet}}((i_\bullet)_*\mathcal{O}_{Y_\bullet}, (i_\bullet)_*\mathcal{O}_{Y_\bullet}) \hookrightarrow \underline{\mathrm{Ext}}^\bullet_{\mathcal{O}_{X_\bullet}}((i_\bullet)_*\mathcal{O}_{Y_\bullet}, (i_\bullet)_*\mathcal{O}_{Y_\bullet})$$

is uniquely extended to an $\mathcal{O}_{X_\bullet}$-algebra map

$$T_\bullet((i_\bullet)_*(i_\bullet^* \mathcal{I})^\vee) \to \underline{\mathrm{Ext}}^\bullet_{\mathcal{O}_{X_\bullet}}((i_\bullet)_*\mathcal{O}_{Y_\bullet}, (i_\bullet)_*\mathcal{O}_{Y_\bullet}),$$

where $T_\bullet$ denotes the tensor algebra. It suffices to prove that this map is an epimorphism, which induces an isomorphism

$$\bigwedge^\bullet ((i_\bullet)_*(i_\bullet^* \mathcal{I})^\vee) \to \underline{\mathrm{Ext}}^\bullet_{\mathcal{O}_{X_\bullet}}((i_\bullet)_*\mathcal{O}_{Y_\bullet}, (i_\bullet)_*\mathcal{O}_{Y_\bullet}).$$

In fact, the exterior algebra is compatible with base change, and

$$\bigwedge{}^{\bullet}((i_{\bullet})_*(i_{\bullet}^*\mathcal{I})^{\vee}) \cong (i_{\bullet})_*i_{\bullet}^* \bigwedge{}^{\bullet}((i_{\bullet})_*(i_{\bullet}^*\mathcal{I})^{\vee})$$
$$\cong (i_{\bullet})_* \bigwedge{}^{\bullet}((i_{\bullet}^*(i_{\bullet})_*)(i_{\bullet}^*\mathcal{I})^{\vee}) \cong (i_{\bullet})_* \bigwedge{}^{\bullet}(i_{\bullet}^*\mathcal{I})^{\vee}.$$

To verify this, we may assume that $i_{\bullet} : Y_{\bullet} \to X_{\bullet}$ is a morphism of single schemes, $X_{\bullet} = \operatorname{Spec} A$ affine, and $\mathcal{I} = \tilde{I}$ generated by an A-sequence. The proof for this case is essentially the same as [19, Lemma IV.1.1.8], and we omit it.

4 Let Z denote the ringed site $(\operatorname{Zar}(X_{\bullet}), (i_{\bullet})_*\mathcal{O}_{Y_{\bullet}})$, and $g : Z \to \operatorname{Zar}(Y_{\bullet})$ the associated admissible ringed continuous functor. By **2–3**, there is a sequence of isomorphisms in $\operatorname{Coh}(X_{\bullet})$

$$(i_{\bullet})_* \bigwedge{}^{d}(i_{\bullet}^*\mathcal{I})^{\vee} \cong \underline{\operatorname{Ext}}_{\mathcal{O}_{Y_{\bullet}}}^{d}((i_{\bullet})_*\mathcal{O}_{Y_{\bullet}}, (i_{\bullet})_*\mathcal{O}_{Y_{\bullet}}) \cong \underline{\operatorname{Ext}}_{\mathcal{O}_{X_{\bullet}}}^{d}((i_{\bullet})_*\mathcal{O}_{Y_{\bullet}}, \mathcal{O}_{X_{\bullet}}).$$

In view of **1**, there is an isomorphism

$$Rg^{\#} \bigwedge{}^{d}(i_{\bullet}^*\mathcal{I})^{\vee} \cong R\underline{\operatorname{Hom}}_{\mathcal{O}_{X_{\bullet}}}^{\bullet}(\mathcal{O}_Z, \mathcal{O}_{X_{\bullet}})[d]$$

in $D_{\operatorname{Coh}}^b(Z)$. Applying $g_{\#}$ to both sides, we get

$$\bigwedge{}^{d}(i_{\bullet}^*\mathcal{I})^{\vee} \cong i_{\bullet}^{\flat}\mathcal{O}_{X_{\bullet}}[d].$$

5 is an immediate consequence of **4** and Lemma 28.4. $\square$

(28.9) Let I and S be as in (28.6). Let $f_{\bullet} : X_{\bullet} \to Y_{\bullet}$ be a morphism in $\mathcal{P}(I, \underline{\operatorname{Sch}}/S)$. Assume that $f_{\bullet}$ is separated so that the diagonal $\Delta_{X_{\bullet}/Y_{\bullet}} : X_{\bullet} \to X_{\bullet} \times_{Y_{\bullet}} X_{\bullet}$ is a closed immersion. Define $\Omega_{X_{\bullet}/Y_{\bullet}} := i_{\bullet}^*\mathcal{I}$, where $\mathcal{I} := \operatorname{Ker}(\eta : \mathcal{O}_{X_{\bullet} \times_{Y_{\bullet}} X_{\bullet}} \to (\Delta_{X_{\bullet}/Y_{\bullet}})_*\mathcal{O}_{X_{\bullet}})$. Note that $(\Delta_{X_{\bullet}/Y_{\bullet}})_*\Omega_{X_{\bullet}/Y_{\bullet}} \cong \mathcal{I}/\mathcal{I}^2$.

Lemma 28.10. *Let the notation be as above. Then we have*

1   $\Omega_{X_{\bullet}/Y_{\bullet}}$ *is locally quasi-coherent.*
2   *If $f_{\bullet}$ is cartesian, then $\Omega_{X_{\bullet}/Y_{\bullet}}$ is quasi-coherent.*
3   *For $i \in \operatorname{ob}(I)$, there is a canonical isomorphism $\Omega_{X_i/Y_i} \cong (\Omega_{X_{\bullet}/Y_{\bullet}})_i$.*

Proof. Easy. $\square$

Theorem 28.11. *Let I be a finite ordered category, and $f_{\bullet} : X_{\bullet} \to Y_{\bullet}$ a morphism in $\mathcal{P}(I, \underline{\operatorname{Sch}})$. Assume that $Y_{\bullet}$ is noetherian with flat arrows, and $f_{\bullet}$ is separated cartesian smooth of finite type. Assume that $f_{\bullet}$ has a constant relative dimension d. Then for any $\mathbb{F} \in D_{\operatorname{Lqc}}^{+}(Y_{\bullet})$, there is a functorial isomorphism*

$$\bigwedge{}^{d} \Omega_{X_{\bullet}/Y_{\bullet}}[d] \otimes_{\mathcal{O}_{X_{\bullet}}}^{\bullet} f_{\bullet}^*\mathbb{F} \cong f_{\bullet}^!\mathbb{F},$$

where $[d]$ denotes the shift of degree.

Proof. In view of Theorem 26.14, it suffices to show that there is an isomorphism $f_{\bullet}^!\mathcal{O}_{Y_{\bullet}} \cong \bigwedge{}^{d} \Omega_{X_{\bullet}/Y_{\bullet}}[d]$. Consider the commutative diagram

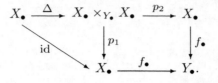

By Lemma 7.17 and Lemma 7.16, the all morphisms in the diagrams are cartesian. As p_1 is smooth of finite type of relative dimension d, Δ is a cartesian regular embedding of the constant codimension d.

By Theorem 21.8, Theorem 27.8, and Proposition 28.8, we have

$$\mathcal{O}_{X_\bullet} \cong f_\bullet^* \mathcal{O}_{Y_\bullet} \cong \Delta^! p_1^! f_\bullet^* \mathcal{O}_{Y_\bullet} \cong \Delta^\natural p_2^* f_\bullet^! \mathcal{O}_{Y_\bullet} \cong$$
$$\textstyle\bigwedge^d \Omega_{X_\bullet/Y_\bullet}^\vee [-d] \otimes_{\mathcal{O}_{X_\bullet}}^{\bullet, L} L\Delta^* p_2^* f_\bullet^! \mathcal{O}_{Y_\bullet} \cong \bigwedge^d \Omega_{X_\bullet/Y_\bullet}^\vee [-d] \otimes_{\mathcal{O}_{X_\bullet}}^{\bullet, L} f_\bullet^! \mathcal{O}_{Y_\bullet}.$$

As $\bigwedge^d \Omega_{X_\bullet/Y_\bullet}$ is an invertible sheaf, we are done.　　　□

Chapter 29
Group Schemes Flat of Finite Type

(29.1) Let S be a scheme.

(29.2) Let $\mathcal{F}$ (resp. $\mathcal{F}_M$) denote the subcategory of $\mathcal{P}((\Delta), \underline{\mathrm{Sch}}/S)$ (resp. subcategory of $\mathcal{P}(\Delta_M, \underline{\mathrm{Sch}}/S)$) consisting of noetherian objects with flat arrows and cartesian morphisms separated of finite type.

Let G be a flat S-group scheme of finite type. Note that G is faithfully flat over S. A G-*scheme* is an S-scheme with a left G-action by definition. Set $\mathcal{A}_G$ to be the category of noetherian G-schemes and G-morphisms separated of finite type.

For $X \in \mathcal{A}_G$, we associate a simplicial scheme $B_G(X)$ by $B_G(X)_n = G^n \times X$. For $n \geq 1$, $d_i(n) : G^n \times X \to G^{n-1} \times X$ is the projection $p \times 1_{G^{n-1} \times X}$ if $i = n$, where $p : G \to S$ is the structure morphism. While $d_i(n) = 1_{G^{n-1}} \times a$ if $i = 0$, where $a : G \times X \to X$ is the action. If $0 < i < n$, then $d_i(n) = 1_{G^{n-i-1}} \times \mu \times 1_{G^{i-1} \times X}$, where $\mu : G \times G \to G$ is the product. For $n \geq 0$, $s_i(n) : G^n \times X \to G^{n+1} \times X$ is given by

$$s_i(n)(g_n, \ldots, g_1, x) = (g_n, \ldots, g_{i+1}, e, g_i, \ldots, g_1, x),$$

where $e : S \to G$ is the unit element. Indeed, $B_G(X)$ satisfies the relations (11^{op}), (12^{op}), and (13^{op}) in [29, (VII.5)].

Note that $(B_G(X)')|_{(\Delta)}$ is canonically isomorphic to $B_G(G \times X)$, where $G \times X$ is viewed as a principal G-action. Note also that there is an isomorphism from $B_G(X)'$ to $\mathrm{Nerve}(p_2 : G \times X \to X)$ given by

$$B_G(X)'_n = G^{n+1} \times X \to (G \times X) \times_X \cdots \times_X (G \times X) = \mathrm{Nerve}(p_2)_n$$

$$(g_n, \ldots, g_0, x) \mapsto ((g_n \cdots g_0, x), (g_{n-1} \cdots g_0, x), \ldots, (g_0, x)).$$

Hence we have

Lemma 29.3. *Let G and X be as above. Then $B_G(X)$ is a simplicial S-groupoid with $d_0(1)$ and $d_1(1)$ faithfully flat of finite type.*

J. Lipman, M. Hashimoto, *Foundations of Grothendieck Duality for Diagrams*   445
of Schemes, Lecture Notes in Mathematics 1960,
© Springer-Verlag Berlin Heidelberg 2009

We denote the restriction $B_G(X)|_{\Delta_M}$ by $B_G^M(X)$. Obviously, $B_G^M(X)$ is an S-groupoid with $d_0(1)$ and $d_1(1)$ faithfully flat of finite type.

For a morphism $f : X \to Y$ in $\mathcal{A}_G$, we define $B_G(f) : B_G(X) \to B_G(Y)$ by $(B_G(f))_n = 1_{G^n} \times f$. It is easy to check that B_G is a functor from $\mathcal{A}_G$ to $\mathcal{F}$. Thus B_G^M is a functor from $\mathcal{A}_G$ to $\mathcal{F}_M$.

We define a $(G, \mathcal{O}_X)$-module to be an $\mathcal{O}_{B_G^M(X)}$-module. That is, an object of $\mathrm{Mod}(B_G^M(X))$. So an equivariant (resp. locally quasi-coherent, quasi-coherent, coherent) $(G, \mathcal{O}_X)$-module is an equivariant (resp. locally quasi-coherent, quasi-coherent, coherent) object of $\mathrm{Mod}(B_G^M(X))$. The category of G-linearized $\mathcal{O}_X$-modules in [32] is equivalent to that of our equivariant $(G, \mathcal{O}_X)$-modules. See also [6] and [19].

We denote the category of $(G, \mathcal{O}_X)$-modules by $\mathrm{Mod}(G, X)$. The category of equivariant (resp. locally quasi-coherent, quasi-coherent, coherent) $(G, \mathcal{O}_X)$-modules is denoted by $\mathrm{EM}(G, X)$ (resp. $\mathrm{Lqc}(G, X)$, $\mathrm{Qch}(G, X)$, $\mathrm{Coh}(G, X)$).

Note that $\mathrm{EM}(B_G(X))$ (resp. $\mathrm{Qch}(B_G(X))$, $\mathrm{Coh}(B_G(X))$) is equivalent to $\mathrm{EM}(G, \mathcal{O}_X)$ (resp. $\mathrm{Qch}(G, X)$, $\mathrm{Coh}(G, X)$) (Lemma 9.4). However, the author does not know whether $\mathrm{Mod}(B_G(X))$ is equivalent to $\mathrm{Mod}(B_G^M(X))$. From our point of view, it seems that it is more convenient to work over Δ_M, which is a finite ordered category, than (Δ).

The discussion on derived categories of categories of sheaves over diagrams of schemes are interpreted to the derived categories of the categories of $(G, \mathcal{O}_X)$-modules.

By Lemma 29.3 and Lemma 12.8, we have

Lemma 29.4. *Let $X \in \mathcal{A}_G$. Then $\mathrm{Qch}(G, X)$ is a locally noetherian abelian category. $\mathcal{M} \in \mathrm{Qch}(G, X)$ is a noetherian object if and only if $\mathcal{M}_0$ is coherent if and only if $\mathcal{M} \in \mathrm{Coh}(G, X)$.*

Let $\mathcal{M}$ be a $(G, \mathcal{O}_X)$-module. If there is no danger of confusion, we may write $\mathcal{M}_0$ instead of $\mathcal{M}$. For example, $\mathcal{O}_X$ sometimes means $\mathcal{O}_{B_G^M(X)}$, since $(\mathcal{O}_{B_G^M(X)})_0 = \mathcal{O}_X$. This abuse of notation is what we always do when $S = X = \mathrm{Spec}\, k$ and G is an affine algebraic group over k. A G-module and its underlying vector space are denoted by the same symbol. Similarly, an object of $D(B_G^M(X))$ and its restriction to $D(B_G^M(X)_0) = D(X)$ are sometimes denoted by the same symbol. Moreover, for a morphism f in $\mathcal{A}_G$, we denote for example $R(B_G^M(f))_*$ by Rf_*, and $B_G^M(f)^!$ by $f^!$.

$D(B_G^M(X))$, or $D(\mathrm{Mod}(G, X))$, is denoted by $D(G, X)$ for short. Thus for example, $D_{\mathrm{Qch}(G,X)}^+(\mathrm{Mod}(G, X))$ is denoted by $D_{\mathrm{Qch}}^+(G, X)$.

Thus, as a corollary to Theorem 25.2, we have

Theorem 29.5 (G-Grothendieck's duality). *Let S be a scheme, and G a flat S-group scheme of finite type. Let X and Y be noetherian S-schemes with G-actions, and $f : X \to Y$ a proper G-morphism. Then the composite*

$$\Theta(f) : Rf_*R\underline{\mathrm{Hom}}^\bullet_{\mathrm{Mod}(G,X)}(\mathbb{F}, f^\times\mathbb{G}) \xrightarrow{H} R\underline{\mathrm{Hom}}^\bullet_{\mathrm{Mod}(G,Y)}(Rf_*\mathbb{F}, Rf_*f^\times\mathbb{G})$$
$$\xrightarrow{\varepsilon} R\underline{\mathrm{Hom}}^\bullet_{\mathrm{Mod}(G,Y)}(Rf_*\mathbb{F}, \mathbb{G})$$

is an isomorphism in $D(G, Y)$ *for* $\mathbb{F} \in D_{\mathrm{Qch}}(G, X)$ *and* $\mathbb{G} \in D^+_{\mathrm{Qch}}(G, Y)$.

Chapter 30
Compatibility with Derived G-Invariance

(30.1) Let S be a scheme, and G a flat S-group scheme. Let X be an S-scheme with a trivial G-action. That is, $a : G \times X \to X$ agrees with the second projection p_2. In other words, $d_0(1) = d_1(1)$ in $B_G^M(X)$.

For an object $\mathcal{M}$ of $\mathrm{Mod}(G, X)$, we define the G-*invariance* of $\mathcal{M}$ to be the kernel of the natural map

$$\beta_{d_0(1)} - \beta_{d_1(1)} : \mathcal{M}_0 \to d_0(1)_* \mathcal{M}_1 = d_1(1)_* \mathcal{M}_1,$$

and we denote it by $\mathcal{M}^G$.

(30.2) Let X be as in (30.1). Define $\tilde{B}_G^M(X)$ to be the augmented diagram

$$G \times_S G \times_S X \underset{p_{23}}{\overset{1_G \times a}{\underset{\mu \times 1_X}{\rightrightarrows}}} G \times_S X \underset{p_2}{\overset{a}{\rightrightarrows}} X \overset{\mathrm{id}}{\to} X.$$

Note that $\tilde{B}_G^M(X)$ is an object of $\mathcal{P}(\Delta_M^+, \underline{\mathrm{Sch}}/S)$. For an S-morphism $f : X \to Y$ between S-schemes with trivial G-actions, $\tilde{B}_G^M(f) : \tilde{B}_G^M(X) \to \tilde{B}_G^M(Y)$ is defined by $\tilde{B}_G^M(f)_n = 1_{G^n} \times f$ for $n \geq 0$ and $\tilde{B}_G^M(f)_{-1} = f$. Thus $\tilde{B}_G^M$ is a functor from the category of S-schemes (with trivial G-actions) to the category $\mathcal{P}(\Delta_M^+, \underline{\mathrm{Sch}}/S)$ such that $(?)|_{\Delta_M} \tilde{B}_G^M = B_G^M$ and $(?)|_{-1} \tilde{B}_G^M = \mathrm{Id}$.

Lemma 30.3. *The functor* $(?)^G : \mathrm{Mod}(G, X) \to \mathrm{Mod}(X)$ *agrees with* $(?)_{-1} R_{\Delta_M}$.

Proof. Follows easily from (6.14). □

(30.4) We say that an object $\mathcal{M}$ of $\mathrm{Mod}(G, X)$ is G-*trivial* if $\mathcal{M}$ is equivariant, and the canonical inclusion $\mathcal{M}^G \to \mathcal{M}_0$ is an isomorphism. Note that $(?)_{\Delta_M} L_{-1}$ is the exact left adjoint of $(?)^G$. Note also that $\mathcal{M}$ is G-trivial if and only if the counit of adjunction $\varepsilon : (?)_{\Delta_M} L_{-1} \mathcal{M}^G \to \mathcal{M}$ is an isomorphism if and only if $\mathcal{M} \cong \mathcal{N}_{\Delta_M}$ for some $\mathcal{N} \in \mathrm{EM}(\tilde{B}_G^M(X))$.

J. Lipman, M. Hashimoto, *Foundations of Grothendieck Duality for Diagrams* 449
of Schemes, Lecture Notes in Mathematics 1960,
© Springer-Verlag Berlin Heidelberg 2009

Let $\mathrm{triv}(G, X)$ denote the full subcategory of $\mathrm{Mod}(B_G^M(X))$ consisting of G-trivial objects. Note that $(?)^G : \mathrm{triv}(G, X) \to \mathrm{Mod}(X)$ is an equivalence, whose quasi-inverse is $(?)_{\Delta_M} L_{-1}$.

Assume that G is concentrated over S. If $\mathcal{M}$ is locally quasi-coherent, then $\mathcal{M}^G$ is quasi-coherent. Thus we get a derived functor $R(?)^G : D_{\mathrm{Lqc}}^+(G, X) \to D_{\mathrm{Qch}}^+(X)$.

Proposition 30.5. *Let G be of finite type over S. Let X and Y be noetherian S-schemes with trivial G-actions, and $f : X \to Y$ an S-morphism, which is automatically a G-morphism, separated of finite type. Then there is a canonical isomorphism*

$$f^! R(?)^G \cong R(?)^G f^!$$

between functors from $D_{\mathrm{Lqc}}^+(G, Y)$ to $D_{\mathrm{Qch}}^+(X)$.

Proof. As $(?)_{-1}$ is exact, we have $R(?)^G \cong (?)_{-1} RR_{\Delta_M}$ by Lemma 30.3. Thus, we have a composite isomorphism

$$f^! R(?)^G \cong f^!(?)_{-1} RR_{\Delta_M} \xrightarrow{\tilde{\xi}^{-1}} (?)_{-1} (\tilde{B}_M^G(f))^! RR_{\Delta_M} \xrightarrow{\bar{c}} (?)_{-1} RR_{\Delta_M} f^! \cong R(?)^G f^!$$

by Lemma 21.3 and Lemma 24.7. $\square$

Chapter 31
Equivariant Dualizing Complexes and Canonical Modules

(31.1) Let $\mathcal{A}$ be a Grothendieck category, and $\mathbb{I} \in D(\mathcal{A})$. We say that $\mathbb{I}$ has a finite injective dimension if $R\operatorname{Hom}_{\mathcal{A}}(?, \mathbb{I})$ is way-out in both directions, see [17, (I.7)]. By definition, an object of $C(\mathcal{A})$ or $K(\mathcal{A})$ has a finite injective dimension if it does in $D(\mathcal{A})$. $\mathbb{F} \in C(\mathcal{A})$ has a finite injective dimension if and only if there is a bounded complex $\mathbb{J}$ of injective objects in $\mathcal{A}$ and a quasi-isomorphism $\mathbb{F} \to \mathbb{J}$.

(31.2) Let I be a finite ordered category, S a scheme, and $X_{\bullet} \in \mathcal{P}(I, \underline{\mathrm{Sch}}/S)$.

Lemma 31.3. *Assume that $X_{\bullet}$ has flat arrows. Let $\mathbb{I} \in D(X_{\bullet})$. Then $\mathbb{I}$ has a finite injective dimension if and only if $\mathbb{I}_i$ has a finite injective dimension for any $i \in \mathrm{ob}(I)$.*

Proof. We prove the 'only if' part. Since $(?)_i$ is exact and has an exact left adjoint L_i, and $\mathbb{I}$ has a finite injective dimension, $\mathbb{I}_i$ has a finite injective dimension for $i \in \mathrm{ob}(I)$.

We prove the converse by induction on the number of objects of I. We may assume that I has at least two objects.

Let i be a maximal element of $\mathrm{ob}(I)$. There is a triangle of the form

$$\mathbb{I} \xrightarrow{u} RR_i(?)_i\mathbb{I} \to \mathbb{C} \to \mathbb{I}[1].$$

Since $\mathbb{I}_i$ has a finite injective dimension and R_i has an exact left adjoint $(?)_i$, it is easy to see that $RR_i(?)_i\mathbb{I}$ has a finite injective dimension. So it suffices to show that $\mathbb{C}$ has a finite injective dimension. Applying $(?)_i$ to the triangle above, it is easy to see that $\mathbb{C}_i = 0$. Let J be the full subcategory of I such that $\mathrm{ob}(J) = \mathrm{ob}(I) \setminus \{i\}$. Then $u : \mathbb{C} \to R_J\mathbb{C}_J$ is an isomorphism by Lemma 18.9. On the other hand, by the only if part, which has already been proved, it is easy to see that $\mathbb{C}_j$ has a finite injective dimension for $j \in \mathrm{ob}(J)$. By induction assumption, $\mathbb{C}_J$ has a finite injective dimension.

J. Lipman, M. Hashimoto, *Foundations of Grothendieck Duality for Diagrams*   451
of Schemes, Lecture Notes in Mathematics 1960,
© Springer-Verlag Berlin Heidelberg 2009

So $\mathbb{C} \cong R_J\mathbb{C}_J$ has a finite injective dimension, since R_J is exact and has an exact left adjoint $(?)_J$. $\qquad\qquad\qquad\qquad\qquad\qquad\qquad\qquad\qquad\qquad\qquad$ $\square$

(31.4) Let the notation be as in Theorem 20.4. Let $X_\bullet$ be an object of $\mathcal{F}$ (i.e., an I^{op}-diagram of noetherian S-schemes with flat arrows). We say that $\mathbb{F} \in D(X_\bullet)$ is a *dualizing complex* of $X_\bullet$ if $\mathbb{F} \in D_{\mathrm{Coh}}(X_\bullet)$, $\mathbb{F}$ has a finite injective dimension, and the canonical map

$$\mathcal{O}_{X_\bullet} \xrightarrow{\mathrm{tr}} R\underline{\mathrm{Hom}}^\bullet_{\mathcal{O}_{X_\bullet}}(\mathbb{F}, \mathbb{F})$$

is an isomorphism. A complex $\mathbb{F} \in C(\mathrm{Mod}(X_\bullet))$ is said to be a dualizing complex if it is as an object of $D(X_\bullet)$.

(31.5) If there is a dualizing complex of $X_\bullet$, then it is represented by a bounded injective complex $\mathbb{F} \in C(\mathrm{Mod}(X_\bullet))$ with coherent cohomology groups such that

$$\mathcal{O}_{X_\bullet} \xrightarrow{\mathrm{tr}} \underline{\mathrm{Hom}}^\bullet_{\mathcal{O}_{X_\bullet}}(\mathbb{F}, \mathbb{F})$$

is a quasi-isomorphism.

    More is true. We may further assume that $\mathbb{F} \in C(\mathrm{Lqc}(X_\bullet))$. Indeed, we may replace $\mathbb{F}$ above by $\mathrm{lqc}\,\mathbb{F}$. Since $\mathbb{F}$ has coherent cohomology groups, it is easy to see that the canonical map $\mathrm{lqc}\,\mathbb{F} \to \mathbb{F}$ is a quasi-isomorphism by Lemma 14.3, **3**. Each term of $\mathrm{lqc}\,\mathbb{F}$ is an injective object of $\mathrm{Lqc}(X_\bullet)$, since lqc has an exact left adjoint. Note that each term of $\mathrm{lqc}\,\mathbb{F}$ is still injective in $\mathrm{Mod}(X_\bullet)$ by Lemma 24.2.

Lemma 31.6. *Let the notation be as in* (31.4)*. An object $\mathbb{F} \in D(X_\bullet)$ is a dualizing complex of $X_\bullet$ if and only if $\mathbb{F}$ has equivariant cohomology groups and $\mathbb{F}_i \in D(X_i)$ is a dualizing complex of X_i for any $i \in \mathrm{ob}(I)$.*

Proof. This is obvious by Lemma 13.9 and Lemma 31.3. $\qquad\qquad\qquad\qquad$ $\square$

Corollary 31.7. *Let the notation be as in* (31.4)*. If $X_\bullet$ is Gorenstein with finite Krull dimension, then $\mathcal{O}_{X_\bullet}$ is a dualizing complex of $X_\bullet$.*

Proof. This is clear by the lemma and [17, (V.10)]. $\qquad\qquad\qquad\qquad\qquad$ $\square$

Lemma 31.8. *Let the notation be as in* (31.4)*. If $X_\bullet$ has a dualizing complex $\mathbb{F}$, then $X_\bullet$ has finite Krull dimensions, and $X_\bullet$ has Gorenstein arrows.*

Proof. As $\mathbb{F}_i$ is a dualizing complex of X_i for each $i \in \mathrm{ob}(I)$, $X_\bullet$ has finite Krull dimensions by [17, Corollary V.7.2].

    Let $\phi : i \to j$ be a morphism of I. As X_ϕ is flat, $\alpha_\phi : X_\phi^*\mathbb{F}_i \to \mathbb{F}_j$ is an isomorphism of $D(X_j)$. As $X_\phi^*\mathbb{F}_i$ is a dualizing complex of X_j, X_ϕ is Gorenstein by [4, (5.1)]. $\qquad\qquad\qquad\qquad\qquad\qquad\qquad\qquad\qquad\qquad$ $\square$

Proposition 31.9. *Let the notation be as above, and $\mathbb{I}$ a dualizing complex of $X_\bullet$. Let $\mathbb{F} \in D_{\mathrm{Coh}}(X_\bullet)$. Then we have*

1 $R\operatorname{\underline{Hom}}^{\bullet}_{\mathcal{O}_{X_{\bullet}}}(\mathbb{F}, \mathbb{I}) \in D_{\mathrm{Coh}}(X_{\bullet})$.

2 *The canonical map*

$$\mathbb{F} \to R\operatorname{\underline{Hom}}^{\bullet}_{\mathcal{O}_{X_{\bullet}}}(R\operatorname{\underline{Hom}}^{\bullet}_{\mathcal{O}_{X_{\bullet}}}(\mathbb{F}, \mathbb{I}), \mathbb{I})$$

is an isomorphism for $\mathbb{F} \in D_{\mathrm{Coh}}(X_{\bullet})$.

Proof. **1** As $\mathbb{I}$ has a finite injective dimension, $R\operatorname{\underline{Hom}}^{\bullet}_{\mathcal{O}_{X_{\bullet}}}(?, \mathbb{I})$ is way-out in both directions. Hence by [17, Proposition I.7.3], we may assume that $\mathbb{F}$ is bounded. This case is trivial by Lemma 13.10.

2 Using Lemma 13.9 twice, we may assume that $X_{\bullet}$ is a single scheme. This case is [17, Proposition V.2.1]. □

Lemma 31.10. *Let X be a noetherian scheme, and $\mathfrak{U} = (U_i)$ a finite open covering of X. Let $\mathbb{I} \in D(X)$. Then $\mathbb{I}$ is dualizing if and only if $\mathbb{I}|_{U_i}$ is dualizing for each i.*

Proof. It is obvious that $\mathbb{I}$ has coherent cohomology groups if and only if $\mathbb{I}|_{U_i}$ has coherent cohomology groups for each i.

Assume that $\mathbb{I}$ is a bounded injective complex. Then $\mathbb{I}|_{U_i}$ is a bounded injective complex, since $(?)|_{U_i}$ preserves injectives. Conversely, assume that $\mathbb{I}$ has coherent cohomology groups and $\mathbb{I}|_{U_i}$ has a finite injective dimension for each i. Then by [17, (II.7.20)], there is an integer n_0 such that for any i, any $G \in \mathrm{Coh}(U_i)$, and any $j > n_0$, we have $\underline{\mathrm{Ext}}^j_{\mathcal{O}_{U_i}}(G, \mathbb{I}|_{U_i}) = 0$. This shows that for any $G \in \mathrm{Coh}(X)$ and any $j > n_0$, $\underline{\mathrm{Ext}}^j_{\mathcal{O}_X}(G, \mathbb{I}) = 0$, and again by [17, (II.7.20)], we have that $\mathbb{I}$ has a finite injective dimension.

Let $\mathbb{I}$ be a bounded injective complex. Let $\mathbb{C}$ be the mapping cone of $\mathrm{tr} : \mathcal{O}_X \to \underline{\mathrm{Hom}}_{\mathcal{O}_X}(\mathbb{I}, \mathbb{I})$. $\mathbb{C}$ is exact (i.e., tr is a quasi-isomorphism) if and only if $\mathbb{C}|_{U_i}$ is exact for each i. On the other hand, $\mathbb{C}|_{U_i}$ is isomorphic to the mapping cone of the trace map $\mathcal{O}_{U_i} \to \underline{\mathrm{Hom}}_{\mathcal{O}_{U_i}}(\mathbb{I}|_{U_i}, \mathbb{I}|_{U_i})$. Thus $\mathrm{tr} ; \mathcal{O}_X \to \underline{\mathrm{Hom}}_{\mathcal{O}_X}(\mathbb{I}, \mathbb{I})$ is a quasi-isomorphism if and only if $\mathcal{O}_{U_i} \to \underline{\mathrm{Hom}}_{\mathcal{O}_{U_i}}(\mathbb{I}|_{U_i}, \mathbb{I}|_{U_i})$ is a quasi-isomorphism for each i. Thus the lemma is obvious now. □

Lemma 31.11. *Let the notation be as in (31.4). Let $f_{\bullet} : X_{\bullet} \to Y_{\bullet}$ be a morphism in $\mathcal{F}$, and let $\mathbb{I}$ be a dualizing complex of $Y_{\bullet}$. Then $f^!_{\bullet}(\mathbb{I})$ is a dualizing complex of $X_{\bullet}$.*

Proof. By Corollary 22.4, $f^!_{\bullet}(\mathbb{I})$ has coherent cohomology groups.

By Lemma 31.6 and Proposition 18.14, we may assume that $f : X \to Y$ is a morphism of single schemes. By Lemma 31.10, the question is local both on Y and X, So we may assume that both Y and X are affine, and f is either an affine n-space or a closed immersion. These cases are done in [17, Chapter V]. □

Lemma 31.12. *Let the notation be as in (31.4), and $\mathbb{I}$ and $\mathbb{J}$ dualizing complexes on $X_{\bullet}$. If $X_{\bullet}$ is d-connected and X_i is non-empty for some $i \in \mathrm{ob}(I)$, then there exist a unique invertible sheaf $\mathcal{L}$ and a unique integer n such that*

$$\mathbb{J} \cong \mathbb{I} \otimes_{\mathcal{O}_{X_\bullet}}^{\bullet, L} \mathcal{L}[n].$$

Such $\mathcal{L}$ and n are determined by

$$\mathcal{L}[n] \cong R\operatorname{\underline{Hom}}_{\mathcal{O}_{X_\bullet}}^\bullet (\mathbb{I}, \mathbb{J}).$$

Proof. Use [17, Theorem V.3.1]. $\qquad \square$

Definition 31.13. Let the notation be as in (31.4), and $\mathbb{I}$ a fixed dualizing complex of $X_\bullet$. For any object $f_\bullet : Y_\bullet \to X_\bullet$ of $\mathcal{F}/X_\bullet$, we define *the* dualizing complex of $Y_\bullet$ (or better, of $f_\bullet$) to be $f_\bullet^! \mathbb{I}$. It is certainly a dualizing complex of $Y_\bullet$ by Lemma 31.11. If $Y_\bullet$ is d-connected and Y_i is non-empty for some $i \in \operatorname{ob}(I)$, then we define *the canonical sheaf* $\omega_{Y_\bullet}$ of $Y_\bullet$ (or better, $f_\bullet$) to be $H^s(f_\bullet^! \mathbb{I})$, where s is the smallest i such that $H^i(f_\bullet^! \mathbb{I}) \neq 0$. If $Y_\bullet$ is not d-connected, then we define $\omega_{Y_\bullet}$ componentwise.

Lemma 31.14. *Let S be a noetherian scheme, and G a flat S-group scheme of finite type. Then $G \to S$ is a (flat) local complete intersection morphism. That is, (it is flat and) all fibers are locally complete intersections.*

Proof. We may assume that $S = \operatorname{Spec} k$, with k a field. Then by [3, Theorem 1], we may assume that k is algebraically closed.

First assume that the characteristic is $p > 0$. Then there is some $r \gg 0$ such that the scheme theoretic image of the Frobenius map $F^r : G \to G^{(r)}$ is reduced (or equivalently, k-smooth) and agrees with $G^{(r)}_{\mathrm{red}}$. Note that the induced morphism $G \to G^{(r)}_{\mathrm{red}}$ is flat, since the flat locus is a G-stable open subset of G [19, Lemma 2.1.10], and the morphism is flat at the generic point.

As the group G_{red} acts on G transitively, it suffices to show that G is locally a complete intersection at the unit element e. So by [3, Theorem 2], it suffices to show that the rth Frobenius kernel G_r is a complete intersection. As G_r is finite connected, this is well known [46, (14.4)].

Now consider the case that G is of characteristic zero. We are to prove that G is k-smooth. Take a finitely generated $\mathbb{Z}$-subalgebra R of k such that G is defined. We may take R so that G_R is R-flat of finite type. Set $H := (G_R)_{\mathrm{red}}$. We may take R so that H is also R-flat. Then H is a closed subgroup scheme over R, since $\operatorname{Spec} R$ and $H \times_R H$ are reduced. As a reduced group scheme over a field of characteristic zero is smooth, we may localize R if necessary, and we may assume that H is R-smooth.

Let $\mathcal{J}$ be the defining ideal sheaf of H in G_R. There exists some $s \geq 0$ such that $\mathcal{J}^{s+1} = 0$. Note that $\mathcal{G} := \bigoplus_{i=0}^s \mathcal{J}^i / \mathcal{J}^{i+1}$ is a coherent $(H, \mathcal{O}_H)$-module. Applying Corollary 10.15 to the case that $Y = \operatorname{Spec} R$ and $X_\bullet = B_H^M(H)$, the coherent $(H, \mathcal{O}_H)$-module $\mathcal{G}$ is of the form $f^*(\tilde{V})$, where V is a finite R-module, and $f : H \to \operatorname{Spec} R$ is the structure map. Replacing R if necessary, we may assume that $V \cong R^u$. Now we want to prove that $u = 1$ so that $H = G_R$, which implies G is k-smooth.

There exists some prime number $p > u$ and a maximal ideal $\mathfrak{m}$ of R such that $R/\mathfrak{m}$ is a finite field of characteristic p. Let κ be the algebraic closure of $R/\mathfrak{m}$, and consider the base change $\overline{(?)} := ? \otimes_R \kappa$. Note that $\bar{\mathcal{G}} = \bigoplus_i \overline{\mathcal{J}^i/\mathcal{J}^{i+1}}$ (recall that R is $\mathbb{Z}$-flat and H is R-flat). Let $\mathcal{I}^{[p^r]}$ denote the defining ideal of the rth Frobenius kernel of $\bar{G}_R$. By [46, (14.4)] again, $\dim_k(\mathcal{O}_{\bar{G}_R}/\mathcal{I}^{[p^r]})_e$ is a power of p, say $p^{v(r)}$. Similarly, the k-dimension of the coordinate ring $(\mathcal{O}_{\bar{G}_R}/(\bar{\mathcal{J}} + \mathcal{I}^{[p^r]}))_e$ of the rth Frobenius kernel of $\bar{H}$ is a power of p, say $p^{w(r)}$. Note that

$$p^{w(r)} \leq p^{v(r)} \leq \dim_k(\mathcal{O}_{\bar{G}_R}/\mathcal{I}^{[p^r]} \otimes_{\mathcal{O}_{\bar{G}_R}} \bar{\mathcal{G}})_e = p^{w(r)}u < p^{w(r)+1}.$$

Hence $w(r) \leq v(r) < w(r) + 1$, and we have $\bar{\mathcal{J}}_e \subset \mathcal{I}_e^{[p^r]}$ for any r. By Krull's intersection theorem, $\bar{\mathcal{J}}_e \subset \bigcap_r (\mathcal{I}_e)^{p^r} = 0$. This shows that $\bar{G}_R$ is reduced at e, which shows that $\bar{G}_R$ is κ-smooth everywhere. So the nilpotent ideal $\bar{\mathcal{J}}$ must be zero, and this shows $u = 1$. $\qquad\square$

(31.15) Let S be a scheme, G a flat S-group scheme of finite type, and X a noetherian G-scheme. By definition, a G-*dualizing complex* of X is a dualizing complex of $B_G^M(X)$. Let us fix X and a G-dualizing complex $\mathbb{I}$. For $(f : Y \to X) \in \mathcal{A}_G/X$, we define *the* G-dualizing complex of Y (or better, of f) to be $f^!(\mathbb{I})$. It is certainly a G-dualizing complex of Y. The canonical sheaf of $B_G^M(Y)$ is called *the* G-*canonical sheaf* of Y, and is denoted by ω_Y.

Lemma 31.16. *Let* $f : X \to Y$ *be a Gorenstein flat morphism of finite type between noetherian schemes. If* $\mathbb{I}$ *is a dualizing complex of* Y, *then* $f^*(\mathbb{I})$ *is a dualizing complex of* X.

Proof. Since Y has a dualizing complex, Y has finite Krull dimension [17, Corollary V.7.2]. Since X is of finite type over Y, X has finite Krull dimension. By [17, Proposition V.8.2], it suffices to show that $f^*(\mathbb{I})$ is pointwise dualizing. So we may assume that $X = \operatorname{Spec} B$ and $Y = \operatorname{Spec} A$ are affine, both A and B are local, and f is induced by a local homomorphism from A to B. Then the assertion follows from [4, (5.1)]. $\qquad\square$

Lemma 31.17. *Let* S, G, *and* X *be as in* (31.15). *Then* $\mathbb{I} \in D(G, X)$ *is a* G-*dualizing complex of* X *if and only if* $\mathbb{I}$ *has equivariant cohomology groups and* $\mathbb{I}_0 \in D(X)$ *is a dualizing complex of* X.

Proof. The 'only if' part is obvious by Lemma 31.6.

To prove the converse, it suffices to show that $\mathbb{I}_i \in D(G^i \times X)$ is dualizing for $i = 1, 2$ by the same lemma. Since $B_G^M(X)$ has flat arrows and $\mathbb{I}$ has equivariant cohomology groups, $\alpha_{\rho_i(i)} : r_i(i)^* \mathbb{I}_0 \to \mathbb{I}_i$ is an isomorphism in $D(G^i \times X)$. Since $r_i(i)$ is Gorenstein flat of finite type, the assertion follows from Lemma 31.16. $\qquad\square$

Lemma 31.18. *Let* R *be a Gorenstein local ring of dimension* d, *and* $S = \operatorname{Spec} R$. *Then* $B_G^M(S)$ *is Gorenstein of finite Krull dimension. In particular,*

$\mathcal{O}_S[d]$ is a G-dualizing complex of S (i.e., $\mathcal{O}_{B_G^M(S)}[d]$ is a dualizing complex of $B_G^M(S)$).

Proof. As $S = \operatorname{Spec} R$ is Gorenstein by assumption and G is Gorenstein over S by Lemma 31.14, the assertions are trivial. $\qquad\square$

(31.19) When R, S and d are as in the lemma, then we usually choose and fix the G-dualizing complex $\mathcal{O}_S[d]$ of S. Thus for an object $X \in \mathcal{A}_G = \mathcal{A}_G/S$, the G-dualizing complex of X is $f^!(\mathcal{O}_S[d])$, where f is the structure morphism $X \to S$ of X. The G-canonical sheaf is defined accordingly.

Lemma 31.20. *Let R, S and d be as in* Lemma 31.18. *Let $X \in \mathcal{A}_G$, and assume that G acts on X trivially. Then the dualizing complex $\mathbb{I}_X := f^!(\mathcal{O}_S[d])$ has G-trivial cohomology groups, where $f : X \to S$ is the structure map. In particular, ω_X is G-trivial.*

Proof. By Proposition 18.14,

$$f^!(\mathcal{O}_S[d]) \cong f^!((\mathcal{O}_{\tilde{B}_G^M(S)})_{\Delta_M})[d] \cong (?)_{\Delta_M}(\tilde{B}_G^M(f)^!(\mathcal{O}_{\tilde{B}_G^M(S)}))[d].$$

By Corollary 22.4, $\tilde{B}_G^M(f)^!(\mathcal{O}_{\tilde{B}_G^M(S)})$ has coherent cohomology groups. Hence, $f^!(\mathcal{O}_S[d])$ has G-trivial cohomology groups. $\qquad\square$

Chapter 32
A Generalization of Watanabe's Theorem

Lemma 32.1. *Let R be a noetherian commutative ring, and G a finite group which acts on R. Set $A = R^G$, and assume that $\operatorname{Spec} A$ is connected. Then G permutes the connected components of $\operatorname{Spec} R$ transitively.*

Proof. Since $\operatorname{Spec} R$ is a noetherian space, $\operatorname{Spec} R$ has only finitely many connected components, say $X_1, \ldots, X_n$. Then $R = R_1 \times \cdots \times R_n$, and each R_i is of the form Re_i, where e_i is a primitive idempotent. Note that $E := \{e_1, \ldots, e_n\}$ is the set of primitive idempotents of R, and G acts on E. Let E_1 be an orbit of this action. Then $e = \sum_{e_i \in E_1} e_i$ is in A. As A does not have any nontrivial idempotent, $e = 1$. This shows that G acts on E transitively, and we are done. $\qquad\square$

Lemma 32.2. *Let R be a noetherian commutative ring, and G a finite group which acts on R. Set $A = R^G$, and assume that the inclusion $A \hookrightarrow R$ is finite. If $\mathfrak{p} \in \operatorname{Spec} A$, then G acts transitively on the set of primes of R lying over $\mathfrak{p}$. Moreover, the going-down theorem holds for the ring extension $A \hookrightarrow R$.*

Proof. Note that A is noetherian by Eakin-Nagata theorem [30, Theorem 3.7]. Let A' be the $\mathfrak{p}A_\mathfrak{p}$-adic completion of $A_\mathfrak{p}$, and set $R' := A' \otimes_A R$. As A' is A-flat, $A' = (R')^G$. It suffices to prove that G acts transitively on the maximal ideals of R'. But R' is the direct product $\prod_i R'_i$ of complete local rings R'_i. Consider the corresponding primitive idempotents. Since A' is a local ring, G permutes these idempotents transitively by Lemma 32.1. It is obvious that this action induces a transitive action on the maximal ideals of R'.

We prove the last assertion. Let $\mathfrak{p} \supset \mathfrak{q}$ be prime ideals of A, and P be a prime ideal of R such that $P \cap A = \mathfrak{p}$. By the lying over theorem [30, Theorem 9.3], there exists some prime ideal Q' of R such that $Q' \cap A = \mathfrak{q}$. By the going-up theorem [30, Theorem 9.4], there exists some prime $P' \supset Q'$ such that $P' \cap A = \mathfrak{p}$. Then there exists some $g \in G$ such that $gP' = P$. Letting $Q := gQ'$, we have that $Q \subset P$, and $Q \cap A = \mathfrak{q}$. $\qquad\square$

J. Lipman, M. Hashimoto, *Foundations of Grothendieck Duality for Diagrams*   457
of Schemes, Lecture Notes in Mathematics 1960,
© Springer-Verlag Berlin Heidelberg 2009

(32.3) Let k be a field, and G a finite k-group scheme. Let $S = \operatorname{Spec} R$ be an affine k-scheme of finite type with a left G-action. It gives a k-algebra automorphism action of G on R. Let $A := R^G$ be the ring of invariants.

Proposition 32.4. *Assume that G is linearly reductive (i.e., any G-module is semisimple). Then the following hold.*

1 *If R satisfies Serre's (S_r) condition, then the A-module R satisfies (S_r), and A satisfies (S_r).*
2 *If R is Cohen-Macaulay, then R is a maximal Cohen-Macaulay A-module, and A is also a Cohen-Macaulay ring.*
3 *If R is Cohen-Macaulay, then $\omega_R^G \cong \omega_A$ as A-modules.*
4 *Assume that R is Gorenstein and $\omega_R \cong R$ as (G, R)-modules. Then $A = R^G$ is Gorenstein and $\omega_A \cong A$.*

Proof. Note that the associated morphism $\pi : S = \operatorname{Spec} R \to \operatorname{Spec} A$ is finite surjective.

To prove the proposition, we may assume that $\operatorname{Spec} A$ is connected.

Set $\bar{G} := G \otimes_k \bar{k}$, and $\bar{R} = R \otimes_k \bar{k}$, where $\bar{k}$ is the algebraic closure of k. Let G_0 be the identity component (or the Frobenius kernel for sufficiently high Frobenius maps, if the characteristic is nonzero) of $\bar{G}$, which is a normal subgroup scheme of $\bar{G}$. Note that $\operatorname{Spec} \bar{R} \to \operatorname{Spec} \bar{R}^{G_0}$ is finite and is a homeomorphism, since G_0 is trivial if the characteristic is zero, and $\bar{R}^{G_0}$ contains some sufficiently high Frobenius power of $\bar{R}$, if the characteristic is positive. On the other hand, the finite group $\bar{G}(\bar{k}) = (\bar{G}/G_0)(\bar{k})$ acts on $\bar{R}^{G_0}$, and the ring of invariants under this action is $A \otimes_k \bar{k}$. By Lemma 32.2, for any prime ideal $\mathfrak{p}$ of $A \otimes_k \bar{k}$, $\bar{G}(\bar{k})$ acts transitively on the set of prime ideals of $\bar{R}$ (or $\bar{R}^{G_0}$) lying over $\mathfrak{p}$. It follows that for any prime ideal $\mathfrak{p}$ of A and a prime ideal $\mathfrak{P}$ of R lying over $\mathfrak{p}$, we have $\operatorname{ht} \mathfrak{p} = \operatorname{ht} \mathfrak{P}$.

Let M be the sum of all non-trivial simple G-submodules of R. As G is linearly reductive, R is the direct sum of M and A as a G-module. It is easy to see that $R = M \oplus A$ is a direct sum decomposition as a (G, A)-module.

1 Since A is a direct summand of R as an A-module, it suffices to prove that the A-module R satisfies the (S_r)-condition. Let $\mathfrak{p} \in \operatorname{Spec} A$ and assume that $\operatorname{depth}_{A_\mathfrak{p}} R_\mathfrak{p} < r$. Note that $\operatorname{depth}_{A_\mathfrak{p}} R_\mathfrak{p} = \inf_{\mathfrak{P}} \operatorname{depth} R_\mathfrak{P}$, where $\mathfrak{P}$ runs through the prime ideals lying over $\mathfrak{p}$. So there exists some $\mathfrak{P}$ such that $\operatorname{depth} R_\mathfrak{P} \leq \operatorname{depth}_{A_\mathfrak{p}} R_\mathfrak{p} < r$. As R satisfies Serre's (S_r)-condition, we have that $R_\mathfrak{P}$ is Cohen-Macaulay. So

$$\operatorname{ht} \mathfrak{p} = \operatorname{ht} \mathfrak{P} = \operatorname{depth} R_\mathfrak{P} \leq \operatorname{depth}_{A_\mathfrak{p}} R_\mathfrak{p} \leq \operatorname{depth} A_\mathfrak{p} \leq \operatorname{ht} \mathfrak{p},$$

and all $\leq$ must be $=$. In particular, $R_\mathfrak{p}$ is a maximal Cohen-Macaulay $A_\mathfrak{p}$-module. This shows that the A-module R satisfies Serre's (S_r)-condition.

2 is obvious by **1**.

We prove **3**. We may assume that $\operatorname{Spec} A$ is connected. Note that $\pi : S = \operatorname{Spec} R \to \operatorname{Spec} A$ is a finite G-morphism. Set $d = \dim R = \dim A$. As A is Cohen-Macaulay and $\operatorname{Spec} A$ is connected, A is equidimensional of dimension d. So $\operatorname{ht}\mathfrak{m} = d$ for all maximal ideals of A. The same is true of R, and hence R is also equidimensional. So $\omega_R[d]$ and $\omega_A[d]$ are the equivariant dualizing complexes of R and A, respectively. In particular, we have $\pi^!\omega_A \cong \omega_R$. By Lemma 31.20, ω_A is G-trivial. By Theorem 29.5, we have isomorphisms in $D(G, \operatorname{Spec} A)$

$$\omega_R \cong R\pi_*R\underline{\operatorname{Hom}}_{\mathcal{O}_{\operatorname{Spec} R}}(\mathcal{O}_{\operatorname{Spec} R}, \pi^!\omega_A) \cong R\underline{\operatorname{Hom}}_{\mathcal{O}_{\operatorname{Spec} A}}(R\pi_*\mathcal{O}_{\operatorname{Spec} R}, \omega_A).$$

As π is affine, $R\pi_*\mathcal{O}_{\operatorname{Spec} R} = R$. As R is a maximal Cohen-Macaulay A-module and ω_A is a finitely generated A-module which is of finite injective dimension, we have that $\operatorname{Ext}_A^i(R, \omega_A) = 0$ $(i > 0)$. Hence

$$\omega_R \cong R\underline{\operatorname{Hom}}_{\mathcal{O}_{\operatorname{Spec} A}}(R, \omega_A) \cong \operatorname{Hom}_A(R, \omega_A)$$

in $D(G, \operatorname{Spec} A)$. As G is linearly reductive, there is a canonical direct sum decomposition $R \cong R^G \oplus U_R$ (as an (G, A)-module), where U_R is the sum of all non-trivial simple G-submodules of R. As ω_A is G-trivial, $\operatorname{Hom}_G(U_R, \omega_A) = 0$. In particular, $\operatorname{Hom}_A(U_R, \omega_A)^G = 0$.

On the other hand, we have that

$$\operatorname{Hom}_A(R^G, \omega_A)^G = \operatorname{Hom}_A(A, \omega_A)^G = \omega_A^G = \omega_A.$$

Hence

$$\omega_R^G \cong \operatorname{Hom}_A(R, \omega_A)^G \cong \operatorname{Hom}_A(U_R, \omega_A)^G \oplus \operatorname{Hom}_A(R^G, \omega_A)^G \cong \omega_A.$$

4 follows from **2** and **3** immediately.   □

Corollary 32.5. *Let k be a field, G a linearly reductive finite k-group scheme, and V a finite dimensional G-module. Assume that the representation $G \to GL(V)$ factors through $SL(V)$. Then the ring of invariants $A := (\operatorname{Sym} V)^G$ is Gorenstein, and $\omega_A \cong A$.*

Proof. Set $R := \operatorname{Sym} V$. As R is k-smooth, we have that $\omega_R \cong \bigwedge^n \Omega_{R/k} \cong R \otimes \bigwedge^n V$, where $n = \dim_k V$. By assumption, $\bigwedge^n V \cong k$, and we have that $\omega_R \cong R$, as (G, R)-modules. By the proposition, A is Gorenstein and $\omega_A \cong A$.   □

Although it has nothing to do with the twisted inverse, we give some normality results on invariant subrings under the action of group schemes. For a ring R, let $R^\star$ denote the set of nonzerodivisors of R.

Lemma 32.6. *Let S be a finite direct product of normal domains, R a commutative ring, and F a set of ring homomorphisms from S to R. Assume that $f(s) \in R^\star$ for any $f \in F$ and $s \in S^\star$. Then*

$$A := \{a \in S \mid f(a) = f'(a) \text{ for } f, f' \in F\}$$

is a subring of S, and is a finite direct product of normal domains.

Proof. We may assume that F has at least two elements. It is obvious that A is closed under subtraction and multiplication, and $1 \in A$. So A is a subring of S.

We prove that A is a finite direct product of normal domains. Let $h : A \to R$ be the restriction of $f \in F$ to A, which is independent of choice of f. Let $e_1, \ldots, e_r$ be the primitive idempotents of A. Replacing S by Se_i, A by Ae_i, R by $R(h(e_i))$, and F by

$$\{f|_{Se_i} \mid f \in F\},$$

we may assume that $A \neq 0$ and that A does not have a nontrivial idempotent. Indeed, if $se_i \in (Se_i)^\star$, then $se_i + (1 - e_i) \in S^\star$ as can be seen easily. So we have $f(se_i) + (1 - f(e_i)) \in R^\star$, and hence $h(e_i)f(se_i) = f(se_i) \in R(h(e_i))^\star$.

Assume that $a \in A \setminus \{0\}$ is a zerodivisor of S. Then there is a non-trivial idempotent e of S such that $ae = a$ and $1 - e + a \in S^\star$. Then for $f \in F$, $h(a)f(e) = h(a)$, and $1 - f(e) + h(a) \in R^\star$. So for any $f, f' \in F$, $f(e)(1 - f'(e)) = 0$, since

$$(1 - f(e) + h(a))f(e)(1 - f'(e)) = h(a)(1 - f'(e)) = h(a) - h(a) = 0.$$

Similarly we have $f'(e)(1 - f(e)) = 0$, and hence

$$f(e) = f(e)(1 - f'(e) + f'(e)) = f(e)f'(e) = (1 - f(e) + f(e))f'(e) = f'(e).$$

This shows that $e \in A$, and this contradicts our additional assumption. Hence any nonzero element a of A is a nonzerodivisor of S. In particular, A is an integral domain, since the product of two nonzero elements of A is a nonzerodivisor of S and cannot be zero.

Let $K = Q(A)$ be the field of fractions of A, and $L = Q(S)$ be the total quotient ring of S. By the argument above, $A \hookrightarrow S \hookrightarrow L$ can be extended to a unique injective homomorphism $K \hookrightarrow L$. We regard K as a subring of L. As $f(S^\star) \subset R^\star$, $f \in F$ is extended to the map $Q(f) : L = Q(S) \to Q(R)$.
Set

$$B := \{\alpha \in L \mid Q(f)(\alpha) = Q(f')(\alpha) \text{ in } Q(R) \text{ for } f, f' \in F\}.$$

Then B is a subring of L. Note that $K \subset B$. As $R \to Q(R)$ is injective, $A = B \cap S$.

If $\alpha \in K$ is integral over A, then it is an element of $B \subset L$ which is integral over S. This shows that $\alpha \in B \cap S = A$, and we are done. □

Corollary 32.7. *Let Γ be an abstract group acting on a finite direct product S of normal domains. Then S^Γ is a finite direct product of normal domains.*

Proof. Set $R = S$ and $F = \Gamma$, and apply the lemma. □

Corollary 32.8. *Let H be an affine algebraic k-group scheme, and S an H-algebra which is a finite direct product of normal domains. Then S^H is also a finite direct product of normal domains.*

Proof. Set $R = S \otimes k[H]$, and $F = \{i, \omega\}$, where $i : S \to R$ is given by $i(s) = s \otimes 1$, and $\omega : S \to R$ is the coaction. Since both i and ω are flat, the lemma is applicable. $\square$

Chapter 33
Other Examples of Diagrams of Schemes

(33.1) We define an ordered finite category $\mathcal{K}$ by $\mathrm{ob}(\mathcal{K}) = \{s, t\}$, and $\mathcal{K}(s, t) = \{u, v\}$. Pictorially, $\mathcal{K}$ looks like $t \underset{v}{\overset{u}{\longleftarrow}} s$.

Let p be a prime number, and X an $\mathbb{F}_p$-scheme. We define the *Lyubeznik diagram* $\mathrm{Ly}(X)$ of X to be an object of $\mathcal{P}(\mathcal{K}, \underline{\mathrm{Sch}}/\mathbb{F}_p)$ given by $(\mathrm{Ly}(X))_s = (\mathrm{Ly}(X))_t = X$, $\mathrm{Ly}(X)_u = \mathrm{id}_X$, and $\mathrm{Ly}(X)_v = F_X$, where F_X denotes the absolute Frobenius morphism of X. Thus $\mathrm{Ly}(X)$ looks like

$$X \underset{F_X}{\overset{\mathrm{id}_X}{\rightrightarrows}} X .$$

We define an *F-sheaf* of X to be a quasi-coherent sheaf over $\mathrm{Ly}(X)$. It can be identified with a pair $(\mathcal{M}, \phi)$ such that $\mathcal{M}$ is a quasi-coherent $\mathcal{O}_X$-module, and $\phi : \mathcal{M} \to \mathcal{F}_X^* \mathcal{M}$ is an isomorphism of $\mathcal{O}_X$-modules. Indeed, if $\mathcal{N} \in \mathrm{Qch}(\mathrm{Ly}(X))$, then letting $\mathcal{M} := \mathcal{N}_s$ and setting ϕ to be the composite

$$\mathcal{M} = \mathcal{N}_s \cong \mathrm{id}_X^* \mathcal{N}_s = \mathrm{Ly}(X)_u^* \mathcal{N}_s \xrightarrow{\alpha_u} \mathcal{N}_t \xrightarrow{\alpha_v^{-1}} \mathrm{Ly}(X)_v^* \mathcal{N}_s = F_X^* \mathcal{N}_s = F_X^* \mathcal{M},$$

$(\mathcal{M}, \phi)$ is such a pair. Thus if $X = \mathrm{Spec}\, R$ is affine, then the category $\mathrm{Qch}(\mathrm{Ly}(X))$ of F-sheaves of X is equivalent to the category of F-modules defined by Lyubeznik [28].

Note that $\mathrm{Ly}(X)$ is noetherian with flat arrows if and only if X is a noetherian regular scheme by Kunz's theorem [24]. Let $f \colon X \to Y$ be a morphism of noetherian $\mathbb{F}_p$-schemes. Then $\mathrm{Ly}(f) : \mathrm{Ly}(X) \to \mathrm{Ly}(Y)$ is defined in an obvious way.

(33.2) For a ring A of characteristic p, the Frobenius map $A \to A$ $(a \mapsto a^p)$ is denoted by $F = F_A$. So $F_A^e(a) = a^{p^e}$. Let k be a perfect field of characteristic p. For a k-algebra $u : k \to A$, we define a k algebra $A^{(r)}$ as follows. As a ring, $A^{(r)} = A$, but the k-algebra structure of $A^{(r)}$ is given by

$$k \xrightarrow{F_k^{-r}} k \xrightarrow{u} A.$$

J. Lipman, M. Hashimoto, *Foundations of Grothendieck Duality for Diagrams of Schemes*, Lecture Notes in Mathematics 1960,

For $e \geq 0$, $F_A^e \colon A^{(r+e)} \to A^{(r)}$ is a k-algebra map. We sometimes denote an element $a \in A$, viewed as an element of $A^{(r)}$, by $a^{(r)}$. Thus $F^e(a^{(r)}) = (a^{p^e})^{(r-e)}$. For a k-scheme X, the k-scheme $X^{(r)}$ is defined similarly, and the Frobenius morphism $F_X^e \colon X^{(r)} \to X^{(r+e)}$ is a k-morphism. This notation is used for $k = \mathbb{F}_p$ for all rings of characteristic p.

Lemma 33.3. *Let k be a field of characteristic p, and K a finitely generated extension field of k. Then the canonical map $\Phi^{\mathrm{RA}} \colon k \otimes_{k^{(1)}} K^{(1)} \to K$ (the Radu-André homomorphism) given by $\Phi^{\mathrm{RA}}(\alpha \otimes \beta^{(1)}) = \alpha\beta^p$ is an isomorphism if and only if K is a separable algebraic extension of k.*

Proof. We prove the 'if' part. Note that $k \otimes_{k^{(1)}} K^{(1)}$ is a field. If $d = [K : k]$, then both $k \otimes_{k^{(1)}} K^{(1)}$ and K have the same k-dimension d. Since Φ^{RA} is an injective k-algebra map, it is an isomorphism.

We prove the 'only if' part. Since $k \otimes_{k^{(1)}} K^{(1)}$ is isomorphic to K, it is a field. So K/k is separable.

Let $x_1, \ldots, x_n$ be a separable basis of K over k. Then K, which is the image of Φ^{RA}, is a finite separable extension of $k(x_1^p, \ldots, x_n^p)$. If $n \geq 1$, then x_1 is both separable and purely inseparable over $k(x_1^p, \ldots, x_n^p)$. Namely, $x_1 \in k(x_1^p, \ldots, x_n^p)$, which is a contradiction. So $n = 0$, that is, K is separable algebraic over k. □

Lemma 33.4. *Let A be a noetherian ring, and $\varphi \colon F \to F'$ an A-linear map between A-flat modules. Then φ is an isomorphism if and only if $\varphi \otimes 1_{\kappa(\mathfrak{p})} \colon F \otimes_A \kappa(\mathfrak{p}) \to F' \otimes_A \kappa(\mathfrak{p})$ is an isomorphism for any $\mathfrak{p} \in \operatorname{Spec} A$.*

Proof. Follows easily from [19, (I.2.1.4) and (I.2.1.5)]. □

Lemma 33.5. *Let $f \colon X \to Y$ be a morphism locally of finite type between locally noetherian $\mathbb{F}_p$-schemes. Then the diagram*

$$(33.6)$$

is cartesian if and only if f is étale.

Proof. Obviously, the question is local on both X and Y, so we may assume that $X = \operatorname{Spec} B$ and $Y = \operatorname{Spec} A$ are affine.

We prove the 'only if' part. By Radu's theorem [38, Corollaire 6], $A \to B$ is regular. In particular, B is A-flat. The canonical map $A \otimes_{A^{(1)}} B^{(1)} \to B$ is an isomorphism. So for any $P \in \operatorname{Spec} B$, $\kappa(\mathfrak{p}) \otimes_{\kappa(\mathfrak{p})^{(1)}} (\kappa(\mathfrak{p}) \otimes_A B_P)^{(1)} \to \kappa(\mathfrak{p}) \otimes_A B_P$ is an isomorphism, where $\mathfrak{p} = P \cap A$. Let K be the field of fractions of the regular local ring $\kappa(\mathfrak{p}) \otimes_A B_P$. Then $\kappa(\mathfrak{p}) \otimes_{\kappa(\mathfrak{p})^{(1)}} K^{(1)} \to K$ is an isomorphism. By Lemma 33.3, K is a separable algebraic extension of $\kappa(\mathfrak{p})$. Since $\kappa(\mathfrak{p}) \subset$

$\kappa(\mathfrak{p}) \otimes_A B_P \subset K$, we have that $\kappa(\mathfrak{p}) \otimes_A B_P$ is a separable algebraic extension field of $\kappa(\mathfrak{p})$. So f is étale at P. As P is arbitrary, B is étale over A.

We prove the 'if' part. By Lemma 33.3, $\kappa(\mathfrak{p}) \otimes_{\kappa(\mathfrak{p})^{(1)}} (\kappa(\mathfrak{p}) \otimes_A B_P)^{(1)} \to \kappa(\mathfrak{p}) \otimes_A B_P$ is an isomorphism for $P \in \operatorname{Spec} B$, where $\mathfrak{p} = P \cap A$. Then it is easy to see that $\kappa(\mathfrak{p}) \otimes_{\kappa(\mathfrak{p})^{(1)}} (\kappa(\mathfrak{p}) \otimes_A B)^{(1)} \to \kappa(\mathfrak{p}) \otimes_A B$ is a isomorphism for $\mathfrak{p} \in \operatorname{Spec} A$. By Lemma 33.4, $A \otimes_{A^{(1)}} B^{(1)} \to B$ is an isomorphism, as desired. $\qquad\square$

By the lemma, for a morphism $f \colon X \to Y$ of noetherian $\mathbb{F}_p$-schemes, $\operatorname{Ly}(f)$ is cartesian of finite type if and only if f is étale.

(33.7) Let I be a small category, and $R_\bullet$ a covariant functor from I to the category of (non-commutative) rings. A left $R_\bullet$-*module* is a collection $\mathcal{M} = ((M_i)_{i \in \operatorname{ob}(I)}, (\beta_\phi)_{\phi \in \operatorname{Mor}(I)})$ such that M_i is a left R_i-module for each $i \in \operatorname{ob}(I)$, and for $\phi \in I(i,j)$, $\beta_\phi \colon M_i \to M_j$ is an R_i-linear map, where M_j is viewed as an R_i-module through the ring homomorphism $R_\phi \colon R_i \to R_j$. Moreover, we require the following conditions.

1 For $i \in \operatorname{ob}(I)$, $\beta_{\operatorname{id}_i} = \operatorname{id}_{M_i}$.
2 For $\phi, \psi \in \operatorname{Mor}(I)$ such that $\psi\phi$ is defined, $\beta_\psi \beta_\phi = \beta_{\psi\phi}$.

For $\phi \in \operatorname{Mor}(I)$, let $s(\phi) = i$ and $t(\phi) = j$ if $\phi \in I(i,j)$. i (resp. j) is the source (resp. target) of ϕ. Set $\mathcal{A}(R_\bullet) := \bigoplus_{\phi \in \operatorname{Mor}(I)} R_{t(\phi)}\phi$. We define $(b\psi)(a\phi) = (b \cdot R_\psi a)(\psi\phi)$ if $\psi\phi$ is defined, and $(b\psi)(a\phi) = 0$ otherwise. Then $\mathcal{A}(R_\bullet)$ is a ring possibly without the identity element. If $\operatorname{ob}(I)$ is finite, then $\sum_{i \in \operatorname{ob}(I)} \operatorname{id}_i$ is the identity element of $\mathcal{A}(R_\bullet)$. We call $\mathcal{A}(R_\bullet)$ the *total ring* of $R_\bullet$.

Let $\mathcal{M} = ((M_i)_{i \in \operatorname{ob}(I)}, (\beta_\phi)_{\phi \in \operatorname{Mor}(I)})$ be an $R_\bullet$-module. Then $M = \bigoplus_i M_i$ is an $\mathcal{A}(R_\bullet)$-module by $(a\phi)(\sum_j m_j) = a\beta_\phi m_{s(\phi)}$ for $\phi \in \operatorname{Mor}(I)$, $a \in R_{t(\phi)}$, and $m_j \in M_j$. It is a unitary module if $\operatorname{ob}(I)$ is finite.

From now on, assume that $\operatorname{ob}(I)$ is finite. Then a (unitary) $\mathcal{A}(R_\bullet)$-module M yields an $R_\bullet$-module. Set $M_i = \operatorname{id}_i M$. Then M_i is an R_i-module via $r(\operatorname{id}_i m) = (r\operatorname{id}_i m) = \operatorname{id}_i(r\operatorname{id}_i m)$. For $\phi \in I(i,j)$, $\beta_\phi \colon M_i \to M_j$ is defined by $\beta_\phi(m) = \phi m$. Thus an $R_\bullet$-module $((M_i), (\beta_\phi))$ is obtained. Note that the category of $R_\bullet$-modules and the category of $\mathcal{A}(R_\bullet)$-modules are equivalent.

Now consider the case that each R_i is commutative. Then $R_\bullet$ yields $X_\bullet = \operatorname{Spec}_\bullet R_\bullet \in \mathcal{P}(I, \underline{\operatorname{Sch}})$. By (4.10), the category of $R_\bullet$-modules is equivalent to $\operatorname{Lqc}(X_\bullet)$.

Lemma 33.8. *Let I be a finite ordered category, and $R_\bullet$ a covariant functor from I to the category of commutative rings. If R_i is regular with finite Krull dimension for each $i \in \operatorname{ob}(I)$, and R_ϕ is flat for each $\phi \in \operatorname{Mor}(I)$, then $\mathcal{A}(R_\bullet)$ has a finite global dimension.*

Proof. Follows easily from Lemma 31.3. $\qquad\square$

Glossary

$[?, -]$	the internal hom, 277
$\heartsuit$	stands for either PA, AB, PM, or Mod, 294
$(?)|_J$	the pull-back associated with the inclusion $J \hookrightarrow I$, 322
$(?)_J^{\heartsuit}$	the abbreviation for $Q(X_\bullet, J)_{\heartsuit}^{\#}$, 323
$(?)_{J_1, J}^{\heartsuit}$	the restriction $\heartsuit(X_\bullet|_J) \to \heartsuit(X_\bullet|_{J_1})$, 323
$(?)_J^{\mathrm{AB}}$	the abbreviation for $Q(X_\bullet, J)_{\mathrm{AB}}^{\#}$, 322
$(?)_J^{\mathrm{PA}}$	the abbreviation for $Q(X_\bullet, J)_{\mathrm{PA}}^{\#}$, 322
$\otimes$	the product structure, 277
$\otimes_{\mathcal{O}_{\mathbb{X}}}$	the sheaf tensor product, 290
$\otimes_{\mathcal{O}_{\mathbb{X}}}^{p}$	the presheaf tensor product, 290
$(?)^!$	the equivariant twisted inverse, 417
$(?)|_x^{\heartsuit}$	the restriction functor, 295
$\mathbb{A}$	the ascent functor, 367
$\mathcal{A}$	the category of noetherian I^{op}-diagrams of schemes and morphisms separated of finite type, 415
$\underline{\mathrm{Ab}}$	the category of abelian groups, 287
$\mathrm{AB}(\mathbb{X})$	the category of sheaves of abelian groups on $\mathbb{X}$, 287
$\mathcal{A}_G$	the category of noetherian G-schemes and G-morphisms separated of finite type, 445
α	the associativity isomorphism, 277
(α)	the canonical map $(d_0)^* \to (?)_{(\Delta)} \circ (?)'$, 364
(α^+)	the canonical map $(d_0^+)^* \to (?)' \circ (?)_{(\Delta)}$, 364
$\alpha_\phi^{\heartsuit}$	the translation map, 322
$\mathcal{A}(R_\bullet)$	the total ring of $R_\bullet$, 465
$a(\mathbb{X}, \mathrm{AB})$	the sheafification functor $\mathrm{PA}(\mathbb{X}) \to \mathrm{AB}(\mathbb{X})$, 287
$a(\mathbb{X}, \mathrm{Mod})$	the sheafification $\mathrm{PM}(\mathbb{X}) \to \mathrm{Mod}(\mathbb{X})$, 289
$B_G^M(X)$	the restriction $B_G(X)|_{\Delta_M}$, 446

$B_G(X)$	the simplicial groupoid associated with the action of G on X, 445
C	the morphism adjoint to η, 281
$\bar{c}$	the canonical isomorphism $f_\bullet^! RR_J \to RR_J(f_\bullet\|_J)^!$, 428
c'	the canonical isomorphism $RR_J(f_\bullet\|_J)^\times \to f_\bullet^\times RR_J$, 428
$C(\mathcal{A})$	the category of complexes in $\mathcal{A}$, 311
$C^b(\mathcal{A})$	the category of bounded complexes in $\mathcal{A}$, 311
$C^-(\mathcal{A})$	the category of complexes in $\mathcal{A}$ bounded above, 311
$C^+(\mathcal{A})$	the category of complexes in $\mathcal{A}$ bounded below, 311
Čech	the Čech complex, 387
$c = c(f)$	the identification $qf^\# = f^\# q$ or its inverse, 294
$c_{f,g}$	the canonical isomorphism $(gf)_\# \xrightarrow{\cong} g_\# f_\#$ of an almost-pseudofunctor, 271
$c = c(gf = f'g')$	the isomorphism $g_* f_* \xrightarrow{c^{-1}} (gf)_* = (f'g')_* \xrightarrow{c} f'_* g'_*$, 271
$\chi(f_\bullet)$	the canonical map $f_\bullet^\times \mathbb{F} \otimes_{\mathcal{O}_{X_\bullet}}^{\bullet, L} Lf_\bullet^* \mathbb{G} \to f_\bullet^\times (\mathbb{F} \otimes_{\mathcal{O}_{Y_\bullet}}^{\bullet, L} \mathbb{G})$, 432
$\bar{\chi} = \bar{\chi}(p_\bullet, i_\bullet)$	see page, 433
$c_{I,J,K}^\heartsuit$	the canonical isomorphism $(?)_{K,I}^\heartsuit \cong (?)_{K,J}^\heartsuit \circ (?)_{J,I}^\heartsuit$, 328
$c_{J,f_\bullet}^\heartsuit$	the canonical isomorphism $(?)_J^\heartsuit \circ (f_\bullet)_*^\heartsuit \cong (f_\bullet\|_J)_*^\heartsuit \circ (?)_J^\heartsuit$, 328
$\mathrm{Coh}(G, X)$	the category of coherent $(G, \mathcal{O}_X)$-modules, 446
$\mathrm{Cone}(\varphi)$	the mapping cone of φ, 313
cosk_J^I	the right adjoint of $(?)\|_J$, 322
$\mathrm{Cos}(\mathcal{M})$	the cosimplicial sheaf associated with $\mathcal{M}$, 365
$\mathrm{Cos}^+(\mathcal{N})$	the augmented cosimplicial sheaf associated with $\mathcal{N}$, 365
$\mathbb{D}$	the descent functor, 368
$(d_0)(X_\bullet)$	the natural map $X_\bullet(\delta_0) : X'_\bullet\|_{(\Delta)} = X_\bullet$ shift $\iota \to X_\bullet$, 363
$(d_0^+)(Y_\bullet)$	the natural map $(Y_\bullet\|_{(\Delta)})' = Y_\bullet \iota$ shift $\xrightarrow{Y_\bullet(\delta_0^+)} Y_\bullet$, 363
$D^?(\mathcal{A})$	the derived category of $\mathcal{A}$ with the boundedness ?, 311
$D_{\mathcal{A}'}^?(\mathcal{A})$	the localization of $K_{\mathcal{A}'}^?(\mathcal{A})$ by the épaisse subcategory of exact complexes, 311
$D_{\mathrm{Coh}}^b(\mathrm{Qch}(X_\bullet))$	a short for $D_{\mathrm{Coh}(X_\bullet)}^b(\mathrm{Qch}(X_\bullet))$, 351
Δ	see page, 281
(Δ)	see page, 359
$(\Delta)_S^{\mathrm{mon}}$	see page, 359
(Δ^+)	see page, 359
$(\Delta^+)^{\mathrm{mon}}$	see page, 359
$(\Delta^+)_S^{\mathrm{mon}}$	see page, 359
(δ_0)	the natural map $\mathrm{Id}_{(\Delta)} \to$ shift ι, 363
(δ_0^+)	the standard natural transformation $\mathrm{Id}_{(\Delta^+)} \to \iota \mathrm{shift}$, 363
Δ_M	$(\Delta)_{\{0,1,2\}}^{\mathrm{mon}}$, 359

Δ_M^+	$(\Delta^+)_{\{-1,0,1,2\}}^{\mathrm{mon}}$, 359
$D_{\mathrm{EM}}^+(X_\bullet)$	a short for $D_{\mathrm{EM}(X_\bullet)}^+(\mathrm{Mod}(X_\bullet))$, 351
$d_{f,g}$	the natural isomorphism $f^\# g^\# \to (gf)^\#$ of a contravariant almost-pseudofunctor, 272
$d = d(gf = f'g')$	the isomorphism $(g')^*(f')^* \xrightarrow{d} (f'g')^* = (gf)^* \xrightarrow{d^{-1}} f^* g^*$, 272
$D(G,X)$	stands for $D(B_G^M(X))$, 446
$d_{I,J,K}^\heartsuit$	the canonical isomorphism $L_{I,J}^\heartsuit \circ L_{J,K}^\heartsuit \cong L_{I,K}^\heartsuit$, 328
$d_{J,f_\bullet}^\heartsuit$	the canonical isomorphism $L_J^\heartsuit \circ (f_\bullet|_J)_\heartsuit^* \cong (f_\bullet)_\heartsuit^* \circ L_J^\heartsuit$, 328
$D_{\mathrm{Qch}}^+(G,X)$	stands for $D_{\mathrm{Qch}(G,X)}^+(\mathrm{Mod}(G,X))$, 446
$D_{\mathrm{Qch}}^+(X)$	a short for $D_{\mathrm{Qch}(X)}^+(\mathrm{Mod}(X))$, 351
$D(X_\bullet)$	a short for $D(\mathrm{Mod}(X_\bullet))$, 351
$\mathcal{D}(X_\bullet)$	stands for $D_{\mathrm{Lqc}}(X_\bullet)$, 415
$\mathcal{D}^-(X_\bullet)$	locally bounded above derived category of $X_\bullet$, 408
$\mathcal{D}^+(X_\bullet)$	locally bounded below derived category of $X_\bullet$, 408
$\mathfrak{D}_\heartsuit(X_\bullet)$	the category of structure data, 331
$\mathrm{EM}(G,X)$	the category of equivariant $(G,\mathcal{O}_X)$-modules, 446
$\mathrm{EM}(X_\bullet)$	the category of equivariant sheaves of $\mathcal{O}_{X_\bullet}$-modules, 324
ε	the counit map of adjunction, 273
$\eta = \eta(f)$	the map $\mathcal{O}_Y \to f_* \mathcal{O}_X$, 278
ev	the evaluation map, 278
$\mathfrak{e}_X$	the isomorphism $\mathrm{Id}_{X_\#} \to (\mathrm{id}_X)_\#$, 271
$\mathcal{F}$	the subcategory of $\mathcal{A}$ consisting of objects with flat arrows and cartesian morphisms, 417
$f_\bullet^\natural$	the twisted inverse for a cartesian finite morphism $f_\bullet$, 437
$(f_\bullet)_\heartsuit^*$	the inverse image functor, 327
$(f_\bullet)_*^\heartsuit$	the direct image functor, 327
$f_\bullet^\times$	the right adjoint of $R(f_\bullet)_*$, 402
$f^\#$	the pull-back associated with f, 289
$f^\times$	the right adjoint of $Rf_* : D_{\mathrm{Qch}}(X) \to D(Y)$, 391
$f_{\mathrm{AB}}^\#$	the pull-back $\mathrm{AB}(\mathbb{X}) \to \mathrm{AB}(\mathbb{Y})$, 289
$f_\#^{\mathrm{AB}}$	the left adjoint of $f_{\mathrm{AB}}^\#$, 290
$f_\#^{\mathcal{C}}$	the left adjoint of $f_{\mathcal{C}}^\#$, 290
$(f_\bullet)_*^{\mathrm{Lqc}}$	the direct image functor for Lqc, 349
$\mathcal{F}_M$	see page, 445
$f_{\mathrm{Mod}}^\#$	the pull-back $\mathrm{Mod}(\mathbb{X}) \to \mathrm{Mod}(\mathbb{Y})$ for a ringed continuous functor $f : (\mathbb{Y}, \mathcal{O}_\mathbb{Y}) \to (\mathbb{X}, \mathcal{O}_\mathbb{X})$, 294
$f_\#^{\mathrm{Mod}}$	the left adjoint of $f_{\mathrm{Mod}}^\#$, 294
$f_{\mathrm{PA}}^\#$	the pull-back $\mathrm{PA}(\mathbb{X}) \to \mathrm{PA}(\mathbb{Y})$ for $f : \mathbb{Y} \to \mathbb{X}$, 289

$f_\#^{\mathrm{PA}}$	the left adjoint of $f_{\mathrm{PA}}^\#$, 289
$f_\flat^{\mathrm{PM}}$	the right adjoint of $f_{\mathrm{PM}}^\#$, 294
$f_{\mathrm{PM}}^\#$	the pull-back $\mathrm{PM}(\mathbb{X}) \to \mathrm{PM}(\mathbb{Y})$ for a ringed functor $f\colon (\mathbb{Y}, \mathcal{O}_\mathbb{Y}) \to (\mathbb{X}, \mathcal{O}_\mathbb{X})$, 294
$f_\#^{\mathrm{PM}}$	the left adjoint of $f_{\mathrm{PM}}^\#$, 294
$F(\mathbb{X})$	the forgetful functor $\mathrm{Mod}(\mathbb{X}) \to \mathrm{AB}(\mathbb{X})$, 289
$\mathfrak{f}_X$	the isomorphism $\mathrm{id}_X^\# \to \mathrm{Id}_{X^\#}$, 272
γ	the twisting (symmetry) isomorphism, 277
Γ_i	$L_{I,J_1} \circ R_{J_1,i}$, 408
H	see page, 279
hocolim	the homotopy colimit, 381
$\operatorname{holim} t_i$	the homotopy limit of (t_i), 381
$\underline{\mathrm{Hom}}_{\heartsuit(\mathbb{X})}(\mathcal{M}, \mathcal{N})$	the sheaf Hom functor, 295
ι	the inclusion $(\Delta) \hookrightarrow (\Delta^+)$, 363
I_x^f	see page, 290
$K^?(\mathcal{A})$	the homotopy category of $\mathcal{A}$ with the boundedness ?, 311
$K_{\mathcal{A}'}^?(\mathcal{A})$	the full subcategory of $K^?(\mathcal{A})$ consisting of complexes whose cohomology groups lie in $\mathcal{A}'$, 311
λ	the left unit isomorphism, 277
$\lambda_{J,i}$	the canonical isomorphism $(L_J^\heartsuit(\mathcal{M}))_i^\heartsuit \cong \varinjlim(X_\phi)_\heartsuit^*(\mathcal{M}_j)$, 334
Lch	the category of locally coherent sheaves, 384
$L_J^\heartsuit$	the left induction functor, 327
$L_{J,J_1}^\heartsuit$	the left adjoint of $(?)_{J_1,J}^\heartsuit$, 327
lqc	the local quasi-coherator for a diagram of schemes, 385
$\mathrm{Lqc}(G, X)$	the category of locally quasi-coherent $(G, \mathcal{O}_X)$-modules, 446
$\mathrm{Lqc}(X_\bullet)$	the full subcategory of locally quasi-coherent sheaves in $\mathrm{Mod}(X_\bullet)$, 346
$L_x^\heartsuit$	the left adjoint of $(?)|_x^\heartsuit$, 295
$\mathrm{Ly}(X)$	the Lyubeznik diagram of X, 463
$\mathcal{M}'$	the pull-back $F_{\mathrm{Mod}}^\#(\mathcal{M})$, 364
$m = m(f)$	the natural map $f_* a \otimes f_* b \to f_*(a \otimes b)$, 278
m_i	the isomorphism $\mathcal{M}_i \otimes_{\mathcal{O}_{X_i}} \mathcal{N}_i \cong (\mathcal{M} \otimes_{\mathcal{O}_{X_\bullet}} \mathcal{N})_i$, 331
$\mathrm{Mod}(G, X)$	the category of $(G, \mathcal{O}_X)$-modules, 446
$\mathrm{Mod}(\mathbb{X})$	the category of sheaves of $\mathcal{O}_\mathbb{X}$-modules, 289
$\mathrm{Mod}(X_\bullet)$	the abbreviation for $\mathrm{Mod}(\mathrm{Zar}(X_\bullet))$, 323
$\mathrm{Mod}(Z)$	the category of $\mathcal{O}_Z$-modules of a scheme Z, 267
$\mu^\heartsuit$	the canonical map $f_\bullet^* R_J \to R_J(f_\bullet|_J)^*$, 341
$\mu(g_\bullet, J)$	the canonical map $g_\bullet^* RR_J \to RR_J(g_\bullet|_J)^*$, 428
$\mathrm{Nerve}(f)$	the Čech nerve of f, 360

ν	the canonical isomorphism $\check{H}^0(\mathcal{U}, f^\# \mathcal{M}) \cong \check{H}^0(f\mathcal{U}, \mathcal{M})$, 295
ν	the canonical isomorphism $\underline{H}^0 f^\# \mathcal{M} \to f^\# \underline{H}^0 \mathcal{M}$, 296
ω_Y	the G-canonical sheaf of Y, 455
$\mathcal{O}_x$	$L_x^{\mathrm{Mod}}(\mathcal{O}_{\mathbb{X}}|_x) \cong a\mathcal{O}_x^p$, 316
P	the canonical map $f^*[a,b] \to [f^*a, f^*b]$, 283
$\mathfrak{P}$	the category of strongly K-flat complexes, 316
$\underset{\rightarrow}{\mathfrak{P}}$	the full subcategory consisting of the direct limits of $\mathfrak{P}$-special direct systems, 316
$\underset{\leftarrow}{\mathfrak{P}}$	the full subcategory consisting of the inverse limits of $\mathfrak{P}$-special inverse systems, 316
$\mathrm{PA}(\mathbb{X})$	the category of presheaves of abelian groups on $\mathbb{X}$, 287
$\phi_\heartsuit^\star$	stands for the pull-back $(\mathfrak{R}_\phi)_\heartsuit^\#: \heartsuit(\mathbb{X}/y) \to \heartsuit(\mathbb{X}/x)$, 295
$\phi_\star^\heartsuit$	stands for $(\mathfrak{R}_\phi)_\#^\heartsuit: \heartsuit(\mathbb{X}/x) \to \heartsuit(\mathbb{X}/y)$, 295
Φ^{RA}	the Radu-André homomorphism, 464
$\mathcal{P}(I, \mathcal{C})$	the category of presheaves over the category I with values in $\mathcal{C}$, 287
$\Pi(f)$	the canonical map (projection morphism) $f_*a \otimes b \to f_*(a \otimes f^*b)$, 431
$\mathrm{PM}(\mathbb{X})$	the category of presheaves of $\mathcal{O}_{\mathbb{X}}$-modules, 289
$\mathrm{PM}(X_\bullet)$	the abbreviation for $\mathrm{PM}(\mathrm{Zar}(X_\bullet))$, 323
$\mathcal{P}(X_\bullet, \mathcal{C})$	the abbreviation for $\mathcal{P}(\mathrm{Zar}(X_\bullet), \mathcal{C})$, 322
Q	the localization $K^?(\mathcal{A}) \to D^?(\mathcal{A})$, 311
$\mathfrak{Q}$	the full subcategory of $C(\mathrm{Mod}(\mathbb{X}))$ consisting of bounded above complexes whose terms are direct sums of copies of $\mathcal{O}_x$, 316
$\mathrm{Qch}(G, X)$	the category of quasi-coherent $(G, \mathcal{O}_X)$-modules, 446
$\mathrm{qch}(X)$	the quasi-coherator on a scheme X, 385
$\mathrm{Qch}(X_\bullet)$	the full subcategory of $\mathrm{Mod}(X_\bullet)$ consisting of quasi-coherent modules, 346
$\mathrm{Qch}(Z)$	the category of quasi-coherent $\mathcal{O}_Z$-modules of a scheme Z, 267
$q(\mathbb{X}, \mathrm{AB})$	the inclusion $\mathrm{AB}(\mathbb{X}) \to \mathrm{PA}(\mathbb{X})$, 287
$Q(X_\bullet, J)$	the inclusion $\mathrm{Zar}((X_\bullet)|_J) \hookrightarrow \mathrm{Zar}(X_\bullet)$, 322
$q(\mathbb{X}, \mathrm{Mod})$	the inclusion $\mathrm{Mod}(\mathbb{X}) \to \mathrm{PM}(\mathbb{X})$, 289
$R^\star$	the set of nonzerodivisors of R, 459
ρ	the right unit isomorphism, 282
$\rho^{J,i}$	the canonical isomorphism $(R_J^\heartsuit(\mathcal{M}))_i^\heartsuit \cong \varprojlim (X_\phi)_*^\heartsuit(\mathcal{M}_j)$, 337
$R_J^\heartsuit$	the right induction functor, 327
$R_{J,J_1}^\heartsuit$	the right adjoint of $(?)_{J_1,J}^\heartsuit$, 327

$\mathfrak{R}_\phi$ the canonical functor $\mathbb{X}/x \to \mathbb{X}/y$ for $\phi : x \to y$, 295

$\mathfrak{R}_x$ the canonical functor $\mathbb{X}/x \to \mathbb{X}$, 295

<u>Sch</u>   the category of schemes, 321

<u>Sch</u>$/S$ the category of S-schemes, 321

<u>Set</u>   the category of small sets, 287

shift   the standard shifting functor $(\Delta^+) \to (\Delta)$, 363

Σ the suspension of a triangulated category, 311

Σ_i the right adjoint of Γ_i, 408

$\Sigma(X_\bullet)$ the simplicial S-scheme associated with $X_\bullet$, 376

$\mathcal{S}(\mathbb{X}, \mathcal{C})$ the category of sheaves over $\mathbb{X}$ with values in $\mathcal{C}$, 287

$\mathcal{S}(X_\bullet, \mathcal{C})$ the abbreviation for $\mathcal{S}(\mathrm{Zar}(X_\bullet), \mathcal{C})$, 322

$\tau_{\geq n}\mathbb{F}$ the truncation of a complex, 317

$\tau_{\leq n}\mathbb{F}$ the truncation of a complex, 317

$\bar{\theta}$ the canonical map $af^\# \to f^\# a$, 296

$\Theta(f)$ the duality isomorphism for schemes, 429

$\Theta(f_\bullet)$ the duality isomorphism, 429

$\theta_\heartsuit(f_\bullet, J)$ the canonical isomorphism $((f_\bullet)|_J)^*_\heartsuit \circ (?)_J \to (?)_J \circ (f_\bullet)^*_\heartsuit$, 340

$\theta(J, f_\bullet)$ the canonical map $L_J(f_\bullet|_J)_* \to (f_\bullet)_* L_J$, 340

$\theta(\sigma)$ Lipman's theta, 276

tr   the trace map, 278

u the unit map of adjunction, 273

Υ the independence isomorphism, 397

$X'_\bullet$ the augmented simplicial scheme $\mathrm{shift}^\#(X_\bullet) = X_\bullet\,\mathrm{shift}$, 363

Ξ the canonical map $QF \to (RF)Q$, 312

$\bar{\xi}$ the canonical map $R(g_\bullet^X)_*(f'_\bullet)^! \to f_\bullet^! R(g_\bullet)_*$, 425

$\xi_\heartsuit(f_\bullet, J)$ the isomorphism $(f_\bullet)_*^\heartsuit R_J \to R_J(f_\bullet|_J)_*^\heartsuit$, 341

$\xi(J, f_\bullet)$ the natural map $(?)_J \circ f_\bullet^\times \to f_J^\times \circ (?)_J$, 405

$\bar{\xi}(J, f_\bullet)$ the canonical map $(?)_J f_\bullet^! \to f_\bullet|_J^! (?)_J$, 419

$\xi(\sigma_2)$ the canonical isomorphism $R(g_\bullet^Z)_*(p'_\bullet)^\times \to p_\bullet^\times R(g_\bullet)_*$, 425

$Y(\mathcal{M})$ the canonical map $\mathcal{M} \to \underline{\check{H}}^0(\mathcal{M})$, 288

$\mathrm{Zar}(X_\bullet)$ the Zariski site of $X_\bullet$, 321

$\zeta(\sigma)$ the canonical map $(g'_\bullet)^* f_\bullet^\times \to (f'_\bullet)^\times g_\bullet^*$, 413

$\bar{\zeta}(\sigma)$ the canonical map $(g_\bullet^X)^* f_\bullet^! \to (f'_\bullet)^! g_\bullet^*$, 420

References

1. L. Alonso Tarrío, A. Jeremías López, and M. J. Souto Salorio, Localization in categories of complexes and unbounded resolutions, *Canad. J. Math.* **52** (2000), 225–247.
2. M. Artin, *Grothendieck Topology,* Mimeographed notes, Harvard University (1962).
3. L. L. Avramov, Flat morphisms of complete intersections, *Dokl. Akad. Nauk SSSR* **225** (1975), *Soviet Math. Dokl.* **16** (1975), 1413–1417.
4. L. L. Avramov and H.-B. Foxby, Locally Gorenstein homomorphisms, *Amer. J. Math.* **114** (1992), 1007–1047.
5. A. A. Beilinson, J. Bernstein, and P. Deligne, Faisceaux pervers, Astérisque **100** (1982), 5–171.
6. J. Bernstein and V. Lunts, *Equivariant sheaves and functors, Lect. Notes Math.* **1578**, Springer Verlag (1994).
7. M. Bökstedt and A. Neeman, Homotopy limits in triangulated categories, *Compositio Math.* **86** (1993), 209–234.
8. A. Bondal and M. van den Bergh, Generators and representability of functors in commutative and noncommutative geometry, *Mosc. Math. J.* **3** (2003), 1–36, 258.
9. J. Franke, On the Brown representability theorem for triangulated categories, *Topology* **40** (2001), 667–680.
10. E. M. Friedlander, *Etale homotopy of simplicial schemes,* Princeton (1982).
11. P. Gabriel, Des categories abeliennes, *Bull. Soc. Math. France* **90** (1962), 323–448.
12. P. Gabriel, Construction de préschémas quotient, in *Schémas en Groupes I (SGA3), Lect. Notes Math.* **151**, Springer Verlag (1970), pp. 251–286.
13. A. Grothendieck, Sur quelques points d'algèbre homologique, *Tôhoku Math. J.* **9** (1957), 119–221.
14. A. Grothendieck, *Eléments de Géométrie Algébrique I,* IHES Publ. Math. **4** (1960).
15. A. Grothendieck, *Eléments de Géométrie Algébrique IV,* IHES Publ. Math. **20** (1964), **24** (1965), **28** (1966), **32** (1967).
16. A. Grothendieck et J.-L. Verdier, Préfaisceaux, in *Théorie des Topos et Cohomologie Etale des Schémas, SGA 4, Lect. Notes Math.* **269**, Springer Verlag (1972), pp. 1–217.
17. R. Hartshorne, *Residues and Duality, Lect. Notes Math.* **20**, Springer Verlag, (1966).

18. R. Hartshorne, *Algebraic Geometry, Graduate Texts in Math.* **52**, Springer Verlag (1977).

19. M. Hashimoto, *Auslander-Buchweitz Approximations of Equivariant Modules, London Mathematical Society Lecture Note Series* **282**, Cambridge (2000).

20. L. Illusie, Existence de résolutions globales, in *Théorie des Intersections et Théorème de Riemann-Roch (SGA 6), Lect. Notes Math.* **225**, Springer Verlag, pp. 160–221.

21. J. C. Jantzen, *Representations of algebraic groups,* Second edition, AMS (2003).

22. B. Keller and D. Vossieck, Sous les catégories dérivées, *C. R. Acad. Sci. Paris Sér. I Math.* **305** (1987), 225–228.

23. G. R. Kempf, Some elementary proofs of basic theorems in the cohomology of quasi-coherent sheaves, *Rocky Mountain J. Math.* **10** (1980), 637–645.

24. E. Kunz, Characterizations of regular local rings of characteristic p, *Amer. J. Math.* **91** (1969), 772–784.

25. G. Lewis, Coherence for a closed functor, in *Coherence in Categories, Lect. Notes Math.* **281**, Springer Verlag (1972), pp. 148–195.

26. J. Lipman, Notes on Derived Functors and Grothendieck Duality, in *Foundations of Grothendieck Duality for Diagrams of Schemes, Lect. Notes Math.* **1960**, Springer Verlag (in this volume) (2009), pp. 1–259.

27. W. Lütkebohmert, On compactification of schemes, *Manuscripta Math.* **80** (1993), 95–111.

28. G. Lyubeznik, F-modules: applications to local cohomology and D-modules in characteristic $p > 0$, *J. reine angew. Math.* **491** (1997), 65–130.

29. S. Mac Lane, *Categories for the Working Mathematician,* 2nd ed. *Graduate Texts in Math.* **52**, Springer Verlag (1998).

30. H. Matsumura, *Commutative Ring Theory,* First paperback edition, Cambridge (1989).

31. J. S. Milne, *Étale cohomology,* Princeton (1980).

32. D. Mumford, J. Fogarty and F. Kirwan, *Geometric Invariant Theory,* third edition, Springer (1994).

33. J. P. Murre, *Lectures on an introduction to Grothendieck's theory of the fundamental group,* Tata Institute, Bombay (1967).

34. M. Nagata, A generalization of the imbedding problem of an abstract variety in a complete variety, *J. Math. Kyoto Univ.* **3** (1963), 89–102.

35. A. Neeman, The Grothendieck duality theorem via Bousfield's techniques and Brown representability, *J. Amer. Math. Soc.* **9** (1996), 205–236.

36. A. Neeman, *Triangulated Categories,* Princeton (2001).

37. N. Popescu, *Abelian Categories with Applications to Rings and Modules,* Academic Press (1973).

38. N. Radu, Une classe d'anneaux noethériens, *Rev. Roumanie Math. Pures Appl.* **37** (1992), 79–82.

39. N. Spaltenstein, Resolutions of unbounded complexes, *Compositio Math.* **65** (1988), 121–154.

40. H. Sumihiro, Equivariant completion. II, *J. Math. Kyoto Univ.* **15** (1975), 573–605.

41. J.-L. Verdier, Base change for twisted inverse images of coherent sheaves, in *Algebraic Geometry (Internat. Colloq.),* Tata Inst. Fund. Res., Bombay (1968), pp. 393–408.

42. J.-L. Verdier, Topologies et faisceaux, in *Théorie des Topos et Cohomologie Etale des Schémas, SGA 4, Lect. Notes Math.* **269**, Springer Verlag (1972), pp. 219–263.

43. J.-L. Verdier, Fonctorialité des catégories de faisceaux, in *Théorie des Topos et Cohomologie Etale des Schémas, SGA 4, Lect. Notes Math.* **269**, Springer Verlag (1972), pp. 265–297.

44. J.-L. Verdier, Catégories dérivées, quelques résultats (etat 0), in *Cohomologie Etale, SGA* $4\frac{1}{2}$, *Lect. Notes Math.* **569**, Springer Verlag (1977), pp. 262–311.
45. K.-i. Watanabe, Certain invariant subrings are Gorenstein I, *Osaka J. Math.* **11** (1974), 1–8.
46. W. C. Waterhouse, *Introduction to Affine Group Schemes, Graduate Texts in Math.* **66**, Springer Verlag (1979).

Index

adjoint pair, 275
admissible functor, 290, 338
admissible subcategory, 337
almost-S-groupoid, 379
almost-pseudofunctor, 271
ascent functor, 368
associated pseudofunctor, 273
augmented simplicial object, 359
augmented simplicial scheme, 360

big, 346

the canonical sheaf, 454
cartesian, 321
Čech complex, 387
Čech nerve, 360
coherent, 346
compact object, 389
compactification, 397
compactly generated, 389
composite, 398
composition data, 395
concentrated, 347, 373
conjugate, 274
connected component, 439
contravariant almost-pseudofunctor, 272

d-connected, 439
defining ideal sheaf, 441
descent functor, 368
direct image, 327
dualizing complex, 452

equivariant, 324
equivariant Grothendieck's duality, 429
equivariant twisted inverse, 417

F-acyclic, 314

F-sheaf, 463
finite projective dimension, 440

the G-canonical sheaf, 455
G-dualizing complex, 455
G-invariance, 449
$(G, \mathcal{O}_X)$-module, 446
G-scheme, 445
Grothendieck, 289

homotopy colimit, 381
homotopy limit, 381
hyperExt, 319
hyperTor, 319

independence diagram, 397
independence isomorphism, 397
inverse image, 327
invertible, 439

K-flat, 316
K-injective, 312
K-injective resolution, 312
K-limp, 316

left conjugate, 274
left induction, 327
Lipman, 283
Lipman's theta, 276
local complete intersection, 441
local quasi-coherator, 385
locally an open immersion, 368
locally coherent, 345
locally free, 439
locally of finite projective dimension, 440
locally quasi-coherent, 345
Lyubeznik diagram, 463

Lecture Notes in Mathematics

For information about earlier volumes
please contact your bookseller or Springer
LNM Online archive: springerlink.com

Vol. 1871: P. Constantin, G. Gallavotti, A.V. Kazhikhov, Y. Meyer, S. Ukai, Mathematical Foundation of Turbulent Viscous Flows, Martina Franca, Italy, 2003. Editors: M. Cannone, T. Miyakawa (2006)

Vol. 1872: A. Friedman (Ed.), Tutorials in Mathematical Biosciences III. Cell Cycle, Proliferation, and Cancer (2006)

Vol. 1873: R. Mansuy, M. Yor, Random Times and Enlargements of Filtrations in a Brownian Setting (2006)

Vol. 1874: M. Yor, M. Émery (Eds.), In Memoriam Paul-André Meyer - Séminaire de Probabilités XXXIX (2006)

Vol. 1875: J. Pitman, Combinatorial Stochastic Processes. Ecole d'Eté de Probabilités de Saint-Flour XXXII-2002. Editor: J. Picard (2006)

Vol. 1876: H. Herrlich, Axiom of Choice (2006)

Vol. 1877: J. Steuding, Value Distributions of L-Functions (2007)

Vol. 1878: R. Cerf, The Wulff Crystal in Ising and Percolation Models, Ecole d'Eté de Probabilités de Saint-Flour XXXIV-2004. Editor: Jean Picard (2006)

Vol. 1879: G. Slade, The Lace Expansion and its Applications, Ecole d'Eté de Probabilités de Saint-Flour XXXIV-2004. Editor: Jean Picard (2006)

Vol. 1880: S. Attal, A. Joye, C.-A. Pillet, Open Quantum Systems I, The Hamiltonian Approach (2006)

Vol. 1881: S. Attal, A. Joye, C.-A. Pillet, Open Quantum Systems II, The Markovian Approach (2006)

Vol. 1882: S. Attal, A. Joye, C.-A. Pillet, Open Quantum Systems III, Recent Developments (2006)

Vol. 1883: W. Van Assche, F. Marcellàn (Eds.), Orthogonal Polynomials and Special Functions, Computation and Application (2006)

Vol. 1884: N. Hayashi, E.I. Kaikina, P.I. Naumkin, I.A. Shishmarev, Asymptotics for Dissipative Nonlinear Equations (2006)

Vol. 1885: A. Telcs, The Art of Random Walks (2006)

Vol. 1886: S. Takamura, Splitting Deformations of Degenerations of Complex Curves (2006)

Vol. 1887: K. Habermann, L. Habermann, Introduction to Symplectic Dirac Operators (2006)

Vol. 1888: J. van der Hoeven, Transseries and Real Differential Algebra (2006)

Vol. 1889: G. Osipenko, Dynamical Systems, Graphs, and Algorithms (2006)

Vol. 1890: M. Bunge, J. Funk, Singular Coverings of Toposes (2006)

Vol. 1891: J.B. Friedlander, D.R. Heath-Brown, H. Iwaniec, J. Kaczorowski, Analytic Number Theory, Cetraro, Italy, 2002. Editors: A. Perelli, C. Viola (2006)

Vol. 1892: A. Baddeley, I. Bárány, R. Schneider, W. Weil, Stochastic Geometry, Martina Franca, Italy, 2004. Editor: W. Weil (2007)

Vol. 1893: H. Hanßmann, Local and Semi-Local Bifurcations in Hamiltonian Dynamical Systems, Results and Examples (2007)

Vol. 1894: C.W. Groetsch, Stable Approximate Evaluation of Unbounded Operators (2007)

Vol. 1895: L. Molnár, Selected Preserver Problems on Algebraic Structures of Linear Operators and on Function Spaces (2007)

Vol. 1896: P. Massart, Concentration Inequalities and Model Selection, Ecole d'Été de Probabilités de Saint-Flour XXXIII-2003. Editor: J. Picard (2007)

Vol. 1897: R. Doney, Fluctuation Theory for Lévy Processes, Ecole d'Été de Probabilités de Saint-Flour XXXV-2005. Editor: J. Picard (2007)

Vol. 1898: H.R. Beyer, Beyond Partial Differential Equations, On linear and Quasi-Linear Abstract Hyperbolic Evolution Equations (2007)

Vol. 1899: Séminaire de Probabilités XL. Editors: C. Donati-Martin, M. Émery, A. Rouault, C. Stricker (2007)

Vol. 1900: E. Bolthausen, A. Bovier (Eds.), Spin Glasses (2007)

Vol. 1901: O. Wittenberg, Intersections de deux quadriques et pinceaux de courbes de genre 1, Intersections of Two Quadrics and Pencils of Curves of Genus 1 (2007)

Vol. 1902: A. Isaev, Lectures on the Automorphism Groups of Kobayashi-Hyperbolic Manifolds (2007)

Vol. 1903: G. Kresin, V. Maz'ya, Sharp Real-Part Theorems (2007)

Vol. 1904: P. Giesl, Construction of Global Lyapunov Functions Using Radial Basis Functions (2007)

Vol. 1905: C. Prévôt, M. Röckner, A Concise Course on Stochastic Partial Differential Equations (2007)

Vol. 1906: T. Schuster, The Method of Approximate Inverse: Theory and Applications (2007)

Vol. 1907: M. Rasmussen, Attractivity and Bifurcation for Nonautonomous Dynamical Systems (2007)

Vol. 1908: T.J. Lyons, M. Caruana, T. Lévy, Differential Equations Driven by Rough Paths, Ecole d'Été de Probabilités de Saint-Flour XXXIV-2004 (2007)

Vol. 1909: H. Akiyoshi, M. Sakuma, M. Wada, Y. Yamashita, Punctured Torus Groups and 2-Bridge Knot Groups (I) (2007)

Vol. 1910: V.D. Milman, G. Schechtman (Eds.), Geometric Aspects of Functional Analysis. Israel Seminar 2004-2005 (2007)

Vol. 1911: A. Bressan, D. Serre, M. Williams, K. Zumbrun, Hyperbolic Systems of Balance Laws. Cetraro, Italy 2003. Editor: P. Marcati (2007)

Vol. 1912: V. Berinde, Iterative Approximation of Fixed Points (2007)

Vol. 1913: J.E. Marsden, G. Misiołek, J.-P. Ortega, M. Perlmutter, T.S. Ratiu, Hamiltonian Reduction by Stages (2007)

Vol. 1914: G. Kutyniok, Affine Density in Wavelet Analysis (2007)

Vol. 1915: T. Bıyıkoğlu, J. Leydold, P.F. Stadler, Laplacian Eigenvectors of Graphs. Perron-Frobenius and Faber-Krahn Type Theorems (2007)

Vol. 1916: C. Villani, F. Rezakhanlou, Entropy Methods for the Boltzmann Equation. Editors: F. Golse, S. Olla (2008)

Vol. 1917: I. Veselić, Existence and Regularity Properties of the Integrated Density of States of Random Schrödinger (2007)

Vol. 1918: B. Roberts, R. Schmidt, Local Newforms for GSp(4) (2007)

Vol. 1919: R.A. Carmona, I. Ekeland, A. Kohatsu-Higa, J.-M. Lasry, P.-L. Lions, H. Pham, E. Taflin, Paris-Princeton Lectures on Mathematical Finance 2004. Editors: R.A. Carmona, E. Çinlar, I. Ekeland, E. Jouini, J.A. Scheinkman, N. Touzi (2007)

Vol. 1920: S.N. Evans, Probability and Real Trees. Ecole d'Été de Probabilités de Saint-Flour XXXV-2005 (2008)

Vol. 1921: J.P. Tian, Evolution Algebras and their Applications (2008)

Vol. 1922: A. Friedman (Ed.), Tutorials in Mathematical BioSciences IV. Evolution and Ecology (2008)

Vol. 1923: J.P.N. Bishwal, Parameter Estimation in Stochastic Differential Equations (2008)

Vol. 1924: M. Wilson, Littlewood-Paley Theory and Exponential-Square Integrability (2008)

Vol. 1925: M. du Sautoy, L. Woodward, Zeta Functions of Groups and Rings (2008)

Vol. 1926: L. Barreira, V. Claudia, Stability of Nonautonomous Differential Equations (2008)

Vol. 1927: L. Ambrosio, L. Caffarelli, M.G. Crandall, L.C. Evans, N. Fusco, Calculus of Variations and Non-Linear Partial Differential Equations. Cetraro, Italy 2005. Editors: B. Dacorogna, P. Marcellini (2008)

Vol. 1928: J. Jonsson, Simplicial Complexes of Graphs (2008)

Vol. 1929: Y. Mishura, Stochastic Calculus for Fractional Brownian Motion and Related Processes (2008)

Vol. 1930: J.M. Urbano, The Method of Intrinsic Scaling. A Systematic Approach to Regularity for Degenerate and Singular PDEs (2008)

Vol. 1931: M. Cowling, E. Frenkel, M. Kashiwara, A. Valette, D.A. Vogan, Jr., N.R. Wallach, Representation Theory and Complex Analysis. Venice, Italy 2004. Editors: E.C. Tarabusi, A. D'Agnolo, M. Picardello (2008)

Vol. 1932: A.A. Agrachev, A.S. Morse, E.D. Sontag, H.J. Sussmann, V.I. Utkin, Nonlinear and Optimal Control Theory. Cetraro, Italy 2004. Editors: P. Nistri, G. Stefani (2008)

Vol. 1933: M. Petkovic, Point Estimation of Root Finding Methods (2008)

Vol. 1934: C. Donati-Martin, M. Émery, A. Rouault, C. Stricker (Eds.), Séminaire de Probabilités XLI (2008)

Vol. 1935: A. Unterberger, Alternative Pseudodifferential Analysis (2008)

Vol. 1936: P. Magal, S. Ruan (Eds.), Structured Population Models in Biology and Epidemiology (2008)

Vol. 1937: G. Capriz, P. Giovine, P.M. Mariano (Eds.), Mathematical Models of Granular Matter (2008)

Vol. 1938: D. Auroux, F. Catanese, M. Manetti, P. Seidel, B. Siebert, I. Smith, G. Tian, Symplectic 4-Manifolds and Algebraic Surfaces. Cetraro, Italy 2003. Editors: F. Catanese, G. Tian (2008)

Vol. 1939: D. Boffi, F. Brezzi, L. Demkowicz, R.G. Durán, R.S. Falk, M. Fortin, Mixed Finite Elements, Compatibility Conditions, and Applications. Cetraro, Italy 2006. Editors: D. Boffi, L. Gastaldi (2008)

Vol. 1940: J. Banasiak, V. Capasso, M.A.J. Chaplain, M. Lachowicz, J. Miękisz, Multiscale Problems in the Life Sciences. From Microscopic to Macroscopic. Będlewo, Poland 2006. Editors: V. Capasso, M. Lachowicz (2008)

Vol. 1941: S.M.J. Haran, Arithmetical Investigations. Representation Theory, Orthogonal Polynomials, and Quantum Interpolations (2008)

Vol. 1942: S. Albeverio, F. Flandoli, Y.G. Sinai, SPDE in Hydrodynamic. Recent Progress and Prospects. Cetraro, Italy 2005. Editors: G. Da Prato, M. Röckner (2008)

Vol. 1943: L.L. Bonilla (Ed.), Inverse Problems and Imaging. Martina Franca, Italy 2002 (2008)

Vol. 1944: A. Di Bartolo, G. Falcone, P. Plaumann, K. Strambach, Algebraic Groups and Lie Groups with Few Factors (2008)

Vol. 1945: F. Brauer, P. van den Driessche, J. Wu (Eds.), Mathematical Epidemiology (2008)

Vol. 1946: G. Allaire, A. Arnold, P. Degond, T.Y. Hou, Quantum Transport. Modelling, Analysis and Asymptotics. Cetraro, Italy 2006. Editors: N.B. Abdallah, G. Frosali (2008)

Vol. 1947: D. Abramovich, M. Mariño, M. Thaddeus, R. Vakil, Enumerative Invariants in Algebraic Geometry and String Theory. Cetraro, Italy 2005. Editors: K. Behrend, M. Manetti (2008)

Vol. 1948: F. Cao, J-L. Lisani, J-M. Morel, P. Musé, F. Sur, A Theory of Shape Identification (2008)

Vol. 1949: H.G. Feichtinger, B. Helffer, M.P. Lamoureux, N. Lerner, J. Toft, Pseudo-Differential Operators. Quantization and Signals. Cetraro, Italy 2006. Editors: L. Rodino, M.W. Wong (2008)

Vol. 1950: M. Bramson, Stability of Queueing Networks, Ecole d'Eté de Probabilités de Saint-Flour XXXVI-2006 (2008)

Vol. 1951: A. Moltó, J. Orihuela, S. Troyanski, M. Valdivia, A Non Linear Transfer Technique for Renorming (2008)

Vol. 1952: R. Mikhailov, I.B.S. Passi, Lower Central and Dimension Series of Groups (2008)

Vol. 1953: K. Arwini, C.T.J. Dodson, Information Geometry (2008)

Vol. 1954: P. Biane, L. Bouten, F. Cipriani, N. Konno, N. Privault, Q. Xu, Quantum Potential Theory. Editors: U. Franz, M. Schuermann (2008)

Vol. 1955: M. Bernot, V. Caselles, J.-M. Morel, Optimal Transportation Networks (2008)

Vol. 1956: C.H. Chu, Matrix Convolution Operators on Groups (2008)

Vol. 1957: A. Guionnet, On Random Matrices: Macroscopic Asymptotics, Ecole d'Eté de Probabilités de Saint-Flour XXXVI-2006 (2008)

Vol. 1958: M.C. Olsson, Compactifying Moduli Spaces for Abelian Varieties (2008)

Vol. 1959: Y. Nakkajima, A. Shiho, Weight Filtrations on Log Crystalline Cohomologies of Families of Open Smooth Varieties (2008)

Vol. 1960: J. Lipman, M. Hashimoto, Foundations of Grothendieck Duality for Diagrams of Schemes (2009)

Recent Reprints and New Editions

Vol. 1702: J. Ma, J. Yong, Forward-Backward Stochastic Differential Equations and their Applications. 1999 – Corr. 3rd printing (2007)

Vol. 830: J.A. Green, Polynomial Representations of GL_n, with an Appendix on Schensted Correspondence and Littelmann Paths by K. Erdmann, J.A. Green and M. Schoker 1980 – 2nd corr. and augmented edition (2007)

Vol. 1693: S. Simons, From Hahn-Banach to Monotonicity (Minimax and Monotonicity 1998) – 2nd exp. edition (2008)

Vol. 470: R.E. Bowen, Equilibrium States and the Ergodic Theory of Anosov Diffeomorphisms. With a preface by D. Ruelle. Edited by J.-R. Chazottes. 1975 – 2nd rev. edition (2008)

Vol. 523: S.A. Albeverio, R.J. Høegh-Krohn, S. Mazzucchi, Mathematical Theory of Feynman Path Integral. 1976 – 2nd corr. and enlarged edition (2008)

Vol. 1764: A. Cannas da Silva, Lectures on Symplectic Geometry 2001 – Corr. 2nd printing (2008)

LECTURE NOTES IN MATHEMATICS   🖎 **Springer**

Edited by J.-M. Morel, F. Takens, B. Teissier, P.K. Maini

Editorial Policy (for the publication of monographs)

1. Lecture Notes aim to report new developments in all areas of mathematics and their applications - quickly, informally and at a high level. Mathematical texts analysing new developments in modelling and numerical simulation are welcome.

   Monograph manuscripts should be reasonably self-contained and rounded off. Thus they may, and often will, present not only results of the author but also related work by other people. They may be based on specialised lecture courses. Furthermore, the manuscripts should provide sufficient motivation, examples and applications. This clearly distinguishes Lecture Notes from journal articles or technical reports which normally are very concise. Articles intended for a journal but too long to be accepted by most journals, usually do not have this "lecture notes" character. For similar reasons it is unusual for doctoral theses to be accepted for the Lecture Notes series, though habilitation theses may be appropriate.

2. Manuscripts should be submitted either to Springer's mathematics editorial in Heidelberg, or to one of the series editors. In general, manuscripts will be sent out to 2 external referees for evaluation. If a decision cannot yet be reached on the basis of the first 2 reports, further referees may be contacted: The author will be informed of this. A final decision to publish can be made only on the basis of the complete manuscript, however a refereeing process leading to a preliminary decision can be based on a pre-final or incomplete manuscript. The strict minimum amount of material that will be considered should include a detailed outline describing the planned contents of each chapter, a bibliography and several sample chapters.

   Authors should be aware that incomplete or insufficiently close to final manuscripts almost always result in longer refereeing times and nevertheless unclear referees' recommendations, making further refereeing of a final draft necessary.

   Authors should also be aware that parallel submission of their manuscript to another publisher while under consideration for LNM will in general lead to immediate rejection.

3. Manuscripts should in general be submitted in English. Final manuscripts should contain at least 100 pages of mathematical text and should always include

   – a table of contents;
   – an informative introduction, with adequate motivation and perhaps some historical remarks: it should be accessible to a reader not intimately familiar with the topic treated;
   – a subject index: as a rule this is genuinely helpful for the reader.

For evaluation purposes, manuscripts may be submitted in print or electronic form, in the latter case preferably as pdf- or zipped ps-files. Lecture Notes volumes are, as a rule, printed digitally from the authors' files. To ensure best results, authors are asked to use the LaTeX2e style files available from Springer's web-server at:

ftp://ftp.springer.de/pub/tex/latex/svmonot1/ (for monographs).

Additional technical instructions, if necessary, are available on request from:
lnm@springer.com.

4. Careful preparation of the manuscripts will help keep production time short besides ensuring satisfactory appearance of the finished book in print and online. After acceptance of the manuscript authors will be asked to prepare the final LaTeX source files (and also the corresponding dvi-, pdf- or zipped ps-file) together with the final printout made from these files. The LaTeX source files are essential for producing the full-text online version of the book (see www.springerlink.com/content/110312 for the existing online volumes of LNM).

   The actual production of a Lecture Notes volume takes approximately 12 weeks.

5. Authors receive a total of 50 free copies of their volume, but no royalties. They are entitled to a discount of 33.3% on the price of Springer books purchased for their personal use, if ordering directly from Springer.

6. Commitment to publish is made by letter of intent rather than by signing a formal contract. Springer-Verlag secures the copyright for each volume. Authors are free to reuse material contained in their LNM volumes in later publications: a brief written (or e-mail) request for formal permission is sufficient.

Addresses:
Professor J.-M. Morel, CMLA,
École Normale Supérieure de Cachan,
61 Avenue du Président Wilson, 94235 Cachan Cedex, France
E-mail: Jean-Michel.Morel@cmla.ens-cachan.fr

Professor F. Takens, Mathematisch Instituut,
Rijksuniversiteit Groningen, Postbus 800,
9700 AV Groningen, The Netherlands
E-mail: F.Takens@math.rug.nl

Professor B. Teissier, Institut Mathématique de Jussieu,
UMR 7586 du CNRS, Équipe "Géométrie et Dynamique",
175 rue du Chevaleret
75013 Paris, France
E-mail: teissier@math.jussieu.fr

For the "Mathematical Biosciences Subseries" of LNM:

Professor P.K. Maini, Center for Mathematical Biology,
Mathematical Institute, 24-29 St Giles,
Oxford OX1 3LP, UK
E-mail: maini@maths.ox.ac.uk

Springer, Mathematics Editorial I, Tiergartenstr. 17
69121 Heidelberg, Germany,
Tel.: +49 (6221) 487-8259
Fax: +49 (6221) 4876-8259
E-mail: lnm@springer.com